I-CAR UNIBODY COLLISION REPAIR

Delmar Publishers Inc.®

NOTICE TO THE READER

Publisher does not warrant or guarantee any of the products described herein or perform any independent analysis in connection with any of the product information contained herein. Publisher does not assume, and expressly disclaims, any obligation to obtain and include information other than that provided to it by the manufacturer.

The reader is expressly warned to consider and adopt all safety precautions that might be indicated by the activities described herein and to avoid all potential hazards. By following the instructions contained herein, the reader willingly assumes all risks in connection with such instructions.

The publisher makes no representations or warranties of any kind, including but not limited to, the warranties of fitness for particular purpose or merchantability, nor are any such representations implied with respect to the material set forth herein, and the publisher takes no responsibility with respect to such material. The publisher should not be liable for any special, consequential or exemplary damages resulting, in whole or in part, from the readers' use of, or reliance upon, this material.

Cover Design: John Lent

For information, address Delmar Publishers Inc.
3 Columbia Circle, Box 15-015
Albany, New York 12212-5015

Delmar Staff
Executive Editor: Michael McDermott
Project Editor: Carol Micheli
Production Coordinator: Teresa Luterbach

Printed in the United States of America
Published simultaneously in Canada
by Nelson Canada,
a division of The Thomson Corporation

10 9 8 7 6 5 4

Library of Congress Cataloging-in-Publication Data

Unibody collision repair / I-CAR
p. cm.
Includes index.
ISBN 0-8273-4671-9
1. Automobiles—Bodies—Maintenance and repair. I. Inter-Industry Conference on Auto Collision Repair.
TL255.U549 1992
629.26—dc20 91-34673
CIP

CONTENTS

PREFACE

The Inter-Industry Conference on Auto Collision Repair (I-CAR) is a nonprofit organization whose purpose is to promote high-quality practices in the collision repair industry. It is made up of vehicle manufacturers, collision repair shops, insurance companies, as well as independent appraisers, recyclers, educators and tool, equipment and supply manufacturers. Since its start is 1979, I-CAR has emerged as the leading training organization in the industry. This development responded to the widening technology gap and shortage of qualified repair personnel, as the industry converted rapidly to unibody designs.

We believe that quality repairs cost less in the long run. The unique designs and materials of unibody vehicles create many special repair considerations. Correct methods and materials are vitally important in producing high-quality, durable repairs. I-CAR's mission is to promote this goal as a standard of performance in the collision repair industry. Actually, the entire automobile industry is changing faster than ever before. This is especially true of the auto body repair and refinishing trade. The role of the body technician and refinisher has changed greatly in the past few years. The reason for this is the major shift by vehicle owners to unibody construction.

In unibody construction, the metal body panels are welded together making a structural unit as opposed to the traditional frame construction where the cosmetic steel body rests on top of a structural steel frame. Virtually all vehicles manufactured today and in the forseeable future will be unibody vehicles. This change in vehicle preference has had major impact on the duties, knowledge, procedures, and responsibilities of body technicians. Before the introduction of unibody vehicles, the body technician's primary concern was to straighten the frame and replace or fix damaged panels or sections. Today the body technician is responsible for a great deal more.

When working on most collision damaged unibody vehicles, it is necessary to have a complete knowledge of the mechanical components, including the operation, diagnosis of problems, and adjustment of the following: strut suspension system, independent rear wheel suspension, rack and pinion steering, and trailing rear axles.

In addition, the modern body technician must have a working knowledge of mechanical and suspension components, such as mechanical component mounting points, engine accessory systems (power steering, fuel systems, and brake systems, for example), removal of mechanical versus partial assembly, emission control systems and certification of repairs, front wheel drive systems, rear wheel drive systems, and heat and air-conditioning systems. Why does the modern body technician need to understand the functions of the vehicle's mechanical components? The answer is simple—most, if not all, are fastened to the body structure. This means that with most major collision damage, these parts must be removed by the technician before any body straightening work is undertaken. While the actual repairs to mechanical components might be done in the body by other persons or it might be "farmed out" to outside vendors, it is usually the responsibility of the body technician to reinstall the mechanical parts back onto the repaired body.

With the introduction of unibody vehicles, the body technician has had to learn about the use of new materials and techniques. The conventional body-over-frame vehicle was primarily made of low carbon or mild steel. These materials were usually welded or cut with an oxyacetylene gas torch. However, in unibody construction, high-strength steel is used and all vehicle manufacturers mandate in their service manuals that MIG welding must be used on all structural parts repair. Only the MIG welding procedures will maintain the structural integrity of the high-strength steel used in today's unibody vehicles. The gas torch should never be used on a unibody constructed vehicle.

All recommendations presented in this book are based upon research programs or upon tests conducted by laboratories, manufacturers, or selected collision repair facilities. If performed as detailed, these recommendations will provide the basis for a thorough, professional repair. For years, most American vehicle manufacturers recommended that replacement parts be installed at factory seams. Thanks to research, primarily by I-CAR, it was discovered that structural parts replacement at factory seams is not always practical. In some cases it requires extensive dismantling of the vehicle. In others, it simply was not an efficient repair operation. This realization led experts in the field to identify a need for the development of additional sound technical information and procedures on how to section structural components when it is not practical to replace them at factory seams.

This book provides general directions for collision damage repair using tested, effective procedures. Following them will help assure the reliability of the repair. I-CAR cannot accept responsibility for any individual repair, nor can it warrant to the quality of such repair. Anyone who departs from the instructions must first establish that neither personal safety nor the integrity of the repair of the vehicle is compromized by the choice of methods, tools, or supplies. I-CAR does not endorse nor recommend any brands or makes of vehicles, repair equipment or other products. However, we would like to thank the manufacturers who loaned us some photographs and information that appears in this book.

Unibody Collision Repair is directed toward knowledge and skills needed by collision repair technicians. An awareness of this same knowledge and skills is just as important for insurance people, recyclers, educators, tool and equipment manufacturers, sales personnel, and students. This knowledge will be useful in estimating repair work that needs to be done, as well as in judging the quality of the work completed. So, while the objectives focus primarily on performance outcomes for repair techicians, these same objectives define the learning that should take place by all who read and study this book.

ACKNOWLEDGMENTS

Grateful acknowledgment is made to the following companies for reference material and information used in the preparation of this book:

American Sikkens, Inc.
Binks Manufacturing Co.
Blackhawk Automotive, Inc.
Carborundum Abrasives Co.
Central Tools, Inc.
Century Manufacturing Co.
Chief Automotive System, Inc.
Chrysler Corp.
Clements National Co.
Continental Manufacturing, Inc.
Dataliner Inc.
Datawelder Inc.
DeVilbiss Company
Du Pont Company
Dur Mor, Inc.
Ford Motor Co.
General Motors
H & S Manufacturing Company, Ltd.
Herkules Equipment Co.
HTP America, Inc.
Kansas Jack, Inc.
Maaco Enterprises, Inc.
Mac Tools, Inc.
Martin-Senour Paint
Nissan Motor Co.
Paint Safe Products
PPG Industries, Inc.
Russel Thrall of Tynel Corp.
Sherwin-Williams Co.
3M
Toyota Motor Corp.

Thanks is extended to the following people for reviewing the manuscript and for their helpful comments:

Tony Jones
Cypress College
Cypress, CA 90706

George Lovejoy
Texas State Technical Inst.
Waco, TX 76705

Tony Passwater
I-CAR Great Lakes Regional Mgr.
Greenfield, IN 46140

Doug Roach
Collision Repair Specialist
San Jose, CA 95111

Ronnie Kendrick
Augusta Technical Inst.
Augusta, GA 30906

Jack D' Armond
Central Piedmont Community College
Charlotte, NC 28217

Jim Johnson
North Dakota State College of Science
Wahpeton, ND 58075

Mike Jund
Scott Community College of Science
Bettendorf, IA 52722

Charles W. Brown
Science Hill Vocational Center
Johnson City, TN 37601

Robert Hodgson
Miami Lakes Tech Ed Center
Miami Lakes, FL 33014

Robert A. Smith
Fox Valley Technical College
Appleton, WI 54913

Scott Berg
Norfolk, NE

Jay Reitz
Dauphin County Vo Tech School
Harrisburg, PA 17109

CHAPTER

1 Collision Repair and Refinishing Shop Procedures

OBJECTIVES

- Explain how the unibody vehicle came into being and what changes it has caused in the automotive industry.
- Know the meaning of a true professional technician.
- Identify the setup and innerworkings of a typical collision repair shop and its various personnel.
- Outline several control measures necessary when working with hazardous substances.
- Describe the various precautions for personnel safety that must be followed in a collision repair shop.
- Describe safety practices designed to avoid fire and explosions.
- Know the sources of professional training and certification available to collision repair shop personnel.

Almost all vehicles registered in the United States in 1977 were of the body-over-frame construction. By 1981, the ratio between body-over-frame and unibody vehicles was almost equal. Now, however, the percentage difference between the two is 98 to 2 in favor of unibodies. Just about every vehicle built in the foreseeable future will be a unibody (Figure 1-1).

The difference in construction between a body-over-frame and unibody vehicle is described in Chapter 2. But for now, it can be stated that the unitized body is constructed of structural panels that are welded together to form a single unit; this is generally done with spot welds. Each panel is dependent upon the panels attached to it. That means that when stress is applied to one of the panels, it will affect all other panels welded to it.

FIGURE 1-1 With the coming of unibody vehicles, the future is now.

Many of the mechanical parts, including the steering and suspension components, are mounted directly to the structural panels. When straightening the vehicle, it is important that these parts and panels are pulled back to their correct dimensions. Positioning of the panels must be precise to allow for steering and suspension systems to function correctly.

The introduction of the unibody vehicle was brought about by many factors. They include:

- Rising cost of gasoline
- Federal mandates for fuel-efficient vehicles
- Changing manufacturing methods
- Manufacturing cost controls
- Consumer demands
- Safety consciousness

The unitized automobile has had a drastic effect on the automotive repair industry. Unibody vehicles require:

- New estimating and damage analysis procedures
- New repair techniques
- More attention to details
- A willingness to keep up with the technical advancement within the industry
- A commitment to professionalism in the collision repair industry. That is, the industry must commit itself to taking greater pride in its work.

Keep in mind that consumers expect more from the vehicles they purchase than ever before. There are many reasons for this, including the following:

- New vehicle financing is averaging more than 48 months.
- The average age of all vehicles on the road is now more than seven years.
- The price of new vehicles keeps rising.
- Vehicle manufacturer warranties are getting longer.

Because of customer expectations, the collision repair shop and technician have a mutual obligation to:

1. Return the damaged vehicle to its pre-accident condition.
   - Restore durability, performance, and structural integrity
   - Perform a quality repair
2. Use full disclosure to inform the customer of the repair procedure (Figure 1-2).
   - How the vehicle will be repaired
   - What specifically will be done
   - Why it will be done
3. Complete the repairs as specified.
   - Use recommended procedures
   - Pay attention to detail

By a professional, we are referring to the attitude, the quality of work, and the image that a business projects to its customers. We are not implying that the repair industry is not being professional, only that the importance of professionalism is ever increasing as competition and the business climate get tougher. It is interesting to note that in Europe a shop cannot open without the presence of a "Meister" or "master" craftsperson. We do not have that in this country yet, but there is a definite trend in that direction.

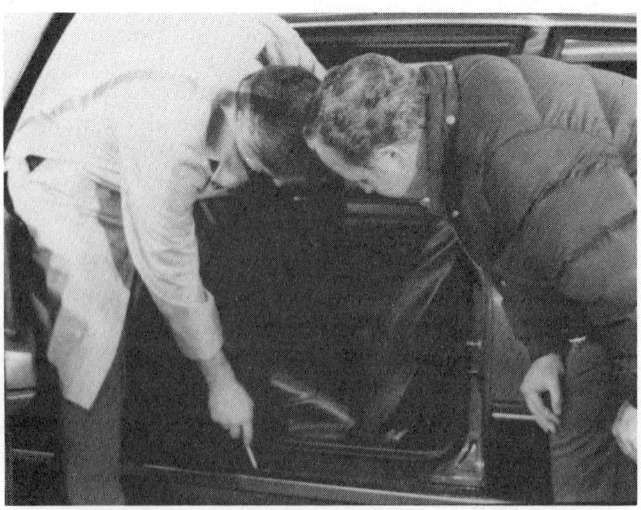

FIGURE 1-2 Communication with the vehicle owner is very important.

It must be kept in mind that the "true" professional (either technician or insurance personnel) is one who:

- Is customer oriented
- Is up-to-date on vehicle manufacturers' developments
- Is willing to keep up with advancements in the repair industry
- Pays attention to detail
- Ensures that his/her work is the best that it can be
- Participates in local, state, and national trade associations

The true professional must be aware of the present liability question; there have been several lawsuits recently in which improper repairs were blamed for personal injury. In fact, two trends within society increase the potential for liability exposure in the collision repair industry:

1. Consumer expectations for products and services are at an all-time high.
2. Intolerance for failure and willingness to sue are also at an all-time high.

At the same time, technicians are dealing with a more sophisticated vehicle that requires advanced technology and better skills and understanding of repair methods. Because the consumer's expectations with regard to quality and durability have increased, the industry must meet these higher expectations.

Collision repair facilities as well as insurance companies have the highest exposure to liability responsibility. The technician must set high standards personally and for the shop in terms of quality. Achieving these standards and employing the latest repair techniques will reduce the chances of being held liable.

There are several principles of liability the collision repair technician should be familiar with.

- *Due care.* Care must be used to do the work thoroughly and correctly.
- *Acts or omissions.* The shop and technician(s) are liable for any faulty work. The shop is also liable for failure to repair damage, which is acknowledged to exist, unless the customer specifically directs that it should not be repaired.
- *Failure to warn.* If knowledge is withheld from the customer about the repair, the shop, technician, and/or insurance personnel can be held liable.
- *Ignorance is no defense.* In the eyes of the law, the collision repair shop, technician, and insur-

ance personnel are expected to be knowledgeable about the latest available techniques for safe, sound repair. Failure to learn and apply those techniques, so as to produce a proper repair, could result in liability.

Pride of workmanship is an important hallmark of a professional craftsperson. Pride of workmanship also means that there will be less customer returns, improved customer relations, and reduced chances of costly liability exposure.

TYPICAL COLLISION REPAIR SHOP OPERATION

There are two basic work areas in most collision repair operations:

- Metalworking shop
- Refinishing or paint shop

Whatever the physical size of the overall collision repair shop, the work flow should be as continuous as on a factory production line. That is, the vehicle should move from its entrance into the shop to the metalworking area and on to the refinishing area. The vehicle should be thoroughly washed to remove mud, dirt, and water-soluble contaminants. Make sure it is completely dry before moving it to the metalworking area.

METAL SHOP

The metalworking area (Figure 1-3) is where the vehicle body is repaired. This damage can be either the result of a collision or deterioration. The repair tasks in this area of the shop are performed by collision repair technicians or mechanics and their helpers.

In most shops, the estimator makes an appraisal of the damage to the vehicle and how much it will cost to repair. Once the owner and/or the insurance company approves the appraisal, the vehicle is turned over to the shop supervisor for the start of the repair work. The supervisor and one of the technicians go over the repair estimate that was filled out by the shop manager or estimator.

In repairing any type of collision damage, the technician must first study and diagnose the damage that has occurred. Once the damage has been evaluated, it must be determined whether it will be cheaper to straighten or repair the damaged section or replace it. Is the cost of repair more or less than the cost of replacement parts plus the cost of installation? In some shops, the estimator, shop manager, or even the owner are involved in these decisions. However, the technician is usually consulted.

FIGURE 1-3 The metalworking area is the heart of the collision repair shop operation.

Today, because of unibody construction design, more and more collision repair shops are offering complete collision services such as wheel alignment, cooling system repairs, electrical system diagnosis, and diagnosing suspension damage (Figure 1-4). While many of these repairs are still done by so-called "specialty shops," the simple fact that these systems are commonly damaged in collisions makes it necessary for the collision repair technician to have a working knowledge of these systems.

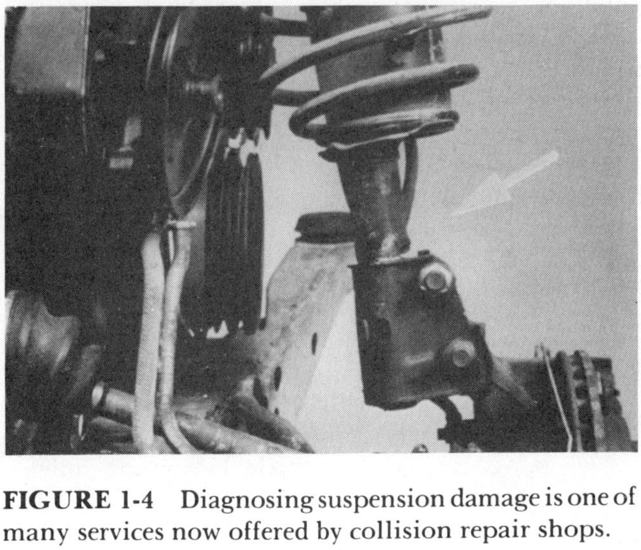

FIGURE 1-4 Diagnosing suspension damage is one of many services now offered by collision repair shops.

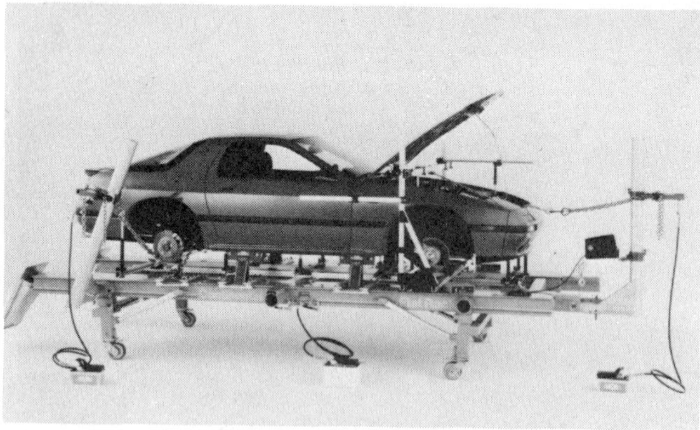

FIGURE 1-5 Typical hydraulic straightening equipment in use

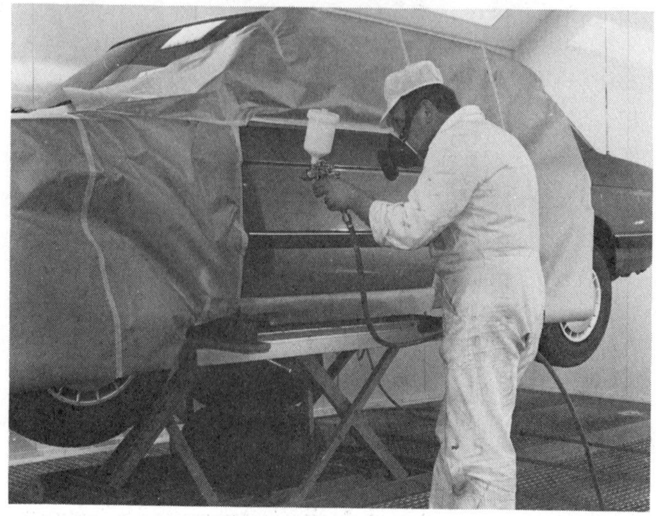

FIGURE 1-6 Masking paper and tape are used to protect areas of the vehicle where paint is unwanted.

It is also important to be able to correct and repair minor defects such as dents and rustouts. More and more collision repair shops are providing rustproofing services to their customers either before rustout occurs or after repairing the problem. Rustproofing is often a joint job of both the collision repair and paint technicians.

Today's professional collision repair technician must have a total knowledge of the tools used in the metalworking area. These include body repair hand tools, air and electric power tools, hydraulic body tools, welding equipment, and various types of hydraulic straightening equipment (Figure 1-5), as well as measuring devices.

PAINT SHOP

Auto refinishing or painting is a very important part of the auto repair business. Not only do collision-damaged vehicles have to be painted, but also many vehicles are repainted to enhance their beauty. New and used auto dealers repaint vehicles to attract buyers. Also, sometimes the owner gets tired of looking at the same old color. Often, the finish on many automobiles needs attention because it has been neglected or damaged by weather conditions. The paint shop area is where the vehicle is refinished. Here, a series of operations is performed on the vehicle as it passes through the following stages:

1. *Preparation.* In the initial preparation, the vehicle is made ready for the spraying operation. The preparation procedure includes:
   - Removing windshield wipers, emblems, nameplates, mirrors, and other trim pieces.
   - Cleaning the vehicle by wiping down each area with a degreaser or wax and grease remover. This will remove grease, road tar, and petroleum-based contaminants from the surface.
   - Machine and/or hand-sanding any chips and scratches.
   - Cleaning both the interior and exterior carefully to remove all dust and dirt.
   - Inspecting all surfaces to be sure that they are properly cleaned before the vehicle is masked.
   - Masking to protect the parts of the vehicle that are not to be painted—windows, chrome, lights, and vinyl tops (Figure 1-6).

   In large shops, the sanding might be handled by one technician, while the masking is performed by another. In small shops, all of the preparation jobs are usually done by the same person.
2. *Spraying.* When the vehicle moves into the spray booth, the painter takes over. Final cleaning is done at this point using tack rags. If a sealer is to be used, it is applied next, followed by the topcoat. Masking is removed after the paint has completely dried.
3. *Drying.* Some shops use drying equipment to speed up the drying. This is especially useful in drying enamel. In some paint shops the drying operation may be done with portable equipment in the spray area (Figure 1-7).
4. *Final Detailing.* After the finish is dried, the vehicle moves to the final detailing area. Compounding or polishing of topcoats is done in the final detailing area, along with other tasks such as polishing chrome, cleaning glass, removing overspray, installing trim, vacuuming the interior, and cleaning the vinyl top and tires.

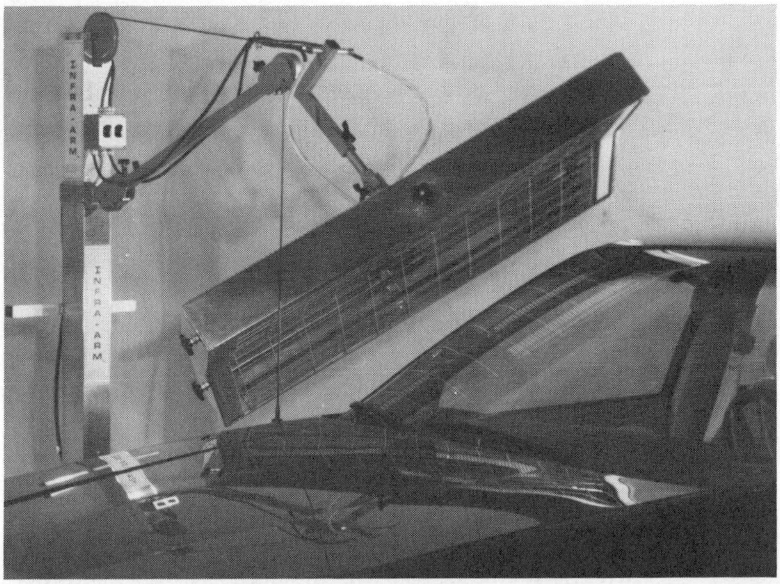

FIGURE 1-7 Examples of portable drying equipment

OTHER COLLISION REPAIR SHOP PERSONNEL

Foremost in any collision repair business are the collision repair and paint technicians and their helpers, but there are other jobs that must be done as well. The following paragraphs describe the personnel that handle these jobs.

Shop Owner and/or Manager. The shop owner must be concerned with all phases of work performed in the shop. In smaller shops, the owner and shop manager are usually the same person. In large operations or dealerships, the owner might hire a shop manager to supervise the operation of the shop. In all cases, the person in charge should understand all of the work done in the shop as well as its business operations.

Parts Manager. The parts manager is in charge of ordering all parts (both new and salvaged), receiving all parts, and seeing that they are delivered to the ordering technician. Since not every collision repair shop has an actual parts manager, the task of ordering parts can fall on each employee at one time or another.

To locate and order the correct parts, it is important that the collision repair shop collect all of the necessary information about the vehicle (Figure 1-8). Not only is accurate ordering of parts vital to the job production, it is equally important to the job cost and profit. For example, the price of a can of paint can vary as much as 200 percent from one color to another.

In order to purchase the correct parts, it is sometimes necessary to consult the vehicle identification number, or VIN, plate. For example, to get a wood grain overlay to match a certain vehicle, the parts department might have to know which assembly plant built the vehicle. The plant can be identified using the VIN plate. The VIN also reveals the model year, body type, engine size, and other data helpful both in making estimates and in ordering parts. The VIN plate is described in more detail in Chapter 2.

There are frequent "running changes" made by the vehicle manufacturers. Vehicles built in the same model year can have one of several individual part applications. The only way to keep track of all of the changes is

FIGURE 1-8 Ordering parts is a vital task in the collision repair shop.

FIGURE 1-9 Many collision repair shops rely on independent towing services.

for the manufacturer to note the build date when a different part is first installed. Use the vehicle certfication label to locate the build date.

Bookkeeper. The bookkeeper keeps the shop's record books, prepares invoices, writes checks, pays bills, makes bank deposits, checks bank statements, and takes care of tax payments. Many shops hire an outside accountant to perform these jobs.

Office Manager. The office manager's duties may include answering the telephone, being a receptionist, typing, and filing letters, estimates, and receipts. In many small shops, the office manager also acts as the parts manager and bookkeeper.

Tow Truck Operator. Larger shops generally have trained tow truck operators to operate their wrecker(s). Rather than own this expensive piece of equipment, many smaller shops depend on independent towing services (Figure 1-9).

Estimator. Estimating the cost of damage repair and refinishing on various vehicles is important to the success of a collision repair shop. Estimates will range from small dents to major wrecks. Refinishing work must be estimated, too. The person who determines the cost value of the damage plus the price required to repair the damaged vehicle is called an estimator or damage appraiser (Figure 1-10).

In addition to determining the cost of repairing and refinishing, the estimator must work with insurance companies' appraisers. The appraiser is the insurance company's representative who estimates a vehicle's damage and authorizes payment to the shop. All estimates or damage appraisals, as they are sometimes

FIGURE 1-10 An estimator at work

called, must be in writing and signed by an insurance company's appraiser, the customer, and the shop's estimator. The estimate then becomes a legal commitment among the parties involved as to work to be done, cost, and method of payment. More details on estimating and the importance of the estimator's function in a collision repair shop is described in Chapter 4.

Except in large operations, only one person in the shop should make estimates and sign them. This person can be the owner, the shop manager, or someone else specifically assigned to this job.

SHOP PUBLICATIONS

There are several publications that shop personnel should become familiar with. The most important of these to the technician is the manufacturer's body dimension manual (Figure 1-11). On the other hand, the

FIGURE 1-11 Typical body dimension manuals

crash estimating guides are vital to the estimator; the replacement catalogs are very important to the estimator and parts manager; and all shop personnel should be interested in the trade publications. Especially for the technician is I-CAR's *Advantage*. All professional handbooks, trade publications, manuals, and catalogs should be available to all shop personnel.

SHOP RECORDS

The most frequently used shop forms are work orders or damage reports. A work order or damage report form is completed before actual work is started. It is the basis of the agreement between the shop and the customer. Even in a small collision repair shop where the owner or manager deals directly with the customer, problems can still arise because information is not written down; for example, parts that are used on the job are not recorded or requested, operations are not performed, or too many operations are performed, which destroys the quote the customer was given. No matter how careful anyone is, problems will occur unless the work order is complete and the information is recorded properly. This problem is compounded as the shop grows in size. The use of computers to store and retrieve shop records is becoming increasingly popular (Figure 1-12).

Various work orders or damage reports are available from different sources. With the inexpensive printing that is available today, it takes little effort to create a work order or damage report that addresses the shop's needs. The best work order not only allows the technician to ask the needed questions, it also has boxes to be checked when the work is completed and when parts are ordered or in stock. Prices can be written beside each operation or part so billing will be accurate.

FIGURE 1-12 Computers provide an excellent means of storing shop records.

Keep in mind that many states have information disclosure laws to protect consumers from fraud. Written job estimates must be furnished to the customer by the collision repair shop. If the cost will be higher than the original estimate, additional approval must be received from the customer. Also any replaced parts might have to be returned to the customer. More information on the use of both estimates and work orders can be found in Chapter 4.

Other forms that are used in the collision repair shop include the following:

- *Parts Requisition.* To order parts, the technician writes the names of what is needed in the requisition section of the work order. Then the technician takes the order to the parts department where it is filled.

  Parts managers must be sure to keep an adequate number of parts in stock. When the stock gets low, new parts are ordered. Occasionally it is necessary to order special parts. To ensure customer and shop satisfaction, these special orders must always be coordinated between the service and parts departments. Parts can be ordered by using standard or computerized catalogs. In some shops, the technician or painter might also have the job of ordering parts.
- *Dispatch Sheet.* A dispatch sheet, or work schedule, keeps track of dates when the work is to be completed. Some dispatch sheets follow the job through each step of the repair and refinish process. Whenever a customer calls, the dispatch sheet is consulted. This sheet is usually posted in the office of the shop or wherever it is convenient for the service manager.

Once all work is completed, the technician turns in the work order to the service manager, or estimator, who adds the labor charges. The bill includes all labor charges, parts charges, sales tax, and any other charges. The billing department totals the amount and receives payment from the customer or insurance company.

COLLISION REPAIR SHOP SAFETY PROCEDURES

True professionalism begins with the way a technician performs his/her given work tasks and how safely they are accomplished. It is very important to remember that safety is a prime concern in all collision repair shops. Carelessness and the lack of safety habits cause accidents. Accidents have a far-reaching effect, not only on the victim, but also on the victim's family and friends. More important, accidents can cause serious injury, or death. Therefore, it is the obligation of all shop employees and the employer to develop a safety program to protect the health and welfare of those involved.

In the following chapters of this book, the text contains special notations labeled **SHOP TALK, CAUTION,** and **WARNING.** Each one is there for a specific purpose. **SHOP TALK** gives added information that will help the collision repair technician to complete a particular procedure or make a task easier. **CAUTION** is given to prevent the technician from making an error that could damage the vehicle. **WARNING** reminds the technician to be especially careful of those areas where carelessness can cause personal injury. The following text contains some general **WARNINGs** that should be followed when working in a collision repair shop.

In addition to the safety procedures given in future chapters, there are certain rules regarding manufacturer's warnings, government regulations, personal safety, work area safety, and tool and machine safety that must be followed at all times.

MANUFACTURER'S WARNINGS AND GOVERNMENT REGULATIONS

Most of the products used in a collision repair shop carry warning and caution information that must be read and understood by all users. Likewise, all federal (including Occupational Safety and Health Administration [OSHA], Mine Safety and Health Administration [MSHA], and National Institute for Occupational Safety and Health [NIOSH]), state, and local safety regulations should not only be fully understood, but also strictly observed.

In collision repair and refinishing shops, hazardous wastes are generated. Every employee in the shop is protected by Right-to-Know laws. These laws started with OSHA's Hazard Communication Standard published in 1983. This document was originally intended for chemical companies and manufacturers that require employees to handle potentially hazardous materials in the workshop. Since then, the majority of states have enacted their own Right-to-Know laws and the federal courts have decided that these regulations should apply to all companies, including the auto collision repair and refinishing professions.

The general intent of the law is for employers to provide their employees with a safe working place as it relates to hazardous materials. Specifically, there are three areas of employer responsibility:

1. *Training/Educating Employees.* All employees must be trained about their rights under the legislation, the nature of the hazardous chemicals in their workplace, the labeling of chemicals, and the information about each chemical posted on material safety data sheets (MSDS). These sheets, available from the product manufacturers, detail chemical composition and precautionary information for all products that can present a health or safety hazard. Employees must be familiarized about the general uses, protective equipment, accident or spill procedures, and any other information regarding the safe handling of the hazardous material. This training must be given to employees annually and provided to new employees as part of their job orientation.
2. *Labeling/Information about Potentially Hazardous Chemicals.* Every hazardous material must be properly labeled to indicate what health, fire, or reactivity hazard it poses and what protective equipment is necessary when handling it. The manufacturer of the hazardous materials must provide all warnings and precautionary information (material safety data sheets), which must be read and understood by the user before application. Attention to all label precautions is essential for the proper use of the chemical and for prevention of hazardous conditions.
3. *Record Keeping.* Shops must maintain documentation on the hazardous chemicals in the workplace, proof of training programs, records of accidents and/or spill incidents, satisfaction of employee requests for specific chemical information via the MSDS, and a general Right-to-Know compliance procedure manual utilized within the shop.

Complete EPA lists of hazardous wastes can be found in the Code of Federal Regulations. Materials and wastes of most concern to the technician are organic solvents: wastes that contain heavy metals, especially lead, and ignitable, corrosive, and/or toxic materials. It should be noted that no material is considered hazardous waste until the shop is finished using it and ready to dispose of it. For instance, a caustic cleaning solution with a heavy concentration of lead in the cleaning tank is not considered hazardous waste until it is ready to be replaced. Now that the shop is ready to dispose of it, it is hazardous waste and must be handled accordingly. The EPA says it is the owner's responsibility to determine whether the waste is hazardous, but the owner must have adequate test results to support the shop's claim.

Testing for hazardous wastes can be done by any qualified laboratory that performs tests on drinking water. They should be capable of performing tests for hazardous materials. The shop's owner(s) should contact them to see how they would like the samples taken and shipped to give accurate results.

Many shops use "full-service" haulers to remove hazardous waste from the property. Besides hauling the hazardous waste away, the hauler will also take care of all the paperwork, deal with the various government agencies, even advise the shop on how to recover the disposal costs. However, contracting with a hauling firm does not relieve the shop from knowing and understanding the laws and regulations. But it can go a long way toward reducing much of the concern.

WARNING

The collision repair shop is ultimately responsible for the safe disposal of hazardous wastes, even after they leave the shop. Be sure that any hauling contract is in writing so as to leave nothing to chance. In the event of an emergency hazardous waste spill, contact the National Response Center (800-424-8802) immediately. Failure to do so can result in a $10,000 fine, a year in jail, or both.

PERSONAL SAFETY

Personal appearance and conduct can help prevent accidents. Guidelines for personal safety are as follows:

- The eyes are sensitive to dust and flying particles from grinding and to mists and vapors from spraying. Such exposure could cause severe pain and possibly the loss of sight. Whenever such external risks exist, a full-face shield and a good pair of safety glasses or chemical splash goggles should be worn (Figure 1-13). Remember that eyes are irreplaceable.

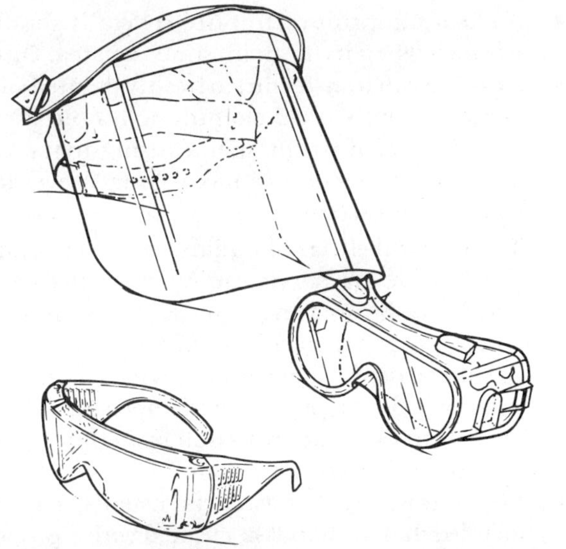

FIGURE 1-13 Recommended eye protection

- Loose clothing, unbuttoned shirt sleeves, dangling ties, loose jewelry, and shirts hanging out are very dangerous. Instead, wear approved shop coveralls or jumpsuits (Figure 1-14). Pants should be long enough to cover the top of the shoes. Keep them mended and tear free.

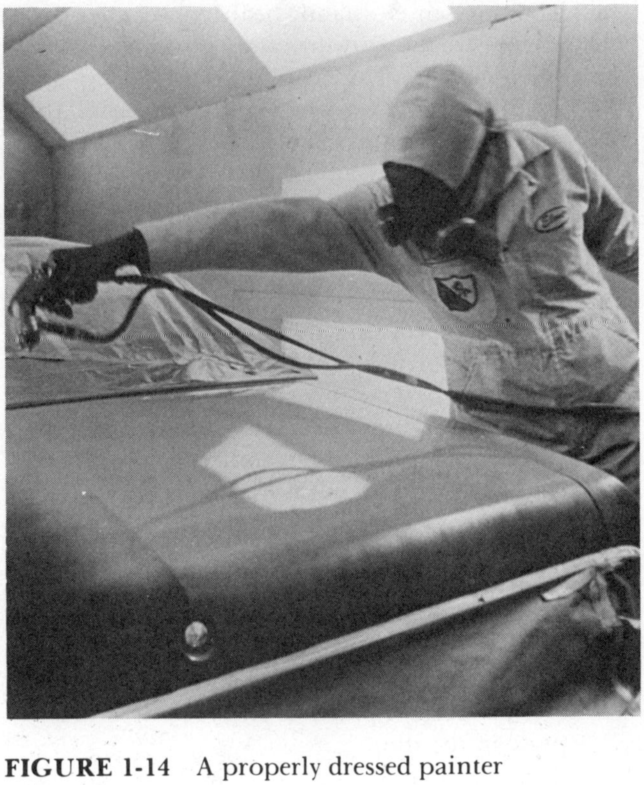

FIGURE 1-14 A properly dressed painter

- A clean jumpsuit or lint-free coveralls should be worn when in the paint spraying area. Dirty clothing can transfer dirt to freshly painted surfaces. Solvent-soaked clothing will hold these chemicals against the skin, causing irritation or a rash. Make sure you wear long sleeves for complete protection.
- The harmful effects of liquids, undercoats, and finishes on the hands can be prevented very effectively by wearing proper gloves. Impervious gloves, such as the nitrile latex type, should be used when working with solvents or two-component primers and topcoats. These gloves offer special protection from the chemicals found in two-component systems.
- When washing hands, it is usually recommended that the hands be cleaned with a proper hand cleaner. At the end of a day's work, it is wise to oil the skin a little by applying a good silicone-free skin cleaner. Do not use thinner as a hand cleaner.
- Wear shoes that have nonslip soles to prevent falls. In addition, good shoes provide support and comfort for someone who is standing for a long time.
- Be sure to tie long hair securely behind the head before beginning to work on a vehicle. The hair also must be protected against dust and sprays. To keep hair clean (and healthy) wear a cap at all times in the work area and a protective painter's stretch hood in the spray booth. Bump caps should be worn when working beneath the hood or under the vehicle.
- Keep clothing away from moving parts when an engine or a machine is running. Any loose or hanging clothing, such as shirttails, ties, lapels, cuffs, or scarves, creates a risk of getting wrapped up in moving parts of the vehicle or machinery, causing serious bodily injury. No type of jewelry should be worn.
- When lifting and carrying objects, bend with the knees, not the back. Also, do not bend the waist when lifting. Remember, heavy objects should be lifted and moved with the proper equipment for the job.
- Never smoke while working on any vehicle or machine in the shop.
- Proper conduct can also help prevent accidents. Horseplay is not fun when it sends someone to the hospital. Such things as air nozzle fights, creeper races, or practical jokes do not have any place in the shop.
- Do not risk injury through the lack of knowledge; use shop machinery or perform work operations only after receiving proper instruction.
- Metalworking, the piercing noise of sanding, and the radio blaring full blast can all make it difficult to hear anything else. It is enough to deafen a person—and that is exactly what will happen if proper precautions are not taken. When in metalworking areas, wear earplugs or earmuffs to protect the eardrums from damaging noise levels.
- A welding helmet or welding goggles with the proper shade lens must be worn when welding. These will protect the eyes and face from flying molten pieces of steel and harmful light rays.
- To prevent serious burns, avoid contact with hot metal parts such as the radiator, exhaust manifold, tailpipe, catalytic converter, and muffler.
- It is most important to use jack stands whenever a procedure requires getting under the vehicle. Never trust jacks alone.
- When it is necessary to drive a vehicle in a shop, watch out for other vehicles and people. It is best to have someone act as a guide. Leave the window open and the radio off, so that the guide's directions can be heard.

WORK AREA SAFETY

It is very important that the work area be kept safe.

- All surfaces should be kept clean, dry, and orderly. Any oil, coolant, or grease on the floor can cause slips that could result in serious injuries. To clean up oil, be sure to use a commercial oil absorbent.
- Keep all water off the floor; remember that water is a conductor of electricity. A serious shock hazard will result if a live wire happens to fall into a puddle in which a person is standing. Be sure the floor is perfectly dry when using electric power tools.
- Make sure that aisles and walkways are kept clean and wide enough for a safe clearance. Cluttered walking areas contain items waiting to cause accidents.
- Use anti-skid floor strips on the floor area where the technician normally stands and mark off each work area. Make certain the work area is well lighted and ventilated. Provide for adequate work space around any machines. The work area should not be readily accessible to anyone except the technician.
- Operate the engine only in a well-ventilated area to avoid the danger of carbon monoxide

(CO). Most shop exhaust systems carry the exhaust away from a vehicle directly to the outside of the shop.

- Space heaters used in some shops can also be a serious source of CO and therefore should be periodically inspected to make sure they are adequately vented and do not become blocked.
- Proper ventilation is very important in areas where caustics, degreasers, undercoats, and finishes are used. Ventilation can be by means of an air-changing system, extraction floors, or central dust extraction combined with spraying area walls. The latter is effective for removing dust from abrasives.
- There should be a list of emergency telephone numbers clearly posted next to the telephone. These numbers should include a doctor, hospital, and fire and police departments. Also the work area should have a first-aid kit for treating minor injuries. This kit should include sterile gauze, bandages, scissors, and other related items. Facilities for flushing the eyes should also be in or near the shop area.
- When diagnosing an engine, set the parking brake when working on the vehicle. If it is an automatic transmission, set it in park unless instructed otherwise for a specific service operation. If it is a manual transmission, it should be in reverse (engine off) or neutral (engine on) unless instructed otherwise for a specific service operation.
- Gasoline is a highly flammable liquid that vaporizes rapidly. Always keep gasoline and diesel fuel in approved safety cans and never use them to wash hands or tools.
- Oily rags should also be stored in an approved metal container. When these oily, greasy, or paint-soaked rags are left lying about or are stored improperly, they are prime candidates for spontaneous combustion, that is, fires that start by themselves.
- Check to be sure that all drain covers are snugly in place. Open drains have caused many toe, ankle, and leg injuries.
- Keep workbenches clean and orderly. When things are stacked on top of benches and tables, they can easily fall. Generally, they cause a cluttered work area, which invites injury.
- Tools should be hung up when not in use. Roll up idle hoses and air lines so they do not create traffic hazards; automatic rewinding wheels can be used.
- Floor jacks, bumper jacks, jack stands, and creepers should be kept in one area, out of aisles and walkways.
- Use only approved explosion-proof equipment in hazardous locations. A UL (Underwriters Laboratories) approved drum transfer pump along with a drum vent should be used when working with drums to transfer chemicals.
- Keep all solvent containers closed, except when pouring.
- Handle all solvents (or any liquids) with care to avoid spillage. Extra caution should also be used when transferring flammable materials from bulk storage. The most important thing to remember is to make sure the drum is grounded and that a bond wire connects the drum to a safety can (Figure 1-15). Otherwise, static electricity can build up enough to create a spark that could cause an explosion.
- Discard or clean all empty solvent containers. Solvent fumes in the bottom of these containers are prime ignition sources. Remember, never use gasoline as a cleaning solvent.

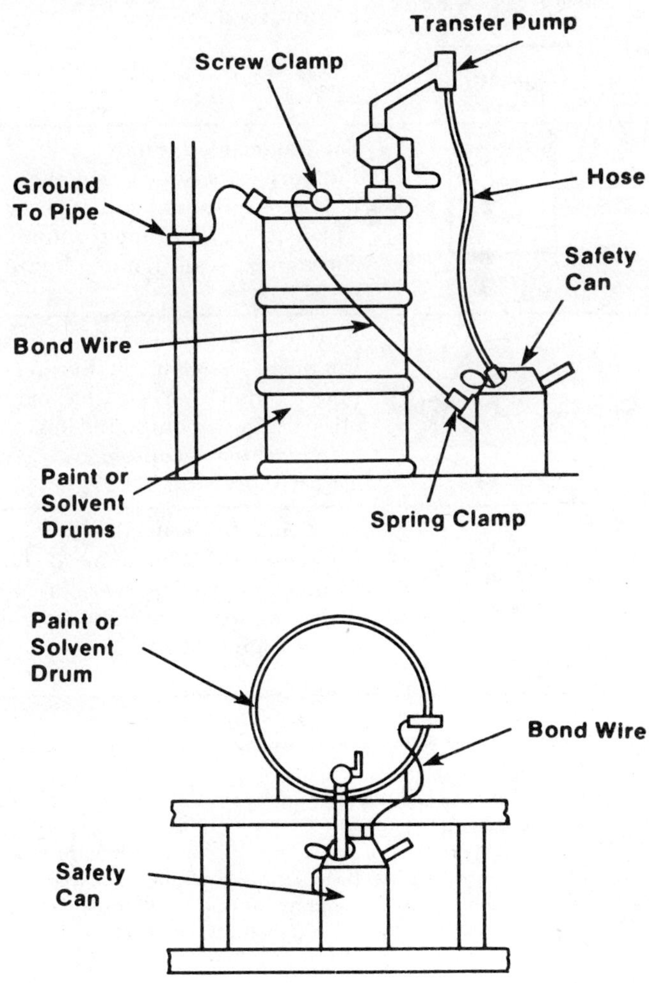

FIGURE 1-15 Safe method of moving flammable liquids from a drum to a portable safety can

- Do not light matches or smoke in the paint spraying area, and make sure that hands and clothing are free from solvent when lighting matches or smoking in other areas of the shop where smoking or an open flame is permitted.
- Paints, thinners, solvents, pressurized containers, and other combustible materials used in the shop must be stored in approved and designated storage cabinets or rooms. Storage rooms should have adequate ventilation. Never have more than one day's supply of paint outside of approved storage areas.
- The connections on all drums containing flammable and combustible liquids must be vapor- and liquid-tight.
- Know where the fire extinguishers are and what types of fires they put out (Table 1-1). A multipurpose dry chemical fire extinguisher will put out ordinary combustibles, flammable liquids, and electrical fires. In case of a gasoline fire, do not put water on it; the water will just spread the fire. Use a fire extinguisher to smother the flames. Remember, during a fire, never open doors or windows unless it is absolutely neces-

TABLE 1-1: GUIDE TO FIRE EXTINGUISHER SELECTION

	Class of Fire	Typical Fuel Involved	Type of Extinguisher
Class A Fires (green)	**For Ordinary Combustibles** Put out a class A fire by lowering its temperature or by coating the burning combustibles.	Wood Paper Cloth Rubber Plastics Rubbish Upholstery	Water*[1] Foam* Multipurpose dry chemical[4]
Class B Fires (red)	**For Flammable Liquids** Put out a class B fire by smothering it. Use an extinguisher that gives a blanketing, flame-interrupting effect; cover whole flaming liquid surface.	Gasoline Oil Grease Paint Lighter fluid	Foam* Carbon dixoide[5] Halogenated agent[6] Standard dry chemical[2] Purple K dry chemical[3] Multipurpose dry chemical[4]
Class C Fires (blue)	**For Electrical Equipment** Put out a class C fire by shutting off power as quickly as possible and by always using a nonconducting extinguishing agent to prevent electric shock.	Motors Appliances Wiring Fuse boxes Switchboards	Carbon dioxide[5] Halogenated agent[6] Standard dry chemical[2] Purple K dry chemical[3] Multipurpose dry chemical[4]
Class D Fires (yellow)	**For Combustible Metals** Put out a class D fire of metal chips, turnings, or shavings by smothering or coating with a specially designed extinguishing agent.	Aluminum Magnesium Potassium Sodium Titanium Zirconium	Dry powder extinguishers and agents only

*Cartridge-operated water, foam, and soda-acid types of extinguishers are no longer manufactured. These extinguishers should be removed from service when they become due for their next hydrostatic pressure test.

Notes:

(1) Freezes in low temperatures unless treated with antifreeze solution, usually weighs over 20 pounds, and is heavier than any other extinguisher mentioned.

(2) Also called ordinary or regular dry chemical (sodium bicarbonate).

(3) Has the greatest initial fire-stopping power of the extinguishers mentioned for class B fires. Be sure to clean residue immediately after using the extinguisher so sprayed surfaces will not be damaged (potassium bicarbonate).

(4) The only extinguishers that fight A, B, and C classes of fires. However, they should not be used on fires in liquefied fat or oil of appreciable depth. Be sure to clean residue immediately after using the extinguisher so sprayed surfaces will not be damaged (ammonium phosphates).

(5) Use with caution in unventilated, confined spaces.

(6) May cause injury to the operator if the extinguishing agent (a gas) or the gases produced when the agent is applied to a fire is inhaled.

sary; the extra draft will only make the fire worse. A good rule is to call the fire department first and then attempt to extinguish the fire. Standing 6 to 10 feet from the fire, hold the extinguisher firmly in an upright position. Aim the nozzle at the base and use a side-to-side motion, sweeping the entire width of the fire. Stay low to avoid inhaling the smoke. If it gets too hot or too smoky, get out. Remember, never go back into a burning building for anything.

- Keep customers' vehicles away from immediate work areas to prevent them from being damaged while waiting to be worked on. If vehicle covers are available, by all means use them.
- Customers should never be allowed into the shop's work areas.

TOOLS AND MACHINES

The collision repair technician and painter must observe the following hand and power tool safety guidelines:

- Hand tools should always be clean and in workable condition. Greasy or oily hand tools can easily slip out of your grasp, possibly causing injury.
- Check all hand tools for cracks, chips, burrs, broken teeth, or other dangerous conditions before using them. If any tools are defective, do not use them.
- Be careful when using sharp or pointed tools that can cause injury. If a tool should be sharp, make sure it is sharp.
- Do not use any hand tools for any job other than that for which they were specifically designed.
- Do not carry screwdrivers, punches, or other sharp hand tools in your pockets. It is possible to injure yourself or damage the vehicle being worked on.
- When using an electric power tool, make sure that it is properly grounded and check the wiring for cracks in the insulation, as well as for bare wires. Also, when using electric power tools, never stand on a wet or damp floor.
- Do not operate a power tool or machine without its safety guard(s).
- Disconnect electrical power before performing any service on the machine or tool (Figure 1-16).
- When doing any power grinding, cutting, sanding, or similar operation, always wear safety glasses. When using power equipment on small parts, never hold the part in your hand. It could slip. Use vise grips or a bench vise instead.

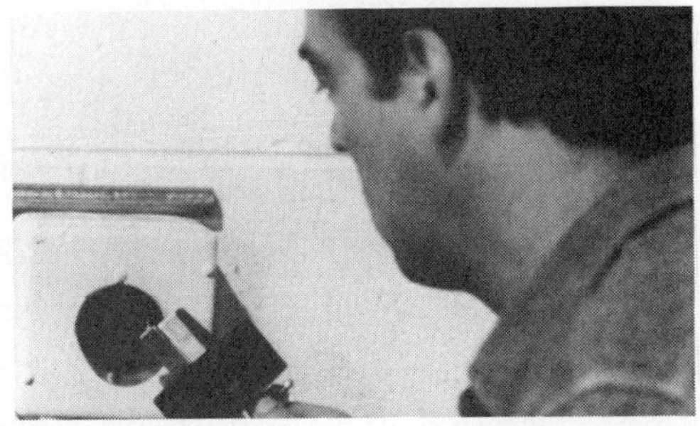

FIGURE 1-16 Always disconnect the power before doing any kind of work on a machine or tool.

- Before plugging in any electric tool, make sure the switch is off to prevent serious injury. When you are through using it, turn it off for the next person.
- Do not attempt to use the machine or tool beyond its stated capacity or for operations requiring more than the rated horsepower of the motor. Never use a tool or machine for operations it was not designed for.
- Keep hands away from moving parts when the machine or tool is under power. Never clear debris when the machine is under power and never use your hands.
- Never overreach. Maintain a balanced stance to avoid slipping.
- Utmost caution should accompany the use of compressed air. Pneumatic tools must be operated at the pressure recommended by their manufacturer. The downstream pressure of compressed air used for cleaning purposes must remain at a pressure level below 30 psi. Do not use compressed air to clean clothes. Even at low pressure, compressed air can cause dirt particles to become embedded in the skin, which can result in infection.
- Store all parts and tools properly by putting them away neatly where people will not trip over them. This practice not only cuts down on injuries, it also reduces time wasted looking for a misplaced part or tool.
- When working with a hydraulic press, make sure that hydraulic pressure is applied in a safe manner. It is generally wise to stand to the side when operating the press. Always wear safety glasses.
- If the shop has a hydraulic lift, be sure to read the instruction manual before using it. Check the pads to see that they are making proper contact with the frame. Then raise the vehicle

about 6 inches and shake it to make sure it is well balanced on the lift. If there are any rattling or scraping sounds, it means that the vehicle is not positioned properly. If this happens, lower the lift and realign the pads to the vehicle. Test it again as previously described. Then, after lifting the vehicle to full height, put the safety catch on before working underneath the vehicle. Never permit anyone, technician or customer, to remain in the vehicle while it is being lifted.

- When welding or cutting near vehicle interiors, remove seats and carpets. Always have a pail of water and a fire extinguisher nearby.

PAINT SHOP SAFETY

Many paints and paint-related products are known to present health hazards unless proper precautions are taken during their use. The issue of paint shop safety has grown more important in recent years, as more and more vehicles are refinished with urethane enamels, which contain isocyanates. So the amount of exposure to chemicals has increased, and some of the chemicals used are more toxic. These chemicals can cause nose, throat, and lung irritation. They can also lead to acute respiratory attacks. Symptoms can include shortness of breath, tightness in the chest, dizziness, nausea, abdominal pain, and vomiting.

Respirators are needed in the paint shop even though adequate ventilation is provided. There are three primary types of respirators available to protect technicians: the hood or air-supplied respirator, cartridge filter respirator, and dust or particle respirator.

AIR-SUPPLIED OR HOOD RESPIRATORS

A NIOSH-approved air-supplied respirator provides protection from sensitization and other dangers of inhaling isocyanate paint vapors and mists, as well as from hazardous solvent vapors.

The fresh air supply respirator is comfortable to wear and does not require fit testing (Figure 1-17). It consists of a half mask, full facepiece, hood, or helmet, to which clean, breathable air is supplied through a hose from a separate air source.

The fresh air supply respirator may include a self-contained oilless pump to supply air to either one hood or two half mask respirators. The pump's air inlet must be located in a clean air area. Some shops mount the pumps on an outside wall, away from the dust and dirt generated by shop operations. If shop-compressed air is used, it must be filtered with a trap and carbon filter to remove oil, water, scale and odor. The air supply must have a valve to match air pressure to respirator equipment and an automatic control to sound an alarm or shut down the compressor in case of overheating and contamination. (Overheating frequently causes carbon monoxide contamination of the air supply.)

FIGURE 1-17 Typical fresh air supply respirator

Remember that the air source for a fresh air supply respirator must be located in a clean environment, outside of the spray area.

CARTRIDGE FILTER RESPIRATORS

If the refinishing system that is sprayed contains no isocyanates, an air-purifying cartridge filter respirator with organic vapor cartridges and prefilters can be used. These respirators protect against vapors and spray mists of nonactivated enamels, lacquers, and other non-isocyanate materials. The cartridge filter respirator consists of a rubber facepiece designed to conform to the face and form an airtight seal. It includes replaceable prefilters and cartridges that remove solvent and other vapors from the air. The respirator also has intake and exhaust valves, which ensure that all incoming air flows through the filters. Because the eyes are very susceptible to irritation, approved safety goggles should also be worn.

It is very important that an air-purifying cartridge filter respirator fits securely around the edge of the face to prevent contaminated air from leaking into the breathing area. To check this, a qualitative fit test should be done prior to using the respirator, performing both negative and positive pressure checks. To check for negative pressure, the wearer should place the palms of his or her hands over the cartridges and inhale. A good fit will be evident if the facepiece collapses onto the wearer's face. To perform a positive pressure check, the wearer covers up the exhalation valve and exhales. A proper fit is evident if the facepiece billows out without air escaping from the mask. Another form of quantitative fit testing consists of exposing amyl acetate (banana oil) near the seal around the face. If no odor is detected, a proper fit is evident.

Cartridge filter respirators are available in several sizes. Users of this type of respirator should be aware that facial hair might prevent an airtight seal, presenting a health hazard. Therefore, refinishers with facial hair should use a fresh air supply respirator. Remember that cartridge filter respirators should be used only in well-ventilated areas. They must not be used in environments containing less than 19.5 percent oxygen.

To maintain the cartridge filter respirator, keep it clean and change the prefilters and cartridges as often as directed by the manufacturer. Here are a few other maintenance tips:

- Replace the prefilters when it becomes difficult to breathe through the respirator.
- Replace the cartridge at proper intervals, and always at the first sign of solvent odor.
- Regularly check the mask to make sure it does not have any cracks or dents.
- Store the respirator in an airtight container.
- Follow the manufacturer's instructions to ensure proper maintenance and fit.

DUST OR PARTICLE RESPIRATORS

To protect against dust from sanding, use a dust respirator (Figure 1-18). Sanding operations create dust that can cause bronchial irritation and possible long-term lung damage if inhaled. Protection from this health hazard is necessary; a NIOSH-approved dust respirator should be worn whenever sanding operations are performed. Follow the instructions provided with the dust respirator to ensure proper fit. Remember that dust masks do not protect against vapors and spray mists.

Many shops are installing so-called "dustless" sanding machine systems. Depending on the system, vacuum pumps, vacuum pullers, brush motors, or turbine motors can be used to provide sufficient air volume and/or velocity to pull airborne sanding dust through either holes in a special sanding pad or a shroud that entirely surrounds the sanding pad. Some systems run constantly, others are started "on-demand" by plugging a vacuum hose into the vacuum outlet or pushing

FIGURE 1-18 Examples of dust respirators

a button at the sanding tool end of the hose. By using a dustless system, the shop can comply with OSHA's airborne dust standards and eliminate costly, nonproductive cleanups. Some dustless system manufacturers claim that their machines can trap over 99 percent of the toxic dust created by sanding lead- or chrome-based automotive paints and primers.

PROFESSIONAL CERTIFICATION

Just as doctors, nurses, accountants, electricians, and other professionals are licensed or certified in order to practice their profession, collision repair and refinishing technicians can also be certified. Certification protects the general public and the practitioner or professional. It assures the general public and the prospective employer that certain minimum standards of performance have been met. Many employers now expect their collision repair and refinishing technicians to be certified. The certified technician is recognized as a professional by the public, employers, and peers. For this reason, the certified technician usually receives higher pay than one who is not certified.

ASE CERTIFICATION

Collision repair and refinishing technicians can get certified in one or more technical areas by taking and passing written certification tests. The National Institute for Automotive Service Excellence (ASE) offers a voluntary certification program that is recommended by the major vehicle manufacturers in the United States. The Body Repair and Painting and Refinishing tests contain 40 questions in various areas.

Craftspeople who pass the written tests are awarded a certificate, shoulder emblem (Figure 1-19), and other credentials attesting to their knowledge. ASE certification is a badge of proven professionalism.

FIGURE 1-19 ASE shoulder emblem certification

To help prepare for the Body Repair and/or Painting and Refinishing tests, some test questions at the end of each chapter in this text are similar in design and content to those used by ASE. For further information on the ASE certification program, write: National Institute for Auto Service Excellence, 1305 Dulles Technology Drive, Herndon, VA 22071-3145.

I-CAR

The Inter-Industry Conference on Auto Collision Repair, or I-CAR, is dedicated to promoting high-quality practices in the collision repair industry. It is made up of vehicle manufacturers, collision repair shops, and insurance companies as well as independent appraisers, recyclers, educators, and tool, equipment, and supply manufacturers. Since its start in 1979, I-CAR has emerged as the leading collision repair training organization in the world.

Courses offered to auto body professionals by I-CAR include Collision Repair (eight parts); Plastic Repair, Retexturing, and Refinishing; Finish Matching to Restore Pre-Accident Appearance; and Advanced Vehicle Systems. The Collision Repair course is made up of the following sections, each of which is normally a four-hour session:

- Identification and Analysis of Damage
- Measuring Principles and Techniques
- Straightening Systems and Techniques
- Welding in Collision Repair
- Replacement of Structural Parts
- Restoring Corrosion Protection
- Suspension, Steering and Alignment
- Working with Mechanical and Electrical Parts

Upon completion of the eight sections, a certificate of achievement is awarded (Figure 1-20).

I-CAR classes are offered in all fifty states, Canada, and overseas. For information on the courses and when they are given, write: I-CAR, Committee Handbook, 3701 Algonquin Road, Suite 400, Rolling Meadows, Illinois 60008.

FIGURE 1-20 I-CAR certificates of achievement exemplify excellence in the field of auto collision repair.

Many equipment and finish manufacturers have training programs designed to educate their customers on the use of their products. These training programs run throughout the course of the year. In most cases they are scheduled in major cities across the United States and Canada, or at their training facilities. Training programs vary in duration and can be tailored to individual requirements. Shops should contact local sales representatives for more information.

REVIEW QUESTIONS

1. Which of the following was not a factor that helped bring about the unibody vehicle?
   a. safety consciousness
   b. changing manufacturing methods
   c. relaxed fuel conservation measures
   d. manufacturing cost controls
2. Who has the highest exposure to liability responsibility?
   a. insurance companies
   b. collision repair shops
   c. both a and b
   d. neither a nor b
3. In most cases, who makes the vehicle damage appraisal?
   a. estimator
   b. vehicle owner
   c. insurance company
   d. technician
4. Which of the following is not found in the vehicle identification number, or VIN?
   a. engine size
   b. gas mileage
   c. body type
   d. assembly plant code
5. Technician A says that an estimate is a legal commitment as long as it is signed by the vehicle owner and the shop's estimator. Technician B says that an insurance company appraiser must also sign it. Who is correct?
   a. Technician A
   b. Technician B
   c. Both A and B
   d. Neither A nor B
6. Which publication is very important to the parts manager?
   a. replacement catalog
   b. crash estimating guide
   c. trade magazines
   d. I-CAR *Advantage*
7. Which form is specifically designed to keep track of when the work is to be completed?
   a. parts requisition
   b. invoice
   c. quote
   d. dispatch sheet
8. Which of the following statements is incorrect?
   a. The Right-to-Know laws have made record keeping an area of employer responsibility.
   b. Once hazardous wastes leave the premises, the collision repair shop is no longer responsible for them.
   c. Material safety data sheets are made available by the product manufacturers.
   d. Both a and b
9. Technician A leaves his shirt sleeves unbuttoned in the shop. Technician B wears loose clothing instead of coveralls. Who is correct?
   a. Technician A
   b. Technician B
   c. Both A and B
   d. Neither A nor B

10. Which of the following is an important shop safety feature?
    a. jack stands
    b. antiskid floor strips
    c. approved metal containers
    d. all of the above

11. Technician A says that space heaters should never be used in the collision repair shop. Technician B says that space heaters are acceptable as long as they are inspected regularly to make sure they are properly vented. Who is correct?
    a. Technician A
    b. Technician B
    c. Both A and B
    d. Neither A nor B

12. When diagnosing an engine in a vehicle with an automatic transmission, set it in ____________.
    a. neutral
    b. reverse
    c. park
    d. drive

13. Which of the following statements is incorrect?
    a. Gasoline is an ideal cleaning solvent.
    b. Never keep more than one day's supply of paint outside of approved storage areas.
    c. Oily rags should be stored in approved metal containers.
    d. Both a and b.

14. Which type of fire involves electrical equipment?
    a. Class A
    b. Class B
    c. Class C
    d. Class D

15. With which type of respirator can facial hair prevent an airtight seal?
    a. cartridge filter
    b. air-supplied
    c. particle
    d. hood

CHAPTER

2 Fundamentals of Unibody Construction

OBJECTIVES

- Explain the major differences between body-over-frame and unibody vehicle design.
- Describe the general evolution of vehicle body design from early body-over-frame to present-day unitized construction.
- List the major design characteristics of modern body-over-frame and modern unibody construction and how they affect repair procedures.
- Identify the major structural components, sections, and assemblies of body-over-frame vehicles.
- Identify the major structural components, sections, and assemblies of unibody vehicles, including front engine/rear drive, front engine/front drive, and mid engine/rear drive vehicles.
- Identify the types of low-carbon and high-strength steels used in modern vehicle design.
- List where the various automotive steels are commonly used, and explain how their characteristics affect repair procedures.
- Know the difference between FRP and SMC.

FIGURE 2-1 A typical unibody shell and a completed vehicle

The goal of collision repair is to restore the vehicle to its pre-accident condition. To fully comprehend the challenges faced by technicians in today's collision repair market, consider the industry's radical changes in the past twenty years.

With passenger cars, many small pickup trucks, and the new minivan models, the conventional body-over-frame construction that served the industry for over sixty years has given way to unitized construction. It is a totally different concept in vehicle design that requires new assembly techniques, new materials, and a completely different approach to collision repairs (Figure 2-1). In unibody designs, heavy-gauge cold-rolled steels have been replaced with lighter, high-strength steel alloys requiring new handling, straightening, and welding techniques. The ever-increasing use of plastics, modular glass, and advanced paint systems is also changing collision repair procedures. And on unibody vehicles, certain mechanical systems, such as suspension and steering, rely on the proper positioning of unibody components for alignment and operation.

But it is not as simple as abandoning an old design concept for a new one. Body-over-frame construction is still being used on some full-size, luxury passenger cars, one-half and three-quarter ton pickup trucks, some small pickups, and most full-size vans. While literally all import passenger cars built during the 1980s featured unitized construction, some larger domestic models maintained traditional coil spring, body-over-frame setups. Following is a brief look at how the two major vehicle designs used today have evolved over the years.

BODY-OVER-FRAME CONSTRUCTION

In the conventional body-over-frame construction, the frame is the vehicle's foundation. The body and all major parts of a vehicle are attached to the frame (Figure 2-2). It must provide the support and strength needed for the assemblies and parts attached to it. The frame must also be strong enough to keep parts in alignment should a collision occur.

The conventional frame is an independent, separate component, because it is not welded to any of the major units of the body shell. The strong steel frame side members are normally made of U-shaped channel sections or box-shaped sections (Figure 2-3). Cross members of the same material reinforce the frame and

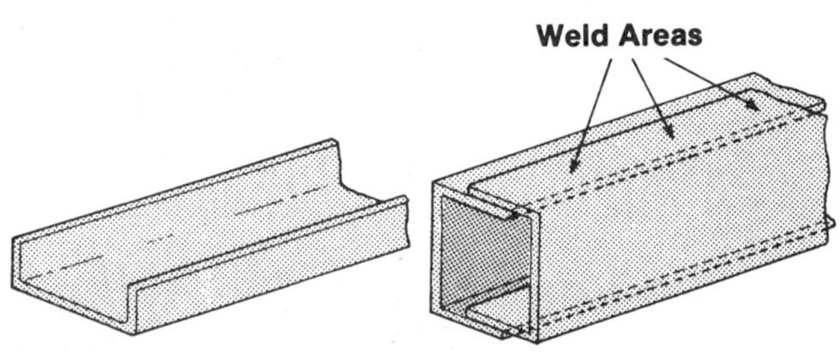

FIGURE 2-3 (Left) U-channel frame section and (right) box channel frame section used in body-over-frame construction

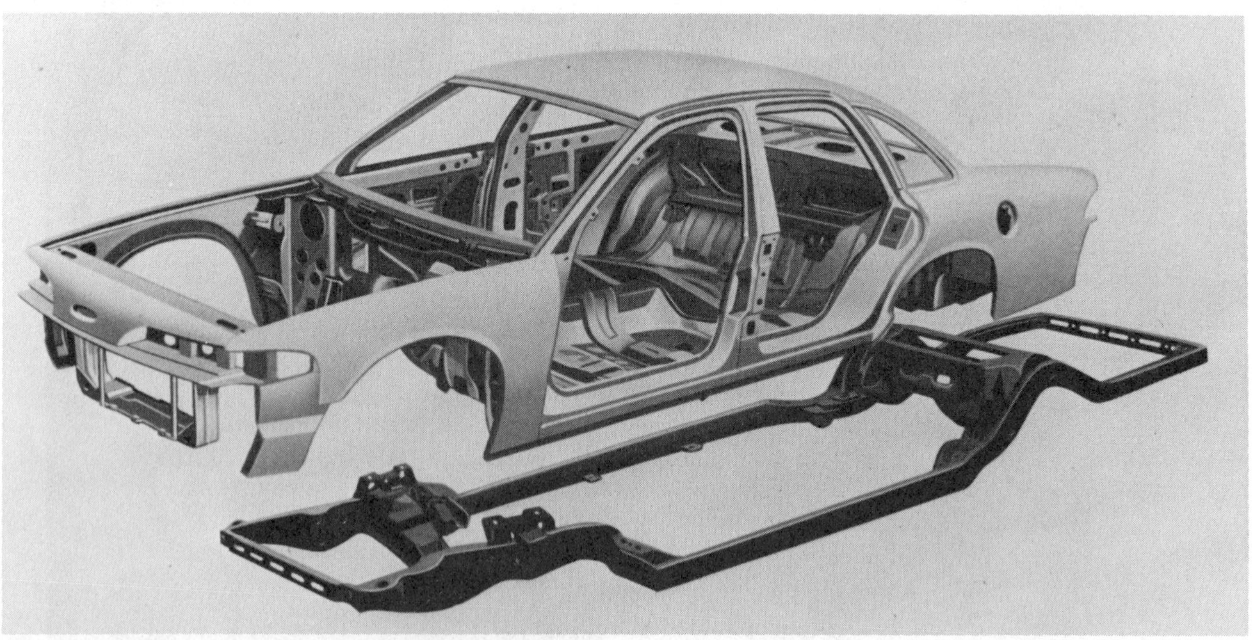

FIGURE 2-2 Body-over-frame construction

provide support for the wheels, engine, and suspension systems. Various brackets, braces, and openings are provided to permit installation of the many parts that make up the automotive chassis. The various cross-members, brackets, and braces are welded, riveted, or bolted to the frame side rails.

Most conventional frames are wide at the rear and narrow at the front. The narrow front construction enables the vehicle to make a shorter turn. A wide frame at the rear provides better support of the body. Other characteristics are as follows:

- Rubber mountings between the body and frame insulate it from vibrations, providing a quiet interior.
- High amounts of energy are absorbed during a collision.
- Suspension and drivetrain parts can be quickly assembled on the basic frame.
- Heavy frame made of thick sheet metal is approximately 3/16 to 1/8 inch.
- The vehicle profile is generally high off the ground.

In a modern body-over-frame vehicle, the front body section is made up of the radiator support, front fenders, and front fender aprons. These components are bolted in place and can be easily disassembled. The radiator support is made up of the upper support, lower support, and left and right side supports, which are welded together to form a single structure. The front fender support panels in the upper inside and rear of the fender are spot welded in place. This not only increases the fender's strength and rigidity, it also works with the fender apron to reduce vibration and noise transmitted to the passenger compartment and helps to minimize damage to the suspension and engine from side impacts.

The main body (Figure 2-4) is made up of the dash panel, floor pan, roof, and so on, to form the passenger and luggage compartments, and is similar in structure to that of a true unibody vehicle. The dash panel is made up of the left and right front body pillars, cowl side panels, and cowl panel.

UNIBODY CONSTRUCTION

The first true domestic unibody vehicle was produced by Nash in 1940. This design featured many of the concepts that still hold true in today's unitized vehicles. It also contained features that caused major repair headaches, some of which still plague modern collision repair shops.

The concept of unibody design had first proved successful in the aircraft industry, where lightweight, high-strength fuselages were needed. Adapted to the auto industry, unitized construction eliminated the need for a separate, rigid frame. The strength and rigidity of the vehicle was integrated into the body shell, which was formed by welding sheet metal into a box- or egg-like configuration. Strength is achieved through shape and design of the individual parts instead of their mass and weight. The entire inner structure acts as the

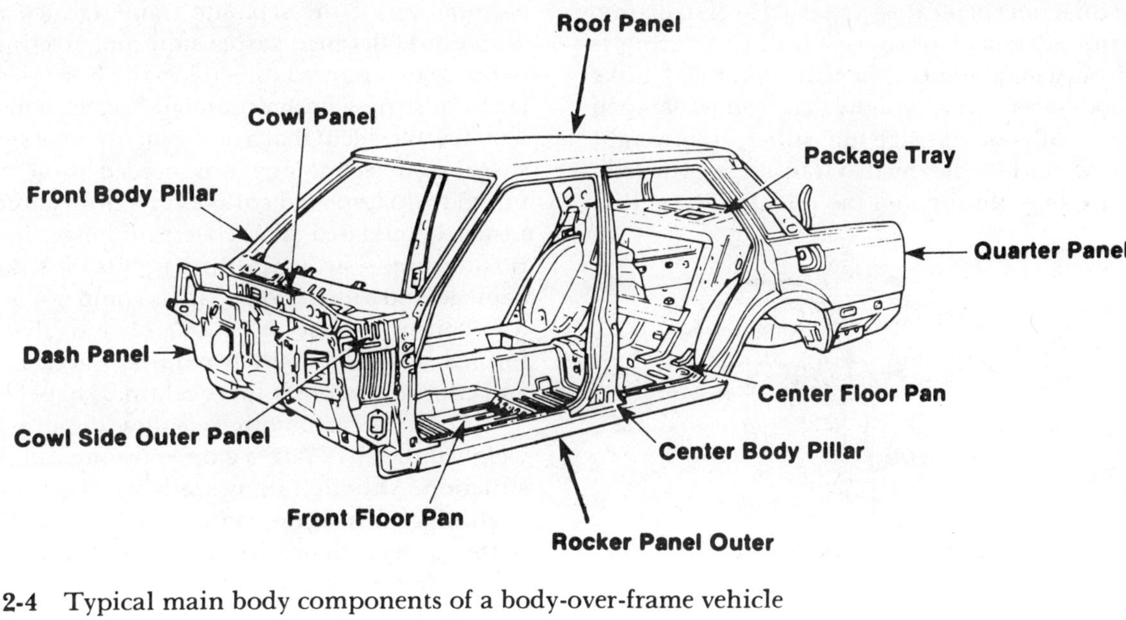

FIGURE 2-4 Typical main body components of a body-over-frame vehicle

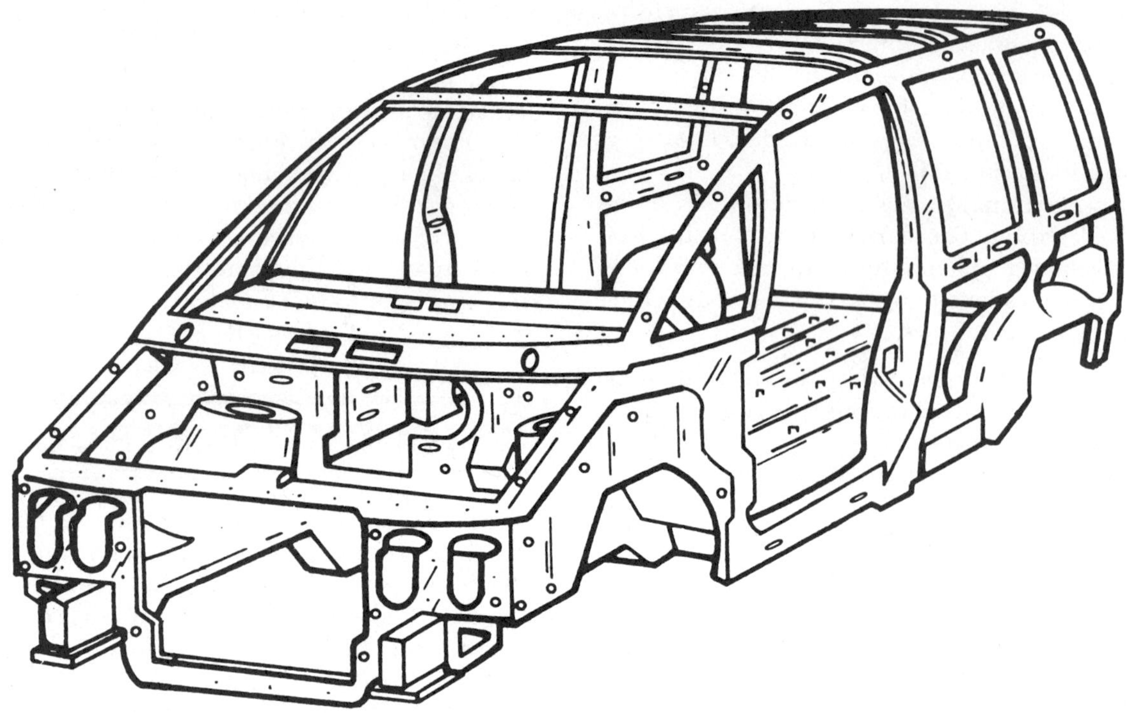

FIGURE 2-5 The inner structure of the modern unibody acts as the support structure for the entire vehicle.

support structure with individual panels depending upon one another for structural integrity (Figure 2-5).

The eggshell concept is an ideal model for understanding the principle of unibody design. Even when pressing hard on an eggshell, it is comparatively difficult to destroy since all the strength applied by the fingers is not concentrated in one place but is dispersed effectively through the entire shell (Figure 2-6). In engineering this concept is called a stressed hull structure.

A major advantage of stressed hull structuring is increased passenger compartment protection. Unlike heavier, body-over-frame designs that tend to dampen and localize collision damage, the stiffer, lightweight sections used in unibodies tend to transmit and distribute impact energy throughout the vehicle. The ability

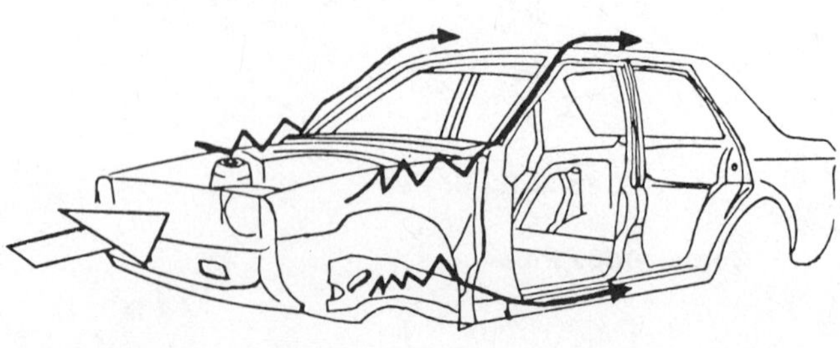

FIGURE 2-6 Unibody designs disperse collision impact energy through the entire body shell, rather than localizing it.

to absorb and dissipate impact energy helps keep passenger compartments intact during many types of collisions. Another advantage is the lighter weight of unibody vehicles. This means that smaller, more fuel-efficient engines are able to adequately power these vehicles.

But Nash's and other first generation unibody vehicles presented several challenges and problems. The elimination of the separate frame created noise and vibrations. Because suspension and steering components were mounted directly to the body shell, setting and adjusting wheel alignment became a major problem. It was evident that a new generation of steering and suspension technology was needed to accommodate unibody design. And collision repair techniques were instantly outdated. There were no longer frame members to supply strong pulling points for straightening damage. Individual body panels could not be unbolted to repair or replace. It became obvious that, once deformed, a unibody vehicle requires a great deal of time and effort to restore it to its original shape.

To American automakers these disadvantages outweighed the advantages gained through unibody construction. Although frames still appeared the best way to attach running gear and support the drivetrain, some of the concepts of unitized design were adopted and are evident in the modern perimeter frame and stub frame designs. While domestic manufacturers resisted a full-scale changeover to unibody designs for over thirty years, European automakers realized unibody design provided a means of producing small, safe, fuel-efficient vehicles for their gas-starved markets. That is why many mechanical systems used on modern unibody vehicles, such as rack-and-pinion steering, MacPherson strut suspensions, and constant velocity driven front wheel drive first gained acceptance through the import market.

In the mid 1970s, a variety of events took place that rocked the foundations of the automobile industry. First of all, the government placed very strict fuel economy and emission control laws and standards on auto manufacturers. The economy standard was known as the Corporate Average Fuel Economy (CAFE) and the emission control standard was set by the Environmental Protection Agency (EPA). This meant that American automakers had to start designing more efficient methods of combustion and emission control systems for engines.

There was also increased pressure for better fuel economy from the public sector. Various Arab oil crises destroyed the American myth of a cheap, endless supply of gasoline for our big, heavy, fuel-hungry vehicles. In addition to being found fuel inefficient, American vehicles were also being proclaimed unsafe in a series of media and consumer protection group reports.

Foreign automakers who had embraced unitized construction years earlier, captured an increased share of the domestic market with their lines of smaller, lighter, more fuel-efficient vehicles. The so-called disadvantages of unitized construction that had kept American automakers entrenched in body-over-frame construction no longer outweighed the advantages, and the wholesale push to unibody was on. At present, most unibodies are constructed of 24 gauge high-strength steel and have an average fleet weight of 900 pounds less than in 1980.

Construction of vehicles has changed, and so has the collision repair profession. In the early days of automobiles, there were no specialized shops for collision repairs. When a damaged vehicle was brought in, the part was usually removed and replaced with a new one that was either forged from steel or cut from wood. This method was expensive and time consuming. Many times there was a long wait for parts. Most of the early body/frame technicians were carpenters or blacksmiths.

When automobile bodies and frames became more complex in design, it became more practical to repair parts instead of replacing them. An early repair procedure involved placing the bent member over a massive structure and forcing it back into shape.

Because more vehicles were manufactured, the repair business needed individuals to specialize as either body or frame technicians. As they became more experienced, their techniques improved. Pushing from within the body compartment became accepted for body and frame alignment.

As procedures were refined, the stationary frame machine evolved. This machine was a more sophisticated version of the railroad iron and mechanical jack methods that blacksmiths used. The stationary frame machine went through some changes to make it efficient and easy to use. By raising the machine off the ground, it was easier to gain access to the various components. Ramps were built so that vehicles could be driven or pulled into place. Later, pits were dug underneath the machine to eliminate the ramps.

In conjunction with the racks, hydraulic jacks were used to push bent frame members back into place. Frame machine manufacturers had to continuously update their machines to keep their units functional. As the automobile frame design became more complex, so did the repair techniques. The job of the frame technician became one for a highly skilled individual.

The basic repair technique of pushing out damaged sections changed to that of pulling out damaged sections. Out of necessity, the portable body and frame straightener was developed and soon became accepted on a worldwide basis.

The manufacturers of frame equipment had to modify their equipment again. The push technique was changed to a pull technique by using adjustable pulling posts (Figure 2-7). These units remained functional but became more massive, complicated, and expensive.

FIGURE 2-7 A pulling post at work

From the technician's standpoint, it is important to know which type of construction is used, since the repair work is different for each type. With unibodies, because the mechanical components (see Chapter 9) are attached directly to the underbody, they require added precision and skill. Since the collision repair shop's responsibility is to return the vehicle to the customer completely repaired, modern technicians must have much more knowledge than their counterparts in the era prior to the advent of the unibody. Without a doubt, the unibody vehicle has brought new challenges to the collision repair industry. It is the purpose of this book to take a look at how a unibody vehicle can be restored to its pre-accident condition.

Unibody vehicles come in two- and four-door sedans, hardtops, convertibles, vans, station wagons, liftbacks, or hatchbacks, as well as light-duty trucks. There are three basic drivetrain designs: front-wheel-drive (FWD), rear-wheel-drive (RWD), and mid-engine rear-drive (MRD). By far the vast majority of unibody vehicles on the road today are FWD.

FWD VEHICLE STRUCTURE

In a FWD vehicle (Figure 2-8), the engine is mounted in the front of the vehicle and drives the front wheels. This unibody construction is characterized by the following:

- The transmission and differential are combined, providing a weight reduction.
- Overall noise and vibration are reduced because they are confined to the front of the vehicle.
- Since the engine and transmission are located in the front, the load on the front suspension and tires is increased.
- The interior of the vehicle is larger because there is no need for a rear drive axle.

The engine of a FWD vehicle can be mounted either longitudinally or transversely. Engine support methods differ between longitudinal mount FWD vehicles and transverse mount FWD vehicles. With longitudinally mounted engines (Figure 2-9) the engine is supported by the front suspension crossmember connected to the front frame rails. With transversely

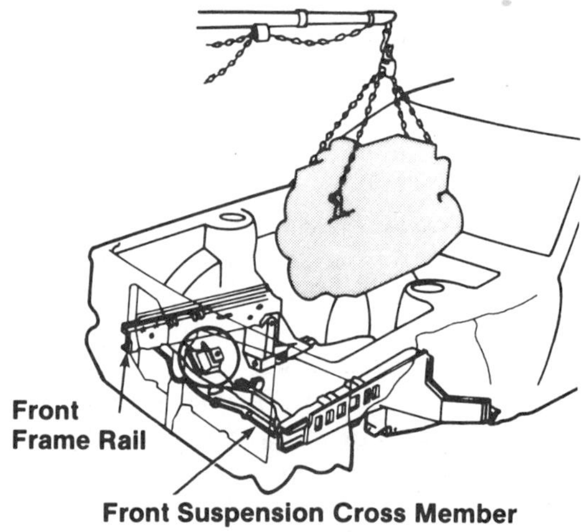

FIGURE 2-9 Longitudinally mounted FWD engine supported by a crossmember

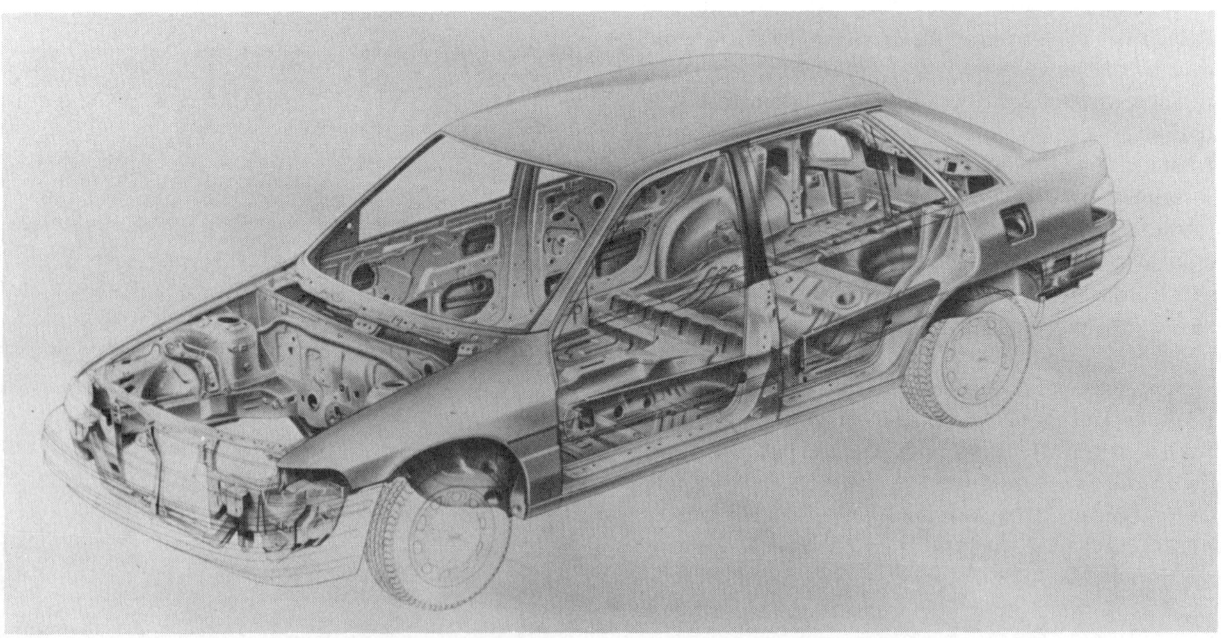

FIGURE 2-8 Typical FWD vehicle main body

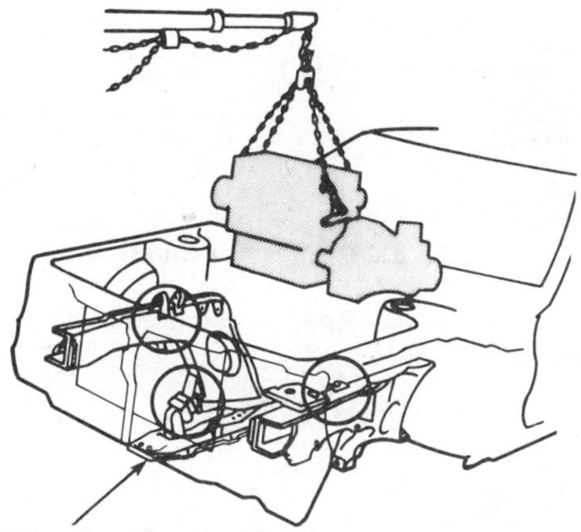

FIGURE 2-10 Transversely mounted FWD engine supports

mounted engines (Figure 2-10), the engine is supported at four points: the front and rear of the engine crossmember (positioned longitudinally through the vehicle's center) and the front frame rails. A typical FWD unibody construction can be divided into sections. The component arrangement of typical FWD vehicles is shown in Figure 2-11.

Front Body Section

The front body components—the engine hood, front fenders, radiator upper support, radiator side support, front crossmember, front frame rails, front fender aprons, and dash panel—are stamped from thin sheet metal. A high-strength structure consisting of reinforced frame rails and motor mounts capable of supporting the engine, transaxle, and suspension is used in the front section of the vehicle. Lightweight plastic bumpers are generally used on these models. The FWD vehicle usually uses an independent strut-type front suspension. The accuracy of the front body has a direct effect on front wheel alignment; therefore, it is important to check the wheel alignment following a collision.

Rear Body Section

The rear body section (Figure 2-12) consists of the hatch, lower rear panel, quarter panel, wheelhouse outer and inner panels, and rear floor pan. Since the entire drivetrain is located in the front of FWD vehicles, the fuel tank is located below the rear floor pan which allows the frame rails to be lower than in FWD vehicles. The rear suspension is then connected to the rear frame rails. Independent rear suspensions are commonly used to improve handling performance and driving stability.

RWD VEHICLE STRUCTURE

The front engine/rear drive (RWD) unibody vehicle has the engine mounted in the front and the driving wheels in the rear. The engine, transmission, front suspension, and steering components are installed in the front body, and the differential and the rear suspension are installed in the rear body.

RWD vehicles are characterized as follows:

- Since the engine, transmission, and differential are in separate positions, weight can be distributed uniformly between the front and rear wheels, lightening the steering force.

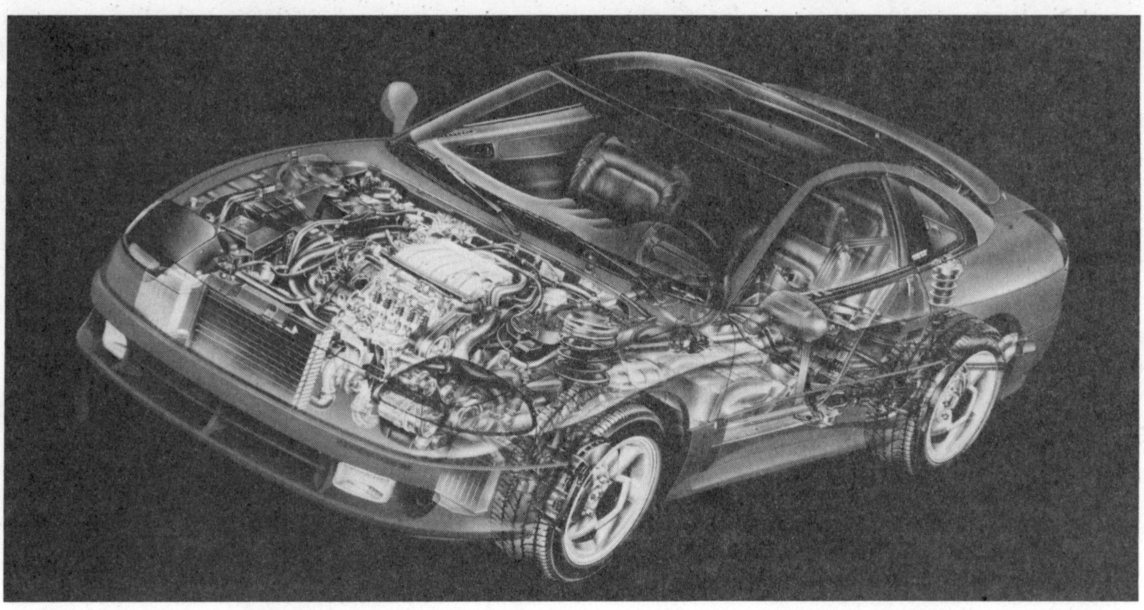

FIGURE 2-11 Component arrangement of a typical FWD vehicle

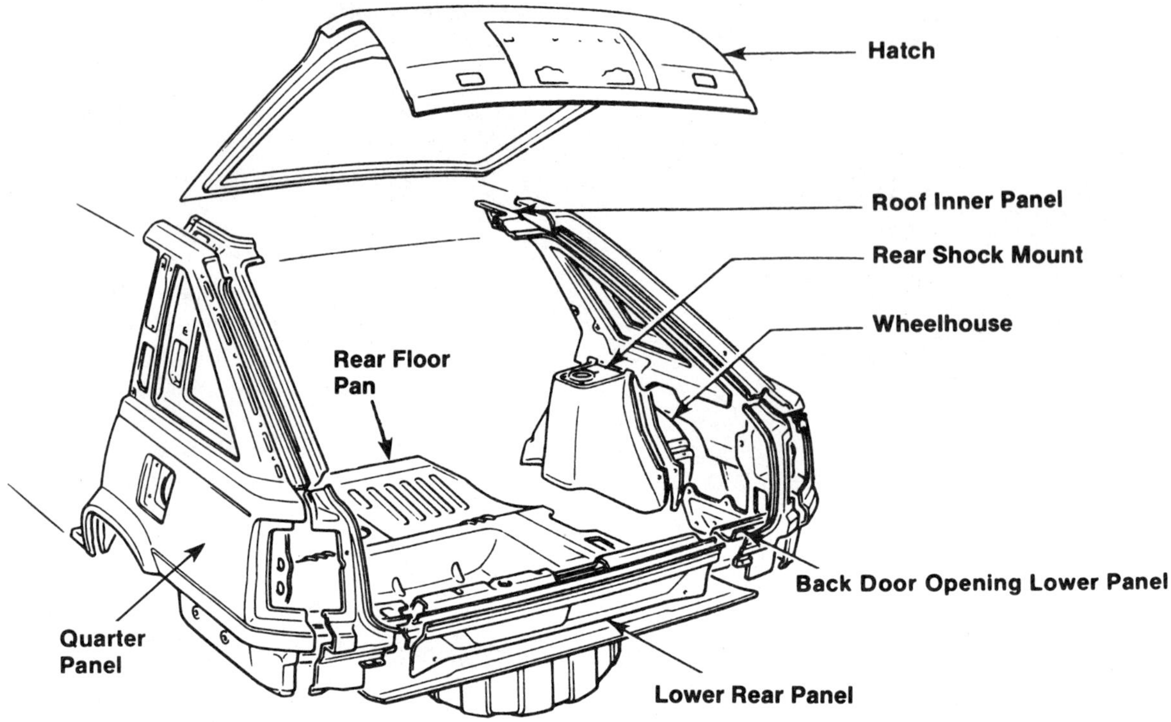

FIGURE 2-12 Rear body structural components of a typical unibody vehicle

- In a RWD vehicle, the engine is placed longitudinally with respect to the vehicle. Most vehicles have a single suspension crossmember placed laterally between the front frame rails to support the engine.
- Because the drive shaft connects the transmission to the differential, a tunnel in the floor is required (Figure 2-13).
- The engine, suspension, and steering components are all mounted to the front fender apron and the front frame rails. These parts are very important because they influence front-wheel alignment and the amount of vibration and noise that is transmitted into the passenger compartment. Therefore, they must be made with great accuracy and strength.

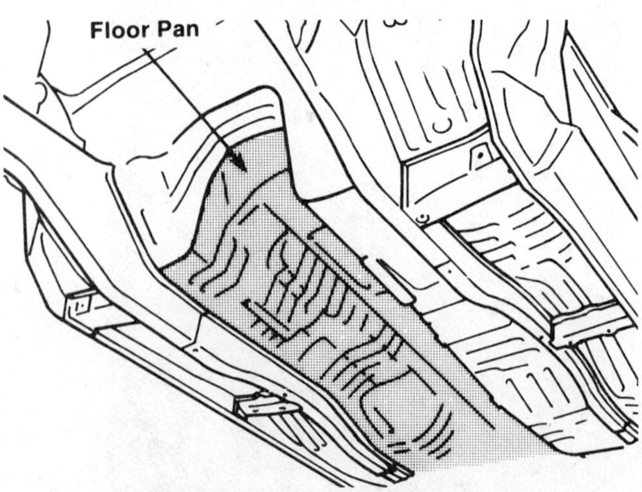

FIGURE 2-13 Tunnel section of RWD vehicle

- With the exception of the outer body parts, such as the engine hood, front fenders, and front valance panel (installed with nuts and bolts), all other exterior parts are welded together.
- The side body (Figure 2-14) is joined to the floor pan, front body, and roof panel to form the passenger compartment. The side body members also serve as door supports and maintain the structural integrity of the passenger compartment if the vehicle should overturn.
- A rear-end collision has an influence on the rear-wheel alignment. As a consequence, it is important that the rear-wheel alignment be checked whenever repairs are performed on a rear body section.

MRD VEHICLE STRUCTURE

MRD designates a mid-engine, rear-wheel-drive vehicle. The term *mid-engine* refers to the central positioning of the engine and drivetrain between the passenger compartment and the rear axle. Due to its unique engine placement, a mid-engine vehicle (Figure 2-15) has a lower profile and hence a lower center of

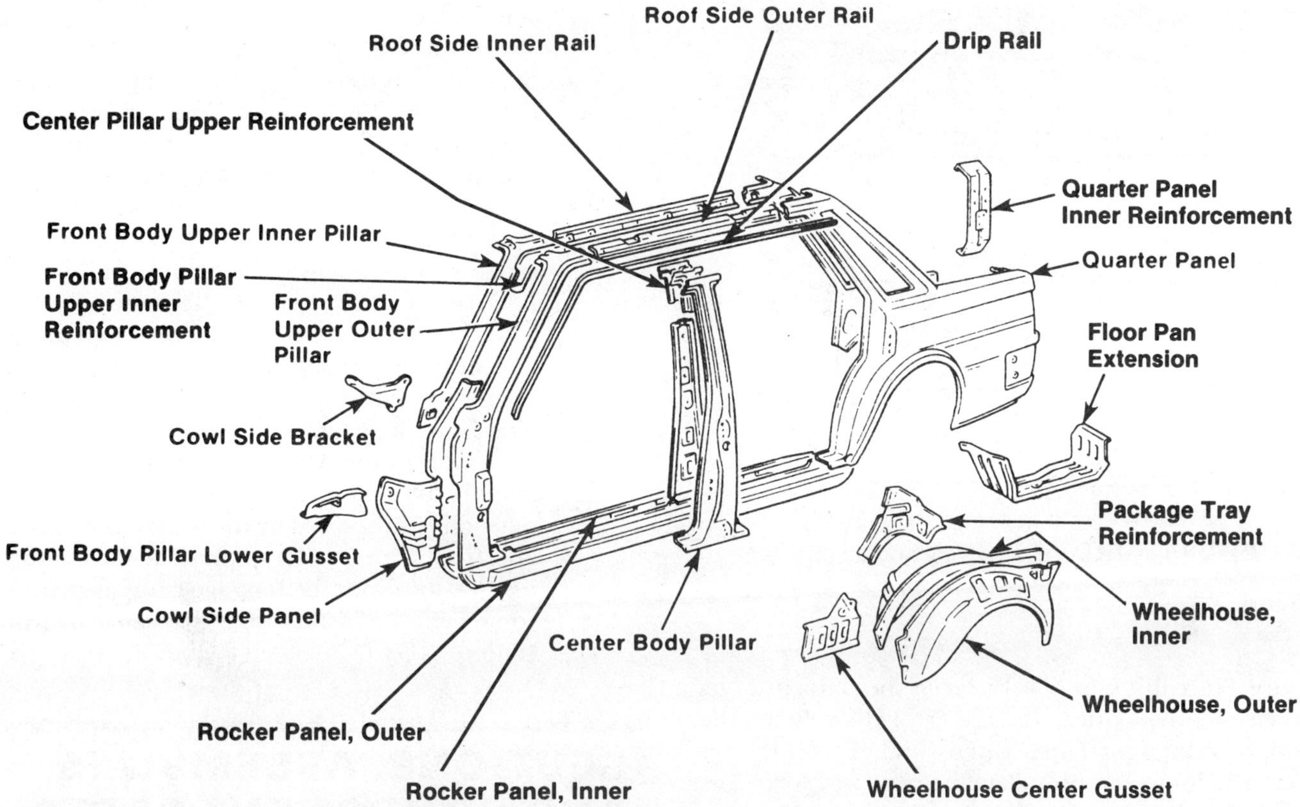

FIGURE 2-14 Side body structural components of a typical unibody vehicle

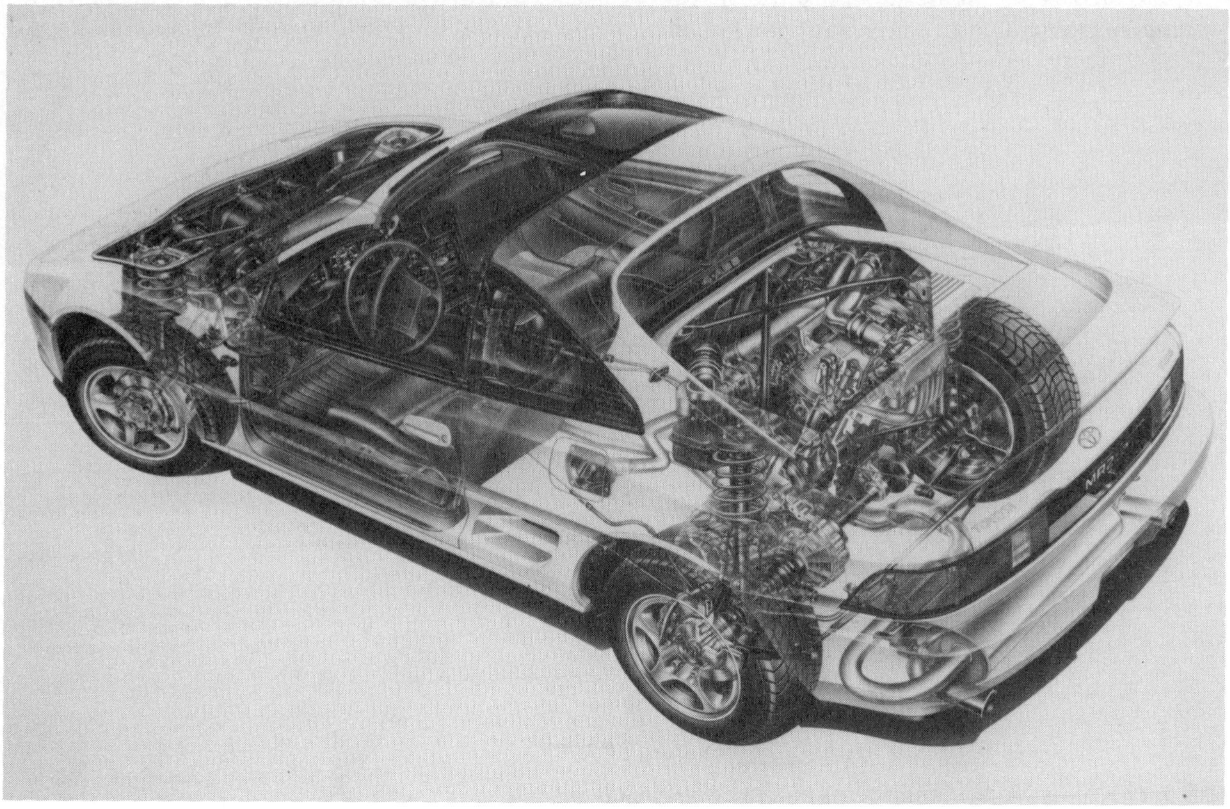

FIGURE 2-15 Typical MRD vehicle

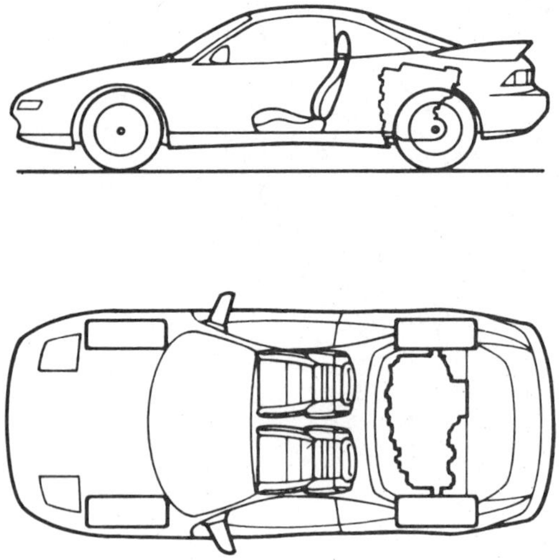

FIGURE 2-16 MRD vehicle weight distribution

gravity. Since this type of vehicle has the majority of its heavy components near the center of the vehicle, the strength of the center structure is higher than that of other vehicles to accommodate the increased load.

MRD vehicles have the following characteristics:

- Due to the central location of the heavy components, such as the engine and transmission, the center of gravity is also concentrated toward the center of the vehicle, which gives it improved steering and handling (Figure 2-16).
- Since the engine is in the rear, the front hood can be sloped downward, improving aerodynamics, lowering the center of gravity, and improving the driver's field of vision.
- Engine access and cooling efficiency are reduced because the engine is between the passenger compartment and the rear axle assembly.
- A barrier is placed between the engine and passenger compartment to reduce noise, vibration, and heat that might otherwise enter the passenger compartment.
- The front suspension, steering, radiator, and air-conditioning condenser are mounted in the front body section. The independent front suspension is supported by the front fender apron and front frame rails. Because of the engine's unique location, the luggage compartment in some vehicles of this type is located in front (Figure 2-17).

SECTIONS, ASSEMBLIES, AND COMPONENT PARTS

On the typical unibody vehicle, the body sections are divided into smaller units known as assemblies.

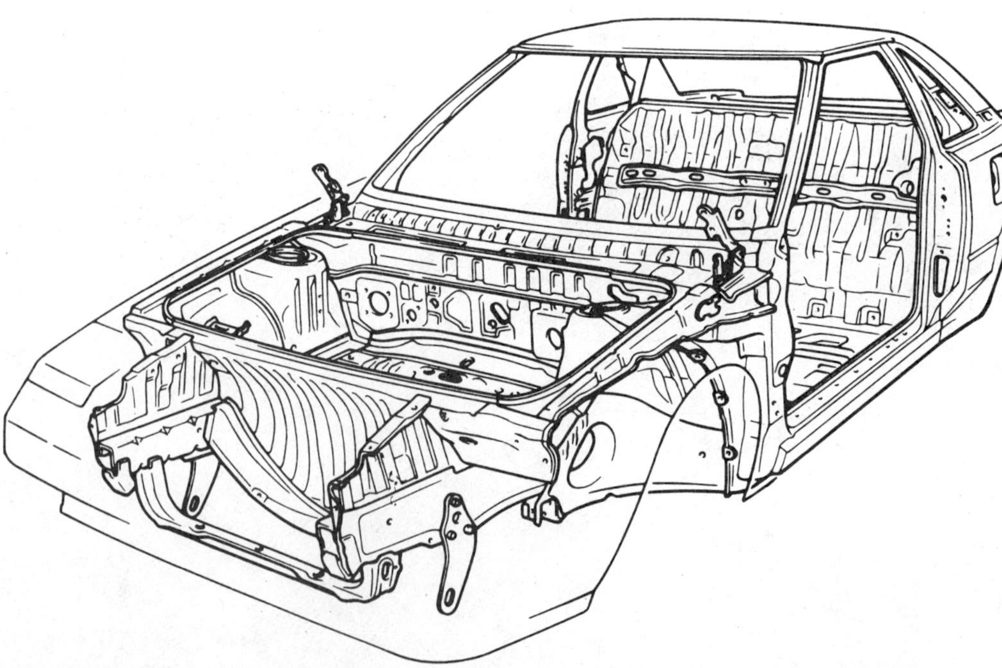

FIGURE 2-17 Main body components of a unitized MRD vehicle

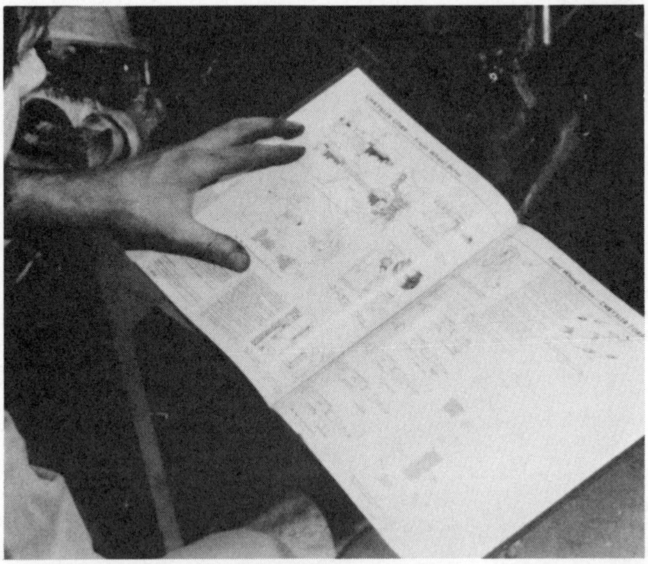

FIGURE 2-18 Typical service manual

These, in turn, are divided into even smaller units called components or parts. For example, the front end section, sometimes called the nose section by veteran technicians, consists of the following assemblies and parts: hood, front right and left fenders and aprons, grille, headlights, radiator support, front bumper and reinforcement, and trim pieces.

PARTS AND VIN NUMBERS

A good collision repair technician must possess a complete understanding of commonly used terms that identify parts, units, components, and assemblies that make up the body structure of a modern passenger vehicle. If the technician is not familiar with this language, it becomes extremely difficult to order parts and read a repair order.

All automobile companies supply yearly service manuals that describe the different makes and models of their vehicles. These manuals (Figure 2-18) give important details on repair procedures and parts. The VIN plate is used to accurately identify the body style, model year, engine, and other pertinent details (Figure 2-19). Since 1981, the VIN plate is riveted to the upper left corner of the instrument panel, visible through the windshield. Prior to 1981 and in foreign vehicles, check the service manual for the location of the VIN, vehicle certification label, or body plate. Service manuals and crash estimating guides also contain all of the necessary VIN number decoding information. Become familiar with each automaker's method of vehicle identification and the specific information it contains. Remember that it is crucial to obtain all of the information possible about the vehicle being worked on.

UNIBODY CONSTRUCTION MATERIALS

Working with unibody vehicles means working with many different construction materials besides just sheet metal. Glass, resins, sealers, and plastic are all important facets of the basic unibody design, and therefore the collision repair technician should be familiar with their unique characteristics. Many late-model vehicles utilize plastic panels glued directly to the frame (Figure 2-20). Remember that collision repair procedures vary not only according to the type of damage and the contour of the damaged panel, but also to the specific properties of the material in question.

AUTOMOTIVE SHEET METAL

Metalworking skills are probably the most important craft a technician can bring to the shop, and one of the most neglected as well. The development of high-quality plastic body fillers in recent years has made it easy to bump the low spots up, bang the high spots down, and coat the damaged area with filler. In many cases, technicians spend more time shaping and sculpting plastic filler than properly reworking the damaged metal. Not only is the inexperienced technician wasting valuable shop time, but in many cases the quality of the repair suffers also. A damaged panel improperly straightened will have tensions that can cause the plastic filler to crack, lose adhesion, and eventually fall off. That, of course, does nothing to build a reputation of customer satisfaction.

But quality work and customer satisfaction are not the only reasons, nor even the best reasons, for properly reworking damaged sheet metal. In today's unitized vehicles, the occupant's safety is at stake. Any kinks and buckles in the vehicle's boxed rail sections, inner braces, and other structural panels affect the strength of the vehicle. The damaged unibody must be restored to its pre-accident condition.

There are two types of sheet metal used in automobile construction—hot-rolled and cold-rolled. Hot-rolled sheet metal is made by rolling at temperatures exceeding 1,472 degrees Fahrenheit. It has a standard manufacturing thickness range of 1/16 to 5/16 inch, and is used for comparatively thick components of auto

FORD - LINCOLN - MERCURY

1981-91 FORD, LINCOLN & MERCURY V.I.N. DEFINED

1st POSITION
COUNTRY

1 = United States
2 = Canada
3 = South America
3 = Mexico
4 = United States
6 = Australia
9 = Brazil
J = Japan
K = Korea
L = Taiwan
W = Germany

2nd POSITION
MAKE

B = Ford
C = Ford
C = Imported Truck
F = Ford
F = Mazda
F = Mercury
L = Lincoln
M = Mercury
N = Continental
N = Ford
Z = Ford

3rd POSITION
TYPE

A = Passenger Car
A = Imported Mercury Tracer
B = Bus
B = Passenger Car
C = Truck, Stripped Chassis
C = Basic, Stripped Chassis
D = Incomplete Vehicle
E = Passenger Car
F = Equiped Without Power Train
F = Imported, Incomplete Truck
H = Incomplete Vehicle
I = Passenger Car
J = Incomplete Vehicle
J = Passenger Car
J = Imported Car, Festiva
M = Multi Purpose Vehicle
N = Passenger Car
P = Passenger Car, Imported
R = Passenger Car
T = Truck, Complete
V = Passenger Car
0 = Imported - Fiesta
1 = Imported - Merker/Scorpio
1 = Passenger Car, Imported
2 = Courier - Complete
3 = Incomplete Vehicle
4 = Multi Purpose Vehicle
4 = Courier - Incomplete
6 = Imported - Festiva

4th POSITION
RESTRAINT SYSTEM

B = Active Belts
C = Air Bags & Active Belts
P = Passive Belts

5th POSITION
IDENTIFICATION

P = Passenger Car (81-89)
P = Ford Make (90-91)
M = Lincoln/Mercury Make
T = Imported & Non-Ford Built

6th & 7th POSITION
LINE, SERIES, BODY TYPE

8th POSITION
ENGINE CODE

A = 2.3-L4, EFI/OHC
B = 3.3-V6, 1 Barrel
C = 2.2-L4, EFI (89-91)
C = 3.8-V6, 2 Barrel (84-86)
C = 3.8-V6, EFI (87-90)
C = 3.8-V6, Supercharged (90)
D = 4.2-V8, 2 Barrel (80-82)
D = 2.3-L4, EFI/Turbo (83)
D = 2.5-L4, CFI (86-90)
E = 5.0-V8, EFI/High Output
F = 5.0-V8, CFI/EFI
G = 5.8-V8, 2 Barrel
H = 1.3-L4, EFI
H = 2.0-L4, Diesel
J = 1.9-L4, EFI
K = 1.3-L4, 2-V, Mazda
L = 2.4-V6, Diesel/Turbo (84-85)
L = 2.2-L4, EFI/TC (89-91)
M = 5.0-V8, CFI or EFI/High Output
M = 2.3-L4, EFI (91)
N = 2.5-L4, EFI
P = 2.3-L4, EFI/TC
R = 2.3-L4, HSC (83-85)
R = 2.2-L4, EFI/Turbo (89-90)
R = 3.8-V6, OHV
S = 2.3-L4, HSC/CFI/High Output/EFI
T = 2.3-L4, EFI/Turbo/IC
T = 5.0-V8, EFI (91)
U = 3.0-V6, EFI
V = 2.9-V6, Fuel Injected
W = 2.3-L4, EFI/Turbo
W = 4.6-V8, EFI (91)
X = 3.3-V6, Fuel Injected
X = 2.3-L4, HSC/CFI/EFI
Y = 3.0-V6, SHO/EFI
Z = 1.6-L4, EFI/Mazda
1 = 1.3-L4, Fuel Injected
2 = 1.6-L4, Fuel Injected
3 = 3.8-V6, CFI
3 = 3.8-V6, 2 Barrel (Canada)
4 = 1.6-L4, FI/High Output (83-85)
4 = 3.8-V6, EFI/SEFI (87-91)
5 = 1.6-L4, EFI/Mazda
6 = 1.6-L4, EFI/Turbo
6 = 2.3-L4, Propane
6 = 1.6-L4, EFI/TC/Mazda (91)
7 = 1.6-L4, Methanol/High Output
8 = 1.6-L4, EFI/Turbo
8 = 1.8-L4, EFI/Mazda
9 = 1.9-L4, CFI

9th POSITION
CHECK DIGIT

10th POSITION
MODEL YEAR

B = 1981
C = 1982
D = 1983
E = 1984
F = 1985
G = 1986
H = 1987
J = 1988
K = 1989
L = 1990
M = 1991

11th POSITION
ASSEMBLY PLANT

A = Atlanta, GA
B = Oakville, Ontario, Canada
E = Mahwah
E = Niehl, West Germany
F = Dearborn, MI
G = Chicago, IL
H = Lorain, OH
J = Los Angeles, CA
J = Monterrey
K = Kansas City, MO
M = West Germany
N = Norfolk, VA
P = Twin Cities, MN
R = Hermosillo, Mexico
R = San Jose, CA
S = Allen Park, MI
T = Metuchen, WI
U = Louisville, KY
W = Wayne, MI
X = St. Thomas, Canada
Y = Wixom, MI
Z = St. Louis, MO
2 = Taiwan
5 = Flat Rock, MI
6 = Korea
8 = Broadmeadows, Australia

12th Thru 17th POSITION
PRODUCTION SEQUENCE NUMBER

FIGURE 2-19 A VIN plate as defined in a typical service manual

FIGURE 2-20 A minivan featuring plastic panels structurally bonded to the frame

bodies such as frames and crossmembers. Cold-rolled sheet metal is hot-rolled sheet metal that is acid rinsed, cold rolled thin, then annealed. Since this sheet metal is worked by cold rolling, it has a dependable thickness accuracy, surface quality, and better workability than hot-rolled steel. Most unibodies are made from cold-rolled steel. In places around the suspension in which corrosion resistance is particularly required, cold-rolled sheet metal subjected to surface treatment is used for its antirust qualities. High-strength steels, in addition to the "mild" steel traditionally used in automotive metals, are becoming more and more common on today's unibody vehicles.

LOW-CARBON STEEL

Much of the sheet metal used in collision repair today is low-carbon or mild steel (MS). Steel with a low level of carbon is relatively soft and easy to work. It can be safely welded, heat shrunk, and cold worked without seriously affecting its strength. But because it is easily deformed and relatively heavy, vehicle manufacturers have begun using high-strength steels in load carrying parts of the vehicle. Mild steel has a yield strength of up to 30,000 psi.

HIGH-STRENGTH STEEL

Today's unitized construction means that a component such as an apron or rail assembly, which is no longer just a splash shield, has to not only be lightweight but also carry a portion of the suspension reaction load. It is also designed with buckling sections to limit damage transmitted to the passenger compartment. In order to successfully achieve these objectives, many manufacturers made this new apron/rail assembly of stronger, lighter weight, high-strength steel. Figure 2-21 shows some typical body parts where high-strength steels are used.

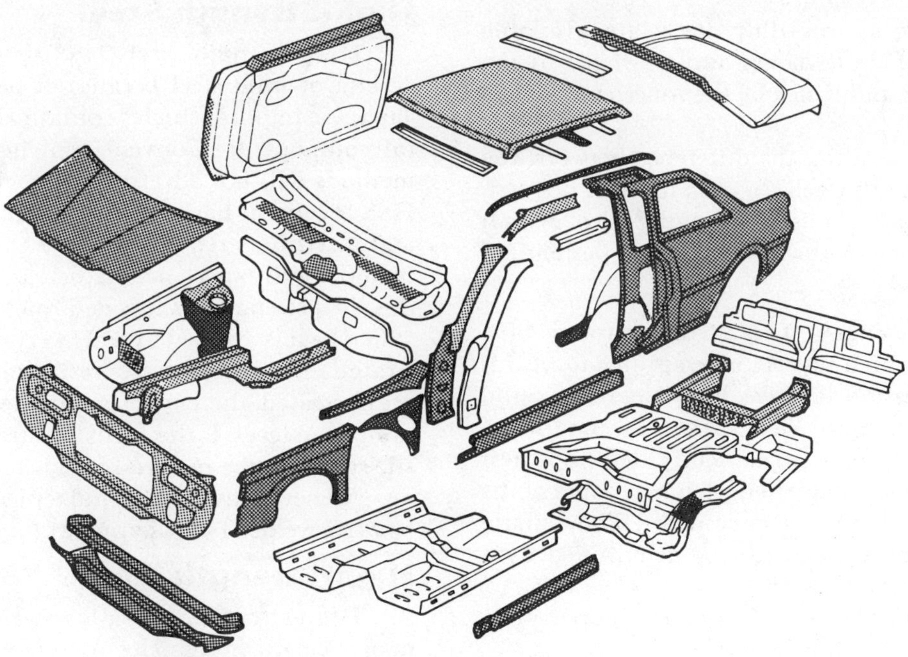

FIGURE 2-21 Typical areas where high-strength steel is used on unibody vehicles

However, these same properties that contribute strength and light weight offer some unique challenges to the collision repair industry. When high-strength steel is deformed on impact, it is more difficult to dimensionally restore than mild steel. Normal repair procedures, which dictate using heat to either aid in relieving the stresses of pulling or to weld in a new part, must be carefully controlled (or not used at all) to avoid structural damage to the part.

Types of High-Strength Steel

In spite of the fact that most types of steel look alike to the naked eye, there can be considerable differences in chemical makeup and crystalline structure that affect strength and sensitivity to heat. There is a variety of high-strength steels, all of which have unique properties that dictate the way in which they can or cannot be repaired.

The following types of steel can be found in this class:

- Chromium alloy
- Carbon molybdenum
- Chromium molybdenum
- Chromium vanadium
- Manganese alloy
- Nickel molybdenum
- Manganese molybdenum
- Nickel chromium
- Nickel chromium molybdenum
- Nickel copper

Other high-strength, low-alloy steels are available under a variety of trade names and are listed in the *Metals Handbook*, published by the American Society for Metals.

Before understanding the difference between the types of high-strength steel, *strength* must be defined. There are four basic kinds of strength, all of which relate to the ability of the steel to resist permanent deformation.

- *Tensile Strength.* This is the property of a material which resists forces applied to pull it apart. Tension includes both yield stress and ultimate strength. Yield stress is the amount of strain needed to permanently deform a test specimen; ultimate strength is a measure of the load that breaks a specimen. The tensile strength of a metal can be determined by a tensile testing machine.
- *Compressive Strength.* This is the property of a material to resist being crushed.
- *Shear Strength.* This is a measure of how well a material can withstand forces acting to cut or slice it apart.
- *Torsional Strength.* This is the property of a material that withstands a twisting force.

Strength is expressed in pounds per square inch (psi) or kilograms per millimeter squared (kg/mm^2). Heat treatment, cold rolling, and chemical additives are among the manufacturing procedures used to increase the strength of steel. Any further heating beyond the set time and temperature limits can alter the strength significantly and permanently. These limits are determined by the type and extent of the heat and chemical treatment.

HIGH-STRENGTH STEEL CLASSIFICATIONS

There are several types of steel within the classification of high strength, all of which must be treated with extreme care when heating. Those most often used in automotive applications are:

- High-strength steel (HSS)
- High-strength, low-alloy steel (HSLA)
- Ultra-high-strength steel (UHSS)
- Martensitic steel

Whenever possible, check with the automobile manufacturer regarding the acceptable heat ranges for specific components. Also, be sure to follow the manufacturer's instructions for restoring corrosion protection after any repairs are completed.

High-Strength Steel

High-strength steel (HSS) is stronger than low-carbon or mild steel because of heat treatment. Most Japanese import vehicles contain HSS in their structural components. Conventional heating and welding methods will not adversely affect the strength level of HSS because it has a yield strength of up to 35,000 psi and a tensile strength of more than 45,000 psi. The material will experience an increase in stress, exceeding the yield strength, as it is deformed during a collision. When heat is applied to an HSS component to assist in straightening, the stresses resulting from the collision are decreased, thereby restoring the strength to a lower or normal level. If the collision stresses exceed the tensile strength, the material will tear or fracture.

All new or used structural replacement parts should be MIG welded (see Chapter 7).

High-Strength, Low-Alloy Steel

High-strength, low-alloy steel (HSLA) is used in many American vehicles for body structural components such as front and rear rails, rocker panels, bumper reinforcements, door hinges, and lock pillars. Its strength is due mainly to the addition of special chemical elements. HSLA steel has a yield strength up to 100,000 psi and can be reworked.

CAUTION

Extreme care must be taken when using heat to relieve stress on HSLA panels. After exposure to a temperature of 900 degrees Fahrenheit, or higher, for several minutes, the special hardening elements are absorbed by larger, softer elements in the heated area, resulting in lower strength. Remember that there are different kinds and grades of high-strength steel and that vehicle manufacturers use different types.

To avoid substantially reducing the ability of the structure to react to normal road loads or collision forces, never exceed the factory recommended temperatures. A good rule of thumb is never heat more than 700 to 900 degrees Fahrenheit and never apply heat for more than 3 minutes. Automobile manufacturers do *not* recommend using oxyacetylene to weld HSLA.

Ultra-High-Strength Steel

UHSS has a yield strength of 110,000 psi or more. This special grade of HSS is so temperature sensitive that no heat should be used during a repair. In fact, damaged parts should not be repaired; replace using only MIG plug welds (see Chapter 7). Check manufacturer's specifications on locations of ultra-high-strength steels.

Martensitic Steel

Martensitic steel is alloy free, with tensile strength almost ten times that of typical mild steel. Door intrusion beams and some bumper reinforcements are martensitic, or ultra-high-strength, steel. The unusually high-strength properties in martensitic steel are the result of a special grain or crystalline composition imparted during forming and fabricating. Any reheating for repair destroys this unique composition and reduces the strength to that of typical mild steel. In addition, these steels are so hard that they cannot be straightened cold or be reworked with typical shop equipment. A damaged martensitic steel member should not be repaired; it must be replaced.

Door intrusion beams and some bumper reinforcements should not be straightened with any amount of heat and should be replaced. Minor damage on door intrusion beams can be ignored if it does not interfere with door alignment or function. If the corrugations in the beam are dented or deformed, the door intrusion beam should be replaced. All new or used replacement parts on a unibody vehicle should be MIG welded, never oxyacetylene welded.

PROPERTIES OF SHEET STEEL

Regardless of the type of steel used in manufacturing the modern unibody, all sheet steel used must be soft enough to be pressed into a variety of shapes that satisfy both safety and styling considerations. The difference in grades of automotive sheet steel is not usually noticeable to the collision repair technician because he or she never works on flat, unworked sheets; the panels have already been shaped and formed. When flat sheet steel is formed to a shape, it takes on certain properties or characteristics that harden the sheet steel.

For example, the roof panel found on most small economy cars is relatively flat. If hit in the center with the palm of the hand, the panel is likely to bend and then pop back to its original shape. On a roof panel that has a corrugated shape, the panel will hardly move at all. Although both roofs might be made of the same grade of steel, the one that has been changed the most is said to be harder and more resistant to deformation. During this stamping process the properties of the metal in the corrugated roof were changed more than the flat roof.

The same is true for panels whose shape has been changed during a collision. The structure of the metal in the affected areas has changed, causing the metal to become harder and more resistant to corrective forces. To repair collision damage, a technician should understand what property changes have taken place in the metal.

Physical Structure of Steel

Steel, just like all matter, is composed of atoms. These very small particles of matter are combined to form grains (Figure 2-22), which are large enough to be

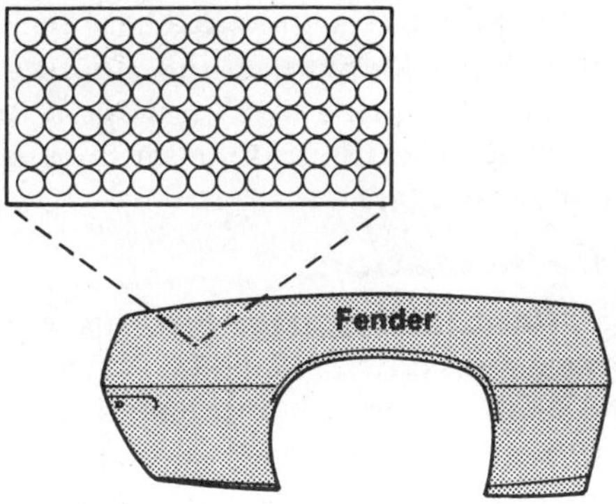

FIGURE 2-22 Granular structure of steel

seen with the aid of a microscope. Grains are formed into patterns called the grain structure. The condition of the grain structure in a piece of steel determines how much it can be bent or shaped. To change the shape of flat sheet steel, it is necessary to change the shape and position of all the individual grains that are located in the area of the creases, folds, or curves.

In mild steels, the individual grains can withstand a considerable amount of change and movement before splitting or breaking occurs. To demonstrate this, take a welding rod and bend it back and forth several times. Notice that in the area of the bend the metal will become very hot. The heat is generated by the internal friction created as the individual grains move against each other in the area of the bend.

EFFECT OF IMPACT FORCES

The atomic structure and grain pattern of a metal will determine how it reacts to force. Sheet metal's resistance to change has three properties:

1. Elastic deformation
2. Plastic deformation
3. Work hardening

All of these properties are closely related in that they are different aspects of the effect of force applied to sheet metal. They are all related to the yield point, which is the amount of force that a piece of metal can resist without tearing or breaking.

Elastic Deformation

Elastic deformation (Figure 2-23) is the ability of metal to stretch and return to its original shape. For example, take a piece of sheet metal and gently bend it to form a slight arc; when released it will spring back to its original shape. This spring-back tendency makes it necessary for the collision repair technician to reorganize elastic deformation in damaged panels, thus enabling the technician to plan his or her work to take advantage of any tendency of the damaged metal to spring back. Spring-back will be found in any area that is still relatively smooth even though it has been carried out of position by buckles formed in adjoining areas. Many such areas will spring back to shape if they are released by relieving the distortion in the buckled areas holding them out of place.

Plastic Deformation

Plastic deformation is the ability of metal to be bent or formed into different shapes. When metal is bent beyond its elastic limit, it will have a tendency to spring back, but not all the way back to its original shape (Figure 2-24). A piece of sheet metal, if bent into a tight U-shape, will spring back when released but not to its original shape. This is because the grain structure

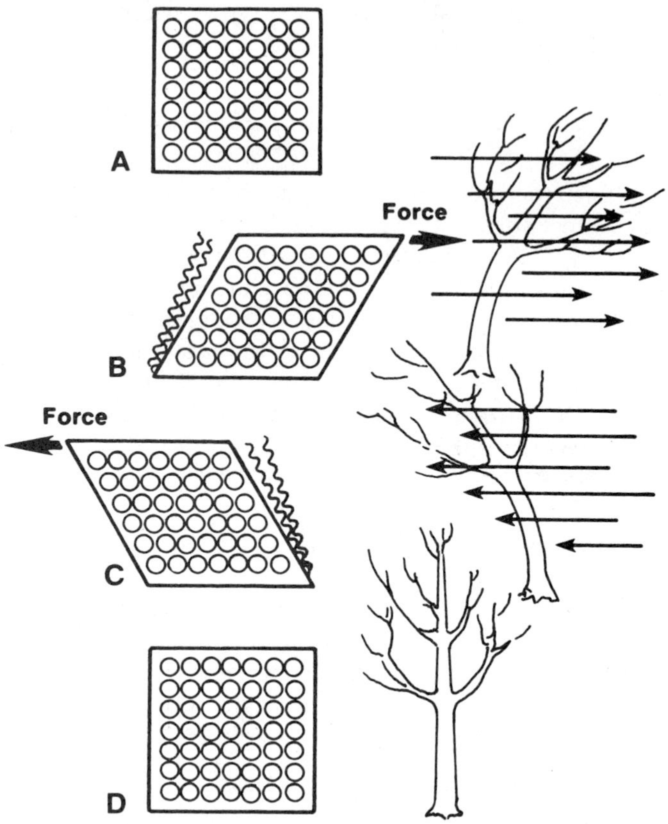

FIGURE 2-23 Under pressure, metal bends; but when the force is removed, the metal returns to its original shape.

has been changed. Plasticity is important to the collision repair technician because both stretching and permanent deformation take place in various areas of most damaged panels.

Elasticity and plasticity are illustrated in Figure 2-25. The graph shows the relationship between the size of the load and the elongation of a piece of sheet metal when a tensile load is applied to the sheet metal. If the load is increased a little at a time, the elongation increases proportionally. However, if the load exceeds a certain limit, internal slipping of the grain pattern occurs, and even if the rate of load increase is kept constant, elongation will suddenly increase. If even more load is applied, elongation will increase and the maximum load will be reached. After that, partial elongation will occur in one portion of the material and it will break.

Point A on the graph is called the elastic limit. If the load is lower than point A, deformation of the sheet metal will disappear when the load is removed, and the metal will return to its original shape. This is called elastic stress. If the load exceeds point A, even if the load is removed, the deformation will remain and the metal will not return to its original shape. This is called

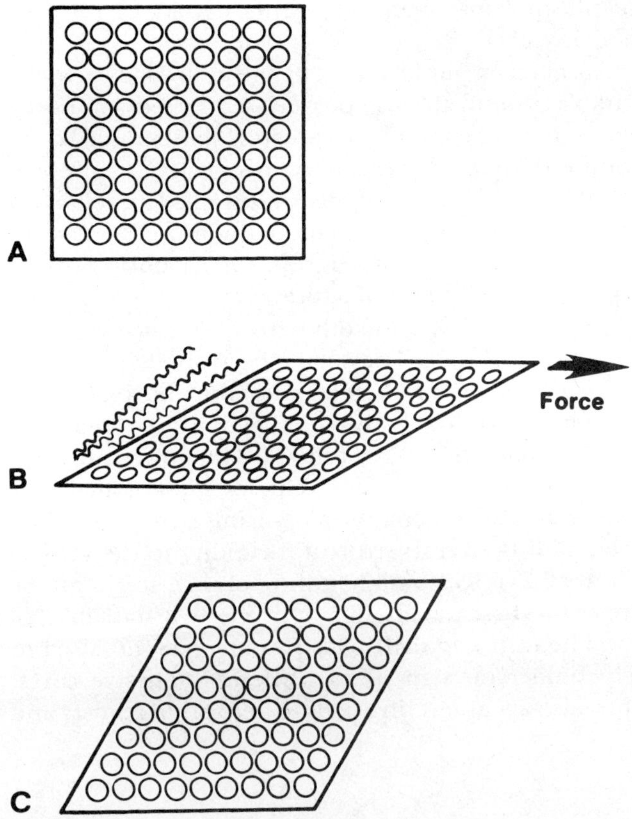

FIGURE 2-24 Plastic deformation occurs when the grain structure is forced beyond its elastic limit and takes on a new set.

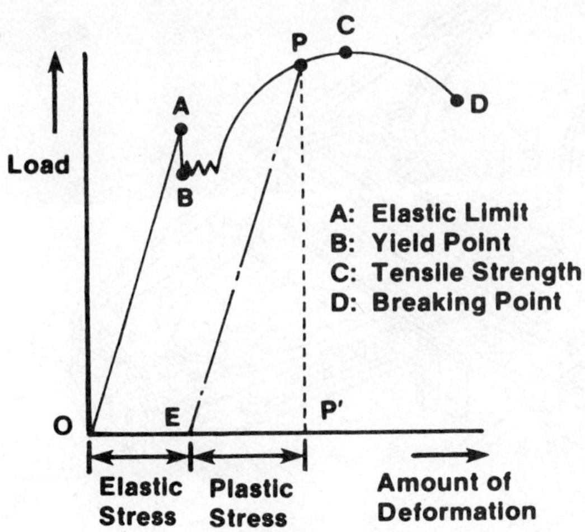

FIGURE 2-25 Load and deformation characteristics

plastic, or permanent, stress. For example, if the load is removed at point P, the elongation of the panel will return to point E but permanent stress O-E will remain. When a vehicle is damaged in a collision, the stress sustained from the impact will remain unless it is taken out. This is a condition in which there is an area with permanent stress surrounded by a neighboring area with elastic stress that cannot be removed. When repairing body panels with this type of damage, first remove the permanent stress that is restricting the elastic stress. The elastic stress will then disappear naturally, and the body panel will return to its original shape by itself.

Work Hardening

Work hardening is the upper limit of plastic deformation that causes the metal to become very hard in the area where it has been bent. For example, if a welding rod is bent back and forth several times, a fold or buckle will appear at the point of the bend. The plastic deformation has been so great at this point that the grain structure has been radically forced out of alignment, causing the metal to become very hard and stiff (Figure 2-26). This increased hardness is called work hardening.

The importance of understanding how metal stiffens, making it stronger in areas that are bent or worked, cannot be overemphasized. It is the basis of practically all collision repair. Some work hardness will be found

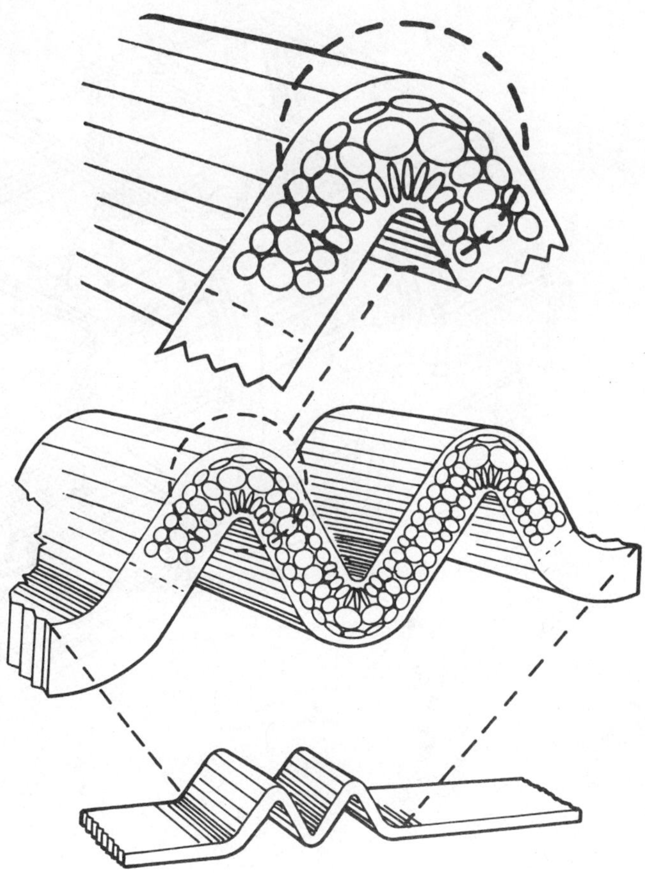

FIGURE 2-26 Plastic deformation of metal can result in work hardening.

in any undamaged body panel as the result of the original forming process. The bending caused by a collision adds still more work hardening in the areas affected. Sometimes much more will be added by the cold working used by the technician to straighten the damaged area. If excessive work hardening is caused by working the metal improperly, the job will be made more difficult.

AUTOMOTIVE PLASTICS

In recent years, more and more plastic has been used in various parts of vehicle bodies, particularly in the front end; for example, bumper and fender extensions, front fascias, fenders, hoods, grille panels, stone shields, instrument panels, trim panels, springs, and elsewhere (see Chapter 13). Because these parts are much lighter in weight than metal, they have become an important part of every auto manufacturer's fuel saving and weight reduction program. And because of the high strength-to-weight ratio of plastic, the weight decrease does not translate into a decrease in strength. Plastic body parts are here to stay, and new applications for plastic components are being developed for each model year.

A good example of the change in unibody vehicles center's around the use of fiber-reinforced composite plastic panels, commonly known as sheet molded compounds (SMC). The reason for using SMC is simple: it is light, corrosion proof, dent resistant, and relatively easy to repair compared to more traditional materials. The use of plastics also significantly reduces tooling expenses for auto manufacturers.

The use of SMC and other fiber-reinforced plastics (FRP) is not new. They have been used in various applications on automobiles since the mid-1950s. The use of large external body panels of reinforced plastic is not unusual either. What is new is that now, unlike the external panels on earlier vehicles, these panels are bonded to a metal space frame using structural adhesives, adding overall structural rigidity to the vehicle (Figure 2-27). Ride and handling of the vehicle can be affected if these panels are not properly installed.

The mill and drill pads (Figure 2-28) are used to help hold panels in place while the adhesive cures. They also act as leveling pads to help align fenders and

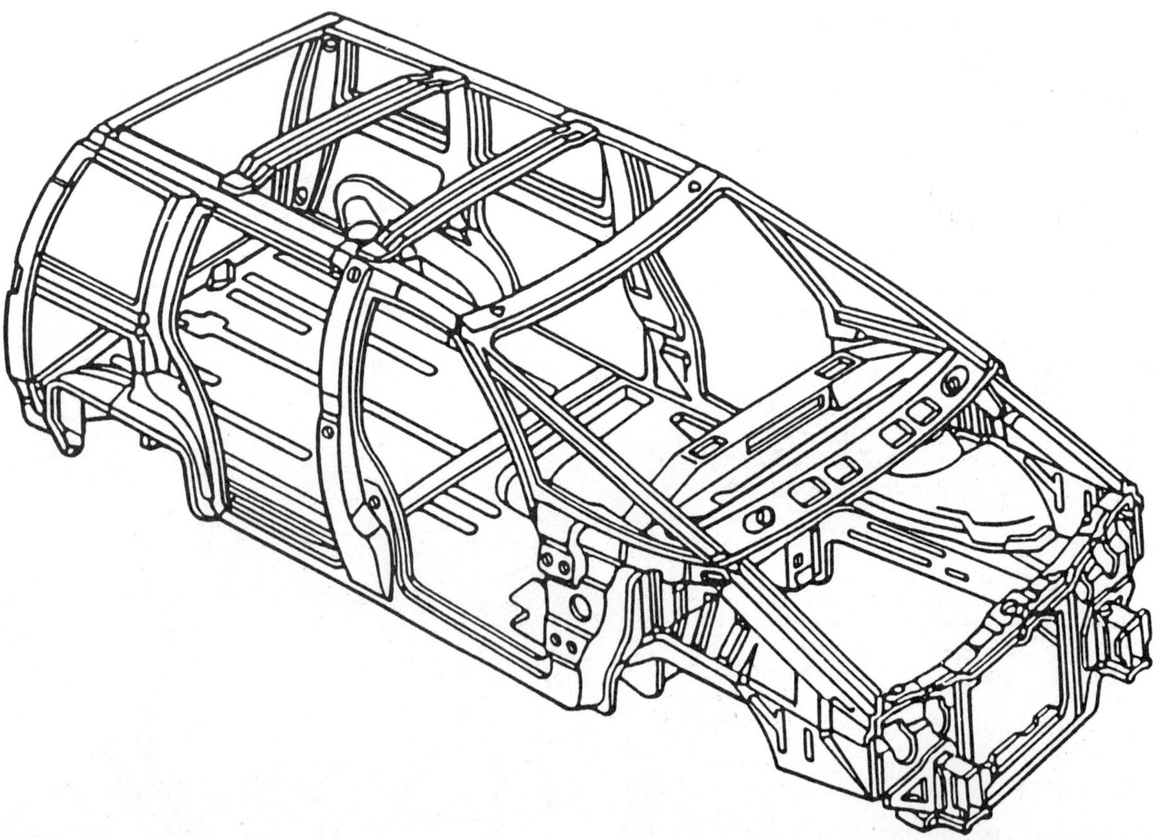

FIGURE 2-27 This multi-purpose vehicle uses SMC panels bonded to a steel space frame, introducing a new type of vehicle construction.

body panels. Whenever a whole or partial panel is replaced, these mill and drill pads will also help the technician to hold, align, and level replacement panels. If a panel is to be sectioned, it should be done between mill and drill pad locations.

SHOP TALK

SMC is very much like FRP (fiberglass) in its composition. Both use glass fibers in a resin base. Both also provide a strong, rigid paintable surface. The big difference between the two materials is their chemistry, how they are manufactured, and the type of resin they use. FRP uses a single polyester resin. FRP is about 70 percent polyester resin. The fiberglass repair resin will adhere well to this. SMC, on the other hand, uses different resins for different reasons such as strength, hardness, appearance. Only 40 percent of SMC is resin.

Conventional fiberglass panels are laid up by hand or are sprayed into a one-piece mold. Because of this, the backside of fiberglass is always rough. SMC uses a sheet of material compressed between the heated halves of both a male and female mold. Pressure and temperature work to liquefy the resins so they will flow and fill the closed mold. The resins then react to the pressure and heat to form a solid of the proper shape. The result is a strong, dense panel of controlled thickness that has a smooth finish on both sides. With fiberglass the mold coating or mold release agent is applied to the mold; with SMC the mold release becomes part of the chemical formula of the material itself.

Types of Adhesives

There are two broad types of adhesive products designed for use with plastics. Each has a specific application:

- *Panel-Bonding Adhesives.* These are two-part epoxy or urethane adhesives approved for bonding structural plastic panels to the space frame.
- *Structural Patching Adhesives.* These are adhesive products that are used to repair the panels but not for attaching them to the space frame. Some of the panel-bonding adhesives can also be used as structural patching adhesives, as well as in filler applications. However, structural patching adhesives *may not* be used to bond panels to the space frame. Further information on these adhesives can be found in Chapter 13.

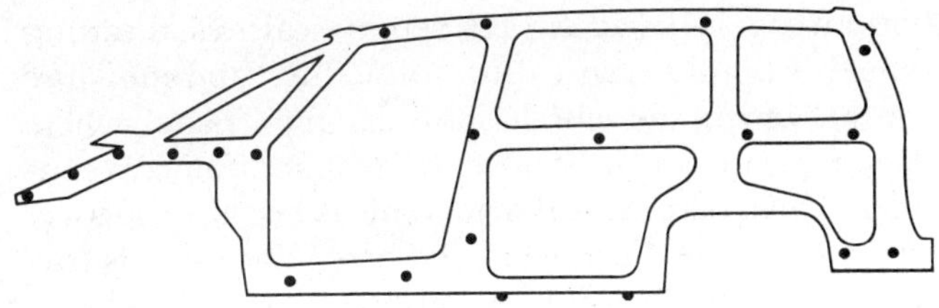

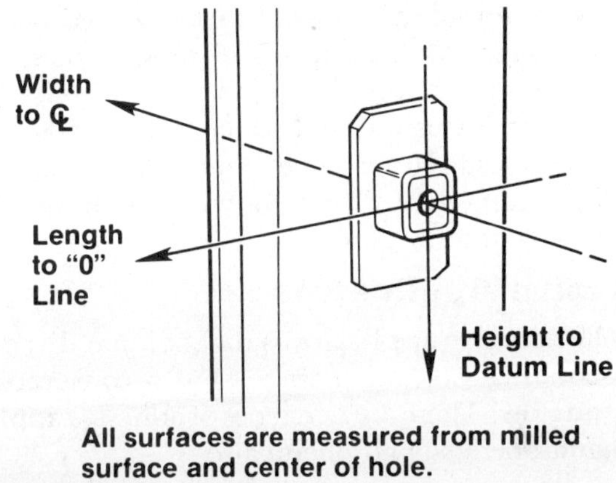

FIGURE 2-28 Typical location of mill and drill pads

The adhesive serves as a seal against moisture. It is also the primary means with which the panels are attached to the space frame and contributes to the structural rigidity of the vehicle. There are areas of the vehicle that are deemed to be 100 percent seal areas, 90 percent seal areas, and 80 percent seal areas. The robot adhesive applicators at the factory must bond at least 85 percent of the projected adhesive in these areas for the vehicle to pass assembly specifications.

Figure 2-29 illustrates the right-hand seal areas. Note that the leading edge of the panel, as well as the

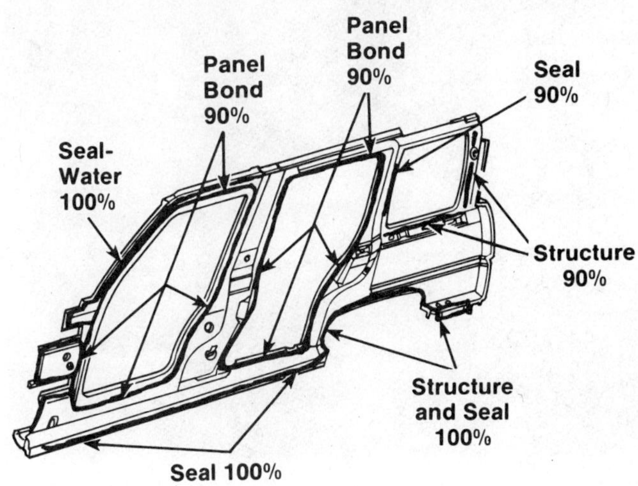

FIGURE 2-29 Right-hand seal areas

bottom edge, is listed as a 100 percent seal area, meaning that this bond is critical to noise, dust, and moisture control within the vehicle. The rear of the panel, where the panel support is bonded to the space frame, is considered a 90 percent seal area. This is because a perfect environment seal is not needed here. The same is true for the front bond point under the fender of the vehicle. All corners must be bonded 100 percent and the adhesive should run a minimum of 8 inches from the corner with no skips. Manufacturers also state that areas marked 100 percent are critical to interior noise level control, dust intrusion, and water leaks. These areas must be replaced 100 percent during a repair. The areas that a collision repair shop deals with are almost all in the 100 percent or 90 percent seal areas.

Adhesive Applicators

The adhesive applicator is the tool that dispenses panel-bonding two-part adhesives at a constant mix ratio and flow. There are two types of adhesive applicators: hand-operated and pneumatic.

Hand-Operated. The hand-operated applicator operates much like a caulking gun. It uses a two-part adhesive, often in a side-by-side tube arrangement (Figure 2-30). When pressure is applied, the adhesive is forced out and through the static mixing nozzle. In the cartridge, the two-part adhesive is separated by a foil barrier that must be broken to mix the components.

FIGURE 2-30 A manually operated two-part adhesive applicator

The material is mixed in the cartridge before inserting the cartridge into a gun. A hand-operated adhesive application is described below; the procedure can vary from one manufacturer to another.

1. The mixing handle is pulled all of the way to the top of the cartridge.
2. The barrier is distorted by squeezing the outside of the tube around the barrier.
3. The mixing rod is pushed through the barrier.
4. Turn the rod slightly clockwise and pull it all of the way up to the top of the cartridge, turn again, and push all of the way to the bottom of the cartridge. Repeat this operation until a uniform mix of adhesive is achieved.
5. Push the mixing rod all of the way to the bottom, remove the mixing rod by turning the handle counterclockwise, and insert a nozzle into the cartridge.

Because these products are reactive, once the two parts have been mixed, the entire tube of adhesive must be used within the given working life of the material.

Pneumatic. The pneumatic adhesive applicator (Figure 2-31) uses compressed air to force a piston or pistons against the bottom of the tube or tubes. This pushes a two-part adhesive through a static mixing nozzle and onto the surface. Before using, take the time to understand the pneumatic dispenser manufacturer's recommendations for use.

1. Turn the air pressure to the gun down to 40 psi or less.
2. Load the cartridge.
3. Point the gun up in the air.
4. Connect the gun to the air supply.

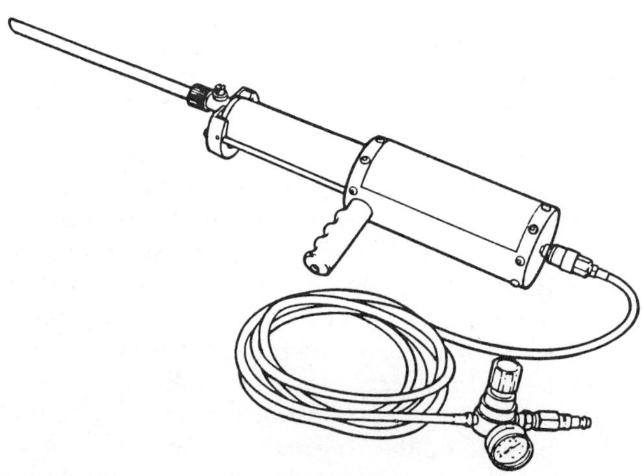

FIGURE 2-31 Pneumatic adhesive applicator

5. Turn the on/off handle to the on position and allow some of the adhesive to run out of the tube ports. Check for any blockage, and make sure that the tube ports are properly filled.
6. Attach the mixing nozzle.

Only with the adhesive cartridge correctly loaded is the air pressure increased to the recommended pressure, usually around 90 to 100 psi. Do not apply this amount of air pressure all at once. Coaxial style cartridges are often used in pneumatic applicators. These cartridges have an inner and an outer tube within the same cartridge. The relative size of the two tubes will depend upon the mixing ratio of the adhesive packaged.

Specific uses of the adhesive applicators can be found in Chapters 11 and 13.

WORKING WITH SMC AND FIBERGLASS

Working with SMC and fiberglass requires thinking about safety at all times. The resin and related ingredients can irritate the skin and stomach lining. The curing agent or hardener is generally a methyl ethyl ketone peroxide, which produces harmful vapors. Read and understand the following safety points before using any of these types of products:

- Read all label instructions and warnings carefully.
- When cutting, sanding, or grinding SMC and FRP, dust control is important.
- Wear rubber gloves when working with fiberglass resin or hardener. Long sleeve shirts, buttoned collar, and cuffs are helpful in preventing sanding dust from getting on the skin. Disposable paint suits will keep dust away from clothes.
- A protective skin cream should be used on any exposed areas of the body.
- If the resin or hardener comes in contact with the skin, wash with borax soap and water or alcohol.
- Safety glasses are always a necessity.
- Always work in a well-ventilated area.
- Wear a respirator to avoid inhaling sanding dust and resin vapors.

It is important for the collision repair technician to be familiar with the properties of sheet metal and automotive plastics (see Chapter 13). It has been estimated that the average technician creates as much damage during the repair as was done in the original impact, due mainly to a lack of knowledge and skill. It is impossible to correct all the damage without creating some during straightening, but the point is to keep it to the absolute minimum. Do not create damage, repair it.

REVIEW QUESTIONS

1. Body-over-frame construction is still being used on ____________.
   a. most imported passenger cars
   b. some full-size, luxury passenger cars
   c. most full-size vans
   d. both b and c
2. The first true domestic unibody vehicle was produced in ____________.
   a. 1930
   b. 1940
   c. 1950
   d. 1973
3. Which of the following is not characteristic of the unibody design?
   a. Once deformed, a good deal of time is needed to restore it to its original shape.
   b. In a collision, impact energy is distributed throughout more of the vehicle.
   c. The structure is lightweight and therefore more fuel efficient.
   d. Individual body panels can be unbolted and repaired or replaced.
4. How many basic unibody structures are there?
   a. five
   b. three
   c. four
   d. two
5. The engine in a FWD vehicle is mounted ____________.
   a. transversely
   b. longitudinally
   c. either a or b
   d. neither a nor b
6. Technician A says that the accuracy of the front body has a direct effect on front wheel alignment. Technician B says that front wheel alignment is not at all affected by the accuracy of the front body. Who is correct?
   a. Technician A
   b. Technician B
   c. Both A and B
   d. Neither A nor B

7. Where are most of the heavy components located in a MRD vehicle?
   a. center
   b. front
   c. sides
   d. rear

8. On the typical unibody vehicle, the body sections are divided into smaller units known as ______________.
   a. support parts
   b. structural panels
   c. structural parts
   d. assemblies

9. Since 1981, where is the VIN plate located on domestic vehicles?
   a. upper left corner of the windshield
   b. inside the trunk
   c. bottom right corner of the rear window
   d. none of the above

10. What is the ability of metal to be bent or formed into different shapes called?
    a. elastic deformation
    b. work hardening
    c. plastic deformation
    d. fatigue resistance

11. Which of the following types of high-strength steel must always be replaced, never repaired?
    a. high-strength, low-alloy steels
    b. high-tensile strength steels
    c. martensitic steels
    d. all of the above

12. Which of the following is not a characteristic of hot-rolled steels?
    a. used for thicker components such as frames and cross members
    b. used to make most of today's unibodies
    c. manufacturing thickness range of 1/16 to 5/16 inch
    d. poorer workability than cold-rolled steels

13. Technician A says that the tensile strength of individual metals cannot be determined. Technician B says it can, using a tensile testing machine. Who is correct?
    a. Technician A
    b. Technician B
    c. Both A and B
    d. Neither A nor B

14. When repairing an HSLA steel structural panel using heat, Technician A applies heat of 700 degrees Fahrenheit, while Technician B uses 1200-degree Fahrenheit heat. Who is correct?
    a. Technician A
    b. Technician B
    c. Both A and B
    d. Neither A nor B

15. Which of the following statements concerning body-over-frame construction is incorrect?
    a. The vehicle profile is generally low to the ground.
    b. The frame is not welded to any of the major units of the body shell.
    c. Rubber mountings help reduce noise and vibration from entering the vehicle's interior.
    d. All of the above

CHAPTER

3 Shop Tools and Equipment

OBJECTIVES

- Identify body repair tools commonly used in the collision repair shop, and explain their uses.
- Identify air tools used in the collision repair shop, and explain their uses.
- Identify electric tools used in the collision repair shop, and explain their uses.
- Describe the basics of hydraulic lift operation.
- Identify the fasteners normally encountered in unibody work.
- Explain the differences between the two types of adhesives used most often with automotive metals.

The importance of having the right tools and equipment for the job cannot be overstated. Even the best collision repair technician cannot do quality work using inferior tools. Of course, knowing how to use them correctly and safely is the true mark of a professional. Although some of the power tools used in collision repair are provided by the shop, every technician should have a complete set of personal hand tools.

GENERAL PURPOSE TOOLS

A good percentage of the tools used in collision repair are common, general purpose hand tools. The apprentice technician will have many of these wrenches, screwdrivers, pliers, etc., in his or her tool collection (Figure 3-1). The less familiar tools are equally important, however, because they are designed for specific types of fasteners often encountered in collision repair work.

FIGURE 3-1 Professional set of automotive hand tools

WRENCHES

The width of the jaw opening determines the size of a wrench. For example, a 1/2-inch wrench has a jaw opening from face-to-face of 1/2 inch. The actual size is slightly larger than its normal size so that the wrench can fit around a nut or bolt head of equal size.

Commonly used wrenches include open-end, box-end, combination, adjustable, pipe, hex, torx, and socket varieties. Most standard (SAE) wrench sets include sizes from 3/16 to 1 inch. Metric sets usually include 6- to 19-millimeter wrenches. Smaller and larger wrenches can be purchased but are rarely used in collision repair. It is important to remember that metric and SAE size wrenches are not interchangeable. For example, a 9/16-inch wrench is 3/10 millimeter larger than a 14-millimeter nut. If the 9/16-inch wrench is used to turn or hold the 14-millimeter nut, the wrench will probably slip, rounding the points on the nut. Having the right tool for the job will avoid unnecessary frustration and wasted time.

SCREWDRIVERS

Many threaded fasteners used in the automotive industry are driven by a screwdriver. Two of the most common are the sheet metal screw and the Torx fastener. Each fastener requires a specific kind of screwdriver, and a well-equipped technician will have several sizes or types of each.

No screwdriver should be used as a chisel, punch, or pry bar. Screwdrivers were not made to withstand blows or bending pressures. When misused in such a fashion, the tips will wear, become rounded, and tend to slip out of the fastener. The screwdriver's usefulness is impaired, and a defective tool is a dangerous tool.

A number of specialty fasteners have started replacing the slot and Phillips head screws. These new breeds are designed to improve the transfer of torque from screwdriver to fastener, slip less, result in less work fatique, and offer some tamper resistance. Figure 3-2 shows a set of specialty screwdriver tips and assorted drivers. Most of these tips will prove useful in unibody repair.

PLIERS

Pliers are used for working with wires, clips, and pins. The collision repair technician must own several types: standard pliers for common parts and wires, needle nose for the very small parts, and large adjustable pliers for heavy-duty work.

MISCELLANEOUS HAND TOOLS

A variety of other miscellaneous hand tools will be useful from time to time. Many are inexpensive and have several uses. A few are very expensive but may be provided by the shop.

A retractable steel rule is a basic ingredient in any toolbox. Its uses in the shop are many—measuring a repair area for a sheet metal patch, finding centerlines when applying woodgrain transfers, measuring frame dimensions, and so on. A 10- or 12-foot self-winding tape measure with a metric scale is usually sufficient.

A common utility knife with a retractable razor blade is handy for cutting or trimming. Most come with extra blades stored in the handle. A wire brush should be used sparingly on bare metal because it can leave scratches in the metal. But wire brushes are ideal for cleaning weld joints. When a third hand is needed, clamps are the answer. Clamps come in various sizes and an assortment is very useful in a collision repair shop. A medium-duty bench vise is normally a shop-provided tool used for holding objects when grinding, bending, or welding. Most have a swivel base and serrated jaws. Clamping the workpiece in a vise frees both hands and keeps the workpiece stationary.

A tap is a tool that cleans and rethreads holes. A die straightens damaged threads on bolts or studs. A tap and die set will perform most rethreading tasks in the body shop. Both SAE and metric sets should be kept on hand in the shop.

A cabinet-type tool chest such as the one shown in Figure 3-1 is standard equipment in the collision repair shop. A chest of drawers organizes wrenches, screwdrivers, hammers, and so forth. The tool chest is on rollers so that it can be conveniently located in the work area and can be locked to safeguard the tools.

For jobs like the installation of locks and glass, there are special hand tools available. They are described in the appropriate chapters in this book.

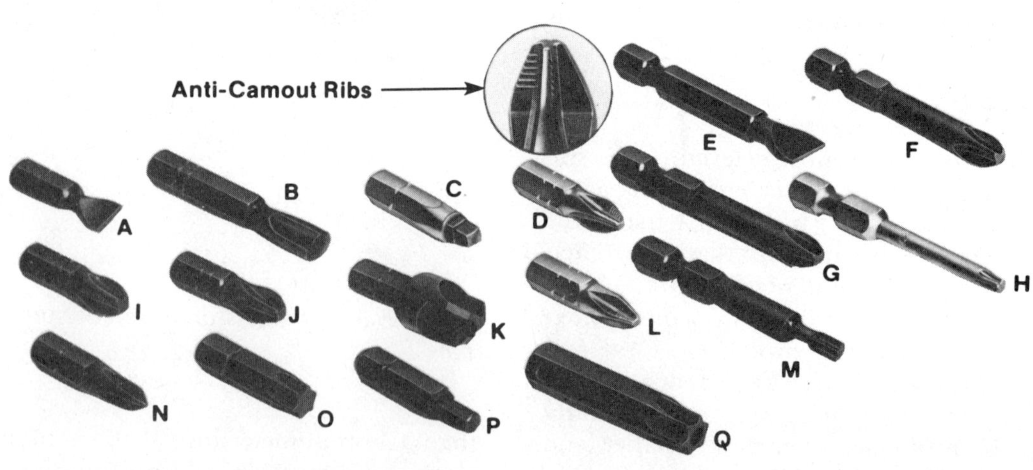

FIGURE 3-2 Specialty screwdriver tips: (A) Flat tip, (B) Clutch A, (C) Scrulox, (D) Anti-Camout ribs Phillips for removal, (E) Flat-tip power bit, (F) Pozidriv power bit, (G) Phillips power bit, (H) Torx power bit, (I) Phillips, (J) Pozidriv, (K) Tri-wing, (L) Anti-Camout ribs Phillips for installation, (M) Clutch A power bit, (N) Reed and Prince, (O) Torx, (P) Hex, and (Q) Tamper-proof Torx

FIGURE 3-3 Typical set of body repair tools

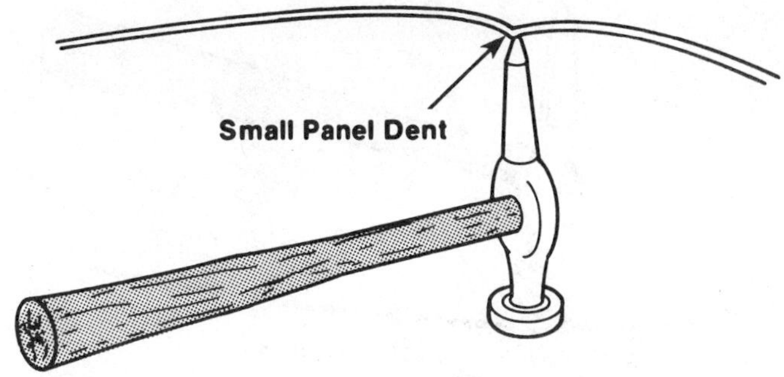

FIGURE 3-4 Raising a dent with a picking hammer

BODY REPAIR TOOLS

Following is a discussion of the most commonly used body repair tools. Included are familiar general purpose metalworking tools such as hammers and tin snips, as well as some designed solely for auto body repair, such as dent pullers and panel cutters. Beyond the typical set of body repair tools shown in Figure 3-3, it is important to have on hand all the tools needed for performing any metal shaping technique as described in Chapter 6. Here are some common body repair tools:

- *Rubber mallets* gently bump sheet metal without damaging the painted finish. Their most frequent use is with a suction cup on soft "cave-in" type dents. While pulling upward on the cup, the mallet is used to tap lightly all around the surrounding high spots. A popping sound occurs as the high spots drop and the low spot springs back to its original contour.
- *Body hammers* are the basic tools for hammering sheet metal back into shape. They come in many different designs. Some have flat, square heads; some have rounded heads; and some, called picking hammers, have pointed heads. Each style is designed for a special use.
- *Picking hammers* will take care of many small dents. The pointed end is used to hammer out dents from the inside (Figure 3-4). The flat end is for hammer-and-dolly work to remove high spots and ripples. Picking hammers come in a variety of shapes and sizes. Some have long picks for reaching behind body panels, some have sharp points, others are blunted.
- *Bumping hammers* are used to bump out large dents. They can have a round face or a square face. The surfaces of the faces are nearly flat and large enough that the force of the blows is spread over a large area. These hammers are used for initial straightening on damaged panels or for working inner panels and reinforced sections that require more force but not a finished appearance.
- *Finishing hammers* are used to achieve the final contour. The face on a finishing hammer is smaller than that of the heavier bumping hammer. The surface of the face is crowned to concentrate the force on top of the ridge or high spot.
- *Shrinking hammers* (Figure 3-5) are finishing hammers with a serrated or cross-grooved face. They are used to shrink spots that have been stretched by excessive hammering.
- *Dollies or dolly blocks* are used like an anvil. They are generally held on the backside of a panel being struck with a hammer. Together the hammer and dolly work high spots down and low spots up (Figure 3-6).
- *Body spoons* are a class of body repair tools that are used sometimes like a hammer and sometimes like a dolly. Available in a variety of shapes and sizes to match various panel shapes, the flat surfaces of a spoon distribute the strik-

FIGURE 3-5 Serrated face of a shrinking hammer

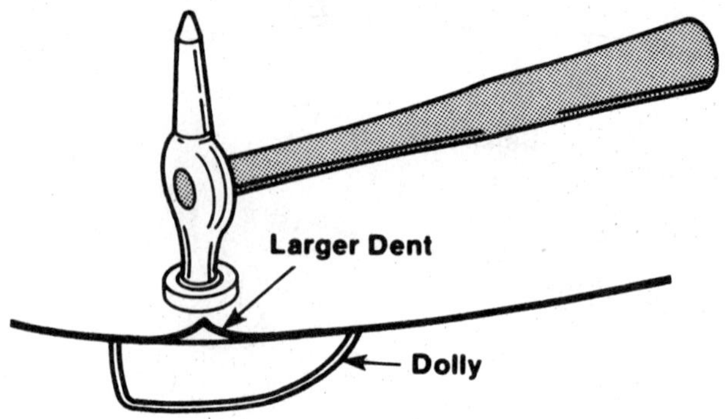

FIGURE 3-6 Bumping a dent with a hammer and dolly

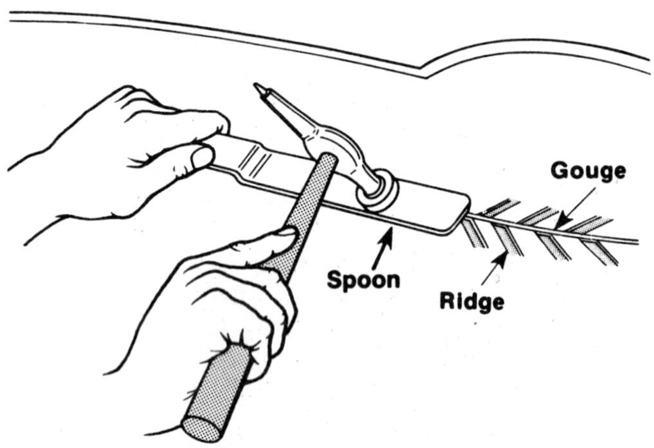

FIGURE 3-7 Lowering a ridge with a hammer and spoon

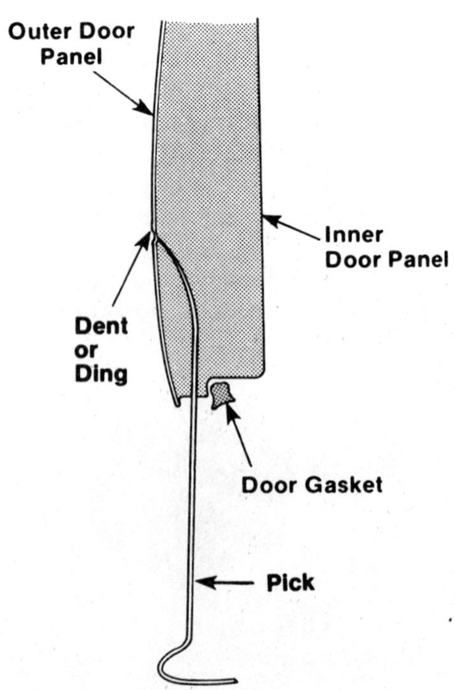

FIGURE 3-8 Prying a dent with a pick

ing force over a wide area (Figure 3-7). They are particularly useful on creases and ridges.

- *Picks* are used to reach into confined spaces. The pick is used only to pry up low spots. This tool varies in length and shape, and most have a U-shaped end that serves as a handle. Picks are commonly used to raise low spots in doors, quarter panels, and other sealed body sections (Figure 3-8). Picks are often preferred to slide-hammers and pull rods because they do not require drilling holes in the sheet metal.
- *Dent pullers* usually come with a threaded tip and a hook tip. The threaded tip is threaded into the drilled hole and the slidehammer is moved along the steel shaft and struck against

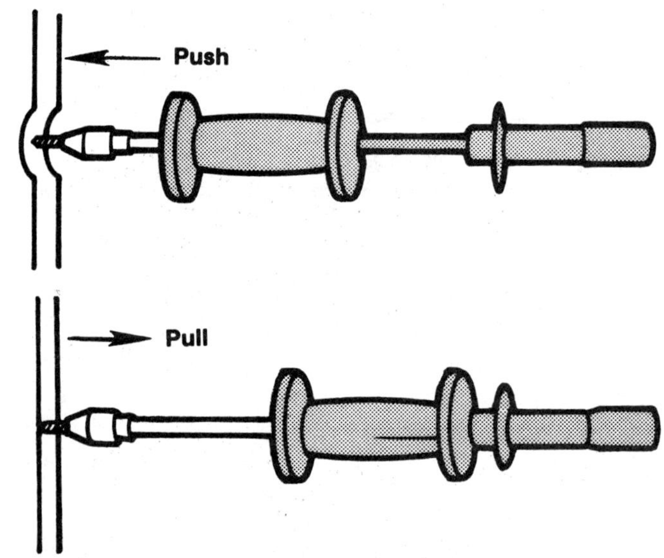

FIGURE 3-9 Using a dent puller

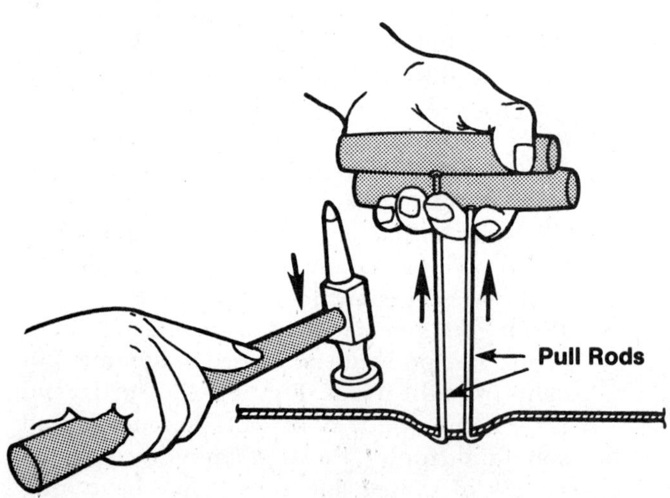

FIGURE 3-10 Bumping and pulling lowers crowns, while raising dings.

the handle. Tapping the slidehammer against the handle slowly pulls up the low spot (Figure 3-9). Creases in sealed body panels or panel sections that cannot be reached from the backside can be pulled out with a dent puller or pull rod. Use of either tool requires that one or more holes be drilled or punched in the crease.

- *Pull rods* have a curved end that is inserted in the drilled hole like the dent puller. A small dent or crease can be pulled up with a single pull rod, but three or four may be used simultaneously to pull up larger dents. A body hammer can also be used with a pull rod. The high crown of a dent can be bumped down, while the low spot is pulled up (Figure 3-10). Simultaneous bumping and pulling returns the panel to its original shape with less danger of stretching the metal.

Straightening tools and techniques are further discussed in Chapter 6.

- *Punches and chisels* (Figure 3-11) are absolutely necessary in every body repair tool chest. Center punches are used to mark the location of a spot for drilling. (The punch mark keeps the drill bit from wandering.) A drift or starter punch has a tapered point with a flat end that is used to drive out rivets, pins, and bolts. A pin punch is similar to the drift, except its shaft is not tapered; thus, it can be used to drive out smaller rivets or bolts. An aligning punch is

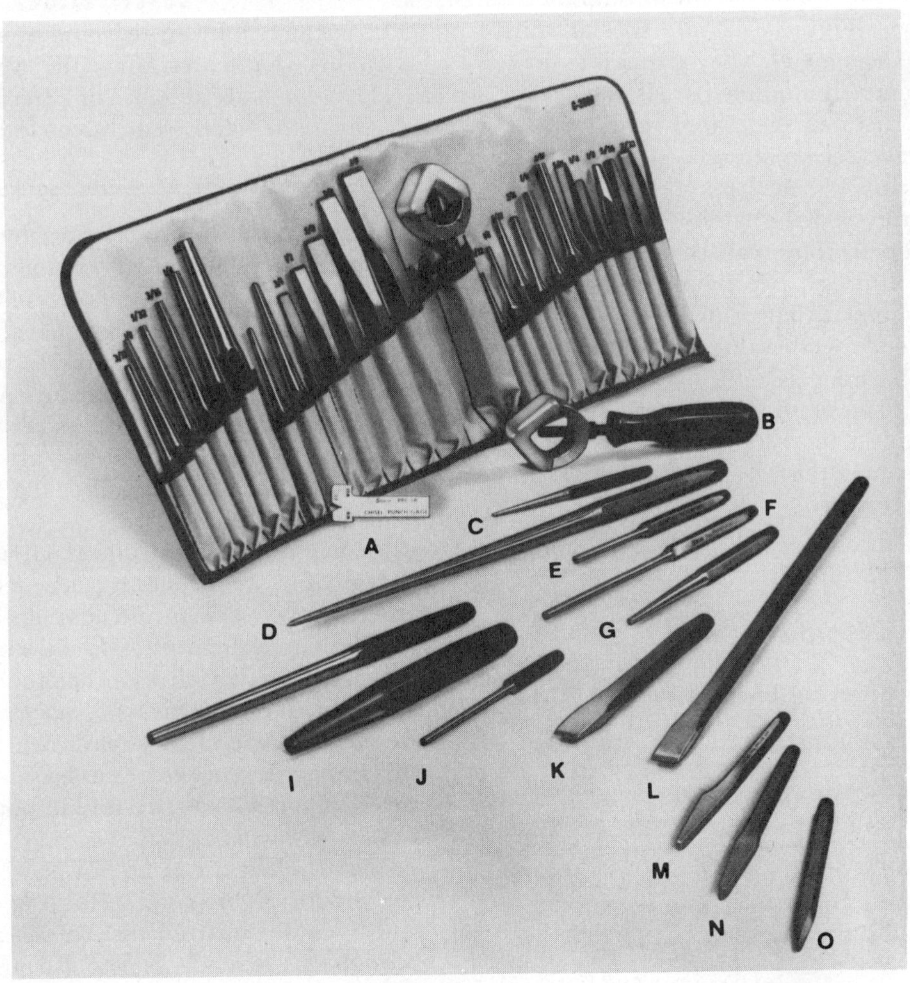

A. Chisel Gauge
B. Punch/Chisel Holder
C. Center Punch
D. Long Center Punch
E. Pin Punch
F. Long Pin Punch
G. Starter Punch
H. Long-Tapered Punch
I. Short-Tapered Punch
J. Roll Pin
K. Flat Chisel
L. Long Flat Chisel
M. Round Nose Cape Chisel
N. Cape Chisel
O. Diamond Point Chisel

FIGURE 3-11 Various punches and chisels

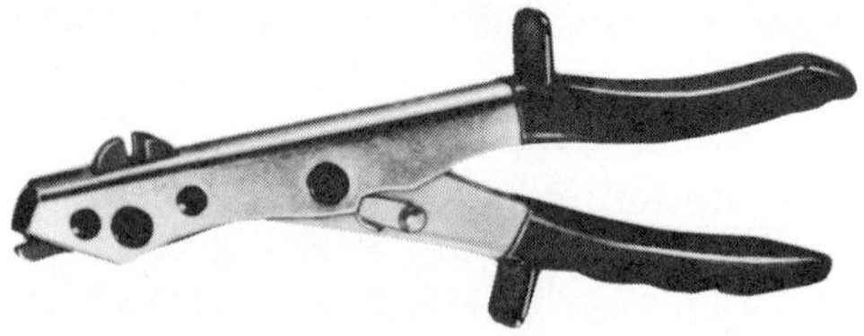

FIGURE 3-12 Panel cutter

long and tapered and used to align body panels. It can be used to line up fender bolt holes and bumpers.

- *Tin snips* are perhaps the most common metal cutting tool. They can be used to cut straight or curved shapes in sheet metal and also to trim panels to size. The narrow profile of the jaws allows the snips to slip between the cut metal, and the jaws are serrated to cut through it.
- *Panel cutters* (Figure 3-12) are special snips used to cut sheet metal. They are used to make straight or curved cutouts in panels that require repairs for rust damage. Panel cutters are designed to leave a clean, straight edge that can be easily welded.
- *Pop rivets* are one of the handiest inventions for collision repair. They can be inserted into a blind hole through two pieces of metal and then drawn up with a riveting tool, locking the pieces of metal together (Figure 3-13). There is no need to have access to the back of the rivets. Pop rivets, however, should not be used in areas that are subject to excessive vibration or for structural panels; the rivets can work loose and weaken the repair.

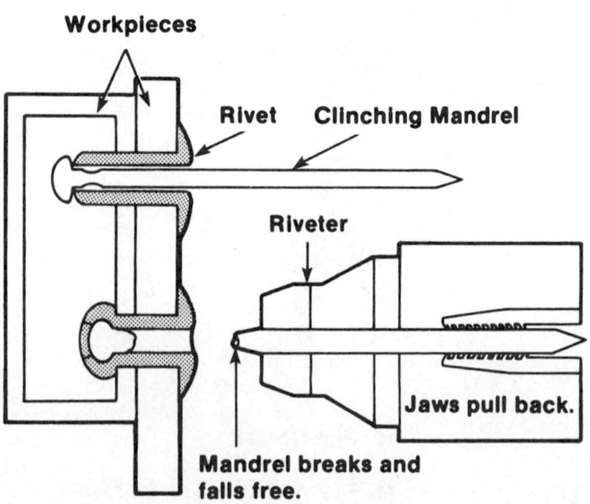

FIGURE 3-13 Riveting two panels together

AIR TOOLS AND COMPRESSORS

The automotive industry recognized early on the advantages of air-powered tools. Those that follow are now considered vital to the professional collision repair technician.

AIR-POWERED WRENCHES

Any job that involves threaded fasteners can be done faster and easier with air-powered wrenches. There are two basic types of wrenches: impact and ratchet.

WARNING

Use of any of the air-driven power tools described in this chapter requires the wearing of safety glasses or a face shield. For hazardous operations, wear both. Never wear loose clothing that might get caught in a tool.

The impact wrench is a portable hand-held reversible wrench. When triggered, the output shaft spins freely at anywhere from 2,000 to 14,000 rpm; the exact amount depends on the wrench's make and model. When the impact wrench meets resistance, a small spring-loaded hammer strikes an anvil attached to the drive shaft, onto which the socket is mounted. Thus each impact moves the socket around a little until torque equilibrium is reached, the fastener breaks, or the trigger is released.

When using an air impact wrench, it is important that only impact sockets and adapters are used with it. Other types of sockets and adapters might shatter and fly off, endangering the safety of the operator and others in the immediate vicinity. Therefore, before using any air-powered tool of this type, make certain that sockets and adapters are clearly marked or labeled "For Use with Impact Wrenches," or at least have the word "Impact." Impact sockets are sold in both SAE and metric sizes.

For all their torquing power, air impact wrenches have practically no recoil. Holding one is rather easy, but this can be misleading. The stream of air blowing through the wrench is very strong. Therefore, the wrench should be held tightly.

The air ratchet wrench (Figure 3-14), like the hand ratchet, has a special ability to work in hard-to-reach places. Its angle drive reaches in and loosens or tightens where other hand or power wrenches cannot fit. The air

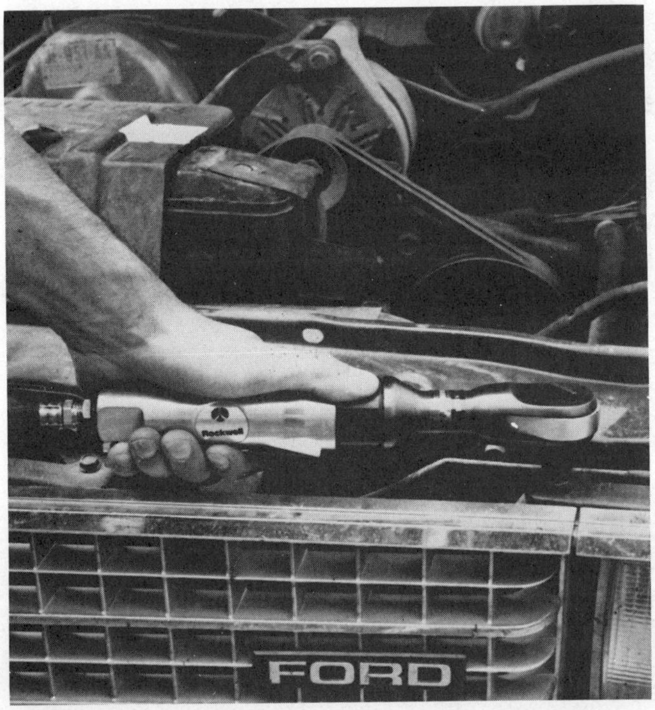

FIGURE 3-14 Loosening a nut using an air ratchet wrench

FIGURE 3-15 Using an air drill

ratchet wrench looks like an ordinary ratchet, but has a fat handgrip that contains the air vane motor and drive mechanism.

Remember that there is no consistently reliable adjustment with an air ratchet or impact wrench. Where accurate, preselected torque adjustments are required, a standard torque wrench should be used. The air regulator on air-powered wrenches can be employed to adjust torque only to an approximate tightness.

SHOP TALK

Actual torque on a fastener is directly related to joint hardness, tool speed, condition of the socket, and the time the tool is allowed to impact.

AIR DRILLS

Air drills are usually available in 1/4-, 3/8-, and 1/2-inch sizes and operate in much the same manner as an electric drill. But, they are smaller and lighter. This compactness makes them a great deal easier to use for drilling operations (Figure 3-15).

Special spot weld removing air drills and/or attachments are available for cutting out spot welds (Figure 3-16). When doing this, the drill can be fastened to the weld area with a clamp attachment that makes the

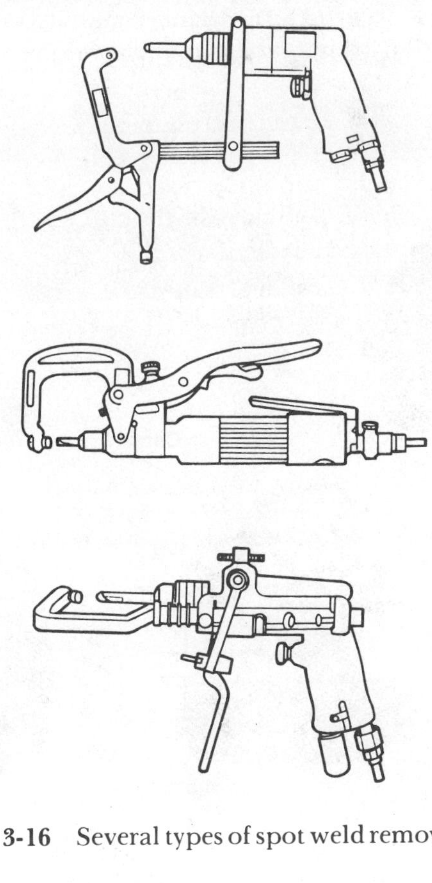

FIGURE 3-16 Several types of spot weld removing air drills

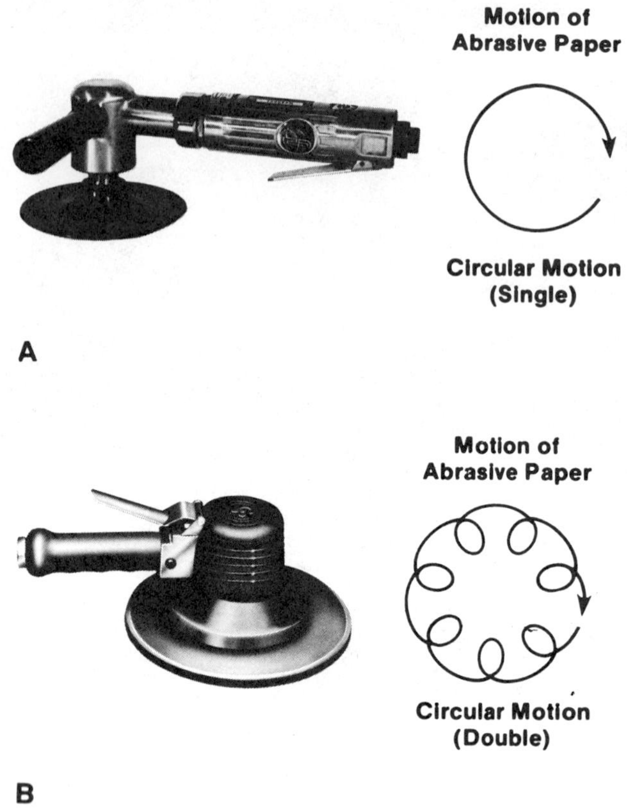

FIGURE 3-17 (A) Disc sander and its motion; (B) dual-action orbital sander and its motion

operation easier. The cutter will not deviate from the weld center during cutting (see Chapters 7 and 8).

AIR SANDERS

In general, air sanders are used in the paint shop. There are two basic types of air sanders: disc and orbital. Most rough sanding in automotive work is done with a disc sander (Figure 3-17A) or its counterpart, the dual-action (DA) orbital sander (Figure 3-17B). The latter oscillates while it rotates, thus creating a buffing pattern rather than the swirls and scratches often caused by the disc sander.

AIR GRINDERS

The most commonly used portable air grinder in collision repair and refinishing shops is the disc-type grinder. It is operated in the same manner as the disc sander. As shown in Figure 3-18, there are several other air grinders available.

AIR CHISELS OR HAMMERS

Of all automotive air tools, the air chisel or hammer is one of the most useful. When using an air chisel or hammer, keep the following points in mind:

- Always use a chisel retainer when operating an air hammer.

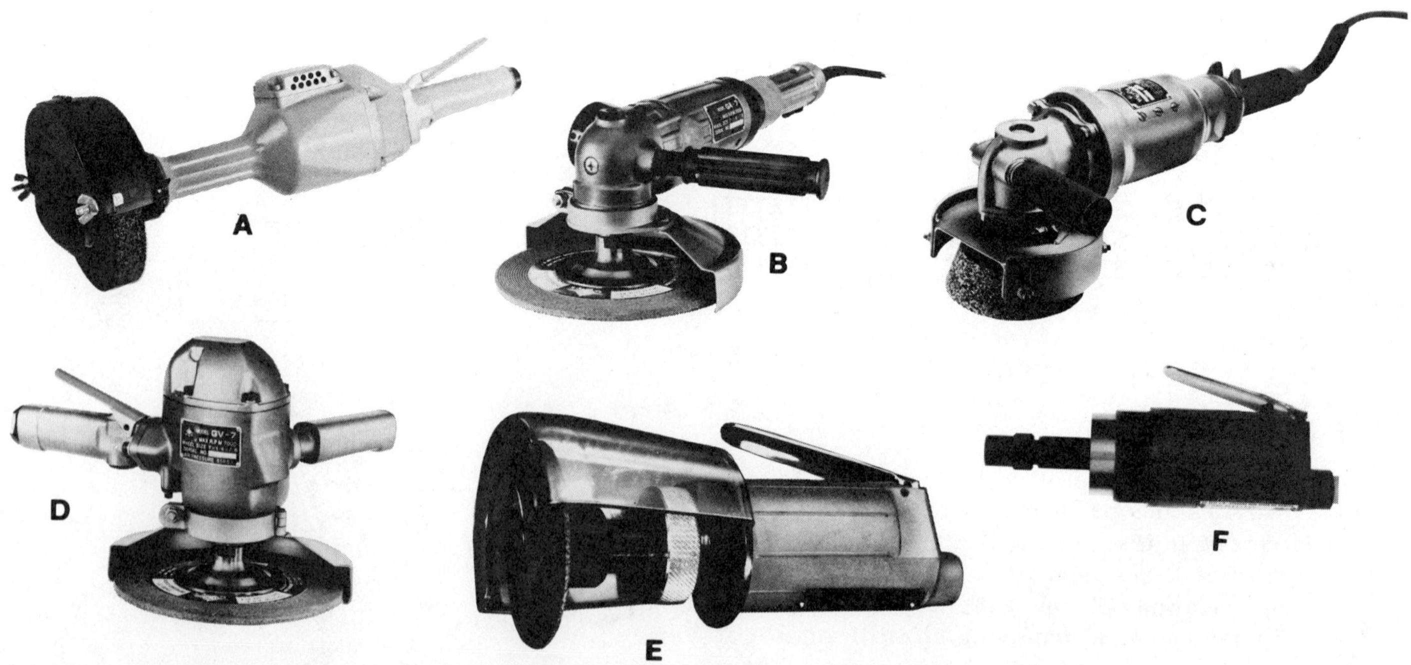

FIGURE 3-18 Various air grinders found in the collision repair shop: (A) horizontal; (B) vertical; (C) vertical with cup wheel; (D) angle; (E) cutoff; (F) die

- Position the tool by starting slowly, then increasing the power. Avoid running into hardware, frames, and so forth with sheet metal cutting tools.
- Check chisel shanks periodically for peening, and grind a new chamfer when required.
- Do not let the chisels ride out of the air hammer.
- Keep the cutters sharp at all times.

PNEUMATIC TOOL MAINTENANCE

Air tools need little upkeep, but could cause problems if the required maintenance is not performed. For instance, moisture gathers in the air lines and is blown into the tools during use. If a tool is stored with water in it, rust will form and the tool will experience reduced efficiency; it will also wear out quickly.

To prevent this from happening, remember that most air tool motors need daily (or as often as used) lubrication with a good grade of air tool oil. If the air line has no in-line oiler or lubricator, run a couple of drops of oil through the tool. The oil can be squirted into the tool's air inlet or into the hose at the nearest connection to the air supply; run the tool after adding the oil. Most air tool manufacturers recommend the use of their special oil; however, when this is not available, standard automatic transmission fluid may be substituted. Do not use in-line oilers in the paint area.

All such tools have recommended air pressures. If the tool is overworked, it will wear out sooner. If something is wrong with the tool, fix it. If the defect is not repaired, a chain reaction might occur and the other parts will become defective as well. For example, if the gearing must be replaced but the job is put off, the rotor and end plate might soon wear out as a result. A tool with worn parts will also use more air pressure. The air compressor, in turn, will become overworked and put out air that is not clean and dry.

COMPRESSED AIR

The compressed air supply system is designed to provide an adequate supply of compressed air at a predetermined pressure to insure efficient operation of all pneumatic equipment. The system can vary in size from small portable units (Figure 3-19) to large stationary installations (Figure 3-20).

FIGURE 3-19 Portable air compressor unit

FIGURE 3-20 Typical stationary air compressor

ELECTRIC TOOLS

In the typical collision repair shop, electric-powered tools are widely used. Drill presses, bench grinders, heat guns, welders, and vacuum cleaners are all vital electric only tools. In addition, items such as drills, polishers, and sanders perform the same tasks as their pneumatic counterparts in some shops.

To protect the operator from electric shock, most power tools are built with an external grounding system. That is, there is a wire that runs from the motor housing, through the power cord, to a third prong on the power plug. When this third prong is connected to a grounded, three-hole electrical outlet, the grounding wire will carry any current that leaks past the electrical insulation of the tool away from the operator and into the ground of the shop's wiring. If the tool is operated at less than 150 volts, it has a plug like that shown in Figure 3-21A. If it is for use on 150 or 250 volts, it has a

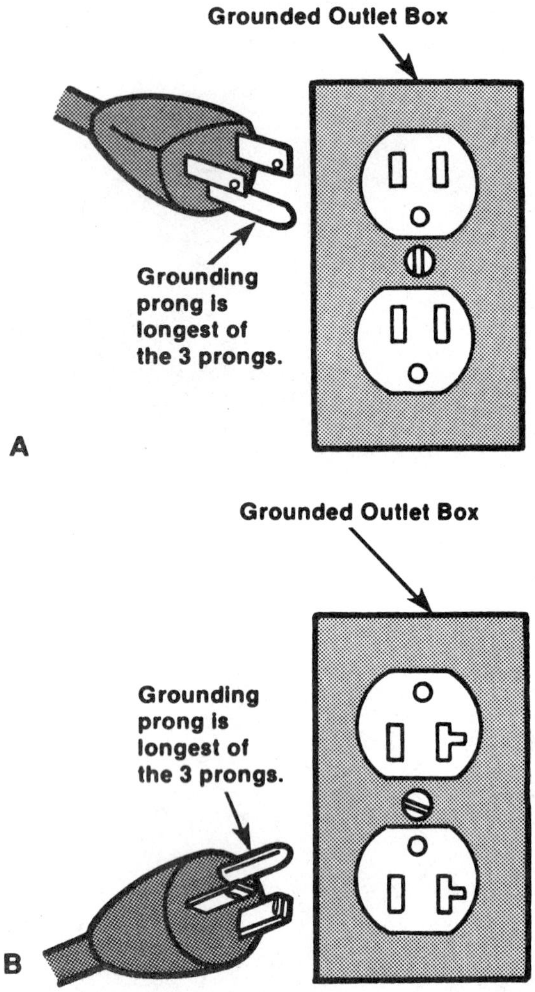

FIGURE 3-21 Approved three-prong grounding plug and outlet box for (A) 115 volts AC and (B) 230 volts AC

plug like that shown in Figure 3-21B. In either type, the green (or green and yellow) conductor in the tool cord is the grounding wire. Never connect the grounding wire to a live terminal.

Some of the new electric power tools are self-insulated and do not require grounding. These tools have only two prongs since they have a nonconducting housing. In shop operations, never use a three-prong adapter plug.

HYDRAULICALLY POWERED EQUIPMENT

Hydraulic equipment uses hydraulic fluid to develop the pressure necessary to operate it. The pressure needed to force the hydraulic fluid into the cylinder is achieved manually (pumping on a handle or lever) or by a small motor (either air or electrically driven).

JACKS AND STRAIGHTENING EQUIPMENT

Hydraulic pressure is used to operate various tools and equipment in the collision repair shop. These range from jacks to large frame/panel pulling and straightening units.

The average shop has approximately a dozen jacks, either air, hydraulic, or a combination of both, depending on preference. (Manual jacks are practically obsolete except for use in confined spaces.)

SHOP TALK

Always use a jack for the tons specified. If a jack is rated for 2 tons, do not attempt to use it for a job requiring any more pressure than that. It is dangerous for the technician and can damage the vehicle.

There are two basic types of frame/panel straighteners on the market: portable and stationary. Portable units are less expensive, but they cannot make as many push and pull actions at one time as can the stationary units.

Body jacks (also known as porta-powers) can be used with frame/panel straighteners or by themselves. To perform the many different straightening operations involving pushing, pulling, or holding a panel to align it, a large assortment of adapters is included in a typical kit. There is also a body repair set, a mechanical set, and even a rescue set, which is used during vehicle accidents.

HYDRAULIC LIFTS

The traditional stationary in-ground lift was usually found only in service stations, muffler shops, transmission shops, and tire dealers for reasons such as oil and lube jobs, brake service, and other underbody repairs. Today, collision repair shops are looking for ways to get the vehicle off the ground to estimate and repair. Damage is easier to evaluate on a vehicle that is up on a lift, and raising a vehicle is a must for many service procedures.

As shown throughout this book, there are several ways to get a vehicle off the ground. Four-post and two-post hoists allow total movement under the vehicle, but they take up more than a work stall in length and width. Side post hoists are great for estimating, but some access to the sides of the vehicle is impaired. Old center post hoists make some areas under the vehicle hard to reach but take up less space.

Raising a vehicle on a lift or a hoist requires special care. Drive-on lifts are fairly safe, but it is important to make sure that the vehicle's exhaust system has enough clearance before driving onto the ramps.

Adapters and hoist plates must be positioned correctly on twin post and rail-type lifts to prevent damage to the underbody of the vehicle. The exhaust system, tie-rod, rod bracket, and shock absorbers could be damaged if the adapters and hoist plates are incorrectly placed.

There are specific contact points to use for lifting the vehicle where the weight is evenly supported by the adapters or hoist plates. The correct lifting points can be found in the vehicle's service manual. Figure 3-22 shows typical locations for a unibody vehicle. This diagram is for example only; always use the manufacturer's instructions for the specific lifting procedure.

FASTENERS

Dozens of types of fasteners are used on a vehicle (Figure 3-23). Each one is engineered for a specific application and for conditions that are encountered in vehicle operation. Always replace fasteners with exactly the same type that was removed from the OEM assembly—do not attempt to re-engineer the vehicle. Keep in mind that using an incorrect fastener or a fastener of inferior quality can result in premature failure and possible injury to the vehicle occupants.

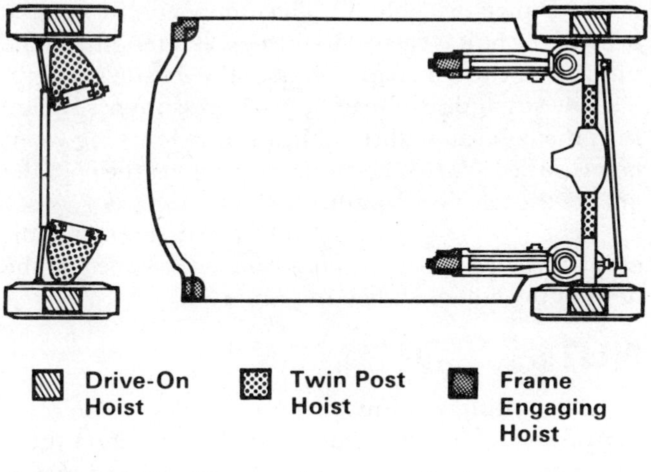

FIGURE 3-22 When raising a unibody vehicle on a lift or hoist, always use the lifting points given in the service manual.

FIGURE 3-23 Many types of fasteners are used on the automobile.

When replacing fasteners, be sure to observe the following precautions:

- Always use the same number of fasteners as used by the OEM.
- Always use the same diameter, length, number of threads (pitch), and type of fasteners as used by the OEM.
- Always observe the OEM's recommendation given in the service manual for tightening sequence, tightening steps (increments), and torque values.
- Always replace a used cotter pin.
- Always replace stretched fasteners or fasteners with damaged threads.
- Always use the correct washers and pins as specified by the OEM.
- Always replace "one-time" fasteners.

As just mentioned, pitch is the term used to denote the number of threads on a screw or bolt. Keep in mind that thread length and total length are two separate measurements. A number of terms have been used over the years to identify the various types of threads. The terms most commonly employed in the automotive trade—the United States Standard (USS), the American National Standard (ANS), and the Society of Automotive Engineers Standard (SAE)—have all been replaced by the Unified National Series. The Unified National Series consists of four basic classifications:

- Unified National Coarse (UNC or NC)
- Unified National Fine (UNF or NF) (SAE)
- Unified National Extrafine (UNEF or NEF)
- Unified National Pipe Thread (UNPT or NPT)

The two common metric threads are coarse and fine and can be identified by the letters SI (System International d'Unites or International System of Units) ISO (International Standards Organization).

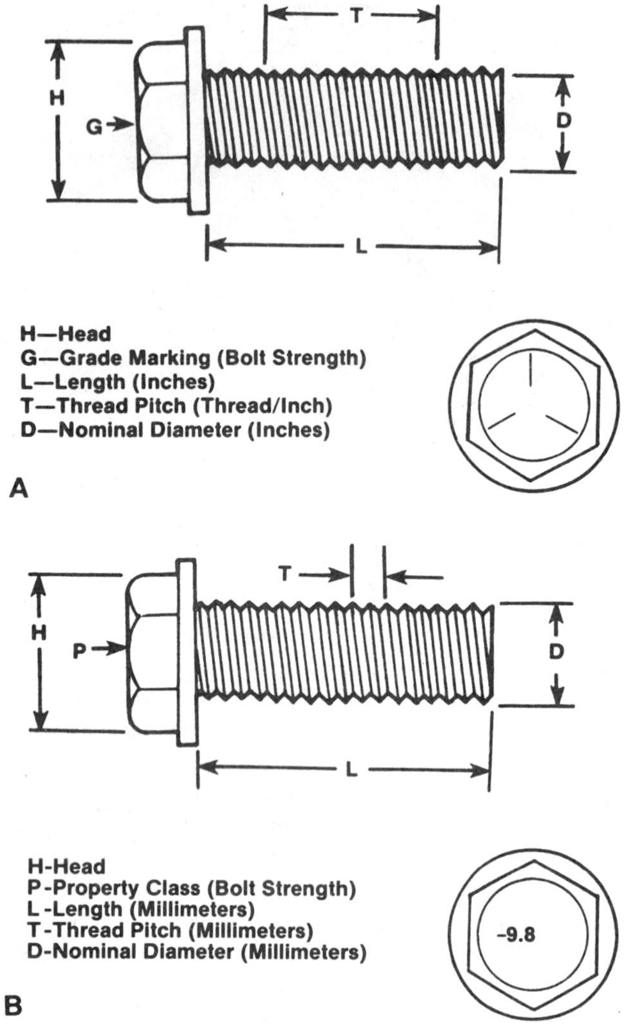

FIGURE 3-24 (A) English and (B) metric bolt measurement

TABLE 3-1: STANDARD BOLT HEAD SIZES

Common English (U.S. Customary) Head Sizes	Common Metric Head Sizes
Wrench Size (inches)*	**Wrench Size (millimeters)***
3/8	9
7/16	10
1/2	11
9/16	12
5/8	13
11/16	14
3/4	15
13/16	16
7/8	17
15/16	18
1	19
1-1/16	20
1-1/8	21
1-3/16	22
1-1/4	23
1-5/16	24
1-3/8	26
7/16	27
1-1/2	29
	30
	32

* The wrench sizes given in this chart are not equivalents, but are standard head sizes found in both inches and millimeters.

BOLTS

To identify bolts properly, bolt terminology should be defined (Figure 3-24). The head is used to torque or tighten the bolt. A socket or wrench fits over the head, which enables the bolt to be tightened. Common USC (United States Customary) and metric sizes are given in Table 3-1. Some USC and metric sockets are very close in size. It is important not to use metric sizes for USC bolts, or USC sizes for metric bolts. The bolt heads can be damaged.

Bolts are made from different materials having various degrees of hardness. Softer metal or harder metal can be used to manufacture the bolts. Under certain conditions, the standard hardness used in a particular situation is not sufficient. Therefore, bolts are made with different hardnesses and strengths for use in different situations. Bolts are graded by SAE and marked with lines on the top of the head to identify bolt hardness as shown in Table 3-2. The number of lines on the head of the bolt is related to the tensile strength. As the number of lines increases so does the tensile strength. Tensile strength is the amount of pressure per square inch the bolt can withstand just before breaking when being pulled apart. The harder or stronger the bolt, the greater the tensile strength.

The hardness or strength of metric bolts is indicated by using a property class number stamped on the head of the bolt as shown in Figure 3-25.

NUTS

Many different nuts are used by the automotive industry (Figure 3-26). But be wary of hexagon (hex) nut rotation on power wrenches. It is deceptively easy to place a bolt into a yield condition within seconds. Im-

TABLE 3-2: STANDARD BOLT STRENGTH MARKINGS

SAE GRADE MARKINGS					
DEFINITION	No lines: unmarked indeterminate quality SAE Grades 0-1-2	3 Lines: common commercial quality Automotive & AN Bolts SAE Grade 5	4 Lines: medium commercial quality Automotive & AN Bolts SAE Grade 6	5 Lines: rarely used SAE Grade 7	6 Lines: best commercial quality NAS & Aircraft Screws SAE Grade 8
MATERIAL	Low Carbon Steel	Med. Carbon Steel Tempered	Med. Carbon Steel Quenched & Tempered	Med. Carbon Alloy Steel	Med. Carbon Alloy Steel Quenched & Tempered
TENSILE STRENGTH	65,000 psi	120,000 psi	140,000 psi	140,000 psi	150,000 psi

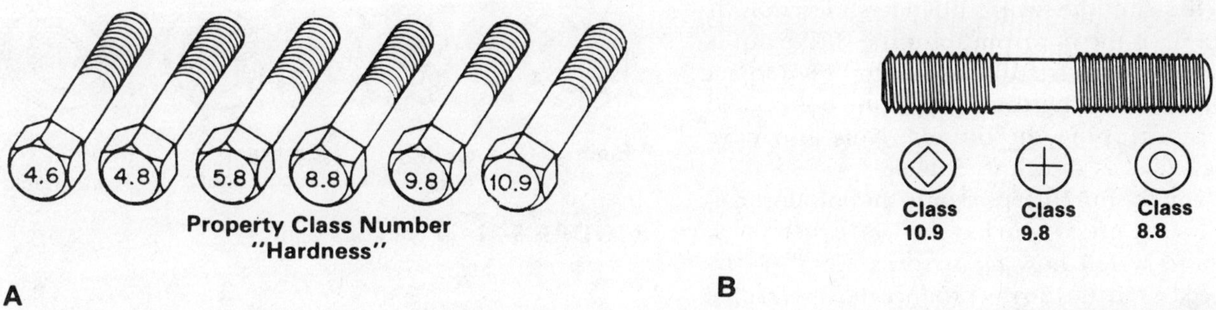

FIGURE 3-25 (A) Metric bolts are rated in hardness by using a property class number. (B) Smaller studs use a geometric code on the end.

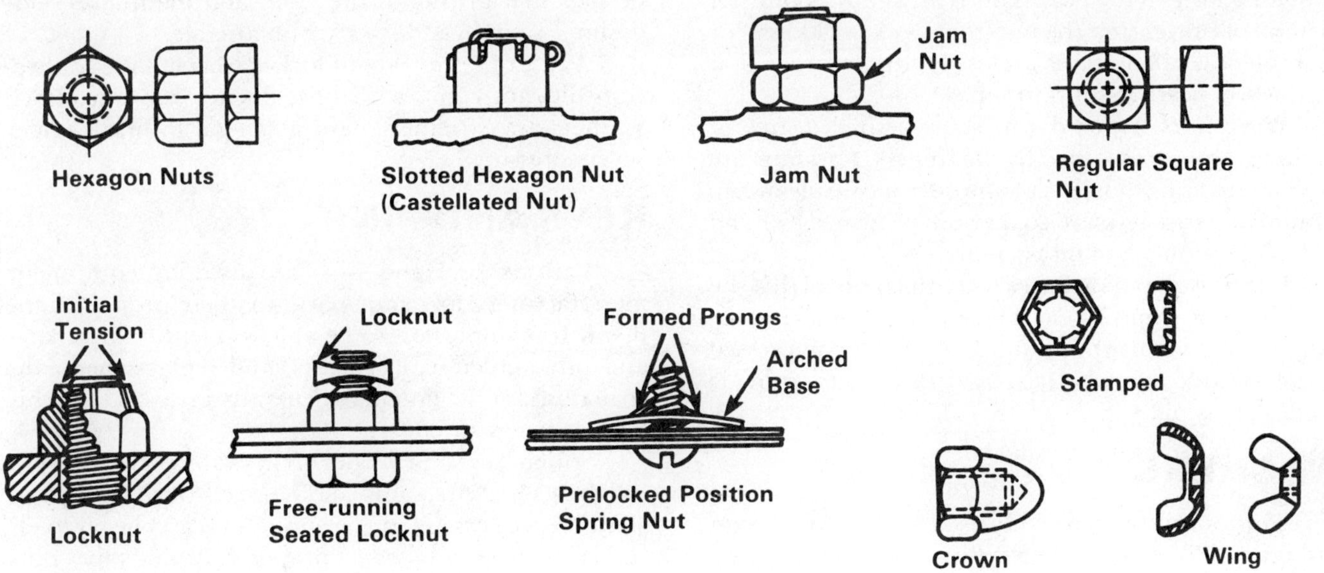

FIGURE 3-26 Each of these nuts has a specific purpose and automotive application.

TABLE 3-3: STANDARD NUT STRENGTH MARKINGS

SAE System		Metric System	
Grade	**Identification**	**Class**	**Identification**
Hex Nut Grade 5	 3 Dots	Hex Nut Property Class 9	 Arabic 9
Hex Nut Grade 8	 6 Dots	Hex Nut Property Class 10	 Arabic 10
Increasing dots represent increasing strength		Can also have blue finish or paint dab on hex flat. Increasing numbers represent increasing strength.	

pact wrenches are the worst offenders. Friction is needed to prevent the nut from spinning. If the nut is lubricated, there is no friction left to stop the impact wrench from hammering the nut past the bolt's yield point and/or stripping the threads. Nuts also have strength markings as shown in Table 3-3.

Do not run the nut full speed onto the bolt. Instead, run it up slowly until it contacts the work, then mark the socket and watch how far it turns. Smaller air-powered speed wrenches do not produce the severe force of impact wrenches and are much safer to use. Follow this procedure with a torque wrench as well.

Castellated nuts are slotted so that a safety wire or cotter pin can be pushed through the slots and into a matching hole in the bolt. This is an almost foolproof method of preventing the nut from working loose. For example, castellated nuts are used with the studs that hold wheel bearings in position.

Wing nuts are used where the desired degree of tightness can be obtained by the fingers. Cap nuts are used where appearance is an important consideration. Thumb nuts are knurled so they can be turned by hand for easy assembly and disassembly.

Elastic stop nuts are used where it is crucial that the nut does not come loose. These nuts have a fiber or composition washer built into them that is compressed automatically against the screw threads to provide holding tension.

WASHERS

The four most commonly used washers are shown in Figure 3-27. Flat washers are used to back up boltheads and nuts and to provide larger load bearing surfaces. They prevent damage to the surfaces of the metal parts.

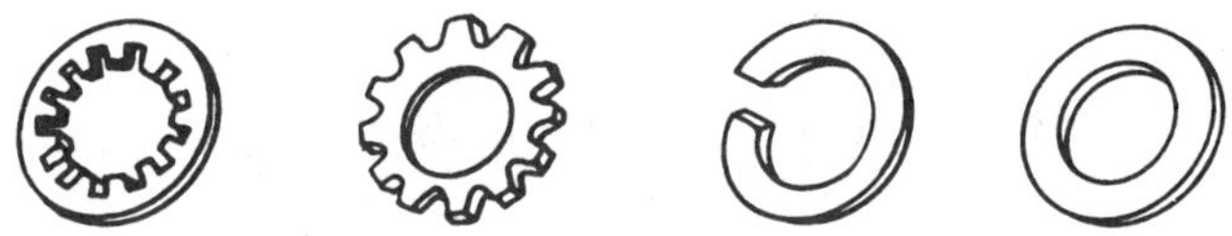

FIGURE 3-27 Types of washers

Split lock washers are used under nuts to prevent loosening by vibration. The ends of these spring-hardened washers dig into both the nut and the work to prevent slippage. Shakeproof lock washers have teeth or lugs that grip both the work and the nut. Several designs, shapes, and sizes are obtainable.

A rule of thumb about lock washers: If the connection did not come with one, do not add one. Lock washers are extremely hard and tend to break under severe pressure.

KEYS AND PINS

Various keys and pins are used by equipment manufacturers to retain parts in their proper alignment. It is important to be able to identify these keys and pins, order replacements, and replace them. Instructions for the latter are generally given in the vehicle's service manual.

Square keys and Woodruff keys are used to prevent handwheels, gears, cams, and pulleys from turning on a shaft. These keys are strong enough to carry heavy loads if they are fitted and seated properly. Round taper pins are used to locate and position matching parts. They can also be used to secure small pulleys and gears to shafts. These pins usually have a taper; holes to accommodate them must be reamed with tapered reamers. If this is not done, the pin will not fit properly.

Dowel pins are used to position and align the units or parts of an assembly. One end of a dowel pin is chamfered, and it is usually 0.001 to 0.002 inch greater in diameter than the size of the hole into which it will be driven. When it is necessary to replace a dowel pin, be sure that it is the same size as the old one.

Cotter pins are used to secure screws, nuts, bolts, and pins. They are also used as stops and holders on shafts and rods. Some cotter pins are more resistant to corrosion than others. Regardless of shape, all cotter pins are used for safety and should never be reused.

The cotter pin should fit neatly into the hole with very little side play. In the preferred installation method (Figure 3-28A), the bent prong above the bolt end should not extend beyond the bolt diameter. Additionally, the bent prong should not rest against the surface of the washer. Cut the prongs down to size if necessary.

If the optional wraparound method is used (Figure 3-28B), the prongs should not extend outward but should be bent over a reasonable radius to the sides of the nut. Sharply angled bends invite breakage. Usually the initial bending of the prongs of a cotter pin is accomplished with needle nose or diagonal pliers; the best tool for final bending of the prongs is a soft-faced mallet.

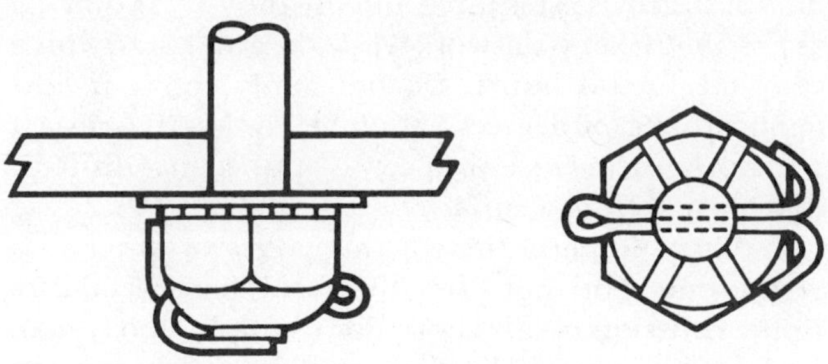

FIGURE 3-28 (A) Preferred and (B) wraparound methods of cotter pin installation

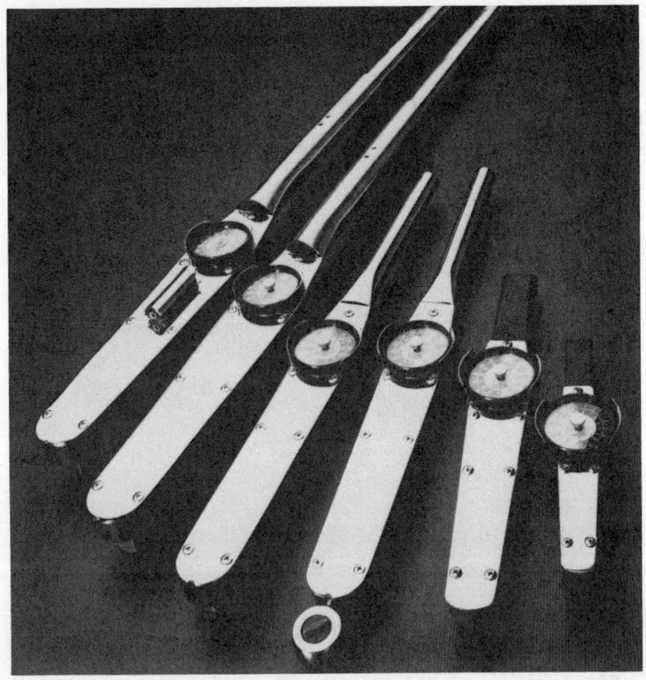

FIGURE 3-29 Dial-type torque wrenches

TORQUING PRINCIPLES

All bolts and nuts used on the automobile should be torqued or tightened to specifications. Many repair manuals list all common torque specifications. A torque wrench (Figure 3-29) is used to measure the tightening force. The correct torque in both inch-pounds and/or foot-pounds, as well as metric measurements, should always be followed when tightening bolts and nuts. If the bolt is tightened to a value less than that given by the specification, the bolt can vibrate loose. The bolt can also be tightened too much. This can cause the threads to strip or the bolt to break. In either case, damage to other automobile parts can result. Always torque all bolts and nuts to the manufacturer's specifications. If there are no specifications listed, use the torque specifications shown in Table 3-4. Similar charts are available for metric torque.

TABLE 3-4: STANDARD BOLT AND NUT TORQUE SPECIFICATIONS

Size Nut or Bolt	Torque (foot-pounds)	Size Nut or Bolt	Torque (foot-pounds)	Size Nut or Bolt	Torque (foot-pounds)
1/4-20	7-9	7/16-20	57-61	3/4-10	240-250
1/4-28	8-10	1/2-13	71-75	3/4-16	290-300
5/16-18	13-17	1/2-20	83-93	7/8-9	410-420
5/16-24	15-19	9/16-12	90-100	7/8-14	475-485
3/8-16	30-35	9/16-18	107-117	1-8	580-590
3/8-24	35-39	5/8-11	137-147	1-14	685-695
7/16-14	46-50	5/8-18	168-178		

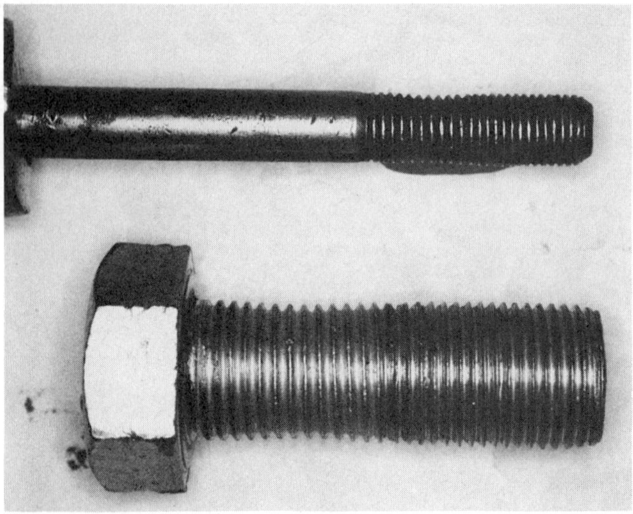

FIGURE 3-30 These bolts have been torqued past their yield points. Notice the "coke bottle" effect.

Proper use of torque will avoid this yield condition. Torque values are calculated with a 25 percent safety factor below the yield point. If a bolt is tightened beyond its yield point, it can no longer safely carry the load it was designed to support. There are some fasteners, however, that are torqued intentionally just barely into a yield condition, although not far enough to create the "coke bottle" shape of a necked out bolt (Figure 3-30). This type of fastener, known as a torque-to-yield (T-T-Y) bolt, will produce 100 percent of its intended strength, compared to 75 percent when torqued to normal values. These fasteners, however, should not be reused, unless otherwise specified in the service manual.

SELF-LOCKING FASTENERS

Several designs of self-locking bolts and nuts are used in automotive applications. These are sometimes referred to as "prevailing torque fasteners." There are many different designs available. As with all fasteners, be sure to replace these bolts and nuts with the same type and torque to specifications.

Prevailing Torque Bolts. These are bolts that incorporate a system to develop an interference between the nut and bolt or the tapped hole threads. Interference is achieved by distorting some of the threads (several methods exist), by applying a nylon patch or strip, or by applying an adhesive coating on the threads.

Prevailing Torque Nuts. These are nuts that incorporate a system to develop an interference between bolt and nut threads. Interference is most commonly achieved by distorting the top of the all-metal nut, but it also may be achieved by distorting at the middle of the hex flat, by a nylon patch on the top nut, and by a nylon insert through the nut.

FASTENING WITH ADHESIVES

The two types of adhesives most often used with automotive metals are acrylic resin and epoxy. Acrylic resin is a two-part adhesive (liquid and powder) that is waterproof and not affected by gas or cracks in objects that hold water. It is most important to follow the manufacturer's directions on the container to get the proper proportions for the job. Drying and setting times are controlled by the amounts mixed; for example, 3 parts of powder to 1 part of liquid will set in about 5 minutes at 70 degrees Fahrenheit. Changing these proportions will allow for faster or slower drying and setting times. Acetone can be used as a cleaning solvent. While acrylic resin provides a very strong bond, it sets too fast for large area work and is generally used for heavy-duty repairs.

Epoxy comes in two separate containers, usually tubes; one contains the epoxy resin and the other contains a hardener. Read the instructions for the proper quantity to use. If both resin and hardener are not in proper proportion, the bond will be weak and might fail. Some epoxy tubes automatically dispense the correct amount of resin and hardener.

Once mixed, the adhesive remains in a workable condition for only a brief time. Therefore, try to mix only as much as is required and use the glue as quickly as possible. Clamp the work while the glue cures, which can take several hours. Do not apply epoxy in low temperatures (50 degrees Fahrenheit or less), because it will not harden. Once an epoxy is applied, it is difficult to remove it from a surface.

When properly mixed, this adhesive provides a very strong bond, not a flexible one. It has several uses in the collision repair shop. For example, epoxy does not shrink when it hardens and is waterproof and heat-resistant.

The use of adhesives for door skin bonding and similar applications is covered in detail in Chapter 11. Chapter 13 discusses the two types of adhesive repair systems used with automotive plastics: two-component and cyanoacrylate.

REVIEW QUESTIONS

1. Technician A uses metric and SAE wrenches interchangeably. Technician B says the two types are not interchangeable. Who is correct?
   a. Technician A
   b. Technician B

c. Both A and B
d. Neither A nor B

2. What is used to straighten damaged threads on bolts or studs?
   a. dolly
   b. tap
   c. die
   d. pull rod

3. Which type of screwdriver is ideal for use as a pry bar?
   a. clutch head
   b. Torx
   c. Pozidriv
   d. none of the above

4. Which of the following is best for removing small dents in automotive sheet metal?
   a. bumping hammer
   b. body hammer
   c. body spoon
   d. picking hammer

5. To cut sheet metal, use ____________.
   a. panel cutters
   b. tin snips
   c. both a and b
   d. neither a nor b

6. Technician A routinely reuses torque-to-yield bolts. Technician B does not. Who is correct?
   a. Technician A
   b. Technician B
   c. Both A and B
   d. Neither A nor B

7. What is the most commonly used portable grinder in collision repair and refinishing shops?
   a. disc-type
   b. cutoff
   c. angle
   d. vertical

8. When operating an air hammer, Technician A uses a chisel retainer. Technician B does not. Who is correct?
   a. Technician A
   b. Technician B
   c. Both A and B
   d. Neither A nor B

9. Body jacks are used ____________.
   a. by themselves
   b. with frame panel straighteners
   c. both a and b
   d. neither a nor b

10. When replacing fasteners, always use ____________.
    a. half the number used by the OEM
    b. 30 percent more than the number used by the OEM
    c. 5 percent more than the number used by the OEM
    d. the same number used by the OEM

11. What type of nuts are slotted to accommodate a safety wire or cotter pin?
    a. wing
    b. thumb
    c. castellated
    d. elastic stop

12. What are the two types of adhesives used most often with automotive metals?
    a. acrylic resin and hardener
    b. liquid and powder
    c. acrylic resin and epoxy
    d. flexible epoxy and rigid epoxy

13. Self-locking fasteners are sometimes referred to as ____________ fasteners.
    a. prevailing torque
    b. interference
    c. power
    d. no-torque

14. For total movement when working under a vehicle, Technician A uses a two-post hoist. Technician B uses a four-post hoist. Who is correct?
    a. Technician A
    b. Technician B
    c. Both A and B
    d. Neither A nor B

15. Which type of washer is used under nuts to prevent loosening by vibration?
    a. flat
    b. shakeproof lock
    c. split lock
    d. hexagon

CHAPTER

4 Estimating Repair Costs

OBJECTIVES

- Explain how damage repair estimates are determined.
- Identify and explain the three most common abbreviations used in crash estimating guides.
- Describe the method of determining the repairability of a damaged vehicle based on observation and by consulting appropriate sources.
- Explain the difference between flat-rate labor time and overlap labor time when making a cost estimation.
- Make a rough estimate of the time required to refinish a given collision repair job.

Nearly any collision damaged automobile can be restored to its pre-accident state if the owner is willing to pay for the repair. Cost is a major consideration for both the owner and the collision repair shop. Insurance companies generally require at least two written estimates from two separate shops. In most cases, the shop that submits the lowest estimate generally gets the work. Therefore, the profit and loss margins for a collision repair shop depend heavily on the accuracy of estimates.

Estimates should be accurately written. In order for an estimate to clearly establish a true cost of repairs, a thorough damage analysis must be performed on the damaged vehicle. Every effort must be made by the estimator and insurance appraiser to locate all the damage present so that the vehicle can be repaired right the first time. The customer is entitled to have his or her vehicle properly repaired to pre-accident state. An estimate that does not indicate all damage present or is poorly written will not produce a profitable or accurate repair for the collision repair shop. However, if the estimate is priced too high, the shop probably will not get to repair the vehicle. An estimate must be fair, accurate, and include all the damage present.

Although it is not the purpose of this chapter to make a collision repair technician into an estimator, it is very important for he or she to understand how estimates are determined and, in some cases, how work orders are prepared from them (Figure 4-1). After all, it is the technician who must do the work. In most successful collision repair, it is a cooperative effort among the estimator, insurance personnel, and technicians that leads to a fair and accurate estimate and a satisfied customer.

FIGURE 4-1 The collision repair technician should always review every line of the work order.

THE ESTIMATE

Before any decision involving the repair of a damaged vehicle can be made, a detailed repair estimate, sometimes called a damage appraisal or physical damage report, must be made (Figure 4-2). Before writing the repair estimate, fill in all preliminary information that identifies the owner of the vehicle and who will pay the bill. It is good business practice to establish the method of payment as well. Also record the make, year, body style or type, license plate number, mileage, and date. The estimate should also give a detailed description of all the labor operations that must be performed and a complete listing of parts and materials needed to make the repairs. This information can help prevent any misunderstanding between the shop and the owner of the vehicle.

At least three copies of the written estimate should be made. One is kept by the shop, one is given to the insurance company, and the other is given to the customer. An estimate is a firm bid for a given period of time—usually for 30 days. The reason for a given time period is obvious: part prices change and damaged parts can deteriorate.

The estimate is also considered the authorization to complete the repair work as listed, but only when it is

ESTIMATE OF REPAIRS

Date Sept 27 19 89

Name Anne Henkel Address RD#1 City Tamaqua Home Phone 467-0089

Make Buick Year 89 Model LeSabre I.D. Number 1G4AZ5AAXKD146792 Body Style 2-Dr Work Phone 678-0098

Mileage 7,987 License No. RN-PD1 **Insurance Co.** IDEAL Adjuster T. Smith

	Re-pair	Re-place	ESTIMATE OF REPAIR COSTS	PAINT	LABOR HRS.	PARTS	NET + SUBLET
1		X	Face Bar w/o bumper guards		2.0	149.00	
2		X	Cover	2.5	Inc.	242.00	
3		X	Moulding, Cover center		Inc.	15.05	
4		X	Cap Moulding		Inc.	1.30	
5		X	Moulding Cover outer	0.3	Inc.	8.95	
6		X	Support, Cover		Inc.	2.40	
7		X	Reinforcement cover outer		Inc.	4.40	
8		X	Reinforcement cover inner		Inc.	2.10	
9		X	Retainer, outer		Inc.	1.45	
10		X	Outer impact bar brace		Inc.	9.65	
11		X	Center impact bar brace		Inc.	12.80	
12		X	Impact absorber		0.3	48.25	
13		X	Absorber plate		Inc.	3.00	
14		X	Absorber Mounting bracket		Inc.	1.45	
15		X	Plate stud		Inc.	1.50	
16		X	Rear combination lamp lens		.5	74.00	
17		X	Rear combination lamp housing		Inc.	56.00	
18		X	Rear combination ornament 2@8.40		Inc.	16.80	
19		X	Bulb		Inc.	.60	
20		X	Stop lamp (high mounted)		0.3	38.50	
21		X	Side marker lamp assembly		0.3	8.75	
22		X	Back up lamp lens		0.2	11.25	
23		X	Rear lower Panel	1.0	5.0	61.25	
24		X	Rear Panel trim finish	0.8	3.0	135.00	
25		X	Moulding Lower		0.2	28.25	
			Continued on 2nd. Sheet **Total**				

THIS ESTIMATE IS BASED ON OUR INSPECTION AND DOES NOT COVER ADDITIONAL PARTS OR LABOR WHICH MAY BE REQUIRED AFTER THE WORK HAS BEEN STARTED. AFTER THE WORK HAS STARTED, WORN OR DAMAGED PARTS WHICH ARE NOT EVIDENT ON FIRST INSPECTION MAY BE DISCOVERED. NATURALLY THIS ESTIMATE CANNOT COVER SUCH CONTINGENCIES. PARTS PRICES SUBJECT TO CHANGE WITHOUT NOTICE. THIS ESTIMATE IS FOR IMMEDIATE ACCEPTANCE.

ESTIMATE GOOD FOR 30 DAYS.

The undersigned authorizes the above repairs for $________. Of this amount the above named insured is to pay $________ insurance deductible

By: ________________

SHOP COPY

LABOR ____ HRS. @ $ ____	$
PAINT ____ HRS. @ $ ____	$
PARTS (subject to invoice)	$
PAINT/MATERIALS	$
SUBLET ITEMS	$
TOWING/STORAGE	$
SUB TOTAL	$
SALES TAX	$
GRAND TOTAL	**$**

FIGURE 4-2 Typical repair estimate

B & J
Collision Estimating Services
20 WEST BROAD ST.
AUBURN, PA 17922
TELEPHONE: (717) 555-7764

ESTIMATE OF REPAIRS
№ 002128

SHEET NO. 1 OF 1 SHEETS

NAME	ADDRESS	PHONE	DATE
KAREN Miller	1143 RAILROAD ST. CRESSONA, PA. 17929	HOME 395-2719 BUS. 623 7347	12-17-89

YEAR	MAKE	MODEL	LICENSE NO.	MILEAGE	SERIAL/V.I. NO.
1989	ford	P.V.	MAE 917	14864	1FTDF15Y5GNA69994

INSURANCE COMPANY	TYPE OF INSURANCE	ADJUSTER	PHONE	CAR LOCATED AT
Amerisure	COL			MFG. 4/89

PARTS NECESSARY AND ESTIMATE OF LABOR REQUIRED	PAINT COST ESTIMATE		PARTS COST ESTIMATE		LABOR COST ESTIMATE	
(1) FRONT FACE BAR Chrome no gds or PADS			205	82		.5
(1) " STONE deflector	1	0	40	50		.5
(1) Left HeadLamp door (with argent GRILL)			52	32		2
(1) " " Shield			3	20		
(1) " front fender	3	1	133	00	1	6
(1) " " " APRON			60	47	1	0
(2) Wheels 15" 55.60 ea.			111	20		6
(2) stems and Balance			2	50		5
(2) HUB CAPS			43	46		—
Repair Radiator Support			—		2	0
Align Front End			—		1	5
(1) Left door trim Panel			73	40		
Stripe left front fender			15	00		.5
LABOR 25.0 HRS @ 23.00			575	00		
(note may be front suspension damage)						
PAINT MAT.			89	10		
UNDERCOAT			15	00		
TOTALS			1,419	.97		

INSURED PAYS $________ INS. CO. PAYS $________ R.O. NO.________

INS. CHECK PAYABLE TO________

The above is an estimate, based on our inspection, and does not cover additional parts or labor which may be required after the work has been opened up. Occasionally, after work has started, worn, broken or damaged parts are discovered which are not evident on first inspection. Quotations on parts and labor are current and subject to change. Not responsible for any delays caused by unavailability of parts or delays in parts shipment by supplier or transporter.

ESTIMATOR________

AUTHORIZATION FOR REPAIRS. You are hereby authorized to make the above specified repairs to the car described herein.

SIGNED X________ DATE________ 19__

GRAND TOTAL	1,419.97
TOWING & STORAGE	
TAX	85.20
TOTAL OF ESTIMATE	$ 1,505.17

FIGURE 4-3 Typical work order

agreed upon and signed by the owner or by the owner and insurance company appraiser. The estimate explains the legal conditions under which the repair work is accepted by the collision repair shop and protects the shop against the possibility of undetected damage that might be revealed later as repairs progress.

Many insurance policies contain a deductible clause. This means that the owner is responsible for the first given amount of the estimate (usually $100 to $500) and the remaining cost is paid by the insurance company. In such cases, both the customer and insurance company should authorize the estimate.

Another important function of the estimate is that it serves as a basis for writing the work order or operational plan. Usually prepared from the damage appraisal of the estimator (using the written estimate) and a visual inspection by the shop supervisor and/or a technician, the work order (Figure 4-3) outlines the procedures that should be taken to put the vehicle back in top condition. It is also a valuable tool to both the estimator and the shop foreman, since it lists the actual times necessary to do the job.

ESTIMATING AIDS

Crash estimating guides or collision damage manuals (Figure 4-4) are essential tools of the estimator. They contain vehicle identification information, the price of new parts, time needed to install the parts, identification of almost any part of the vehicle, and refinishing data such as a paint code reference. These guides are updated at different times of the year to reflect manufacturers' parts price changes and other revisions.

ABBREVIATIONS USED IN ESTIMATING GUIDES

When using crash estimating guides, it is important to be familiar with the abbreviations that are used. The three most common ones are:

- *R & I: Remove and Reinstall.* The item must be removed as an assembly, set aside, and later reinstalled and aligned for a proper fit. This is generally done to gain access to another part. For example, R & I bumper would mean that the bumper assembly would have to be removed to install a new fender or quarter panel.
- *R & R: Remove and Replace.* Remove the old parts, transfer the necessary items to the new part, install, and align.
- *O/H: Overhaul.* Remove an assembly, disassemble, clean, inspect, and replace parts as needed; then reassemble, install, and align. Overhaul should be used only if the time for the individual parts (less overlap) is more than the overhaul time.

FIGURE 4-4 Crash estimating guides are vital estimating tools.

MICROFILM AND COMPUTER ESTIMATORS

Two of the newest developments in estimating are the microfilm estimator and the computer estimator. Basically, a microfilm estimator is a projector in which a very small film card is inserted and its image is displayed on the estimator screen. Generally, the image is an enlarged page of the crash estimating guide, which makes the information easier to store.

Computers are revolutionizing the art of estimating by making enormous amounts of data readily accessible and by automating the process of producing a written estimate. There are basically two types of computer estimating systems: dial-up systems and resident or in-shop systems.

The dial-up system relies on data stored in a large central storage facility and usually accessed via the telephone. For a fee, estimators enter the necessary data relating to damaged parts, and the computer calculates part prices and labor charges and prints out a completed estimate. Insurance companies prefer this method because it gives them access to statistical data developed as a by-product of the estimating process.

Most collision repair shops, however, prefer resident computer systems that can be used in the shop, are controlled by the estimator, and do not require a fee for each estimate. Some computerized estimating systems rely on the estimator to enter damage information on the shop's computer, which then calculates and prints a completed estimate. But newer systems store an electronic data base containing up-to-date part prices and labor times that can be retrieved on the shop's computer (Figure 4-5). With the use of these sophisticated programs, estimators can now use the stored data to make virtually error-free estimates rapidly at a very low cost. As mentioned in Chapter 1, with the countless other tasks that can be done on a computer—letter writing, payroll, etc.—it is easy to see why it has become such an important management tool.

ESTIMATING SEQUENCE

When estimating any type of damage to a vehicle, a logical sequence must be followed. Before making a written estimate, the estimator should visually inspect the entire vehicle, paying special attention to damaged subassemblies and parts that are mounted to (or part of) a damaged component. It is a good idea at this time to take a few snapshots of the damage to serve as a permanent record of the work to be done. The estimator must consider the same points as the technician does before making any decision on repair work. Most estimators start from the outside of the vehicle and work inward, listing everything on paper—by vehicle section—that is bent, broken, crushed, or missing. For example, if the front grille and some of the related parts are damaged, list the needed repairs or replacements as follows:

FIGURE 4-5 Generating a damage report at a PC workstation using crash estimating guides

Front Grille	Replace
Opening Panel	Straighten and Replace
Deflector (or Valance Panel)	Replace
Headlight Door	Replace
Grille Opening Panel	Refinish

Notice that the parts are listed in a definite sequence according to factory disassembly operations or exploded views as provided in service manuals (Figure 4-6) or crash estimating guides.

The estimator must be on constant guard for primary and secondary damage. Damage that happens to a vehicle at the point of impact is referred to as primary damage.

Another type of damage is hidden or secondary damage which is not at or near the impact zone but is caused by it. This can occur to any part or component on a vehicle that has been involved in a collision. Naturally, if the damage is only minor—creased fender, gouge in a door, headlight assembly smashed, and so forth—the chances of hidden damage occurring are somewhat remote. However, due to the tremendous forces involved in a severe collision, the estimator must always suspect some form of hidden damage whenever a vehicle sustains more than a minor impact.

Severe collisions from any direction often cause the frame or unitized body to distort. At one time the estima-

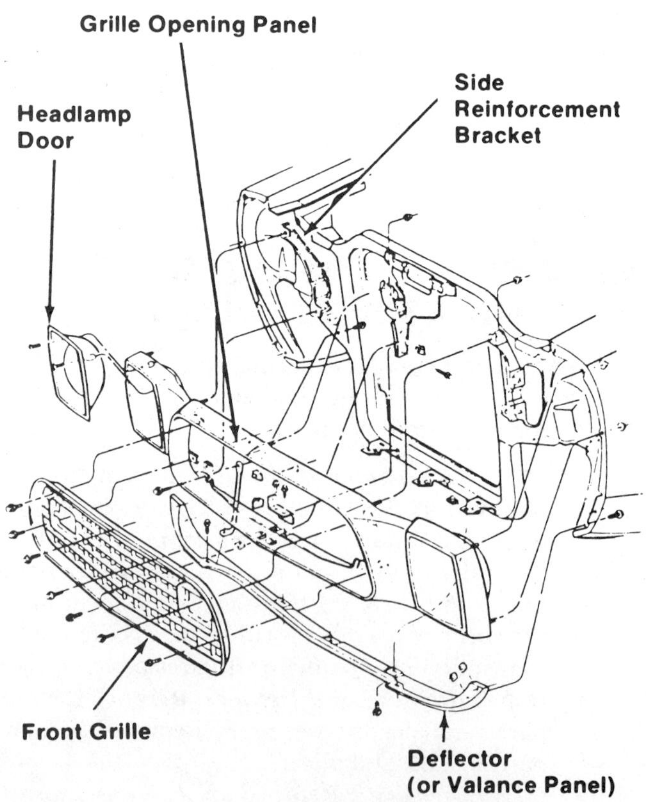

FIGURE 4-6 Typical exploded view found in service manuals and crash estimating guides

tor could check this damage with the naked eye and would be fairly close in giving an estimate. But with the unibody, the "eyeballing" technique is sometimes not enough to detect misalignment. It is far better to use measuring gauges or devices to check for misalignment. (The use of these tools is a must when it comes to checking for misalignment.) However, unless the estimator is thoroughly skilled in the use of such tools, it is wise to consult with an experienced technician to determine the time necessary to get the vehicle back into proper alignment.

Always raise a badly damaged vehicle off the floor so that a good visual inspection can be made of all underbody and drivetrain components (Figure 4-7). In some unibody vehicles, it might even be necessary to remove the drivetrain and suspension components to make a thorough damage inspection.

WARNING

Never support the vehicle on a hydraulic jack to inspect underbody damage; always use jack stands.

Doors out of alignment and cracked stationary glass are often a good clue to hidden damage in the frame or body structure. The operation of instrument panel gauges and lights, as well as the heater, air conditioner, radio, and all other comfort and convenience components, must be carefully checked.

In other words, the estimator must give the entire vehicle—top to bottom and front to rear—a thorough inspection and take nothing for granted. If something is missed on the original estimate, it often becomes difficult to reopen it for further negotiations later. On the other hand, some estimates do include a so-called hidden damage clause that permits added charges to the original estimate if hidden damages are discovered later.

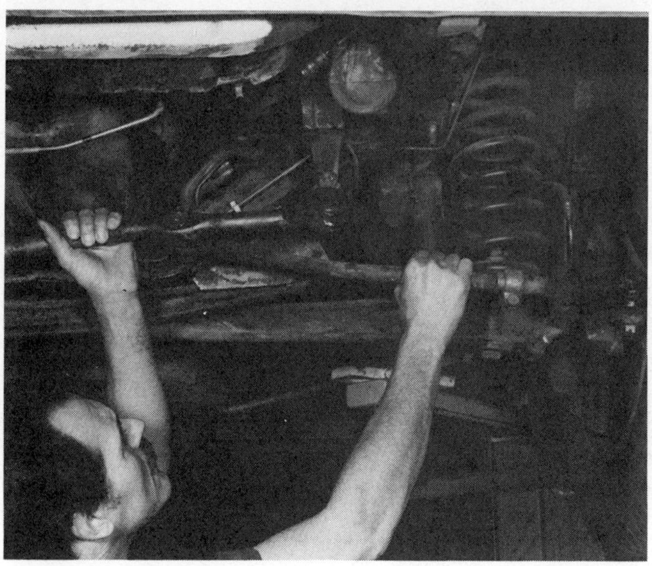

FIGURE 4-7 Determining the extent of underbody damage is much easier if the vehicle is raised off the floor.

FLAT RATE OPERATIONS

Repair operations fall into three categories in the crash estimating guides: remove and replace (R&R), remove and reinstall (R&I), and overhaul (O/H). In addition to these major categories, operation time is also given for drilling holes and for refinishing. However, unless specifically stated in the text or footnotes, the times given in the crash estimating guides are for R&R. It is important that the estimator understands which category each specific repair falls into and that he or she estimates the repair accordingly.

REMOVE AND REPLACE

The first critical decision that the collision repair shop must make involves the repairability of the vehicle. If it was involved in a severe collision, there is a possibility that the total repair costs can exceed the vehicle's market value. For example, a vehicle with a market value of $2,150 and an estimate that totals $2,710 is a situation where repairs are not practical. More than likely the insurance company appraiser will agree with such a decision and would authorize "totaling" the vehicle.

If the vehicle is judged repairable, the next decision is whether or not the damage requires replacement parts or if the parts can be repaired. From an economical (and practical) standpoint, it might be unwise to call for all new parts to replace those that have become bent, twisted, creased, torn, buckled, or broken. As a result, the estimator must have the ability to judge the seriousness of damage that a specific part or unit has received and balance the cost of repairs against the cost of replacement.

As a rule of thumb, repair costs should never exceed replacement costs. If there is some doubt that repairs and straightening will not produce a quality job, then the estimate should list the required new parts. Sheet metal parts usually offer the most opportunities for repair and straightening. As a result, sheet metal repair, replacement, and panel refinishing generally account for the largest number of estimate dollars.

To reduce parts costs, many insurance appraisers and some customers might want to use salvage parts. These are parts that were removed from wrecked vehicles by salvage yards (Figure 4-8) and are usually in good condition. There is always a chance that such parts have been previously damaged and repaired. For

FIGURE 4-8 Salvage yards are excellent sources of used parts.

this reason, it is important to carefully inspect all salvage parts before they are installed to be sure that they are in usable condition.

Each damaged vehicle poses different problems that must be solved in order to arrive at a final decision concerning repair versus replacement. The most difficult questions arise when a vehicle is involved in a major collision. If the frame rails received extensive damage, for instance, the estimator might want to consider listing their replacement rather than the more costly job of installing a complete new frame assembly. Another option is to list the replacement of a front side rail section or a complete front frame section, both of which are available for a number of models. These solutions might be more practical (from a time and cost approach) than making repairs with heat and frame straightening equipment.

On the other hand, estimating damage to the body shell is often a much more difficult job. Knowing what would be needed to restore the body shell to its original dimensions and configuration requires a good deal of experience. If body shell distortion is suspected, it can be checked by careful measurement prior to listing any necessary repair or straightening procedures. Generally, a clue to minor body distortion (twist, sag, sidesway) is if any cracking of stationary glass is noted, yet there is no visible sign of a direct impact with any object.

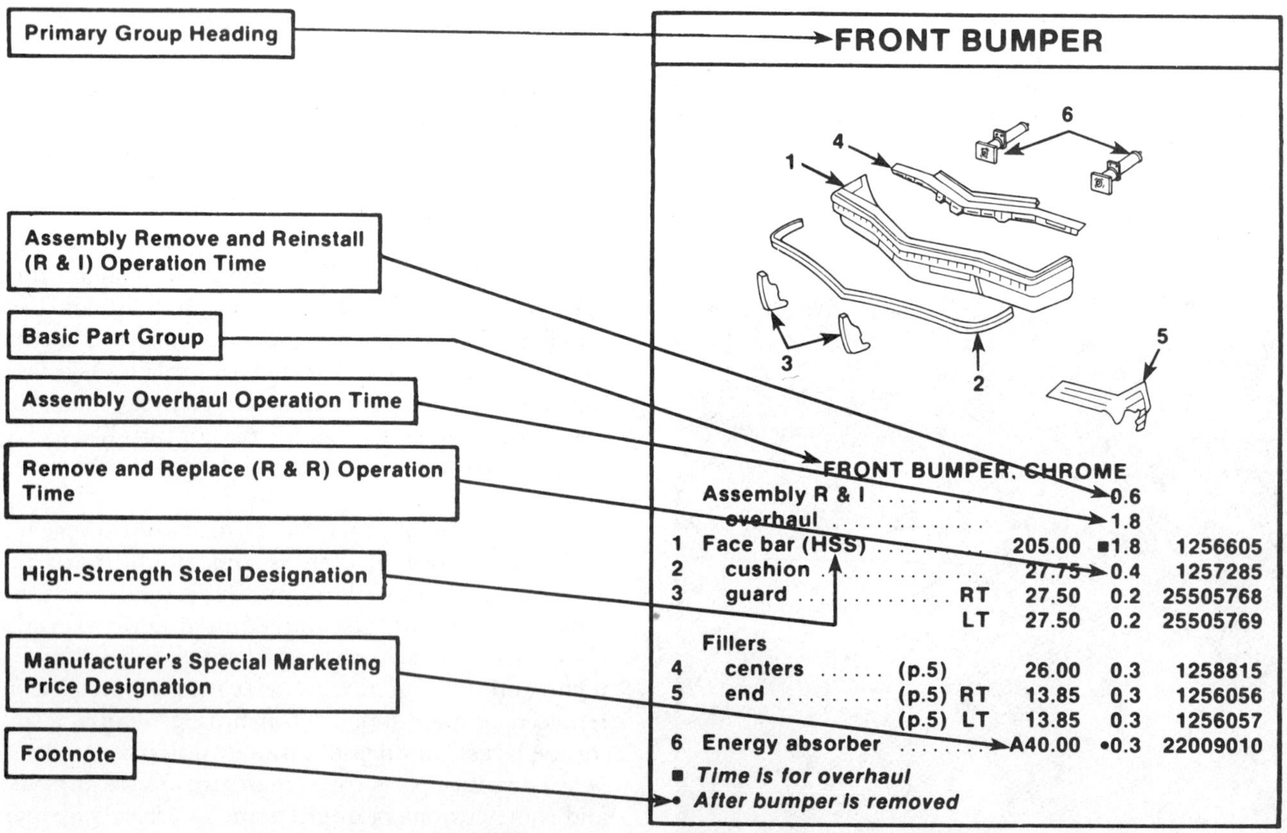

FIGURE 4-9 Typical crash estimating guide information

In some situations, such as a severely damaged front end, an estimator might want to consider the purchase and installation of a used front end assembly. In the body repair trade this is referred to as a "front clip." This assembly generally includes the front bumper and supports, grilles and baffles, the radiator and its supporting members, the hood and its hinges (or hood torsion bars), the front fenders and skirts, all front lights, the wiring, and all other related parts. Often, this method of making collision repairs to the front end will decrease the total time the vehicle is in the shop.

Crash estimating guides can be used as a reference for pricing parts. However, never use them to determine the absolute price. Usually these guides will list the name of the part, the year of the vehicle it will fit, its part number, the estimated time required to replace, and the current price (Figure 4-9). The current prices are the factory-suggested list prices. Parts that have been discontinued are usually listed; the price that appears in the guide is the last available one.

OVERHAUL

Overhaul (O/H) times allow for removal of a subassembly such as a bumper or suspension assembly; disassembly, inspection, and replacement of damaged parts; and reinstallation and adjustment of the subassembly on the vehicle.

Overhaul time is actually a short cut for the estimator. The crash estimating guides list overhaul time most often in connection with bumpers. It is very common for a vehicle involved in a collision to have bumper damage. But, in most cases, many of the individual parts can be salvaged, so that an overhaul of the assembly is in order. To correctly estimate the disassembly, inspection, replacement, reassembly, and adjustment of a bumper and its individual parts, an estimator would have to add up all the operation times for each part, plus figure out overlap for each one. Crash estimating guides do that for the estimator by listing the overhaul time.

SECTIONING

Another operation that may be flat rated in the crash estimating guides is sectioning. Oftentimes it is less expensive and time consuming to section a damaged panel and weld part of a new panel in place than to replace the entire panel. As discussed earlier, rails and other parts may be sectioned (Figure 4-10). Sectioning disturbs less OEM corrosion protection and spot welds. In some cases, it is the best repair procedure to choose. However, the crash estimating guides do not always provide flat rate time for sectioning.

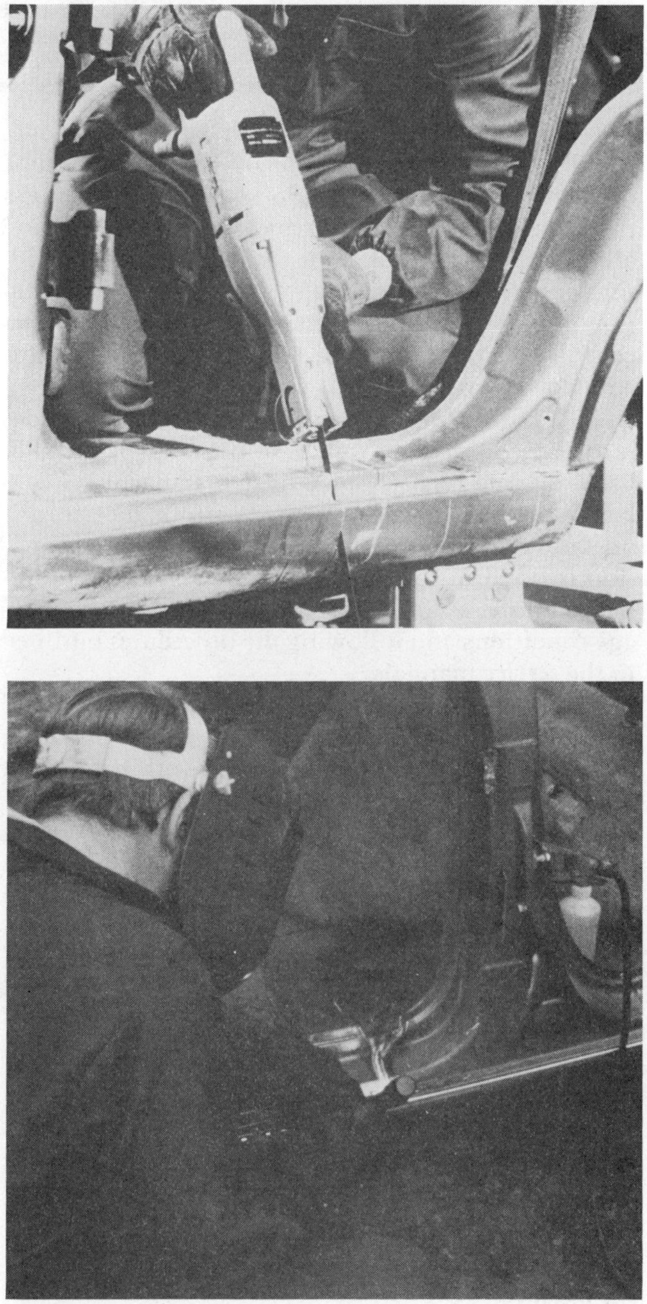

FIGURE 4-10 Sectioning, shown in the examples above, is often the preferred repair procedure.

OTHER FLAT RATE OPERATIONS

The crash estimating guides also contain flat rate times for drilling operations and painting. The allowances for drilling holes are for drilling round holes only. If square holes or holes of another shape are required, they should be estimated separately.

LABOR COSTS

The procedure pages of crash estimating guides usually provide an explanation of what the labor time does and does not include. For example, replacing a panel or fender includes removing, installing, and aligning; it does not include the installation of moldings or antennas, refinishing, pin striping, or application of decals. Also not considered are rusted or inaccessible bolts, undercoating, and alignment or straightening of damaged adjacent parts (Figure 4-11).

The labor time reported in crash estimating guides is to be used as a guide only. The times are principally based on data reported by vehicle manufacturers who have arrived at them by repeated performance of each operation under normal shop conditions. An explanation is listed in the guide of the established requirements for the average technician, working under average conditions and following the procedures outlined in the service manuals.

The times reported apply only to standard models listed in identification sections of the crash estimating guide. They do not apply to vehicles with equipment other than that supplied by the vehicle manufacturer as standard or regular production options. If other equipment is used, the time might have to be adjusted to compensate for the variables.

All labor times given in crash estimating guides include the time necessary to insure proper fit of the new part. The times are based on new, undamaged parts installed as an individual operation. Additional time has not been added to compensate for collision damage to the vehicle. Removal and replacement of exchanged or used parts is also not considered. If additional aligning or repairs must be made, such factors should be considered when making an estimate.

When jobs overlap, reductions in the labor times must be considered. This occurs when replacement of one component duplicates some labor operations required to replace an adjacent component. For example, when replacing a quarter panel and rear body panel on the same vehicle, the area where the two components join is considered overlap. Where a labor overlap condition exists, less time is required to replace adjoining components collectively than is required when they are replaced individually.

Another labor cost reduction is noted as included operations. These are jobs that can be performed individually but are also part of another operation. As an example, when replacing a door, the suggested time would include the replacement of all parts attached to the door, except for the ornamentation. It would be impossible to replace the door without transferring these parts. Consequently, the time involved in transferring them would be included operations and should be disregarded because the times for the individual items are already included in the door replacement time.

Following are additional factors to be considered when writing an estimate. Such added time is usually negotiated between the estimator and the insurance company or customer.

- Time for the setup of the vehicle on a frame machine and damage diagnosis
- Time for pushing, pulling, cutting, and so on to remove collision-damaged parts
- Time to straighten or align related parts
- Time to remove undercoating, tar, grease, and similar materials
- Time to repair rust or corrosion damage to adjacent parts
- Time for the free-up of rusted or frozen parts
- Time for drilling for ornamentation or mounting holes
- Time for filling or plugging unneeded holes in new parts

1. BUMPER

FRONT OR REAR

NOTE: Disconnect at energy absorber (mounting bracket) or frame mounting.

ASSEMBLY REMOVE AND REINSTALL (R & I)

Included:
- R & I unit as assembly
- Alignment to vehicle

ASSEMBLY OVERHAUL

Included:
- R & I assembly
- Disassemble
- Replace damaged parts
- Reassemble unit
- Align to vehicle

FACE BAR REMOVE AND REPLACE (R & R)

Included:
- R & R face bar
- R & I guards and face bar cushions (unless otherwise noted in text)

ALL BUMPER OPERATIONS

Does Not Include:
- Additional time for frozen or broken fasteners
- Refinishing
- Optional moldings, name plates, emblems, and ornamentation; air bags and lamps
- Stripe tape, decals, overlays
- Removal of hydraulic energy absorbers
- Aim headlamps

FIGURE 4-11 Information found in the procedure pages of a crash estimating guide

- Time to repair damaged replacement parts prior to installation
- Time to check suspension and steering alignment/toe-in
- Time for removal of shattered glass
- Time to rebuild, recondition, and install aftermarket components, not including refinishing time
- Time for application of sound-deadening material, undercoating, caulking, and painting of the inner areas
- Time to replace anti-corrosion material
- Time to R&I main computer module when excessive temperatures (above 176 degrees Fahrenheit) are necessary in repair operations
- Time to R&I wheel or hub cap locks
- Time to replace accessories such as trailer hitches, sun roofs, and fender flares

It is easy to forget about the removal of exterior trim and body sheet metal hardware. These items must often be removed prior to bumping or painting operations and then replaced, creating a labor cost that should not be overlooked. Another thing to keep in mind is that any repairs or replacement of front end parts should include a headlight alignment. The bottom line is that the estimator must charge out every job function needed to restore the vehicle to its pre-accident condition.

Estimators must also consider the materials used in vehicle manufacturing that require special handling by the technician. For example, the use of lightweight, high-strength steel and plastic in various locations requires specific repair procedures. It is not always possible to accurately identify the correct locations on a vehicle where HSS or HSLA steel is used. However, crash estimating guides generally will identify the type of steel with the letter designation next to the description. When in doubt, contact the automobile manufacturer for the specific location(s) of high-strength steel and plastic and the recommended repair procedures.

TOTAL ESTIMATED LABOR COSTS

Once all the repairs and labor times have been entered on the estimate form, refer to the estimating guide's time/dollar conversion table (Figure 4-12). This conversion table is used to convert labor time into dollars to fit local labor or operating rates per hour. When

TIME/DOLLAR CONVERSION TABLE

	Dollar Per Hour Rates										
Time	.50	$1.00	$10.00	$15.00	$20.00	$25.00	$30.00	$35.00	$40.00	$45.00	$50.00
0.6	.30	.60	6.00	9.00	12.00	15.00	18.00	21.00	24.00	27.00	30.00
0.7	.35	.70	7.00	10.50	14.00	17.50	21.00	24.50	28.00	31.50	35.00
0.8	.40	.80	8.00	12.00	16.00	20.00	24.00	28.00	32.00	36.00	40.00
0.9	.45	.90	9.00	13.50	18.00	22.50	27.00	31.50	36.00	40.50	45.00
1.0	.50	1.00	10.00	15.00	20.00	25.00	30.00	35.00	40.00	45.00	50.00
1.1	.55	1.10	11.00	16.50	22.00	27.50	33.00	38.50	44.00	49.50	55.00
1.2	.60	1.20	12.00	18.00	24.00	30.00	36.00	42.00	48.00	54.00	60.00
1.3	.65	1.30	13.00	19.50	26.00	32.50	39.00	45.50	52.00	58.50	65.00
1.4	.70	1.40	14.00	21.00	28.00	35.00	42.00	49.00	56.00	63.00	70.00
1.5	.75	1.50	15.00	22.50	30.00	37.50	45.00	52.50	60.00	67.50	75.00
1.6	.80	1.60	16.00	24.00	32.00	40.00	48.00	56.00	64.00	72.00	80.00
1.7	.85	1.70	17.00	25.50	34.00	42.50	51.00	59.50	68.00	76.50	85.00
1.8	.90	1.80	18.00	27.00	36.00	45.00	54.00	63.00	72.00	81.00	90.00
1.9	.95	1.90	19.00	28.50	38.00	47.50	57.00	66.50	76.00	85.50	95.00
2.0	1.00	2.00	20.00	30.00	40.00	50.00	60.00	70.00	80.00	90.00	100.00
2.1	1.05	2.10	21.00	31.50	42.00	52.50	63.00	73.50	84.00	94.50	105.00
2.2	1.10	2.20	22.00	33.00	44.00	55.00	66.00	77.00	88.00	99.00	110.00
2.3	1.15	2.30	23.00	34.50	46.00	57.50	69.00	80.50	92.00	103.50	115.00
2.4	1.20	2.40	24.00	36.00	48.00	60.00	72.00	84.00	96.00	108.00	120.00
2.5	1.25	2.50	25.00	37.50	50.00	62.50	75.00	87.50	100.00	112.50	125.00

FIGURE 4-12 A portion of a typical time/dollar conversion table, details of which can be found in most crash estimating guides

establishing labor rates, the shop overhead (including such items as rent, management and supervision, supplies, and depreciation on equipment) must be determined. Then the actual labor cost of all employees and the profit required to keep the business operating must be added to the shop overhead to obtain a dollar rate for repairs. This rate is usually figured on an hourly basis. Keep in mind that labor times shown in crash estimating guides are listed in tenths of an hour.

SHOP TALK

When using a time/dollar conversion table, read across from the labor time column to the appropriate flat rate column. For time or dollar rates not listed, use a combination of columns. For dollar rates ending with 50 cents, add 50 from the cent column to the appropriate rate column.

On the whole, flat-rated operations comprise all work in which components must be removed and replaced (Figure 4-13). Estimated work is work done to straighten and/or repair body parts. However, it is not always that simple. A good example of this is when body parts must be straightened first before cutting them off to make sure the new member will line up for welding. The procedure for this operation, even though it is a flat rated R&R, must allow for some extra time (which must be estimated) added on to the total.

The repair estimate (see Figure 4-2) is used for determining and totaling all of the flat-rated damage. This information is then transferred to a modified crash estimating guide in order to combine the flat-rated time with the estimated time.

The modified crash estimating guide includes damage identification and the flat-rated times. The bottom section does not include bolt-on flat-rated R&R; this is taken into account on a separate sheet. By combining the information from these two worksheets, it is possible to get a much more accurate picture of:

- First, exactly what must be done to the vehicle to correctly repair it
- Second, how to estimate and charge a fair amount for the work that must be performed

These are, of course, the objectives of good estimating in the first place.

FIGURE 4-13 Flat-rated operations include the time needed to install replacement panels.

REFINISHING TIME

Making a correct estimate of the amount of time required to refinish panels, doors, hoods, and so forth is a vital part of an estimator's job. Although the wide range of materials and conditions sometimes makes it difficult to arrive at a precise refinishing cost, there are a number of generally approved concepts that will help an estimator arrive at a fair judgment of the amount of materials that will be needed for the job.

Flat rate manuals published by the major United States automobile manufacturers list a labor time plus a materials allowance (in dollars) for many different refinishing operations. Independently published crash estimating guides list the labor times, but not the dollar value of the materials required.

In most crash estimating guides on the market today, the time required for refinishing is shown in the parentheses adjacent to the part name. The refinishing time generally includes:

- Cleaning/light sanding
- Masking adjacent panels
- Priming/scuff sanding
- Final sanding/cleaning
- Mixing paint/filling spray equipment
- Applying color coat
- Removing masking
- Buffing or compounding (if required)
- Cleaning equipment

The refinishing time generally does not include:

- Cost of paints or materials
- Matching and/or tinting
- Grinding, filling, and smoothing welded seams
- Blending into adjacent panels
- Removal of protective coatings
- Custom painting
- Undercoating
- Anti-corrosion materials

- Sound deadening
- Refinishing underside of hood or trunk lids
- Covering entire vehicle prior to refinishing
- Additional time to produce custom, non-OEM finishes

The refinishing times given in most estimates are for one color on new replacement parts—outer surfaces only. Additions to refinishing times are usually made for the following operations:

- Refinishing underside of hood
- Refinishing underside of trunk lid
- Edging new part
  - First panel
  - Each additional panel
- Anti-corrosion coating
- Two-tone operations
  - First panel
  - Each adjacent panel
- Stone chip (protective material)
  - First panel
  - Each additional panel
- Clear coat (basecoat/clear coat) after deduction for overlap
  - First major panel
  - Each additional panel

SHOP TALK

A good test to determine if a clear coat was used is to lightly sand the painted surface in a damaged or hidden area. If the sandpaper shows a whitish dust, it is clear coated. If the sandpaper shows color, it is a single-stage color coat.

As previously noted, the labor allowance given in crash estimating guides does not include any material costs. These must be estimated using a locally compiled refinishing materials list or one that is accepted on a national or regional basis. Such guides permit estimators to place a fair value on the refinish materials actually used.

TOTALING THE ESTIMATE

Once all the columns of the estimating form—parts, labor, and refinishing costs—are filled out, they can be added together for a subtotal. To this figure, add any extra charges such as wrecker and towing fee, storage fee, and state and local taxes. These figures, added to the subtotal, will give the grand total of the estimate.

While more and more large collision repair shops are doing repair jobs such as wheel alignment, rustproofing, and tire replacement, smaller operations still "farm out" or "sublet" these jobs to specialty shops. When the work is done, the specialty shop bills the collision repair shop for the work. Generally, this is done at a rate less than the normal retail labor cost. In this way, the collision repair shop can charge the normal retail cost and still make a small profit. Shops that sublet work usually have a "sublet" column on the estimate where the retail labor cost is marked. This figure is added to the rest to obtain the grand total of the estimate.

If the customer wants extra work performed (damage that occurred prior to the collision), this should be noted as customer-requested, or C/R, repairs. As a general rule, the insurance company will not pay for customer-requested repairs. Often a separate estimate must be made in such cases.

REVIEW QUESTIONS

1. How many copies of the written estimate should be made?
   a. one
   b. two
   c. three
   d. four

2. How long does an estimate usually remain a firm bid?
   a. 1 year
   b. 6 months
   c. 90 days
   d. 30 days

3. Which of the following abbreviations found in crash estimating guides means that the item in question should be removed as an assembly, set aside, and later reinstalled?
   a. R&R
   b. R&I
   c. O&H
   d. O/H

4. Which computer estimating system is preferred by most collision repair shops?
   a. microfilm
   b. resident
   c. dial-up
   d. CARDA guide

5. When inspecting a collision-damaged unibody vehicle, Technician A looks for damage only near the area of impact. Technician B looks for dam-

age both in the area of impact and some distance away. Who is correct?
   a. Technician A
   b. Technician B
   c. Both A and B
   d. Neither A nor B

6. Which of the following statements is incorrect?
   a. Estimating damage to the body shell is easier than estimating damage to the frame.
   b. Generally, repair costs should never exceed replacement costs.
   c. Discontinued parts are usually specially marked in crash estimating guides.
   d. Both b and c

7. Technician A supports a vehicle on jack stands to inspect for underbody damage. Technician B uses a hydraulic jack. Who is correct?
   a. Technician A
   b. Technician B
   c. Both A and B
   d. Neither A nor B

8. Which of the following conditions is a good sign of minor body distortion?
   a. cracking of stationary glass
   b. front clip
   c. side clip
   d. all of the above

9. Unless otherwise stated, the times given in crash estimating guides are for ______________.
   a. O/H
   b. R&I
   c. I&R
   d. R&R

10. What is the name given to jobs that can be performed individually but are also part of another operation?
   a. R&R
   b. overlapping procedures
   c. included operations
   d. flat rate

11. What is the smallest increment in which labor times are listed in crash estimating guides?
   a. hours
   b. half hours
   c. quarter hours
   d. tenths of an hour

12. Which of the following is considered in crash estimating guides to be part of the basic color coat application?
   a. matching and/or tinting
   b. masking adjacent panels
   c. cleaning/light sanding
   d. both b and c

13. Lightly sanding a painted surface yields a whitish dust. Technician A says this means that it is a clear coat, but Technician B says it is a single-stage color coat. Who is correct?
   a. Technician A
   b. Technician B
   c. Both A and B
   d. Neither A nor B

14. In the procedure pages of a crash estimating guide, which of the following is not included in a bumper assembly overhaul?
   a. removal of hydraulic energy absorbers
   b. reassemble unit
   c. align to vehicle
   d. all of the above

15. The refinishing times given in most crash estimating guides are for ______________.
   a. one color on new replacement parts, outer surfaces only
   b. two colors on new replacement parts, outer surfaces only
   c. one color on new replacement parts, inner and outer surfaces
   d. two colors on new replacement parts, inner and outer surfaces

CHAPTER

5 Measuring Principles Used in Damage Analysis

OBJECTIVES

- Explain the importance of the datum plane and centerline concepts as related to unibody repair.
- Interpret body dimension information and locate key reference points on a vehicle, using body dimension manuals.
- Discuss the use of tram bars, self-centering gauges, and strut tower gauges.
- Diagnose various types of body damage, including twist, mash, sag, and sidesway.
- Given a damaged vehicle and a body specification manual, locate and measure key points using a tape measure, tram bar, and self-centering gauges.
- Describe the two types of universal measuring systems.
- Explain the operation of the dedicated fixture system, including the use of bolt-on, pin-type, and MacPherson fixtures and bench extensions.

Proper measuring is at the heart of all unibody repair. In fact, thorough and accurate measurement is the most important aspect of unibody repair. It should be remembered that the modern unibody vehicle is precision manufactured using the latest technology. This includes the use of robotic assembly and laser technology, thus creating tighter tolerances than ever before. Repair tolerances are generally no more than ± 3 millimeters, and in some cases even less.

There is good reason for such close body structure tolerances. Steering and suspension systems, for instance, are mounted to the unibody structure. Some of the suspension geometry is built into the mounting design. As a result, the angles of caster and camber have a fixed (nonadjustable) value on some vehicles. Thus, body damage often seriously affects suspension geometry. The rack-and-pinion steering assembly is also generally mounted to a panel, resulting in a fixed relationship to the steering arms. The mechanical components, engine, transmission, and differential are also mounted directly to body members or to cradles supported by the body members (panel or integral rails). In the past, repairs to the sheet metal were primarily cosmetic. When a unibody vehicle is involved in an accident, the resulting damage is often structural as well as cosmetic (Figure 5-1).

A distortion of any of these measuring points will change steering or suspension geometry or misalign mechanical components. This can result in improper steering and handling, vibration and noise in the drivetrain, and excessive wear of tie-rod ends, tires, rack-and-pinion assemblies, universal joints, or other suspension or steering components. Thus, to maintain proper steering, handling, and driveability, the tolerances of critical manufacturing dimensions must be held to within a maximum value of 3 millimeters, or less than 1/8 inch.

The challenge faced by the collision repair industry is to find out which panels are out of place and in which direction and how far they have moved. Only then can a proper plan be devised to bring the vehicle back to pre-accident specifications. This process involves accurate measuring and monitoring of the entire vehicle. The dimensional control points should be monitored throughout the pulling and straightening process. To correctly analyze damage on a unibody vehicle, the entire structure must be considered. To do this, it is necessary to be able to take proper measure-

FIGURE 5-1 Damage to a unibody vehicle is usually both cosmetic and structural.

ments to locate the damage as well as to determine where the vehicle is not damaged. It will also help to plan where to pull.

To successfully repair a unibody vehicle, it is necessary to restore the control points in the damaged area of the vehicle to their specified dimensions, while the control points in the undamaged area remain in their correct locations. Therefore, the collision repair technician must work with the whole vehicle and monitor the progress throughout the course of the repair. This is done by measuring and recording dimensional changes. The most widely accepted method of checking body dimensions is to use the charts supplied in the body dimension manuals.

METRIC SYSTEM

All dimensions given in the body dimension manuals use the metric system of measurement. The dimension manuals use the metric system for the following reasons:

1. In all nations except the United States, metric measurement is standard. Since automobiles are an international product, it makes sense to use metric measurements in their repair manuals. A few manufacturers of measuring equipment give body dimensions in both metric and English systems.

SHOP TALK

In 1975 the United States passed the Metric Conversion Act. This committed the United States to a voluntary and gradual transition from its customary English system of measurement to the metric system. To date only a few industries—including automotive—have made any movement toward adopting the metric system.

2. The metric system is far easier for computation than the English system. This is because multiples of metric units are related to each other by the factor 10. Every metric unit can be multiplied or divided by a factor of 10 to get larger units (multiples) or smaller units (submultiples). There is less chance for a math error with the metric system.
3. Metric measurements are easier to read. Table 5-1 gives an exact English and metric equiva-

TABLE 5-1 METRIC CONVERSION FACTORS

Category	USC Unit and Symbol	Metric Unit and Symbol	English (USC) to Metric	Metric to English (USC)
Length	Inch (in.)	Millimeter (mm)	1 in. = 25.4 mm*	1 mm = 0.0394 in.
		Centimeter (cm)	1 in. = 2.54 cm*	1 cm = 0.394 in.
	Foot (ft.)		1 ft. = 30.5 cm	
	Yard (yd.)	Meter (m)	1 yd. = 0.914 m	1 m = 3.28 ft.
	Mile (mi.)	Kilometer (km)	1 mi. = 1.61 km	1 km = 0.62 mi.
Weight	Ounce (oz.)	Gram (g)	1 oz. = 28.3 g	1 g = 0.0353 oz.
(mass)	Pound (lb.)	Kilogram (kg)	1 lb. = 454 g	1 kg = 2.2 lb.
Area	Square inch (in.^{2})	Square centimeter (cm^2)	1 in.^{2} = 6.45 cm^2	1 cm^2 = 0.155 in.^{2}
	Square foot (ft.^{2})	Square meter (m^2)	1 ft.^{2} = 929 cm^2	1 m^2 = 10.8 ft.^{2}
	Square yard (yd.^{2})		1 yd.^{2} = 0.836 m^2	1 m^2 = 1.2 yd.^{2}
Volume	Cubic inch (in.^{3})	Cubic centimeter (cm^3)	1 in.^{3} = 16.4 cm^3	1 cm^3 = 0.061 in.^{3}
	Cubic foot (ft.^{3})	Cubic meter (m^3)	1 ft.^{3} = 28,300 cm^3	1 m^3 = 35.3 ft.^{3}
	Fluid ounce (fl. oz.)	Millimeter (ml)	1 fl. oz. = 29.6 mL	1 mL = 0.033 8 fl. oz.
	Liquid quart (liq. qt.)	Liter (L or l)	1 liq. qt. = 0.946 L	1 L = 1.057 liq. qt.
Time interval	Second (s)	Second (s)		
	Minute (min)	Minute (min)		
	Hour (h)	Hour (h)		
Speed	Miles per hour (mph)	Kilometer per hour (km/h)	1 mph = 1.61 km/h	1 km/h = 0.621 mph
Pressure	Pounds per square inch (psi)	Kilopascal (kPa)	1 psi = 6.89 kPa	1 kPa = 0.145 psi
Power	Horsepower (hp)	Kilowatt (kW)	1 hp = 0.746 kW	1 kW = 1.34 hp
Temperature	Degrees Fahrenheit (°F)	Degrees Celsius (°C)	°C = 5/9 (°F - 32)	°F = 9/5 (°C + 32)

*Accurate

lent chart. With body dimensions, it usually is not necessary to convert English to metric.

4. Using the metric system is quicker than the English system and looks more professional to the shop's customers.

DIMENSIONAL REFERENCE POINTS

Two major dimensional references are indicated in all body dimension manuals: the datum plane and centerline.

DATUM PLANE

A datum plane, or datum line, is a flat surface parallel to the underbody of the vehicle at some fixed distance from the underbody (Figure 5-2). It is the plane from which all height dimensions are taken by the vehicle manufacturer. It is also the plane that is used to measure the vehicle during a repair.

CENTERLINE

The datum plane is divided by an imaginary center plane. This center plane or centerline (℄) divides the vehicle into two equal halves: the passenger side and the driver side (Figure 5-3). All width or lateral dimensions of symmetrical vehicles are measured from the center. That is, the measurement from the centerline to a specific point on the right side will be exactly the same as the measurement from the centerline to the same point on the left side. One side of the structure would be a perfect mirror image of the other.

Most vehicles are built symmetrically. But if the vehicle is not symmetrical (asymmetrical), the self-centering gauges will not align and will not indicate a true center reference. Symmetrical, in this case, says that the dimensions on the right side of the vehicle are equal to the dimensions on the left side of the vehicle. If the vehicle is asymmetrical, these dimensions are not the same. In such a case, use gauges that compensate for the asymmetry.

It is usually necessary to think of the vehicle as a rectangular structure divided into three zero plane sections (Figure 5-4). That is, break the vehicle into three sections—front, center, and rear—using the torque box location as the dividing lines. This three-section principle, which is a carryover from the body-over-frame days, is a result of the vehicle's design and the way it reacts during a collision. In unibody construction, each section should be checked for diagonal squareness by comparing diagonal lengths. Length and width should also be compared. The center section should be used as a base when reading structural alignment. All measurements and alignment readings will be taken relative to the center section.

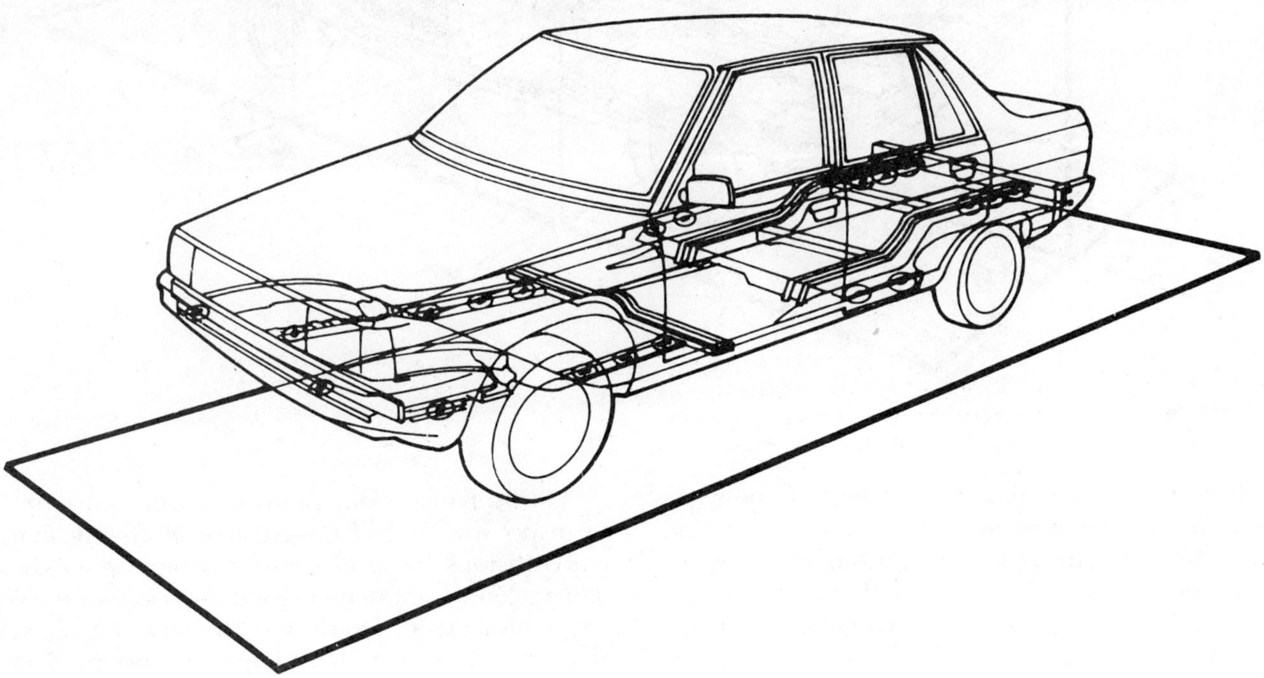

FIGURE 5-2 The datum plane is a reference for all height measurements.

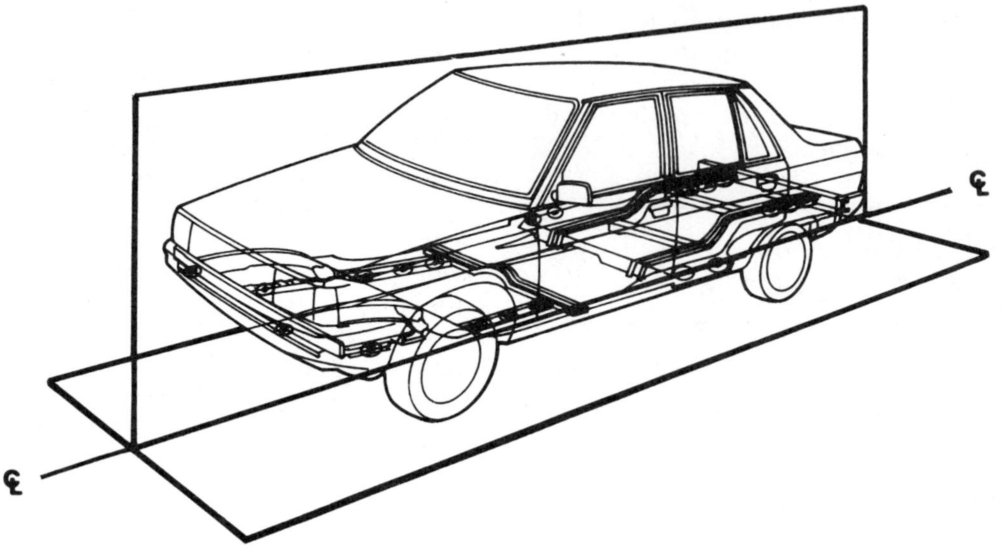

FIGURE 5-3 The centerline divides the datum plane in half.

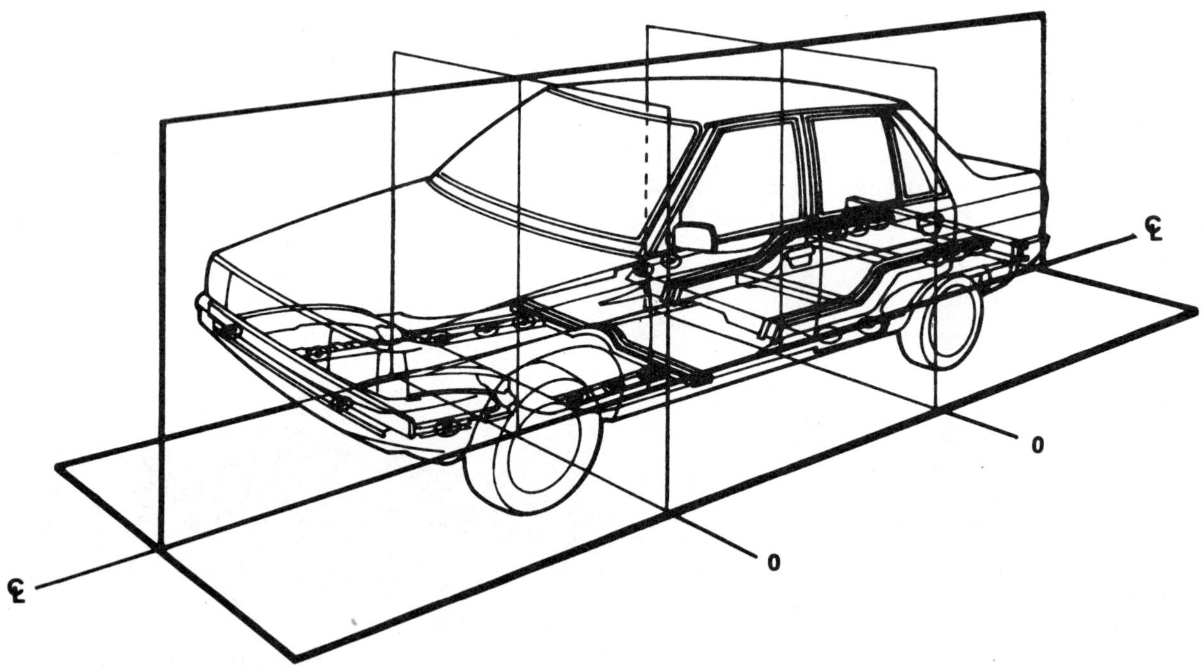

FIGURE 5-4 The three zero vehicle planes

Start measuring in the center or middle section; if it is not square, then move to the undamaged end of the vehicle to find three correctly positioned reference points. Keep in mind that to accurately measure a vehicle, the technician must start with at least three dimensions he/she knows are right. The way to do this is to check the squareness of the vehicle. If the vehicle is not symmetrical, refer to the dimension chart for correct measurements (Figure 5-5).

Measurement points and tolerances are determined by inspections of the damaged area. Normally, in front end collisions that cause slight amounts of door sag, the damage does not extend beyond the center of the vehicle, so measurement in the rear section is not necessary. In a situation where a large impact has occurred, many measurements must be taken to assure proper alignment. However, taking too many measurements might cause unnecessary confusion.

FIGURE 5-5 Dimension charts are critical for working with asymmetrical vehicles.

SHOP TALK

The terms control point *and* reference point *have different meanings. The control points used in manufacturing are not necessarily the same as the reference points the collision repair technician uses to measure the vehicle. Reference points refer to the points, bolts, holes, etc., used to give unibody dimensions in the body specification manuals. The distance between reference points can be measured with either a tram bar or a tape measure (Figure 5-6).*

In the entire repair process of a unibody vehicle it is not possible to overemphasize the importance of measuring. A vehicle cannot be satisfactorily repaired unless all of the reference points in the damaged area are returned to pre-accident specifications. To achieve this, the collision repair technician must:

- Measure accurately
- Measure often
- Recheck all measurements

Because of the importance of measuring in unibody repairs, many kinds of special equipment have been developed and marketed by a large cross section of automotive equipment manufacturers strictly for the purpose of providing the capability to measure very quickly and accurately. Although a number of styles of measuring equipment can be found in collision repair shops, most of it can be divided into three basic systems:

- Gauge measuring systems
- Universal measuring systems
- Dedicated fixture systems

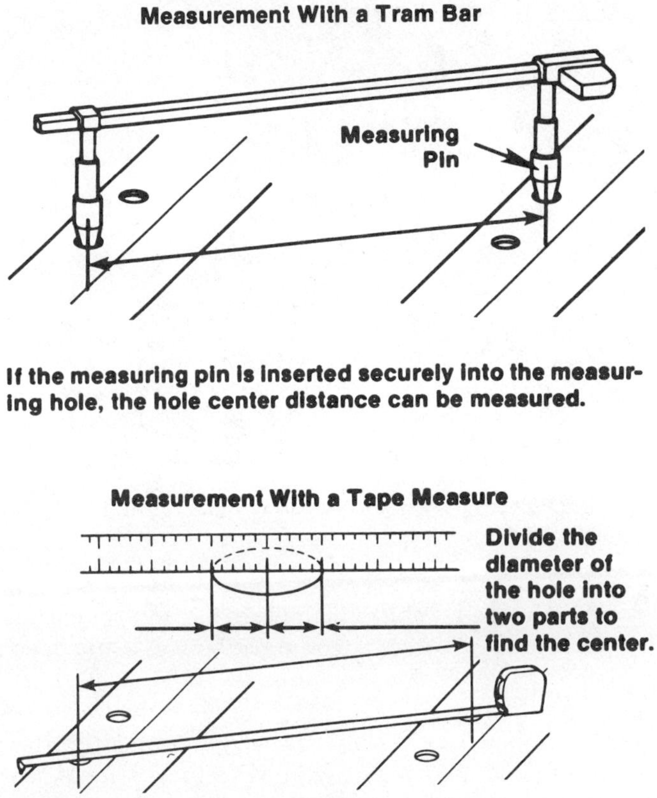

FIGURE 5-6 Methods of measuring the distance between reference points

SHOP TALK

The modern unibody will flex a bit, even when securely mounted to pinch weld clamps. For this reason, the weight of the mechanical and drivetrain components must be taken into account when measuring. Most equipment manufacturers will provide measuring specifications for the vehicle with drivetrain weight compensated for.

GAUGE MEASURING SYSTEMS

The tram bar, the self-centering gauge, and the strut tower gauge are three individual measuring systems that can be used separately or in conjunction with one another. Tram bars are used for measurement; self-centering gauges check for misalignment. Supported by MacPherson strut tower domes, the strut tower gauge allows visual alignment of the critical reference

FIGURE 5-7 Making an engine measurement with a tram bar

points of unibody vehicles. The tram bar, self-centering gauge, and strut tower gauge are available as a unit or as separate vehicle diagnostic tools.

Another popular measuring gauge is the tracking gauge. This gauge is used to check alignment of the front and rear wheels. If the front and rear wheels are not in alignment, the vehicle will not handle properly.

TRAM BARS

The tram bar is a bar with two adjustable pointers attached to it (Figure 5-7). The pointers slide along the bar's length and are adjustable for height. Since the tram bar measures one dimension at a time, each must be recorded and cross-checked from two additional reference points—at least one being a diagonal measurement. The best areas to select for tram bar measurements are the attachment points for suspension and mechanical components since these are critical to alignment. Throughout the pulling and straightening process, critical reference points must be measured (and recorded) repeatedly with the tram bar to monitor progress and to prevent overpulling.

Since these reference point measurements must be taken and written down several times in a repair operation, a method of tabulation must be devised. One way to accomplish this is to use a data or tabulation chart similar to the one shown in Figure 5-8.

In the first column of the tabulation chart, write the location of what is being measured. Numbers or letters (Figure 5-9) can be used. The second and third column contains the manufacturer's specifications taken from the body dimension manual and actual distance as it exists on the damaged vehicle. The 1-2-3 are the readings taken at measurement step 1, measurement step 2, and so on. That is, as each pull is made, the measurements should be recorded, including those dimensions that have just been corrected. This chart tells the collision repair technician at a glance how the job stands in restoring the vehicle to its pre-accident state.

The tram bar might have a scale superimposed on it. However, since almost all dimension charts list measurements in metric, use a steel tape with both English and metric scales to set up the tram bar. The tape can also be used to take quick measurements between reference points. Be sure that the tape has been checked for accuracy before using it.

SHOP TALK

More accurate measurements can be taken if the front end of a tape measure is machined according to Figure 5-10 and is inserted into the reference point hole.

Reference point holes are frequently larger in diameter than the tram bar tip. To measure accurately with the tram bar (when the holes are the same diameter), measure like edge to like edge. When the holes are

TRAM BAR MEASUREMENT CHART

	Mfg. Spec.	1	2	3	4	5	6	7	8	9	10	11	12
A													
B													
C													
D													
E													
F													
G													
H													
I													
J													
K													
L													
M													
N													
O													
P													
Q													
R													
S													
T													
U													
V													
W													

FIGURE 5-8 Typical tabulation chart

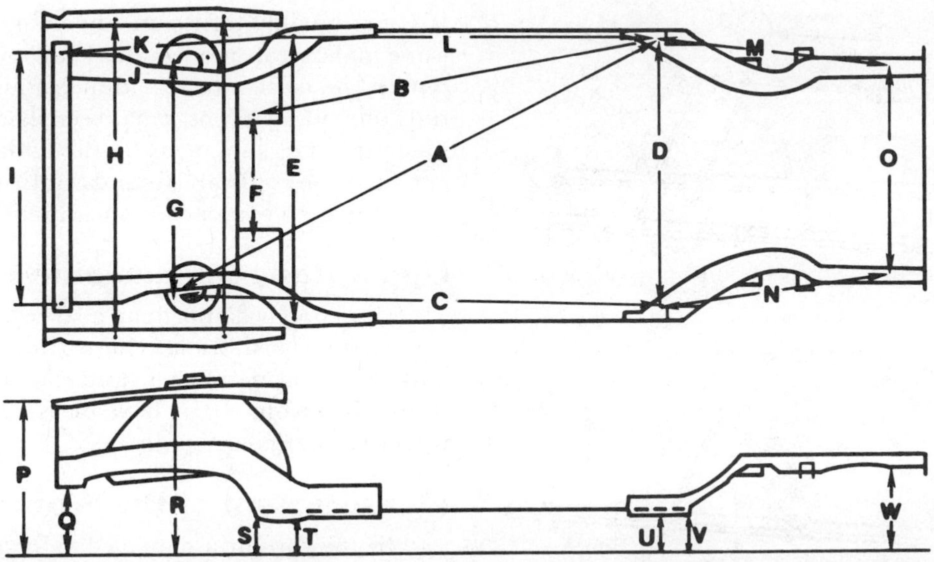

FIGURE 5-9 Many collision repair technicians prefer using letters instead of numbers when making up a tabulation chart.

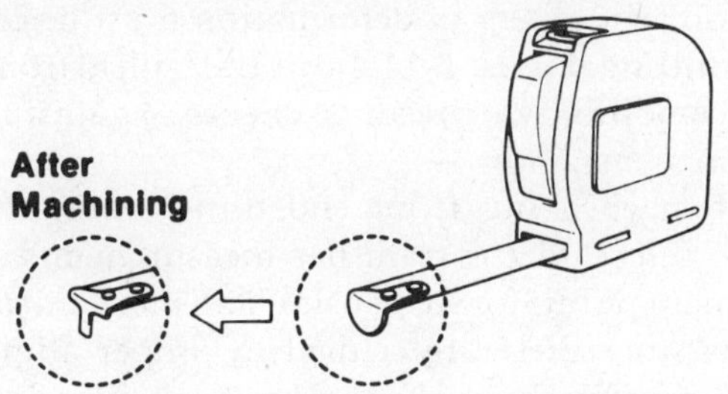

FIGURE 5-10 Machining the tip of a tape measure yields more accurate measurements.

not the same size, they will usually be the same type of hole: round, square, oblong, and so on. In this case, to find the center-to-center measurement, measure inside edge to inside edge, then outside edge to outside edge (Figure 5-11). Add the results of the two measurements and divide by 2. For example, two round holes, one 12.5 mm in diameter, the other 38 mm in diameter, have an inside measurement of 762 mm and an outside measurement of 812 mm. The center-to-center dimension is (762 mm + 812 mm) ÷ 2 = 787 mm (Figure 5-12). The 787 mm is the dimension for the tram bar. Center at

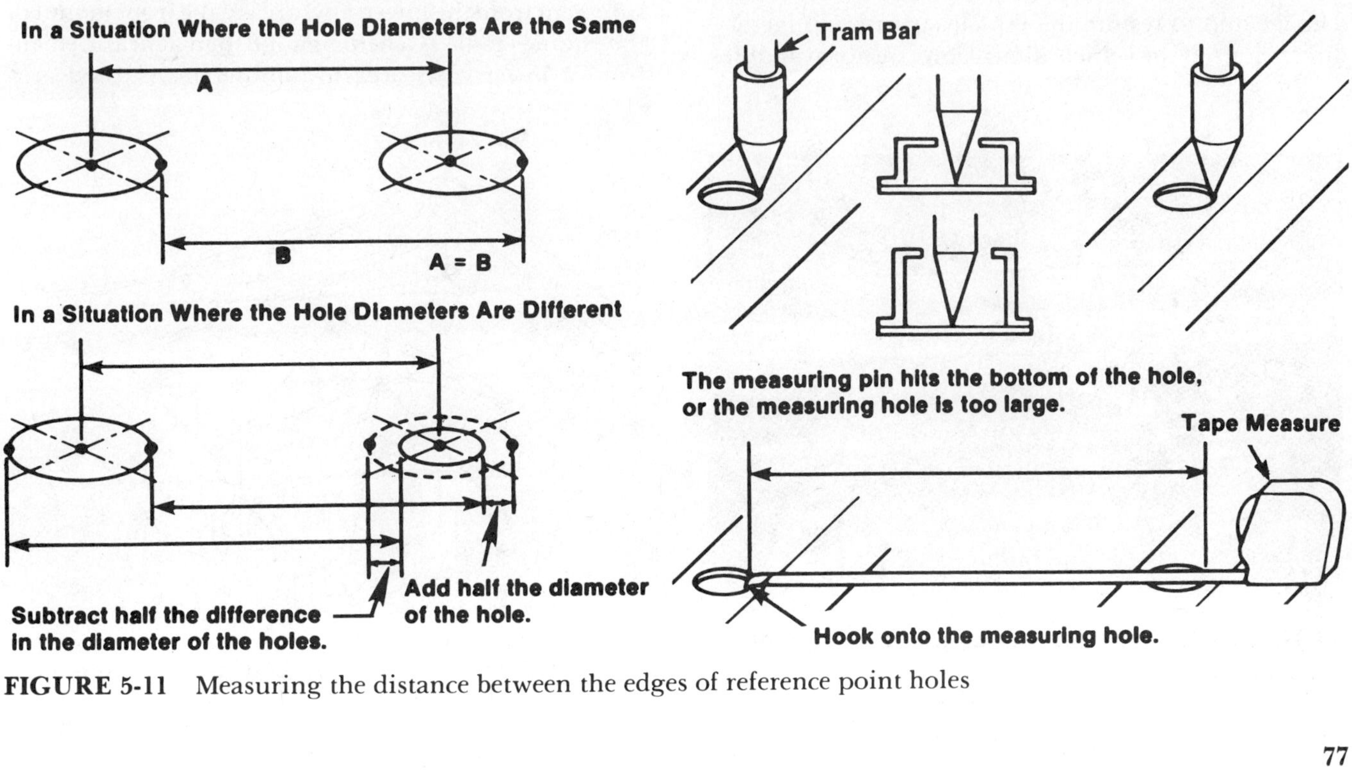

FIGURE 5-11 Measuring the distance between the edges of reference point holes

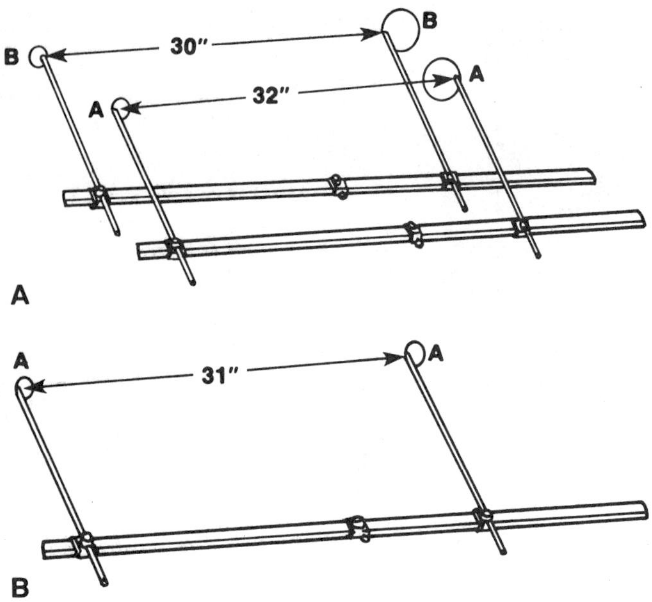

FIGURE 5-12 Measuring when the holes are not the same size

the smallest hole and the other end should center at the larger hole. If not, there is damage to the unibody.

A few body dimension manuals give the measurements based on the bar's length. Here the bar of the tram becomes the datum plane beneath the vehicle (Figure 5-13). The pointers are set at the horizontal height dimensions specified on the dimension chart. It is important to always check the method used on the dimension chart because the two distances are not the same.

In using a tram bar for measuring, the vehicle dimensions are needed in order to accurately assess the damage and to restore the vehicle structure to factory dimensions. If the vehicle dimensions are not available, it is permissible to use an undamaged vehicle of the same make, year, model, and body style as a source to obtain the correct factory dimensions. Frequently, if only one side of a vehicle has been damaged, it is possible to take measurements on the undamaged side that can be set on the tram bar and applied to the damaged side to make a comparison measurement.

Upper Body Dimensioning

Upper body damage can also be determined by the use of a tram bar and a steel measuring tape. Their use is basically the same as when doing an underbody evaluation. Dimension charts have measurements for many upper body reference points.

Measurement of the Front Body

In the case of a damaged vehicle that needs the hood and front side member replaced, it is reasonable to take measurements along with the repair. Even if only the front right side of the vehicle body received the impact, the left side will usually be damaged also. Therefore, the extent of deformation must be checked before pulling. Figure 5-14 shows the typical front body reference points, which can be checked against the dimension chart.

When checking front end dimensions, the best areas to select for the tram bar measurements are the attachment points for suspension and mechanical components since these are critical to proper alignment. Each dimension should be checked from two additional reference points with at least one reference point being a diagonal measurement. The longer the dimension, the more accurate the measurement. For example, a measurement from the lower cowl area to the front mount of the engine cradle is a better gauge than a measurement from a lower cowl area to another lower cowl area

FIGURE 5-13 The tram bar used as the datum plane beneath the vehicle

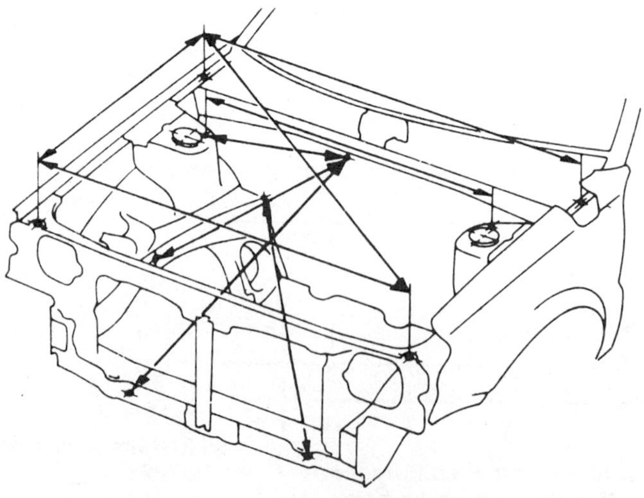

FIGURE 5-14 Typical front body reference points

because the longer dimension takes in a larger area of the vehicle. The use of two or more measurements from each reference point assures greater accuracy and helps identify the extent and direction of any panel damage.

Measurement of the Body Side Panel

Any deformation of the body side structure can be determined by the appearance and irregularities in the door when it is opened and closed. Depending on where the deformation is located, attention should be given to possible water leakage. Thus, accurate measurements must be taken. The tram bar is used to measure the body side panel (Figure 5-15).

Warping can generally be detected if the left-to-right symmetry of the body is used for measuring diagonal lines, as shown in Figure 5-16A. Use this measuring method if the data on the engine compartment and underbody are missing, if there is no data available in the body dimension chart, or if the vehicle has been severely damaged in a rollover.

The diagonal line measurement method is not adequate when inspecting damage to both sides of the vehicle or in the case of twisting, since the left-to-right difference in the diagonal lines cannot be measured (Figure 5-16B). If deformation is the same on the left and the right, left to right difference will not be apparent (Figure 5-16C).

In Figure 5-17, the measurement and comparison of the left and right lengths between yz and YZ will give an even better indication of damage conditions (this method should be used in conjunction with the diagonal line measurement method). This method can be applied where there are parts that are symmetrical on the left and right sides.

Measurement of the Rear Body

Any deformation of the rear body can be roughly estimated by appearance and irregularities evident when the trunk lid is opened and closed. Because of the location of the deformation and the possibility of water leakage, it is essential to take accurate measurements (Figure 5-18). Furthermore, any wrinkle in the rear floor is usually due to buckling of the rear frame rail. Thus, measure the rear body together with the underbody. In this way, the straightening work can be performed effectively.

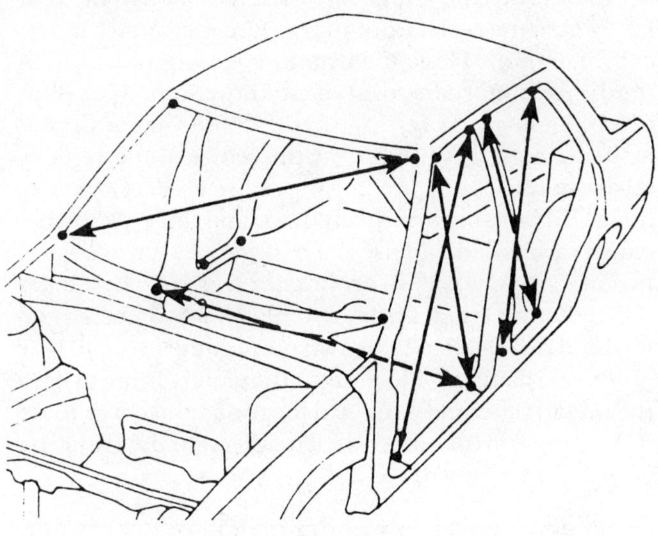

FIGURE 5-15 Typical body side reference points

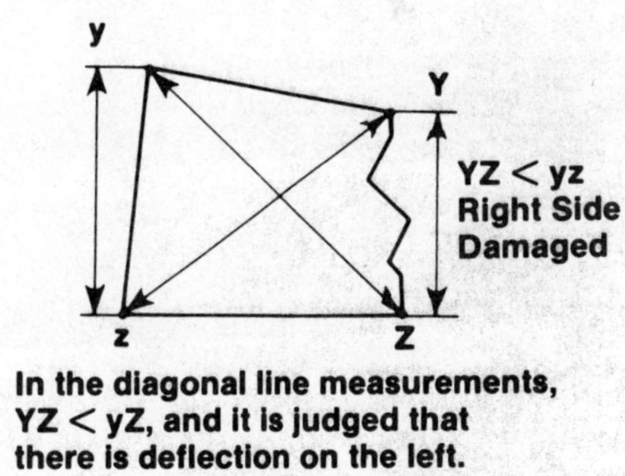

FIGURE 5-17 Comparison of length dimensions

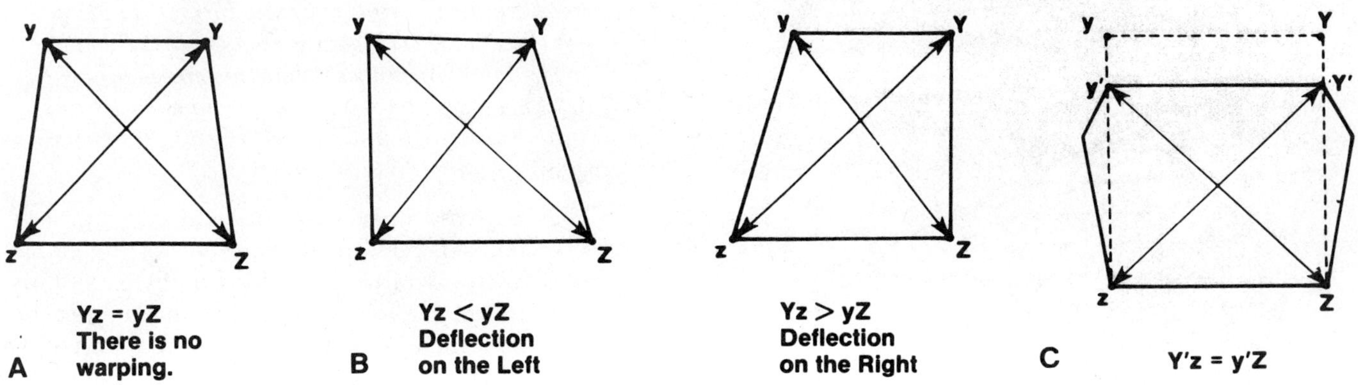

FIGURE 5-16 Using the diagonal line measurement method

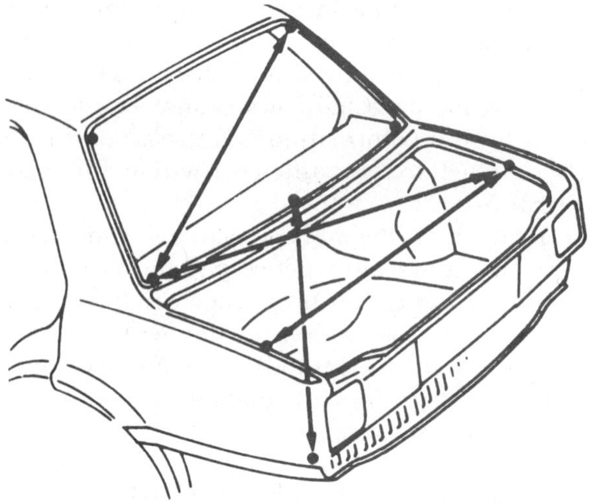

FIGURE 5-18 Typical rear body reference points

FIGURE 5-20 Hanging self-centering gauges to establish the datum plane

SHOP TALK

Keep in mind that all reference points need to be checked from two or more reference points. Diagonal measurements are a good way to cross-check dimensional reference points. Correct use of self-centering gauges will also reveal the kind of misalignment that would be shown by diagonal tram bar measurements.

SELF-CENTERING GAUGES

Although self-centering gauges (Figure 5-19) are closely related to the tram bar, as mentioned earlier, they do not measure. They show alignment or misalignment by projecting points on the vehicle's structure into the technician's line of sight. They are installed at various control areas on the vehicle. Self-centering gauges have two sliding horizontal bars that remain parallel as they move inward and outward. This action permits adjustment to any width for installation on various areas of the vehicle.

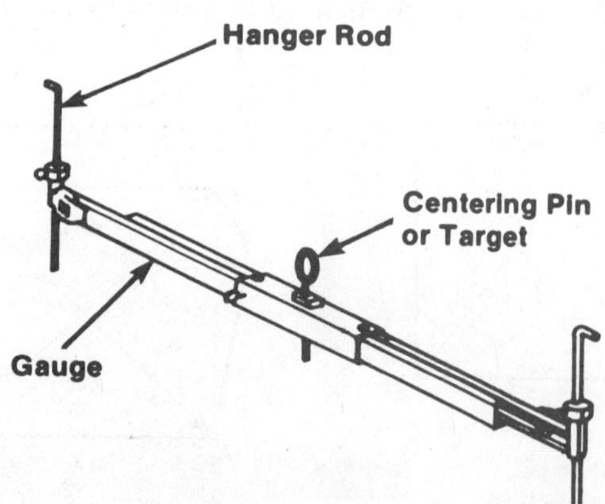

FIGURE 5-19 Typical self-centering gauge

Self-centering gauges are used to establish the vehicle centerline and datum plane. Use the gauges in sets of four or more. Hang them at the locations indicated in the dimensional data to establish the datum plane (Figure 5-20). Begin by setting up the "base" gauges—that is, the two gauges hung near the center of the vehicle. (Usually the torque box locations are used.) Generally these will be gauges numbered two and three. Add other gauges as needed and use the base gauges for reference. In some cases, when inspecting for warpage, first hang self-centering gauges from two places where there is no visible damage, then hang two more gauges where there is obvious damage. Then, sight the gauges hung at both the undamaged and damaged locations and check to see if the gauges are parallel to each other or if the centering pins are misaligned.

SHOP TALK

There should not be any deformation at the point the self-centering gauge is installed. That is, there are many instances where the reference point holes have been deformed by previous collisions or other causes. Do not use deformed holes unless they are repaired first (Figure 5-21).

These gauges are also equipped with center pins or sights, which will remain in the center of the gauge regardless of the width of the horizontal bars. This allows the technician to read centerline throughout the length of the vehicle. When sighting gauges for parallel, the technician always stands directly in the middle, scanning with both eyes. To ensure accuracy, readings

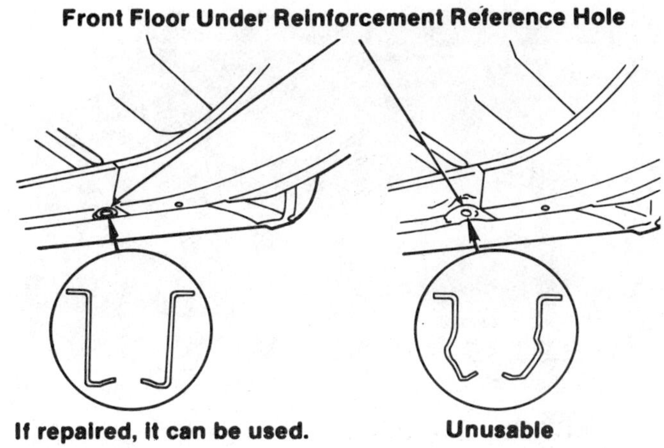

FIGURE 5-21 Do not use deformed reference point holes without repairing them first.

should be made at the outer edge of the self-centering gauge, not in the middle. The farther one stands from the gauges while reading, the more accurate the reading will be. Standing close changes the line of sight to the front gauges so drastically that an accurate reading is nearly impossible. Self-centering gauges should always be set at the same height or plane. Different heights will change the angle of sight and give a false reading. It is sometimes beneficial to sight over one gauge and under another. Going to the end of the vehicle opposite the damage to make readings will sometimes result in a more accurate reading. This is true because the technician is able to read the base gauges before sighting into the damaged area. With practice and a certain amount of experimentation, the technician can improve the damage analysis.

The sighting of centerline pins must be done with one eye. Since the center section is always the base for gauging, the line of sight must always project through the pins of the base gauges. Observing pins in other sections of the body will then reveal how much they are out of alignment. Never attach self-centering gauges to any movable parts such as control arms or springs.

Each self-centering gauge accommodates two vertical scales—one on the left side, one on the right. These scales can be adjusted vertically to assure that the horizontal bars accurately reflect the true positions of the parts to which they are attached (Figure 5-22). Once hung in specific locations, the gauges generally remain on the vehicle throughout the entire repair operation, unless one or all of them interfere with straightening, anchoring, or with the tram bar.

Special centering gauges are available that can be used to check such items as body pillar damage (Figure 5-23). The same system of alignment is employed when using gauges to check underbody damage. Some self-centering gauges can be adjusted for asymmetrical vehicles as well (Figure 5-24). Again, check the dimensional data for these asymmetrical vehicles and follow directions provided with the gauges.

Datum Plane

Using the datum line as a reference, self-centering gauges can be strategically suspended under the vehicle from side to side at varying distances along the length of

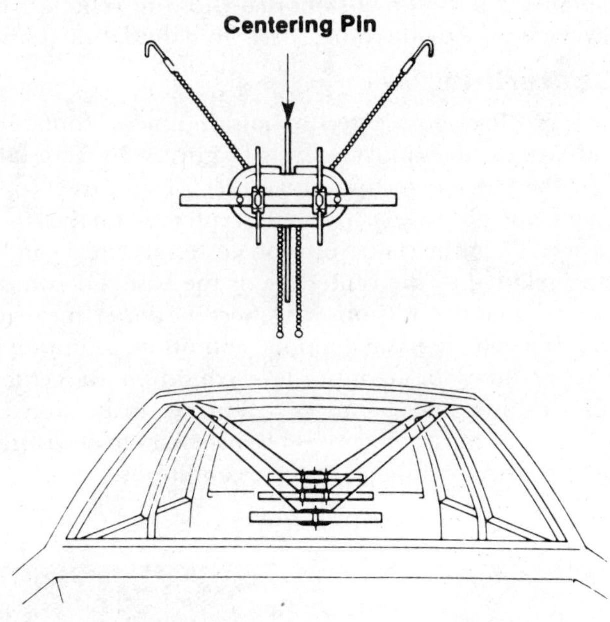

FIGURE 5-23 Special centering gauges can be used to check body pillar damage.

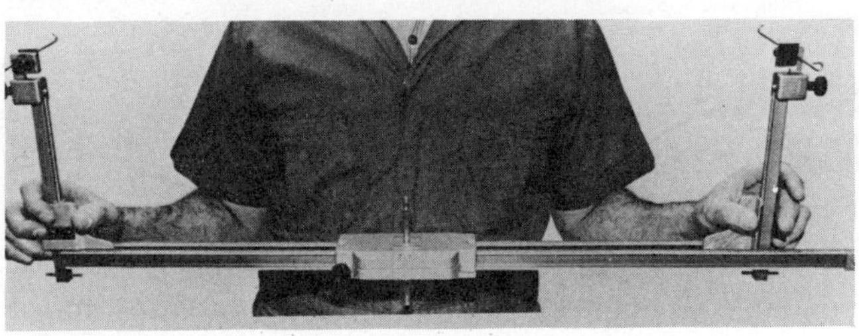

FIGURE 5-22 Self-centering gauges can be adjusted for vertical measurement.

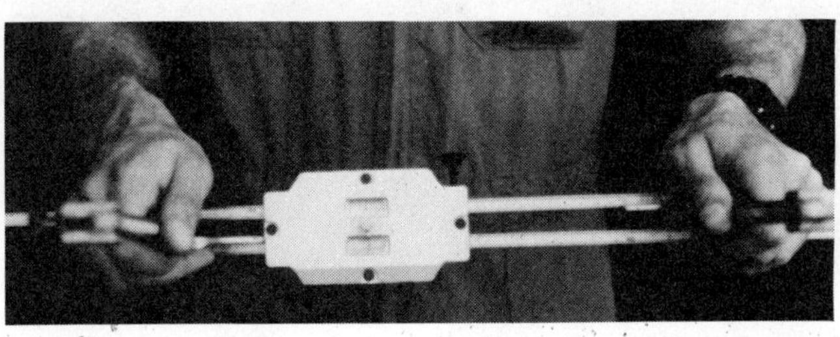

FIGURE 5-24 Some self-centering gauges can be adjusted for asymmetrical vehicles.

the unibody. First, place the base gauges at the main platform: one across the vehicle beneath the rear seat and another under the cowl area. Add two more gauges before and after the base gauges: one located at the main front crossmember and a second at the rear-most crossmember. Additional gauging of the front crossmember area and/or strut tower completes the picture.

To read for datum, all gauges must be on the same plane. After hanging all four gauges, read across the top to determine if datum is correct. If all four gauges are parallel, the vehicle is on datum. If they are not parallel, the vehicle is off datum.

Since the datum line is an imaginary plane, datum heights can be raised or lowered to facilitate gauge readings. Just remember that if the datum height is changed at one gauge location, all the gauges must be adjusted an equal amount to maintain accuracy.

While datum readings are usually obtained from self-centering gauges, there are individual gauges available for measuring datum heights. These datum gauges are usually held in position by magnetic holders. Remember that the dimensions that allow the vehicle to be level with the road are measured from the datum plane.

Centerline

To check for centerline misalignment, four self-centering gauges must be hung (Figure 5-25). To establish the true centerline, the center pin on the No. 2 gauge must be lined up with the center pin on the No. 3 gauge. Then the center pins of No. 1 and No. 4 can be read relative to the centerline of the base. Of course, other damage conditions will affect the centerline reading. If a vehicle has a diamond condition, a shortened rail or subrail, or an out-of-level condition, the centerline reading may be affected. Further inspection by gauging or measuring might be necessary to determine the presence or absence of these conditions.

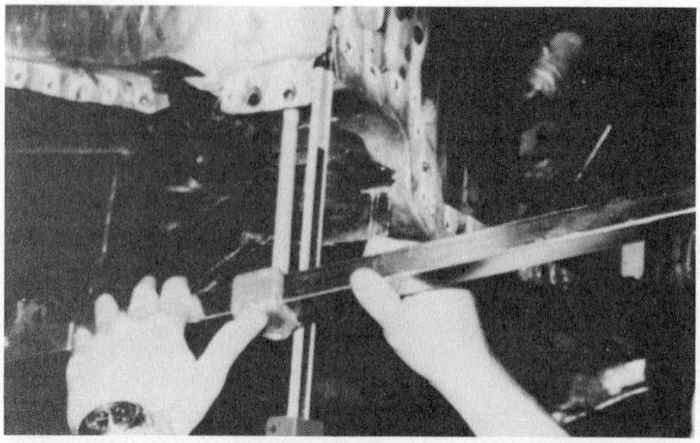

FIGURE 5-25 Four gauges are needed to check for centerline misalignment.

FIGURE 5-26 Typical strut tower gauge

STRUT TOWER GAUGES

Supported by the MacPherson tower domes, the strut tower gauge (Figure 5-26) allows visual alignment of the upper body area. It shows misalignment of the strut tower/upper body parts in relation to the vehicle's centerline plane and datum plane.

The strut tower gauge (Figure 5-27) features an upper and lower horizontal bar, each with a center pin. The upper bar is usually calibrated from the center out. Pointers, which are positioned in an adjustable housing on the upper horizontal bar, are used to mount the gauge to the strut tower/upper body locations. Two types of pointers are provided: "cone" and "reverse cone." The 4-inch reverse cone is notched to provide additional means of mounting on the vehicle, for example, ridged surfaces. The pointers are usually held in the housing by means of thumbscrews.

The vertical scales that link the upper and lower horizontal bars are used to set the lower bar at the datum plane of the parts the gauge mounts to. The scales fasten in housings at the ends of the horizontal bars. Height adjustments are made at the housings of the upper horizontal bar.

The strut tower gauge is used most often to detect misalignment of strut towers; however, it can also be used to detect misalignment of a radiator core support, cowl, quarter panels, and so on.

DIAGNOSING DAMAGE USING GAUGE MEASURING SYSTEMS

To correct collision damage, pull or push the damaged area in the opposite direction of impact. The repair must be made in the reverse sequence that it happened. Therefore, the damage must be measured in the reverse sequence.

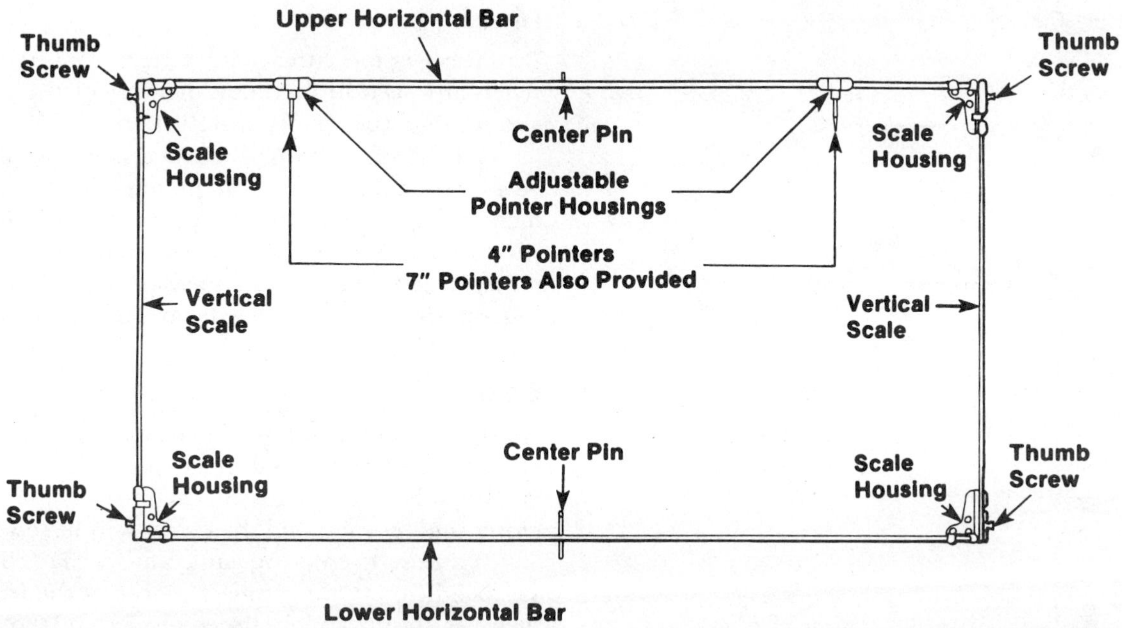

FIGURE 5-27 The parts of a strut tower gauge

When proceeding through the measurement of the damage, keep in mind that a vehicle is similar to a building. If the foundation is not square and level, the rest of the structure will be uneven also. The vehicle's foundation, which is the center section of the vehicle, is measured for twist and diamond first. These two measurements will tell the collision repair technician if the foundation is square and level. The remaining measurements use the foundation as a reference.

Twist

The first damage condition to look for is twist. Twist is the last damage condition to occur to the vehicle and, therefore, the first to measure. Twist is a condition that exists throughout the entire vehicle. True twist exists when one side of the vehicle is pushed either low or high on one end (front or rear) and then moves in the opposite direction (low or high) toward the other end. At the same time, the opposite side would have exactly the opposite damage.

Twist can only be checked in the center section; otherwise, other misalignment in the front or rear might give an inaccurate reading of twist. To check for twist, two base gauges must be hung. These base gauges are also referred to as the No. 2 (front center) and No. 3 (rear center) gauges. The No. 2 should be hung as far forward of the center section as possible. The No. 3 would then be hung as far rearward of the center section as possible. The No. 2 gauge is then read relative to No. 3. If the gauges are parallel, no twist exists. If the gauges are not parallel, then a twist could exist. Remember that a true twist must exist throughout the entire structure. To check for a true twist versus an out-of-parallel condition in the center section, hang another gauge. Go to the undamaged section of the vehicle and hang either the No. 1 or No. 4 (rear) gauge.

These gauges will be read relative to the nearest base gauge. The No. 1 will be read relative to No. 2 and the No. 4 will be read relative to No. 3. If the front or rear gauge reads parallel to the nearest base gauge, true twist cannot exist and the condition is out-of-parallel in the center section. If a true twist exists, the gauges would read like those shown in Figure 5-28.

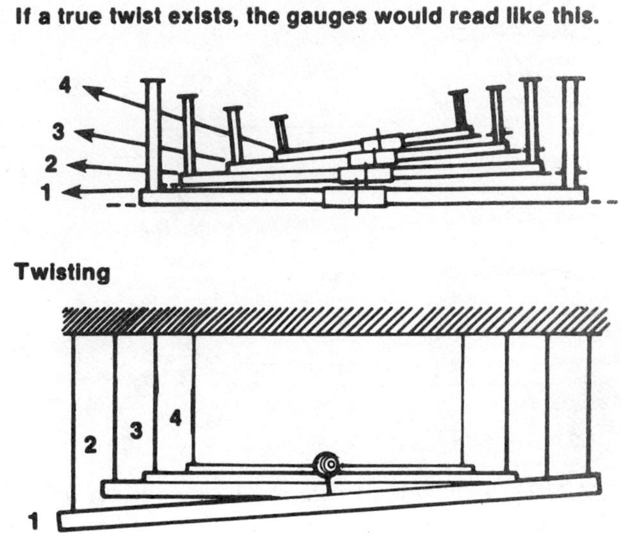

FIGURE 5-28 Gauges indicating twist

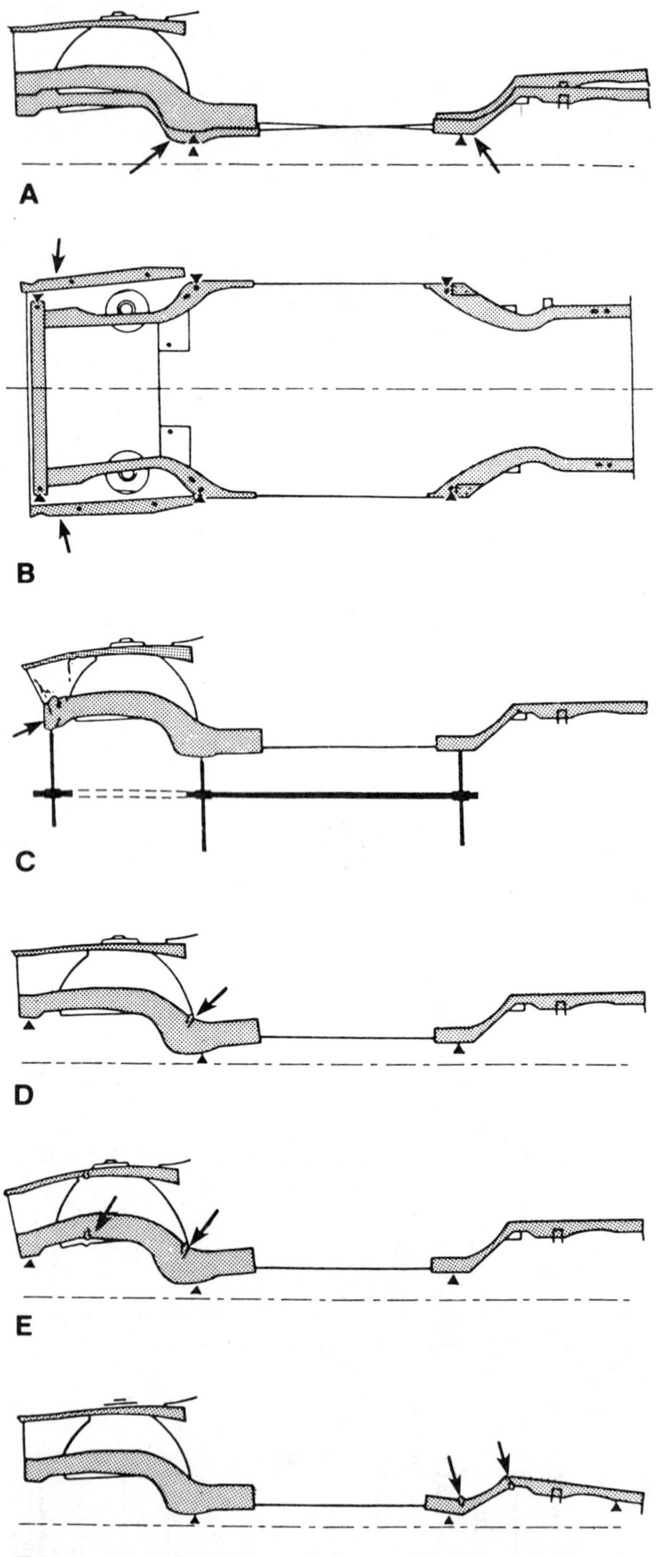

FIGURE 5-29 Measuring mash damage on unibody vehicles: (A) High front impact with secondary damage in the rear; (B) right front corner impact; (C) direct front impact; (D) low front impact; (E) high front impact; and (F) high rear impact

Mash

Mash is measured with a tram bar. It is present when any section or frame member of the vehicle is shorter than the factory specifications.

When using a tram bar on a mash damaged vehicle, be sure to make the measurement called for in the dimension chart. The amount of mash is determined by subtracting the actual measurement from the specified measurement. The proper methods of measuring various mash impacts with a tram bar are illustrated in Figure 5-29.

Sag

Sag is a condition where the cowl area of the vehicle is lower than normal. It can also occur at the front crossmember. The ends of the crossmember will be closer than normal and the center will be too low.

Three self-centering gauges are used to check for a sag condition. One gauge is placed at the front cross member, the next one at the cowl area, and the third one at the rear torque box area. The gauges are on-center and parallel with each other, but the center gauge is lower than the others (Figure 5-30). This indicates a sag condition occurring at the cowl area.

Sidesway (Widening)

Sidesway is present when the front, center, or rear portion of the vehicle is out of lateral alignment. Three self-centering gauges are used to check for sidesway.

If the vehicle was hit in the front, base gauges No. 2 (cowl area) and No. 3 (rear torque box area) are hung and gauge No. 1 is hung in the front crossmember area. If the centering pin on the No. 1 gauge does not line up with the other two, front sidesway is present (Figure 5-31); if it does line up, there is no front sidesway.

If the vehicle has been struck in the rear, the misalignment would appear on the self-centering gauges in a way similar to front sidesway, except the rear centering pin would be out of alignment.

A center hit on a vehicle causes a misalignment known as double sidesway. While this results from a severe impact in the center section, it affects the entire vehicle. The dimensions of both front and rear sections

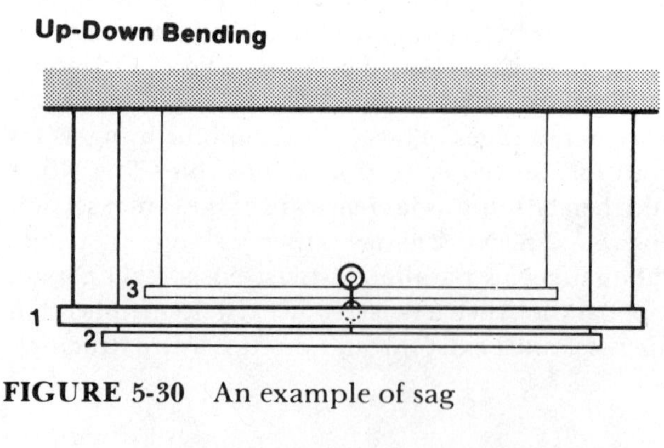

FIGURE 5-30 An example of sag

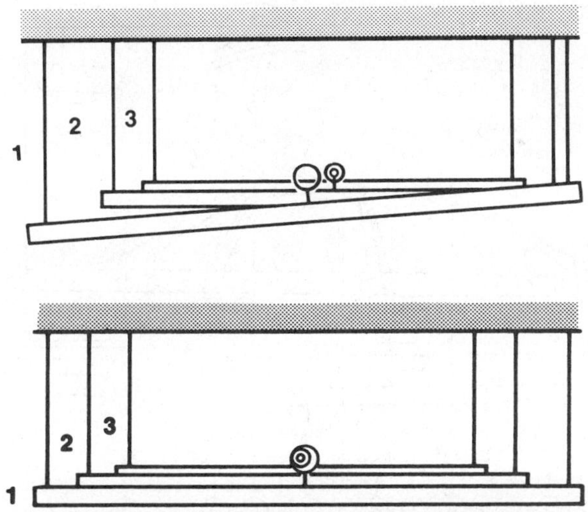

FIGURE 5-31 An example of front sidesway

must be checked during the pulling of double sidesway damage.

UNIVERSAL MEASURING SYSTEMS

Universal measuring systems are the most efficient application of measuring system technology. They make some parts of the measuring job much easier and more accurate, but still require a degree of skill and attention to detail to operate properly. These systems have the ability to measure several reference points at the same time. But to get the proper measurement reading, the equipment must be set using the manufacturer's specifications.

Universal measuring systems are of two types:

- Mechanical
- Laser

Some are a combination of the two.

In practice, a universal measuring system offers the advantage of visually inspecting all the reference points by just moving around the vehicle and determining where each reference point on the vehicle is in comparison to where the reference point on the measuring system is (Figure 5-32). If the reference point on the vehicle is not in the same horizontal and vertical position as the dimension chart says it should be, the reference point on the vehicle is wrong. When the system is set up properly, the technician can monitor the key points by simply looking at the pointers. If the pointers are out of position, then the vehicle is not dimensionally correct. Thus, that reference point is out of position and must be brought back to pre-accident specifications.

FIGURE 5-32 Inspecting the reference points

FIGURE 5-33 Lowering pointers to make room for a pull

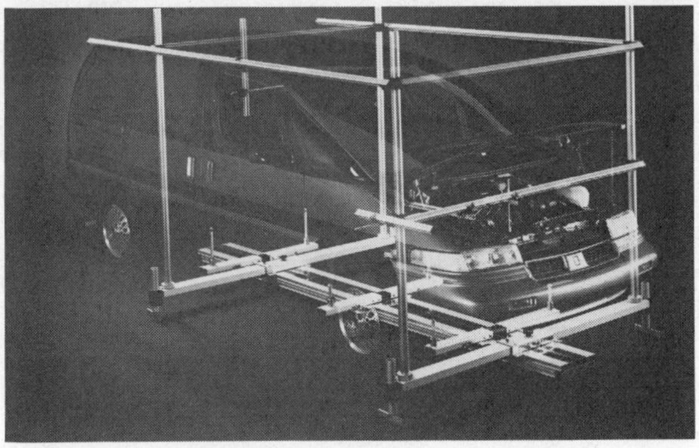

FIGURE 5-34 Typical freestanding bridge

CAUTION

The pointers can be damaged during pulling. Lower the pointers, if necessary, to clear the way for the pull and to avoid damaging or misaligning the measuring system (Figure 5-33).

Universal measuring systems vary in degree of operational difficulty. Figure 5-34 shows a freestanding

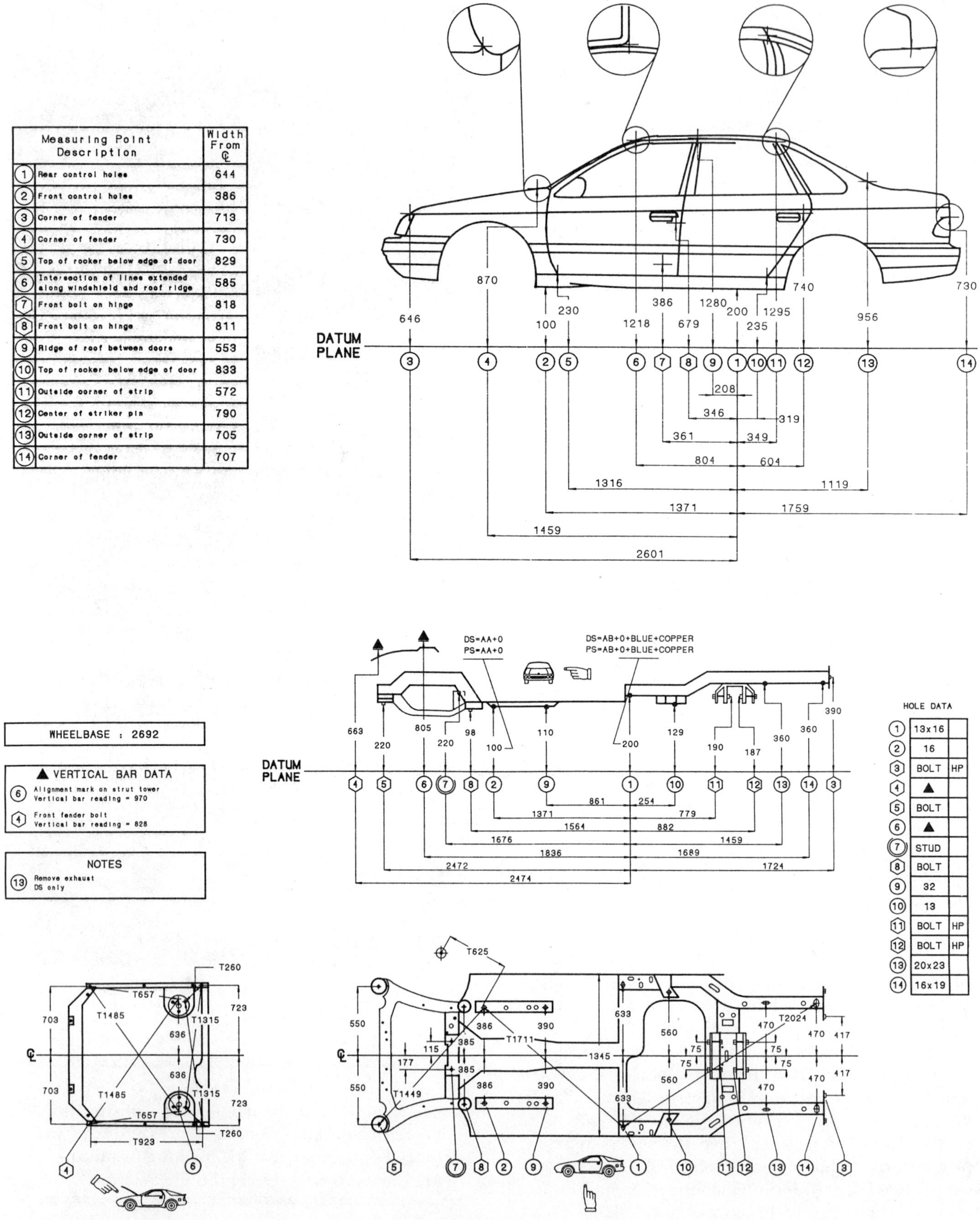

	Measuring Point Description	Width From ℄
1	Rear control holes	644
2	Front control holes	386
3	Corner of fender	713
4	Corner of fender	730
5	Top of rocker below edge of door	829
6	Intersection of lines extended along windshield and roof ridge	585
7	Front bolt on hinge	818
8	Front bolt on hinge	811
9	Ridge of roof between doors	553
10	Top of rocker below edge of door	833
11	Outside corner of strip	572
12	Center of striker pin	790
13	Outside corner of strip	705
14	Corner of fender	707

HOLE DATA

1	13x16	
2	16	
3	BOLT	HP
4	▲	
5	BOLT	
6	▲	
7	STUD	
8	BOLT	
9	32	
10	13	
11	BOLT	HP
12	BOLT	HP
13	20x23	
14	16x19	

FIGURE 5-35 Typical manufacturer's dimension chart

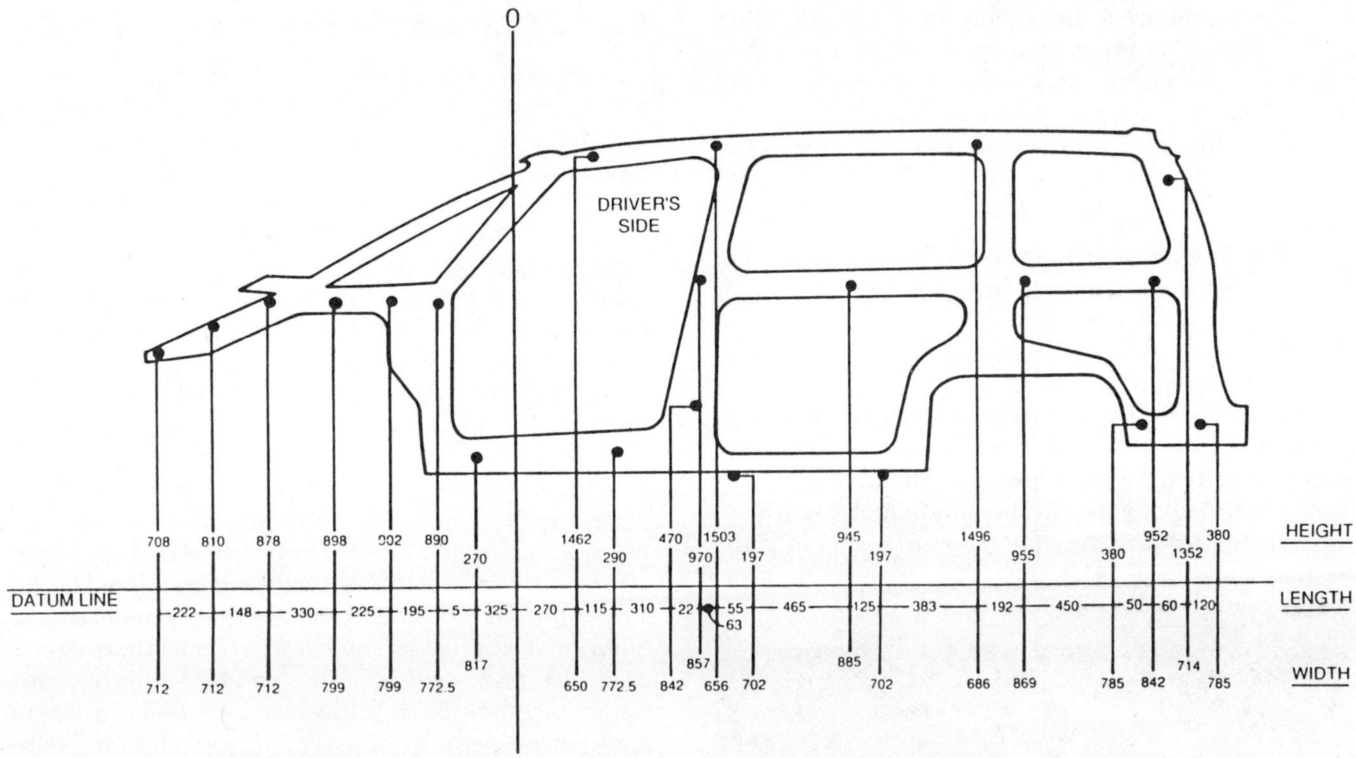

FIGURE 5-36 Typical mill and drill pad measurement specifications

type of bridge, which allows simultaneous observation of a number of reference points when the pointers are in position. While the important vehicle dimensions can be found in body dimension manuals, most universal measuring equipment manufacturers have specific dimension charts for their equipment. These charts, one for each vehicle model manufactured, serve as guides to use before and during the repair (Figure 5-35). The dimension chart usually illustrates two views of the vehicle underbody. The upper illustration is a side view; the lower is the underbody plan view, which identifies the passenger and driver sides. Some charts also give underhood and upper body dimensions. The latter is most important with the plastic paneled mill and drill pad dimensions described in Chapter 13 (Figure 5-36). It is important to remember that many equipment manufacturer's dimension charts are intended only for that specific piece of equipment. Because of this variation between systems, it would be rather difficult to explain how each measures a vehicle; read the owner's manual for these details.

Before beginning any universal measuring operations, be sure to:

- Remove detachable damaged body parts, both mechanical and sheet metal body panels.
- If the damage is severe, perform rough straightening to the center section or foundation of the vehicle.
- If the mechanical parts are left in the vehicle and an overhang condition exists, this must be compensated for.

MECHANICAL MEASURING SYSTEMS

The mechanical measuring system (Figure 5-37) consists of:

- A bridge (or bridges) that runs the length of the vehicle from front to back.

FIGURE 5-37 Typical mechanical measuring system

- Sliding arms that mount to the bridge. They can be moved from front to back on the bridge, for length measurements, and outward or inward, for width measurements.
- Pointers which are mounted on the arms. They can be adjusted up or down for height measurements (Figure 5-38).

On some systems the end of the pointer will accept a special adapter designed to fit over a bolt head or into a reference point hole (Figure 5-39). The equipment manufacturer will provide instructions for this type of setup.

Mechanical measuring systems are generally organized according to guidelines found in dimension charts supplied by the equipment manufacturer for each family of body styles. They might look complicated, but after reading the charts carefully, especially the explanation of the setup, the task is easy.

FIGURE 5-38 The pointers are adjustable for height measurements.

FIGURE 5-39 Some mechanical systems utilize adapters that fit over a bolt head.

LASER MEASURING SYSTEMS

The laser measuring system is extremely accurate when properly installed. The word LASER is an acronym for Light Amplification by Stimulated Emission Reaction.

All laser measuring systems operate basically the same way. The laser unit is aimed at a target or measuring scale that is either hung from or attached to the vehicle. Some systems even use parts of the vehicle as targets. There are several types of systems available, and it does not matter where the laser is set up, or how it is set up. The thing to remember is that the laser system works either from the centerline or from the datum plane reference.

Measurements are taken by observing the laser beam on the targets. Some targets (Figure 5-40) are clear, allowing the laser beam to pass through; thus, several clear targets can be used with one light source. Beam splitters are capable of reflecting the laser beam to additional targets. (In some laser systems these targets are called *laser guides.*) Using combinations of transparent targets and beam splitters, it is often possible to simultaneously measure several dimensions on a vehicle using a single laser source.

Some laser systems (often called three-dimensional systems) use up to three laser units to give length/width/height coordinates anywhere on the vehicle—deck lid openings, cowls, door openings, hinges, pillars, and roof lines (Figure 5-41). With such a system, a single laser unit can be used to make measurements in conjunction with measuring devices such as a tape measure or calibrated bar (Figure 5-42).

Laser measuring systems are made up of both optical and mechanical parts. For example, in the clear target system, the optical parts are:

FIGURE 5-40 Some laser measuring systems use clear targets that allow the beam to pass through.

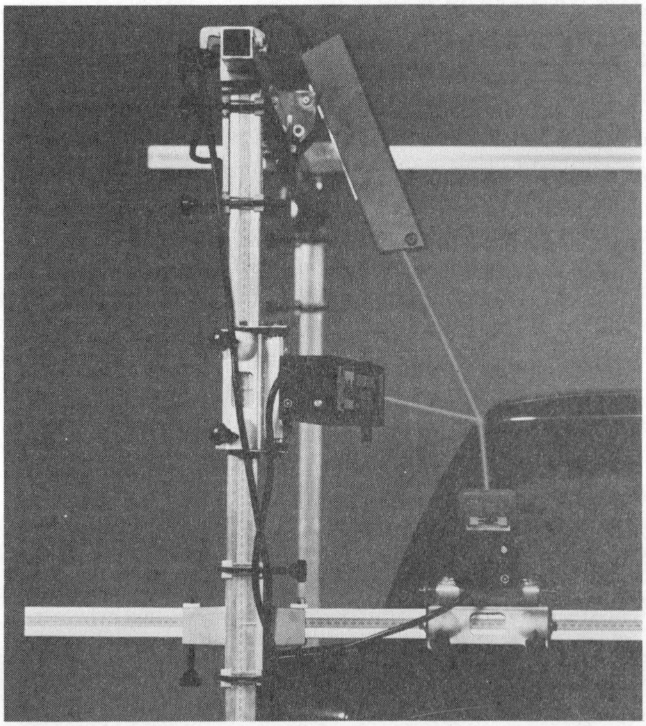

FIGURE 5-41 The three-dimensional system permits length, width, and height measurements.

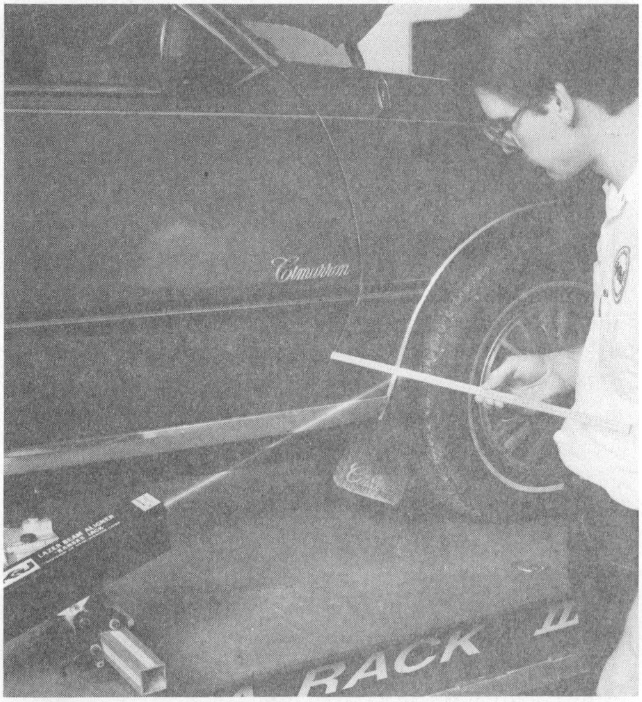

FIGURE 5-42 Taking a measurement using a laser unit and calibrated bar

- Laser power unit, which emits a safe, low-powered laser beam.
- Beam splitters that project beams at a precise right angle so height, width, and length can be measured simultaneously.
- Laser guide that reflects the laser beam at exactly 90 degrees.

Other items used with laser measuring devices include calibrated bars that attach to the vehicle or act as support devices for the laser unit itself, depending on the manufacturer. Transparent scales create a datum plane that brings the reference points on the vehicle to a level that can be illuminated with the laser beam. Scales are hung from reference points on the underside of the vehicle. The laser beam passes through the center of the scale on the target when the measuring point is in its correct position.

After the laser measuring system has been set up and targets hung under the vehicle, measuring can be started by using two sets of undamaged reference points under the vehicle, vertically adjusting the scales, and calibrating the laser beam using the control on the two laser guides (Figure 5-43). The calibrations are complete when the laser beam passes through or hits the two target scales.

Some laser measuring systems permit the technician to monitor upper body dimensions, for instance, pillar locations and window openings. It assures that the front door just installed will fit, or that the windshield to be installed is snug. Also, some laser systems offer an integral four-wheel alignment capability (Figure 5-44). Wheel alignment on the measuring system is

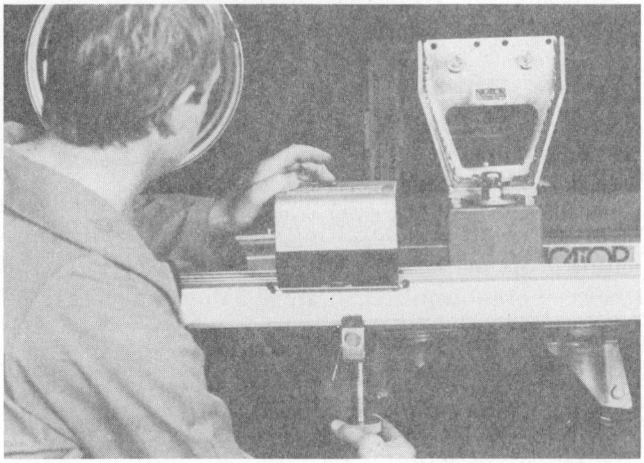

FIGURE 5-43 Calibrating the laser beam

FIGURE 5-44 Four-wheel alignment can be performed with some laser systems.

beneficial because suspension problems can be measured and corrected during repairs.

SHOP TALK

The accuracy of laser measurement systems depends on their proper installation by a trained technician, as well as the accuracy of the targets and calibration of the system to the targets. When a laser beam is projected through one target to another, the targets must be optically perfect, otherwise the laser beam will be deflected as it passes through the targets. The deflection will be magnified by the distance between the targets. This effect can result in serious error. Laser targets that become scratched or warped should be discarded.

When properly set up, the laser system can generally stay with the vehicle during the pulling and straightening process, like the mechanical systems. However, if the laser or mounting bars are in danger of being damaged during the pull, they must be moved.

Laser measuring systems provide direct, instantaneous dimensional readings so that the reference points in both the damaged and undamaged areas of the vehicle can be monitored continually during the pulling and straightening operation.

COMPUTERIZED MEASURING SYSTEMS

There are several systems on the market today that combine laser measuring with a computer for fast entry and checking of measurements. In one such system (Figure 5-45), measurements are made electronically and are displayed digitally on the laser guide. With the hand unit connected, measurements appear immediately on the computer screen. The hand unit has storage capacity for data and specifications of up to four different vehicle models. The vehicle measurements can be quickly compared with specifications, and deviations are displayed. If the measurements and deviations need to be documented, the hand unit can be connected to a printer to provide a printout. Measurements are available either on disk or via modem from the manufacturer's main computer. All data and specifications can be stored in the shop's personal computer and are easily transferred to and from the hand unit via an adapter.

Figure 5-46 shows another computerized measuring system in which the diagnostic computer can be used for direct reading of vehicle dimensions. A command transmitter provides convenient handling and directional transmission to the receiving computer from all measurable points, and a floppy disk stores the operational program.

DEDICATED BENCH AND FIXTURE MEASURING SYSTEMS

The dedicated bench and fixture system is a completely different type of measuring method. Instead of taking actual measurements, the bench acts as a "go–no go" gauge. The bench consists of a flat work surface to which fixtures are attached (Figure 5-47).

Fixtures are designed from the vehicle manufacturer's drawings to duplicate the vehicle body as closely as possible. The fixtures measure suspension mountings or other key locations of the underbody. Fixtures bolt to the transverse beams and mate with the underbody of the vehicle. If the fixtures fit the vehicle properly, the technician knows that the vehicle is in perfect alignment. All that is necessary is to straighten the vehicle until the reference points match the fixtures. No other underbody measurements are usually required.

The dedicated bench and fixture measuring system requires a specific set of fixtures for each family of body styles. If the collision repair shop does not own the required fixtures, they can be rented. Some systems require the entire fixture to be changed (Figure 5-48A)

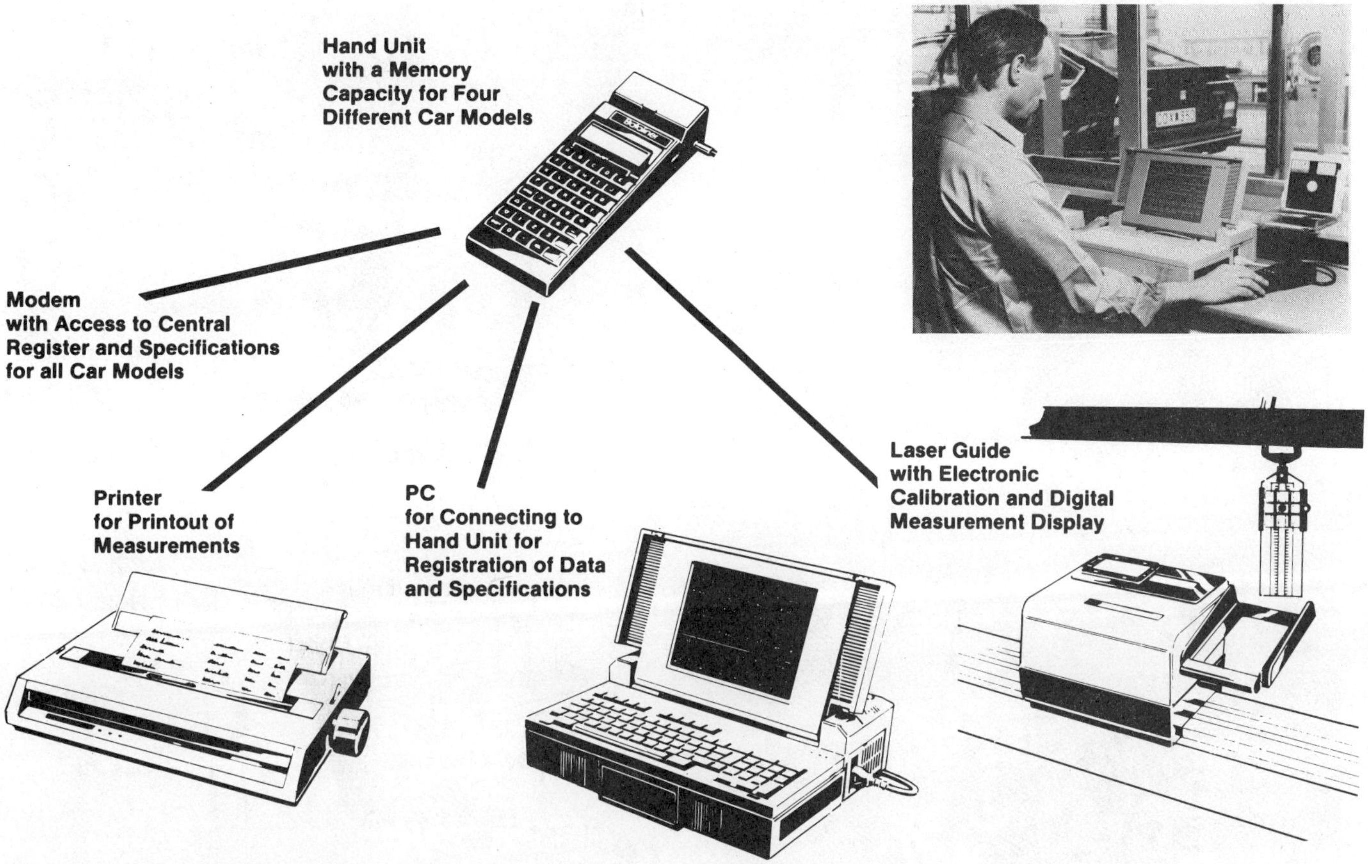

FIGURE 5-45 Hand-held type of computerized measuring system

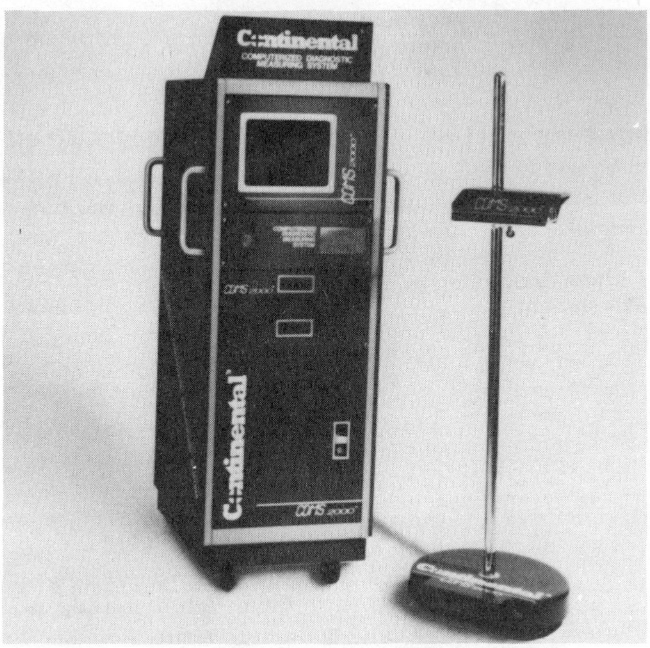

FIGURE 5-46 This diagnostic computer is used for direct reading of vehicle dimensions.

FIGURE 5-47 Typical dedicated bench and fixture system

and others have common fixture bases that require only a different fixture head for each body type (Figure 5-48B). The bench has built-in reference positions, and the fixtures are positioned to the specific references according to instructions supplied by the manufactur-

A

B

FIGURE 5-48 (A) The entire fixture must be changed on some systems, (B) while others require only a different fixture head for each body type.

FIGURE 5-49 A minimum of three fixtures must be in place on the bench prior to mounting the vehicle.

er. At least three fixtures must be positioned on the bench before the vehicle is mounted (Figure 5-49). Generally these are the torque-box fixtures. However, the damage to a specific vehicle will dictate which fixtures should be used.

Figure 5-50 shows four of the more common types of fixtures. They are:

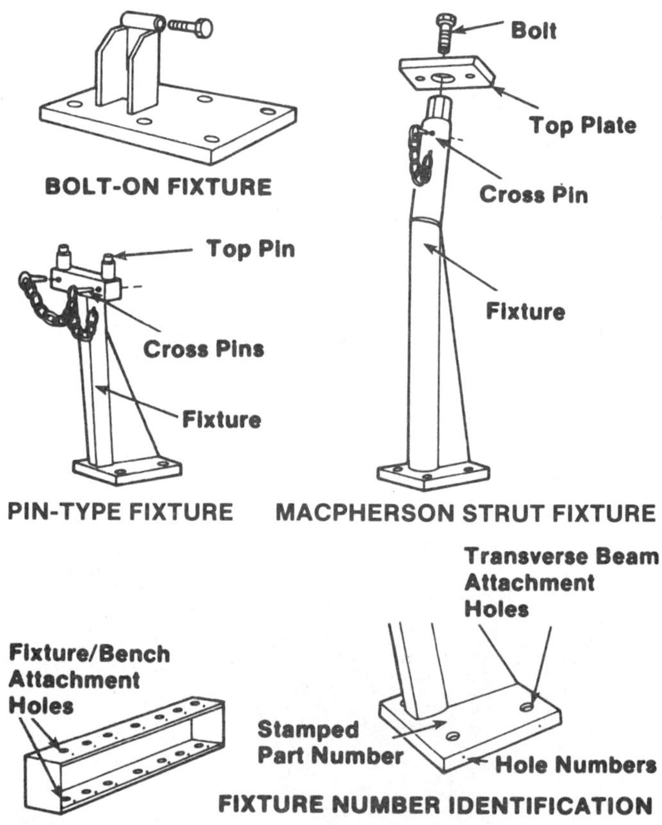

FIGURE 5-50 Common fixture types

- *Bolt-on Fixtures.* These are used when attachment is required to steering or suspension mountings. The studs or bolts normally used to attach these parts to the vehicle are also used to attach the fixtures. Depending on the damage, the fixtures can either be bolted to the vehicle first and lined up with the bench during the repair or attached to the bench first and lined up with the vehicle during repair.
- *Pin-type Fixtures.* These are used most often to mate with reference point holes in the underbody. They can also be used to mate with suspension mounting holes.

- *MacPherson Strut Fixtures.* These are used in the same manner as a pin-type fixture. A typical MacPherson strut fixture consists of a bottom plate assembly as with the other fixtures, an adjustable shaft with a cross pin, and a bolt-on top plate.
- *Bench Extensions.* These are included with certain fixture sets where the length of the vehicle requires that fixtures be positioned beyond the bench surface. The extensions are always used at the rear of the vehicle. Each extension is drilled with holes on top and on the bottom.

Most fixtures share a common attachment method to the transverse beams. The base plate for each fixture is drilled with four, eight, or twelve holes and is marked with a part number and the location on the transverse beam. Stamped at the rear of each fixture base are the numbers that correspond to the numbers on the transverse beam. Make sure these numbers are on the same side as the numbers on the transverse beam. When there are more than four holes on the fixture base, minor length differences in some models require two or three sets of four holes. They are stamped in the base of the fixture and noted on the dimension chart.

Once the vehicle is mounted on the bench, fixtures should be set up in the undamaged area of the vehicle. Then place as many fixtures as possible in the damaged area of the vehicle. The fixtures perform a number of functions:

- They show visually where a reference point should be located. If the reference points on the vehicle are not positioned where the fixtures say they should be, the reference points on the vehicle are out of alignment and need to be returned to their proper positions.
- They provide gauging of all the reference points at the same time. No measuring is required. If all the points line up, the steering, suspension, engine mounts, and so on are all in the exact position they were in when the vehicle was manufactured.
- Once the damaged parts are properly lined up with the fixtures they can hold these parts in position while further straightening is done. This eliminates the "pull-measure," "pull-measure" sequence required with universal measuring systems.
- They allow accurate assembly of parts on the vehicle before actually welding those pieces together (Figure 5-51). A good example of this would be a lower rail and strut tower assembly on some unibody vehicles. The sequence would go like this:

FIGURE 5-51 Dedicated fixtures allow assembly of parts before the actual welding is done.

—Position and hold the lower rail pieces on the fixtures.
—Weld them together.
—Position and hold the strut tower pieces on top of the rail.
—Weld them in their correct position.

SHOP TALK

Manufacturers of measuring systems are constantly furnishing informational updates and bulletins on their products. Be sure to read and study them because they will help to make the repair procedure easier. All of these systems can be used to properly repair unibody vehicles, if the collision repair technician knows how to use them. Remember, measuring the vehicle is a critical step in the repair process. The best measuring system in the world will not help if technicians do not take the time to learn how to use it correctly. Damage cannot be repaired if it cannot be found; and damage cannot be found without precise, careful measurements.

FINAL MEASUREMENTS

The dimensions that are shown in most body dimension manuals are based on the engineering blueprints for a given vehicle rather than on actual measurements from the vehicle itself. The blueprint specifications are, after all, supposedly the dimensions to which the vehicle is built on the assembly line. Robot welding fixtures and laser body inspection has tightened up the assembly specs considerably over what they used to be. Yet there is still quite a margin for error depending on which part of the body is being measured. Consequently, most aftermarket body dimension manual publishers and pulling equipment manufacturers base their dimension charts on actual measurements taken from vehicles.

To compensate for assembly line variations, a number of identical year, make, and model vehicles must be carefully measured to figure out how much variation is common in a given dimension. The numbers are then averaged so the most likely value can be printed as the one to use when restoring a collision-damaged vehicle. For that reason, aftermarket body dimension manuals and those from pulling equipment manufacturers may be more accurate than body dimension manuals from vehicle manufacturers.

The body dimension manuals are a must because without them, you have no way of knowing whether or not the strut towers are supposed to be symmetrical or asymmetrical. On a vehicle with symmetrical strut towers, the tops of the towers should be an equal distance from the vehicle's centerline (Figure 5-52). The same is true for the location of the lower ball joints. But the position of the strut towers can often vary up to 3/8 inch (10 mm) on many vehicles from the factory. They are not supposed to vary that much, but they do—and that can make a big difference when it comes to determining whether or not a strut tower's position is the result of a collision.

A 3/8 inch (10 mm) difference in the strut tower location in or out with respect to vehicle centerline will change SAI and camber by about 1 degree (Figure 5-53). The same is true for the position of a lower ball joint. A

FIGURE 5-52 Adjusting the upper horizontal bar between two symmetrical strut towers

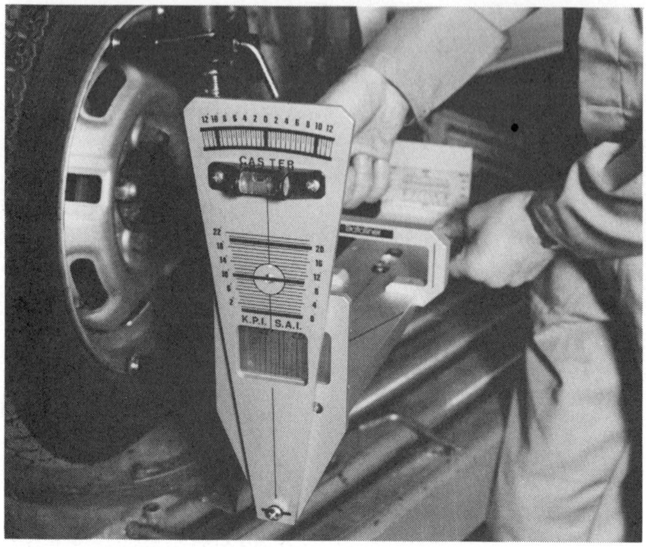

FIGURE 5-53 Checking the SAI measurement with a laser measuring system

1-degree difference in SAI side-to-side is about the acceptable limit before it causes steering problems. But half a degree is the limit for camber. So in this instance, the strut tower is right on the edge as far as SAI is concerned, but it is too far out of position with respect to camber. But since camber can be adjusted on most vehicles, the easiest "fix" (though not necessarily the most desirable one) would be to leave the strut tower as is and simply readjust camber to restore proper wheel alignment. Ideally, the strut tower should be within 3 mm of its specified position for best wheel alignment. But if the strut tower is out of position as a result of a manufacturing defect, trying to correct the factory problem might end up creating more headaches—like welded seams coming apart or alignment problems between the fender and hood, and so on. It does not make as much difference where the strut towers are with respect to the rest of the body as it does where they are with respect to each other and the lower ball joints.

Other problems common to all makes include centerline misalignment, where the left or right rail might be as much as 10 mm closer or further from centerline than it should be, and datum line problems where one side of the vehicle is higher or lower than the other.

The locations of some components are obviously more critical than others. In addition to the strut towers, other key points that influence wheel alignment, steering, and tracking include: the attachment points for the engine cradle subframe on front-wheel-drive vehicles, the attachment points for the lower control arms on the front suspension; the locations where the rack-and-pinion bolts to the subframe or dash panel; and the attachment points for the rear control arms or leaf springs.

The exact position of these components is not as important as their position in space with respect to the wheels and chassis. The rack-and-pinion, for example, must be parallel to the ground and parallel to the wheel axis; otherwise, it can cause a problem known as bump steer. This happens when the tie-rods are not an equal height from the road or an equal distance from the wheel axis. This produces uneven toe changes when the suspension undergoes jounce and rebound, causing a twitching or jerking to one side.

The rear axle must likewise be square to the chassis and perpendicular to the vehicle centerline. If cocked to the right or left, it creates a thrust angle that causes the vehicle to pull to the opposite direction. A difference of only a few millimeters in wheel base side-to-side on a vehicle that is supposed to have a symmetrical chassis can be enough to cause a thrust angle problem.

Twist or sag in the chassis can cause variations in ride height. Only a 1-inch change in ride height at the rear of a vehicle, for example, can alter front caster readings nearly a full degree. (Additional details on wheel alignment can be found in Chaper 9.)

When restoring a vehicle to its original dimensions, a certain amount of common sense must be used in deciding whether or not a given reference point's location is "close enough" to those published in dimension charts.

REVIEW QUESTIONS

1. On some unibody designs, what has a fixed (nonadjustable) value?
   a. camber
   b. caster
   c. both a and b
   d. neither a nor b
2. Which of the following statements is incorrect?
   a. The dimensions given in manufacturers' body dimension manuals use the English system of measurement.
   b. The repair tolerances of critical manufacturing dimensions on unibody vehicles must be held to no more than $\pm$ 3 millimeters.
   c. The metric system is easier for computation than the English system.
   d. When a unibody vehicle is involved in an accident, the damage is often structural as well as cosmetic.
3. From what part of a unibody vehicle are all height dimensions taken?
   a. datum plane
   b. center plane
   c. centerline
   d. reference point
4. All unibody measurements and alignment readings should be taken relative to what section?
   a. front
   b. rear
   c. center
   d. both a and b
5. To accurately measure a vehicle, how many correct dimensions are required as a starting point?
   a. at least two
   b. at least three
   c. at least six
   d. one
6. What measuring system is used to check alignment of the front and rear wheels?
   a. tram bar
   b. self-centering gauge
   c. tracking gauge
   d. strut tower gauge
7. Technician A says that the weight of the mechanical and drivetrain components must be taken into account when measuring a unibody vehicle. Technician B says that these weights are irrelevant and need not be considered. Who is correct?
   a. Technician A
   b. Technician B
   c. Both A and B
   d. Neither A nor B
8. Which of the following statements is incorrect?
   a. Upper body damage can be measured by the use of tram bars and a steel measuring tape.
   b. Reference point holes are frequently larger in diameter than the tram bar pointer.
   c. When checking front end dimensions, the best areas to use for the tram bar measurements are the attachment points for suspension and mechanical components.
   d. None of the above
9. When checking front end dimensions, Technician A measures from the lower cowl area to the front mount of the engine cradle. Technician B measures from one lower cowl area to another. Whose measurement will provide a better reading?
   a. Technician A
   b. Technician B
   c. Both A and B
   d. Neither A nor B

10. Any deformation of the body side structure can be ascertained by irregularities in the ____________ when it is opened and closed.
    a. hood
    b. trunk
    c. door
    d. none of the above

11. Technician A uses the diagonal line measurement method to inspect damage to both sides of a vehicle. Technician B uses the same method when twist has occurred. Who is correct?
    a. Technician A
    b. Technician B
    c. Both A and B
    d. Neither A nor B

12. What are self-centering gauges used to establish?
    a. datum plane
    b. vehicle centerline
    c. both a and b
    d. neither a nor b

13. Technician A stands 3 feet away from a self-centering gauge when taking a reading, while Technician B stands 6 feet away. Who will get the more accurate reading?
    a. Technician A
    b. Technician B
    c. Both A and B
    d. Neither A nor B

14. To check for centerline misalignment, at least ____________ gauges must be hung?
    a. four
    b. three
    c. two
    d. one

15. What type of fixture is used when attaching to a suspension mounting location?
    a. bolt-on
    b. bench extension
    c. pin-type
    d. MacPherson strut

CHAPTER

6 Straightening Systems and Techniques

OBJECTIVES

- Identify the various types of straightening equipment and explain how they are used.
- Describe the usual sequence for a total unibody realignment procedure.
- List important safety considerations for working with alignment equipment.
- Explain why the multiple-pull approach is required for unibody vehicles.
- Plan and execute a pulling sequence.
- Perform basic stress relieving and metalworking techniques.
- Identify signs of stress/deformation on a unibody vehicle.
- Mix, apply, and shape plastic body filler.
- Perform a final check of alignment procedures.

Until recently the straightening process was a simple matter of applying brute force until the damage in the impact area bore a visual resemblance to the proper structure. But with today's unibody construction, precision is the key factor in collision repair. It is not enough just to make the damaged or replacement panels fit by shimming and stretching, then turning the alignment and mechanical problems over to a mechanic. The collision repair technician now has the responsibility of restoring the vehicle structure back to OEM specifications; only the fine-tuning of the suspension and mechanical components is left to the specialists in those fields.

If, after a vehicle has been repaired following a collision, a customer complains of abnormal tire wear and pulling to one side, a check might find an enlarged slotted hole for the installation of the front fender or door hinge. Examples such as these are indicative of improper structural alignment techniques, which are the major causes of the vehicle not being restored to its original dimensions. Structural alignment is an extremely important operation that affects safety, repair time, and quality, not to mention the confidence of the customer. In unibody repair, the technician has a twofold responsibility: to return the vehicle to its pre-accident shape and pre-accident state.

STRUCTURAL ALIGNMENT BASICS

It is a fundamental principle in structural alignment that force must be applied to pull the damaged area in the direction opposite of the direction of impact. This works well when the impact is small and damage is simple. However, where there is sectional deformation such as folding, or where the damage is severe, the original shape cannot be restored simply by a basic standard pulling operation. When body panels are deformed by the force of impact into complex shapes, their strength is altered. So to simply pull in the opposite direction would not work because of the differences in strength and recovery rates of each panel. Therefore, it is necessary to apply force according to the recovery rate of the individual panel and to change the direction in which the force is applied to meet the changes that occur during the repair operation.

Length is the most important body dimension and thus must be corrected first during the straightening procedure. When determining the direction of pull, set the equipment at an angle where the pulling force will be applied perpendicular to the damage. One way to alter the direction while pulling is to divide the pulling force into two or more directions. Since applying force in only one place will not result in proper repairs, it is recommended to exert pulling force on many places at the same time. For convenience, the term *direction opposite to the impact* will be used to describe the effective pulling direction.

STRAIGHTENING EQUIPMENT

Every collision repair facility must have equipment that performs the basic straightening functions. However, unless this equipment has been upgraded to the required level, repairs should not be attempted.

More specifically, auto manufacturers believe that their front-wheel-drive vehicles can be competently repaired and original design safety restored only with equipment that provides:

- Precision and full structural alignment
- Multiple anchoring locations
- Multiple and full range pulling

The relatively fragile unibody structure requires multiple anchoring points to offset the pulling forces. Multiple pulls, and the ability to pull in more than one direction at the same time, are needed to return the vehicle to its correct dimensions. Single, hard pulls in one direction are fairly effective for straightening body-over-frame vehicles. Overpulling and tearing metal is not a very big issue when body metal is 1/8- to 1/4-inch thick. However, this is seldom the case when repairing the 24-gauge steel used on unibody vehicles.

It is important to keep in mind that no one part of a unibody vehicle can withstand the full force of a pull. Since these vehicles do not have frames to help hold the vehicle, some type of stabilizing is required. This can be accomplished by anchoring the vehicle firmly to the floor, platform, bench, or rack. By stabilizing the vehicle and distributing the pulling force, it is possible to make a pull without distorting the anchoring points.

Because of the unibody's thin gauge structure, it is very important that the pull setup is right the first time. Therefore, a measuring system that only indicates where there is misalignment is not sufficient. For the unibody repair, the measuring system must also show the amount of misalignment at each reference point and the direction of the misalignment. Only with such equipment can the collision repair technician plan exactly how the total pull sequence should be set up and how to monitor the progress and effectiveness of each pull.

Pulling and anchoring systems come in a variety of styles from many manufacturers. Regardless of the make of equipment being used, it has to accomplish two basic tasks:

- Hold the vehicle securely in place.
- Apply controlled force in controlled directions.

The pulling equipment may be portable, used with a floor system, or attached to a bench or rack (see Chapter 5). It must have the capability to properly repair unibody vehicles. The most popular systems are described in the following pages.

IN-FLOOR SYSTEMS

The two types of in-floor straightening systems presently available are:

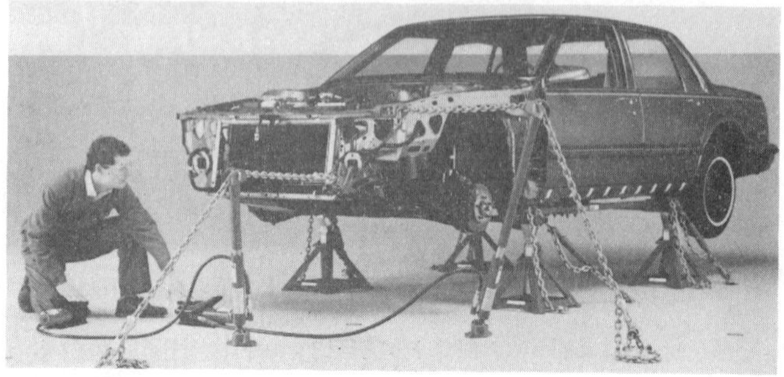

A

B

FIGURE 6-1 Typical in-floor straightening systems: (A) anchor pot and (B) floor grid

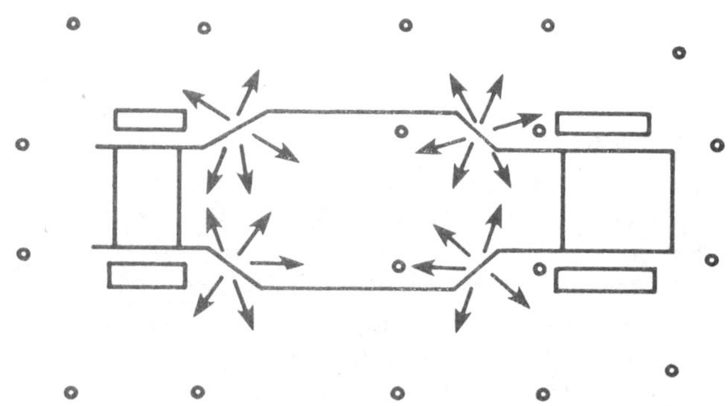

FIGURE 6-2 The anchoring setup must be balanced in both the direction and force of the pull.

- Anchor pot (Figure 6-1A)
- Floor grid (Figure 6-1B)

Both systems use an anchoring setup (Figure 6-2) that must be balanced in both the direction and force of the pull. Each system has a pump, either hydraulic or air operated, for supplying pressure and a ram for putting that pressure to work. There is a chain for connecting

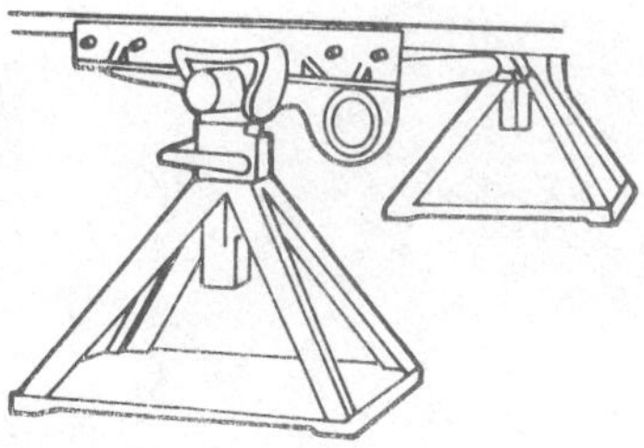

FIGURE 6-3 Proper positioning of the cross tube pinchweld anchoring clamps is important.

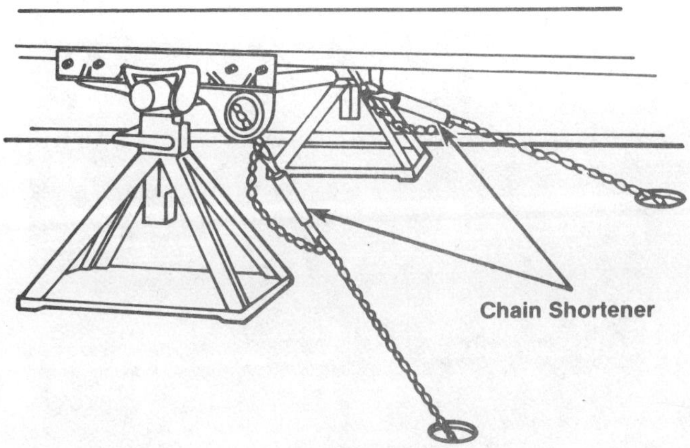

FIGURE 6-4 Use chain shorteners to remove slack.

the ram to the vehicle and anchors for securing the ram and chain to the grid system. The vehicle is usually supported on jack stands using cross tube pinchweld anchoring clamps, as shown in Figure 6-3. Chains are attached to the clamps and hooked to the anchor pots or floor grid. Pull the chains tight for secure anchoring. Chain shorteners should be used to remove slack from the chain (Figure 6-4).

Position the ram in the ram foot so it will exert force in the desired direction, and build the ram up to the desired height using extensions. Pull the chain tight and lock it in the chain head with the cross pin. Hook the chain onto the anchor. The anchor, ram foot, and attachment point on the vehicle should be in a straight line in the direction of the pull. Hook the pump to the ram and connect an air hose to the pump. After engaging the pump to take up slack in the chain, pulling can be started.

An in-floor system is ideal for a small collision repair shop. Both the anchor pot and floor grid systems provide fast hookup, single or multiple pulls, and positive anchoring without sacrificing space. It is possible with most in-floor systems to use a pulling post to provide the pulling power (Figure 6-5).

FIGURE 6-5 Typical pulling post

FIGURE 6-6 Typical above-floor platform alignment system

While in-floor alignment systems are built right into the concrete floor, the above-floor platform arrangement shown in Figure 6-6 sits above ground. It offers all the same advantages of the in-floor system, and since it is only bolted to the floor, it can be moved about the shop without a lot of time and effort.

CHAINLESS ANCHORING SYSTEMS

The low-profile design of the chainless anchoring system (Figure 6-7) keeps it close enough to the floor to load a vehicle quickly, yet high enough to allow easy access to the vehicle's underbody. The unit's adjustable length and width allows adjustment to virtually all vehicle sizes.

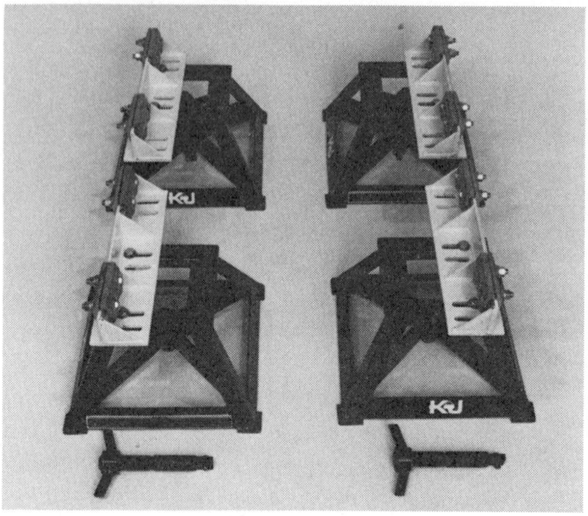

FIGURE 6-7 Chainless anchoring system stands

The following steps are required to set up a vehicle for pulling with a chainless anchoring system:

1. Raise one end of the vehicle and attach the underbody clamps (Figure 6-8A).
2. Insert the support tube through the underbody clamp and the stand (Figure 6-8B). Lower the vehicle onto the stand.
3. Secure the system to the in-floor anchors by positioning the lock arm. Hammer the wedges in place to lock the vehicle to the system (Figure 6-8C).
4. Repeat the operation on the opposite end of the vehicle (Figure 6-8D).

Once the vehicle is secured, it is locked to stay. It is possible to pull from any angle, 360 degrees around the entire vehicle. A pulling post is often used to provide the pulling force.

PORTABLE BODY AND FRAME PULLERS

This type of pulling equipment (Figure 6-9) uses a hydraulic pressure system installed between the removable main frame and the mast to extract the damaged portion by means of chains and clamps. It is often used for minor damage. Being easily movable, portable body and frame pullers can set the pulling direction to the damage input (Figure 6-10). Many units of this type, however, are able to pull only in one direction.

RACK SYSTEMS

Most rack systems have a pulling post or articulating arms that give the technician infinite positioning with complete freedom to pull from any angle and

A

B

C

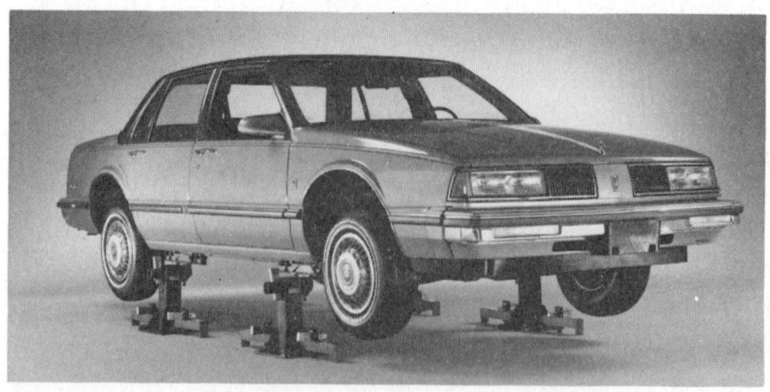

D

FIGURE 6-8 Installing a vehicle on a chainless anchoring system

FIGURE 6-9 Typical portable body and frame puller

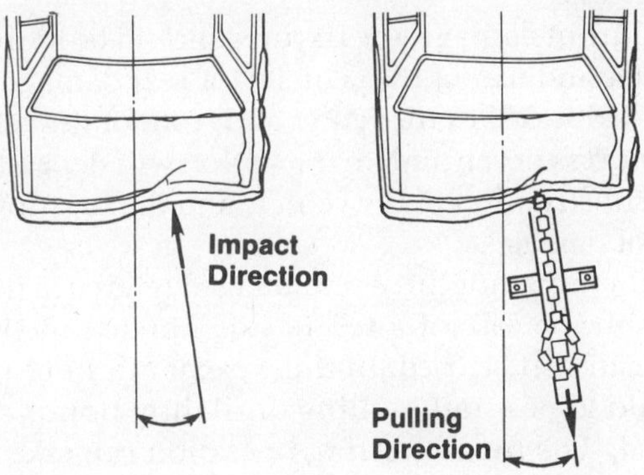

FIGURE 6-10 Portable body and frame pullers can set the pulling direction to the damage input.

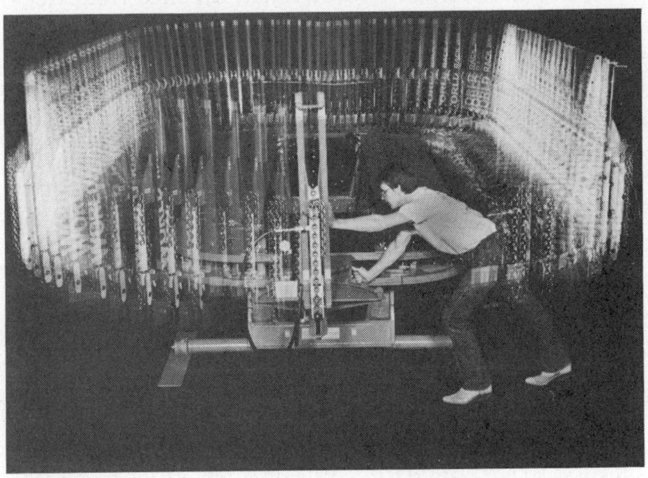

FIGURE 6-11 Most rack systems provide 360 degrees of pulling power around the entire vehicle.

FIGURE 6-12 Some racks can be positioned at a height flush to the floor.

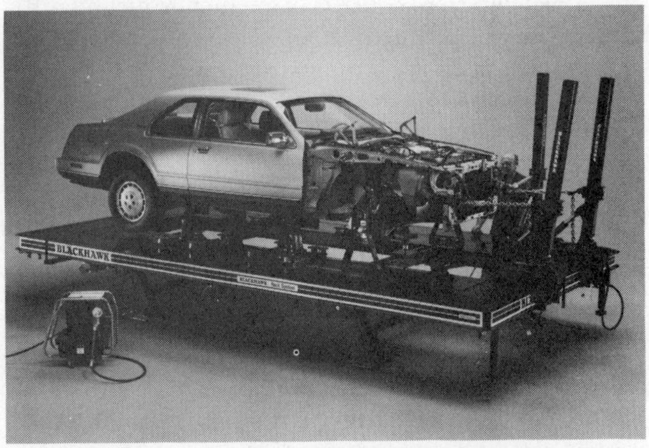

FIGURE 6-13 Typical stationary alignment bench

height, 360 degrees around the vehicle (Figure 6-11). It is possible to make upward pulls or downward pulls; pulls can also be made with the rack positioned at a height flush to the floor (Figure 6-12). Most racks tilt hydraulically so that a vehicle can either be driven on or pulled on with the optional power winch.

In most cases, a rack system provides either a mechanical or laser measuring system, both of which are described in detail in Chapter 5.

BENCH SYSTEMS

This type of equipment is used to straighten the body and frame. It is also capable of performing multidirectional pulls and measuring the vehicle simultaneously. Alignment benches are currently available in two types: stationary (Figure 6-13) and portable (Figure 6-14).

FIGURE 6-14 Typical portable alignment bench

FIGURE 6-15 Fixtures in place on a bench

A bench system consists of an anchoring and pulling system, plus a bench. Often called a dedicated bench and fixture system, it is discussed in Chapter 5.

Before putting the vehicle on any bench, there are some things that should be done to make the most effective use of the bench. One of the most important things is to visually determine the degree of damage and begin making a plan for the best way to repair it. As a general rule, a bench system should be used whenever the damage involves the suspension, steering, or power train mounting points. This would include situations such as a side collision where the suspension components and their mountings are not damaged directly, but because of deformation in the center section of the vehicle, the whole body is out of alignment. Determine whether a particular collision meets this criterion by making some general measurements with a tape measure or tram bar. These would include diagonal measurements to check for sidesway, and length measurements to check for mash. Try to get as good an idea as possible of where the damage begins and ends. Use all the data available, including body dimension manuals and the vehicle manufacturer's service manuals. Remember that these are only general measurements to determine whether to use the bench system; they do not have to be made as accurately as when actually straightening the vehicle.

Fixtures (Figure 6-15) are an important component of many bench systems. It is necessary to have at least three reference points on the undamaged part of the vehicle to set it up properly on the bench. These three locations will form the datum plane on which all of the other measurements will be based. If there are more than three locations that are undamaged, they can also be used for setup. The fixtures chosen should require the minimum amount of disassembly of undamaged components.

For front damage, the fixtures should be located in the center and rear of the vehicle; for rear damage, they should be located in the center and front of the vehicle. The fixtures chosen and their location will depend both on the construction of the vehicle and the location and degree of damage.

If there are not three undamaged reference points, such as in the case of a severe side impact, the bench system can still be used. In this case, it might be necessary to do some rough pulling until three points can be obtained. The rough pulling operation can take place on the bench and might require some shifting of the vehicle on the bench once it is completed. It can also be done using conventional measuring methods prior to loading the vehicle on the bench.

Once the vehicle is positioned on the cross tubes and the bench slid under the vehicle, survey the damaged area and the positions of the fixtures to determine exactly the location, direction, and extent of the damage. Loosely attach the outermost fixtures to the proper location on the bench if they are the bolt-on variety. Raise the bench until the fixtures mate with the proper reference point on the vehicle and on the bench. Continue raising the bench until the weight of the vehicle is off the cross tubes.

Remove the cross tubes, stands, and underbody clamps (if they were used), and lower the vehicle and bench. The next step is the installation of the pinchweld anchoring clamps as shown in Figure 6-16. Repeat the procedure at the opposite end and side of the vehicle, then install the remaining fixtures. Once all of the clamps are fastened in place on the bench, the vehicle is ready for pulling.

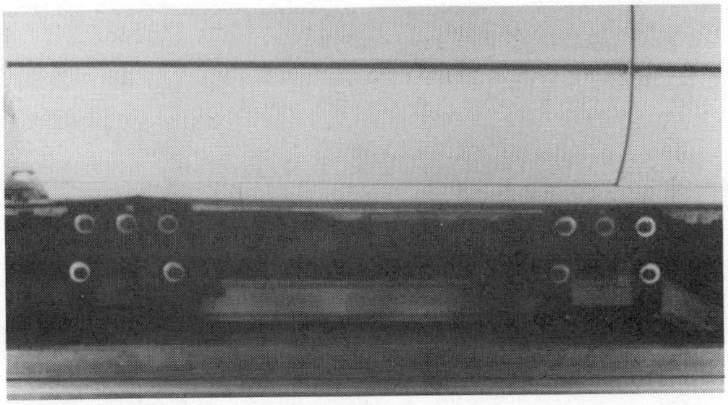

FIGURE 6-16 Pinchweld anchoring clamps

OTHER STRAIGHTENING ACCESSORIES

Two accessories that are often used with alignment systems are the engine stand and a portable pulling/pushing arm. Whenever it is necessary to remove the engine or transmission mounts, the engine stand can be used to support the engine. It rests on the inner fenders and is adjustable in width. The center contains an adjustable chain hook that is used to hold the chain attached to the engine.

A pulling/pushing arm can pivot completely around the end of the bench from the center position on the special flange. In the other positions, it can reach the end and one side of the bench. The unit can also be used anywhere along the side of the bench by hooking the inner clamp on the outer flange on the opposite side.

Portable hydraulic rams are very handy tools, since they can be used to push, spread, clamp, pull, and stretch. Figure 6-17 summarizes the various applications of portable hydraulic equipment.

STRAIGHTENING AND REALIGNING TECHNIQUES

The usual sequence for a total unibody realignment procedure is as follows:

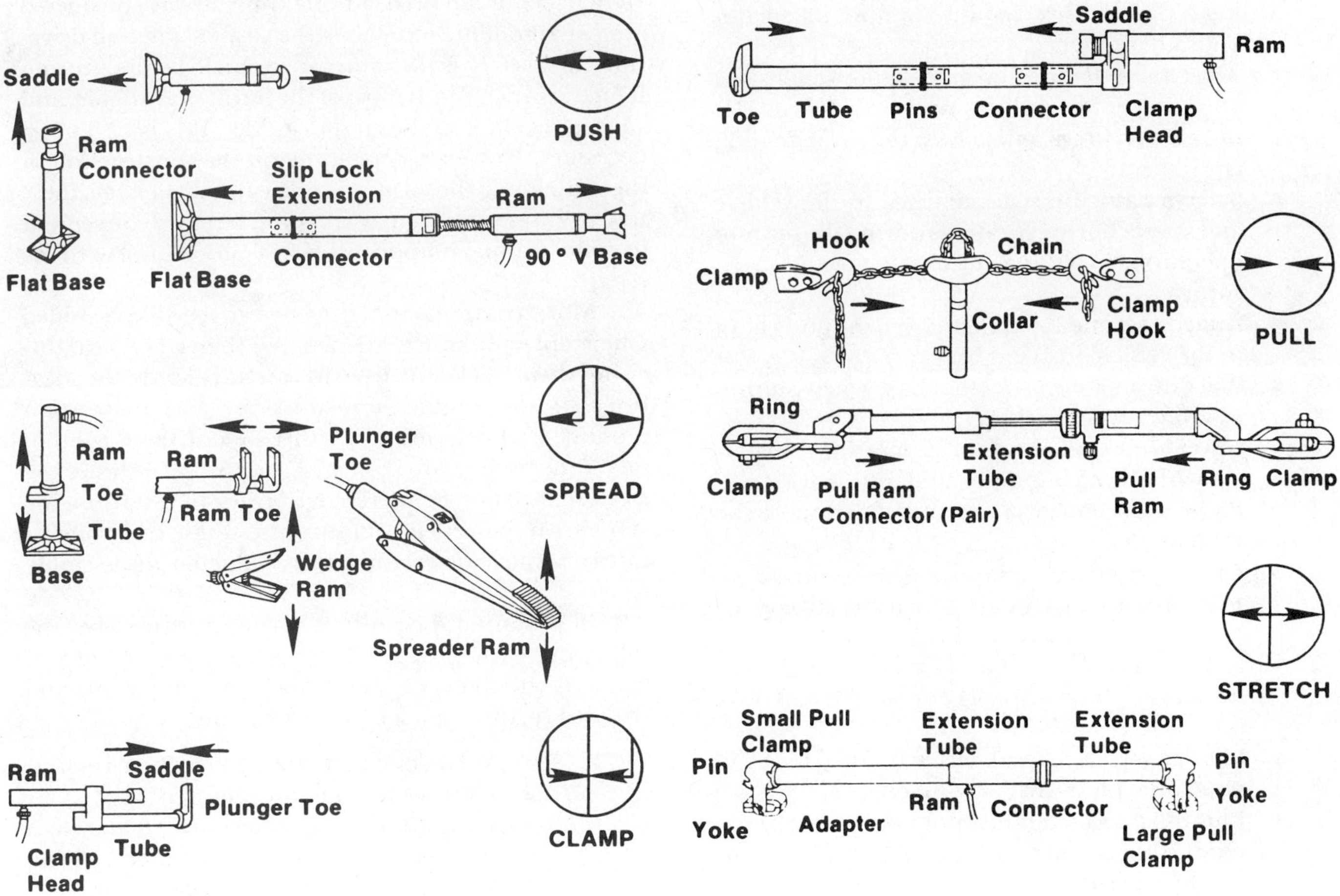

FIGURE 6-17 Basic hydraulic ram setup utilizing the standard push, spread, clamp, pull, and stretch symbols

1. Understanding the safety considerations of the equipment used
2. Damage analysis (this is fully covered in Chapter 5)
3. Initial clamping and alignment checks
4. Executing the pulling sequences with additional clamping and alignment checks

ALIGNMENT SAFETY CONSIDERATIONS

When using alignment equipment, inadequate attention to any of the rules given here can result in material and/or serious physical damage:

- Be sure to use all equipment correctly according to the instruction manual prepared by the manufacturer. Watch the equipment carefully during the pull and learn to look for hazardous situations.
- Never allow unskilled or improperly trained personnel to operate the equipment.
- Always anchor the vehicle securely before making a pull. Check that the clamps and anchor bolts are tightened.
- Always use the size and grade (alloy) chain recommended for pulling and anchoring. Use only the chain supplied with the straightening equipment.
- Safety chains must be attached to the vehicle and/or anchoring locations during the pulling operation. Avoid placing chains around sharp corners.
- Attach a second loose chain from the pull chain to the vehicle to avoid backlash.
- Place throw covers over the chain when pulling to protect from backlash.
- Never use a service jack for supporting the vehicle while working on or under it. Always use the recommended jack stands to support the vehicle.
- Clamps can slip and cause the sheet metal to tear. Prevent bodily harm and material damage by always using safety wires.
- Never stand in line of a chain or clamp. Chain breakage, clamp slippage, or sheet metal tearing could cause injury.
- It is dangerous to work inside the vehicle at the same time pulls are being made.
- Pumps and rams should not have to work very hard. If there is an overload, back off, rethink the pull, and try again. Change the direction of the pull, add more pulls, or stress relieve as described later in this chapter.

WARNING

Failure to observe any of these cautions or other unsafe use might result in property damage and/or serious personal injury.

PLANNING COLLISION REPAIR PROCEDURES

Before attempting any repair work, determine exactly the procedure that should be taken. The first step is damage analysis as discussed in Chapter 5. A little extra time spent on an operational plan can save many hours in the overall repair procedure.

Unless the damage is minor and limited to cosmetic sheet metal, chances are good that parts will have to be removed before pulling can be started. This is because of the unibody's structure, which tends to spread damage into remote and often unexpected places, some of which might be hidden behind these parts.

A general rule is to disassemble only those components that must be removed to get to the area of the vehicle being repaired. At one time, it was considered almost automatic to remove the suspension and driveline completely from a unibody vehicle before putting it on the bench. With most of the fixtures available, and such accessories as the engine holder, this is no longer necessary. However, depending on the construction of the vehicle and the location and degree of damage, there will be cases where it may very well be more convenient to remove some components before proceeding with the repair.

More components will have to be disassembled when replacing structural members than when straightening them. Take the time to carefully study the locations of the engine, transmission, and suspension mounts, and determine whether or not these components themselves are damaged. Figure 6-18 shows the front suspension being removed prior to installing the vehicle on the bench. In most cases these components can be removed as a unit, thus reducing disassembly time.

In planning the pulling sequence, remember the following basic rules to assure that misalignment and damage will be corrected with minimum work and without further damage to the structure.

- Repair the damage in the reverse (first-in, last-out) sequence in which it occurred.
- The pulling forces must never be greater than the combined anchoring force. When pulling with a two-ton force, for example, the anchoring system must be able to hold at least two tons.

FIGURE 6-18 Front suspension removal

- No single anchoring point on a unibody vehicle can withstand all the pulling force. The pulling forces must be distributed throughout the vehicle. A multiple anchoring system and multiple pulls accomplish this.

INITIAL CLAMPING WITH ALIGNMENT CHECKS

Multiple pulls (Figure 6-19) accomplish the following objectives:

- The desired direction of pull can be achieved from three or four points at one time.

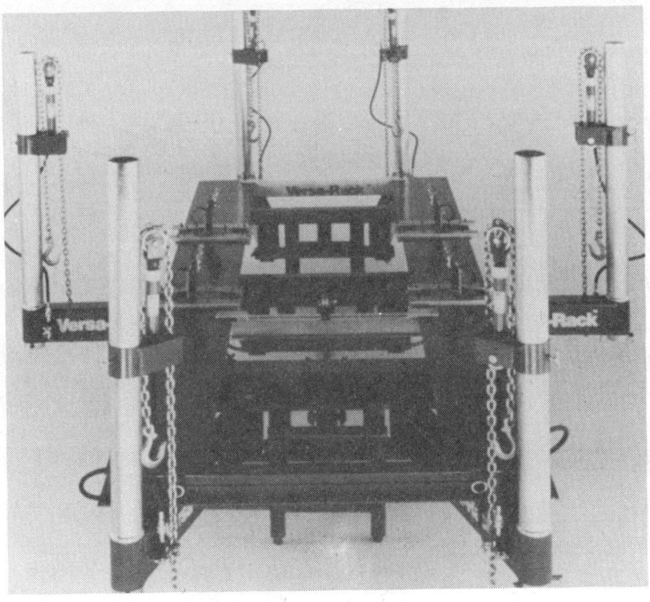

FIGURE 6-19 The multiple-pull approach

- The amount of force required at any single point is reduced, thus lessening the risk of tearing the thin sheet metal.

WARNING

If a vehicle is not anchored properly before pulling, there is a risk of causing additional damage to the vehicle, as well as personal injury.

There are two commonly used anchoring methods:

- Clamping onto pinchwelds.
- Bolting onto mechanical or suspension mounting points.

If working on a vehicle which has no pinchwelds, check the equipment manufacturer's instructions. There may be specific recommendations for anchoring those types of vehicles.

With the use of thinner sheet metals, pinchwelds are getting weaker. For this reason, large pinchweld clamps are recommended. One manufacturer, for instance, suggests 10- to 12-inch clamp jaws. Before attaching the clamps to the vehicle, clean any sealer or undercoating from the area, and clean the jaws of the clamps. Look out for any lines or hoses in the area to avoid damaging them. Keep an eye on the pinchwelds during the pulling operation to make sure they do not fold or buckle under pulling stress. Check the pinchweld clamps after the first pulls and periodically throughout the repair process.

When anchoring the vehicle, lean toward "over anchoring" or "over clamping." An extra anchor point or two takes very little time and surely cannot hurt anything. In those cases where a clamp cannot be fastened to the exact area of deformation, a piece of steel can be temporarily welded to the section (Figure 6-20). After repairs are completed, remove the steel.

Set the pulling clamp so that the line extending along the path of the pulling force passes through the middle of the teeth of the clamp. If this is not done, rotational force on the clamp will pull it off, further damaging the section (Figure 6-21). It is also important to treat each damaged area as individually as possible, since unibody vehicles are manufactured for isolated collapse upon impact.

SHOP TALK

Always make the first pulls before cutting away any damaged metal. For the initial pulls, this could be the best attachment point. Do not attach to any of the mechanical or suspension parts, only to their mounting parts.

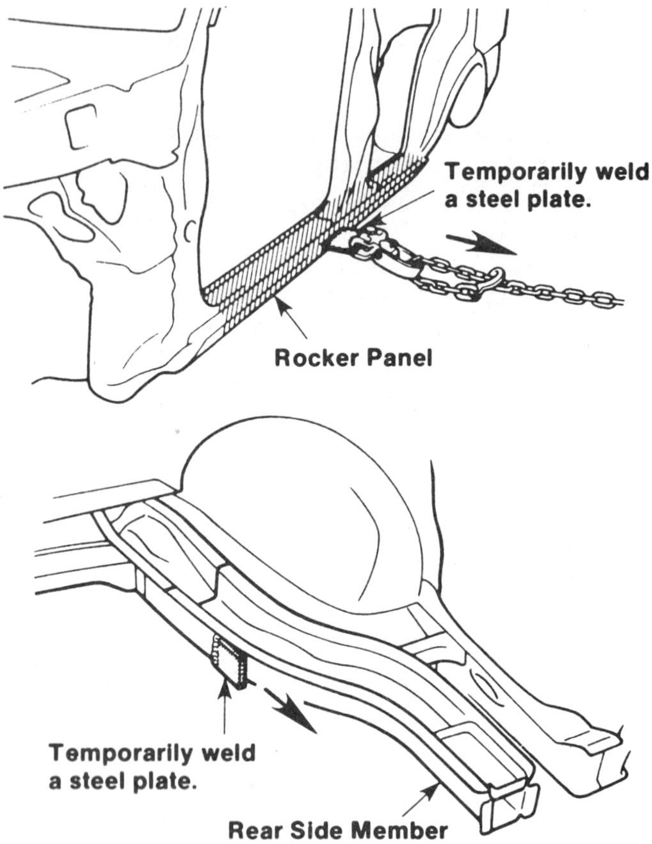

FIGURE 6-20 Two methods of welding a temporary steel pull plate in place

EXECUTING A PLANNED PULLING SEQUENCE

The pulling force will travel throughout the entire unibody vehicle. That is, a pull applied anywhere can affect the whole vehicle, so the technician must think through the entire pulling process. Regardless of how complex the pulling setup is, the angle of the chain—from the attachment point to the ram or tower—is the angle of the pull.

The correct pulling sequence consists of solving a variety of small problems rolled into one. Find the first problem, work on it, then move to the next problem (Figure 6-22). Make the pulls a little at a time, relieve the stress, take a measurement (or, in the case of the bench, check how close the fixtures are to their corresponding reference points on the body), and start the sequence again. Work from the center section outward, achieving correct length, then sidesway removal, and finally correct height.

The most effective way to perform a pull procedure is to approach the operation as though it was going to be done with your bare hands. That is, determine how the metal should be worked to move it back into shape if no tools were available. How many areas can be moved at one time and in which directions?

There are a number of possible setups for pulling or pushing. The pulling arrangement with the vector system is determined by a simple triangle formed by the ram, the anchored portion of the pull chain, and the

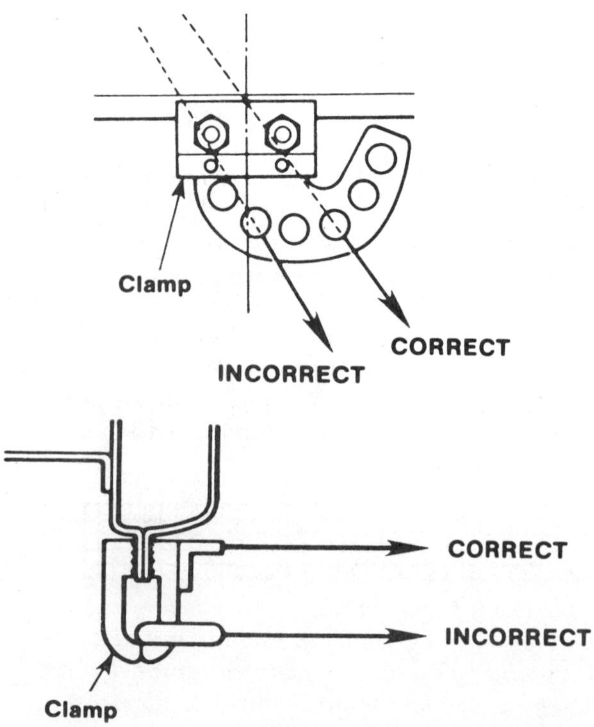

FIGURE 6-21 The right and wrong way to set clamps

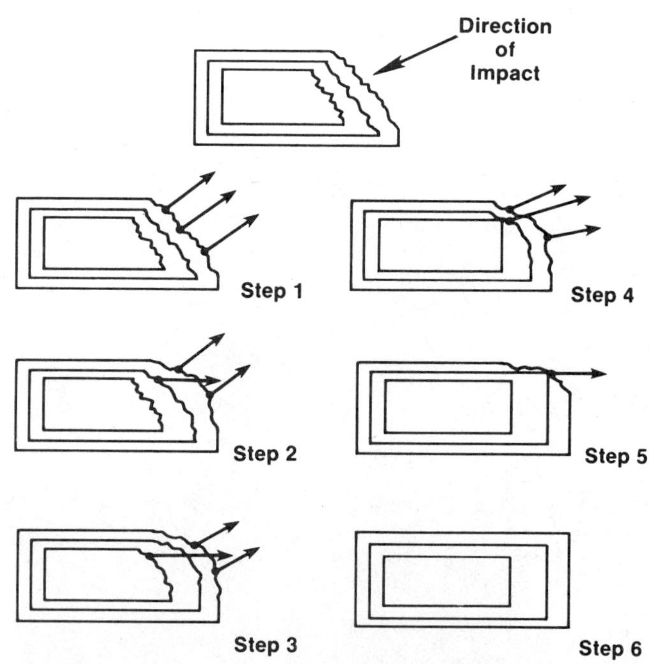

FIGURE 6-22 A correct pulling sequence is best accomplished through a series of steps.

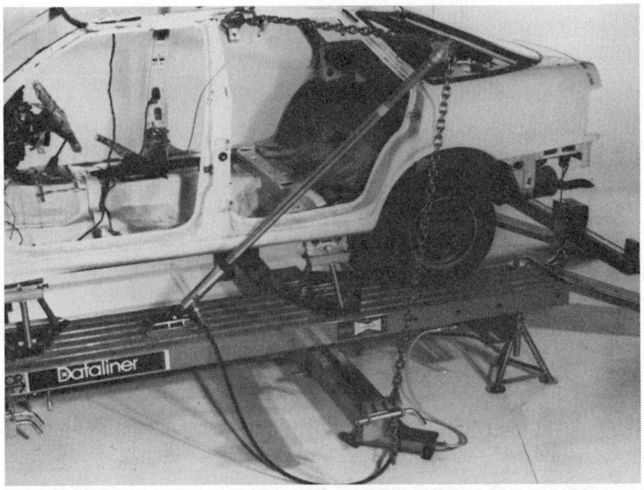

FIGURE 6-23 Pulling upward and outward

FIGURE 6-24 Setting up for a straight out pull

section of the frame between the ram foot and the chain anchor. The setup shown in Figure 6-23 is used for pulling up and out. As the ram is extended, one side of the triangle becomes longer; this causes the ram to swing to the right. As the triangle's shape changes, it brings the damaged area of the vehicle with it.

Figure 6-24 shows a triangular arrangement that will provide a straight out pull. Note that the ram is placed at an angle to the right of true vertical. As force is applied, the ram will swing to the right, pulling the damage with it. It is important that the ram is at the proper height. This can be achieved by adding a length of tubing to the ram. At no time should pulling continue if the chain between the ram and the anchor goes beyond perpendicular; overloading the chain is a distinct possibility in this case because of the added stress placed on the anchored end of the chain. To avoid this, be sure that the chain lockhead is not placed behind the chain anchor.

By setting up the equipment as designed, the vector system will provide strong pull capabilities at frame horn height and allow 10 inches of chain travel in one continuous pull. As higher, longer pulls are needed, the vector system automatically trades power for motion and allows as much as 44 inches of chain travel in one pull. By controlling the angle of the ram and adjusting the length of the tubing, this system can pull up, out, or down.

In a typical pull setup at frame rail height, the ram is set so that the angle between the ram and the chain is equal to the angle between the ram and the anchor (Figure 6-25). With approximately 10 tons of force and 10 inches of chain travel available, there is ample power for tough lower structure pulls.

As the pulls move higher, part of the available power is converted to longer chain travel. For example, at roof rail height 10 tons of force are not needed to position a top section, but more than 10 inches of continuous chain travel might be needed to get the vehicle into proper alignment with one setup. In a typical roof rail height pull, approximately 33 inches of continuous pull is usually available. This saves setup time and speeds the repair procedure. Other basic pulling setups are illustrated in Figure 6-26.

Because of the high-strength (and, in some cases, heat-sensitive) steel in the unibody structure, it is best not to attempt to make a pull in one step. Instead, use a sequence that consists of a pull . . . hold . . . pull . . . hold, and so on. This will allow more time for working the metal and for checking the progress. Start the hydraulics slowly and carefully. Watch the movement closely.

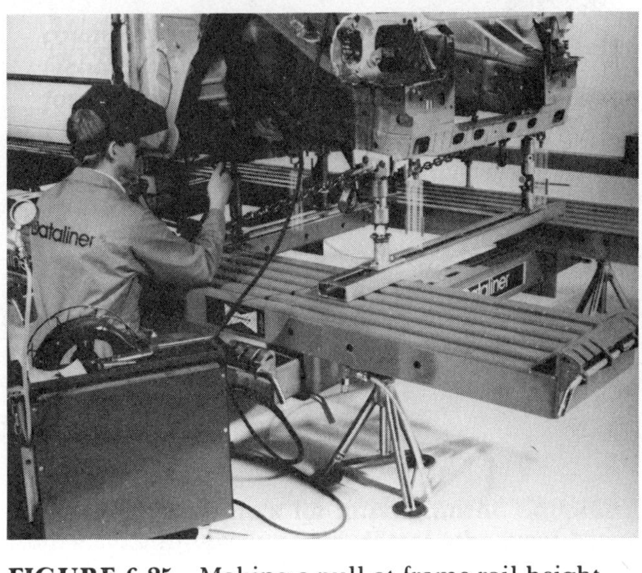

FIGURE 6-25 Making a pull at frame rail height

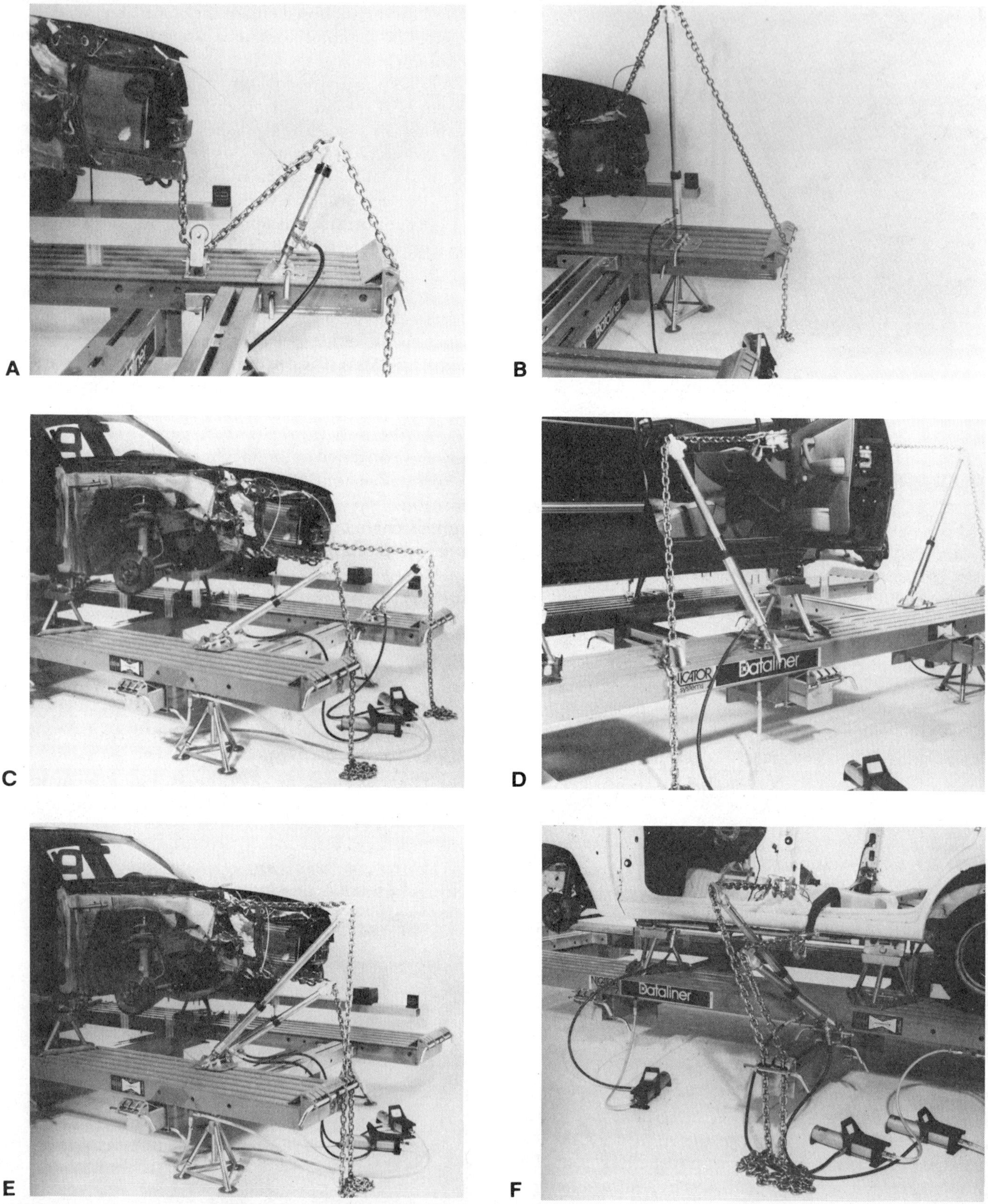

FIGURE 6-26 Various pulling setups: (A) downward pull using a guide roller; (B) up and outward pull; (C) stretching a vehicle lengthwise; (D) pulling the windshield (A-pillar); (E) using two rams (high and low) to make a forward pull; and (F) pulling a center section

Is it doing what it is supposed to do? If so, keep going. If not, determine why and make the required angle or directional adjustment, then try again. Relieve the stressed or locked-up metal by pulling it to apply tension, then loosen it up with hammering.

Although unibody structures are extremely strong in the direction of their design loads, a thin-gauge pulling clamp can still overstress and damage the area where it is attached. It might be a good idea to distribute the load over more of the structure with additional pulling clamps.

The repair of a damaged closed cross-sectional structure such as a frame rail is done by clamping the surface on the damaged side and pulling. The pulling direction should be such that the force is applied in the direction of an imaginary straight line extending through the original position of the part. The damaged portion can be repaired by welding studs in place and pulling with a dent puller or other pulling device.

If some of the buckles are folded so tightly that they threaten to tear, it might be necessary to use a little heat to soften the metal. Use heat sparingly and only on the corners and double panels.

To summarize, when planning a pulling sequence it is important to keep some basic ideas in mind:

- Make the first pull a multiple pull, opposite the direction of impact.
- Make as many pulls as practical in the area of direct impact.
- Repair as much damage as possible at one time.
- Look out for secondary damage.
- Remember that the damage that occurred last during the collision should be corrected first. If the secondary damage is not coming out, make corrections in the pulls, or add a pull.

Overpulling

Uncontrolled overpulling (Figure 6-27) is most often the result of failing to measure accurately and often. A small amount of controlled overpulling is usually necessary to overcome springback. But, to prevent overpull damage, measure the progress periodically when pulling the damaged area. There is no way to push an overpull back, and it is impossible to shrink or compress metal that has been stretched beyond its critical control dimensions. In most instances, an overpulled panel must be replaced.

There is some evidence to suggest that bonded SMC panels make the pulling process slightly more complicated. This is due to the bracing effect of SMC panels. Normally, when pulling a conventional unibody vehicle, the metal is overpulled somewhat to allow for springback. This can be tricky with SMC panels which are bonded to the frame; they do not "give" in the same way sheet metal panels would. It may be necessary to remove damaged or separated SMC panels before pulling the space frame. In most cases, these will be panels which will need to be replaced or rebonded to the space frame anyway.

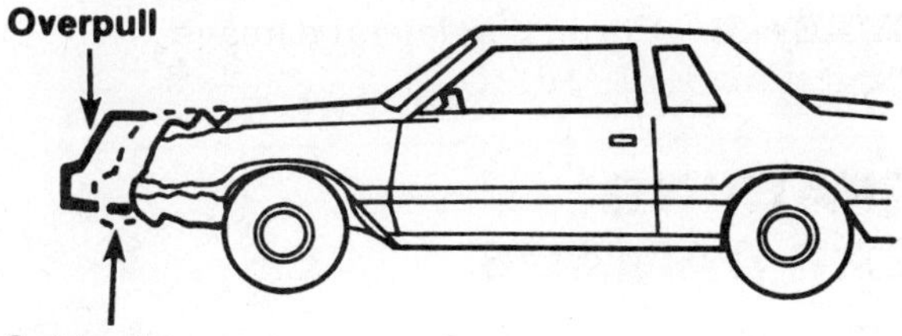

FIGURE 6-27 Uncontrolled overpulling must be avoided.

REPAIRING FRONT END DAMAGE

The general repair method for front end damage is best illustrated by going through a typical example of front end collision damage. This will include the replacement of the front crossmember, front fender apron, and frame rail, as well as repairs to the front fender, front fender apron, and frame rail on the opposite side of the vehicle. It is important to begin the repair by restoring the front fender apron and frame rail to their original condition and to repair the support structure on the replacement side.

First, pull the frame rail on the replacement side in the direction opposite to the impact. Then repair the fender apron and frame rail on the repair side, followed by the front fender apron and frame rail installation areas on the replacement side. There are many cases in which the entire fender apron or frame rail on the repair side is deflected left or right only. Since there is practically no warping in the lengthwise direction, repairs involve measuring the diagonal dimensions (points A and B in Figure 6-28) and correcting that distance while

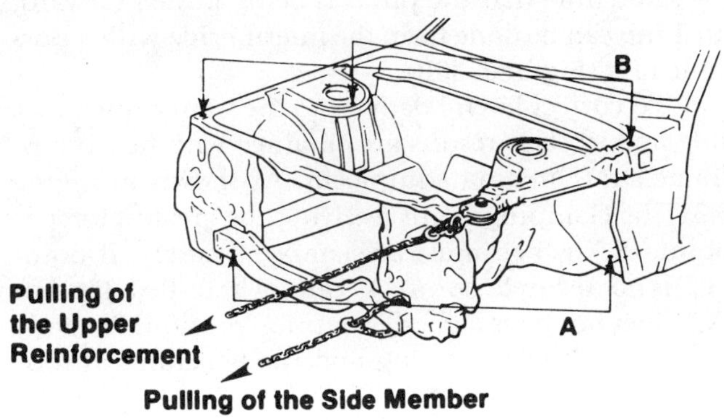

FIGURE 6-28 Checking the underhood dimensions

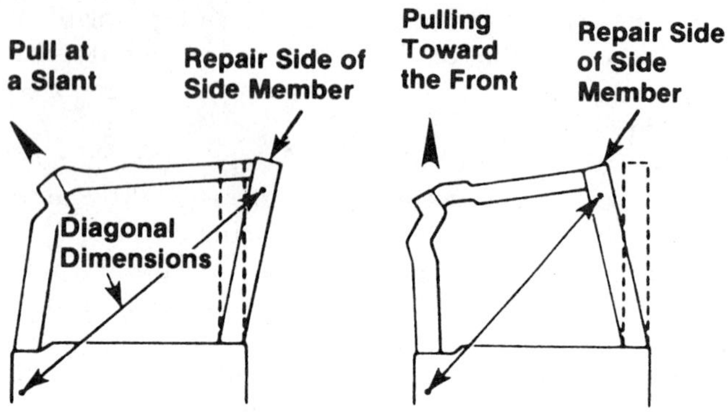

FIGURE 6-29 Frontal pulling

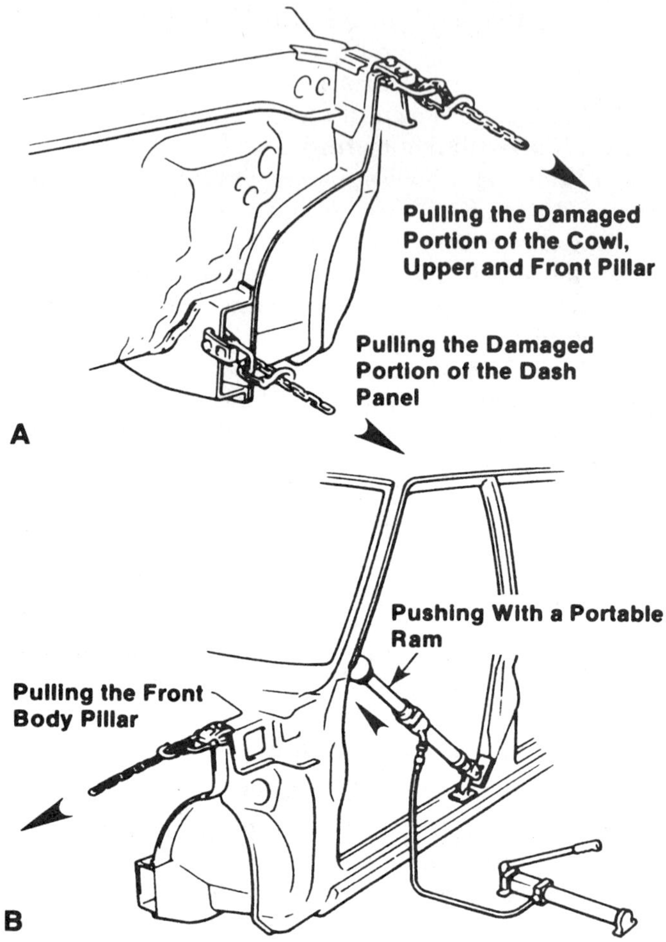

FIGURE 6-30 Front pillar and dash panel repairs

keeping an eye on the repair condition. The operation can be done efficiently if the fender apron upper reinforcement is pulled at the same time as the frame rail.

If the frame rail on the repair side is deflected toward the outside, angle the pull toward the front, keeping an eye on the diagonal dimensions; if it is deflected toward the inside, pull directly toward the front (Figure 6-29). If there is severe damage to the frame rail on the repair side, separate the front crossmember and radiator upper support at the point where the diagonal dimension is correct and repair them separately. Grip the inside broken face of the frame rail, and while pulling it forward, pull the broken piece from the inside or push it from the outside. After repairing the damaged portion, match up the dimensions to the standard diagonal dimensions.

To repair the replacement side front fender apron and frame rail installation area, the main repairs will be near the dash panel and the cowl panel. If the impact was severe, the damage will extend into the front body pillar; the door will fit poorly in this case. Simply holding the front edge of the fender apron and pulling will not repair the major damage to the front body pillar or the dash panel (Figure 6-30A). Instead, cut the fender apron and frame rail near the installation area, clamp near the major damage, and make the pull. At the same time that the pillar is being pulled forward, pushing can be done from the interior side with a portable ram (Figure 6-30B).

To correct lateral damage to the front caused by a side impact, best results are obtained with bench type frame straightening equipment. As shown in Figure 6-31, the clamping point receiving the greatest force is point "B," which must be clamped securely. If point "C" is not secured, point "A" cannot be pulled. If point "C" does not have a good clamping location, the body still can be fixed by using underbody clamps at four points.

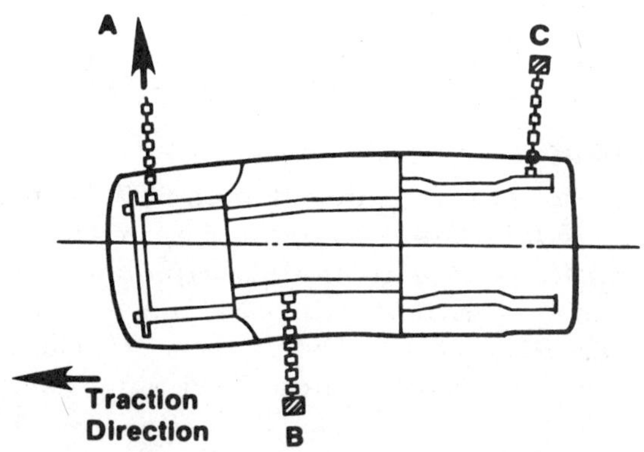

FIGURE 6-31 Correcting lateral damage

REPAIRING REAR DAMAGE

Since panel construction of the rear body is more complex than the front body, the damage from a rear

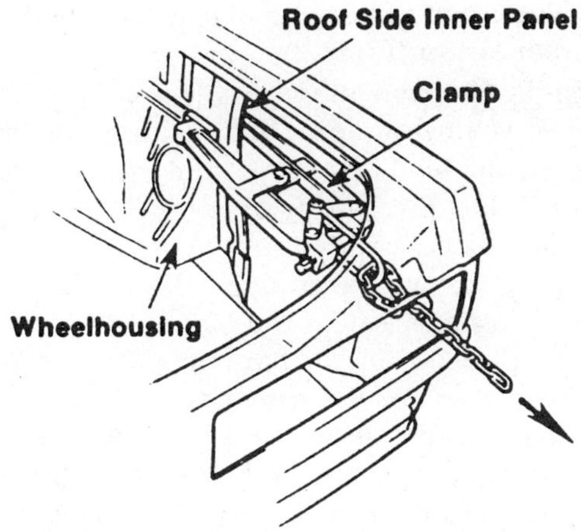

FIGURE 6-32 Pulling to correct rear damage

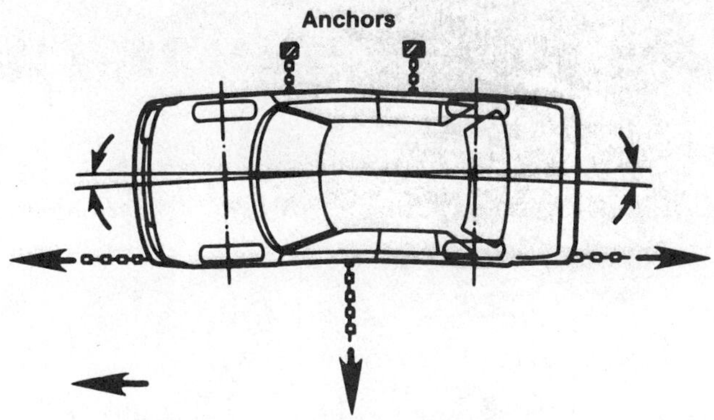

FIGURE 6-33 Three-way pulling

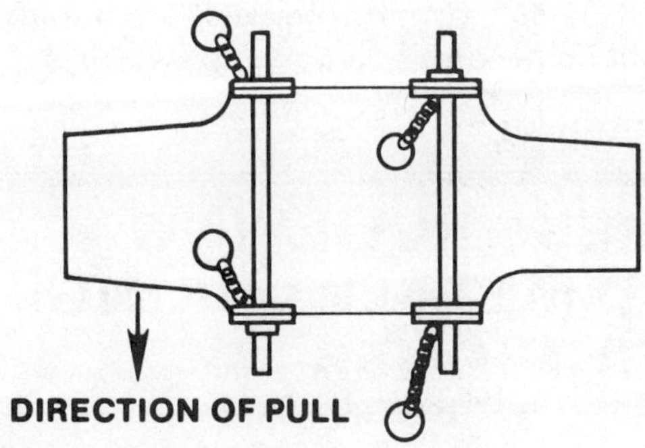

FIGURE 6-34 Anchoring a side pull

impact can be more extensive. Therefore, the damage diagnosis must be done accurately. In most cases the bumper is impacted during rear end collisions, with the force of the impact traveling through the rear frame rails or nearby panels and causing damage to the kick-up area. Next, the wheelhousings will deform and the entire quarter panel will move forward, causing clearance problems between other components. If the impact is severe enough, it will also have an effect on the roof, door panels, and center body pillar.

Attach clamps or hooks to the rear portion of the rear frame rail, rear floor pan, or rear quarter panel. Pull while measuring the dimensions of each part of the underbody. Do not clamp and pull a quarter panel that has little or no strain on it when the rear frame rail is pushed into the wheelhousing or there are clearance problems at the rear door. Relieve stress in the quarter panel by pulling on the frame rail only.

If the wheelhousing or roof side inner panel is clamped and pulled along with the rear frame rail, the door panel clearances can be maintained properly (Figure 6-32). Where there is rear structure twist from a front hit, the lower rear structure should be clamped to the bench. The preliminary pulling will restore some of the lower alignment points, and clamps can be positioned to preserve the alignment. While moving forward with subsequent pulls, the alignment will, of course, move as well.

Install upper fixtures as soon as the upper damage is reduced enough to permit their positioning. Also, excessively damaged sections requiring replacement can be cut away at this time.

REPAIRING SIDE DAMAGE

If there is a severe impact to the center of the rocker panel, the floor panel will deform and the entire body will take on a curved shape like a banana. To align this type of damage, the two ends of the body are pulled apart and the side that is caved in is pulled outward. This is known as three-way pulling and is illustrated in Figure 6-33.

Anchoring unibody vehicles for side pulls can be very difficult due to the limited hook-up areas. Figure 6-34 shows a recommended anchoring method. Keep in mind that when making a side pull on the end of a vehicle, the center section can be anchored by passing a chain around the pinchweld clamp and hooking it to the edge of the bench. Tension can be applied by attaching the pull chain to the pinchweld clamp.

The portable beam and knee can be used as a side anchor with either inside or outside contact. By attaching the pulling chain to the beam and knee, it can be used as a pulling attachment. It is advisable to make an end-to-end stretch pull whenever pulling outward on the center section of a vehicle. If pulling on the body, it will be necessary to tie the vehicle down on the opposite side. Pulling outward on the center section of the vehicle can also be done with the portable beam and knee attached to the pulling post or ram (Figure 6-35).

FIGURE 6-35 Using a portable beam and knee to pull out on the center section

STRESS RELIEVING AND METALWORKING TECHNIQUES

Stress is defined as the internal resistance a material offers to being deformed when subjected to a specific load. In the collision repair industry, stress can be defined as the internal resistance a material offers to cor-caused by factors such as deformation, overheating, improper welding techniques, and undesirable stress concentrations.

Signs of stress/deformation on unibody vehicles include

- Misaligned door, hood, and trunk
- Dents and buckles in aprons and rails
- Misaligned steering, suspension, and motor mounts
  mounts
- Damaged floor pans
- Cracked paint and undercoating
- Pulled or broken spot welds
- Split seams and seam sealer

Since the damaged area will frequently exhibit greater resistance to alignment than adjacent areas, additional clamping is sometimes required; for example, a damaged rail with greater resistance to tension than usual where it attaches to the front cowl/bulkhead. By using a clamp, movement at the front cowl/bulkhead is prevented and the corrective forces are applied directly to the rail. During the pulling operation, all critical reference points must be measured to monitor direction and prevent overpulling.

If the strain from an impact does not lead to buckling, effective pulling force during aligning will alleviate the problem. The buckles and creases in a panel can be unlocked in a variety of ways. On panels where the backside is accessible, body hammers, dollies, or spoons are used for the initial roughing out. On sealed panels, or in areas where the backside is difficult to reach, dent pullers, picking hammers, and welded studs can be used to reverse the damage.

Damage to an exterior panel can be locked in by indirect damage to structural panels or inner reinforcements. Before the repair techniques described in this chapter are used, structural damage must be repaired by pushing or pulling the damaged understructure into alignment. Usually when this is necessary, it is best to maintain the push or pull until the exterior panels have been straightened.

BODY HAMMERS

The body hammer is designed to strike the sheet metal and rebound off the surface rather than be driven down as in driving a nail. A driving action would create additional damage in the sheet metal. The secret of metal straightening is to hit the right spot at the right time, with the right amount of force. Swing in a circular motion, using the wrist rather than the whole arm and shoulder. Hit squarely and let the hammer rebound off the metal. Space each blow 3/8 to 1/2 inch apart until the damaged metal is level.

The face of the hammer must fit the contour of the panel. Use a flat face on flat or low-crown panels. Use a convex-shaped or high-crown face when bumping inside curves. Heavy body hammers should be used for roughing out the damage. Finishing, or dinging, hammers should be used for final shaping. The secret to finish hammering is light, rapid taps. It is also important to hit squarely; hitting with the edge of the hammer will put additional nicks in the metal.

SHOP TALK

Before hammering sheet metal, make sure that both sides of the panel are clean to ensure that the tool comes in direct contact with the metal.

BUMPING DENTS WITH SPOONS

Spoons can be used in a number of ways, depending on their design. They can be used to pry out dam-

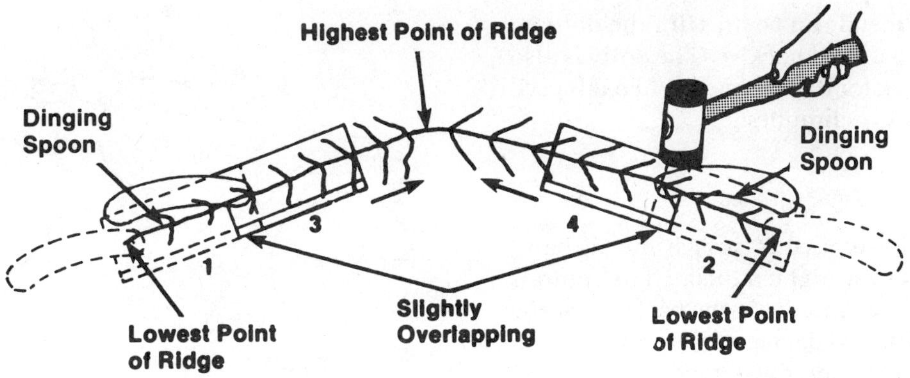

FIGURE 6-36 Spring hammering a ridge

age, certain kinds can be struck with a hammer to drive out damage, and others are used as a dolly in hard-to-reach areas. Some are even designed to be used in place of a body hammer.

Spring hammering is a method of bumping damage with a hammer and a dinging spoon. The dinging spoon is lightweight and has a low crown. When used, it is held firmly against a ridge or crease. The spoon is then struck with a ball peen or bumping hammer (Figure 6-36). The force of the blow is distributed by the spoon over a large area of the ridge. This reduces the likelihood of stretching the metal. Always keep firm pressure on the spoon when spring hammering; it must never be allowed to bounce. Part of the corrective force is the pressure of the spoon. Begin at the ends of the ridge and work toward the high point, alternating from side to side. Remember that hammer blows on top of a panel can be a corrective force only when they are placed on high areas of the damage. A bumping file can also be used to "slap" down ridges.

SHOP TALK

The effect of hammering is increased when pressure is kept on the underside of the panel. This can be done with a dolly, spoon, or pick.

Spoons can be used to back up the hammer or in combination with a slapper spoon. With a long body spoon, the technician can often reach into places inaccessible to a hammer or dolly. Pressure can be applied to tension areas with the spoon, while high areas are bumped down (Figure 6-37).

Spoons can also be used to pry up metal in the rough-out stage or to drive out deep dents. Figure 6-38 shows a double end spoon being used to pry out a dent in a door panel. The door is supported on blocks of wood to provide clearance for the panel to move. Care must be taken not to stretch the metal by prying it beyond the original contour. Once the dent is roughed out by prying with a spoon or dolly, a body hammer can be used to finish the area.

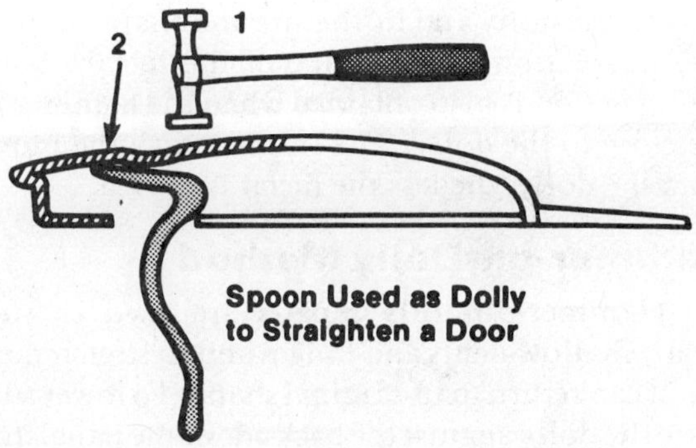

FIGURE 6-37 Bumping a dent with a spoon

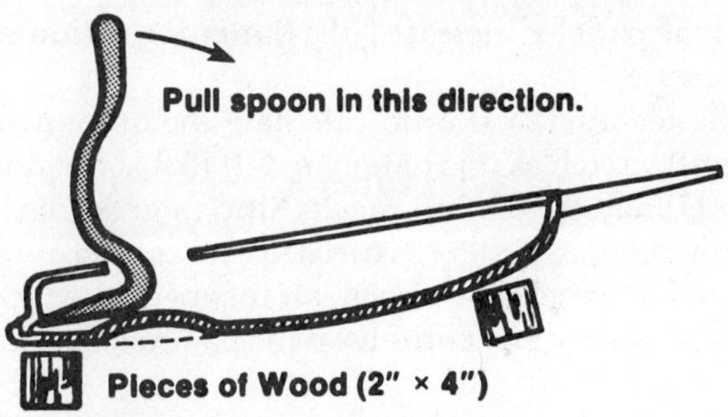

FIGURE 6-38 Using a spoon to pry out a door panel dent

BUMPING DENTS WITH DOLLIES

The dolly block is used throughout the repair process. In the rough-out phase, it is used as an impact tool.

The backside of the metal can be hit with the dolly to raise low areas and to unroll buckles. The dolly is also used as a backing block for the hammer. When it is used this way, there are two techniques:

- Hammer-on-dolly
- Hammer-off-dolly

Hammer-on-dolly is used to stretch metal; hammer-off-dolly is used to straighten metal. The contour of the dolly must fit the contour of the backside of the damaged area or additional damage will result. As each hammer blow is struck, the dolly tends to rebound slightly from its position against the metal, creating somewhat of a secondary lifting action. The amount of rebound force is determined by the pressure exerted against the dolly and by the size and shape of the dolly face in relation to the shape of the metal. A greater lifting action is also obtained when the hammer blows are close to the dolly. The further away the blows are from the dolly, the less the metal is moved.

Hammer-on-Dolly Method

Hammer-on-dolly repairs are used to smooth small, shallow dents and bulges and to stretch metal so that it can return to its original shape. To lower a bulge, place the dolly against the backside of the panel directly behind the bulge and use a hammer from the front side (Figure 6-39). There will be a slight rebound as a result of the hammer hitting the dolly. The dolly will hit the backside of the panel; as the force of the dolly pressing against the panel is increased, the flattening action will also increase.

The technician should calculate the amount the panel will stretch as it is hammered. If this is not done, the panel may stretch too much. Since a great deal of collision repair experience is needed to accurately judge how much a panel will stretch, an inexperienced technician should not rely entirely on the hammer-on-dolly repair method.

Any shrunken metal on a crown can be stretched to its normal level faster and easier by the hammer-on-dolly method than by any other method. Hammer-on-dolly is used only if there is access to the backside of the panel. If not, dent pullers and filling are used. Figure 6-40 shows a typical low area.

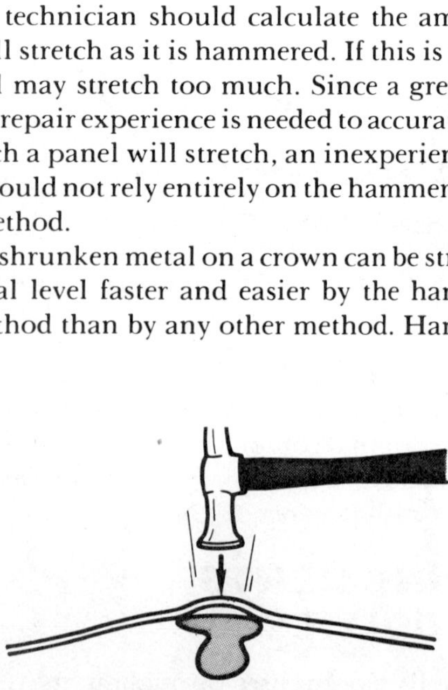

FIGURE 6-39 Using the hammer-on-dolly method

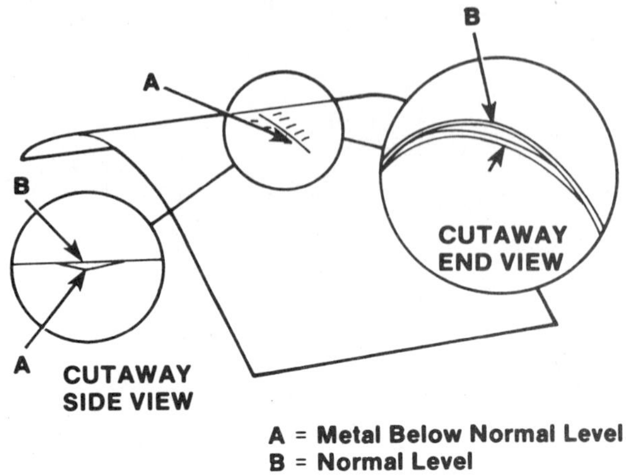

FIGURE 6-40 Low areas must be raised using the hammer-on-dolly method, pull rods, or dent pullers.

This panel was damaged and has been roughed out but is still low in the area of the buckle. Because it shows no evidence of a pressure area adjacent to the buckle, the panel is low at point A as shown in the drawing. The only way point A can be raised to the original level (point B) is by stretching. The stretching has to be along the line of the rolled buckle. Picking up the low area will not correct the condition. The hammer-on-dolly method should be used at point A.

In the hammer-on-dolly procedure, there are actually two actions:

- When the hammer strikes the metal
- When the dolly rebounds upward and strikes the backside of the metal.

Tension is kept on the dolly at all times. The greater the tension, the greater the rebound action. When hammering fast, the tension is increased on the dolly to ensure a quick rebound. When low metal is being raised, each rebound action should be deliberately driven up by hand action.

Note that improper hammer blows placed on a crown will shrink rather than stretch. All blows that are designed to stretch should be hard and accurate. An inaccurate hard blow can damage the panel. Keep in mind that light hammer blows are for straightening, not stretching. In other words, when using the hammer-on-dolly technique, hit hard and do not miss! Be sure to choose the proper shaped dolly. When the contour of the dolly is larger than the contour of the panel, additional damage will result.

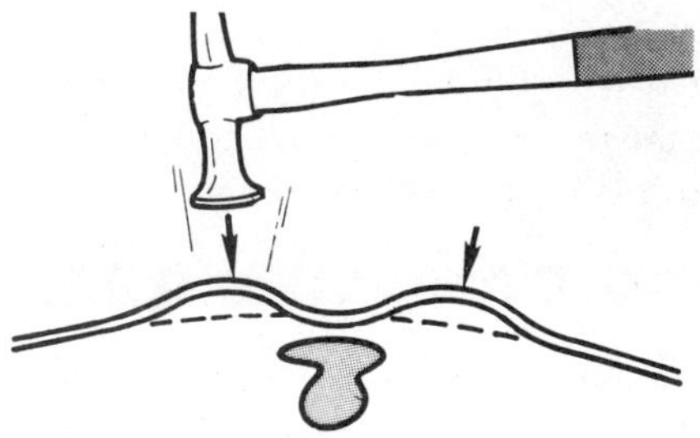

FIGURE 6-41 Using the hammer-off-dolly method

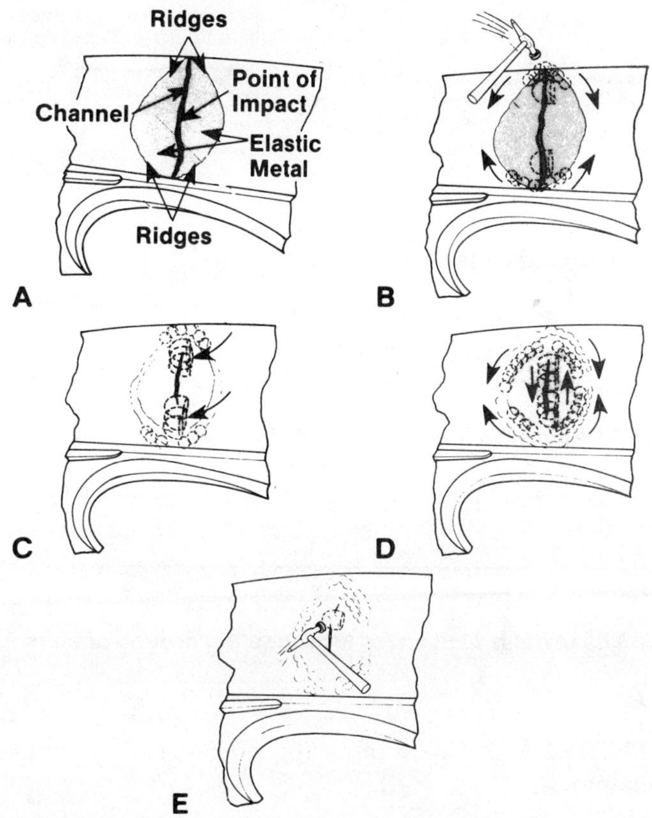

FIGURE 6-42 Steps in repairing a damaged panel with hammer and dolly

Hammer-off-Dolly Method

The hammer-off-dolly method is used to straighten metal just before the finishing stage. In this procedure, the dolly is held under the lowest area, and the hammer hits the high area nearby (Figure 6-41). The dolly, like the hammer, is designed to correct damage in that it must strike only in tension areas to be effective as an impact tool.

REPAIRING DAMAGE WITH A HAMMER AND DOLLY

The damaged metal in a minor dent as shown in Figure 6-42A is straightened by using the hammer and dolly to "roll out" the metal in the reverse order it went in. As the metal is pushed in, a channel is gradually formed on either side of the point of impact. This channel is usually deepest next to the point of impact, and decreases in size toward the outer ends of the channel. At the same time as the channels are being pushed in, ridges are being formed around the outer area of the dent. They also have the greatest degree of bend in their center, gradually decreasing the size toward each end. Both the ridges and channels have been work hardened, the amount of which depends on the degree of bend.

To remove the dent, the damage must be rolled outward from the outside, working toward the center in a reverse order to which it went in. The dolly is held tightly under the channel at the outer end where the least degree of bend is present (Figure 6-42B). A flat-faced dinging hammer is used to direct light to medium blows at the outer ends of the ridge closest to the dolly. The blows from the hammer gradually force down the ends of the ridges, while the arm pressure held against the dolly forces the end of the channel upward. The same procedure is then repeated on the other end of the channel and the adjacent ridges (Figure 6-42B).

The hammer-off-dolly method is gradually worked toward the center or greatest degree of bend in the ridges and channels. As the pressure is released in the ridges and channels, the surrounding elastic metal tends to move back to its original position. The dolly can also be used as a driving tool to help work the channel (Figure 6-42C). However, if the dolly does not move when the channel is hit upward, there is still too much pressure on the ridges and/or channel. More dollying must be done to relieve the tension (Figure 6-42D).

Once the area has been brought back to its basic shape, the hammer-on-dolly method is used to smooth and level (Figure 6-42E). It is then ready for either metal finishing or filling procedures.

PICKING HAMMERS

Picking hammers, picks, the edge of a dolly, and even a scratch awl can be used to pick up metal. When picking up a small dent, it is better to use several light blows rather than one or two hard ones. After an area has been raised, use a file or grinder to level the damaged area. Figure 6-43 shows a low spot being raised with a picking hammer. Some stretching has occurred, so the metal must be filed after the low spot is eliminated. Filing will heat the metal, causing it to shrink partially back to shape.

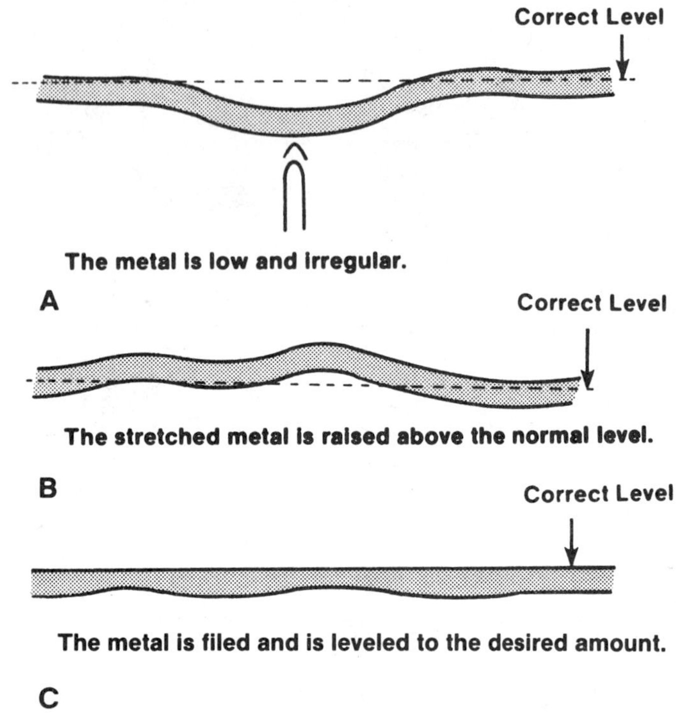

FIGURE 6-43 Steps in raising a dent with a picking hammer

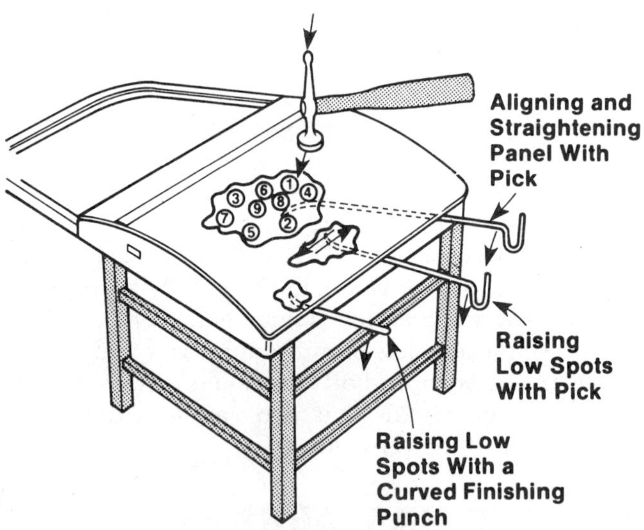

FIGURE 6-44 Raising dents on a vehicle door

Picks can be used to pry up metal in areas that cannot be reached with a dolly or spoon. A vehicle door is a good example (Figure 6-44). A pick can sometimes be inserted through a drainage hole. This eliminates the need to remove the inside door trim or to drill holes in the outer panel for pulling the dent.

When prying with a pick, be careful not to stretch the metal by exerting too much pressure. Start with the point of impact or the lowest point and slowly pry up the damaged area. On larger areas, use a flat blade pick rather than a pointed one. Tap down pressure areas, while prying up low tension areas.

PULLING DAMAGED AREAS

Damage can be pulled out with a number of tools: suction cups, pull rods, dent pullers, even a sheet metal screw and vise grips. The slide hammer is one of the collision repair technician's most used tools. One reason is the growing complexity of automobile body construction. Access to the inside of many panels is blocked by panel reinforcements and window mechanisms. Using a dent puller, the technician can repair a simple dent in a minimal amount of time.

After scrutinizing the damage to determine the point and angle of the initial impact, drill or punch holes 1 inch apart along the length of the crease. Punching is preferred to drilling because punching leaves more metal for the screw tip to grab. Beginning at the point of impact, thread the screw tip on the dent puller into the first hole. Hold the handle of the dent puller in one hand and forcefully slide the weight straight back against the handle.

CAUTION

Be careful not to pinch your fingers between the weight and the handle.

To gradually work the damage out, pull the first hole slightly. Then, go on to the next hole; work from the lowest spot to the damage in both directions. After pulling out each hole slightly, repeat the process. Work the surface as close to the finished contour as possible without pulling the metal beyond the original surface. Tap high spots down with a body hammer to relieve pressure in ridge areas. Holes must be welded closed and low spots can be filled with body filler.

Shallow, small dents in enclosed panels can be pulled out with one or more pull rods. Again, a body hammer should be used to tap high spots down while pulling. Keep in mind that it is senseless to pierce a large number of holes in the surface of a panel that could have been straightened by other means. The result is often an unsatisfactory job. Worse, it might be the case that more time was required to do it wrong than would have been required to do it right.

Pulling with Studs

The most advanced way to pull damage is with a stud spot welded to the damaged area. The stud can be a washer, a pull tab, or a pin. Regardless of the system used, pulling with spot-welded studs avoids drilling or

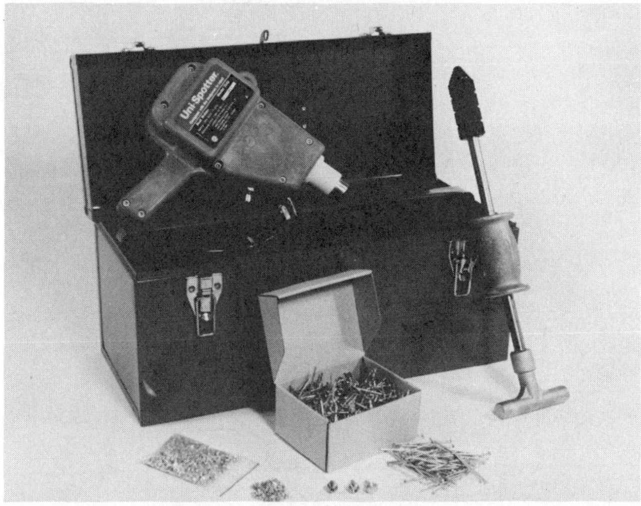

FIGURE 6-45 Typical stud pulling kit

punching through the metal and undercoating, which can lead to corrosion.

Figure 6-45 shows a complete set of stud pulling equipment. A specially designed spot welder fuses metal pins to the damaged area (Figure 6-46A). Fusing a pin to the panel takes only a fraction of a second. Then, the pins can be pulled with either a dent puller (Figure 6-46B) or pulling power (Figure 6-46C). When the damage is corrected, the pins can be snipped off with cutters and ground flush with the panel (Figure 6-46D). The whole process is very quick, with no damage to the panel. It is especially useful in pulling out small door dings where the backside of the panel is inaccessible.

SHRINKING METAL

It is safe to assume that the panel has been stretched if the strain on it cannot be removed even after it has been worked properly. The damaged area can be easily filled with plastic body filler to give the proper appearance, but road vibrations can cause the panel to make a flapping noise as the vehicle is driven. In addition, the owner can see and feel the surface irregularities when he or she waxes the vehicle. Eventually, the collision repair shop will be confronted by the dissatisfied owner and be required to spend extra time correcting the work that should have been done properly in the first place.

Identifying Stretched Metal

When metal is severely damaged in a collision, it is often stretched in the badly buckled areas. These same areas are also sometimes stretched slightly during the straightening process. Most of the stretched metal will be found along the ridges, channels, and buckles in the direct damage area. When there are stretched areas of metal present in the panel, it is impossible to correctly

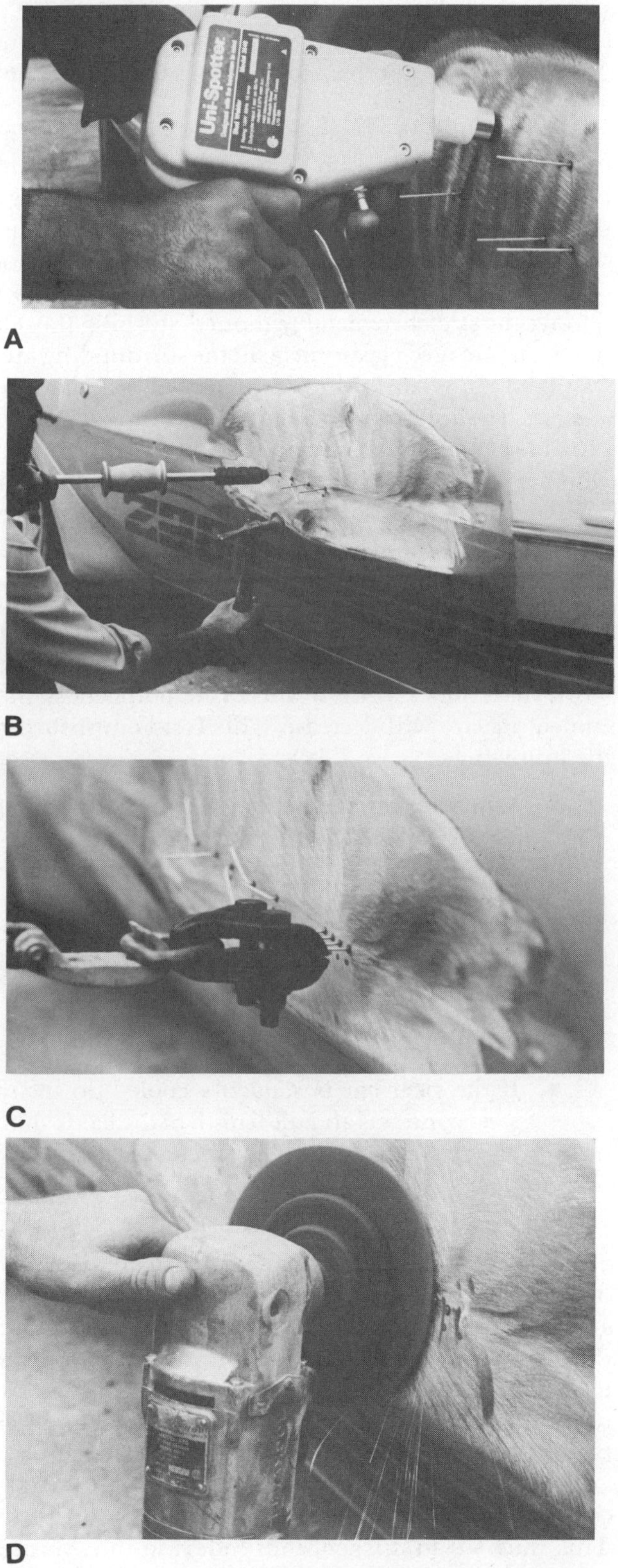

FIGURE 6-46 (A) Welding the metal pins to the damaged area, (B) pulling the damage with a dent puller or (C) pulling power, and (D) grinding the pins flush with the panel

straighten the area back to its original contour. The stretched areas can be compared to a bulge on a tire. There is no place for them to fit within the correct panel contour.

When an area is stretched, the grains of metal are moved further away from each other. The result is that the metal is thinned and work hardened. Shrinking is the method used to bring the molecules back to their original position and to restore the metal to its proper contour and thickness. The object of shrinking is to remove the stretched metal without disturbing the relatively undamaged elastic metal in the surrounding area.

Before any shrinking is attempted, the damaged area must be dollied back as close to its original shape as possible. Then the technician can accurately determine whether or not there is any stretched metal in the damaged area and, if so, where shrinking should be done.

Principles of Shrinking

A steel bar, with both ends free to expand or contract, will expand when heated and contract to its original length when cooled. If the same steel bar is heated while it is blocked or restricted at both ends, then cooled, its size will decrease. This is accomplished by the following:

- When heated, the steel bar tries to expand (Figure 6-47A). Since it is prevented from expanding at both ends, a strong compression load is generated inside the bar.
- When the temperature is increased, and the steel becomes red hot and soft, the compression load concentrates in the red hot area and is relieved as the diameter of the red hot area increases (Figure 6-47B).
- If the steel bar is suddenly cooled down, the steel contracts and the length of the bar is shortened by the amount of the increased diameter of the red hot area (Figure 6-47C)

The above stated principle of shrinking a steel bar also applies to the shrinking of a warped area in a piece of sheet metal. A small spot in the center of the warped area is heated to a dull red. When the temperature rises, the heated area swells and attempts to expand outward toward the edges of the heated circle. Since the surrounding area is cool and hard, the panel cannot expand, so a strong compression load is generated.

If heating continues, the stretching of the metal is centered in the dull red hot portion, pressing it out. This causes it to thicken, thus relieving the compression load. If the red hot area is suddenly cooled while in this state, the steel will contract and the surface area will shrink. As the panel contracts from the cooling, a tensile load is generated in the panel, as opposed to the compression load generated during heating.

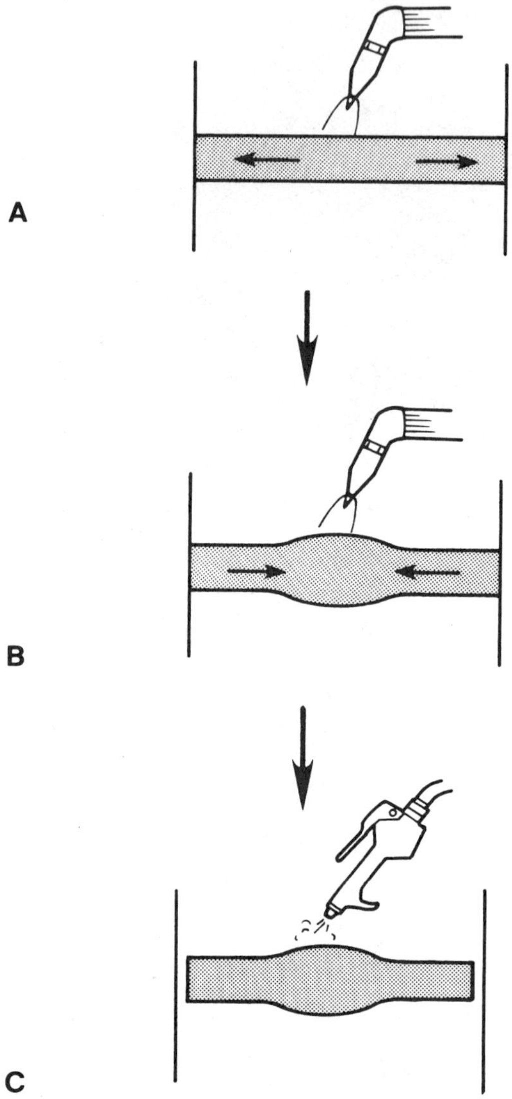

FIGURE 6-47 (A) Heat is applied, (B) causing the softened metal to expand and thicken in the red hot area. (C) When the metal cools, the panel contracts to an area smaller than its original size.

A variety of welding equipment can be used to heat metal for shrinking. Attachments are available for spot and MIG welding equipment to transform them into shrinking tools. But the most commonly used tool is the oxyacetylene torch with a No. 1 or No. 2 tip.

SHRINKING OPERATION WITH A TORCH

To shrink an area with a torch, a small spot of the bulge is heated to a cherry red. The "shrink" is done in the highest spot of the stretched area, then in the next highest spot, and so on, until the area has been shrunk back to its proper position (Figure 6-48).

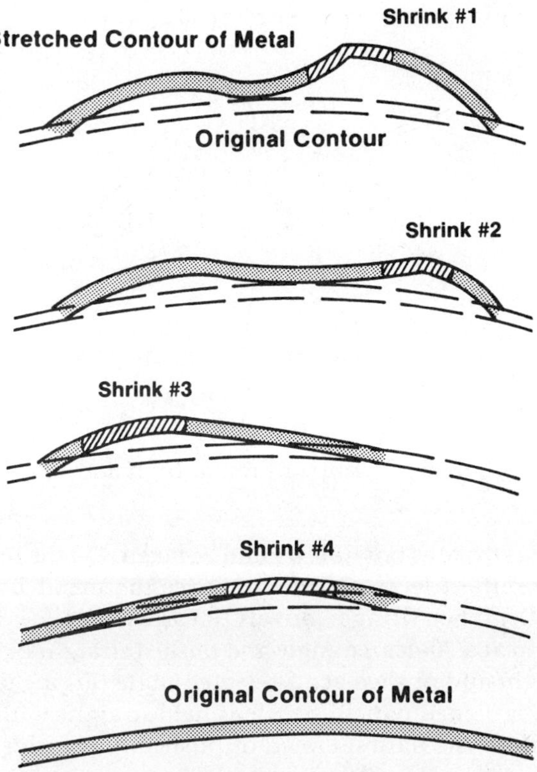

FIGURE 6-48 Shrinking stretched metal

The size of the shrink or "hot spot" is determined by the amount of excess metal. The larger the shrink, the harder the heat is to control. An average-sized shrink is usually about the size of a quarter. Small shrinks should be used on flat panels, as such panels tend to warp easily. The area should never be larger than the size of the hammer face being used.

A neutral flame and a No. 1 or No. 2 tip used for welding sheet metal are used to heat the shrinks. The point of the cone is brought straight down to within 1/8 inch of the metal and is held steady until the metal starts to get red. The torch is then slowly moved outward in a circular motion until the complete shrink is cherry red.

SHOP TALK

Do not heat the metal past a cherry red. It will start to melt, and a hole might be burned through the metal.

As the heat from the torch enters the panel, the metal expands. The cooler metal surrounding the shrink resists the expansion forces. As the temperature increases, the heated metal becomes softer. This soft metal piles up and forms a bulge.

The metal usually will bulge up instead of down because the top of it is heated first. When it starts to bulge, the rest of the metal in the shrink follows. When the shrink has been heated, tap around it with several sharp hammer blows to drive the molecules of metal closer together.

During this procedure, it may be necessary to support the panel with a dolly to prevent the metal from collapsing. If it is necessary to support the panel, the dolly should only be held lightly under the metal. As soon as the redness disappears, hammer-off-dolly and light hammer-on-dolly techniques are used to smooth and level the area around the shrink.

SHOP TALK

Never use hard, hammer-on-dolly blows to level the area. This will restretch the metal.

Once the redness has disappeared and the area has been smoothed, the shrink can be cooled with a wet rag or sponge. When this is done, a greater degree of contraction occurs, and a slight amount of distortion could result. The warpage should be straightened before the next shrink is attempted. It is very hard to determine accurately the amount that each area will shrink. One shrink can remove far more excess metal in one area than the same size would in another.

It is not uncommon to find that an area has been overshrunk. When this occurs, the metal in the area of the last shrink is usually collapsed or pulled flat; sometimes the metal surrounding the shrink area can even be pulled out of the proper contour. Overshrinking is corrected by using hard, hammer-on-dolly blows to stretch the affected area.

SHRINKING A GOUGE

A gouge is an example of metal that has been stretched by impact. Gouges must be shrunk to their original size to repair the damage. Simply picking up the low area would distort the panel, and filling the gouge with filler without restoring the panel's original contour will leave tensions in the panel that could cause the filler to fail.

Follow this procedure for shrinking gouges:

1. Heat the lowest point of the gouge with an oxyacetylene torch until the metal is cherry red.
2. Use a dolly to hammer up the shrink. This will increase the tension on the soft area, forcing it to swell and return to its original position.

3. While the metal is still hot, hold the dolly directly under the groove and tap down the ridges that will have developed on either side of the groove. This will not only drive down the ridges, but will also bump up the gouged metal.
4. If the gouge is long, this process will have to be repeated several times to raise the whole length of the gouge. Only heat as much of the gouge that can be worked before the metal cools.

KINKING

Kinking is another way to deal with stretched metal. Instead of using heat to shrink the metal, a hammer and dolly are used to create pleats, or kinks, in the stretched area. This is done by striking the stretched area with a picking hammer slightly off-dolly (Figure 6-49). Kinking the metal will lower the area slightly below the rest of the panel. The low area should be filled with body filler, then filed and sanded level with the panel.

FILING THE REPAIR AREA

When the damaged area has been bumped and pulled as level and smooth as possible, the body file must be used to locate any remaining high and low spots (Figure 6-50). Begin filing on one side of the undamaged area and progress across the damaged area to the undamaged metal on the opposite side. This way the correct plane can be maintained from undamaged metal to the damaged area.

The file should be pushed forward by the handle for the cutting stroke. Downward pressure and direction is controlled by holding the front of the file. As long a stroke as possible should be taken. On the return stroke, the file is pulled back over the metal by the handle. When filing relatively flat areas, the file is either held at a 30-degree angle and pushed straight or held straight and pushed at a 30-degree angle (Figure 6-51). On a crowned panel, the file is held straight with the length of the flattest crown or pushed to one side at a 30-degree angle or less (Figure 6-52).

The scratch pattern created by the file identifies any high and low spots. The technician then "picks up" the low spots and bumps down the high spots. This process is repeated until all the low spots disappear and the area has been filed smooth.

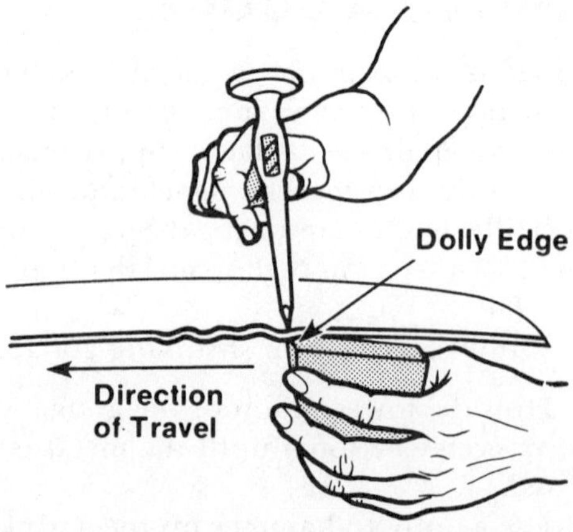

FIGURE 6-49 Shrinking a gouge by kinking the high spot

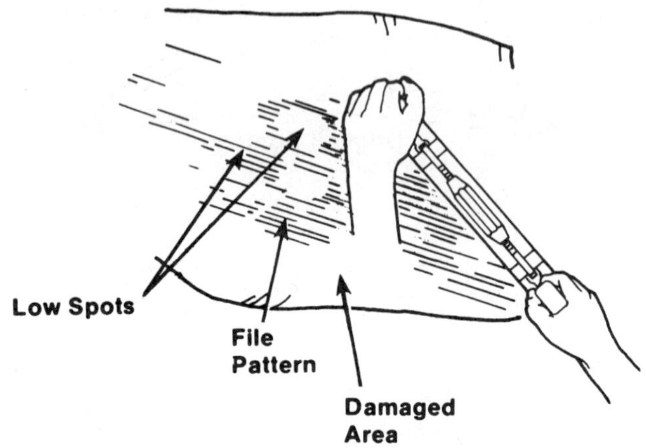

FIGURE 6-50 Filing will reveal high and low spots.

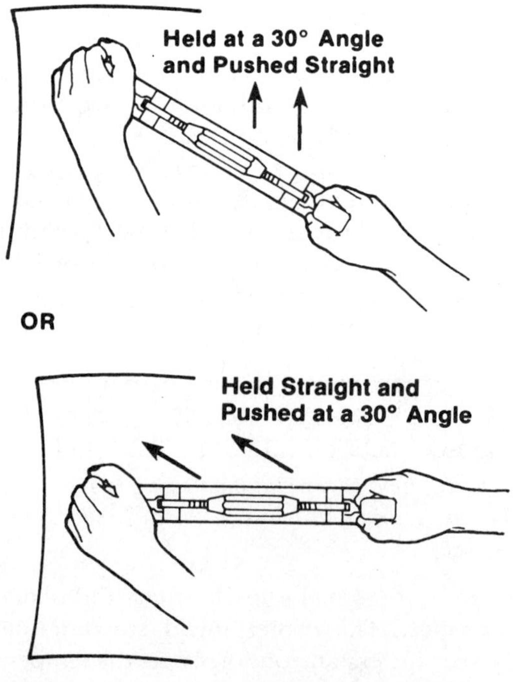

FIGURE 6-51 Methods of filing flat surfaces

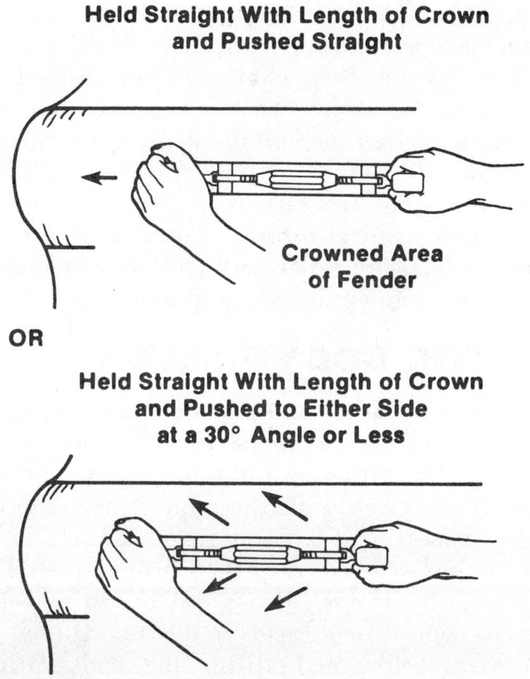

FIGURE 6-52 Methods of filing crowned panels

WORKING ALUMINUM

Aluminum is used for a variety of automotive panels, such as hoods and deck lids. The repair of aluminum requires much more care than working steel panels. Aluminum is much softer than steel, yet it is more difficult to shape once it becomes work hardened. It also melts at a lower temperature and distorts readily when heated.

Hammer-On-Dolly Work

Because aluminum is soft, it responds very readily to the hammer-on-dolly method. If hammered too hard or too much, the panel will stretch. Use several light strokes rather than a few heavy blows.

Hammer-Off-Dolly Work

Aluminum does not readily bend back to its original shape after being buckled by an impact. Therefore, it does not respond well to the hammer-off-dolly method. Care must be exercised not to create additional damage when attempting to lower ridges with hammer and dolly blows.

Picking

Raising small dents with a picking hammer or pry bar is an excellent way to repair aluminum panels. However, be careful not to raise the panel too far and stretch the soft aluminum.

Spring Hammering

Spring hammering with a hammer and spoon is an excellent way to unlock stress in high-pressure areas. The spoon distributes the force of the blow over a wider area, minimizing the possibility of creating additional damage in the unyielding buckle.

Filing

Because aluminum is soft, reduce hand pressure on the body file. Use a file with rounded edges to avoid scratching and gouging the metal.

Grinding

Grinding must be done very carefully on aluminum. Not only will a coarse grit grinding disc on a high-speed grinder quickly burn through aluminum, but the heat from the grinding operation can quickly warp the panel. A No. 36 grit open coat disc can be used, but grind carefully in order to remove only paint and primer, and not the metal. Make two or three passes, then quench the area with a wet rag to cool the metal and minimize heat gain in the panel. Grinding and featheredging should be done with a dual-action sander or an electric sander that rotates at less than 2500 rpm. Use No. 80 or No. 100 grit paper and a soft, flexible backing pad.

Heat Shrinking

Heat shrinking of stretched aluminum must be done slowly to avoid distorting the panel. Do not heat the spot over 1200 degrees Fahrenheit; use an appropriate temperature-sensitive heat crayon to monitor the heat application. Mark the part with the crayon when it is still cold. When the stated temperature has been reached, the mark will liquefy.

Unlike steel, aluminum does not turn red in color as it is heated. Instead, it turns ash gray just before it melts. Thus, a lack of caution will result in a melted panel. Also, the heated area must be very slowly quenched to avoid distorting the panel by excessive contraction.

CRUSH ZONES

Certain built-in conditions of a metal structure can reduce its strength. These are called crush zones or stress concentrators (Figure 6-53). Stress concentrators, as the name implies, result in a localized concentration of stress as a load is applied. Crush zones are designed into unibody vehicles to control and absorb collision forces, minimize structural damage, and increase occupant protection. Do not remove any of these crush zones. In addition, always follow the vehicle manufacturer's recommendations for straightening or replacing parts that have crush zones.

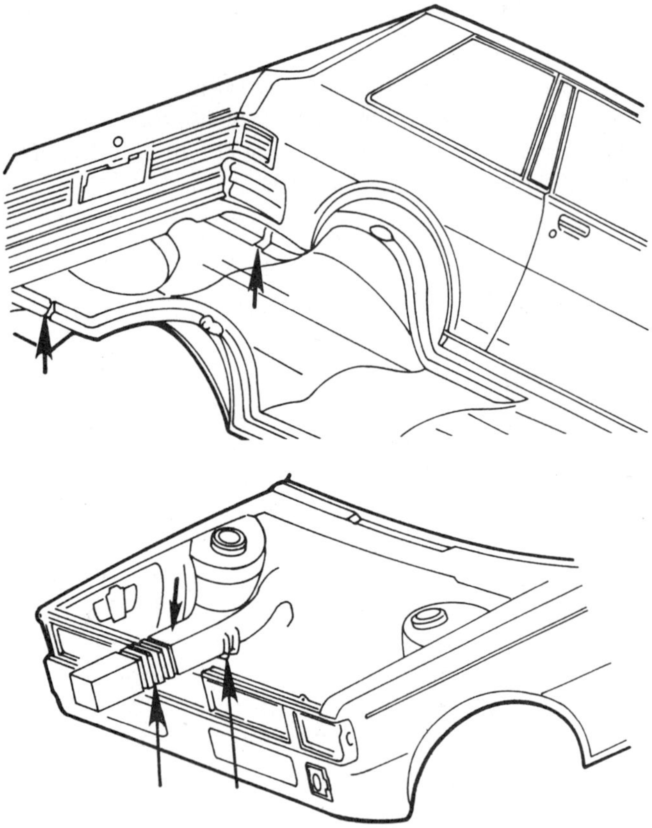

FIGURE 6-53 Typical crush zones

A quality repair can be achieved only if function, durability, and appearance have been restored. When stress is not removed, the following can occur:

- Fatigue caused by the loading and unloading of suspension and steering components.
- Dimensional distortion, which can cause handling problems.
- In the event of a second similar collision, less of an impact can cause the same or greater damage to the vehicle and could jeopardize the safety of the occupants.

APPLYING BODY FILLER

The repair techniques discussed previously are fundamental to any repair, but they are not the end of the job. After the damaged metal has been bumped, pulled, pried, and dinged, the finishing touch must be applied; namely, an application of body filler. Body fillers are designed to cover up the minor defects that remain after the metal has been straightened. Some are made with specific applications in mind, while others are more general in use. Keep in mind that the permanence of the repair and the quality of the finish is adversely affected by the wrong choice of filler (Table 6-1).

Taking shortcuts when mixing and applying body fillers might save time, but it will affect the quality of the repair. Fillers that are improperly mixed and applied will eventually crack, lose adhesion, permit rust to form on the metal substrate, and fall off the panel. Sooner or later, the repair will have to be redone. The loss in time, money, and reputation is great.

PLASTIC BODY FILLERS

One of the most important steps in applying plastic body fillers is surface preparation. Begin by washing the repair area with soap and water to remove dirt and grime. Then clean with wax and grease remover to eliminate wax, road tar, and grease. Be sure to use a remover that will take away the silicones often present in automotive waxes. Also, wash brazed or soldered joints to neutralize the acids in the flux. Do not grind these areas before neutralizing the acids. Grinding simply drives the acids deep into the metal.

Grind the area to remove the old paint 3 to 4 inches around the area to be filled. If filler overlaps any of the existing finish, the paint film will absorb solvents from the new primer and paint, destroying the adhesion of the filler. The filler will lift, cracking the paint and allowing moisture to seep under the filler. Rust will then form on the metal.

Use a No. 24 or No. 36 grit grinding disc to remove the paint. This also etches the metal to provide better adhesion. If applying filler over a metal patch, do not hammer down the excess weld bead. Grind it level with the surface. As shown in Figure 6-54, hammering the weld distorts the metal, creates stress in the panel, and increases the area to be filled. After grinding away the finish from the repair area, blow away the sanding dust

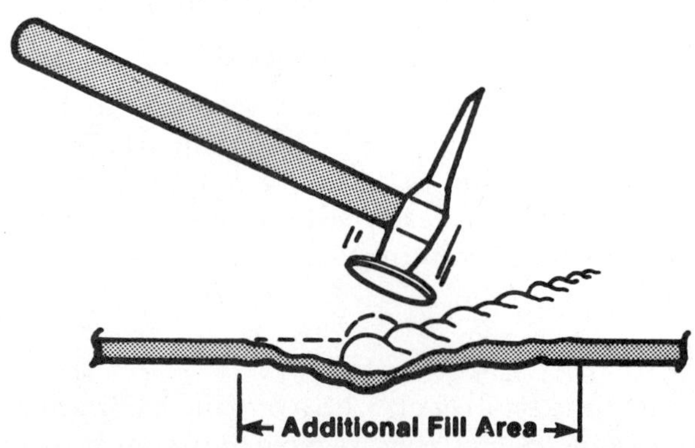

FIGURE 6-54 Hammering the weld bead down increases the fill area, and thus should not be done.

TABLE 6-1: COMPARING BODY FILLERS

Filler	Composition	Characteristics	Application
Conventional Fillers			
Heavyweight fillers	Polyester resins and talc particles	Smooth sanding; fine featheredging; nonsagging; less pinholing than lightweight fillers	Dents, dings, and gouges in metal panels
Lightweight fillers	Microsphere glass bubbles; fine grain talc; polyester resins	Spreads easily; nonshrinking; homogenous; no settling	Dings, dents, and gouges in metal panels
Premium fillers	Microspheres; talc; polyester resins; special chemical additives	Sands fast and easy; spreads creamy and moist; spreads smooth without pinholes; dries tack-free; will not sag	Dings, dents, and gouges in metal panels
Fiberglass-Reinforced Fillers			
Short strand	Small fiberglass strands; polyester resins	Waterproof; stronger than regular fillers	Fills small rustouts and holes. Used with fiberglass cloth to bridge larger rustouts.
Long strand	Long fiberglass strands; polyester resins	Waterproof; stronger than short strand fiberglass fillers; bridges small holes without matte or cloth	Cracked or shattered fiberglass. Repairing rustouts, holes, and tears.
Specialty Fillers			
Aluminum filler	Aluminum flakes and powders; polyester resins	Waterproof; spreads smoothly; high level of quality and durability	Restoring classic and exotic vehicles
Finishing filler/ polyester putty	High-resin content; fine talc particles; microsphere glass bubbles	Ultra-smooth and creamy; tack-free; nonshrinking; eliminates need for air dry type glazing putty	Fills pinholes and sand scratches in metal, filler, fiberglass, and old finishes.
Sprayable filler/ polyester primer-surfacer	High-viscosity polyester resins; talc particles; liquid hardener	Virtually nonshrinking; prevents bleed-through; eliminates primer/glazing/ primer procedure	Fills file marks, sand scratches, mildly cracked or crazed paint films, and pinholes. Seals fillers and old finishes against bleed-through.

with compressed air and wipe the surface with a tack rag to remove any remaining dust particles.

CAUTION

Do not apply a metal conditioner to the repair area before applying the filler. The acids in the conditioner will prevent the filler from bonding to the metal substrate.

Mixing the Filler

Mix the can of filler to a uniform and smooth consistency free of lumps (Figure 6-55). Heavyweight fillers should be shaken on a paint shaker for several minutes. If they are not stirred thoroughly to a smooth and uniform consistency before use, the filler in the upper portion of the can will be too thin. This results in runs and sags, slow and poor curing, a gummy condition when sanded, poor featheredging, a very tacky surface, and blistering and lifting when coated with

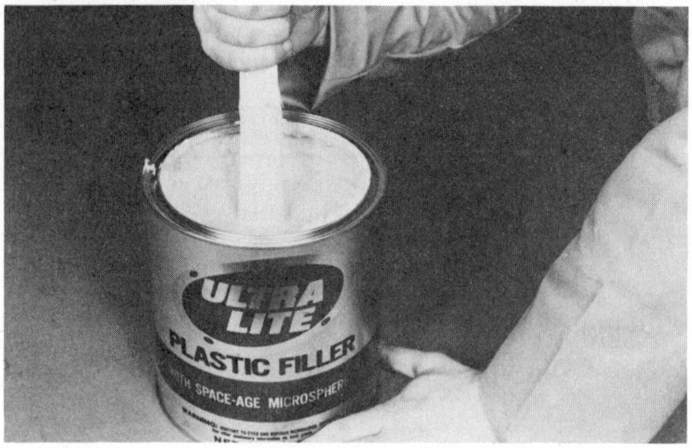

FIGURE 6-55 Mixing filler with a paint paddle

primers and refinishing materials. If the filler in the lower portion of the can is too thick, the final results will be very coarse and grainy. There will be poor adhesion, excessive pinholing, poor featheredging, and poor color holdout when the area is coated with primers and refinishing materials.

Kneading the Hardener

Loosen the cap of the hardener tube to release the air. Knead the tube thoroughly to assure a smooth, pastelike consistency (like toothpaste) when squeezed out. If this is not done, the result can be slow and ineffective curing, a tacky surface, poor adhesion, poor featheredging, blistering, and peeling after finishing. If the hardener is kneaded thoroughly and remains thin and watery, it is spoiled and should not be used.

Mixing Filler and Hardener

Open the can slowly and remove the desired amount of filler. Using a clean putty knife or spreader, place the filler on a smooth, clean surface such as sheet metal, glass, or a nonporous mixing board (Figure 6-

FIGURE 6-56 A nonporous mixing board is good for mixing the filler and hardener.

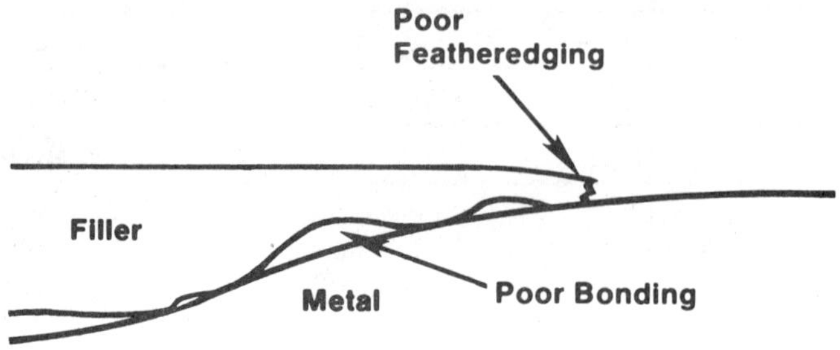

FIGURE 6-57 Too little hardener causes adhesion difficulties and poor featheredging.

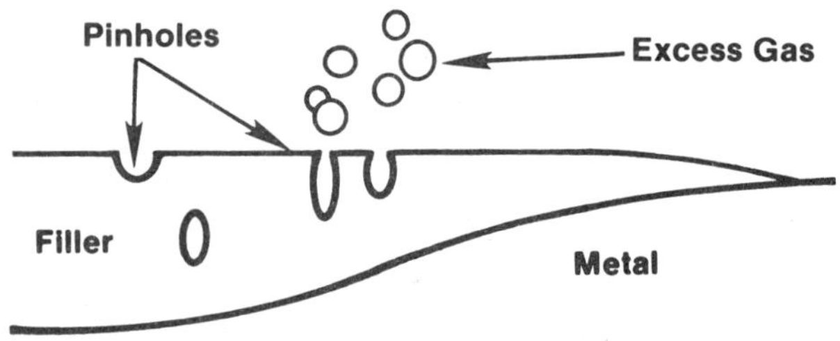

FIGURE 6-58 Too much hardener creates excess gas and pinholes.

56). Add the hardener according to the proportion indicated on the can.

Too little hardener will result in a soft, gummy filler that will not adhere properly to the metal and will not sand or featheredge cleanly (Figure 6-57). Too much hardener will produce excessive gases, resulting in pinholing (Figure 6-58).

SHOP TALK

Cardboard should not be used as a mixing surface for the filler. It is porous and contains waxes for waterproofing. These waxes will be dissolved in the mixed filler and cause poor bonding to both the metal repair area and to earlier cured filler. Cardboard also absorbs some of the chemicals in the filler and hardener, reducing the quality of the job. Use glass, plastic, or any nonporous surface. Plastic mixing boards are also available for this purpose.

With a clean putty knife or spreader, use a scraping motion (back and forth) to mix the filler and hardener together thoroughly and achieve a uniform color (Figure 6-59). Do not stir the filler. Stirring whips air into the filler, producing air pockets and pinholes in the application. Do not return the knife, spreader, or mixed filler to the can. Replace the cover on the can and close it tightly.

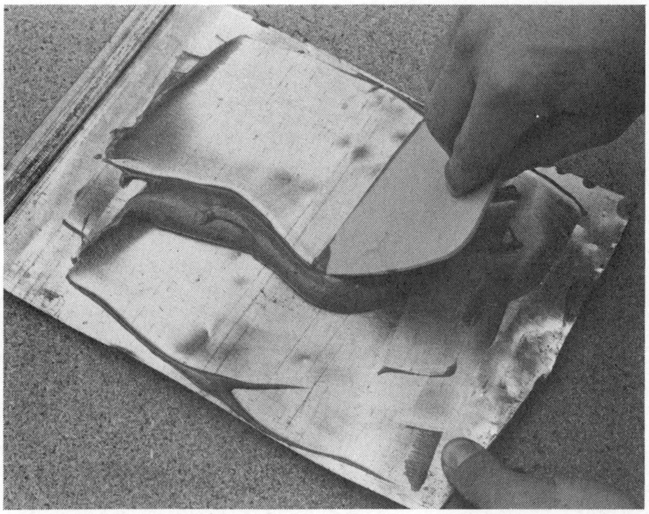

FIGURE 6-59 Use a scraping motion when mixing the filler and hardener.

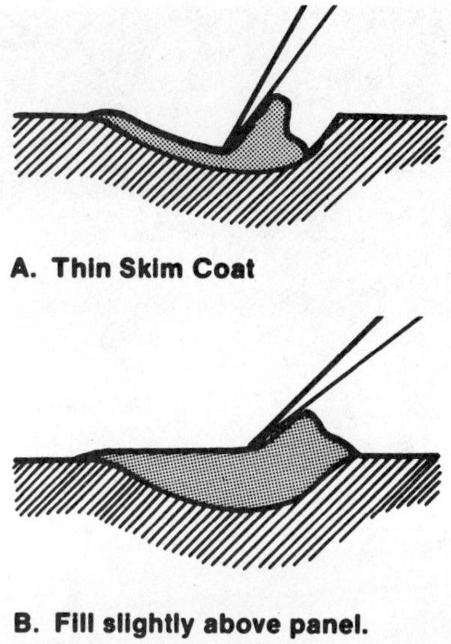

FIGURE 6-60 Build up the filler thickness in several applications.

If the filler and hardener are not thoroughly mixed to a uniform color, soft spots will form in the cured filler. The result is an uneven cure, poor adhesion, lifting, and blistering.

Always use clean, uncontaminated tools when removing the filler from the can and mixing the filler and hardener together. Otherwise, hard lumps of filler might appear in the can and/or the applied filler, which will cause poor adhesion and an uneven cure.

Applying the Filler

Apply the mixed filler promptly to a thoroughly clean and well-sanded surface. A tight, thin application is recommended first. Press firmly to force filler into the sand scratches to maximize the bond. When cured, apply additional thicker applications as needed to build up the repair area to proper contour (Figure 6-60). Allow each application to set up before applying the next coat of filler. Conventional body fillers should be built up slightly so that the waxy film that curing produces on the surface of the filler can be removed with a surform (cheese grater) file.

SHOP TALK

Always use a clean plastic spreader or putty knife to apply the filler, and use a different spreader to mix the filler. A small amount of unmixed filler will always remain on the spreader. If any of this is transferred to the metal, soft spots will be present in the cured filler and the paint finish will peel.

Applying the mixed filler thickly without first applying a thin, tight application causes poor adhesion and pinholing. Wiping over the repair area with solvents before applying the filler also causes pinholes and poor adhesion.

Avoid use in cold temperatures. When the filler, shop, and/or panel temperatures are cold, the filler will not cure properly; it will have a tacky surface and poor sanding properties. Tremendous pinholes could also be created. Filler should be stored at room temperature (65 to 70 degrees Fahrenheit) and a heat lamp used to warm cold surfaces.

In winter, moisture will form on metal surfaces when a cold vehicle is brought inside. In summer, condensation might form on the metal. Use a heat lamp to dry damp surfaces before applying filler. If the repair area is not first warmed to remove moisture accumulation, poor adhesion, a poor featheredge, pinholing, and lifting when refinishing could result.

Shaping the Filler

Allow the filler to cure to a semihard consistency. This usually takes 15 to 20 minutes. If the filler leaves a firm white track when scratched with a fingernail, it is ready to be filed. Filing is perhaps the most important factor in achieving a quality surface and controlling material cost and labor. The cheese grater file is used to cut the excess filler to contour quickly. It produces an even, level surface. The teeth in the file are open enough to prevent the tool from clogging. Grinders, sanders, and air files do not level well; they become loaded quickly, create too much dust, and waste a great deal of sandpaper.

To use the cheese grater file, hold it at a 30- to 40-degree angle and pull it lightly across the filler (Fig-

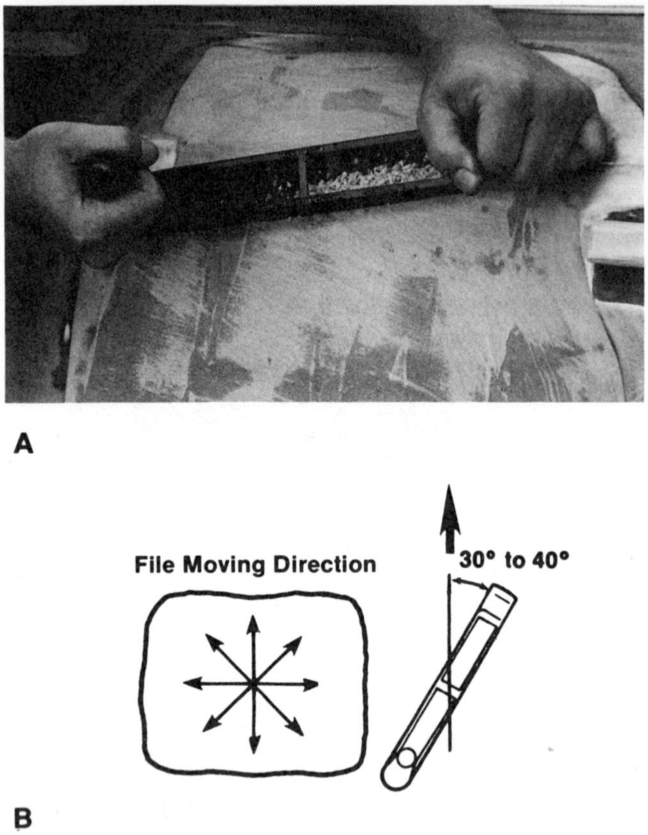

FIGURE 6-61 (A) File the semihard filler with a cheese grater file. (B) Hold the file at a 30° to 40° angle and work it in several directions.

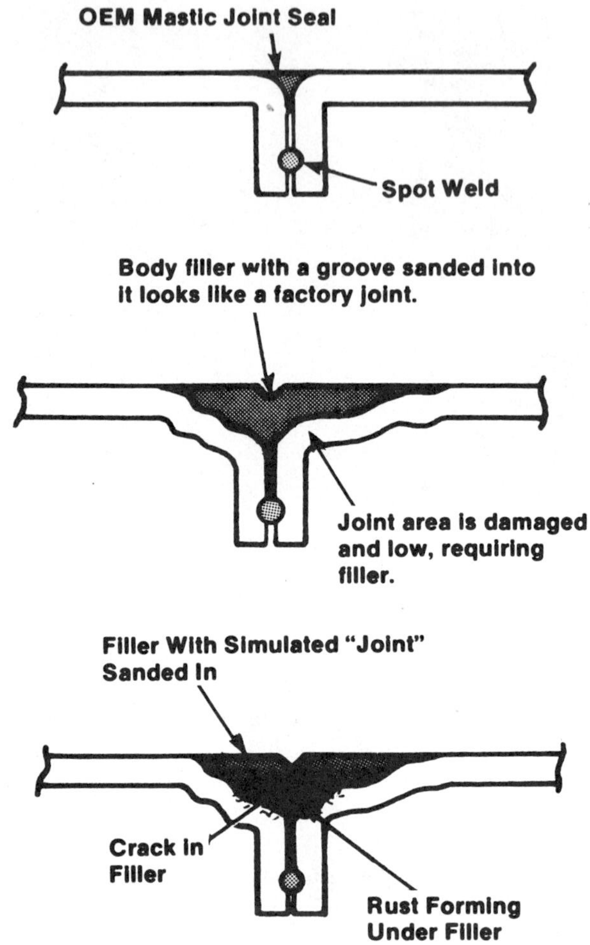

FIGURE 6-62 Normal road conditions cause inflexible filler to crack.

ure 6-61). Work the file in several directions. Stop filing when the filler is slightly above the desired level. This will be sufficient for sanding out the file marks and for feathering the edges. If the filler is undercut, additional filler must be applied.

Applying Filler to Body Lines

Many vehicles today have sharp body lines in doors, quarter panels, hoods, and so forth. Maintaining the sharpness of these lines when doing filler work is difficult, especially in recessed areas. The best way to get straight, clean lines is to fill each plane, angle, or corner separately.

Masking tape is applied along one edge, then filler is applied to the adjacent surface. Before the filler sets up, the tape is pulled off, removing the excess filler from the body line. After the first application is dry and sanded, the opposite edge is taped. Masking tape is applied along the body line and over the filler, and the adjacent surface is coated with filler. When the tape is removed and the filler sanded, the result is a straight, even line or corner.

Applying Filler to Panel Joints

Many panels on unibody vehicles have joints that are factory finished with a seam sealer to allow the panel to flex and move. Often both halves of the joint suffer damage and require filling. Many collision repair technicians make the mistake of covering the seam sealer with body filler. Figure 6-62 shows what happens when the inflexible filler is subjected to the twisting action of the panels under normal road conditions. A crack develops that allows moisture to seep under the filler, which causes rust to form on the metal surface. This eventually results in the failure of the filler-to-metal bond and a weakened sheet metal joint.

The original flexibility of the joint can be preserved by taping its alternate sides. As shown in Figure 6-63, tape is applied to one panel. Filler is then applied to the other panel and the tape is pulled up, removing the excess filler. The other panel is filled in the same way. After the filler in both panels is cured and shaped, a seam sealer is forced into the joint.

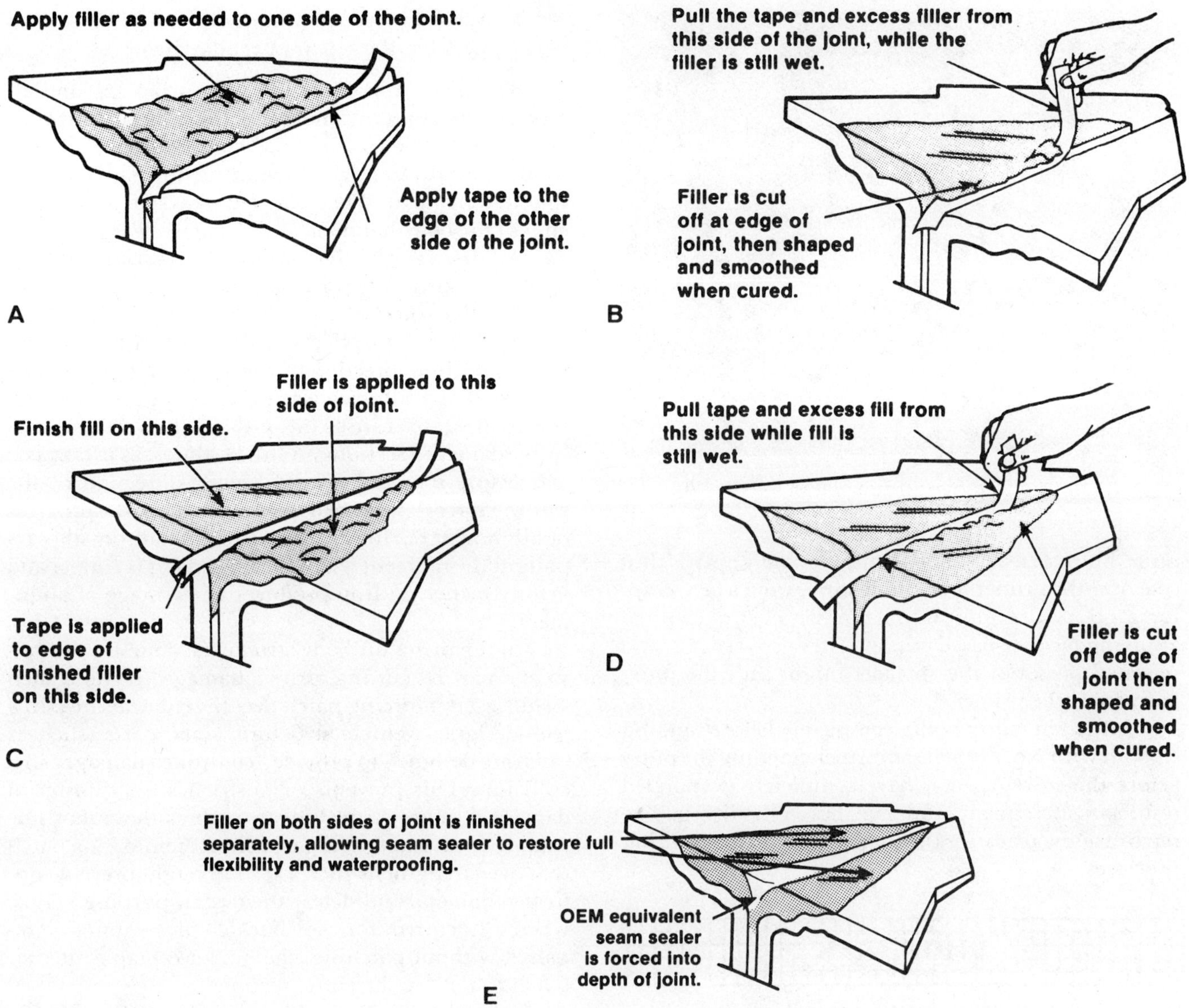

FIGURE 6-63 Using masking tape to create a flexible filler-to-metal joint

Sanding the Filler

After filing, sand out all file marks (Figure 6-64). Use a No. 36 or No. 40 grit disc on a sanding board or block. An air file can also be used on large flat areas. Then, follow with No. 80 grit paper until all scratches are removed.

Final smoothing should be done using No. 180 grit paper. A sanding block or air file can again be used, as well as a long sanding board. Be careful not to oversand—this results in the filled area being below the desired level, which makes it necessary to apply more filler. After final sanding, blow with an air gun and wipe with a tack cloth. This removes any fine sanding dust that might be hiding surface pinholes and also exposes holes lying just below the surface. These holes and remaining sand scratches must be filled.

Run a hand or straightedge over the surface often to check for evenness. Do not trust eyeing for accuracy. Remember, paint does not hide imperfections; it highlights them. Do not be satisfied until the repaired surface feels perfectly smooth. When satisfied with the smoothness, clean the area with a tack cloth. The repair area must then be primed.

Applying Glazing Putty

Once the primer is dry, small pinholes and scratches can be filled with glazing putty. If using a polyester putty, mix the putty and hardener according to the manufacturer's instructions. Place a small amount of

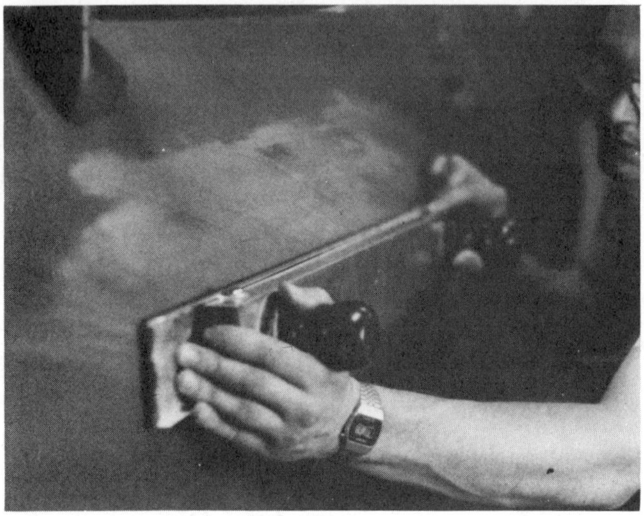

FIGURE 6-64 Remove all file marks by sanding.

putty onto a clean rubber squeegee and apply a thin coat over the primer. Use single strokes and a fast scraping motion. Use a minimum number of strokes when applying lacquer-based putties. They dry very fast, and repeated passes of the spreader might pull the putty away from the primer.

Allow the putty to dry completely before sanding smooth with No. 240 grit sandpaper. Sanding the putty before the solvents in it have completely evaporated results in subsequent sand scratches in the finish. The repair is now ready for the final priming and painting processes.

SHOP TALK

When properly mixed, applied, sanded, and primed, a quality body filler will have very few imperfections. Excessive use of glazing putties is usually an indication of a lack of expertise on the part of the collision repair technician.

STRAIGHTENING SUMMARY

The structural qualities of unibody construction bring up questions about repairing or replacing severely damaged sections that cannot be restored by pulling. The basic questions are:

- When should the repair or replacement be made in relation to the pull—before, after, or during?
- What procedures and tools should be used to perform the repair or replacement?

The basic rule of thumb is to make any and all repairs when parts alignment permits. In other words, when the torn edges become butted or sheared spot welds move back and line up during pulling, that's the time to weld them. When buckled areas start to straighten, use the body hammer.

At first, apply the pulling force intermittently while checking the movement of the panel and confirming that the force is working effectively on the damaged area. If no effect can be seen, consider changing the pulling direction or the area being pulled. Remember that pulling generates forces opposite in direction to those that caused the initial damage. Accordingly, the damaged section can still be almost as strong as it was before it was damaged—in the direction of the pulling force. It is crucial to visually check the progress of alignment during the pull, making all possible repairs at the moment when the alignment is correct and before further pulling produces a new stage of alignment.

In repairing unibody structures, do not try to salvage a part by cutting away a damaged section, then welding a reinforcing patch over the cutaway area. In a well-designed vehicle structure, some parts (such as rails) are designed to provide "controlled damage" in a collision. This prevents or postpones the transfer of damage to more critical areas (such as those that surround the passenger compartment). A reinforcing patch that overstrengthens such a section might prevent controlled damage and defeat the design purpose. Thus, when a fractured, torn, or buckled area cannot be repaired without patching, the only acceptable alternative is to replace the entire section.

Once the parts are welded into place as described in Chapters 7 and 8, follow the vehicle manufacturer's recommendations for sealing, caulking, and restoring corrosion protection to all the structural parts. As mentioned in Chapter 10, this is extremely important on a unibody to assure lasting strength and quality of the repair.

When the repair is completed, including all straightening and welding operations, the alignment procedures are ready for a final check. If the vehicle was repaired without using fixtures, final measurements should be made and compared to the body dimension manuals. All measurements must be checked against factory dimensions. This is admittedly time consuming, but it must be done. Of course, if the vehicle was repaired with fixtures, the underbody of the vehicle is within specifications.

Begin the final checks by slowly walking around the vehicle looking for obvious signs of misalignment.

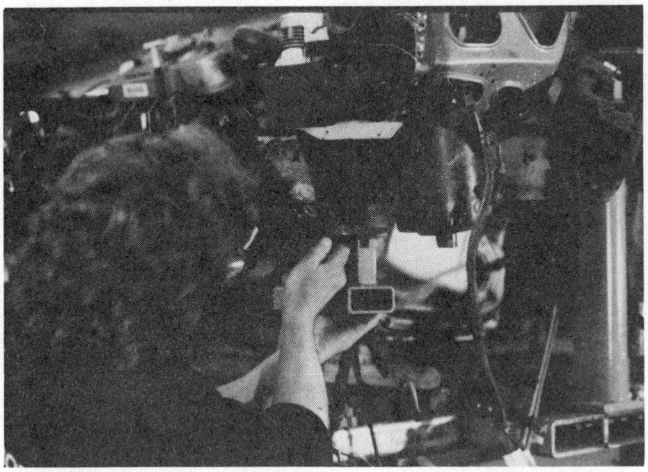

FIGURE 6-65 Checking the fixture reference points

Too large of gaps between roof lines and doors is a good sign that some damage is still present. If fixtures were used, go back and check all the fixture reference points (Figure 6-65). Do they still line up true, without having to force or hammer the fixture? Go back over the work order or estimate to be sure everything that was supposed to be done was done. Remember that it is much easier to set up and make additional pulls now rather than wait until more steps of the repair procedure have been completed and then find additional damage.

Other items that should be carefully inspected are:

- The alignment between the door and rocker sill. This should be a straight and narrow gap.
- The general alignment of all the upper body areas. Make sure everything looks as though nothing was ever out of alignment.
- The doors and deck lid. Do they feel tight and secure when latched? Make sure they close smoothly and open easily.

Once the final inspection is made, the vehicle can be left on the bench for replacement of components removed earlier, or it can be removed from the bench. The vehicle can be reassembled by simply reversing the sequence of steps used when removing the components from the vehicle originally.

REVIEW QUESTIONS

1. When determining the direction of a pull, Technician A sets the equipment at an angle where the pulling force will be applied parallel to the damage. Technician B sets the equipment at an angle where the pulling force will be applied perpendicular to the damage. Who is correct?
   a. Technician A
   b. Technician B
   c. Both A and B
   d. Neither A nor B

2. For a unibody repair, the measuring system must show the ____________.
   a. direction of the misalignment
   b. amount of misalignment at each reference point
   c. both a and b
   d. neither a nor b

3. Anchor pots and floor grids are examples of which straightening systems?
   a. bench
   b. in-floor
   c. rack
   d. both a and b

4. Which of the following statements concerning bench systems is incorrect?
   a. There must be at least five reference points on the undamaged part of the vehicle to set it up properly on a bench.
   b. The bench system should be used whenever the damage involves the suspension, steering, or power train mounting points.
   c. Benches are available in both stationary and portable varieties.
   d. None of the above.

5. Technician A says that any amount of overpulling, no matter how small, is detrimental. Technician B says that a small amount of controlled overpulling can be helpful. Who is correct?
   a. Technician A
   b. Technician B
   c. Both A and B
   d. Neither A nor B

6. Portable hydraulic rams have the capability to ____________.
   a. pull
   b. push
   c. spread
   d. all of the above

7. For front damage, Technician A locates the fixtures in the front of the vehicle. Technician B locates them in the front and center of the vehicle. Who is correct?
   a. Technician A
   b. Technician B

c. Both A and B
d. Neither A nor B

8. Which of the following statements is incorrect?
   a. More components must be disassembled when straightening structural members than when replacing them.
   b. With today's unibody vehicles, it is considered almost automatic to remove the suspension and driveline before putting it on the bench.
   c. Damage should be repaired in the sequence in which it occurred.
   d. All of the above.

9. When anchoring a unibody vehicle, Technician A favors "over anchoring," while Technician B favors "under anchoring." Who is correct?
   a. Technician A
   b. Technician B
   c. Both A and B
   d. Neither A nor B

10. In a pulling sequence, what is the correct order of problem solving?
    a. length, sidesway removal, correct height
    b. correct height, sidesway removal, length
    c. sidesway removal, correct height, length
    d. length, correct height, sidesway removal

11. The hammer-on-dolly metalworking technique is used to ______________.
    a. straighten metal
    b. stretch metal
    c. both a and b
    d. neither a nor b

12. In a typical roof rail height pull, approximately how many inches of continuous pull are usually available automatically?
    a. 10
    b. 24
    c. 33
    d. 44

13. In most instances, an overpulled panel ______________.
    a. results from failure to measure accurately and often
    b. must be replaced
    c. can be repaired
    d. both a and b

14. Technician A uses an air file to shape and level excess plastic body filler. Technician B uses a surform (cheese grater) file to do this. Who is correct?
    a. Technician A
    b. Technician B
    c. Both A and B
    d. Neither A nor B

15. Which of the following statements is incorrect?
    a. Stress can be caused by overheating.
    b. Cracked paint and undercoating is a sign of stress/deformation on a unibody vehicle.
    c. Crush zones can be removed from a vehicle if necessary.
    d. None of the above.

CHAPTER 7

Welding in Unibody Collision Repair

OBJECTIVES

- Name some of the advantages of MIG welding.
- Explain how to use a MIG welding machine.
- Explain how factors such as welding current, voltage, wire stick-out distance, gun angle, shielding gas flow rate, travel speed, wire speed, gun nozzle adjustment, and welding position affect a MIG weld.
- Describe how tack, butt, lap, flange, plug, spot, and stitch welds are formed.
- Test welds using both destructive and nondestructive methods.
- Explain the basic processes for flux-cored arc welding, resistance spot welding, pneumatic spot welding, plasma arc cutting, oxyacetylene welding, and brazing.

As defined by the American Welding Society, a weld is formed when separate pieces of material are fused together through the application of heat. The heat applied is high enough in temperature to cause the softening or melting of the pieces to be joined. Once the pieces are heated, pressure may or may not be used to force them together. In some cases, filler material is added to the weld joint when needed. For practical purposes, welding can be divided into three main categories:

- *Pressure Welding.* The pieces of metal are heated to a softened state by electrodes; pressure is applied; and the pieces are joined. Of the various types of pressure welding, resistance spot welding is an indispensable welding method used in automobile manufacturing and to a lesser degree in collision repair operations. In fact, resistance spot welding is generally acceptable for cosmetic panel repairs.
- *Braze Welding.* A metal with a melting point that is lower than the pieces of base metal to be joined is melted into the weld joint (without melting the pieces of base metal). Braze welding is classified as either soft or hard brazing, depending on the temperature at which the brazing material melts. Soft brazing or soldering is done with brazing materials that melt at temperatures below 900 degrees Fahrenheit and hard brazing is done with brazing materials that melt at temperatures above 900 degrees Fahrenheit.
- *Fusion Welding.* The pieces of metal are heated to their melting point, joined together (usually with a filler rod), and allowed to cool. The once popular fusion welding methods of shielded arc welding (Figure 7-1) and oxyacetylene welding (Figure 7-2) have been completely replaced in unibody repair work by metal inert gas (MIG) welding. The reason is simply that the new steel alloys used in today's unibody vehicles cannot be welded properly with the two previous processes. Today's steels demand a technique that leaves a narrow heat-affected zone. This is especially true with coated steels; oxyacetylene welding would burn off too much of the galvanized zinc coating in the weld area. Also, unibody vehicle designs require 100 percent fusion for most repairs, normally impossible with oxyacetylene or shielded arc welding.

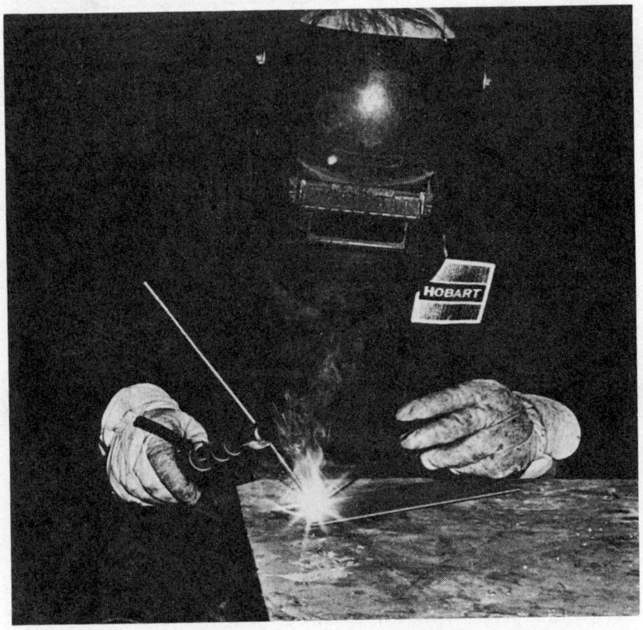

FIGURE 7-1 The shielded arc welding process

FIGURE 7-2 Oxyacetylene welding was once a popular repair method, but is not recommended for unibody work.

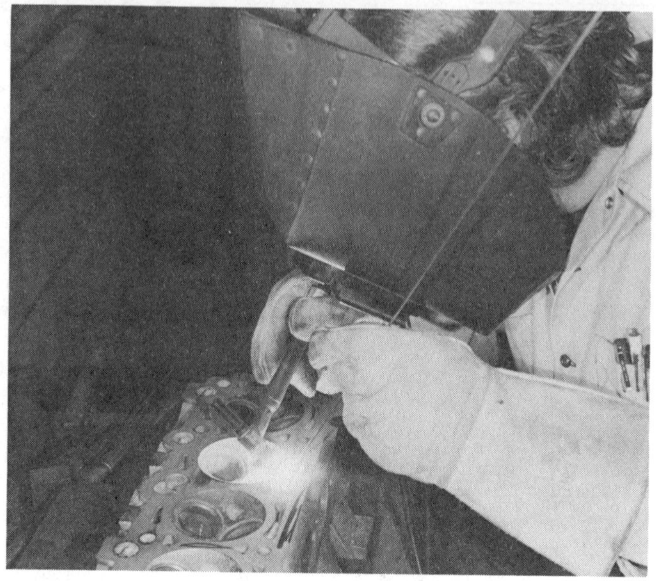

FIGURE 7-3 TIG welding is used by the automotive engine rebuilding trade for repairing cracks in aluminum cylinder heads.

There are two general types of gas metal arc welding methods: metal inert gas (MIG) and tungsten inert gas (TIG). TIG is generally used in automotive engine rebuilding shops to repair cracks in aluminum cylinder heads (Figure 7-3) and in reconstructing combustion chambers, but has only very limited applications for collision repair.

The advantages of MIG welding (Figure 7-4) over the other fusion methods used in collision repair are so numerous that vehicle manufacturers now recommend that it be used not only for HSS, but for all structural collision repair. The advantages of MIG welding are as follows:

- MIG welding is easy to learn. The typical welder can learn to use MIG equipment and actually reach a peak level of proficiency with just a few hours of instruction and practice. Moreover, experience shows that even an average MIG welder can produce higher quality welds faster and more consistently than a highly skilled welder using stick electrode welds.
- MIG produces 100 percent fusion in the base metals (Figure 7-5). This means MIG welds can be ground down flush with the surface (for cosmetic reasons) without loss of strength.
- Low current can be used to MIG weld thin metals. This prevents heat damage that can cause strength loss and warping.

FIGURE 7-4 MIG is now the number one welding repair method for unibody vehicles.

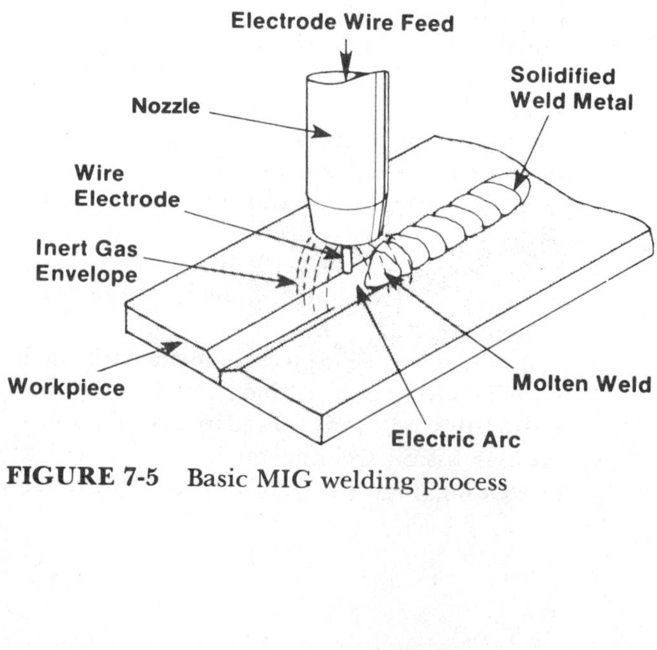

FIGURE 7-5 Basic MIG welding process

- The arc is smooth and the weld puddle small. This ensures maximum metal deposit with minimum spatter.
- Gaps can be spot welded by making several spot welds on top of each other—immediately, with no slag to remove.
- Almost all steels can be MIG welded with one common type of weld wire.
- Metals of different thicknesses can be MIG welded.
- The welder can control the temperature of the weld and the time the weld takes.
- MIG welding produces a hotter, more concentrated heat, which means the technician stays on the workpiece for a much shorter time period (Figure 7-6). This minimizes the heat-affected zone, therefore reducing metal fatigue, warpage, and distortion of the panel. Vertical and/or overhead welding is possible because the metal is molten for a very short time.
- MIG welding is safer to use when making repairs under the vehicle next to brake lines, fuel lines, and fuel tanks because the wire is not energized until the gun trigger is depressed.
- With MIG welding, there is minimum waste of welding consumables.

The other type of welding that is recommended in collision repair work of newer vehicles is squeeze-type resistance spot welding (Figure 7-7). To use squeeze-type resistance spot welding equipment, install the proper extensions and electrodes on the welder to provide access to the area being welded. The clamping force on the squeeze-type resistance spot welder must be properly adjusted. On some equipment, the current flow and weld time are made with one adjustment. After the adjustments are made, the squeeze-type resistance spot welder is positioned, making sure the electrodes are directly opposite each other. The trigger is squeezed and the weld is made.

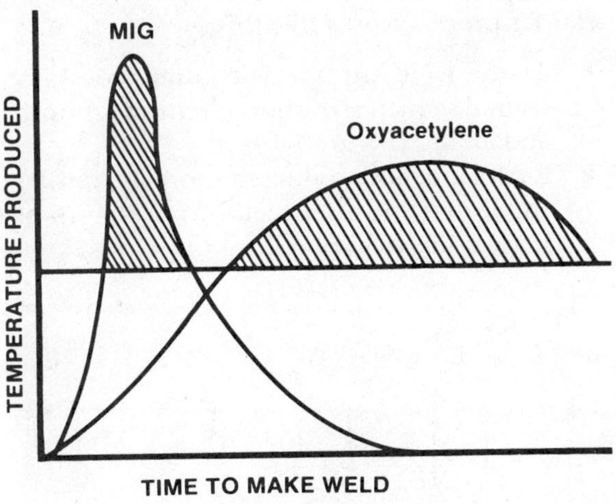

FIGURE 7-6 Although the MIG welding process produces a hotter and more concentrated heat than oxyacetylene welding, it is not necessary to stay on the piece as long, so the heat-affected zone is smaller.

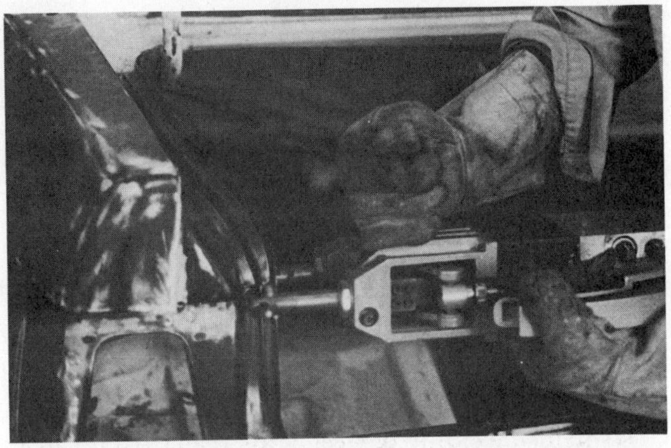

FIGURE 7-7 Using a portable squeeze-type resistance spot welder to weld a left rear quarter panel to the lower rear panel

MIG WELDING

The basic components used in MIG welding are simple. MIG welding uses a welding wire electrode that is fed automatically at a constant speed. A short arc is generated between the base metal and the wire, and the resulting heat from the arc melts the welding wire and joins the base metals together. Since the wire is fed automatically at a constant rate, this method is often mistakenly called semiautomatic arc welding. During the welding process, either an inert or active gas shields the weld from the atmosphere and prevents oxidation of the base metal. The type of inert or active gas used depends on the base metal to be welded.

Several types of shielding gases and shielding gas mixtures are used with the MIG process. Argon and helium are primarily used for welding nonferrous metals and can also be used in combination. Carbon dioxide and argon/carbon dioxide mixtures are other popular shielding gases.

Recently, MIG welders have been made available that run on low voltage. These units use a flux-cored wire that does not require a shielding gas. While low-voltage MIG welders do produce more weld spatter, they are favored by smaller shops as a means of cutting down on overhead expenses.

TYPES OF METAL TRANSFER

There are a number of methods used in MIG welding for transferring the filler, or welding wire, to the base metals. Spray transfer is accomplished by the movement of a stream of tiny droplets across the arc from the electrode to the base metal (Figure 7-8A). The arc has a characteristic sizzling sound or, in some cases, a buzzing sound. Spray transfer usually employs relatively high voltages and current.

As the name implies, globular transfer across the arc occurs in large, irregularly shaped drops or "globs" (Figure 7-8B). This type of metal transfer is indicative of CO_2 shielded welding of steel. It also can exist because of low arc voltage, or low amperage, when using other shielding gases.

The short-circuiting arc method of metal transfer is unique in that no metal transfers across the arc. Instead, the metal transfer occurs when the electrode wire makes contact with the base metal. Welding thin automobile sheet metal can cause strain, blow holes, and warped panels. To prevent these problems, it is necessary to limit the amount of heat near the weld. The short-circuiting arc method uses very thin welding wires, a low current, and low voltage. With this technique, the amount of heat introduced into the panels is kept to a minimum and penetration of the base metal is quite shallow.

As shown in Figure 7-9, the end of the wire is melted by the heat of the arc and forms into a drop, which then contacts the base metal and creates a short circuit. When this happens, current flows through the metal and the shorted portion of the welding wire is torn away by the pinch force or burnback, which reestablishes the arc. That is, the electrode wire is fed continuously into the weld puddle at a controlled, constant rate, where it short circuits, and the arc goes out. When the arc is out, the puddle flattens and cools; but the wire continues to feed, shorting to the workpiece again. This heating and cooling occurs on an average of 100 times a second. The filler metal is transferred to the workpiece with each short circuit. Generally, if current is flowing through a cylindrically shaped fluid (in this case molten metal) or through an arc, the current is pulled toward the weld. This works as a constricting force in the direction of the center of the cylinder. This action is known as the *pinch effect,* and the size of the force is called the *pinch force.* In summary, the short-circuiting arc transfer process works like this:

- At the weld site the wire undergoes a split-second sequence of short circuiting, burnback, and arcing (Figure 7-10).
- Each sequence produces a short arc transfer of a small drop of filler metal from the tip of the wire to the weld puddle.

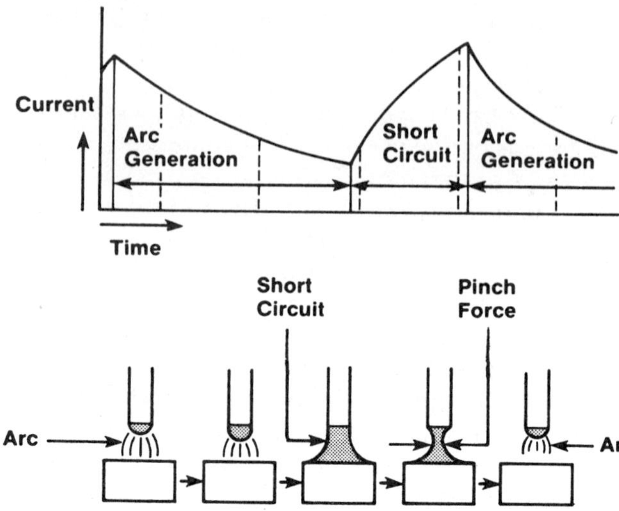

FIGURE 7-9 Short-circuiting arc method of metal transfer

A

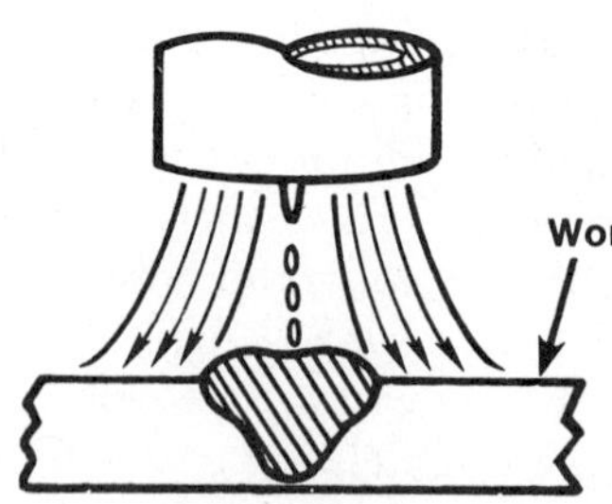

B

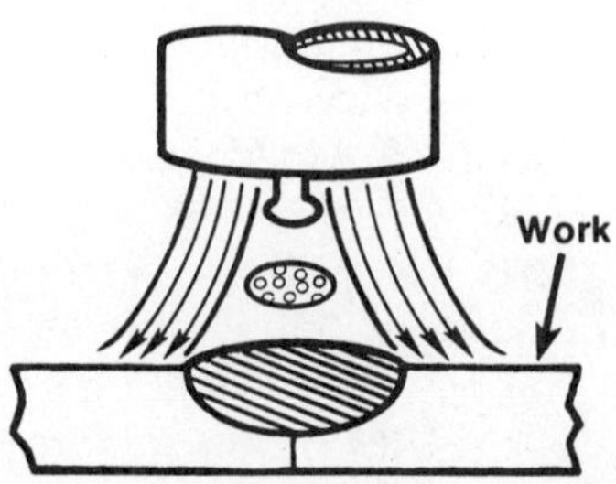

FIGURE 7-8 Types of metal transfer: (A) spray and (B) globular

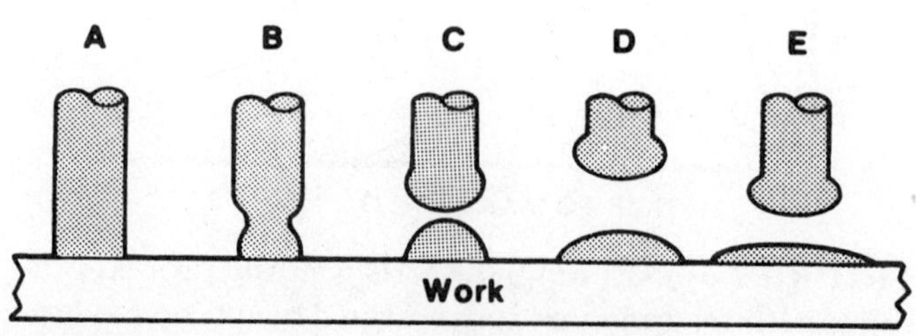

FIGURE 7-10 Typical action of the welding wire as it burns back from the work during the MIG short-circuiting arc transfer process

- A gas shield surrounds the wire electrode. This shielding gas prevents contamination from the atmosphere and helps to stabilize the arc.
- The continuously fed electrode wire contacts the metal and sets up a short circuit, and resistance heats the wire and the base metal.
- As the heating continues, the wire begins to melt and thin out.
- Increasing resistance in the molten neck accelerates the heating in this area.
- The molten neck burns through, depositing a puddle on the workpiece and starting the arc.
- The arc tends to flatten the puddle and burn back the electrode.
- With the arc gap at its widest, it cools; this allows the wire feed to move the electrode closer to the work.
- The short end starts to heat up again, enough to further flatten the puddle but not enough to keep the electrode from recontacting the workpiece. This extinguishes the arc, reestablishes the short circuit, and restarts the process. The complete cycle occurs automatically at a frequency ranging from 50 to 200 cycles a second.

SHOP TALK

The MIG welder can be set up for either reverse or straight polarity. Reverse polarity, electrode positive (REP) puts more of the heat into the weld puddle, making a better weld. The manual that comes with the welder will explain how to set the polarity.

MIG WELDING EQUIPMENT

Most MIG welding equipment that is used in and designed for collision repair work is considered semiautomatic. This means that the machine's operation is automatic, but the gun is controlled by hand. Before starting to weld, the operator sets the following:

- Voltage
- Wire speed
- Shielding gas flow rate

Once the power button is pressed, the operator has complete freedom to concentrate entirely on the weld site, the molten puddle, and the welding technique. The burnback control, which is found on some MIG welders, provides an adjustable burnback of the electrode to prevent it from sticking in the puddle at the end of a weld.

Regardless of the brand of MIG welding equipment used, it will consist of the following basic components (Figure 7-11):

- Power supply
- Electrode (wire) feed unit
- Electrode (wire) supply of a specific type and diameter
- Flow meter or regulator
- Work cable and clamp assembly
- Welding gun and cable assembly
- Supply of shielding gas (Figure 7-12)

The most common shielding gas combinations are: for sheet metal use—75 percent argon/25 percent carbon dioxide; for stainless steel use—90 percent heli-

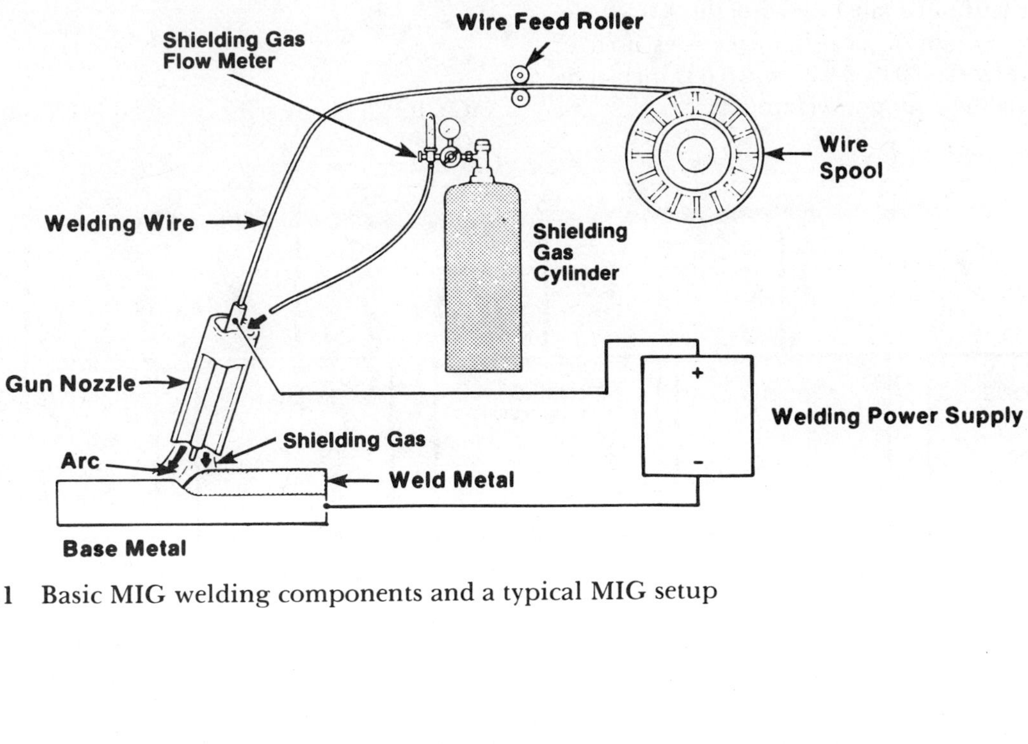

FIGURE 7-11 Basic MIG welding components and a typical MIG setup

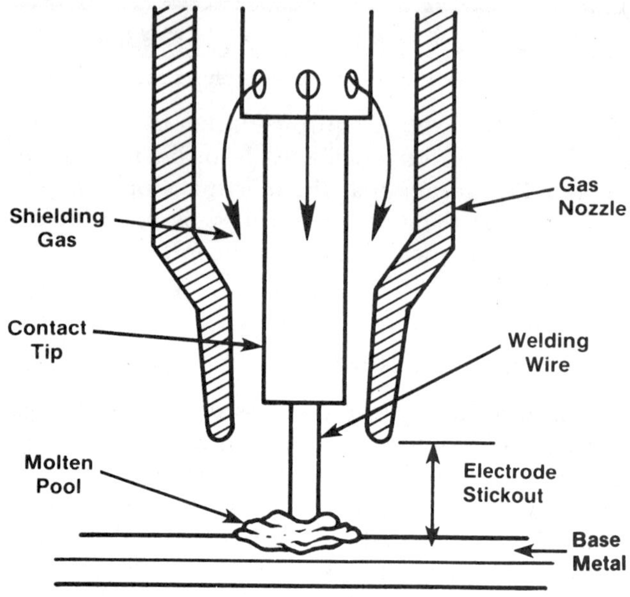

FIGURE 7-12 Shielding gas flow

um/7-1/2 percent argon/2-1/2 percent carbon dioxide; and for aluminum use—100 percent argon. There are also tri-mix gases that contain mostly argon, with 5 to 10 percent carbon dioxide and 1 to 3 percent oxygen. The addition of oxygen further settles the welding arc at low current and seems to improve welding on coated steel. The flow meter or regulator controls the flow of shielding gas. Shielding gas improves the mechanical strength properties of weld metal.

Welding wire with an AWS code rating (Figure 7-13) must meet certain standards. Wire offered by one manufacturer may differ from wire with the same code offered by another. If the AWS ratings are the same, however, both at least meet the same set of standards. Never use wire without a label showing the AWS code. There are three common wire diameter sizes used in collision repair work: 0.035, 0.030, and 0.023 inch. The latter is preferred for unibody welding.

Automakers, Tech-Cor, I-CAR, and most other testing organizations recommend AWS ER70S-6 wire for MIG welding HSS. Using the recommended wire, or one that exceeds the standards, can help avoid comebacks and possible legal problems.

Most MIG welders used for collision repair are self-contained machines, meaning the power source, wire feed drive motor, controls, and electrode wire are contained in one unit (Figure 7-14).

FIGURE 7-14 A self-contained MIG unit

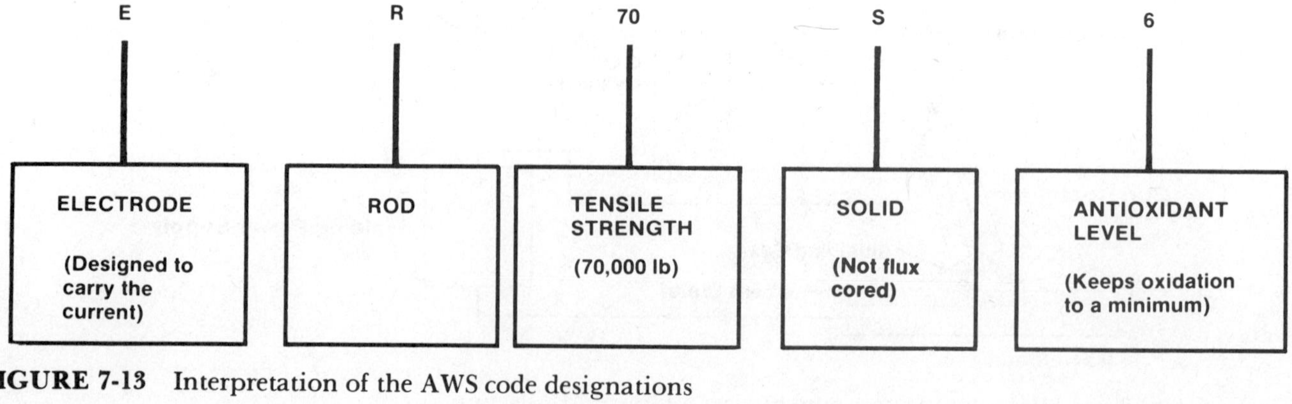

FIGURE 7-13 Interpretation of the AWS code designations

FIGURE 7-15 Removing coatings with a scraper and torch

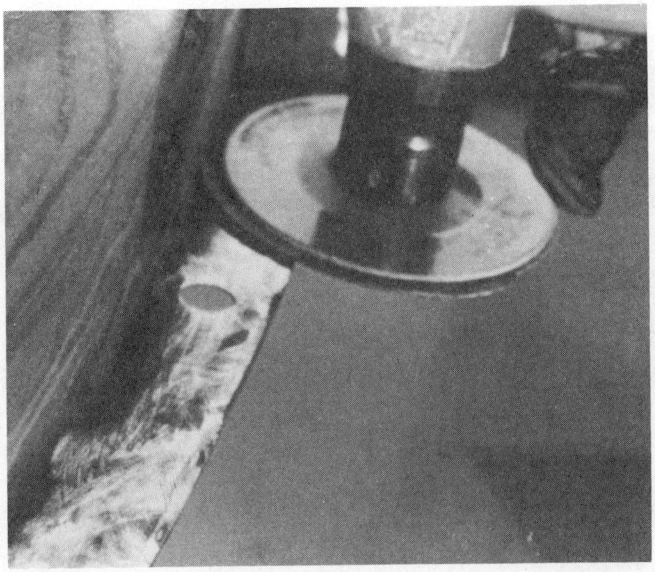

FIGURE 7-16 It is important to make sure that the surfaces to be welded are completely free of rust and scale before starting to weld. One of the most effective ways of doing this is by grinding.

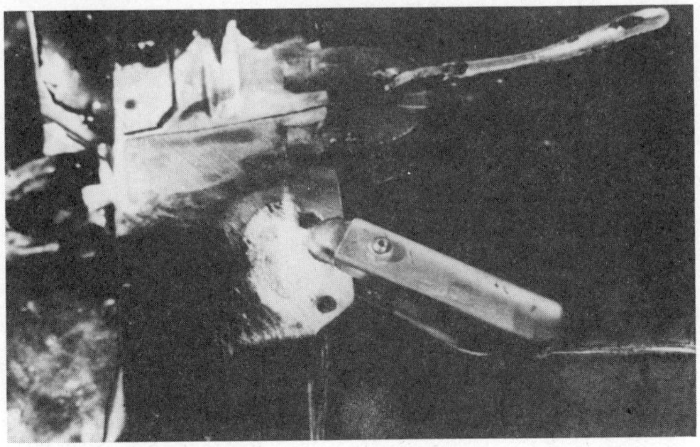

FIGURE 7-17 Attach the work cable clamp to a clean bare metal surface.

WELDING PREPARATION

Before starting to weld, be sure to thoroughly clean the surfaces to be joined. The weld site must be completely free of any foreign material that might contaminate the weld. Otherwise a brittle, porous weld of poor strength will be produced. Use a scraper and a torch (Figure 7-15) to remove heavy undercoatings, rustproofing, tar, caulking and sealants, road dirt and oil, primers, and the like.

SHOP TALK

A small, inexpensive propane torch is the best choice for removing paint and undercoatings from the weld site. The propane torch provides a cooler and broader heat.

Do the finish cleaning with a wire brush and torch. Keep the metal at a very low heat—just enough to get the job done.

Make sure the surfaces to be welded are completely free of rust and scale. This is best removed by grinding (Figure 7-16), sanding, or sandblasting until there is a clean metal welding surface. In addition, attach the work cable clamp to a clean bare metal surface for a trouble-free welding circuit (Figure 7-17).

WARNING

Whenever possible, grind so that sparks fly down and away. Always wear proper eye protection when grinding. Do not grind off galvanized coatings.

Locking jaw pliers (vise grips), C-clamps, sheet metal screws, and special clamps are all necessary tools for welding. Clamping panels together correctly will require close attention to detail (Figure 7-18). A hammer can often be used to fit panels closely together in places that cannot be suitably held in position with clamps (Figure 7-19).

Clamping both sides of a panel is not always possible. In these cases, a simple technique using self-tapping sheet metal screws or rivets can be employed. To hold panels together with sheet metal screws or rivets, the panels must have holes punched or drilled through them. When using the plug weld technique as described later in this chapter, the temporary panels are held in place by setting a screw in every other hole. The empty holes are then plug welded. After the original holes are plug welded, the screws are removed and the remaining holes are then plug welded.

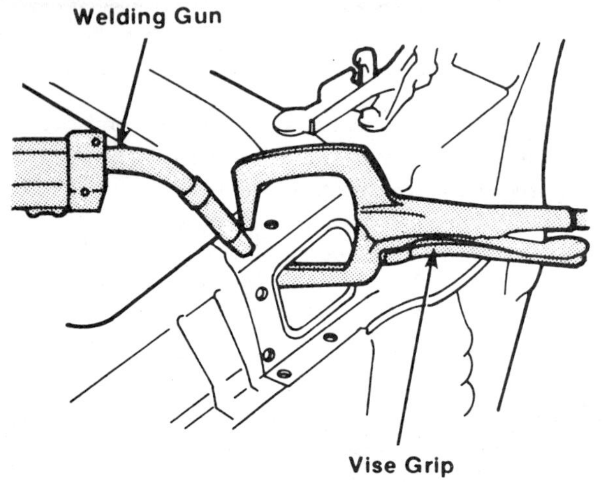

FIGURE 7-18 Clamping a fender apron in preparation for welding

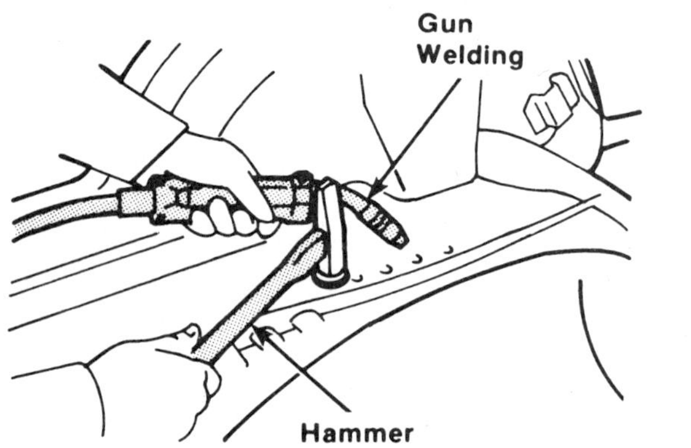

FIGURE 7-19 Using a hammer to hold a floor pan in place for welding

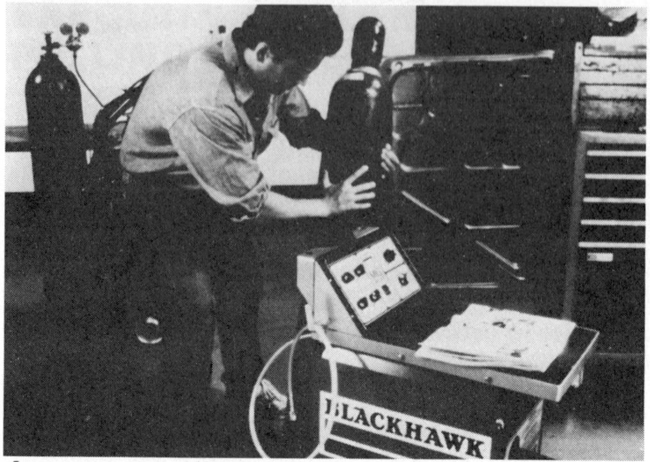

A

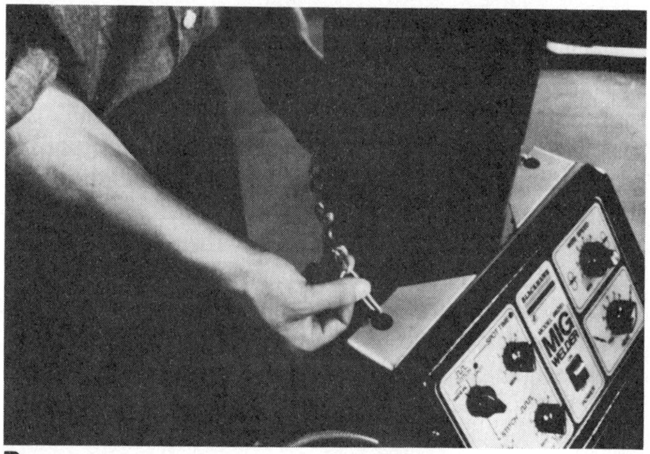

B

FIGURE 7-20 (A) Install the shielding gas cylinder with care and (B) chain or strap it in place.

Fixtures can also be used in some cases to hold panels in proper alignment. Fixtures alone, however, should not be depended upon to maintain tight clamping force at the welded joint. Some additional clamping will be required to make sure that the panels are tight together and not just held in proper alignment.

SHOP TALK

Apply weld-through primer (as an anti-corrosion treatment) to the welding surfaces where the paint film was removed. Remove any excess primer after welding.

MIG WELDER PREPARATION

Match the welding power unit in the MIG machine to the available input voltage, following the procedure outlined on the machine or in the owner's manual. Due to the fact that the input voltage varies in many shops, some manufacturers recommend use of a separate circuit for MIG welding machines.

Handle the shielding gas with care; the cylinder might be pressurized to more than 2000 psi. Chain or strap it to a support sturdy enough to hold it securely to the MIG welder (Figure 7-20) or to a wall, post, or the like. Standing to one side of the cylinder, quickly crack the valve to blow out any dirt before the flow meter or regulator is attached. Install the regulator, making sure to observe the recommended safety cautions. Keep in mind that the regulator must be matched to the type of gas being used. Attach the hose from the regulator to the "gas-in" connection on the welder.

Install the spool of electrode wire on the holder and secure it as directed in the owner's manual (Figure 7-21A). Check the roller size to ensure that it matches the wire size. The liner size should also be checked to be

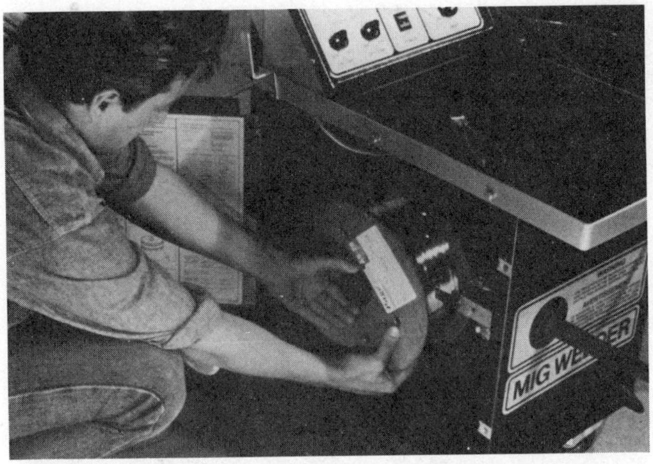

A

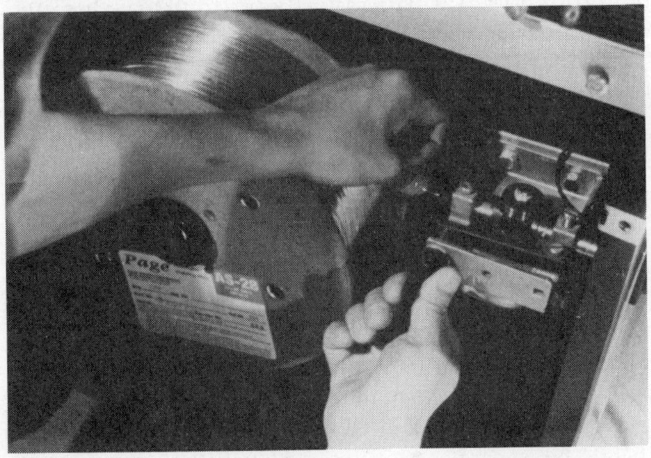

B

FIGURE 7-21 (A) Install the spool of electrode wire, (B) making sure the drive roller grooves, wire guides, cable liner, and gun contact tube correspond with the size of wire being used.

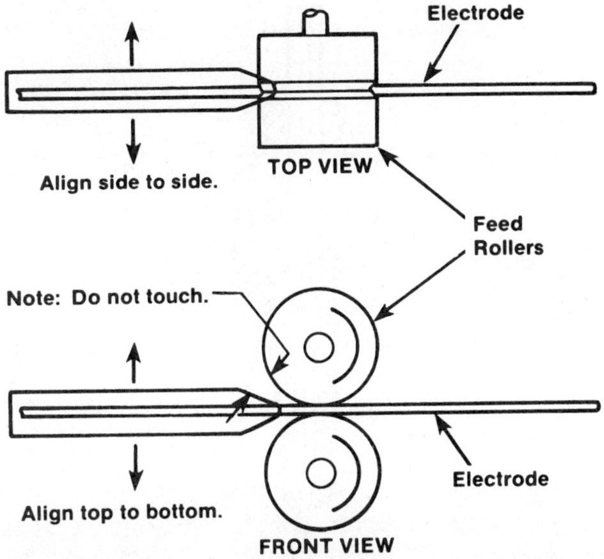

FIGURE 7-22 Feed roller and conduit alignment

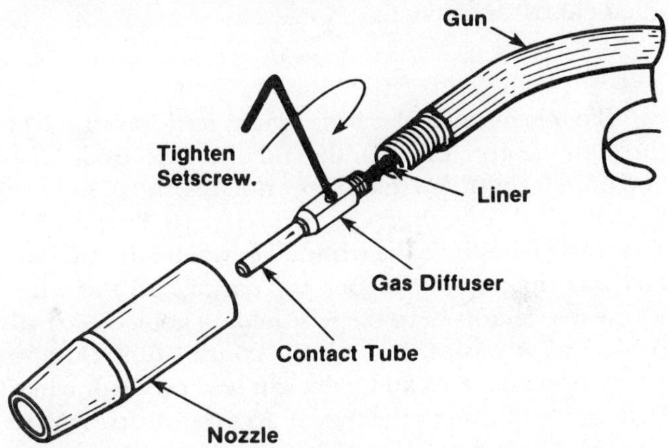

FIGURE 7-23 The welding gun is attached to the main lead cable and liner.

sure that it is compatible with the wire diameter(Figure 7-21B). Connect the conduit to the feed unit. The conduit or an extension should be aligned with the groove in the roller and set as close to the roller as possible without touching (Figure 7-22). Misalignment at this point can contribute to birdnesting.

By attaching the work clamp to clean metal on the vehicle near the weld site, the welding circuit is completed. The work clamp is not referred to as a ground cable or ground clamp. The ground connection is for safety purposes and is usually made from the machine's case to the building ground through the third wire in the electric input cable. Attach the shielding "gas-out" side of the solenoid to the gun lead. If a separate splice is required from the gun switch circuit to the feed unit, it should be connected at this time. Be sure the welding contactor circuit is connected from the feed unit to the power source. The welding gun should be securely attached to the main lead cable and liner (Figure 7-23). There should be a gas diffuser attached to the end of the liner to ensure proper alignment. Install a contact tube of the correct diameter to match the electrode wire diameter, and attach a shielding gas nozzle.

Recheck all fittings and connections for tightness. Loose fittings can leak; loose connections can cause added resistance, thus reducing the welding efficiency.

Next, the MIG wire must be fed through and the wire feeder adjusted. Consult the owner's manual to see that the unit is assembled correctly according to the manufacturer's specifications. Turn on the power and check the gun switch circuit by depressing the switch. The power source relays, feed relays, gas solenoid, and feed motor should all activate.

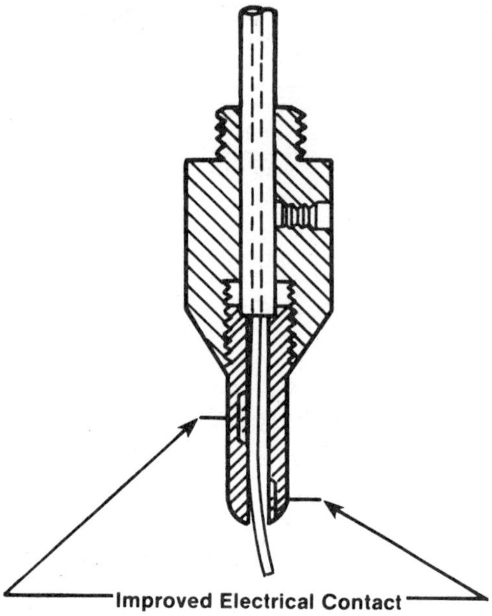

FIGURE 7-24 Cast and helix help the wire make good electrical contact.

To check the electrode wire, feed several feet through the gun, then cut the end of the electrode wire and hold it tightly so that it does not unwind. The wire has a natural curve that is known as its cast. Put another way, cast is the circle the wire makes when it is laid flat. Helix is the amount of flex or "spring" in the wire. These two factors help the wire make a good electrical contact as it passes through the contact tube (Figure 7-24). However, cast and helix can be a problem when feeding wire through the system. As a test, drop the wire on the floor. It should form a circular loop, 15 inches or more across, and lay relatively flat. If the wire forms a tight loop, or looks like a spring, it could cause a jamming problem. To make feeding easier, straighten about 12 inches of the end of the wire and remove any kinks by cutting.

Separate the wire feed rollers and push the wire first through the guides, then between the rollers and into the liner. Reset the rollers so there is a slight amount of compression on the wire by adjusting the wire feed tension. Set the wire speed control to slow speed. Hold the welding gun so that the cable is as straight as possible.

Press the trigger. The wire should start feeding into the liner. Watch to make certain that the wire feeds smoothly, and release the trigger as soon as the end comes through the contact tube. If the wire stops feeding before it reaches the end of the contact tube, stop and check the system. If no obvious problem can be found, mark the wire with tape and remove it from the gun. It then can be held next to the system to determine the location of the problem.

FIGURE 7-25 The tension of the spool spindle should be tight enough to stop the spool from freewheeling when the trigger is released.

Adjust the drive rollers so that they apply just enough pressure on the wire to pull it off the spool and through the gun/cable assembly. The tension must be such that the wire will slip on the rollers when it is stopped at the nozzle, but still tight enough to withstand a 30- to 40-degree deflection. If too much pressure is applied, the wire will be deformed, creating a spiral effect through the liner and an erratic feed.

Stopping the wire at the tip with excessive pressure will also cause the wire to bird-nest between the rollers and cable entrance. The tension on the spool spindle should be set so that the wire can be pulled off easily, yet still be tight enough to stop the spool from freewheeling when the trigger is released (Figure 7-25).

The recommendations for welder maintenance, including cleaning the drive rollers, provided in the owner's manual should be followed.

WELDING CURRENT

The welding current directly affects the base metal penetration depth (Figure 7-26), the speed at which the

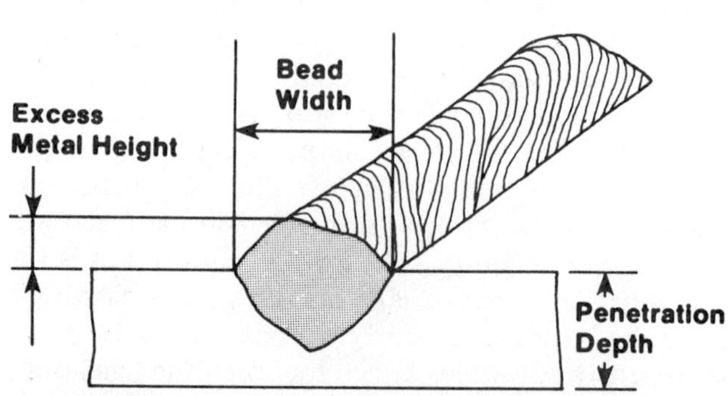

FIGURE 7-26 The base metal penetration depth, bead height and bead width are all affected by the welding current.

TABLE 7-1: RELATIONSHIP BETWEEN WELDING CURRENT, WIRE DIAMETER, AND PANEL THICKNESS

	Panel Thickness						
Wire Diameter	1/64"	1/32"	Less Than 3/64"	3/64"	1/16"	3/32"	1/8"
1/64"	20-30A	30-40A	40-50A	50-60A	—	—	—
1/32"	—	—	40-50A	50-60A	60-90A	100-120A	—
More Than 1/32"	—	—	—	—	60-90A	100-120A	120-150A

wire is melted, arc stability, and the amount of weld spatter. As shown in Table 7-1, as the panel thickness increases, the wire diameter and welding current must likewise be increased.

VOLTAGE

Good welding results depend on a proper arc length. The length of the arc is determined by the voltage. When the voltage is set properly, a continuous light hissing or cracking sound is emitted from the welding area. When the voltage is low, the arc length decreases, penetration is deep, and the bead is narrow and dome shaped.

WIRE STICK-OUT DISTANCE

The wire stick-out distance (Figure 7-27) is also an important factor in obtaining good welding results. The standard distance is approximately 1/4 to 5/8 inch.

If the wire stick-out distance is too long, the length of wire protruding from the end of the gun increases and becomes preheated, which increases the melting speed of the wire. Also, the shielding gas effect will be reduced. If the wire stick-out distance is too short, it is difficult to see the progress of the weld because it will be hidden behind the tip of the gun.

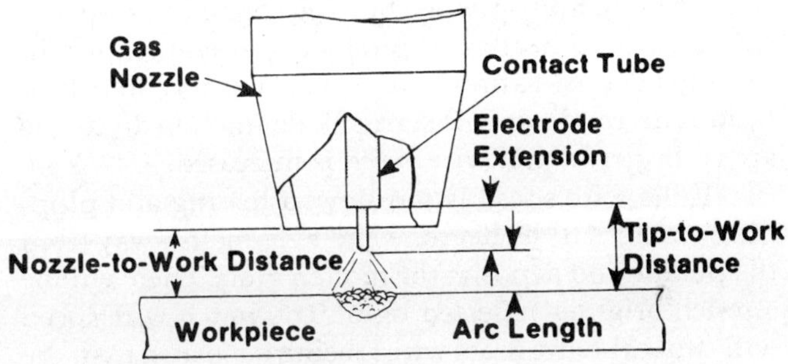

FIGURE 7-27 Wire stick-out distance

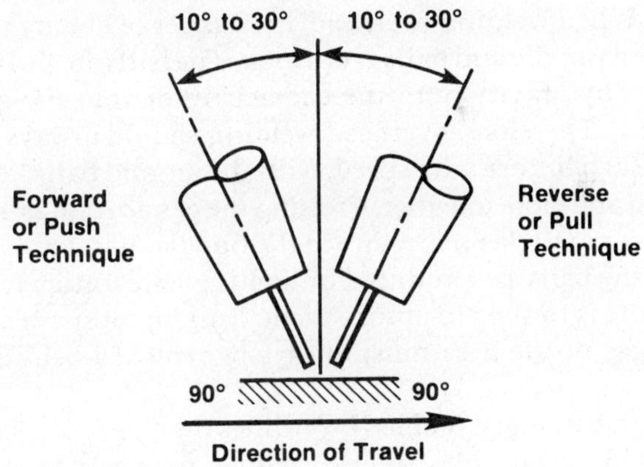

FIGURE 7-28 Proper gun angle and travel direction for both the push and pull methods

GUN ANGLE

There are two basic MIG welding methods: the forward or push method and the reverse or pull method. With the push method, the penetration depth is shallow and the bead is flat. With the pull method, the penetration is deep and a larger amount of metal is deposited. The gun angle for both methods should be between 10 and 30 degrees (Figure 7-28).

SHIELDING GAS FLOW RATE

Precise gas flow is essential for a good weld. If the flow of gas is too high, it will flow in eddies and reduce the shielding effect. If there is not enough gas, the shielding effect will also be reduced. Adjustment is made in accordance with the distance between the nozzle and the base metal, the welding current, travel speed, and welding environment (nearby air currents). The standard flow rate is 25 cubic feet per hour.

TRAVEL SPEED

If the operator welds at a rapid pace, the penetration depth and bead width will decrease, and the bead will be dome shaped. On the other hand, welding at too slow of a travel speed can cause burn-through. Ordinarily, travel speed is determined by the base metal thickness and/or the voltage of the welder (Table 7-2).

TABLE 7-2: WELDING SPEED

Panel Thickness	Welding Speed (in./min.)
1/32"	41-11/32–45-9/32
More Than 1/32"	39-3/8
3/64"	35-7/16–39-3/8
1/16"	31-1/2-33–15/32

WIRE SPEED

An even, high-pitched buzzing sound indicates the correct wire-to-heat ratio producing a temperature in the 9000-degree-Fahrenheit range. A steady, reflected light is desirable; it will start to fade in intensity as the arc is shortened and wire speed is increased.

If the wire speed is too slow, a hissing and plopping sound will be heard as the wire melts away from the puddle and deposits the molten glob. There will be a much brighter reflected light. Too much wire speed will choke the arc; more wire is being deposited than the heat and puddle can absorb. The result is that the wire melts into tiny balls of molten metal that fly away from the weld, creating a strobe light arc effect.

When welding overhead, the danger of having too large a puddle and balls is obvious. The balls are pulled down by gravity onto the contact tip or into the gas nozzle. Therefore, overhead welding should always be done at a higher wire speed, with the arc and balls kept small and close together. Pressing the gas nozzle against the work will keep the wire in the puddle. If it is moved out, the balls are produced by melting wire until a new puddle is formed to absorb them. Buildup of spatter in the gas nozzle area must always be removed before it causes a short circuit.

In summary, Table 7-3 outlines the effects of several welding variables and the changes necessary to alter the characteristics of a MIG weld.

GUN ADJUSTMENT

The nozzles used on MIG welders serve two main functions:

- Provide proper gas protection
- Apply some gun pressure to the work area to prevent the wire from being moved out of the puddle

The basic procedure for adjusting the contact tube to the nozzle is as follows:

- *Arc Generation.* Position the contact tip of the gun near, but not touching, the base metal. When the gun switch is activated, the wire is fed at the same time as the shielding gas. Bring the end of the wire in contact with the base metal to create an arc. If the distance between the tip and the base metal is shortened a little, it will be easier to create the arc (Figure 7-29). If the end of the wire forms a large ball, it will be difficult to generate an arc, so cut off the end of the wire with a pair of wire cutters.

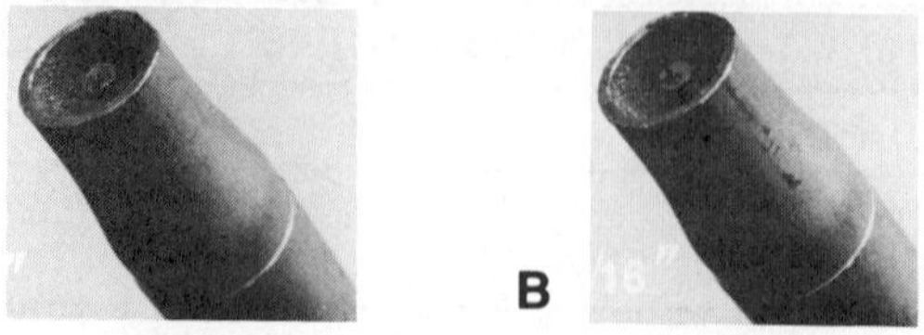

FIGURE 7-29 (A) Typically, the contact tip should be flush with the nozzle to 1/8 inch inside it. (B) The wire should extend 3/16 to 5/16 inch beyond the contact tip.

CAUTION

Hold the tip of the gun away from the face when cutting off the end of the wire.

- *Spatter Treatment.* Remove weld spatter promptly. If it adheres to the nozzle, the shielding gas will not flow properly and a poor weld will result. Weld spatter on the tip will prevent the wire from moving freely. As a general rule, the technician will not need antispatter compounds. The paste type tends to clog the nozzle, which prevents the gas from protecting the weld site. The spray-on type burns off. Even if the spray is reapplied during the welding process, it vaporizes on contact with the hot welding gun and does not have much positive effect. If the wire feed switch is turned on and the wire is not able to move freely through the tip, the wire will become twisted inside the welder. Use a suitable tool, such as a file, to remove spatter from the tip, then check to see that the wire comes out smoothly.
- *Contact Tip Conditions.* To ensure a stable arc, the tip should be replaced if it has become worn. For a good current flow and stable arc, keep the tip properly tightened.

WELDING POSITION

In collision repair, the welding position is usually dictated by the location of the weld on the vehicle. Both the heat and wire speed parameters can be affected by the welding position.

Flat welding is generally easier and faster and allows for the best penetration (Figure 7-30). When weld-

TABLE 7-3: ADJUSTMENTS IN MIG WELDING VARIABLES AND TECHNIQUES

Welding Variables to Change	Desired Changes							
	Penetration		Deposition Rate		Bead Size		Bead Width	
	Increase	Decrease	Increase	Decrease	Increase	Decrease	Increase	Decrease
Current and Wire Feed Speed	Increase	Decrease	Increase	Decrease	Increase	Decrease	No effect	No effect
Voltage	Decrease	Increase	No effect	No effect	Increase	Decrease	Increase	Decrease
Travel Speed	Little effect	Little effect	No effect	No effect	Decrease	Increase	Increase	Decrease
Stickout	Decrease	Increase	Increase	Decrease	Increase	Decrease	Decrease	Increase
Wire Diameter	Decrease	Increase	Decrease	Increase	No effect	No effect	No effect	No effect
Shielding Gas Percent CO_2	Increase	Decrease	No effect	No effect	No effect	No effect	Increase	Decrease
Gun Angle	Backhand to 25°	Forehand	No effect	No effect	No effect	No effect	Backhand	Forehand

FIGURE 7-30 Flat welding position

FIGURE 7-31 Horizontal welding position

ing a part that is off the vehicle, try to place it so that it can be welded in the flat position.

When welding a horizontal joint (Figure 7-31), angle the gun upward to hold the weld puddle in place against the pull of gravity. To weld a vertical joint (Figure 7-32), the best procedure is usually to start the arc at the top of the joint and pull downward.

Overhead welding, as mentioned previously, is the most difficult. In this position (Figure 7-33), the danger of having too large a puddle is obvious; some of the molten metal can fall down into the nozzle, where it can create problems. It can also fall on the operator, so be sure to wear the full complement of safety gear. Always do overhead welding at a lower voltage, while keeping the arc as short as possible and the weld puddle as small

FIGURE 7-32 Vertical welding position

FIGURE 7-33 Overhead welding position

as possible. Press the nozzle against the work to stabilize the gun, and pull the gun along the joint at a steady speed.

BEFORE OPERATING THE WELDER

Be sure the following safety rules are completely understood and practiced while welding:

- *Electric Shock.* Always be aware of the possibility of electric shock. The hazard of electric shock can be reduced by installing the machine according to the manufacturer's instructions, making sure that the cables and wires are in good condition, and by ensuring that all of the connections are in good shape. Never place the machine in a wet place, nor stand in a wet area when welding.
- *Arc Rays.* Arc rays are ultraviolet rays, like those from the sun. Therefore, they have several hazards associated with them. They will produce a burn similar to sunburn on exposed skin. Arc rays can also affect the eyes if a face shield is not worn.
- *Clothing.* Clothing should be tightly woven to prevent sparks from penetrating. Shirt collars and cuffs should be buttoned to prevent sparks from burning the skin. Cuffed pants are not advisable, since the cuffs will catch sparks. Wear the proper protective equipment such as helmet, leather gloves, and sleeve protectors.
- *Eye Protection.* Protecting the eyes is essential. "Flash burn," an eye condition caused by looking at the arc without protective lenses, can be very painful. Severe flash burns can cause permanent eye damage. Safety glasses are an absolute necessity under the welding helmet. Side shields should be used to keep sparks from entering the eye. A welding helmet with the appropriate shaded lens is also a must. The lens will help to protect the entire face from the arc rays and sparks. The choice of shade number will depend upon the welder and the current used. The higher the number, the darker the shade and the greater the protection. Shade numbers range from 2 for torch soldering up to 14 for heavy plasma arc cutting.
- *Fumes.* All welding operations generate fumes. Zinc coatings are especially troublesome. Zinc coatings vaporize during the welding operation and are dangerous to breathe. Adequate ventilation must be provided by a fan or other air circulator, and a source of fresh air is necessary.
- *Gas Cylinders.* Gas cylinders used for welding and cutting are under extreme pressure. Handle them with care. If they are damaged, there is a real danger of explosion. Secure them before use.
- *Matches and Lighters.* Never carry a lighter or matches in pockets while welding. The sparks could cause the lighter to explode or the matches to ignite.

Before operating the MIG welder, be sure to:

- Read the owner's manual carefully. Remember that when in doubt, follow directions.
- Use the proper power source, ideally on a separate circuit.
- Use proper wire and shielding gas.
- Set all drive rollers, cable liners, and tensions to manufacturer's specifications.
- Set the machine to the proper voltage and wire speed suggested for the thickness of the metal.
- Adjust the shielding gas flow rate to the recommended settings.
- Cover upholstery and glass to prevent spark damage. Sparks can be lessened by making sure the welding area is clean, reducing the voltage setting, and making sure the wire stick-out distance is correct.

SHOP TALK

Always disconnect the battery terminals before doing any welding on the vehicle. Also, remove computers and store them outside the immediate welding area. The magnetic fields created by welding machines have been known to affect on-board computers.

To make sure that the MIG machine is correctly adjusted for the specific joint being welded, always do a test weld. Especially when welding a closed section where the backside of the weld cannot be checked, a test weld is the only way to ensure that the welding techniques and machine adjustments will restore the original strength, integrity, and alignment of the panel. Make the test welds on metal that is the same as that of the area of the vehicle to be welded.

SHOP TALK

Allow the weld to cool naturally. Do not use water or compressed air to cool the weld. Forced cooling can cause cracking. If the weld can be touched, it is cool enough.

MIG WELDING TECHNIQUES

There are six basic welding techniques employed with MIG equipment, as shown in Figure 7-34.

TACK WELDS

The tack weld is exactly that: a tack, or relatively small, temporary MIG spot weld that is used instead of a clamp or sheet metal screw to hold the fit-up in place while proceeding to make a permanent weld. And like the clamp or sheet metal screw, a tack weld is always and only a temporary device (Figure 7-35). Tack welds are very important in maintaining proper panel alignment and therefore must be done accurately.

BUTT WELDS

Butt welds are formed by fitting two edges of adjacent panels together and welding along the mating or butting edges. When butt welding, especially on thin panels, it is wise not to weld more than 3/4 of an inch at one time. Closely watch the melting of the panel and welding wire, and the continuity of the bead. At the same time, be sure the end of the wire does not wander away from the butted portion of the panels.

In a continuous butt weld, an uninterrupted seam or bead is laid down in a slow, steady, ongoing movement. The gun must be supported so it does not wobble. Move the gun at a constant speed, looking frequently at

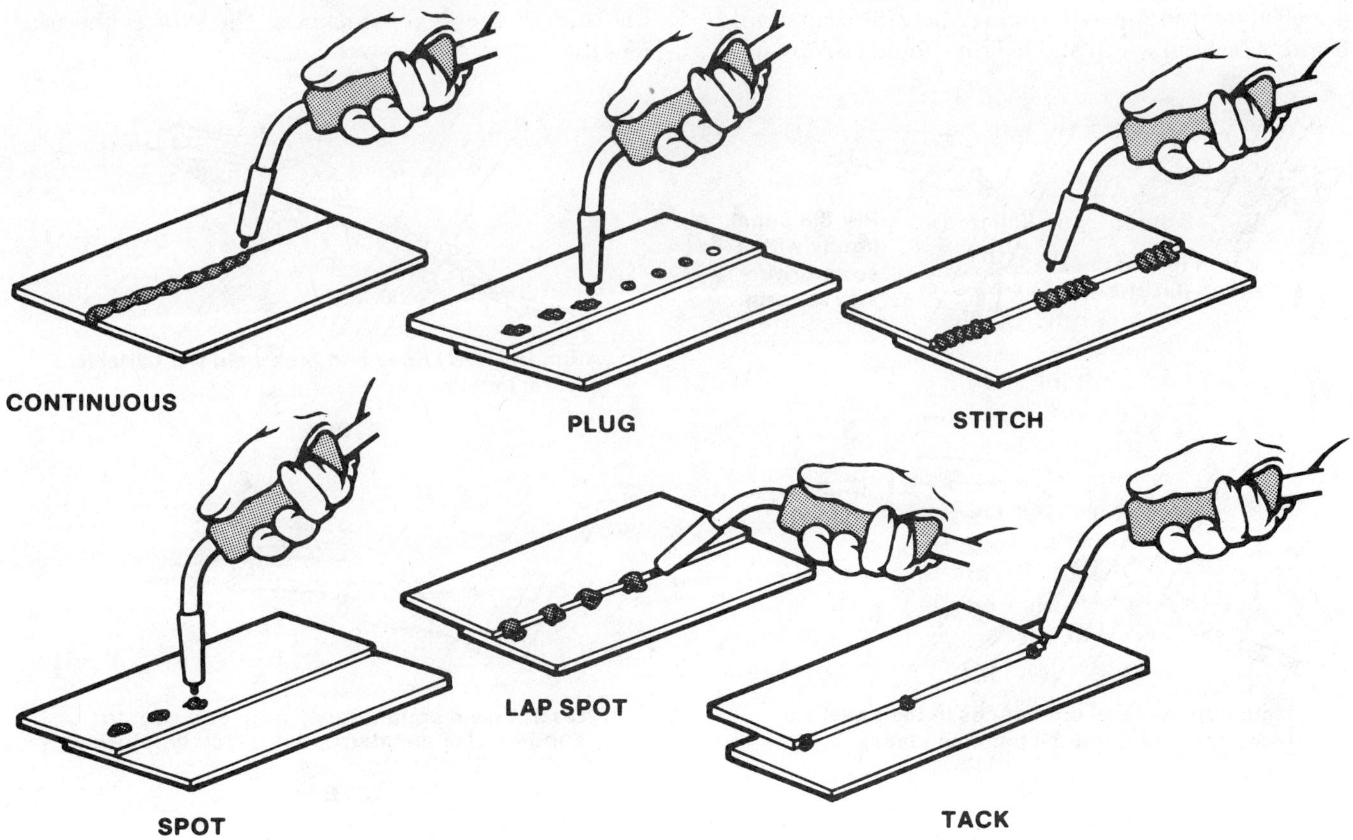

FIGURE 7-34 Basic MIG welding techniques

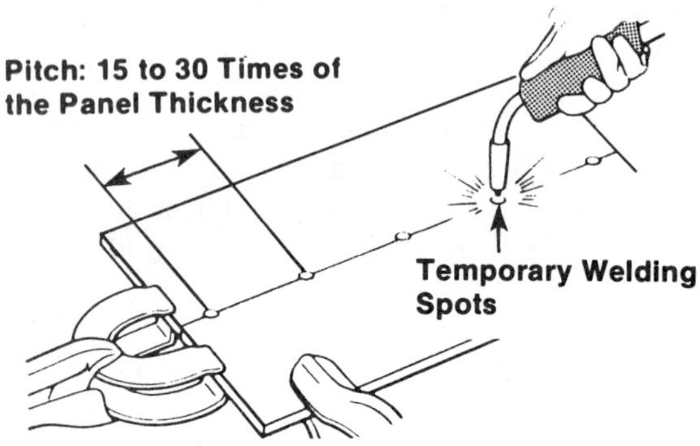

FIGURE 7-35 Temporary or tack welding

the welding bead. The gun should be inclined between 10 to 15 degrees to obtain the optimum bead shape, welding line, and shielding gas effect.

Maintain proper wire stick-out distance and correct gun angle when making a continuous weld. If the gun handling is smooth and even, the bead will be of consistent height and width, with a uniform, closely spaced ripple.

If the butt weld is to be long, it is a good idea to tack weld the panels in several locations to prevent panel warpage (Figure 7-36). Generate an arc a short distance ahead of the point where the weld ends and then immediately move the gun to the point where the bead should begin. The bead width and height should be consistent at this time, with a uniform, closely spaced ripple, provided the gun handling is smooth and even—not too fast and not too slow (Figure 7-37).

When butt welding, a sequence should be established to allow the weld area to cool naturally before the next area is welded (Figure 7-38). To fill the spaces between intermittently placed beads, first grind the beads along the surface of the panel using a sander or grinder, then fill the spaces with weld metal (Figure 7-39). If weld metal is placed without grinding the surface of the beads, blowholes can be produced.

When butt welding panels that are 1/32 of an inch thick or less, an intermittent or stitch welding technique is a must to prevent burn-through. The combination of the proper gun angle and correct cycling techniques will achieve a satisfactory weld bead.

Moving the gun too fast or too slowly will yield poor results (even if the speed of the wire feed is constant). A travel speed that is too slow will cause melt-through. Conversely, a travel speed that is too fast will cause shallow penetration and inferior weld strength.

Even if a proper bead is formed during butt welding, panel warpage can result if the weld is started at or near the edge of the metal. Therefore, to prevent panel warpage, disperse the heat into the base metal by starting the weld in the center and frequently changing the location of the weld area. A good rule of thumb is this: the thinner the panel thickness, the shorter the bead length.

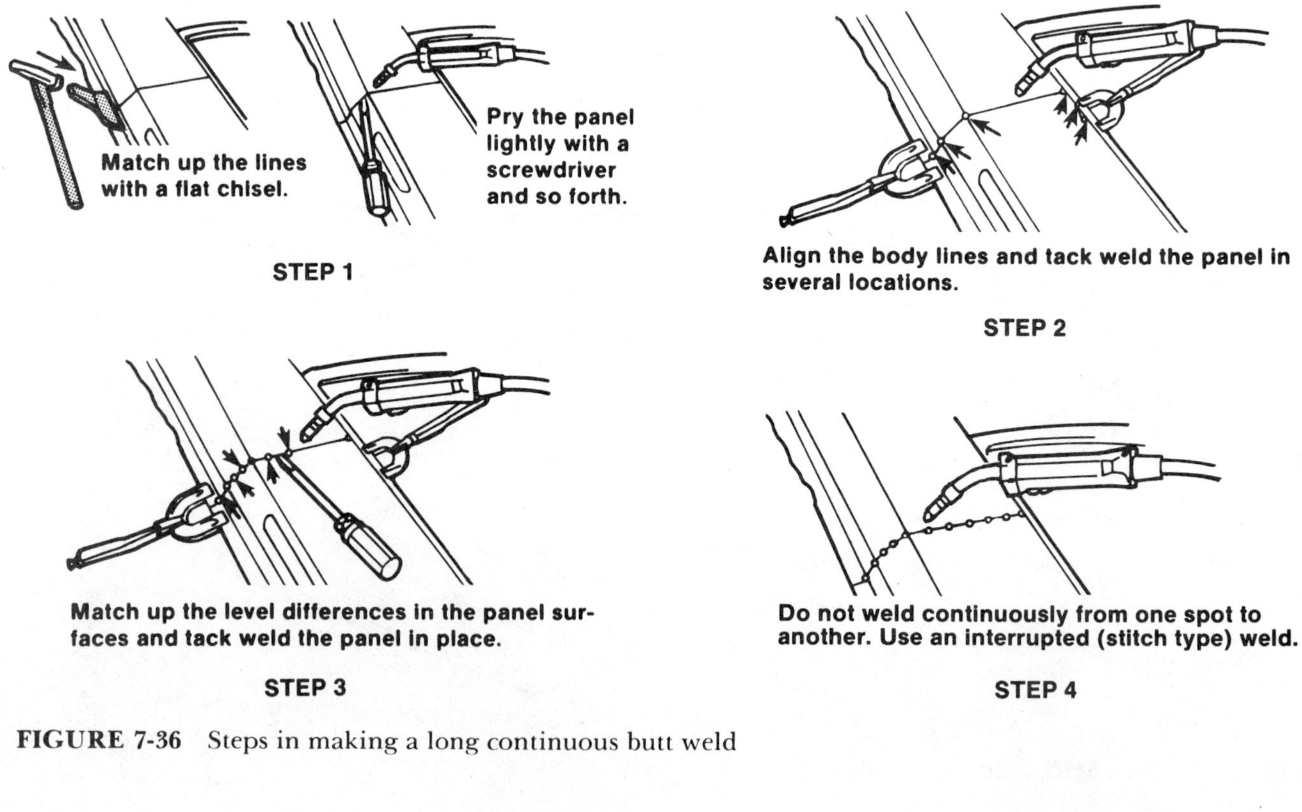

FIGURE 7-36 Steps in making a long continuous butt weld

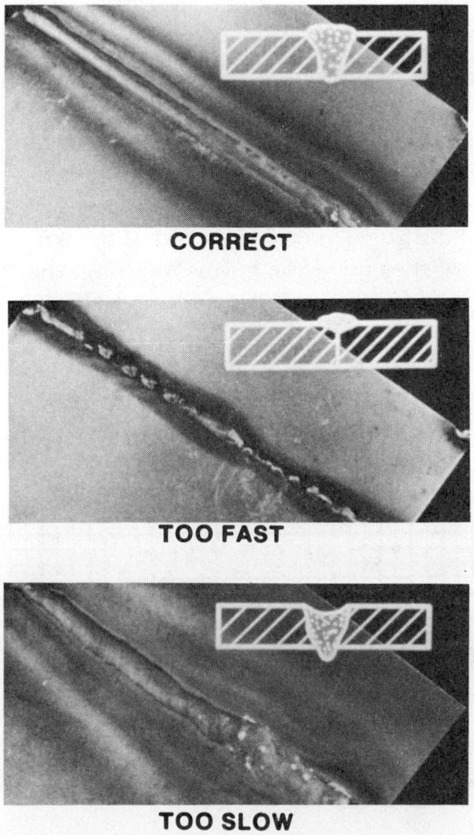

FIGURE 7-37 Gun speed and bead shape

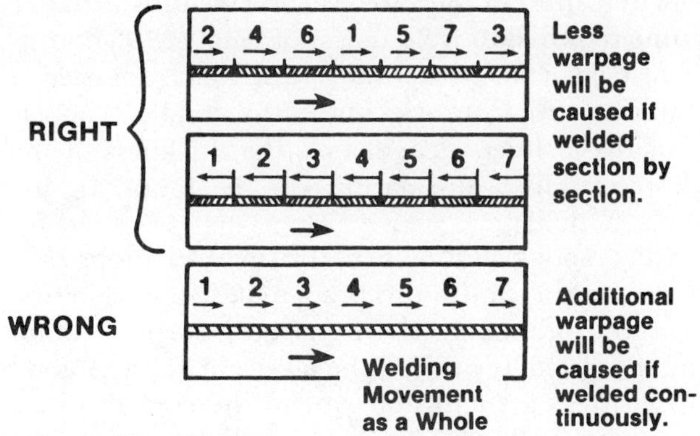

FIGURE 7-38 Right and wrong butt welding sequences

LAP AND FLANGE WELDS

Both lap and flange welds (Figure 7-40) are made with identical MIG techniques. They are formed by melting the two surfaces to be joined at the edge of the top of the two overlapping surfaces. This is similar to butt welding, except only the top surface has an edge. Lap and flange welds should only be made in repairs where they replace identical original factory welds, or

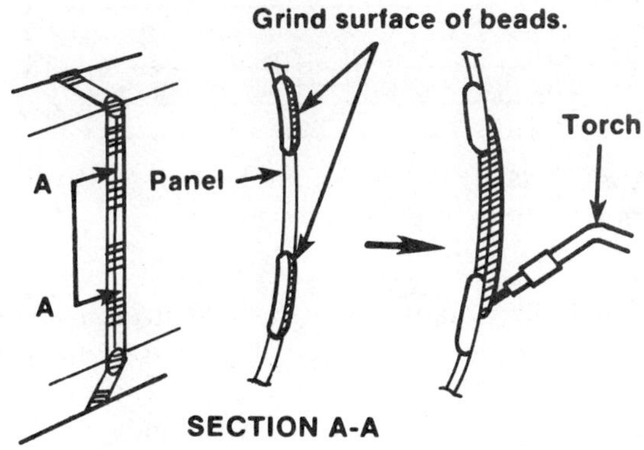

FIGURE 7-39 Filling the spaces between intermittently spaced beads

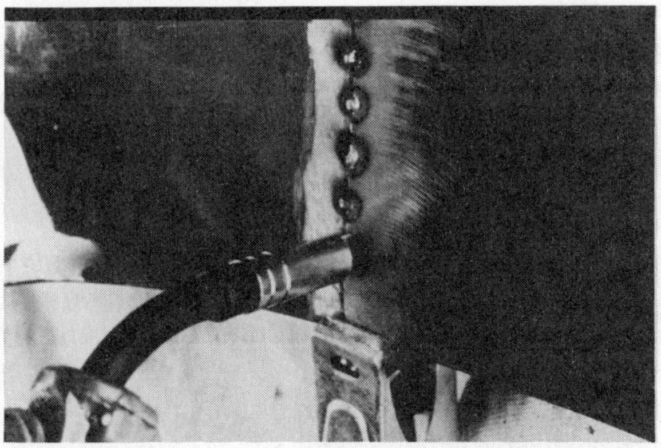

FIGURE 7-40 Welding a lap or flange joint

where outer panels and not structural panels are involved. These welds should not be used to join more than two thicknesses of material together. The same technique used for temperature control in butt welding should be followed for lap and flange welding. Welds should never be made continuously, but should be sequenced to allow for natural cooling; this will prevent temperature buildup in the weld area.

A flat lap or flange joint can be easily welded if some basic measures are taken. When heating the two pieces, caution must be exercised to ensure that they start melting at the same time. Otherwise, the heat is not distributed uniformly in the joint.

Because of this difference in heating rate, the gun must be directed on the bottom piece. The filler metal should be added to the top piece. Gravity will pull the molten puddle down, so it is not necessary to put filler metal on the bottom piece.

PLUG WELDS

The plug weld is the collision repair shop alternative to the OEM resistance spot welds made at the facto-

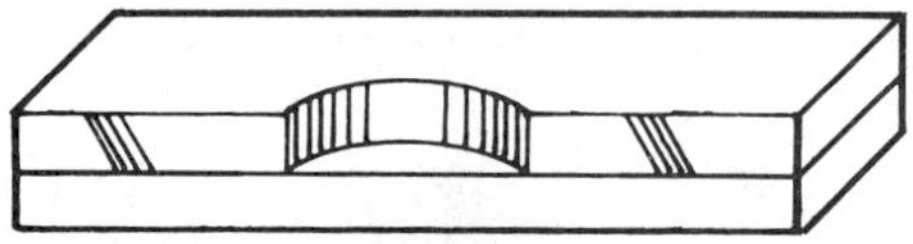

FIGURE 7-41 Plug welds are formed by drilling or punching a hole in the outer panel.

ry. It can be used anywhere in the body structure where a resistance spot weld was used. A plug weld has ample strength for welding load bearing structural members, and it can also be used on cosmetic body panels and other thin-gauge sheet metal.

Plug welding is a form of spot welding; it is basically spot welding through a hole. That is, a plug weld is formed by drilling or punching (Figure 7-41) a hole in the outer panel being joined. The panels should be tightly clamped together. Holding the gun at right angles to the surface, put the electrode wire in the hole, trigger the arc briefly, then release the trigger. The puddle fills the hole and solidifies.

When plug welding, duplicate both the number and the size of the original factory spot welds. The hole that is punched or drilled should not be larger in diameter than the factory weld nugget. The 3/16-inch hole customarily used for plug welding cosmetic panels is not sufficient for plug welding structural members. Most structural members require a 5/16- to 3/8-inch hole to achieve acceptable weld strength. When filling a larger hole, move the gun slowly in a circular motion around the edges of the hole (Figure 7-42A), filling in the cavity. For smaller holes, it is best to aim the gun at the center of the hole and keep it stationary (Figure 7-42B). When the situation calls for plug welding of multi-layered panels, it is a good idea to step out the hole sizes (Figure 7-42C).

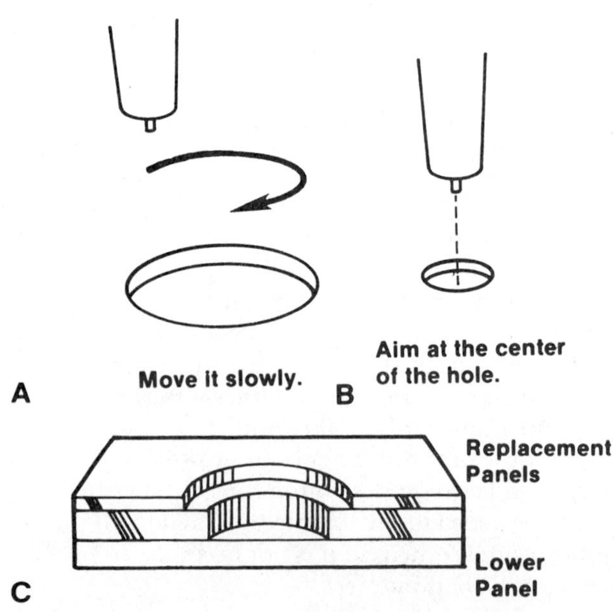

FIGURE 7-42 (A) When making a large plug weld, begin along the edge and work toward the center. (B) When making a small plug weld, gun motion is not usually necessary. (C) For multi-layered panels, the hole sizes should be stepped out.

Proper wire stick-out distance is an important factor in obtaining a good plug weld. If the wire protruding out of the end of the gun is too long, the wire will not melt properly, causing inferior weld penetration. The weld will improve if the gun is held closer to the base metal. Be sure the weld penetrates into the lower panel; round, dome-shaped protrusions on the backside of the metal are good indicators of proper weld penetration.

SHOP TALK

Intermittent plug welding leads to the generation of oxide film on the surface, which in turn causes blowholes. If this occurs, remove the oxide film with a wire brush.

SPOT WELDS

It is possible to spot weld with MIG equipment. In fact, most of the better MIG machines now available and designed for collision repair work have built-in timers that shut off the wire feed and welding arc after the time required to weld one spot. Some MIG equipment also has a burnback time setting that is adjusted to prevent the wire from sticking in the puddle. The setting of these timers depends on the thickness of the workpiece. This information can be found in the owner's manual.

Since spot welding usually involves more heat than running a continuous bead, make a test weld first to select the right welder settings. Check the penetration of the spot weld by pulling the test weld apart. A good weld will tear a small hole out of the bottom piece, while a weak weld will break off at the surface. To get more penetration, increase the weld time and/or heat. To reduce penetration, reduce the weld time and/or heat.

For MIG spot welding, a special nozzle (Figure 7-43) must replace the standard nozzle. Once the gun is in place and the spot weld time, welding heat, and burnback time are set for the given situation, the spot nozzle is held against the weld site and the gun triggered. For a very brief period of time, the timed pulses of wire feed and welding current are activated, and the arc melts through the outer layer and penetrates the inner layer. Then the automatic shutoff goes into action and no matter how long the trigger is squeezed, nothing

FIGURE 7-43 Special MIG spot welding nozzle

FIGURE 7-44 The intermittent cooling and solidifying of the stitch weld makes it preferable to continuous welding on vertical joints.

will happen. However, when the trigger is released and then squeezed again, the next spot pulse is obtained.

STITCH WELDS

In MIG stitch welding, the standard nozzle is used, not the spot nozzle. To make a stitch weld, combine the spot welding process with the continuous welding technique and gun travel. Set the automatic timer—either a shutoff or pulsed interval timer, depending on the MIG machine. The spot weld pulses and shutoffs recur with automatic regularity: weld-stop-weld-stop-weld-stop as long as the trigger is held in.

Another way to look at this is weld-cool-weld-cool-weld-cool and so on, because the arc off period allows the last spot to cool slightly and start to solidify before the next spot is deposited. This intermittent technique means less distortion and less melt-through or burn-through. These characteristics make the stitch weld preferable to the continuous weld for working thin-gauge cosmetic panels.

The intermittent cooling and solidifying of the stitch weld also makes it preferable to continuous welding on vertical joints where distortion is a problem (Figure 7-44). The operator does not have to contend with a continuous puddle that gravity is trying to pull down the joint ahead of the arc. By the same token, stitch welding is also preferable in the overhead position, because there is virtually no puddle for gravity to pull.

MIG WELDING GALVANIZED METALS

Since there is little question as to the superior durability of coated steels when compared to uncoated steels, it is very important that the collision repair technician is able to weld them. But although the name might change (galvanized, zinc-metallized, zinc-coated), the basic problem remains the same: coated steels just do not weld very well. In fact, the weldability is so poor that some technicians grind off the zinc near the weld seam to avoid dealing with it. This results in premature corrosion of the steel unless a weld-through primer is used. For this reason, the galvanized coating should never be removed prior to welding.

DEALING WITH FUMES AND SPATTER

The fumes given off while welding galvanized metals are due to the fact that zinc melts at 786 degrees Fahrenheit and has a boiling point of 1665 degrees Fahrenheit. The melting point of steel is about 2600 degrees Fahrenheit. For this reason, it is easy to understand why welding galvanized metals yield fumes. The zinc melts at a temperature so much lower than that of steel that as soon as the welding arc is initiated, zinc fumes appear in the form of white powdered smoke. Besides being hazardous, these fumes cause the weld puddle to become very erratic, which results in porosity, spatter, and reduced penetration.

The first considerations for anyone welding galvanized metals are safety and health. Inhaling substantial amounts of zinc fumes will make a person sick. Always provide adequate ventilation and wear a respirator. A small fan set off to one side will help clear out the fumes. Do not direct the fan right at the work area or a porous, weak weld could result.

ERRATIC ARCS

As already noted, zinc melts at a lower temperature than steel, and this causes white smoke or residue. When the zinc fumes become vaporized in the heat of the arc, the weld puddle becomes quite erratic. It is very hard trying to weld thin-gauge metals with an arc that seems to have a mind of its own, but there are some things the technician can do to stabilize the arc.

First, increase the amperage by five or ten amps and increase the voltage by one volt. On most MIG machines, the increased amperage means an increase in wire speed of two or three marks on the wire speed adjustment dial. This change in machine setting to a "hotter" arc will help vaporize the zinc fumes somewhat.

Next, move the gun a bit slower than normal. This will give the leading edge of the weld puddle time to vaporize the zinc before the bead actually passes over that area. Pulling the gun concentrates the heat of the arc, which aids in burning out the zinc. Pushing the gun seems to spread the heat of the arc sideways, which vaporizes more zinc than necessary.

HEAT INPUT

Because extra heat is recommended for welding galvanized metals, heat input is very important. Heat input is directly related to the machine settings, travel speed, and how quickly the welds must be made to complete a repair. Because the hotter settings and slow travel speed work better on galvanized metals, use a stitch welding technique, weld short beads (no more than 1-1/2 inches at a time), and pause several minutes between welds to allow the heat to dissipate before continuing.

Penetration is reduced slightly when welding galvanized metals. Wider gaps are usually required for butt joints, which makes penetration control difficult. For this reason, the I-CAR procedure, which uses a backup piece when making butt joints, is the ideal way to weld these joints.

FIGURE 7-45 When welding galvanized metals, use a slower gun travel speed.

DEALING WITH POROSITY

One of the biggest problems with welding coated steels is porosity. If the holes are no larger than 1/32 inch in diameter and are no more frequent than one hole for every 1 inch of weld, porosity will probably not impair the quality of the weld or the strength of the joint. But if there are three or four holes for every inch of weld, grind them out and reweld the bad portion. The extent of porosity will depend on amperage, voltage, travel speed, shielding gas, and thickness of the coating. Remember to increase amperage slightly, increase the voltage by one volt, and travel slower when welding coated steels (Figure 7-45).

MIG WELDING ALUMINUM

It was once thought that aluminum would become widely used in automotive construction because of its light weight and corrosion resistance. However, the cost of aluminum has prevented this. Only a few production passenger cars currently feature all-aluminum construction.

Still, collision repair shops are seeing more aluminum transmission cases, bell housings, alternator and starter housings, and mounting brackets. Often, an ear or lug on such parts will be broken or cracked. Because shops do not often get to practice welding aluminum, they automatically either replace the damaged part or send it out to a welding specialist instead of doing it themselves. However, most shops can MIG weld aluminum. All it takes is a gun and liner designed for aluminum wire, a spool of aluminum wire, argon shielding gas, and a little practice.

First, it is important to be able to tell the difference between magnesium and aluminum. Although there is only a small chance of running into a magnesium part, the risk of starting a magnesium fire is serious enough to make it crucial to be absolutely sure. Just brush the part with a stainless steel wire brush. Aluminum will turn shiny; magnesium will turn a dull gray.

There are some major differences to keep in mind when working with aluminum as opposed to steel, particularly when it comes to welding. Pure aluminum is lightweight and useful more for its ability to be formed than for its strength. When used on vehicles, it is alloyed with other elements and heat-treated for additional strength. Pound for pound, aluminum is a better conductor of electricity. It is also an excellent conductor of heat—about three times that of steel—and is resistant to low temperatures. Even in extreme cold, it gets stronger, rather than becoming brittle. Aluminum cannot be magnetized, and most aluminum alloy combinations are easy to weld.

Aluminum's natural resistance to corrosion is a special advantage. When first exposed to air, a thin oxide film forms on the surface; aluminum oxide serves as protection and prevents further corrosion. This property varies depending on the type of alloy and whether or not it is heat-treated.

Because aluminum conducts heat so well, it is most suitable for MIG welding. With MIG, it is easier to produce good results than with other welding methods. MIG also causes its own cleaning action. When the gun is pushed, the arc cleans the area in front of the gun. Pushing, in effect, scrubs away the aluminum oxide on the surface at the site of the weld.

Use the following guidelines when MIG welding aluminum as compared to steel:

- Use aluminum wire and 100 percent argon shielding gas.
- Set the wire speed faster than with steel.
- Hold the gun closer to vertical when welding aluminum. Tilt it only about 5 to 10 degrees from the vertical in the direction of the weld.
- Use only the forward welding method with aluminum. Always push—never pull. When making a vertical weld, start at the bottom and work up.
- Set the tension of the wire drive roller lower to prevent twisting. But do not lower the tension too much or the wire speed will not be constant.
- Use about 50 percent more shielding gas with aluminum.
- Because there tends to be more spatter with aluminum, use an antispatter compound to control buildup at the end of the nozzle and contact tip.

A noticeable difference with welding aluminum is that it does not change color when heated, so it is nearly impossible to judge the exact melting point. Cleaning the weld site is important. Use a stainless steel brush to remove the aluminum oxide. All dirt must be removed before welding, along with any oil or grease. Tools such as stainless steel brushes and sanding discs must be kept separate from those used on steel. If the same tools are used for both, they could carry contaminants from one piece to another and imbed the contaminants in the aluminum.

Besides the regular welding safety gear, extra protection is needed for aluminum. This is because the lightweight spatter can fly farther than steel. Use long-sleeved overalls, foot protectors, a dustproof mask, goggles, and apron. It is wise to cover exposed areas of the vehicle as well.

The typical procedure for MIG welding aluminum is as follows:

1. Clean the weld area completely, both front and back, using wax and grease remover and a clean rag.
2. If the pieces to be welded are coated with a paint film, sand a strip about 3/4 inch wide to the bare metal, using a disc sander and No. 80 disc. Do not press too hard, or the sander will heat up and peel off aluminum particles, clogging the paper.
3. Clean the metal until shiny with a stainless steel wire brush.
4. Load 0.030 aluminum wire into the welder. Trigger it to extend about an inch beyond the nozzle.
5. Set the voltage and wire speed according to the instructions supplied with the welding machine. Remember that the wire speed must be faster than for steel.
6. Nip off the end of the wire to remove the meltdown.
7. Position the two pieces together and lay a bead along the entire joint. The distance between the contact tip and the weld should be 5/16 to 9/16 inch.
8. If the arc is too large, turn down the voltage and increase the wire speed. The bead should be uniform on top, with even penetration on the backside.

When doing continuous welding on any material, the heat can cause warping of the base metal. The high heat conduction of aluminum means that the technician must protect against warpage. There are two methods for doing this: stitch welding and center out welding (Figure 7-46). In the latter, a small distance at a time is welded in each direction.

Parts made of aluminum are usually 1-1/2 to 2 times as thick as steel parts. When damaged, aluminum feels harder or stiffer to the touch because of work hardening. There is one major difference with straightening aluminum. With steel, the use of heat is avoided whenever possible to prevent reducing the strength of the metal. With aluminum, heat must be used to restore flexibility caused by work hardening. If not, it will crack when straightening force is applied.

Before straightening, heat the damaged aluminum. It is easy to apply too much heat since aluminum does not change color with high temperatures. It melts at a relatively low 1220 degrees Fahrenheit, so careful heat control is very important. A temperature-sensitive heat crayon comes in handy for this.

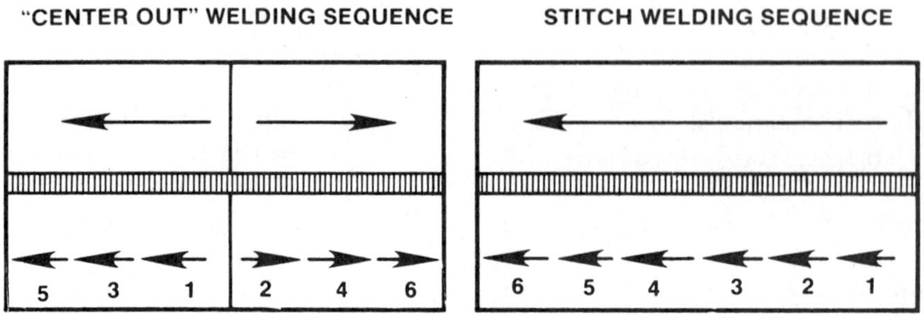

FIGURE 7-46 Two methods of preventing warpage of aluminum

Straightening using a hammer and dolly is basically the same with aluminum as for steel, with a few exceptions. Aluminum alloys bend too quickly when the panel is sandwiched between the hammer and dolly, as with the hammer-on-dolly method. For this reason, use only the hammer-off-dolly method. Also, the large body hammers used for steel should not be used because they could cause cracking.

FLUX-CORED ARC WELDING

Flux-cored arc welding (FCAW) is an electric arc welding process that uses a tubular wire with flux inside. With the development of 0.030 self-shielded flux-cored wire, this process has proven to be valuable for work on high-strength steel (coated or uncoated). FCAW uses the same power source as MIG. It also uses the electrode feed system, contact tip, electrode conduit, welding gun, and many other pieces of MIG equipment (Figure 7-47). Nevertheless, the process itself differs somewhat from MIG (Figure 7-48).

There is no external shielding gas in FCAW. As the flux within the wire melts in the heat of the arc, the created gases shield the weld puddle, stabilize the arc, help to control penetration, and reduce porosity. The melted flux also mixes with the impurities on the metal surface and brings them to the top of the weld, where they solidify as slag. The slag can then be chipped or brushed away.

Two very important advantages of the FCAW process over MIG are its abilities to tolerate surface impurities (thus, it requires less precleaning) and to stabilize the arc. Other beneficial characteristics of the process include the following:

- High deposition rate
- Low voltage requirement
- Efficient electrode use
- Requires little edge preparation

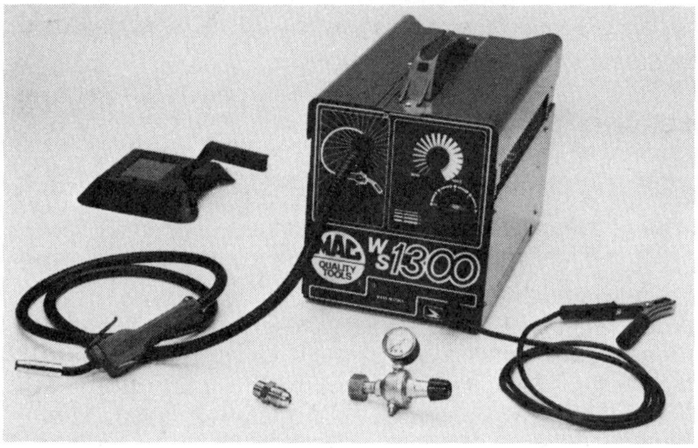

FIGURE 7-47 Small welders are suitable for flux-cored arc welding.

- Welds in any position
- Welds a wide range of metal thicknesses with one size of electrode
- Produces high-quality welds
- Weld puddle is easily controlled and its surface appearance is smooth and uniform, even with minimal operator skill
- Produces a weld with less porosity than MIG when welding galvanized metals

SHOP TALK

If the nozzle is removed when using self-shielded wires, visibility is improved.

While the FCAW process has a number of advantages over MIG, it has the following drawbacks:

- FCAW wires are more expensive per pound than MIG wires. However, the cost is quickly recovered through higher productivity.
- The 0.030 flux-cored wire contains fluoride compounds that require adequate ventilation at all times during the welding process.

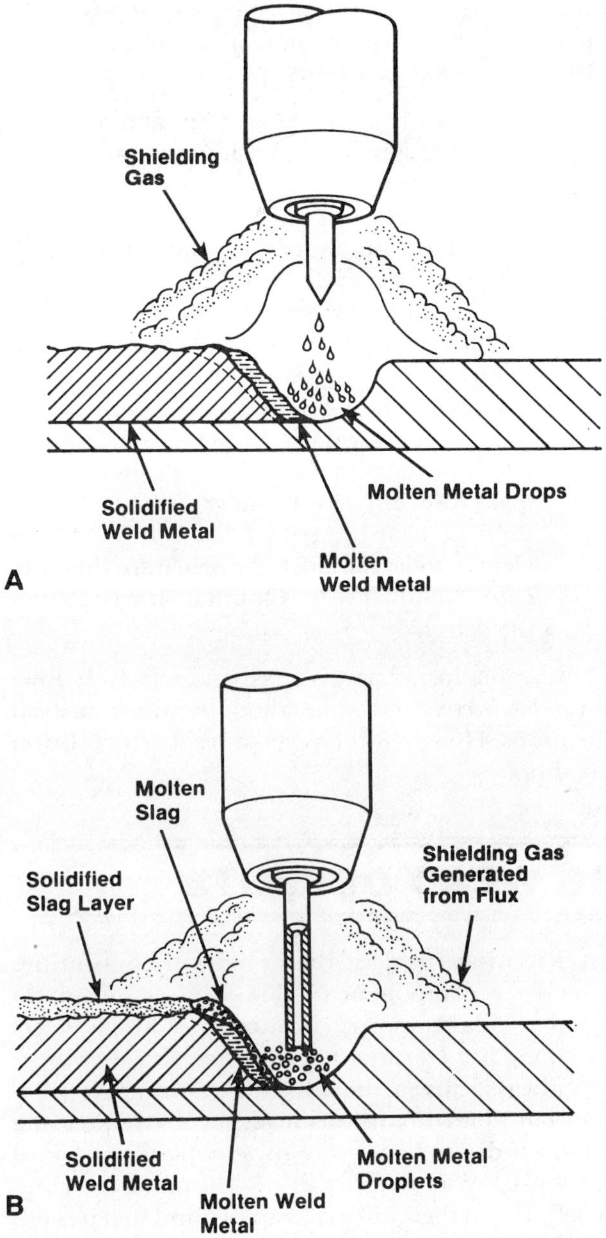

FIGURE 7-48 Differences between (A) MIG and (B) FCAW

- The flux from the wire changes to slag as it cools. Until it does cool, the slag is sharp and hot and should be considered an eye and skin hazard. Once it cools, this slag must be removed prior to the application of fillers, seam sealers, primers, or paint.
- Spatter is worse when using flux-cored wires. Use nozzle gel and keep the nozzle scraped clean. Spatter buildup in the gun nozzle can jam the wire in the contact tip; it can also fall off during welding and mix with the molten puddle, diminishing the quality of the weld.
- Only ferrous metals can be welded with FCAW.
- Wire feed problems for FCAW are similar to those encountered with MIG, but with one major difference. Because flux-cored wires are not solid but tubular, excessive tension on the drive rollers or using the incorrect style of drive rollers might collapse the wire; this leads to feeding problems. Check the owner's manual for correct drive rollers and tension requirements for flux-cored wire.

If a machine is used for both MIG and FCAW, the welder must have polarity switching capabilities. FCAW with 0.030 or 0.035 inch wire uses straight polarity, while 0.023, 0.030, and 0.035 inch wires for MIG use DC reverse polarity. Many of the MIG welding machines sold over the past few years were originally designed to run DC reverse. Without going inside the machine to change polarities, which is difficult and time consuming, this type of machine will not run straight polarity. Check the owner's manual for polarity reversing capabilities.

PERFORM TEST WELDS

For all of the welding procedures described in this chapter, some test welds should be fabricated to determine weld quality and strength before welding on the vehicle. Test welds should be made using scraps of similar metals. This should be done several times during the course of a repair. Testing will help determine if adjustments in the equipment are required, such as voltage, wire feed rate, wire size and type, or dressing of the MIG gun nozzle.

DESTRUCTIVE TESTING

Most of the destructive testing methods used in industrial welding situations are simply not feasible in an automotive application. Tensile, fatigue, and impact testing are more suited to evaluating welds on small, individual specimens; they are of little use on vehicles weighing more than 2000 pounds. Probably the simplest, yet most effective, destructive test is to try to break the weld apart using a hammer and chisel. If reasonable force does not harm it, the weld integrity is good (Figure 7-49).

Other destructive tests include trying to bend or twist the pieces apart and checking for cracks or voids in the weld. When attempting to tear the pieces apart, keep in mind that good welds will be revealed by the fact that the sheet metal around the weld tears away. The weld itself should not tear apart. If it does, the weld is not a good one, and the welder needs to be readjusted.

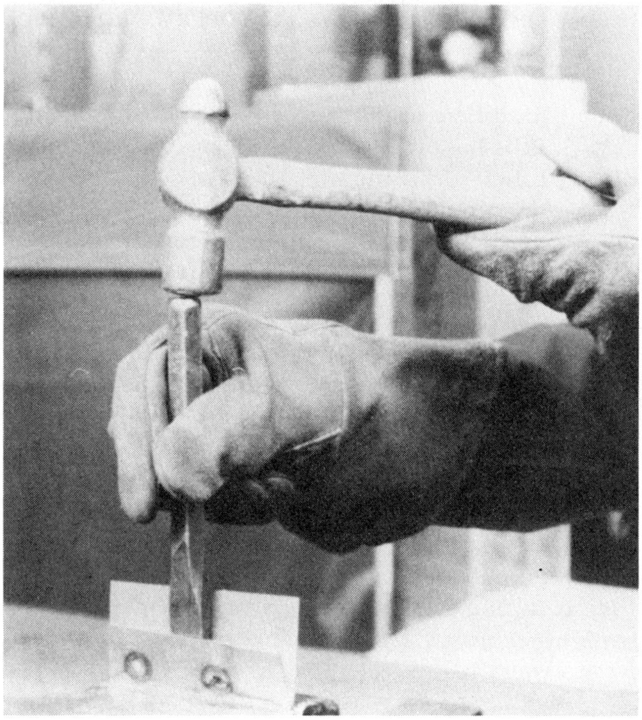

FIGURE 7-49 Destructive testing of a MIG weld

NONDESTRUCTIVE TESTING

Nondestructive testing is an effective method of checking for surface defects, including cracks, undercuts, arc burns, and lack of penetration. The methods of nondestructive testing that will be discussed here are visual inspection and penetrant inspection.

Visual Inspection

Visual inspection is the most frequently used nondestructive testing method. Many times, a weld receives only a visual inspection. If the weld looks good, it passes; if it looks bad, it is rejected. This procedure can be easily overlooked when more sophisticated nondestructive testing methods are used, but it should not be. In fact, a visual inspection should be done before any other tests to eliminate the obvious problem welds. A visual inspection can reduce the finished weld rejection rate by more than 75 percent. Minor problems can be identified and corrected before the weld is completed, thus eliminating costly repairs or rejection.

Penetrant Inspection

Penetrant inspection is used to locate very small surface cracks, as well as porosity. There are two types of penetrants now in use: the color-contrast and the fluorescent versions. Color-contrast penetrants contain a colored dye that is visible under ordinary white light. Fluorescent penetrants contain a more effective fluorescent dye that is only visible under black light.

Use a penetrant as follows:

1. The test surface must be clean and dry. Any flaws must be free of oil, water, and other contaminants.
2. Cover the test surface with a film of penetrant. This can be accomplished by dipping, bathing, spraying, or brushing.
3. Wipe, wash, or rinse the test surface free of excess penetrant. Dry with cloths or hot air.
4. Apply developing powder to the test surface. It will act like a blotter and speed the penetrant out of any flaws on the test surface.
5. Depending on the type of penetrant applied, make the inspection under ordinary white light or near-ultraviolet black light. When viewed under the latter, the penetrant fluoresces to a yellow-green color that clearly defines the defect.

Ultrasonic inspection is a fast and relatively low-cost nondestructive testing method used in industrial applications. However, it has yet to reach the collision repair shop.

MIG WELD DEFECTS

A defect is defined as "one or more discontinuities that render a component unable to meet minimum accepted standards or specifications." While a defect may or may not be cause for rejection of a weld, it is almost always cause for repair. It is most important to become familiar with the various types of MIG welding defects in order to better recognize, repair, and, most importantly, avoid them. Table 7-4 summarizes common defects and their causes. Keep in mind that it is not intended to be all-inclusive; there are others not mentioned here, and those that are mentioned will not always conform to these exact descriptions. When defects do occur, always think of ways that the method of operation can be changed to prevent making the same mistake again.

SQUEEZE-TYPE RESISTANCE SPOT WELDING

Although squeeze-type resistance spot welding has been used in the European and Japanese unibody colli-

TABLE 7-4: MIG WELDING DEFECTS AND THEIR PROBABLE CAUSES

Defect	Defect Condition	Remarks	Probable Causes
Pores/Pits (Porosity)	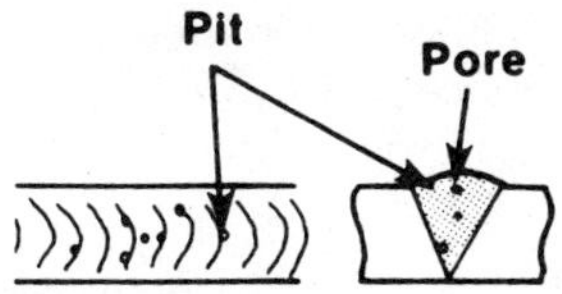	Holes form when gas is trapped in the weld metal.	Rust or dirt on the base metal. Rust or moisture adhering to the wire. Improper shielding action (the nozzle is blocked or the gas flow volume is low). Weld is cooling off too fast. Arc length is too long. Wrong wire used. Gas is sealed improperly. Weld joint surface is not clean. Electrode contamination.
Undercut	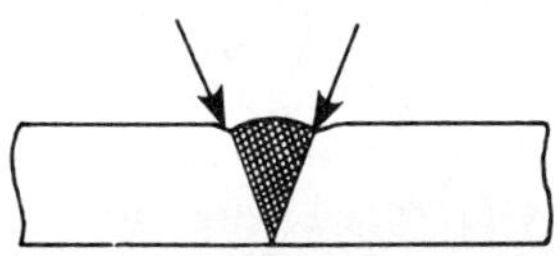	The overmelted base metal creates grooves or an indentation. The base metal's section is made smaller and, therefore, the weld zone's strength is severely lowered.	Arc length is too long. Improper gun angle. Welding speed is too fast. Current is too high. Torch feed is too fast. Torch angle is tilted.
Improper Fusion	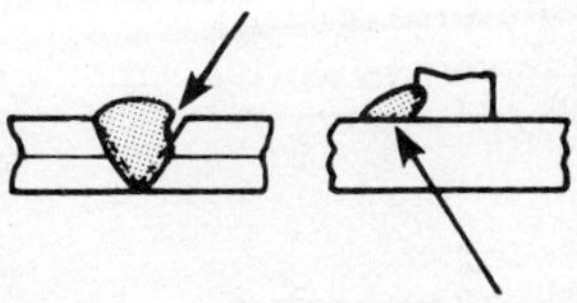	An absence of fusion between weld metal and base metal or between deposited metals.	Improper torch feed operation. Voltage is too low. Weld area is not clean. Improper gun angle.
Overlap	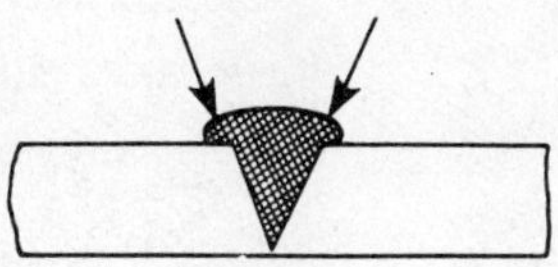	Apt to occur in a fillet weld rather than a butt weld. Causes stress concentration and results in premature corrosion.	Welding speed is too slow. Arc length is too short. Torch feed is too slow. Current is too low.
Insufficient Penetration	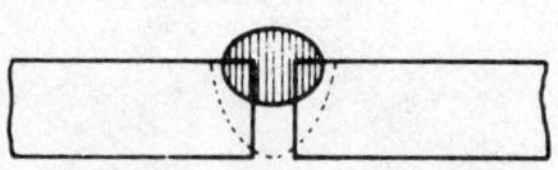	Insufficient deposition made under the panel.	Welding current is too low. Arc length is too long. The end of the wire is not aligned with the butted portion of the panels. Groove face is too small. Gun speed is too fast. Improper electrode extension.
Excess Weld Spatter	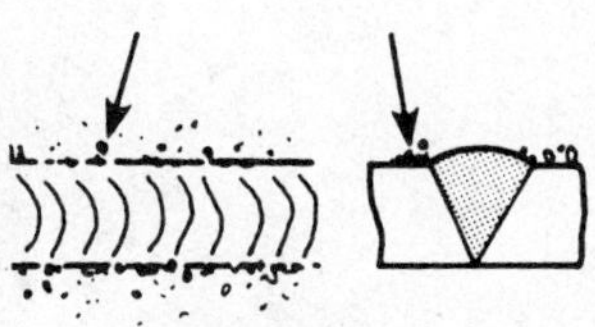	Shows up as speckles and and bumps along either side of the weld bead.	Arc length is too long. Rust on the base metal. Gun angle is too severe. Worn tube or drive rolls. Wire spool too tight. Voltage is too high. Wire speed too slow.
Spatter (short throat)		Prone to occur most often in fillet welds.	Current is too high. Wrong wire used.
Waviness of Bead		Uneven bead pattern usually caused by torch mishandling.	Wavering hands. Electrode extension too long.

TABLE 7-4: MIG WELDING DEFECTS AND THEIR PROBABLE CAUSES (CONTINUED)

Defect	Defect Condition	Remarks	Probable Causes
Cracks	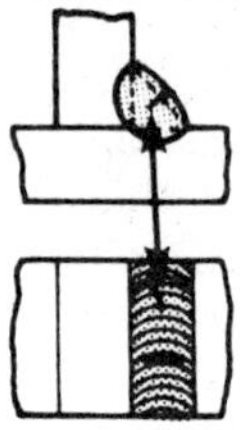	Usually occur on the top surface only.	Stains on welded surface (paint, oil, rust, etc.). Voltage is too high. Wire speed is too fast. Gun speed is too slow.
The Bead Is Not Uniform.		The weld bead is misshapen and uneven rather than streamlined and even.	The contact tip hole is worn or deformed and the wire is oscillating as it comes out of the tip. The gun is not held steady during welding.
Burn Through	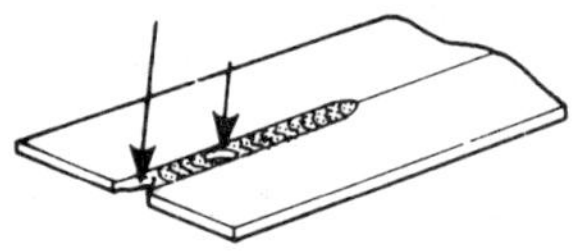	Holes in the weld bead.	The welding current is too high. The gap between the metal is too wide. Gun speed is too slow. The gun-to-base metal distance is too short. Reverse polarity to straight polarity (gun negative, work positive).

sion repair industry for more than 25 years, it is a relatively new process in this country. This is regrettable since the squeeze-type resistance spot welder (Figure 7-50) is ideal for welding many of the unibody's thin-gauge sections that call for good strength and no distortion. Since the early 1980s, a growing number of manufacturers have published service manuals that specify the use of squeeze-type resistance spot welding to repair many areas of their unibody vehicles. These areas include rocker panels, radiator core supports, window and door opening flanges, roofs, many exterior panels, and pinchweld areas (Figure 7-51).

On thin material like the 22- and 24-gauge sheet metal used on today's vehicles, squeeze-type resistance spot welding has the following advantages:

- Uses half as much heat as MIG, which means less finishing time because of less heat distortion
- Finishing time is further reduced because there is no weld metal to grind down
- Duplicates OEM factory weld appearance
- Clean; no smoke or fumes
- Allows use of weld-through conductive zinc primers to restore corrosion protection
- Fast weld times of 1/2 second or less, which produce strong welds on HSS steel, HSLA steel, and mild steels
- Low initial cost and no consumables such as wire, rods, or gas are required

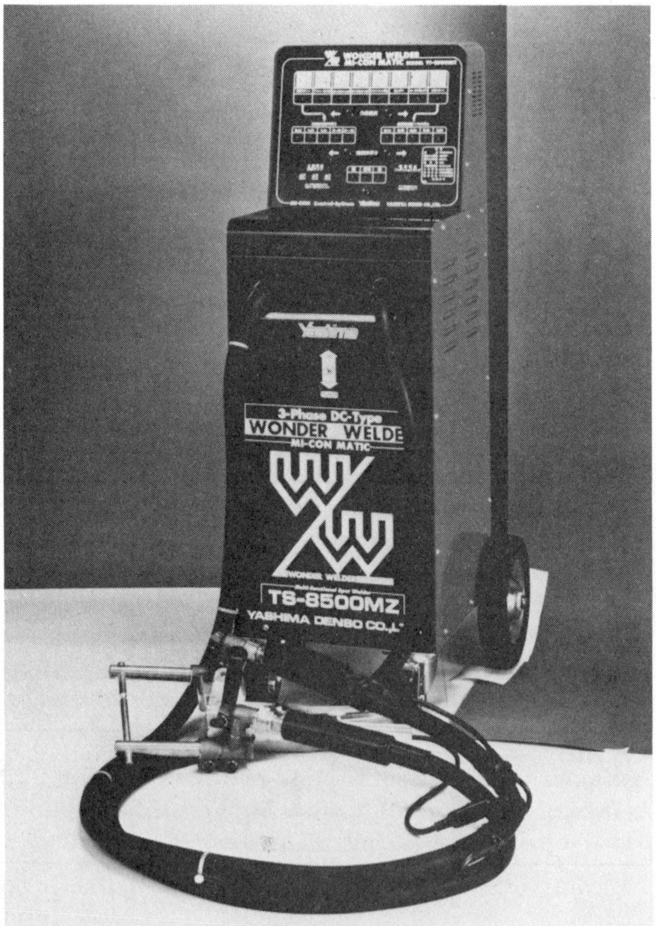

FIGURE 7-50 Typical squeeze-type resistance spot welding unit

Keep in mind that squeeze-type resistance spot welding and MIG welding, should not be viewed as

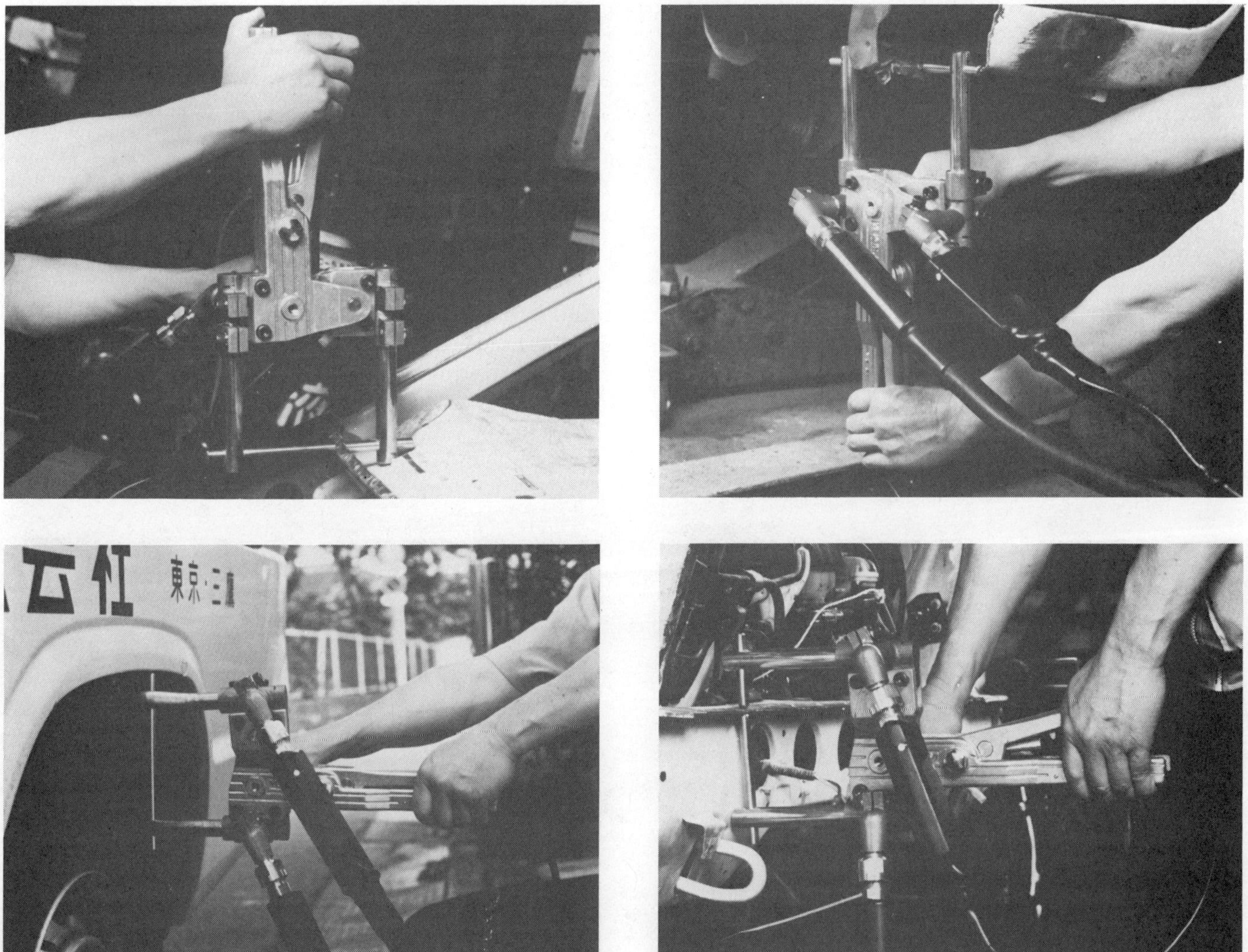

FIGURE 7-51 Typical squeeze-type resistance spot welding operations

competitive, but as complementing each other. The properly equipped shop uses both squeeze-type resistance spot welders and MIG welders.

A squeeze-type resistance spot welder must have the proper heat, pressure, and time to make strong welds. The heat of the weld comes from the resistance of the metal to the passage of heavy current. In all cases, the current must flow or the weld cannot be made. The pressure of the electrode tip on the workpiece holds the parts together in close contact. The right amount of pressure is needed to duplicate the OEM weld. Excessive pressure will cause distortion, while insufficient pressure can lead to cracking and porosity in the weld. The force applied before, during, and after the current flow forges the heated parts together so that coalescence will occur. The welding time is very short, usually 1/2 second (30 cycles) or less. Of course, the correct combination of welding current intensity, weld force, and weld time must be selected to make consistently strong welds.

SQUEEZE-TYPE RESISTANCE SPOT WELDING COMPONENTS

The components of a squeeze-type resistance spot welder are the transformer, the welder control, and the gun (with interchangeable arm sets and electrode tips). Those used for unibody repair welding are available with a full range of interchangeable arm sets (Figure 7-52). Standard arm sets are designed to reach difficult areas such as wheel well flanges, drip rails, taillight openings, and other tight pinchweld areas, as well as

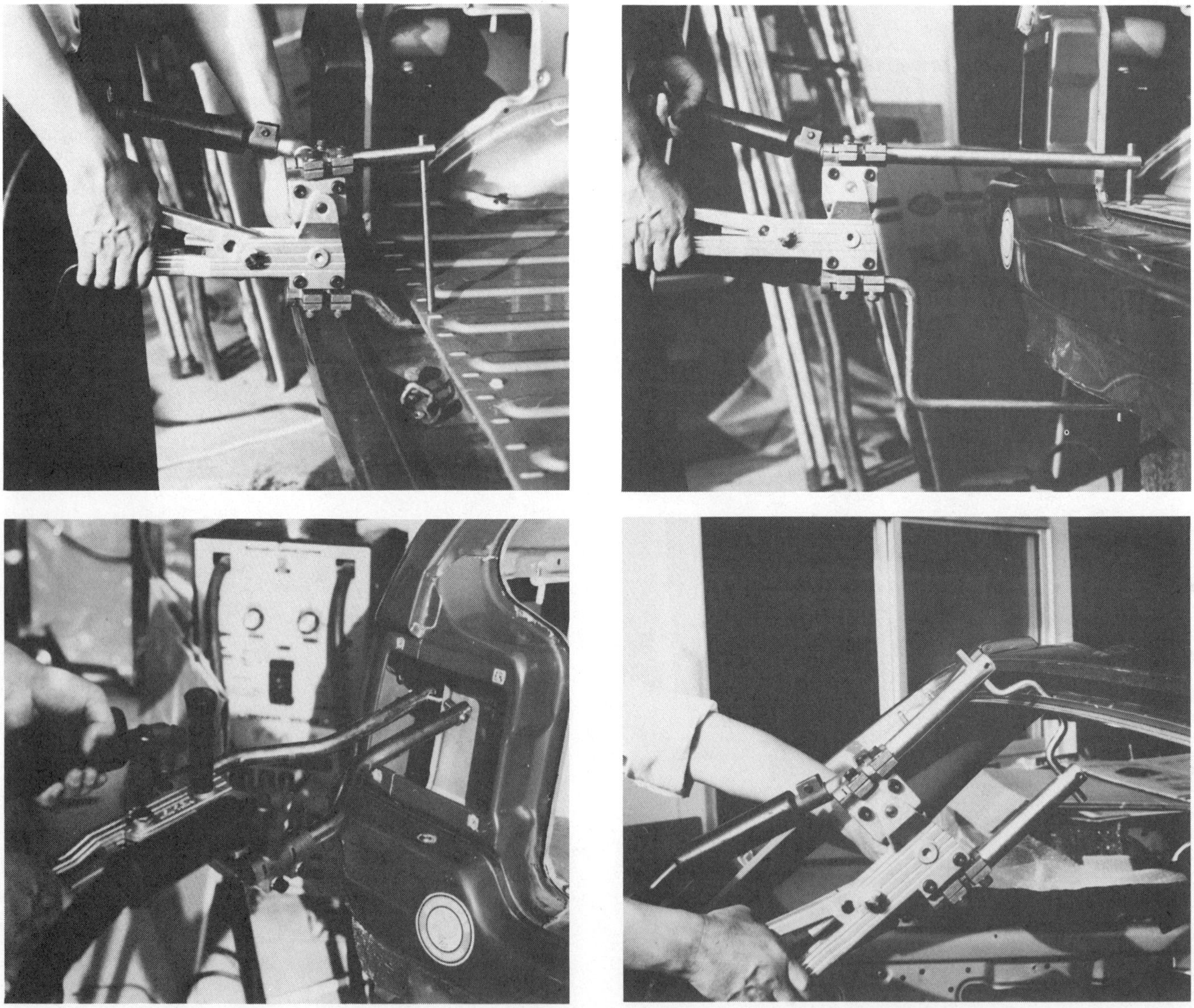

FIGURE 7-52 Select the proper type of arm for the job.

floor pan sections, rocker panels, and window and door openings.

Copper is the base metal that is normally used for squeeze-type resistance spot welding tongs and tips. The purpose of the electrode tips is to conduct the welding current to the workpiece, to be the focal point of the pressure applied to the weld joint, to conduct heat from the work surface, and to maintain integrity of shape.

SQUEEZE-TYPE RESISTANCE SPOT WELDER ADJUSTMENTS

In order to obtain sufficiently strong welds, perform the following checks and adjustments on the squeeze-type resistance spot welder prior to beginning the operation:

- *Arm Set Selection.* It is important to select the appropriate arm set for the specific area to be welded.
- *Arm Adjustment.* Keep the arms as short as possible to obtain the maximum pressure for welding (Figure 7-53). Tighten the arm and tip securely so that they will not come loose during the welding operation.
- *Electrode Tip Alignment.* Align the upper and lower electrode tips on the same axis. Poor alignment of the tips causes insufficient pressure, which results in insufficient current and reduced strength of the welds.
- *Electrode Tip Diameter.* The diameter of the weld decreases as the diameter of the electrode tip increases. However, if the electrode tip is too small, the weld will not increase in size. The tip diameter must be properly controlled to obtain the desired weld strength. Before starting the operation, make sure that the tip diameter is the proper size and file it cleanly to remove burnt or foreign matter from its surface.
- *Electrical Current Flow Time.* When the electrical current flow time increases, the resulting heat increases the weld diameter and penetration. The amount of heat that is dissipated at the weld increases as the current flow time increases. Since the weld temperature will not rise after a certain amount of time, even if current flows longer than that, the weld size will not increase. However, tip pressure marks and heat warping might result.

In addition, keep in mind that the pressurization force and welding current of many squeeze-type resistance spot welders cannot be adjusted and the current value might be low. However, welding strength can be assured by increasing the current flow time (letting low current flow for a long time). The best welding results are obtained by adjusting the welding time according to the thickness of the panels.

OPERATING A SQUEEZE-TYPE RESISTANCE SPOT WELDER

The collision repair technician must hold the squeeze-type resistance spot welding gun and position it so that the electrodes contact the parts to be welded. The technician then uses the squeeze mechanism to apply force to both sides of the metal. As force is applied and maintained on the metal, the force mechanism initiates an electrical signal to the welder control; this signal then switches on the flow of weld current for a preset time and then switches it off. Since the weld time is usually less than 1/2 second, the entire process is very fast.

Other important operational procedures that should be considered when using a squeeze-type resistance spot welder are:

- *Clearance between Welding Surfaces.* Any clearance between the surfaces to be welded causes poor current flow (Figure 7-54). Even if welding can be performed without removing such a gap, the welded area would become smaller,

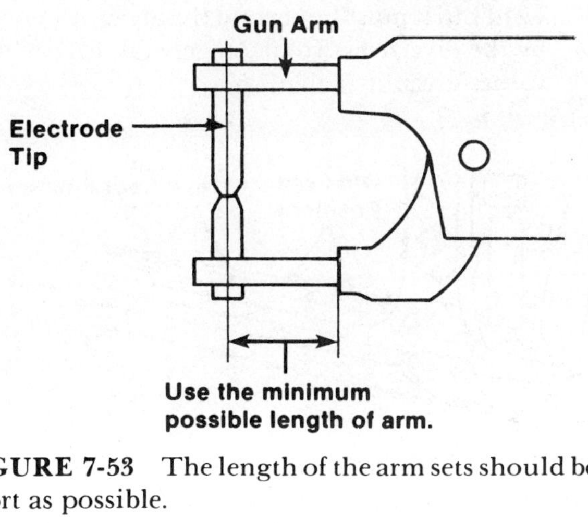

FIGURE 7-53 The length of the arm sets should be as short as possible.

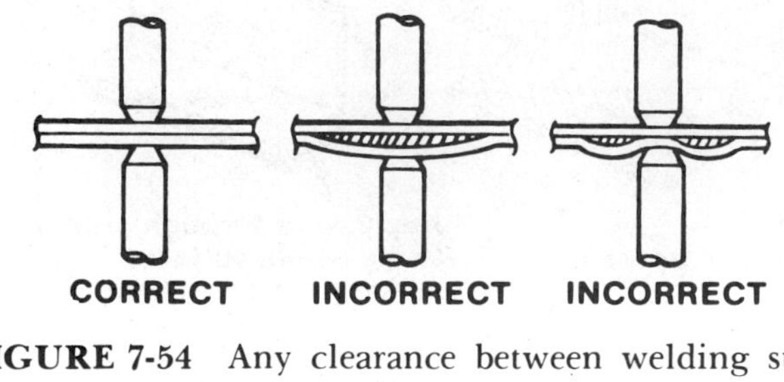

FIGURE 7-54 Any clearance between welding surfaces causes poor current flow.

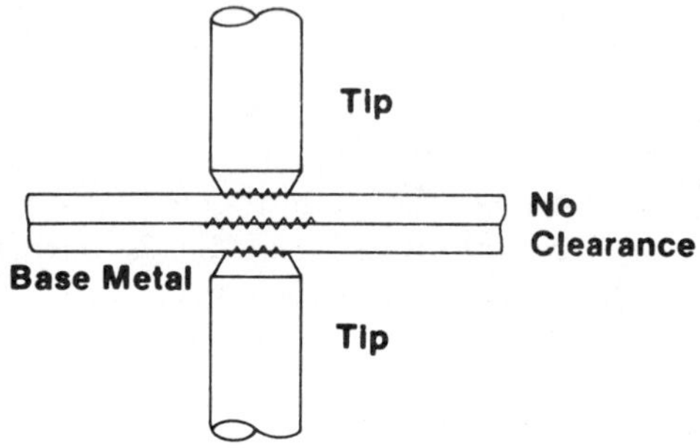

FIGURE 7-55 These surfaces have been flattened sufficiently to remove any gaps.

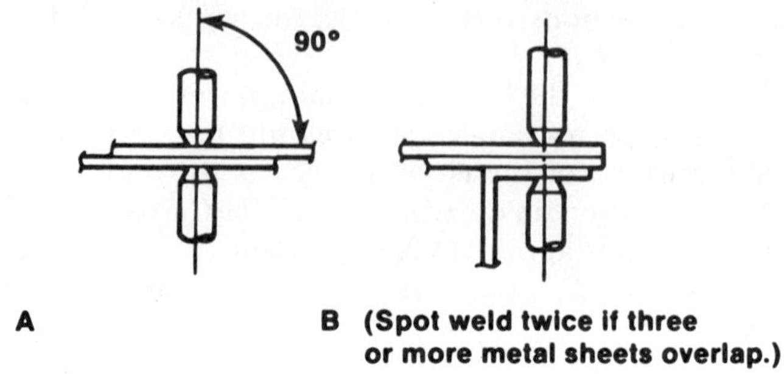

FIGURE 7-57 (A) Align squeeze-type resistance spot welding electrodes at a right angle to the panel. (B) Weld twice if three or more sheets overlap.

resulting in insufficient strength. Flatten the two surfaces to remove the gaps (Figure 7-55), and secure them tightly with a clamp before welding.

- *Metal Surfaces to be Welded.* Paint film, rust, dust, or any other contamination on the metal surfaces to be welded cause poor results. Remove foreign materials from the surfaces to be welded to promote sufficient current flow.
- *Corrosion Prevents Processing on Metal Surfaces.* Coat the surfaces to be welded with a weld-through primer (see Chapter 10) that has higher conductivity. It is important to apply the primer uniformly even to the end face of the panel (Figure 7-56).
- *Performance of Welding Operations.* When performing squeeze-type resistance spot welding operations, be sure to:
  —Use the forward welding method. For the portions to which forward welding cannot be applied, use MIG plug welds.
  —Apply the electrodes at a right angle to the panel (Figure 7-57A). If the electrodes are not applied at right angles, the current flow will be low, resulting in insufficient weld strength.
  —For portions where three or more layers of metal are overlapping, welding should be done twice (Figure 7-57B).
- *Number of Welding Points.* Generally, the capacity of squeeze-type resistance spot welding machines available in a collision repair shop is smaller than that of welding machines at the factory. Accordingly, the number of welds should be increased by 30 percent in a collision repair shop compared to factory welds (Figure 7-58).
- *Minimum Welding Pitch.* The strength of individual squeeze-type resistance spot welds is determined by the weld pitch (the distance between welds) and edge distance (the distance of the welds from the panel edge). The bond between panels becomes stronger as the weld pitch is shortened. However, past a certain point, the metal becomes saturated and further shortening of the pitch will not increase the strength of the bond because the current will flow to the spots that have previously been welded. This diverting of the current increases as the number of welds increases, and the diverted current does not raise the temperature at the welds (Figure 7-59). The distance of the weld pitch must be beyond the area influenced by the diverted current. In general, follow the values given in Table 7-5.

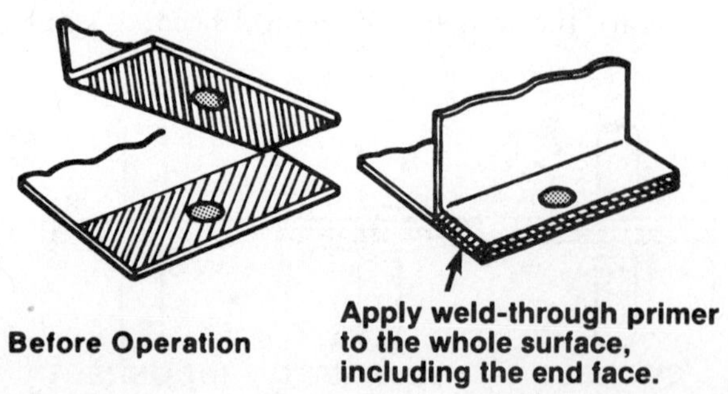

FIGURE 7-56 Areas to be protected with a weld-through primer

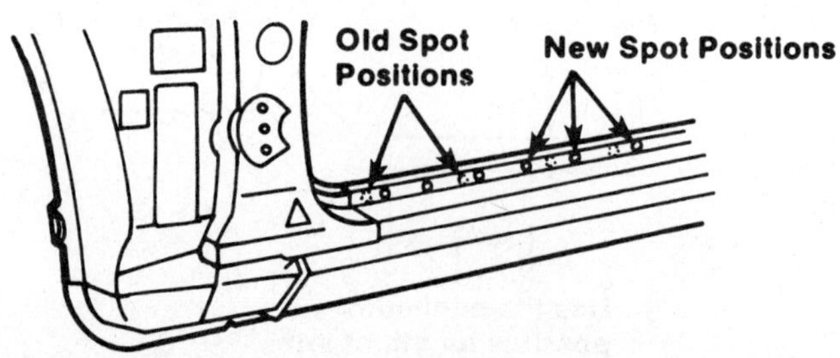

FIGURE 7-58 Use 30 percent more spot welds than the number of original factory welds.

TABLE 7-5: SQUEEZE-TYPE RESISTANCE SPOT WELDING POSITIONS

Panel Thickness	Pitch S	Edge Distance P
1/64″	7/16″ or more	13/64″ or more
1/32″	9/16″ or more	13/64″ or more
Less Than 3/64″	11/16″ or more	1/4″ or more
3/64″	7/8″ or more	9/32″ or more
1/16″	1-9/64″ or more	5/16″ or more

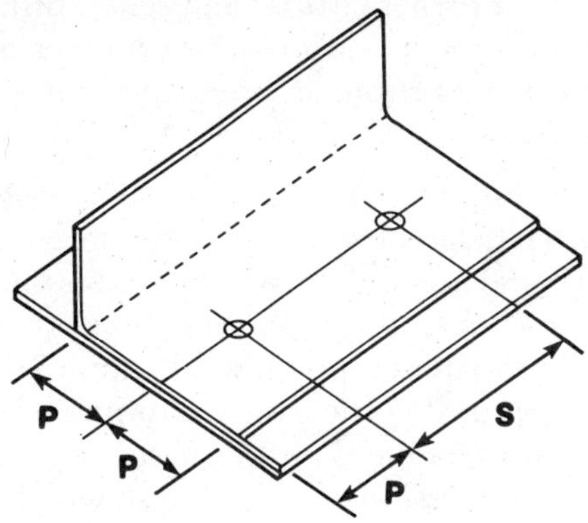

TABLE 7-6: POSITION OF WELDING SPOT FROM THE END OF PANEL

Thickness (t)	Minimum pitch (†)
1/64″	7/16″ or over
1/32″	7/16″ or over
Less than 3/64″	15/32″ or over
3/64″	9/16″ or over
1/16″	5/8″ or over
5/64″	11/16″ or over

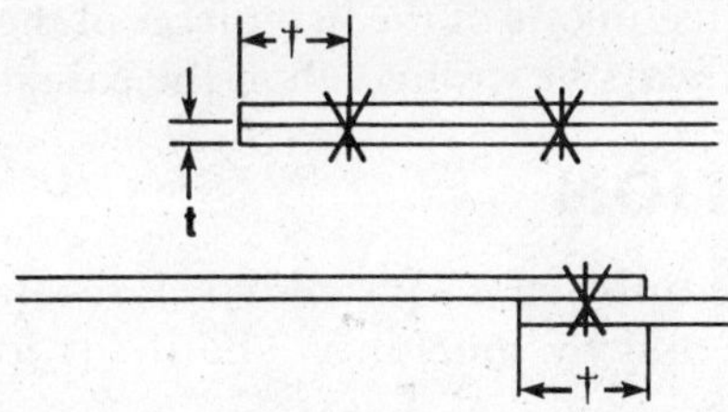

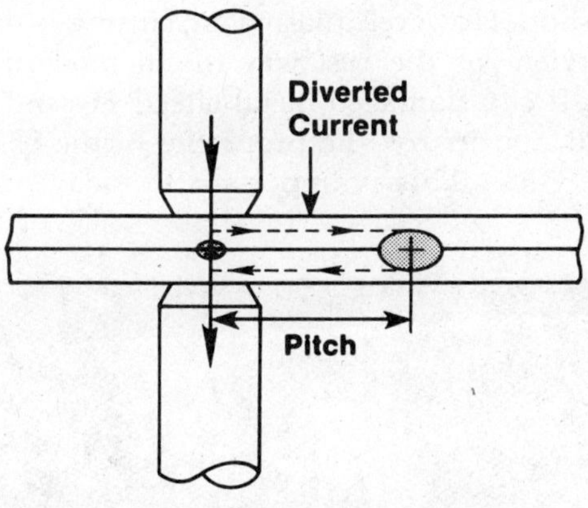

FIGURE 7-59 Minimum welding pitch

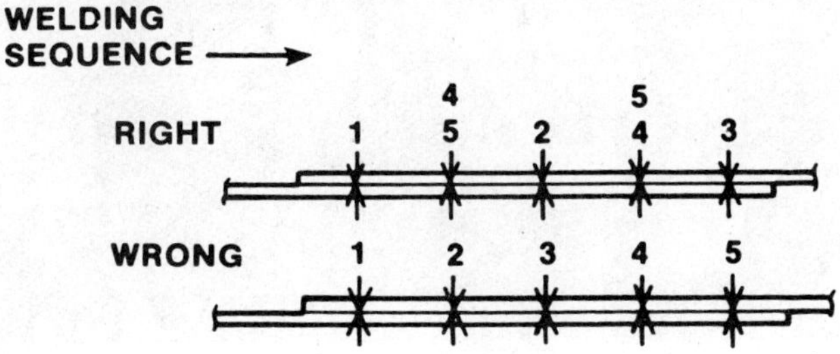

FIGURE 7-60 Proper squeeze-type resistance spot welding sequence

- *Position of Welds from the Edge and End of the Panel.* The edge distance is also determined by the position of the welding tip. Even if the welds are normal, they will not have sufficient strength if the edge distance is insufficient. When welding near the end of a panel, observe the values given in Table 7-6. If the distance is too small, it results in insufficient strength and also in a strained panel.
- *Spotting Sequence.* Do not spot continuously in one direction only; this method provides weak welding (Figure 7-60). If the welding tips become hot and change their color, stop welding and allow the tips to cool.
- *Welding Corners.* Do not weld the radius portion of a corner (Figure 7-61). Welding this portion results in concentration of stress that

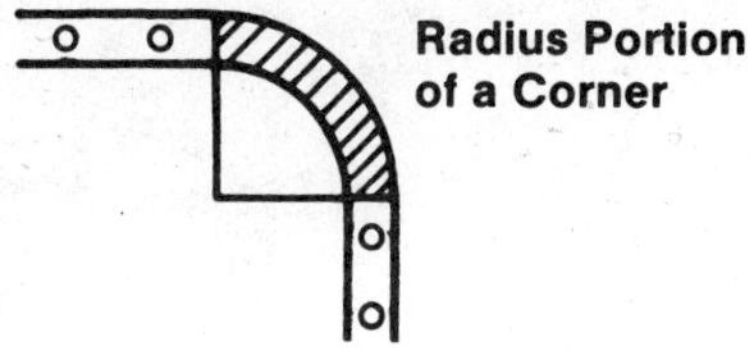

FIGURE 7-61 Proper method of welding corners

leads to cracks. For example, the following locations require special consideration:
—Upper corner of front and center pillars
—Front upper portion of rear quarter panels
—Corner portion of front and rear window openings

Test the sample weld for strength. The easiest and best method is the peel test. Simply peel the ends of the top sheet of metal upward, away from the bottom sheet. Then, peel the ends of the bottom sheet downward, away from the top sheet until the weld sample looks like two letter U's joined at the base. Continue to rock the pieces apart until a hole is pulled in one of them.

An alternate version of the peel test, which is similar to the test for MIG welding given earlier, can be made using a vise, hammer, and chisel. Clamp the welded sample upright in a vise, gripping just below the weld. Place the chisel alongside the weld (never over the weld) and force the two welded pieces apart. A hole should be pulled in one of the two pieces of metal; the weld should always be stronger than the base metal.

INSPECTION

Check the finish of a squeeze-type resistance spot weld visually and by touching. The items to check include:

- *Spot Weld Position.* The weld should be in the center of the flange, with no tip holes and no welds overriding the edge. Avoid the original spot weld position.
- *Number of Spot Welds.* There should be at least 30 percent more than the amount made by the manufacturer. (For example, four original factory welds require roughly six new repair welds.)
- *Pitch.* It should be a little shorter than that of the manufacturer, and the spot welds should be uniformly spaced. The minimum pitch should be at a distance where diverted current will not affect it.
- *Dents.* There should be no dents, or tip bruises, on the surfaces that exceed half the thickness of the panel.
- *Pinholes.* There should be no pinholes large enough to see.
- *Spatter.* Spatter is excessive if a glove can catch on the surface when rubbed across it.

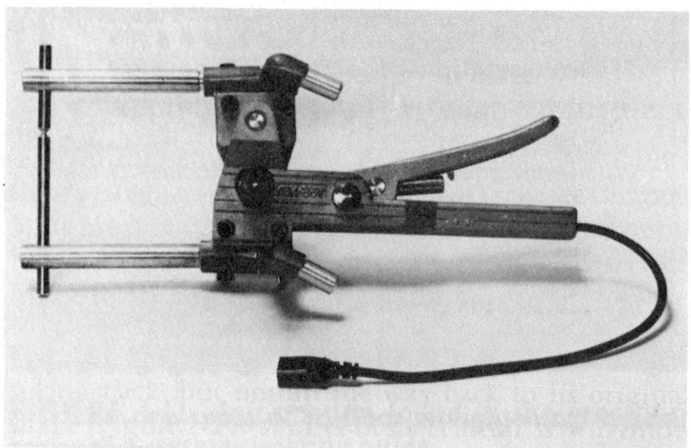

FIGURE 7-62 Typical pneumatic spot welder

PNEUMATIC SPOT WELDING

The pneumatic spot welder (Figure 7-62) operates in the same manner as the conventional squeeze-type resistance spot welder, except that compressed air pressure is used to pinch the panels together instead of a lever. This means that the operator does not have to use force to obtain the necessary pinching action.

PLASMA ARC CUTTING

Plasma welding is seldom used in the collision repair shop. However, plasma arc cutting is replacing oxyacetylene as the best way to cut modern vehicle metals. It cuts damaged metal effectively and quickly but will not destroy the properties of the base metal (Figure 7-63). This is important to today's collision

FIGURE 7-63 Plasma arc is considered the best method for cutting damaged metal.

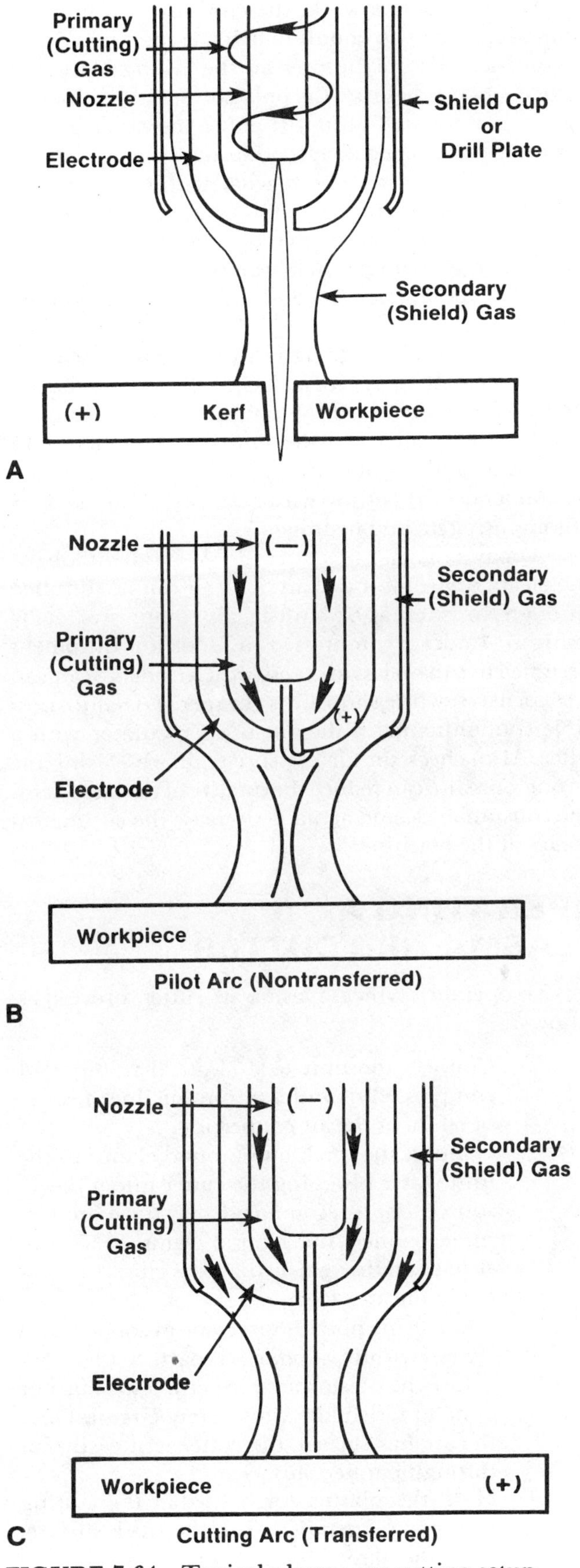

FIGURE 7-64 Typical plasma arc cutting setup

repair shop, which sees more unibody vehicles with high-strength steel or high-strength alloy steel components. The old method of flame cutting just does not work that well anymore. The high heat, fast travel speed, low-heat input qualities, coupled with the fact that plasma arc cutting will cut rusted, painted, or coated metal with little difficulty, make it an ideal process for collision repair.

Basically, plasma arc cutting creates an intensely hot air stream over a very small area, which melts and removes metal. Extremely clean cuts are possible with plasma arc cutting, and, because of the tight focus of the heat, there is no warpage, even when cutting thin sheet metal.

Figure 7-64A shows that there are two areas for gas flow. In plasma arc cutting, compressed air is used for both shielding and cutting. As a shielding gas, air covers the outside area of the torch nozzle, cooling the area so the torch does not overheat. Air also becomes the cutting gas; it swirls around the electrode as it heads toward the nozzle opening. The swirling action helps to constrict and narrow the gas. When the machine is turned on, a pilot arc is formed between the nozzle and the inner electrode (Figure 7-64B). When the cutting gas reaches this pilot arc, it is super heated—up to 60,000 degrees Fahrenheit.

The gas is now so hot it ionizes and becomes capable of carrying an electrical current (the ionized gas is actually the plasma). The small, narrow opening of the nozzle accelerates the expanding plasma toward the workpiece. When the workpiece is close enough, the arc crosses the gap, with the electrical current being carried by the plasma (Figure 7-64C). This is the cutting arc.

The extreme heat and force of the cutting arc melt a narrow path through the metal. This serves to dissipate the metal into gas and tiny particles. The force of the plasma literally blows away the metal particles, leaving a clean cut.

SHOP TALK

Air ordinarily will not conduct electricity. But with a very high voltage, the air molecules ionize and become electrically conductive. The air is super heated and forms a path along which voltage can easily travel.

PLASMA ARC CUTTERS

For plasma arc cutting with air, power sources can range in output from about 10 to 100 amperes. The output of the power source helps determine maximum thicknesses that can be cut. A 10- to 15-amp plasma arc cutter is generally adequate for mild steel up to 3/16

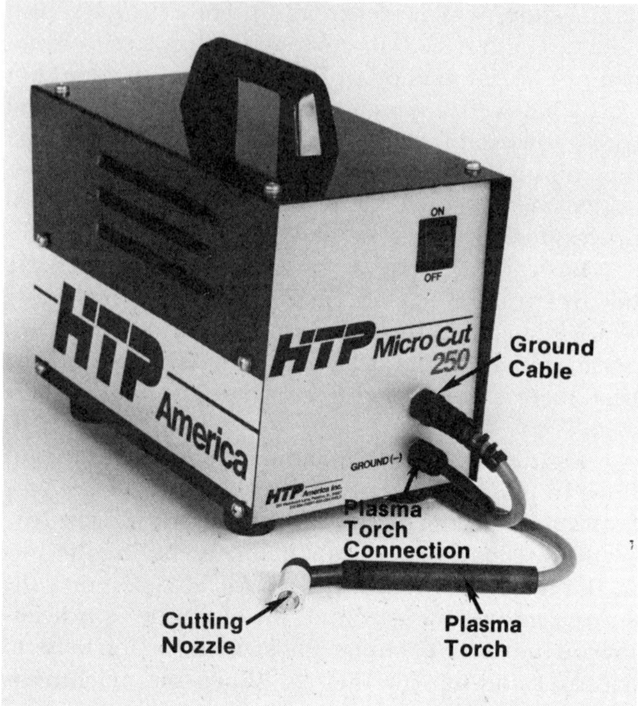

FIGURE 7-65 Controls on a typical plasma arc cutter

inch thick, a 30-amp unit can cut metal up to 1/4 inch thick, and a 60-amp unit will slice through metal up to 1/2 inch thick.

Controls are usually quite simple (Figure 7-65). Plasma arc cutters made specifically for thinner metals might only have an on/off switch and a ready light to show when conditions are proper for cutting. More elaborate equipment can include a built-in air compressor, variable output control, on-board coolant, and other features. Some equipment has a built-in safety system to protect the operator. This type of system cuts output power automatically if the safety cup is removed from the torch, if the tip and electrode are accidentally short circuited because of insufficient air pressure, or if the duty cycle is exceeded. The open circuit voltage of plasma arc cutting equipment can be very high (in the range of 250 to 300 volts), so insulated torches and internally connected terminals are also essential.

On some units, a switch is provided that allows the operator to alter the current mode depending on the surface being cut. When cutting painted or rusty metal, a continuous high-frequency arc is best for punching through the nonconductive surface layer and for keeping the arc going while cutting. When cutting bare metal, on the other hand, a high-frequency arc is only needed to start the cutting. Once the torch starts to cut, a direct current pilot arc is all that is needed to keep things going. This position gives the longest electrode and nozzle life.

For automotive work, the cutting torch must be compact and easy to handle to help get into tight areas. Two critical parts of the torch are the cutting nozzle and the electrode. These are the only consumables (besides air) in plasma arc cutting. If either the nozzle or the electrode is worn or damaged, the quality of the cut will be affected. They wear somewhat with each cut, but such things as moisture in the air supply, cutting excessively thick materials, or poor operator technique will make them deteriorate prematurely. Keep a supply of electrodes and nozzles on hand and replace them when wear appears.

The electrode is actually quite small and is an insert held in place by a metal jacket. Plasma arc cutting electrodes are often made of hafnium or zirconium because of the hardness and durability of these metals. In heavy-duty industrial cutting, tungsten electrodes are recommended for cutting with gases besides air, such as argon, nitrogen, and hydrogen.

Today's plasma arc cutters do an excellent job using clean, dry compressed air. The air can be supplied through an external or built-in air compressor or by using a cylinder of compressed air. Look for equipment designed to run on less air pressure. Cylinders of air can be expensive, while shop air is cheaper. To reduce possible contaminants in the air, use a regulator with a filter. Also, check the air pressure regularly. Using the wrong pressure can reduce the quality of the cuts, damage consumables, and actually decrease the cutting capacity of the machine.

OPERATING A PLASMA ARC CUTTER

To operate a typical plasma arc cutter, proceed as follows:

1. Connect the unit to a clean, dry source of compressed air with a minimum line pressure of 60 psi at the air connection.
2. Connect the torch and ground clamp to the unit. After plugging the cutter into a power source that is recommended by the manufacturer, connect the ground clamp to a clean surface on the vehicle that is as close as possible to the area to be cut.
3. The cutting nozzle must come in contact with an electrically conductive part of the work before the plasma arc can start. This must be done to satisfy the work safety circuit. Once the arc has started, the cutter will easily cut through painted surfaces.
4. Hold the plasma torch so that the cutting nozzle is perpendicular to the work surface. Push the plasma torch down. This will force

the cutting nozzle down until it comes in contact with the electrode, at which time the plasma arc will start. Immediately release the downward force on the torch to let the cutting nozzle return to its normal position. It is not necessary to keep the cutting nozzle in contact with the work once the plasma arc has started; however, it makes cutting easier. While keeping the cutting nozzle in contact with the work, put very little, if any, downward force on the torch. It is only necessary to drag it lightly on the work surface.

5. Begin to move the plasma torch where the metal is to be cut. The speed of the cut will depend on the thickness of the metal. If the torch is moved too fast, it will not cut all the way through the workpiece. If moved too slowly, it will put too much heat into the workpiece and might also extinguish the plasma arc.

Other pointers that should be remembered when using a plasma arc cutter are:

- When piercing materials that are 1/8-inch thick or more, it is a good idea to angle the torch at 45 degrees until the plasma arc pierces the material. This will allow the stream of sparks to shoot off at a 45 degree angle away from the gas diffuser. If the torch is held perpendicular to the work when piercing heavy-gauge material, then the sparks will shoot back up at the gas diffuser. The molten metal will then collect on the diffuser, plugging the air holes and greatly shortening the life of the diffuser.
- Cooling of the torch is important to extend the life of the electrode and nozzle. At the end of a cut, the air continues to flow for several seconds after the torch switch has been released to prevent the nozzle and electrode from overheating until the next cut is started. Some equipment suppliers also recommend "idling" the unit for a couple of minutes after the cut is made.

CAUTION

When angling the torch, be aware of the fact that the sparks will shoot off as far as 20 feet away. Be sure that there are no combustibles or bystanders in the area that might be ignited or hurt by the sparks.

- When making long straight cuts, it might be easier to use a metal straightedge as a guide. Simply clamp it to the work to be cut. For elaborate cuts, make a template out of thin wood and guide the tip along that edge.
- When cutting 1/4-inch materials, it is beneficial to start the cut at the edge of the material.
- When making rust repairs on cosmetic panels, it is possible to piece the new metal over the rusted area and then cut the patch panel at the same time the rust is cut out. This process also works when splicing in a quarter panel.
- Be aware of the fact that the sparks from the arc can damage painted surfaces and can also pit glass. Use a welding blanket to protect these surfaces.
- Remember that these variables will have a bearing on cut quality:
  —Travel speed. The thicker the material, the slower the travel speed.
  —Parts wear. The tip and electrode will wear with use. The more wear, the poorer the quality of the cut.
  —Air quality. Moist or oil-contaminated air will lead to a poor quality cut.

OXYACETYLENE WELDING

Oxyacetylene is a form of fusion welding in which oxygen and acetylene are used in combination. The process is simple; the two are mixed in a chamber, ignited at the tip of the torch, and used to melt and join a welding filler rod and base metal together. Oxyacetylene is no longer used for repairs; it is predominantly used for cleaning and heating on unibody vehicles.

A typical oxyacetylene welding outfit (Figure 7-66) consists of the following:

- Steel tanks (cylinders) hold the oxygen and acetylene.
- Regulators reduce the pressure coming from the tanks to the desired level and maintain a constant flow rate. Each regulator has a cylinder pressure gauge and a working pressure gauge.
  —Working oxygen pressure: 5 to 100 psi
  —Working acetylene pressure: 1 to 12 psi
- Hoses from the regulators and cylinders connect the oxygen and acetylene to the torch.
- The torch mixes the oxygen and acetylene from the cylinders in the proper proportions and produces a flame capable of melting steel. There are two main types of torches: welding and cutting.

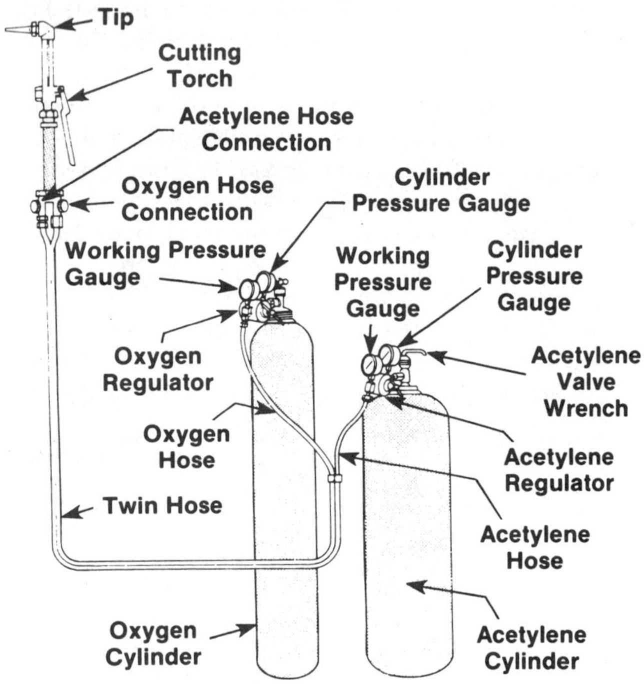

FIGURE 7-66 Typical oxyacetylene welding outfit

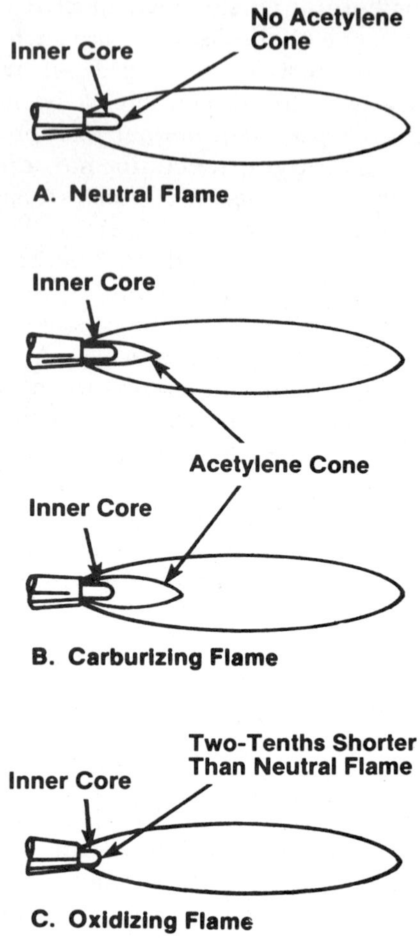

FIGURE 7-67 Types of cutting flames

WARNING

The acetylene line pressure must never exceed 15 psi. Free acetylene has a tendency to dissociate at pressure above 15 psi and could cause an explosion.

All welding should be done with a number 4, 5, or 6 tinted filter lens. A spark lighter is a necessity for lighting oxyacetylene welders.

FLAMES AND ADJUSTMENT

When acetylene and oxygen are mixed and burned, the condition of the flame varies depending on the ratio of oxygen and acetylene used.

There are three types of flames:

1. *Neutral Flame.* The standard flame is said to be a neutral flame (Figure 7-67A). It is produced by mixing acetylene and oxygen in a 1 to 1 ratio. A neutral flame has a brilliant white core surrounded by a clear blue outer flame.
2. *Carburizing Flame.* This flame, also called a reducing flame, is obtained by mixing slightly more acetylene than oxygen. Figure 7-67B shows that this flame has three parts. The core and the outer flame are the same as the neutral flame, but between them there is a light-colored acetylene cone enveloping the core. The length of the acetylene cone varies according to the amount of surplus acetylene in the gas mixture. For a carburizing flame, the oxygen-acetylene mixing ratio is about 1 to 1.4. A carburizing flame is used for heating and welding aluminum, nickel, and other alloys.
3. *Oxidizing Flame.* The oxidizing flame (Figure 7-67C) is obtained by mixing slightly more oxygen than acetylene. It resembles the neutral flame in appearance, but the inner core is shorter and its color is a little more violet, while the outer flame is shorter and fuzzy at the end. Ordinarily, this flame oxidizes melted metal, so it is not used in the welding of mild steel, but is used in the welding of brass and bronze.

Oxyacetylene Torch Flame Adjustment

When using an oxyacetylene welding torch, proceed as follows:

1. Attach the appropriate size tip to the end of the torch. Use the standard size tip for mild steel, keeping in mind that each torch manufacturer has a different system for measuring the size of the tip orifice.
2. Set the oxygen and acetylene regulators to the proper pressure.
3. Open the acetylene hand valve about half a turn and ignite the gas. Continue to open the hand valve until the black smoke disappears and a reddish yellow flame appears. Slowly open the oxygen hand valve until a blue flame with a yellowish white cone appears. Further open the oxygen hand valve until the center cone becomes sharp and well defined. This is a neutral flame and is used for welding mild steel.

If acetylene is added to the flame or oxygen is removed from the flame, a carburizing flame will result. If oxygen is added to the flame or acetylene is removed from the flame, an oxidizing flame will result.

CLEANING WITH AN OXYACETYLENE TORCH

It is important before starting any welding operation that the surfaces to be joined are thoroughly cleaned. The weld site must be completely free of any foreign material that might cause contamination. If this is not done, the finished weld is quite likely to be brittle, porous, and of poor integrity. To remove heavy undercoating, rustproofing, tars, caulking and sealants, road dirt, primers, and the like, first use a scraper and an oxyacetylene torch with a rosebud tip. Do a thorough cleaning with a wire brush and the torch, using a carburizing flame. In any event, keep the torch at a very low, controlled heat—just enough to get the job done. It should be noted that a grinder can also be used to remove primers and paints prior to welding.

SHUTTING DOWN THE EQUIPMENT

Always turn off the oxyacetylene torch when not in use; never lay down a lit torch. Standard practice is to turn off the torch whenever it is not being held in the hand. Do this by first closing the acetylene hand valve, then the oxygen hand valve. Shutting off the acetylene valve first offers less chance of unburned acetylene escaping in the work area; that is, shutting off the flow of acetylene will immediately extinguish the flame. If the oxygen is shut off first, the acetylene will continue to burn, throwing off a great deal of smoke and soot. These instructions are equally valid for both low-pressure and medium-pressure torches.

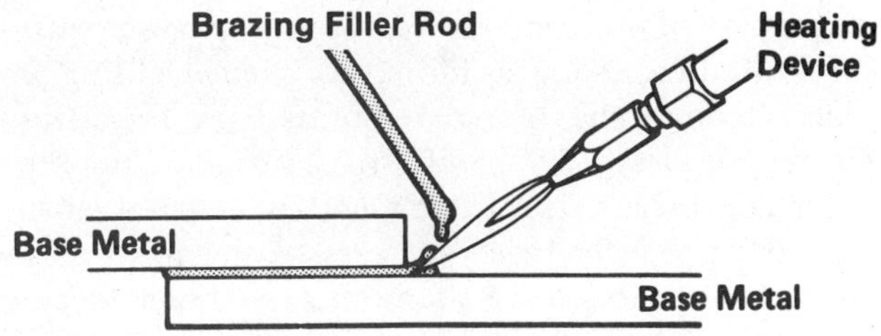

FIGURE 7-68 The basic brazing process

BRAZING

Brazing is applied only for sealing purposes. This is a method of welding whereby a filler rod, whose melting point is lower than that of the base metal, is melted without melting the base metal (Figure 7-68). Brass brazing is frequently applied to automotive bodies.

Brazing is similar to joining two objects with adhesives or solder; melted brass sufficiently spreads between the base metals to form a strong bond. The bending intensity against impact is less than that of the base metal, but the same as the melted brass. Therefore, under no circumstances should brazing be applied to any portions other than those to which it was applied at the factory.

There are two types of brazing:

- Soft brazing (lead soldering)
- Hard brazing

Ordinarily, the term *brazing* refers to hard brazing. The basic characteristics of brazing are:

- The pieces of metal are joined together at a relatively low temperature where the base metal does not melt. Therefore, there is a lower risk of distortion and stress in the base metal.
- Brazing metal has excellent flow characteristics; it penetrates well into narrow gaps and it is convenient for filling gaps in body seams.
- Because the base metal does not melt, it is possible to join metals that would otherwise be incompatible.
- Since there is no penetration and the metals are joined only at the surface, it has very low strength to resist repeated loads or impacts.

More often than not, brazing utilizes brass or bronze filler metals. However, silver solder, aluminum, copper, and many other materials fit just as well into the category of brazing. The remaining brazing equipment is about the same as for oxyacetylene welding: an oxyacetylene torch, flux-coated brass filler rods, flux, welding goggles, gloves, and a spark lighter. Under the proper conditions, the filler metal is distributed between the properly fitted surfaces of the joint to be brazed. This process is very similar to that of soldering copper pipes with wire solder.

Although many brazed joints are as strong as welded steel, there is one notable exception: a butt joint on light-gauge steel. Because of its high incidence of failure, this joint should be avoided if at all possible. To repair a brazed butt joint, the brass deposited on the base metal must be completely removed and rewelded with steel filler rod. Many inexperienced welders fail to do this when trying to repair old fenders with cracks caused by stress or vibrations. Because of the constant expansion and contraction due to extreme heat, brazed butt joints should also be avoided when repairing exhaust systems; use the lap joint technique instead.

When working with sheet metal or other light-gauge steel, heat can cause warping because of the large surface area of the panels. Because the metal is very thin, it reacts to any significant amount of heat; the temperature for mild steel oxyacetylene fusion welding is relatively high, in the 2200- to 2400-degree-Fahrenheit range. This is why the brazing method is preferred over fusion welding methods. The controlled temperature of brazing is sufficiently low as to practically eliminate warping. When brazing, use a slightly carburized flame.

Automobile assembly plants use arc brazing to join the roof and quarter panels together. Arc brazing uses the same principles as MIG welding. However, argon is used with the brazing metal instead of welding wire. Since the heat applied to the base metal is low, overheating is minimized and there is little distortion or warpage of the base metal. When compared to the previously used methods of depositing brass on the base metal, arc brazing shortens both the time for making the weld and for finishing. Also, there is no danger of lead poisoning that is associated with lead fillers.

BRAZING OPERATIONS

In discussing the general brazing technique, the procedure will be geared toward collision repairs on 20-to 24-gauge sheet metal. It is wise to remember that the great majority of brazing jobs will not take place under ideal conditions; the metal will likely be covered with paint, rust, undercoating, galvanizing, and other assorted materials. And while it might be tempting not to bother stripping the surface, the time and effort should be taken to ensure a high-quality brazing job.

The general brazing procedure is as follows:

1. *Cleaning the Base Metal.* If there is oxidation, oil, paint, or dirt on the surface of the base metal, the brazing material will not flow properly over the base metal. These contaminants, if allowed to remain on the surface, can cause eventual joint failures. The surface, therefore, must first be cleaned with a wire brush, grinder, or sand blaster.
2. *Flux Application.* After the base metal is thoroughly cleaned, flux is applied uniformly to the brazing surface. (If a brazing rod with flux in it is used, this operation is not necessary.) Flux neutralizes the oxidation on the metal and prevents further oxidation as the brazing is performed. The importance of flux is illustrated in Figure 7-69. If brazing material is melted over a surface with oxidized film and other foreign matter adhering to it, there will not be adequate bonding to the base metal. This is because the surface tension causes the brazing material to ball up and not stick.

CAUTION

Some fluxes give off very toxic fumes when heated. Use them in a well-ventilated place. When removing flux residue, wear safety glasses to prevent eye infection or other injuries.

3. *Base Metal Heating.* The joining area of the base metal is heated to a uniform temperature

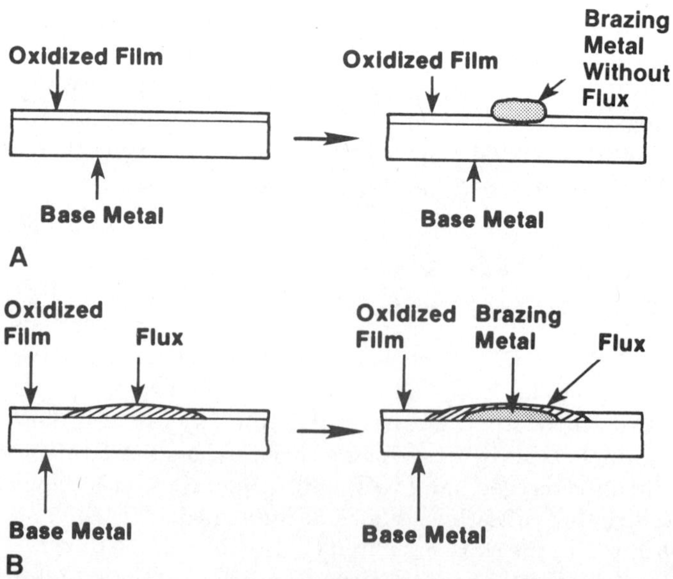

FIGURE 7-69 Comparing the effects on brazing metal (A) when flux is not used and (B) when flux is used

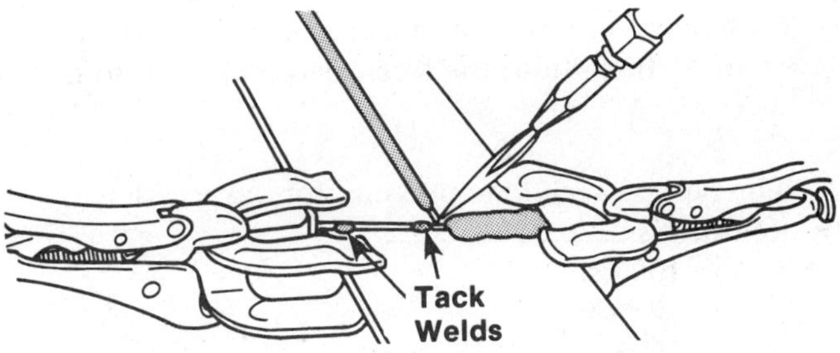

FIGURE 7-70 Base metal brazing operation

capable of accepting the brazing material. When the flux begins to melt, the metal has reached the proper temperature for the brazing material.

4. *Base Metal Brazing.* When the base metal has reached the proper temperature, the brazing filler rod is held in the flame and melted onto the base metal (Figure 7-70). Heat and filler rod are applied to the joining area until the brazing filler rod has filled all of the gaps in the base metal.

   Stop heating when the brazing filler rod has flowed into all of the gaps. If the surface temperature of the base metal is allowed to get too high, the flux will not clean it properly; the result will be a poor brazing bond and inferior joint strength.
5. *Flux Removal.* Once the brazed portion has cooled down sufficiently, the remaining flux sediment is rinsed off with water and the surface is scrubbed with a stiff wire brush. Baked and blackened flux can be removed with a sander or a sharp-pointed tool. If the remaining flux sediment is not adequately removed, the new finish will not adhere properly, and corrosion and cracks might form in the joint.

REVIEW QUESTIONS

1. Technician A says that normal MIG welding procedures can be used to restore HSS components. Technician B says that the MIG process should not be used on HSS steel. Who is correct?
   a. Technician A
   b. Technician B
   c. Both A and B
   d. Neither A nor B
2. Which type of metal transfer is indicative of CO_2 shielded welding of steel?
   a. spray
   b. arc
   c. short-circuiting
   d. globular
3. Welding current affects which of the following?
   a. base metal penetration depth
   b. arc stability
   c. amount of weld spatter
   d. all of the above
4. When MIG welding, what happens if the wire stick-out distance is too long?
   a. The shielding gas effect is reduced.
   b. The wire protruding from the end of the gun increases and becomes preheated.
   c. The melting speed of the wire increases.
   d. All of the above
5. Technician A starts a butt weld in the center of the metal. Technician B starts a butt weld at the edge of the metal. Who is correct?
   a. Technician A
   b. Technician B
   c. Both A and B
   d. Neither A nor B
6. Which of the following welds is the collision repair shop alternative to the OEM resistance spot welds made at the factory?
   a. spot
   b. plug
   c. stitch
   d. all of the above
7. The most common shielding gas for welding sheet metal is ______________.
   a. 75 percent argon and 25 percent CO_2
   b. 25 percent argon and 75 percent CO_2
   c. 50 percent argon and 50 percent CO_2
   d. 50 percent argon and 50 percent helium
8. In addition to very small surface cracks, what is penetrant inspection used to locate?
   a. porosity
   b. undercut
   c. overlap
   d. burn-through
9. When working with a squeeze-type resistance spot welder, Technician A installs a larger diameter electrode tip to increase the diameter of the spot weld. Technician B installs a smaller diameter electrode tip to increase the diameter of the spot weld. Who is correct?
   a. Technician A
   b. Technician B
   c. Both A and B
   d. Neither A nor B

10. How many squeeze-type resistance spot welds should be made in a panel?
    a. a number equal to the original factory welds
    b. 50 percent more than the number of original factory welds
    c. 30 percent more than the number of original factory welds
    d. none of the above

11. Flux-cored arc welding does not utilize a(n) ______________.
    a. electrode feed system
    b. electrode conduit
    c. external shielding gas
    d. contact tube

12. Which of the following statements concerning plasma arc cutting is incorrect?
    a. The plasma arc process cuts damaged metal effectively.
    b. Extremely clean cuts are possible with plasma arc cutting.
    c. The cutting nozzle must come in contact with an electrically conductive part of the work before the arc can start.
    d. When piercing material that is more than 1/8-inch thick, hold the torch perpendicular to the work.

13. The typical acetylene pressure for oxyacetylene welding is ______________.
    a. 5 to 100 psi
    b. 1 to 12 psi
    c. 3 to 25 psi
    d. 30 to 120 psi

14. Mixing slightly more acetylene than oxygen will obtain what type of flame?
    a. neutral
    b. standard
    c. carburizing
    d. oxidizing

15. Which of the following is not characteristic of brazing?
    a. excellent flow characteristics
    b. low strength to resist repeated loads or impacts
    c. greater risk of distortion in the base metal
    d. can join otherwise incompatible metals

CHAPTER

8 Repairing or Replacing Structural Components

OBJECTIVES

- Explain the difference between a kink and a bend.
- List the factors that determine whether to repair or replace a structural component.
- Describe the methods by which spot welds are separated.
- Explain the two ways new body panels can be positioned on a vehicle body.
- List the steps in the general procedure for welding new body panels in place.
- Describe how to install foam panel fillers.
- Section rails, rocker panels, A- and B-pillars, floor pans, and trunk floors.

The question of whether to repair or to replace structural unibody parts is an issue that must be addressed almost daily in collision repair shops (Figure 8-1). Actually, the decision comes down to a judgment call. To help the technician decide, I-CAR has, after a great deal of research, come up with the following simple rule for deciding when to repair and when to replace damaged structural components: If it is *bent,* repair it; if it is *kinked,* replace it.

KINK VS. BEND

To define the terms *kink* and *bend* more closely, keep the following in mind. A part is bent when:

- The change in the shape of the part between the damaged and undamaged areas is smooth and continuous.
- Straightening the part by pulling restores its shape to pre-accident condition without any areas of permanent deformation (Figure 8-2).

A part is kinked when:

- It has a sharp bend of a small radius, usually more than 90 degrees over a short distance (Figure 8-3).
- After straightening, there is a visible crack or tear in the metal or there is permanent deformation that cannot be straightened to its pre-accident shape without the use of excessive heat.

FIGURE 8-1 This is the result of a frontal collision; the vehicle is an ideal candidate for the repair methods covered in this chapter.

FIGURE 8-2 This badly bent and distorted upper rail shows no serious kinking. Straightening brought the part back to pre-accident shape without permanent deformation.

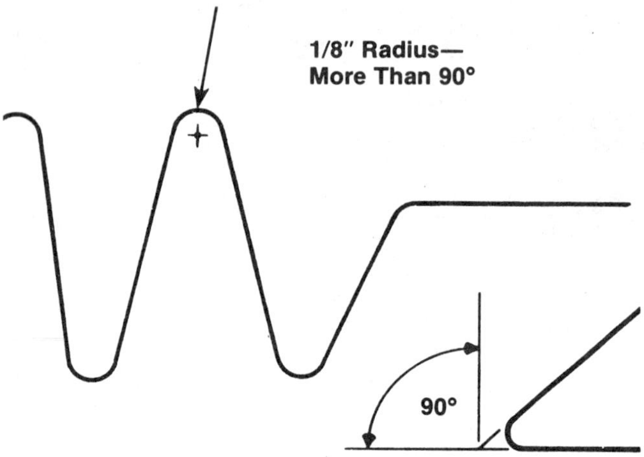

FIGURE 8-3 The radius of bend determines whether or not the damage should be called a kink.

While I-CAR's "kink versus bend" rule is a very useful one, it has raised some questions among collision repair technicians for the following reasons:

- The need to understand the changes that occur inside the steel during kinking and then straightening.
- The need to understand why a large structural part, which has only a few small kinks in it, should be sectioned or replaced, rather than just straightening and welding the small cracks that might form (Figure 8-4).

Actually, there are many factors that must be considered before the decision to repair or replace a structural component can be made. Some of the more important concerns are:

1. *Change in the Structure of the Metal.* Before considering the structure of the metal in a bend/kink situation, be sure to reread the *Unibody Construction Materials* portion of Chapter 2. Although high-strength steel has two or three times the yield strength of mild steel, it is also more brittle. Instead of stretching or compressing smoothly, it tends to kink and crack.

SHOP TALK

With structural components, it is best to treat them all like high-strength steel—just as is done in MIG welding. That way it is not necessary to identify every piece of high-strength steel in an assembly.

As mentioned in Chapter 2, steel is made of microscopic particles called *grains.* Figure 8-5 shows this clearly. Much of the strength of HSS comes from the strong (but somewhat brittle) bond where the grains meet each other. When this is deformed, the grains and the grain boundaries do not move as easily as in mild steel. When the metal does flow, it flows differently.

Yield strength (tensile strength) is the amount of stress the metal will withstand be-

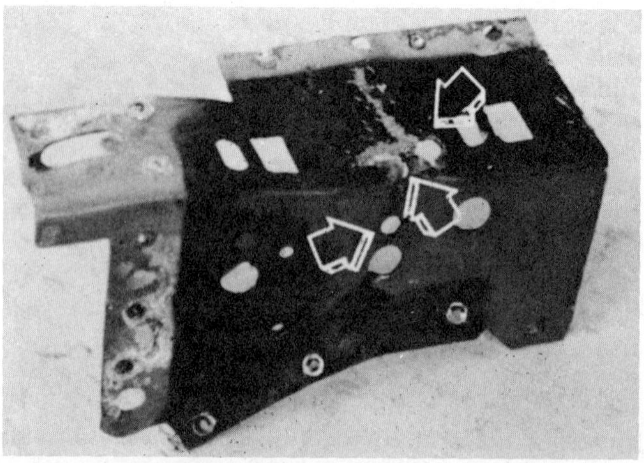

FIGURE 8-4 This rail extension was fully straightened, but the metal cracked at the corner bend and at two holes, so the part was replaced.

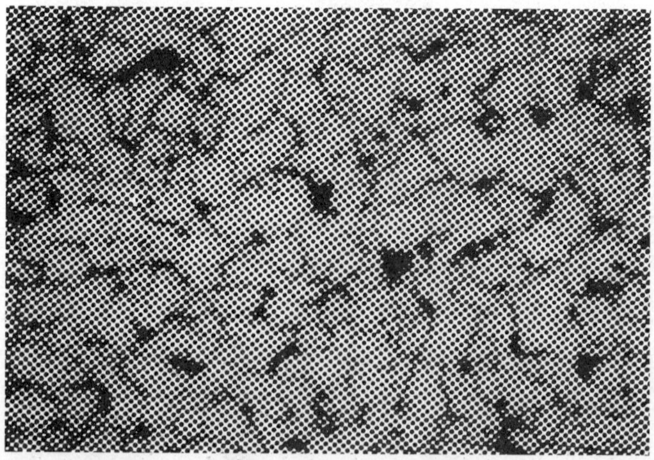

FIGURE 8-5 In this typical grain arrangement of a structural part made of high-strength steel, note the fine, even grain structure.

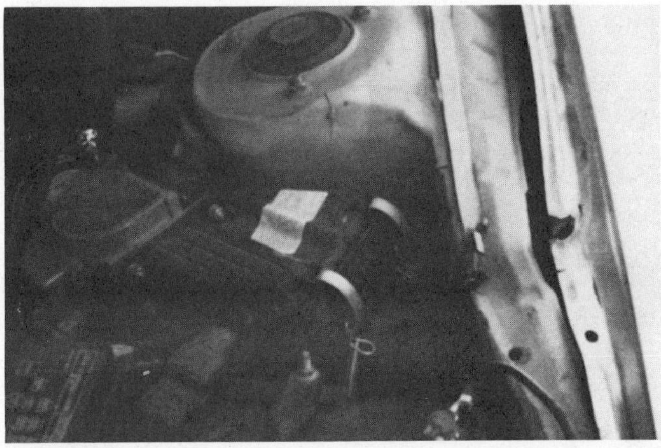

FIGURE 8-6 Damage to this upper rail consists of gradual distortion, with no sharp kinks, so straightening can restore its pre-accident shape and state.

FIGURE 8-7 This lower rail buckled and folded sharply onto itself. This kind of sharp kink causes severe work hardening of the metal and disruption of its grain structure.

fore it takes on permanent deformation; that is, it will not return to its original shape when the load is released. For mild steel this is generally around 30,000 psi or less. For high-strength steel it can be as high as 100,000 psi. When the metal finally "yields" in mild steel, the deformation tends to be an even, plastic deformation. For high-strength steel, it is a more brittle deformation that can begin to create microscopic breaks in the grain structure. This can occur with much less deformation than with mild steel (Figure 8-6).

2. *Why Changes Cause Permanent Damage.* When an HSS part is kinked, there may or may not be visible tears in the metal. But there are almost certainly varying degrees of invisible cracks in its structure. This is further complicated when you try to straighten the damaged part. During the deformation, work hardening occurs as a result of the "cold working" taking place. Work hardening further increases the strength, but also the hardness and brittleness of steel. Interestingly, even a mild steel wire can be strengthened to double its normal tensile strength by the process of cold working. This is a principle that makes it possible to use mild steel to make products like steel siding, and vehicle body panels. These cold-formed products have greater strength than the mild steel has in its unworked state. Work hardening is found in the areas of greatest deformation (Figure 8-7). In a rail that has a gradual bend as well as a sharp fold or kink, the greatest work hardening occurs in the kinked area. When pulling this rail, it will straighten easily in the areas of the gradual bend, but it will move very slowly—if at all—at the kink. The kinked area is now even harder and more brittle than it was to begin with. So when applying the muscle to get it back to where it should be, it is likely to crack even if it did not during the collision.

3. *Repairing/Sectioning Permanent Damage.* Even if the pulling does not produce a visible break, the microscopic fractures in the grain structure will likely be there. These microscopic fractures can result in fatigue failure of the metal, which are worsened by road vibrations. Or they can result in a different response in a later collision.

Using too much heat to remove kinks is a common mistake that can cause more problems. Be sure to follow the stress-relieving with heat guidelines as recommended by the vehicle manufacturer. Limiting the heat to the recommended temperature and time limits will probably not help to remove the kink. And using sufficient heat to soften the metal can cause permanent damage to the strength of the structural member. Even if the appearance is restored, the microscopic fractures mentioned before will still be there. In addition, the heat might reduce the strength of the HSS in the heat-affected area. So while the part might look the same, it will not perform the same.

CAUTION

Although it might seem that small cracks can be welded shut, this will not produce a satisfactory repair. Work hardening and damage to the structure can result in fatigue failure later. Replacing the kinked part by structural sectioning or full panel replacement are the best solutions.

4. *Crush Zones.* Certain structural components have crush zones designed into them for absorbing the impact in a collision. This is particularly true of the front and rear rails because they take the brunt of the impact in most collisions. Crush zones are found in all front and rear rails (Figure 8-8). Some are in the form of convoluted or crinkled areas, some are dents or dimples, and others are holes or slots—all put in deliberately so the rail will collapse at these points. In so doing, the crush zones help preserve the integrity of the passenger compartment and promote a safer ride. In virtually all cases, crush zones are ahead of the front suspension and behind the rear suspension.

   When sectioning, avoid crush zones. Sectioning procedures can change the designed collapsibility if improperly located. If an apron has suffered major damage, it will normally already be buckled in the crush zone, so the crush zone will be easy to locate (Figure 8-9). If only moderate damage has occurred, be very careful; the hit might not have used up the entire crush zone. Be aware of other potential areas where designed-in collapse might occur. Replacing the damaged crush zone ensures that the structure will react the same in another collision.

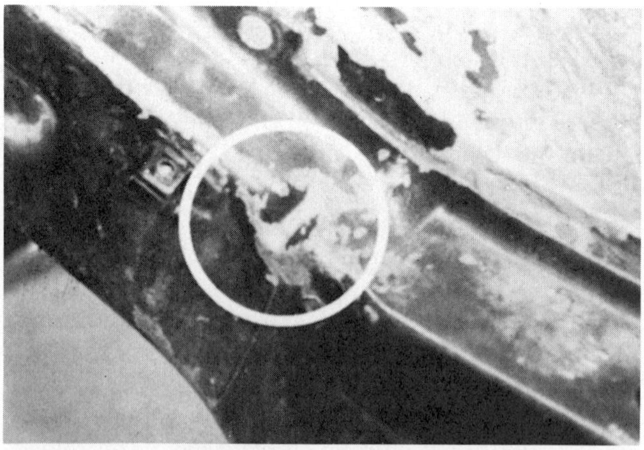

FIGURE 8-8 The dimple on this lower rail is part of an engineered crush zone. The rail folded inward at the dimple—as designed. When pulled back, its shape was restored, but a small crack developed in the center of the dimple. The rail should be sectioned behind the crush zone.

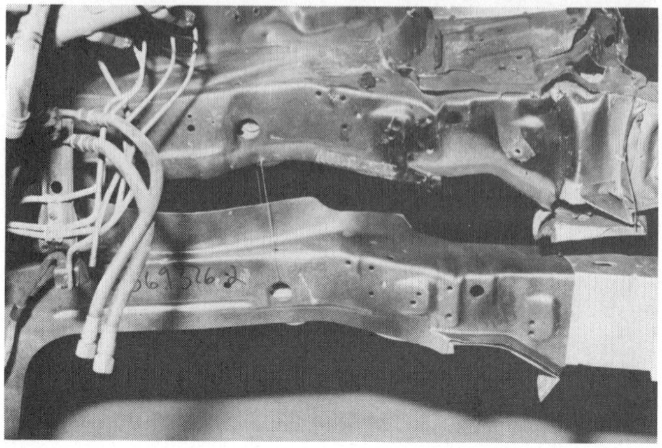

FIGURE 8-9 In the case of this severely damaged vehicle, the crush zone compressed as it absorbed the collision force. The damage calls for sectioning the apron at the area directly behind the damaged portion.

5. *Other Factors to Consider.* There are several other factors that can play a role in a decision on whether to repair or to replace: type of surface, location of damage, and extent of damage. The following are some useful guidelines:
   - If the damage is in a flat surface, chances are it will straighten much more easily than a sharp fold or buckle along a corner. Kinks along sharply formed edges almost certainly call for replacement (Figure 8-10).
   - If the damage is located near the end of a rail, with the crush zone unaffected, replacement is not as critical as when the damage is in the crush zone or even further into the rail. A kink in the crush zone calls for replacement.
   - If the damage is located near engine or steering mounting areas, repeated stress loading can create fatigue failure. Kinks at these mounting points require replacement rather than repair.
   - Tightly folded metal that is "locked" due to severe work hardening requires replacement (Figure 8-11).

FIGURE 8-10 This underbody view of a lower rail shows a sharp kink. Such distortion should not be repaired without partial or full panel replacement.

FIGURE 8-11 A hard hit in the front corner of the vehicle pushed the upper rail down, folding it into sharp kinks in both the front and middle. Replacement is the only solution in such cases.

- If there is just one small kink that was not completely removed, the decision might be different than if the area is large, with multiple kinks.
- If part of a structural member is being replaced anyway, it is easier to decide to go back a little farther and also replace a section that might be borderline.
- Straighten the surrounding areas to the proper dimensions before removing or sectioning. That is, try straightening. Then, if the kinks do not straighten properly, replace the components.

As can be seen, there are many variables that can affect the repair/replace decision. The best decisions are made by having a good understanding of all the variables. Before it is possible to make any decision, the following must be done:

1. Begin by removing all damaged exterior sheet metal and trim needed to gain proper access to the damaged area.
2. Remove the engine, transmission, and front cradle as an assembly by lifting the body up and away from the assembly. (Follow the guidelines provided in Chapter 9.)
3. Pull the damaged area to restore proper body dimensions. Use multiple pulls whenever possible, following the guidelines given in Chapter 6. This must be done prior to cutting and removing any damaged structural member.

TYPES OF STRUCTURAL REPLACEMENTS

Structural replacement can take one of two forms: replacement at factory seams and sectioning. Replacement of panels along factory seams is a common practice in unibody repair and should be done whenever it makes practical and economical sense. When removal of the damaged panel and attachment of the replacement part are carried out according to procedures, the result is almost identical to factory production in both strength and appearance.

On the other hand, sectioning might or might not be a factory-recommended practice by some manufacturers. Special care must be taken when sectioning any part due to the wide use of high-strength steels. Repairs of this type should only be performed after the technician has analyzed the vehicle and determined that sectioning will not jeopardize the structural integrity.

The procedures given here are considered proper techniques for sectioning structural components. Various manufacturers and independent testing agencies have tested these procedures and found that they provide structurally adequate repairs for the specified areas. However, be sure to follow the automaker's recommendations.

Major auto manufacturers are continually investigating the sectioning of structural components and are publishing repair procedures as they are developed and verified. Always give careful attention to the techniques described for specific vehicle makes and models in technical manuals or notices published by the manufacturers. This is very exacting work which has major impact on the strength and serviceability of the vehicle—and it should be treated as such.

REPLACEMENT AT FACTORY SEAMS

Replacing panels at factory seams is a very straightforward repair, one that leaves little room for judgmental error. It should always be considered as the first option when repairing a damaged vehicle. Since new parts are generally all that are used in this technique, none of the considerations exist that are so common when working with salvage parts. For instance, care must be taken with salvage parts not to alter the strength of the metal by cutting too close with an oxyacetylene torch. However, two possible pitfalls should be kept in mind. First, it is easy to destroy more of the factory welds than is necessary to do the job. Second, it is easy to disturb an excessive amount of factory corrosion protection.

The entire replacement procedure will be made easier if the damaged rails and panels are returned to factory dimensions before they are removed from the vehicle. The preliminary work will progress a lot smoother and it will avoid the prospect of having to work around kinked and twisted metal. Most vehicle manufacturers have specific recommendations for replacing body panels at factory seams; adhere to these recommendations as closely as possible.

DETERMINING SPOT WELD POSITIONS

Structural body panels are joined in the factory by spot welding. Therefore, removing body panels mainly involves the separation of spot welds. It is usually necessary to remove the paint film, sealer, or other coatings covering the joint area to find the spot welds (Figure 8-12). To do this, scorch the paint film with a propane torch and remove it with a wire brush. Because propane has a cooler flame than oxyacetylene and subjects the metal to less heat stress, it is the best choice for loosening the paint film. A coarse wire wheel or brush attached to a drill can be used to remove the paint, sealers, and other coatings.

Scrape off thick portions of undercoating or seam sealer before scorching the paint. Do not burn through the paint film so that the sheet metal panel begins to turn color. Heat the area only enough to soften the paint and then brush or scrape it off. It is not necessary to remove paint from areas where the spot welds are visible through the paint film.

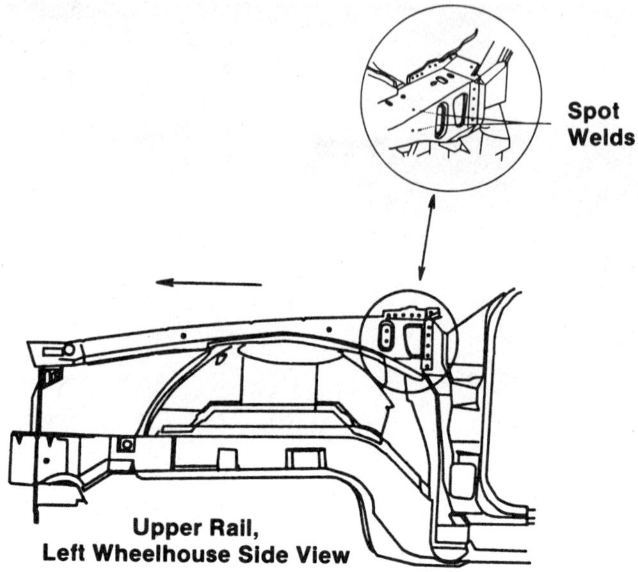

FIGURE 8-12 Paint and other coatings were removed to locate these hidden spot welds.

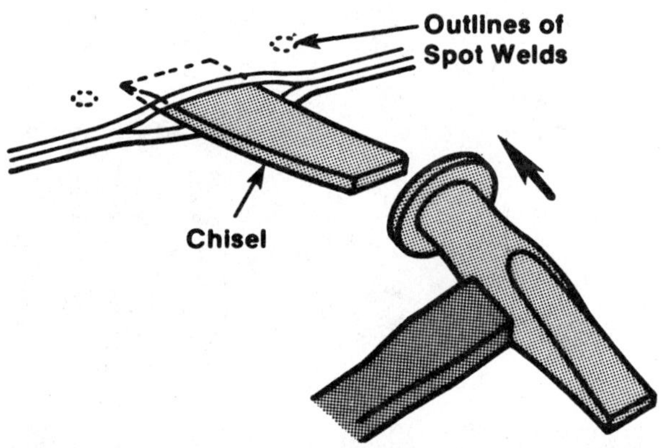

FIGURE 8-13 Determining the location of spot welds with a chisel

In areas where the spot weld locations are not visible after the paint is removed, drive a chisel between the panels as shown in Figure 8-13. Doing so will cause an outline of the spot welds to appear.

SEPARATING SPOT WELDS

After the spot welds have been located, there are several methods of separating them (Figure 8-14). To drill out spot welds, use either of two types of bit: a compound drill type or a hole saw type (Figure 8-15). Regardless of which is used, be careful not to cut into the lower panel and be sure to cut out the spot welds without creating an excessively large hole.

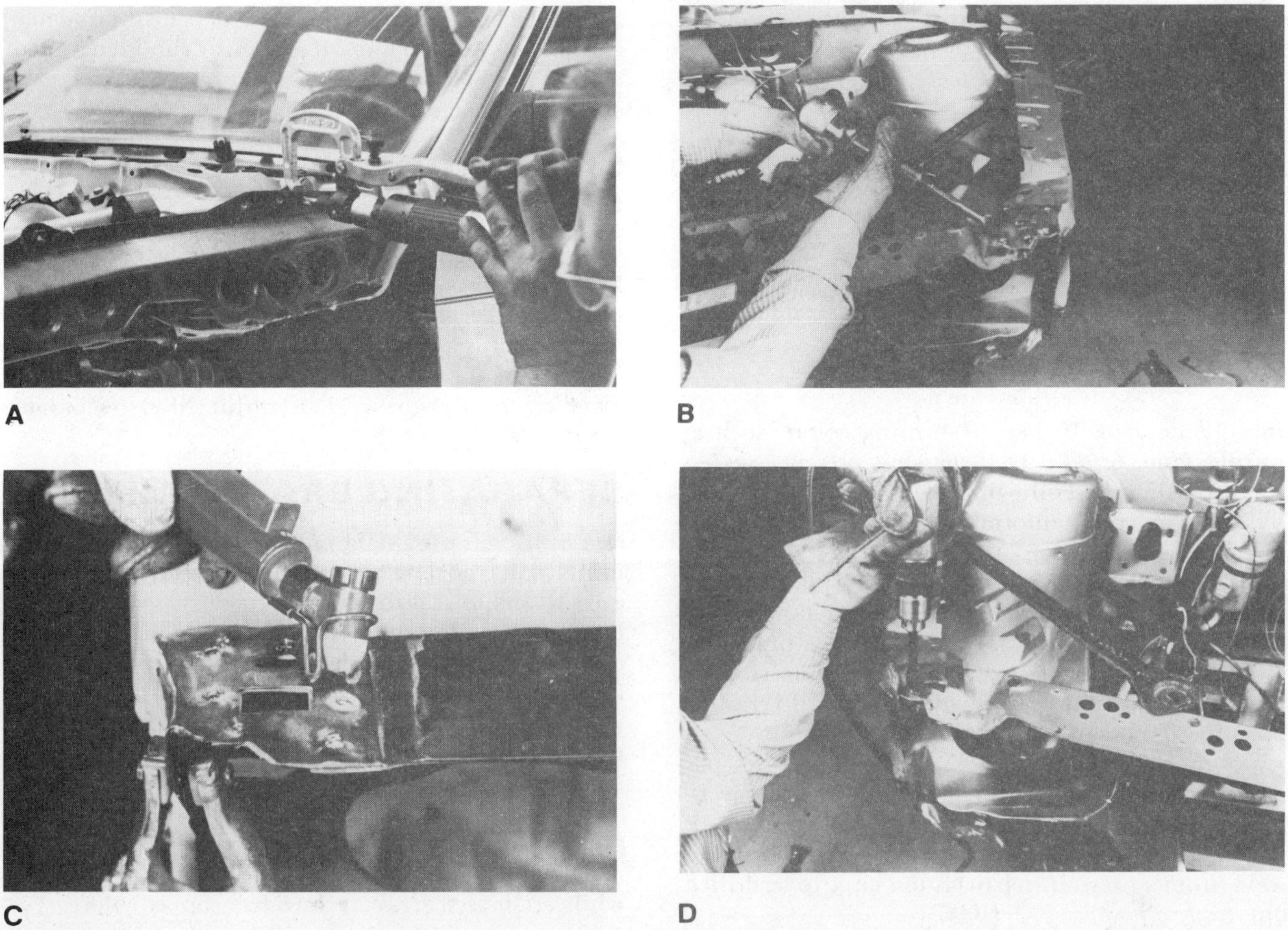

FIGURE 8-14 Methods of separating spot welds: (A) with a spot weld removing tool; (B) with an air chisel after the spot welds have been drilled or cut out; (C) with a plasma torch; (D) with a spot weld cutter

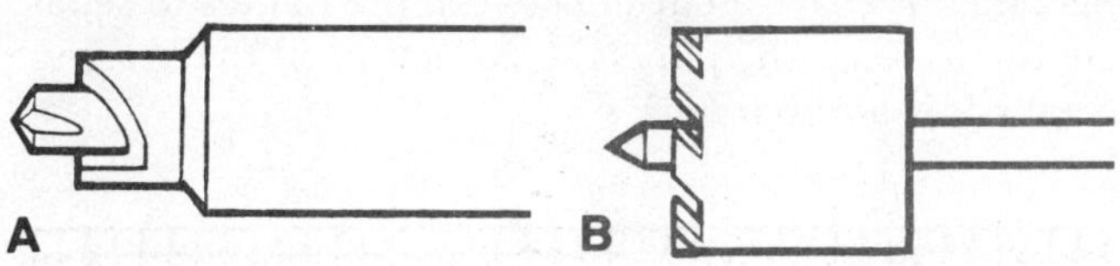

FIGURE 8-15 Spot weld cutters: (A) compound drill type and (B) hole saw type

---

CAUTION

Never use only an air chisel and chisel bit to separate spot welds.

Another drill type spot weld separating tool is the clamp saw. It is a shallow bit that cuts a smooth hole through the metal. The clamping action, plus a spur in the center of the bit, keep the bit from wandering, so center punching and predrilling are not needed. The main advantage of the clamp saw over the hole saw is that the bit can only go straight in and straight out, which lessens the danger of breaking the bit. Also, because most tools of this kind can be adjusted so the bit cuts only through the top panel, there is no remaining metal to grind off. Some types come attached to an air drill designed for removing spot welds (Figure 8-16). They are set up to drill at the right speed (rpm) and pressure for best performance and longest tool life. A disadvantage with the clamp saw is accessibility. It is designed primarily for spot welds along an outside edge, since both sides must be accessible.

A standard drill bit can also be used, but if there are numerous spot welds to cut out, the job can be tedious. To make the job easier, use a variable speed drill and special HSS bits. Make sure the bits are sharp, and use the same pressure on the drill as if you were drilling mild steel. The speed at which the bit turns should be slower than for mild steel; this will keep the heat from affecting the metal. Observe the color of the chips com-

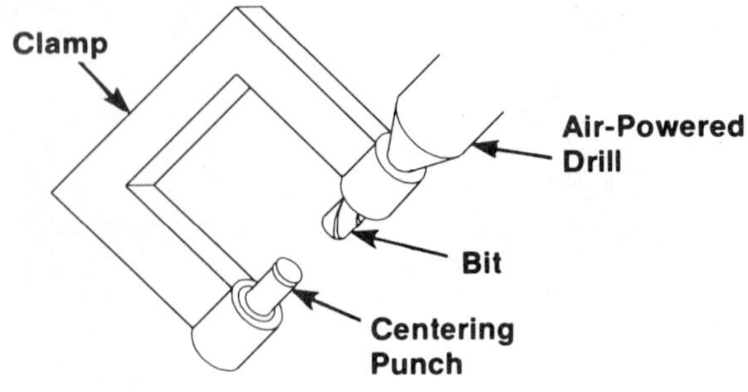

FIGURE 8-16 Clamp saw drill bit attached to an air drill

ing off the drill. If they are turning colors, such as purple, gold, or blue, or if smoke is present, use less speed on the cut. Following these simple guidelines will yield an even, uniform cut.

The removal of spot welds can also be accomplished using a plasma torch cutter. The plasma torch works a little like an oxyacetylene torch; it blows a hole in all the thicknesses of metal at the same time. Obviously, the use of a plasma torch does not preserve the integrity of the underlying panels; it can also create a potential fire hazard.

A high-speed cut-off tool can also be used to separate spot-welded panels (Figure 8-17). Use this technique only when the weld is not accessible with a drill, when the replacement panel is on top, or when a plug weld (from a previous repair) is too large to be drilled out.

After the spot welds have been drilled out, cut out, or ground down, drive a chisel between the panels to separate them. Be careful not to cut or bend the undamaged panel. Keep in mind that a chisel should never be used by itself to remove spot welds because it can create excessive damage to adjacent panels.

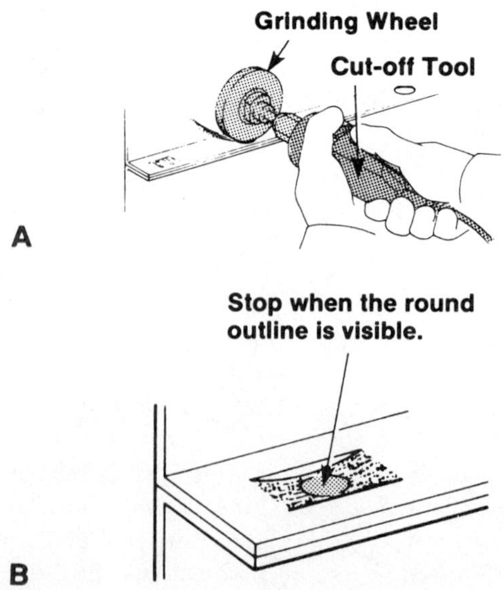

FIGURE 8-17 Removing spot welds with a cut-off tool

SEPARATING CONTINUOUS WELDS

In some vehicles, panels are joined by continuous MIG welding. Since the welding bead is long, use a grinding wheel or cut-off tool to cut through the weld without cutting into or through the panels. Hold the cut-off tool at a 45-degree angle to the lap joint. After grinding through the weld, use an air chisel to separate the panels.

SEPARATING BRAZED AREAS

Brazing is used at the ends of outer panels or at the joints of the roof and body pillars to improve the finish quality and to seal the body. Generally, separation of brazed areas is done by melting the brazing metal with a torch. However, in areas where arc brazing is used, the fusion temperature of the brazing metal is higher than with ordinary brazing, and melting the brazing metal would result in damaging the panels underneath. Therefore, areas that are arc brazed are normally separated by grinding. Ordinary brazing can be distinguished from arc brazing by the color of the brazing metal. Ordinary brazed areas are the color of brass, while arc-brazed areas are a reddish copper color.

First, soften the paint with a propane torch and remove it with a wire brush or scraper. Heat the brazing metal until it starts to melt and puddle, then quickly brush it off. Be careful not to overheat the surrounding sheet metal. Drive a chisel between the panels to separate them. Keep the panels separated until the brazing metal cools and hardens.

SHOP TALK

It is easier to separate the brazed areas after all the other welded parts have been separated.

If after removing the paint, the brazed joint is determined to be arc brazing, use a cut-off tool and grinding wheel to cut through it (Figure 8-18). If replacing the top panel, do not cut through the panel below it. After grinding through the brazing, separate the lapped panels with a chisel and hammer.

VEHICLE PREPARATION

After removal of the damaged panels, prepare the vehicle for installation of the new panels using the

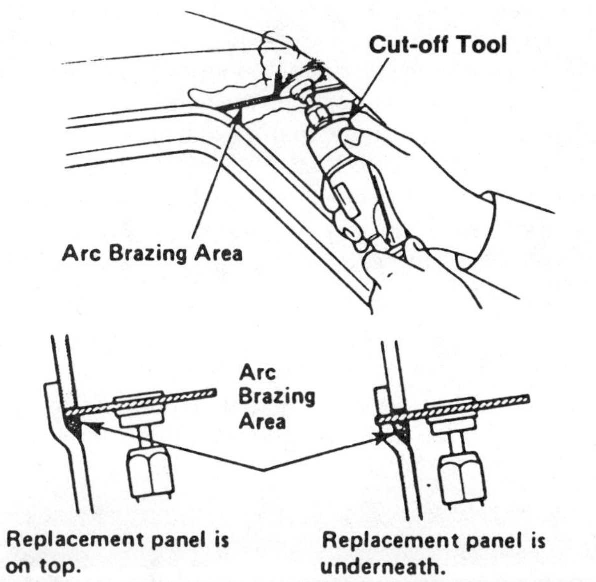

FIGURE 8-18 Separating body panels joined by arc brazing

following procedure. While the steps that follow are typical of many panel replacement operations, always refer to the appropriate manufacturer's body dimension manual for type and placement of welds.

1. Grind off the welding marks and use a wire brush to remove dirt, rust, paint, sealers, and so on from the joint surfaces. (Zinc coatings are not to be removed.) Do not grind the flanges of structural panels. Grinding will thin the section and weaken the joint.
2. Remove paint and undercoating from the backsides of the panel joining surfaces on parts that will be spot welded during installation.
3. Smooth the mating flanges with a hammer and dolly.
4. Apply weld-through primer to areas where bare metal is exposed after the paint film and rust have been removed from the joining surfaces. It is very important to apply the weld-through primer to joining surfaces or to areas where painting cannot be done in later processes.

REPLACEMENT PANEL PREPARATION

Since all new parts are coated with primer, it is important that this coating be removed from the mating flanges in order to allow the welding current to flow properly. Also, holes for the plug welds must be drilled carefully and precisely. To prepare a replacement panel for welding, follow these steps:

1. Use a disc sander to remove the paint from both sides of the spot weld area. Do not grind into the panel and do not heat it so that it turns blue or begins to warp.

CAUTION

Wherever possible, grind so that sparks fly down and away. Always wear proper eye protection when grinding.

2. Apply weld-through primer (as an anti-corrosion treatment) to the welding surfaces where the paint film was removed.
3. Drill holes for plug welding. Always refer to the body dimension manual for the size of plug weld holes. Always be sure to duplicate the location and number of spot welds used in the factory.
4. Be sure all plug weld holes are of the proper diameter. If they are too large or too small for the thickness of the panel, either the metal will melt through or the weld will be a poor one.
5. If the new panel is sectioned to overlap any of the existing panels, rough cut the new panel to size using an air saw or a cut-off wheel. The edges should overlap the portion of the remaining panel by 3/4 to 1 inch. If the overlap portion is too large, it will make matching the position of the panel more difficult.

POSITIONING NEW PANELS

Aligning new panels is a very important step in repairing unibody vehicles. Improperly aligned panels will affect both the appearance and the driveability of the repaired vehicle. Use a measuring system to determine the installation position. Remember that there is a definite relationship between the fit of the new and old parts and the finish appearance. Whether structural or cosmetic panels are being replaced, the emphasis is on proper measurement and fit.

CAUTION

All measurements must be accurate before welding body panels in position. Because there are no shims on unibodies, no adjustments can be made to the outer panels. So, each panel must be precisely positioned before welding.

Positioning by Dimensional Measurement Methods

When using a dedicated bench or a universal measuring system, the vehicle must be properly positioned on the bench before the new panel can be correctly aligned (Figure 8-19). This will usually have already been done in order to pull and straighten damaged panels that do not require replacing. All straightening must be done before replacing panels; otherwise, proper alignment of the new panels will be impossible.

So, when using a dedicated bench system for panel replacement, place the fixtures on the bench in their correct locations and bolt them in place. Then, place the new panel in position on the fixtures and see how it lines up with the good panels on the vehicle. Make any necessary adjustments, clamp the panel in place (Figure 8-20), and weld it to the mating panels.

In the procedure that follows, a vehicle is being fitted with a front fender apron assembly, a front crossmember, and a radiator core support using a tram bar and self-centering gauge.

1. Match the assembly reference marks on the installation areas of the front fender apron and the rail, and secure them with vise grips (Figure 8-21). Parts that have no assembly reference marks should be positioned in the same location as the old parts.

A

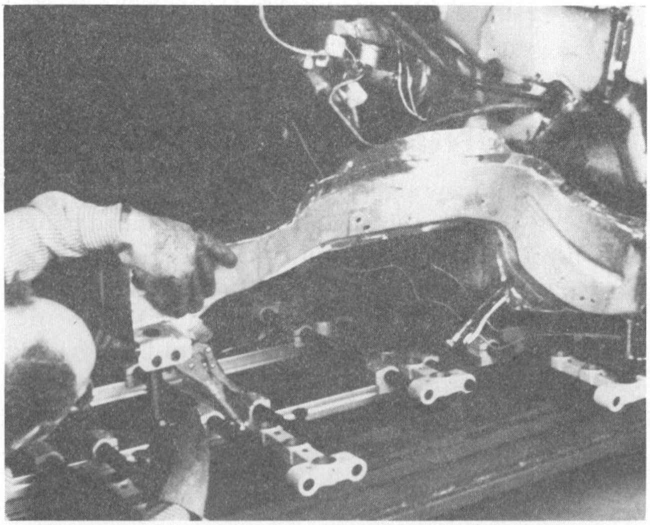

B

FIGURE 8-19 Using the dimensional measurement method: (A) aligning a new panel on a dedicated bench; (B) aligning a new panel on a universal measuring system

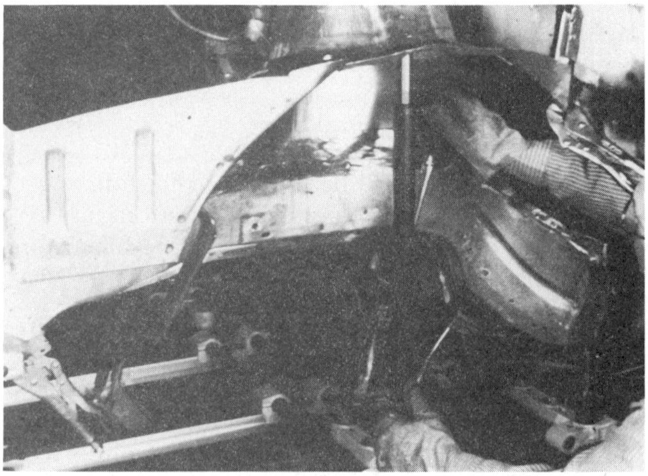

FIGURE 8-20 Clamp the new panel in place.

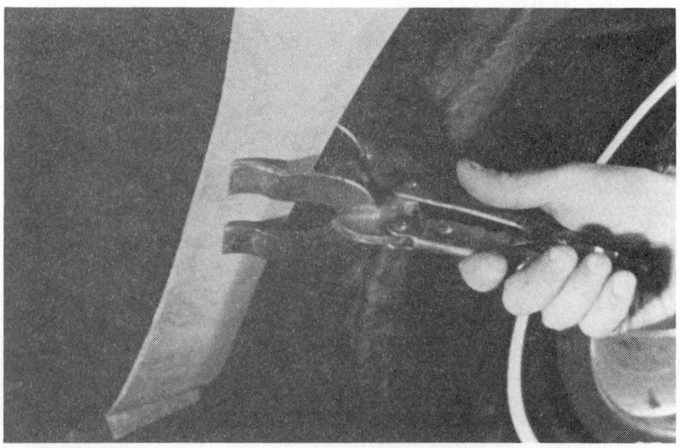

FIGURE 8-21 Vise grips are used to align the body panels.

2. Align the parts by measuring the length between reference points with a tram bar. Temporarily install the front crossmember by tack welding one spot. Make any length adjustments by tapping a hammer against a wood block that, in turn, strikes the panel and moves it in the desired direction (Figure 8-22).
3. Scribe a positioning line at the end of the part that is not welded; drill holes and fasten the parts together with sheet metal screws. Scribe a line on the apron area but do not weld the panels together.
4. Use self-centering gauges to match the height of the new parts to the parts on the opposite side of the vehicle (Figure 8-23). Support the new parts with a hydraulic jack to make sure that the height does not change at all.
5. Measure the lower diagonal and width dimensions (Figure 8-24). Support the parts with a hydraulic jack so that the height does not change, then adjust the rail as needed to obtain the correct dimensions. Confirm the height dimensions again.
6. Take care to position the front crossmember so that both the left and right ends are uniform.
7. Once the dimensions of the rail match the dimensions found on the dimension chart, secure the part in place. The suspension crossmember can also be installed with fixtures. Use a sufficient number of plug welds to fasten the joining area of the rail to the front crossmember.
8. Make sure that the apron's upper length has not changed. Confirm by checking for shifting of the scribed line.
9. Measure the diagonal dimensions between the fender rear installation hole and the strut tower hole or fender front installation hole (Figure 8-25).

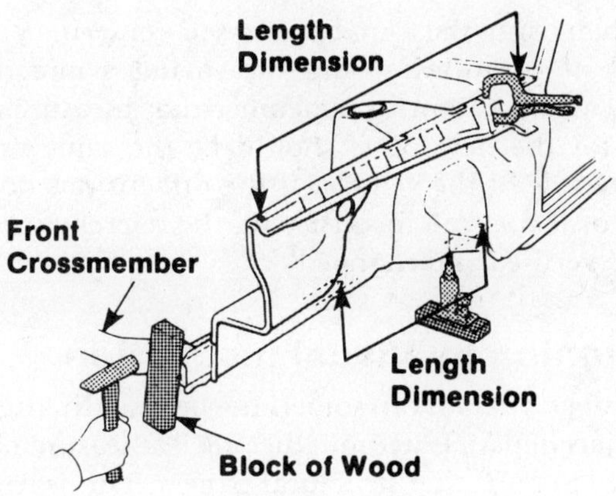

FIGURE 8-22 Making length adjustments

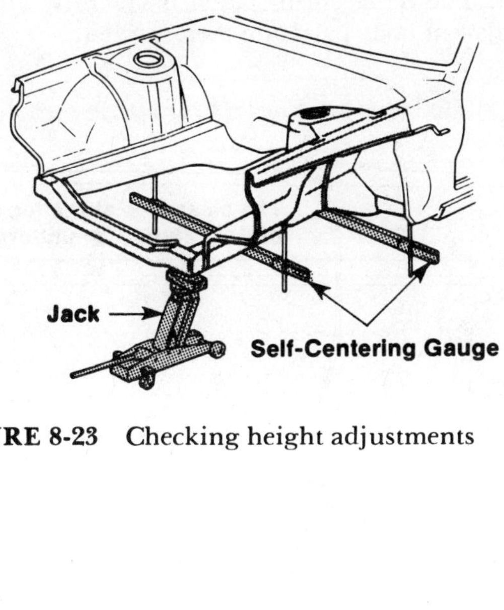

FIGURE 8-23 Checking height adjustments

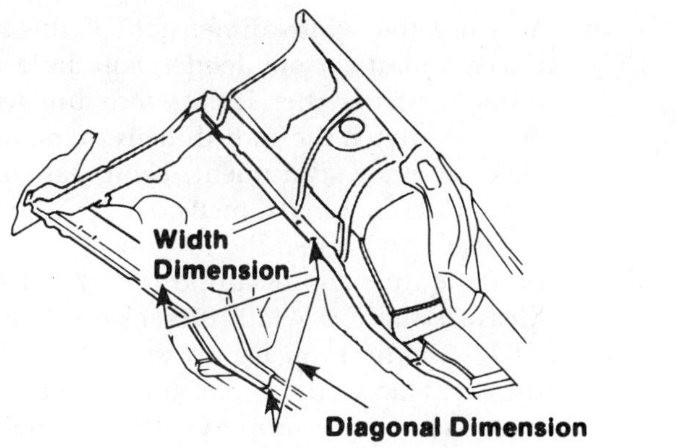

FIGURE 8-24 Checking lower diagonal and width dimensions

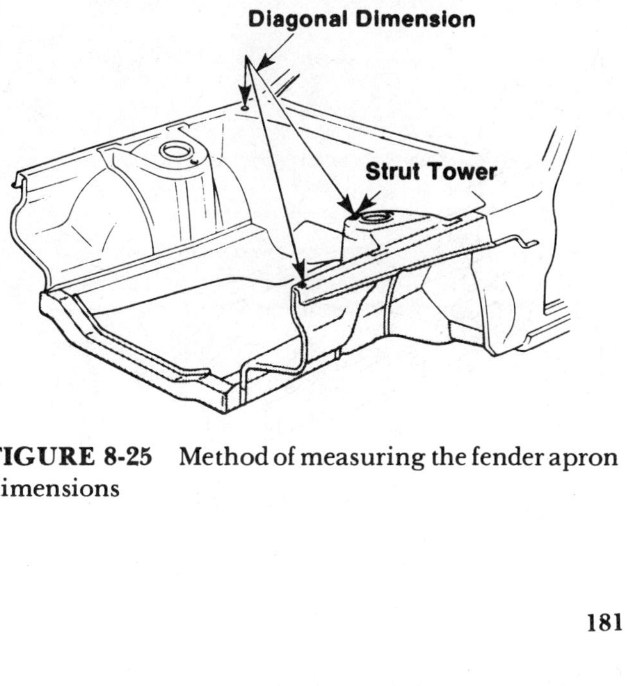

FIGURE 8-25 Method of measuring the fender apron dimensions

10. Measure the width dimension of the strut tower and the front fender bolt hole and fasten them together. If the width dimension does not match the body dimension manual, make a small adjustment, being careful of changes in diagonal dimensions. Temporarily install and fasten the radiator upper support and the radiator support (Figure 8-26).
11. Measure the rail width dimensions (Figure 8-27). Set the tram bar to the proper measurement and adjust the apron as needed. Lightly fasten the support with vise grips and tap it softly by hand to move it in place.
12. Measure the diagonal dimensions for the side supports (Figure 8-28). Be sure these dimensions match.
13. Temporarily install the front fender and inspect it for proper fit with the door (Figure 8-29). If the clearance is not correct, it might be because the fender apron or the rail height is not correct.
14. Measure as previously described, and verify the overall dimensions once more before welding.

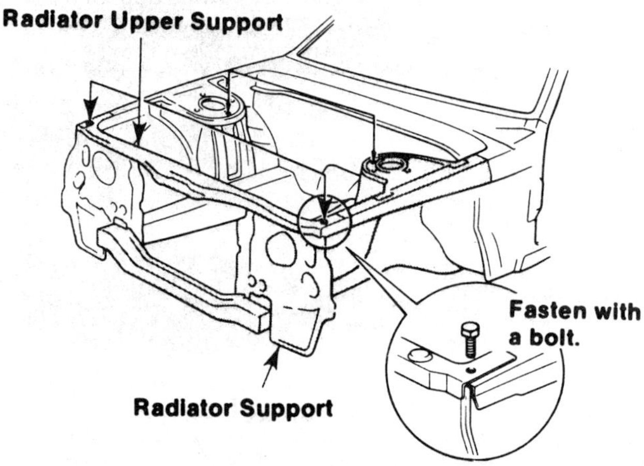

FIGURE 8-26 Installing the radiator upper support and the radiator support

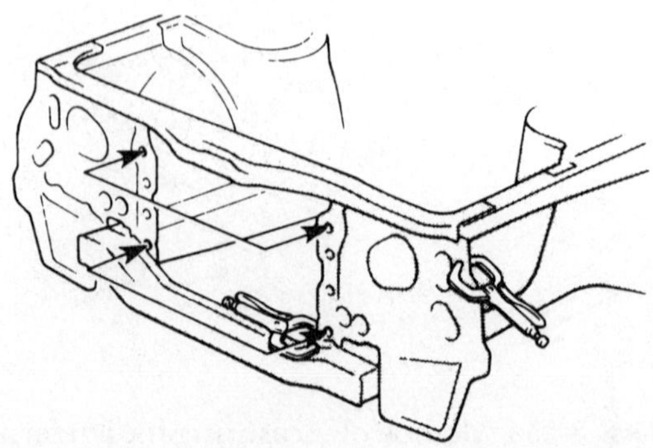

FIGURE 8-27 Measuring the rail width dimensions for proper alignment

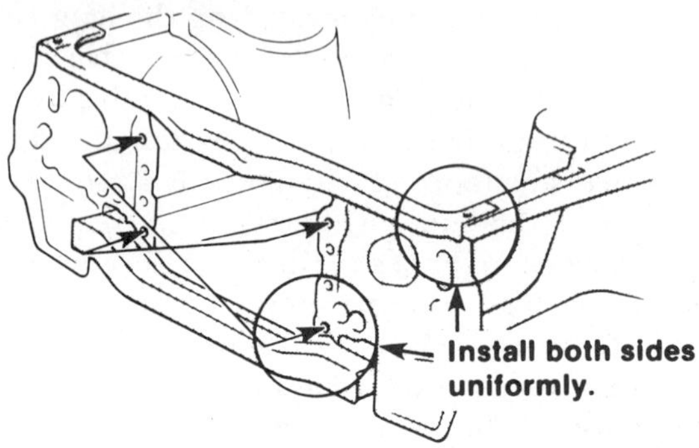

FIGURE 8-28 Measuring the radiator support diagonal dimensions

When using the tram bar and self-centering gauge method of positioning and measuring symmetrical parts, it is important to remember that measurement points for the new parts should be the same as the opposite side of the vehicle. If the dimensions do not match, or if they are inconsistent, the reference points must be verified and changed.

Positioning by Visual Inspection

Cosmetic panels can sometimes be visually aligned with adjacent panels without the precise measurements necessary in replacing structural panels. This is true of both mechanically fastened and welded panels. The emphasis here is on appearance. Body lines must be aligned and gaps between panels must be even, not tapered. The replacement panel is carefully aligned with adjacent body parts and then secured.

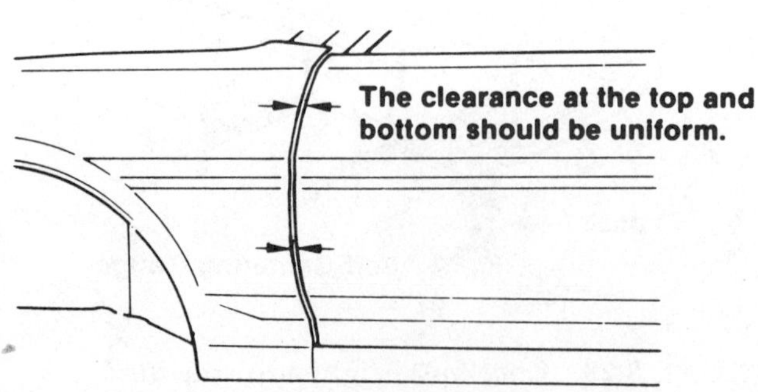

FIGURE 8-29 Inspecting the fender-to-door gap

When installing an overlapping panel, position it carefully in place and then cut the overlapping portion of the joining area with an air saw or cutoff wheel. Be precise when making cuts in the sectioning area; if a gap opens up at the cut or if the panels overlap, welding will be difficult. It is important that the cutting be done accurately.

Cutting the overlapping section can be accomplished in one of two ways: If there is sufficient overlap, both panels can be cut simultaneously. If the overlap is small, a line can be scribed at the end of the overlapping panel. When cut along the scribed line, the panels should fit snugly together with little or no gap.

After cutting the overlap to fit, remove the replacement panel and clean any metal chips and other foreign material from the inside of the panel before proceeding. Apply body sealer around the inside perimeter of the panel. Secure the panel and other parts with self-tapping screws in the same screw holes as before. Verify the fit once more.

WELDING NEW PANELS

When satisfied with the position and dimensions of the new panel, it can be permanently welded in place (Figure 8-30). Under ideal circumstances, the general welding procedure should be:

1. Corrosion protection
2. Clamp
3. Tack weld in position
4. Remeasure
5. Plug weld in final position
6. Grind cosmetic weld surfaces

Remember that it is very important to duplicate exactly the location and number of original factory spot welds. Further details on welding operations can be found in Chapter 7. Repair of plastic panels, such as those made of RRIM and SMC, are discussed in detail in Chapter 13.

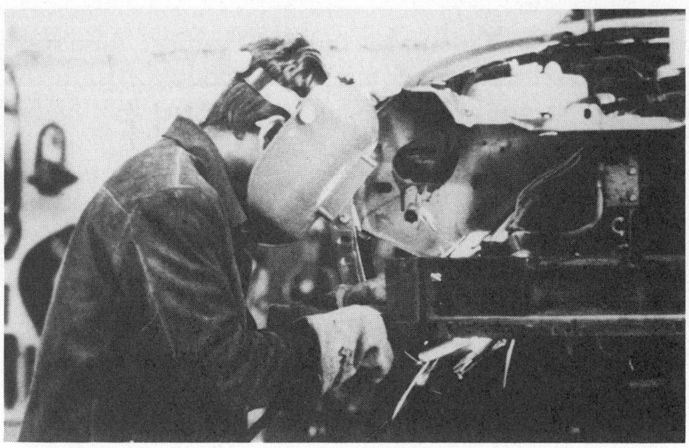

FIGURE 8-30 Begin the welding only when satisfied with the position of the new panel.

SECTIONING

When damaged unibody components need to be replaced, replacing them at factory seams is the logical first choice. However, this is impractical when many seams have to be separated in undamaged areas of the vehicle. In some repairs, sectioning of components such as rails, pillars, and rocker panels may be required to make their repair economically feasible. But, it is important to remember that sectioning requires a high degree of workmanship, as well as strict adherence to procedure. When done correctly, it can be an effective and economical alternative to replacing the panels along factory seams. As a first step, always check the body dimension manual for the vehicle manufacturer's recommended procedure for the specific sectioning location. Because import manufacturers have accepted sectioning for years, their manuals are very specific as to procedures. If such recommendations do not exist, use the general guidelines that follow. As a collision repair technician, never forget that sectioning finally comes down to sound judgment; it is the technician who must assure the quality of the repair through the proper application of tested and proven procedures.

The structural parts of a vehicle that can be sectioned include the following components: rocker panels, floor pans, front rails, rear rails, quarter panels, trunk floors, B-pillars, and A-pillars (Figure 8-31). These unibody components involve the following three basic types of construction:

- Closed sections (Figure 8-32A), such as rocker panels and A- and B-pillars
- Hat or open U-channels (Figure 8-32B), such as rear rails
- Single layer or flat components (Figure 8-32C), such as floor pans and trunk floors

There are certain areas of the vehicle that should always be avoided when sectioning. They include the following:

- Crush zones (Figure 8-33)
- Suspension mounts
- Dimensional reference holes
- Compound shapes/reinforcements/double layered metal
- Compound structures
- Engine or drivetrain mounts

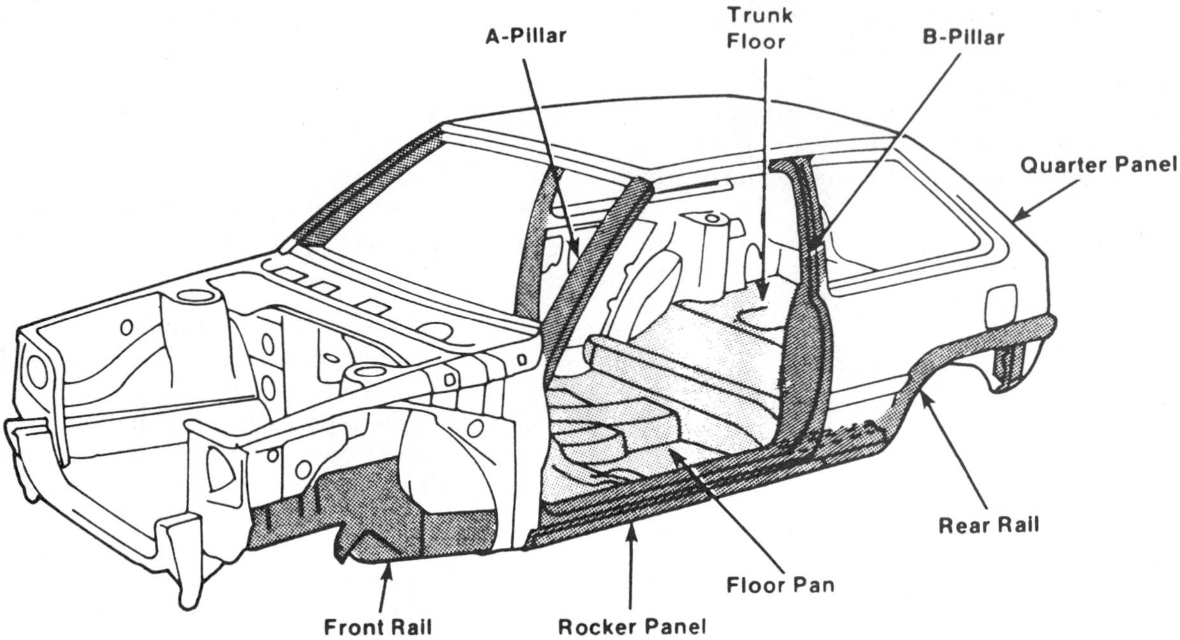

FIGURE 8-31 Sectioning areas of a typical unibody vehicle

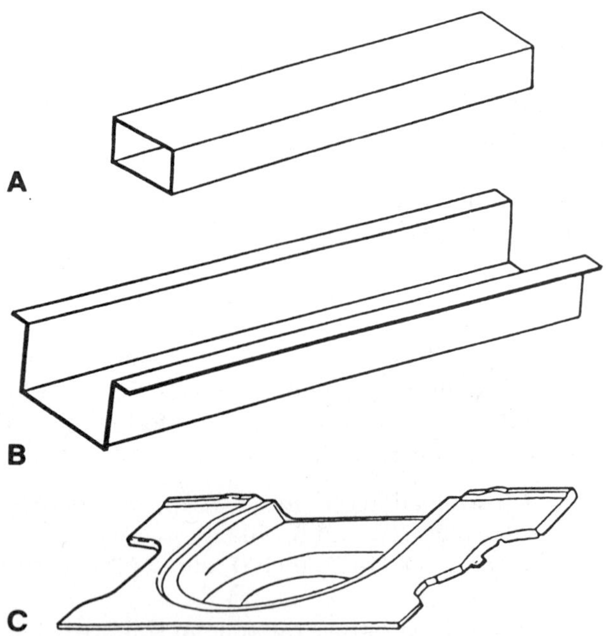

FIGURE 8-32 Unibody construction types: (A) closed sections, (B) hat or open U-channels, and (C) single layer or flat components

FIGURE 8-33 Crush zones should not be sectioned under any circumstances.

The techniques given here involve the use of recycled assemblies consisting of the upper rail, partial lower rail, apron extension panel, radiator support, and opposite side rail extension. The panel assembly is installed at factory seams, except for the sectioning of the lower front rail at a designated location. Actually, the use of recycled assemblies in unibody collision repair makes sense for a number of reasons, including the following:

- Fewer welds need to be made when using recycled assemblies, than are required when separate new parts are used.
- Less factory corrosion protection is disturbed.
- More measuring is required when welding separate new parts and attaching them to the vehicle.
- There is an abundance of recyclable assemblies available in most areas.

SHOP TALK

When deciding where to section, look for an area with a uniform cross section. Before beginning, check the body dimension manuals and bulletins provided by the vehicle manufacturers. Much of this literature provides specific instructions on how and where to section. If there are no such recommendations, follow the general guidelines presented in this chapter.

BASIC TYPES OF SECTIONING JOINTS

Proper sectioning techniques involve three basic types of joints, plus certain variations and combinations of them. A butt joint with insert (Figure 8-34) is used mainly on closed sections, such as rocker panels, A- and B-pillars, and rails. Inserts make it easy to fit and align the joints correctly and also help make the welding process easier.

Another basic joint is an offset butt joint without an insert, also known as a staggered butt joint (Figure 8-35). This type of joint is used on A- and B-pillars and front rails. The third type is a lap joint (Figure 8-36), which is used on rear rails, floor pans, trunk floors, and B-pillars.

The configuration and makeup of the component being sectioned might call for a combination of joint types. Sectioning a B-pillar, for instance, might require the use of an offset cut with a butt joint in the outside piece and a lap joint in the inside piece.

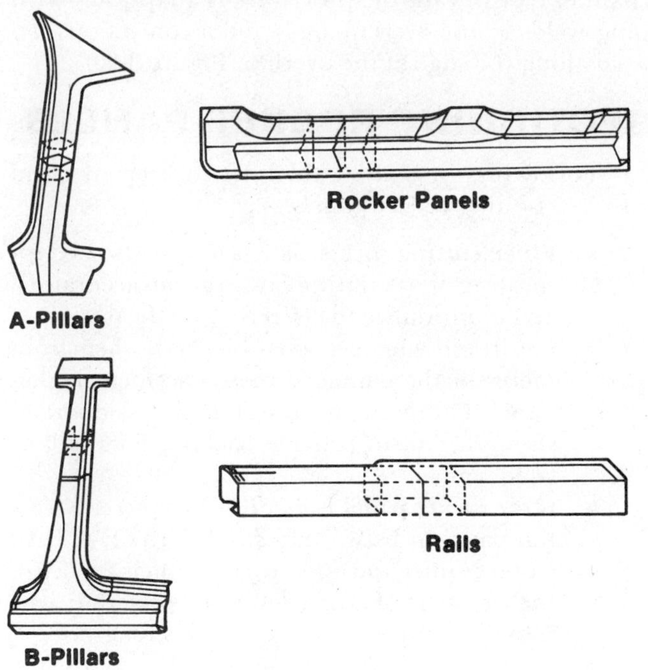

FIGURE 8-34 Examples of butt joints with an insert

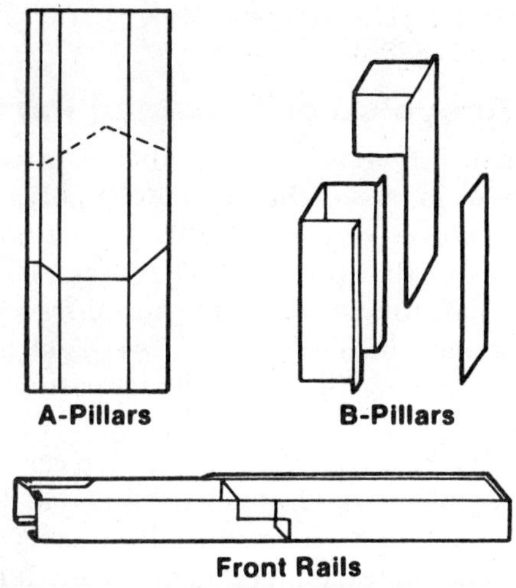

FIGURE 8-35 Examples of offset butt joints without inserts

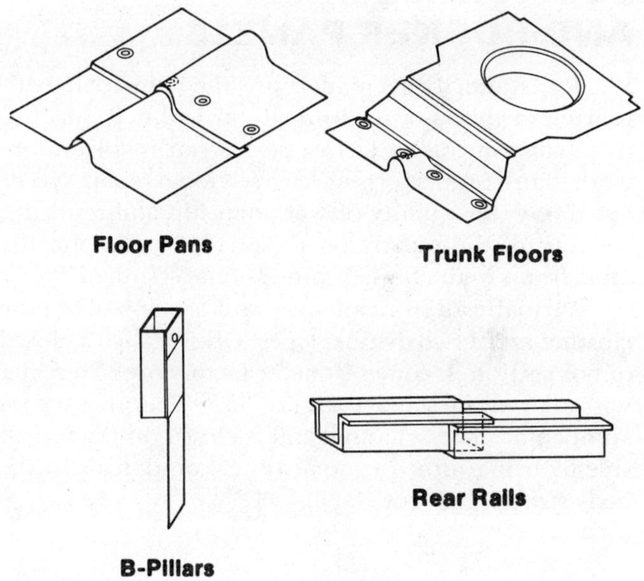

FIGURE 8-36 Examples of lap joints

PREPARING TO SECTION

When preparing to section and replace a structural part of a damaged vehicle, certain steps must be taken to ensure the quality of the repair. The first of these is the sectioning of the replacement part by the collision repair technician. Specific instructions must be provided as to the placement of the section and the method of sectioning to be used. Other important considerations are the welding techniques used (see Chapter 7) and the cleanliness of the joint.

Using Recycled or Salvaged Parts

When using recycled parts, the technician must know where to make the cuts. Preferably, the part should be removed with a metal saw. If a cutting torch is used, make sure that at least 2 inches of extra length are left on the part to insure that the heat dispersion from the cut does not invade the joint area. Make the cut so that the reinforcing pieces inside the part are not cut through.

When a recycled or salvaged part is received, examine it for corrosion. If it has a lot of rust on it, do not use it; ask for another one. Before installing a recycled part, check it for possible damage and make sure it is dimensionally accurate. Remember: quality material is a must to achieve a quality repair.

SECTIONING RAILS AND ROCKER PANELS

Rails and rocker panels are the principal load-bearing members of the unibody structure. Replacing these components at factory seams can involve more work than a properly completed sectional repair. Needless to say, the quality of workmanship and resulting strength of the repairs on these critical components must be up to the highest professional standards.

Virtually all front and rear rails are closed, but the closures are of two distinct types. One is called a closed or box section. It comes from the factory or the recycler with all four sides intact (Figure 8-37). The other type is an open or "hat" channel and is closed on the fourth side by being joined to some other component in the body structure (Figure 8-38).

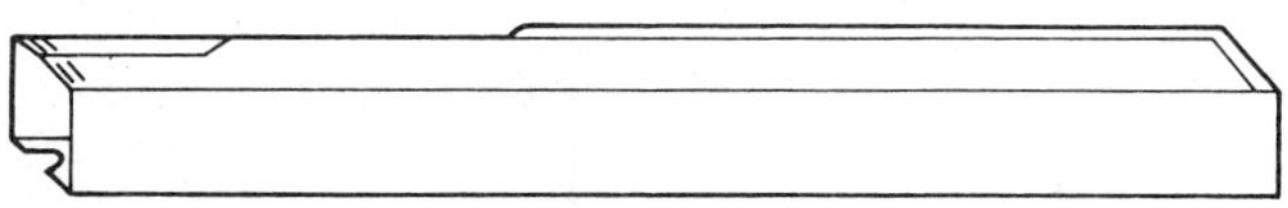

FIGURE 8-37 Typical closed or box rail

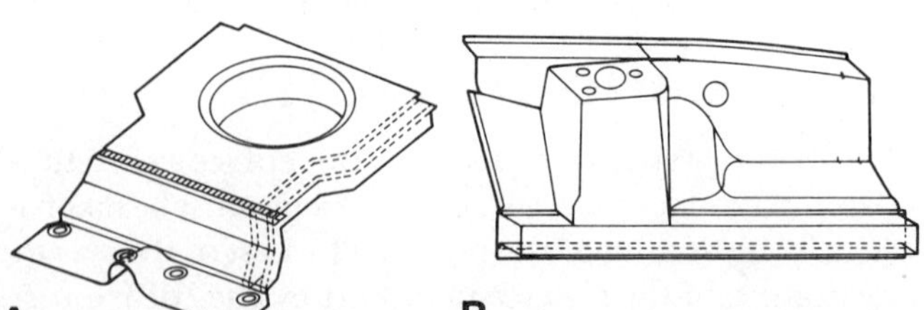

FIGURE 8-38 Typical hat channels: (A) rear rail and (B) front rail

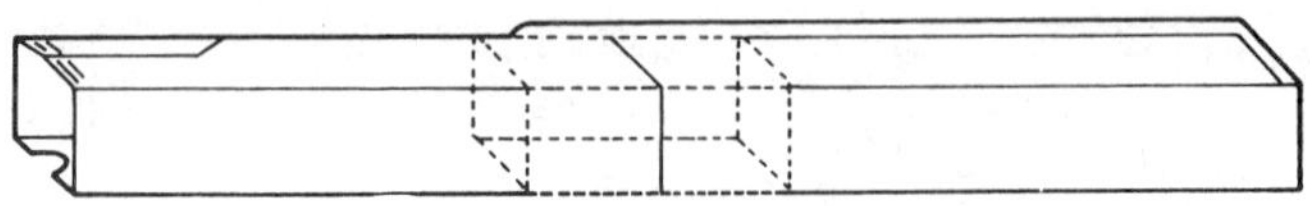

FIGURE 8-39 A closed section rail with butt joint and insert

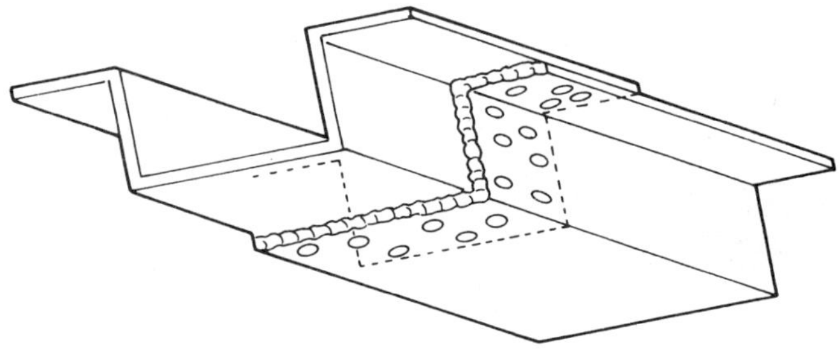

FIGURE 8-40 Joining hat channel rails

The butt joint with insert is the procedure for repairing a closed section rail (Figure 8-39). When sectioning front or rear rails, always remember that they contain crush zones that must be avoided when making the cuts. And remember to stay away from any holes and reinforcements in the rails as well.

Most rear rails, plus various makes of front rails, are of the hat channel type. Some hat channel closures are vertical, such as a front rail joined to a side apron. Others are horizontal, such as a rear rail joined to a trunk floor. In most cases when sectioning the hat channel type of rail, the procedure is a lap joint with plug welds in the overlap areas and a continuous lap weld along the edge of the overlap (Figure 8-40).

SECTIONING ROCKER PANELS

Following are some guidelines to keep in mind when sectioning rocker panels:

- When cutting rocker panels to size, use a reciprocating or oscillating saw to get an accurate fit and to minimize the effects of the heat.
- Use inserts whenever possible. Make them from pieces of the damaged or replacement rocker panel. The main purpose of inserts is not to add strength, but to provide backing for the butt weld seams.
- If the rocker panel is made of two pieces with a flat panel in-between, section with offset cuts. Cut the inner and outer rocker panels 2-1/2 to 3 inches apart (Figure 8-41). The overlap will provide continuity and increase strength.
- Be sure to weld all butt seams fully to close the joints all the way around (Figure 8-42). Lack of

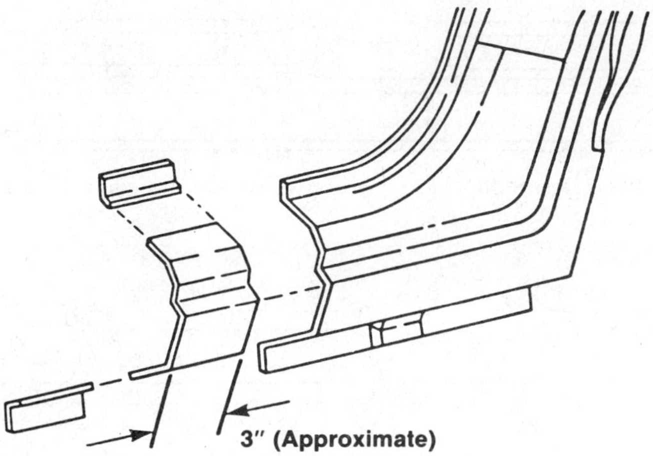

FIGURE 8-41 Cutting inner and outer rocker panels

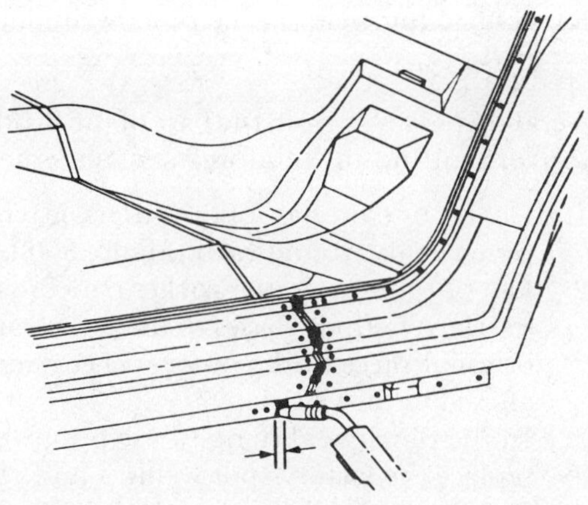

FIGURE 8-42 Welding the rocker panel butt joint

continuity over even a small portion of the seam will reduce the strength of the joint severely and may cause it to break apart later.

- Use plug welds to secure any inserts used.
- Plan the joints so that the internal reinforcement provides continuity between the two sides. Secure them with plug or seam welds.

Rocker panels can come in two-piece or three-piece designs, depending on the make and model of the vehicle. In both cases, the rocker panel might contain reinforcements, which can be either intermittent or continuous. Depending on the nature of the damage, the rocker panel can be replaced with the B-pillar or without it. If possible, rocker panel replacement should be made at the original factory seams.

To section a rocker panel, a straight-cut butt joint with insert can be used (Figure 8-43), or the outside piece of the panel can be cut and the replacement piece installed with overlap joints (Figure 8-44). Generally speaking, the butt joint with insert is used when installing a recycled rocker panel with the B-pillar attached.

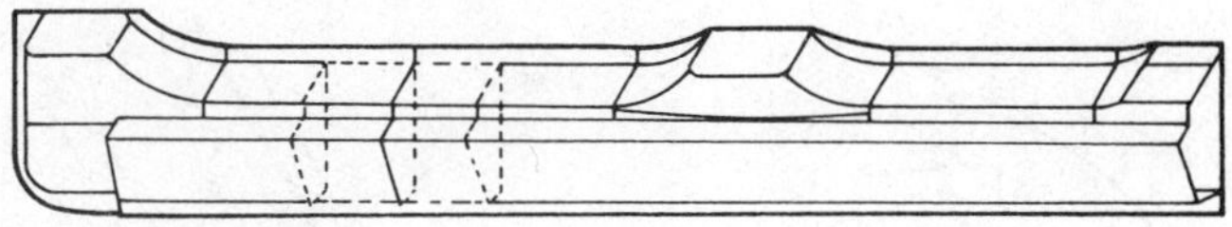

FIGURE 8-43 Joining rocker panel sections with a straight-out butt joint and insert

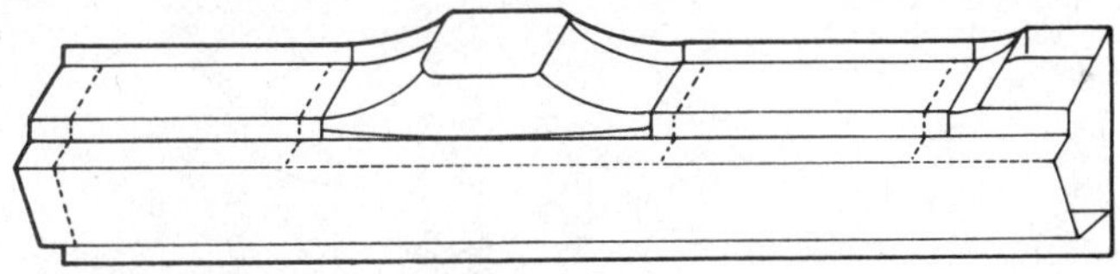

FIGURE 8-44 Replacing the outer rocker panel pieces using lap joints

To do a butt joint with insert, cut straight across the panel. The insert is fashioned out of one or more pieces cut from the excess length on the repair panel or from the end of the damaged panel. The insert should be 6 to 12 inches long and cut lengthwise into two to four pieces, depending on the rocker panel configuration (Figure 8-45). Remove the pinchweld flange so that the insert will fit inside the panel. With the insert in place, secure it with plug welds. For structural sectioning, 5/16-inch plug weld holes are required to achieve an adequate nugget and acceptable weld strength. Making a 5/16-inch weld requires a circular motion of the gun to properly fuse the edge of the hole to the base metal (see Chapter 7).

When installing an insert in a closed section, whether it is a rocker panel, A- or B-pillar, or rail, make sure the closing weld fully penetrates the insert. When closing the joint with a butt weld (Figure 8-46), leave a gap that is wide enough to allow thorough penetration of the insert. The width of the gap depends on the thickness of the metal, but ideally it should not be less than 1/16 inch or more than 1/8 inch. Be careful to remove the burrs from the cut edges before welding; otherwise the weld metal tends to travel around and up under the burr. This can result in stress concentration that can cause cracks and weaken the joint (Figure 8-47).

In general, use the lap joint procedure on a rocker panel when installing only the outer rocker panel or a portion of it. Leave the inner piece intact and cut only the outer piece. One way to make a lap joint is to make the cut in the front door opening and allow for an overlap there when measuring. When making this cut,

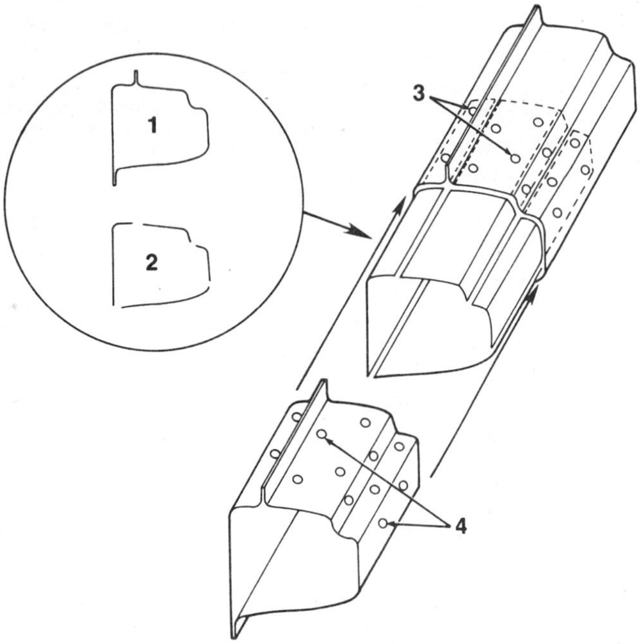

1. **Cross Section of Rocker Panel Insert Material Before Cutting Lengthwise**
2. **Insert Cut Lengthwise into Sections**
3. **Insert Inside Rocker Panel, Secured With Plug Welds or Sheet Metal Screws**
4. **5/16" Holes for Plug Welds**

FIGURE 8-45 Cutting an insert to fit a rocker panel

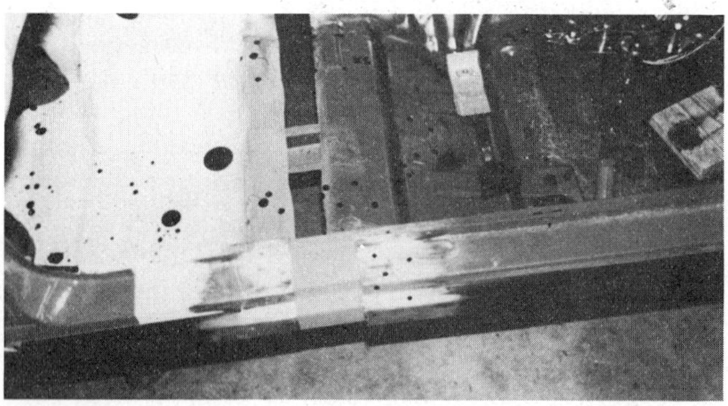

FIGURE 8-46 Rocker panel butt joint

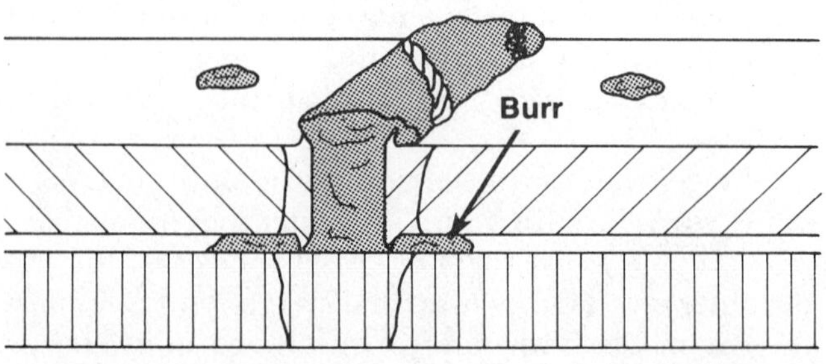

FIGURE 8-47 Burrs in the joint will weaken the weld.

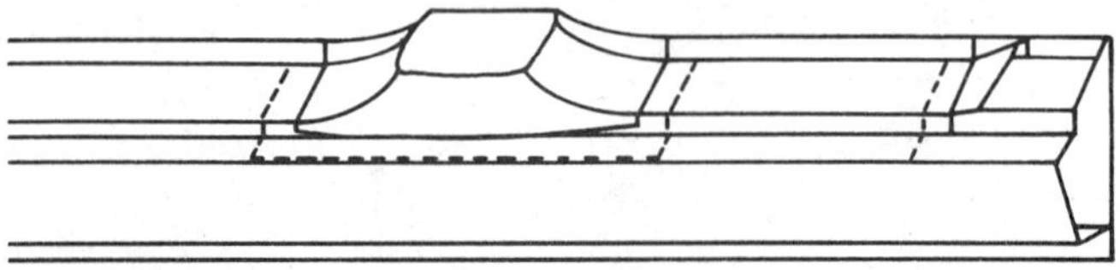

FIGURE 8-48 Leave pinchwelds around each pillar.

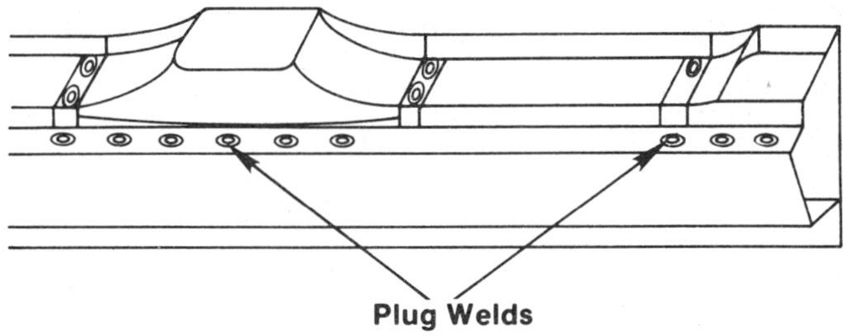

FIGURE 8-49 Plug welding the pinchweld flanges

stay several inches away from the base of the B-pillar in order to avoid cutting any reinforcement underneath it.

1. Cut around the base of the pillars, leaving the pinchwelds around each (Figure 8-48).
2. Cut out the new outer rocker panel so that it overlaps both the bases of the pillars and the original piece of the outer rocker panel still affixed to the vehicle.
3. In the pinchweld flanges, use plug welds to replace the factory spot welds (Figure 8-49).
4. Plug weld the flanges around the B- and C-pillars, using approximately the same spacing as in the pinchweld flanges (Figure 8-50).
5. Lap weld the edges with about a 30 percent intermittent seam; about 1/2 inch of weld in every 1-1/2 inches of overlap edge (Figure 8-51).
6. Plug weld the pinchweld flanges in the door opening, and lap weld around the edges to close the joint (Figure 8-52).

To reverse this procedure (depending on the nature of the hit), make the overlap cut in the rear door open-

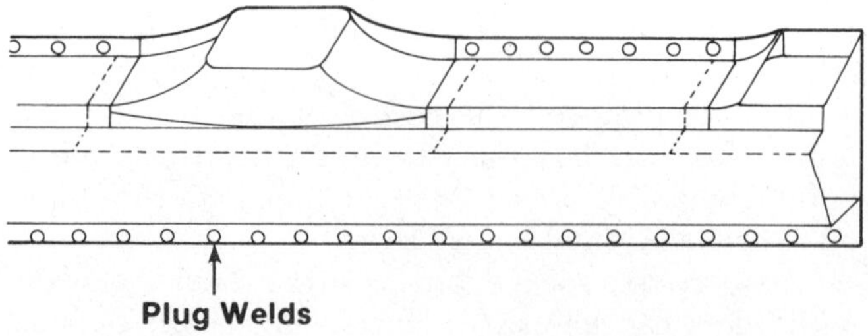

FIGURE 8-50 Plug welding the flanges around the B- and C-pillars

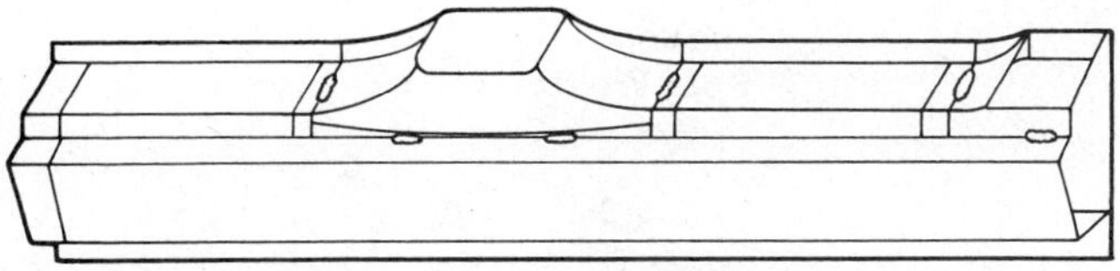

FIGURE 8-51 Intermittent lap welds

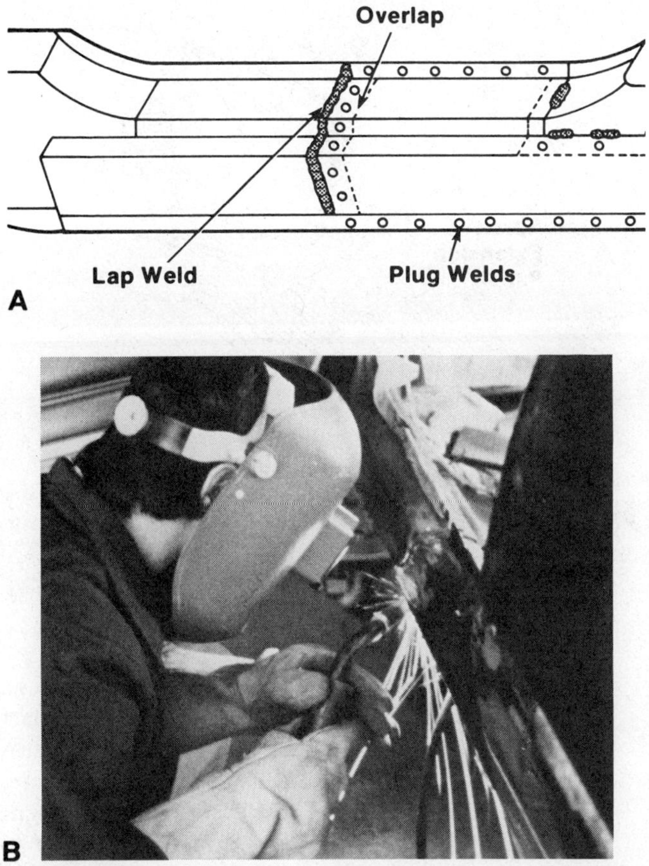

FIGURE 8-52 (A) Plug and lap welding the flanges in the door opening and (B) welding to close the joint

ing and cut out and overlap around the bases of the A-and B-pillars. Use this same basic technique to replace the entire outer rocker panel. In this version, cut around the bases of all three pillars and overlap all three bases in the same way as before.

SECTIONING RAILS

To fully understand the use of lap welds, consider the following example, which involves a vehicle that has sustained damage to the left side, with the left rail to be sectioned. These guidelines apply to damage on the right side or on both sides of the vehicle as well.

1. Locate and drill out the factory spot welds that attach the upper rail to the cowl at the base of the windshield.

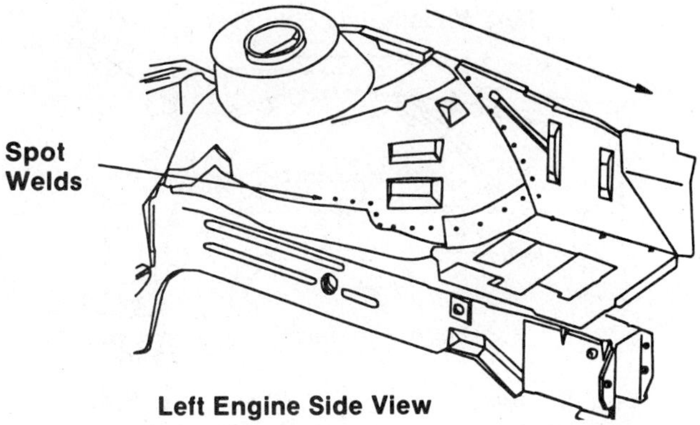

FIGURE 8-53 General location of the spot welds in the strut tower area

2. Remove the spot welds that attach the strut tower to the rail extension panel at the base of the strut tower (Figure 8-53). Remove any seam sealer covering these welds inside the engine compartment. The wheelhouse side is likely coated with sealers and sound-deadening materials.
3. The lower rail sectioning is performed forward of the center of the strut tower. The sectioning procedure uses a staggered cut of the inner and outer lower rail, with a lap joint at both cutlines. There is an inner rail reinforcement that is located in this area. Because of this reinforcement, this part of the rail is ideal for sectioning.
4. There are two spot welds that attach the reinforcement to the inside of the lower rail. These must be removed before any cuts are made; they are visible on the wheelhouse side of the rail.
5. The location of the cut on the engine side is about 12 inches from the cowl (Figure 8-54).

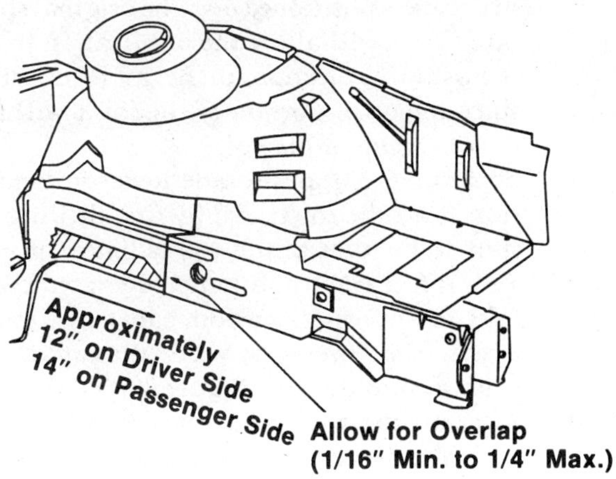

FIGURE 8-54 Sectioning cut on the engine side

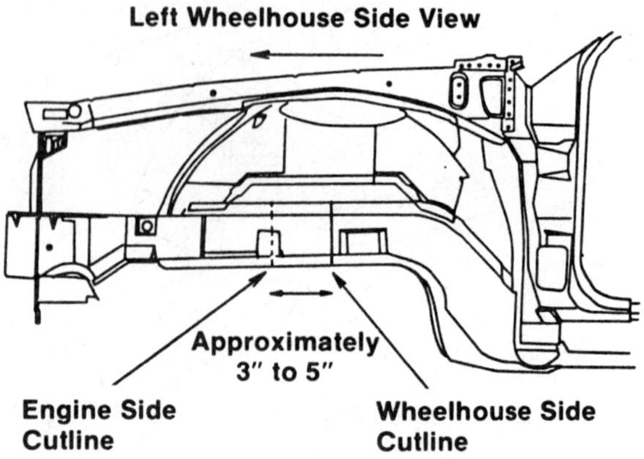

FIGURE 8-55 Sectioning cut on the outer rail

This will position the cutline near the end of the inner reinforcement.

CAUTION

If the reinforcement is accidentally damaged with a small cut, it must be welded. If the cut is larger than 1/4 inch in length and completely through the reinforcement, the sectioning process may not be successful even if the reinforcement is repaired. For this reason, it is important that extreme care is used while cutting around an internal reinforcement.

6. The outer rail cutline (wheelhouse side) is made 3 to 5 inches rearward of the engine side cutline (Figure 8-55).
7. To achieve the correct overlap, carefully "split" the corners on the original structure at the exposed end. These splits should not exceed 1/4 inch. Any part of the splits that is exposed after fit-up must be welded closed.
8. It is very important that the replacement structure is positioned *over* the original structure. This will allow the application of the corrosion protection to be more effective, since the open portion of the joint will face the open end of the rail.
9. Separate the opposite side lower rail extension from the lower rail by first drilling the spot welds that secure the radiator support and apron extension panel. Next, carefully fold the apron extension panel upward to expose the other spot welds that attach the extension to the rail (Figure 8-56).
10. After all the spot welds have been removed and the offset cuts made, the damaged assembly can be removed from the vehicle.

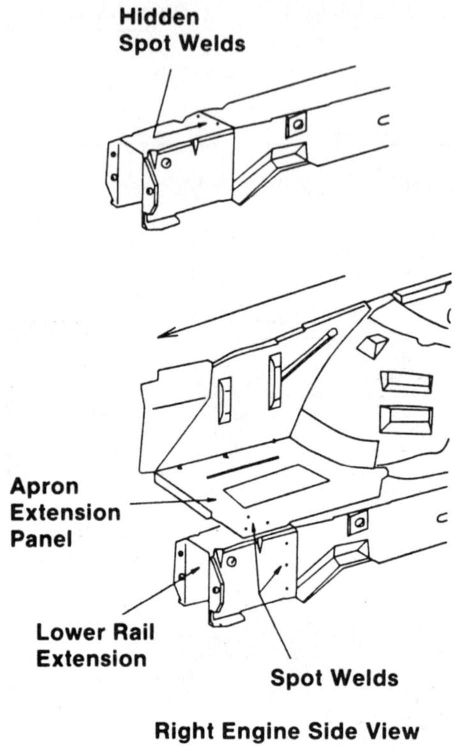

FIGURE 8-56 Fold the apron extension panel up to expose the spot welds.

11. Preparation of the replacement assembly (i.e., spot weld removal and the lower rail offset cut) is identical to that of the damaged assembly. It is very important to inspect, measure, and, if necessary, straighten the replacement assembly to the proper dimensions prior to installing it on the vehicle. Remember to add to the length measurement on the replacement rail to allow for the overlap.
12. All adjoining flanges and weld areas must be thoroughly cleaned using a propane torch and wire brush. Do not grind or burn off any galvanized coatings. After cleaning and before welding, a weld-through primer must be applied to all bare metal mating surfaces.
13. After the replacement assembly is securely clamped in place, measuring equipment can be used to check for dimensional accuracy (Figure 8-57).
14. After checking that all dimensions are within tolerance, the assembly can be welded. Continuous welds should be made in alternating segments of 1/2 to 3/4 inch to minimize distortion. Be sure to completely close all seams and do not leave any gaps.

FIGURE 8-57 Check dimensional accuracy before the new assembly is welded.

15. Corrosion protection, including refinishing of the replacement pieces, should be completed as detailed in Chapter 10.

SECTIONING A-PILLARS

Repairing A-pillars by sectioning can result in significant savings over working strictly along factory seams. These repairs can be readily made using either butt joints with inserts, offset butt joints, or a modified lap and butt joint.

A-pillars can be either two-piece or three-piece components, and can be reinforced at the upper end, the lower end, or both; they are not likely to be reinforced in the middle. Therefore, A-pillars should be cut near the middle to avoid cutting through any reinforcing pieces. It is also the easiest place to work.

To section an A-pillar, use a straight-cut butt joint with insert or an offset butt joint without an insert. The butt joint with insert repair is made in the same manner as already described for the rocker panel. The A-pillar insert should be 4 to 6 inches in length. After cutting the insert lengthwise and removing any flanges, tap the pieces into place. Secure the insert with plug welds and close all around the pillar with a continuous butt weld (Figure 8-58).

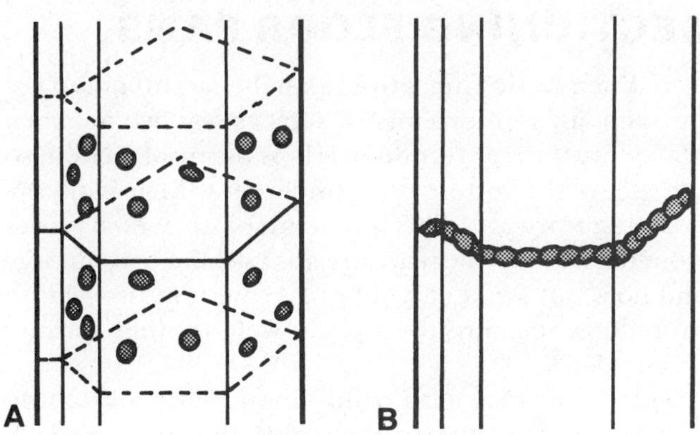

FIGURE 8-58 Welding a butt joint with insert in an A-pillar using (A) plug welds and (B) butt welds

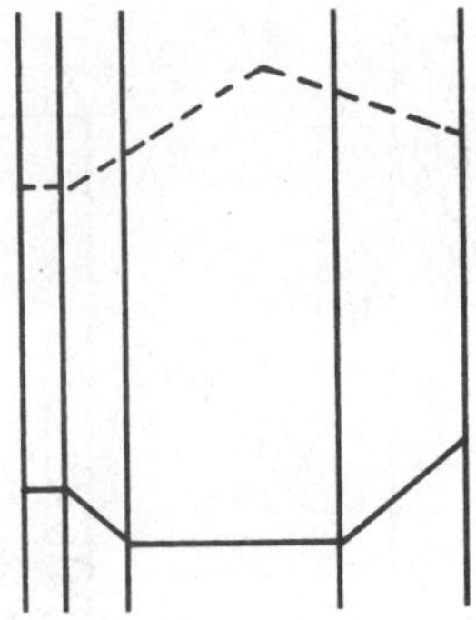

FIGURE 8-59 An offset butt joint

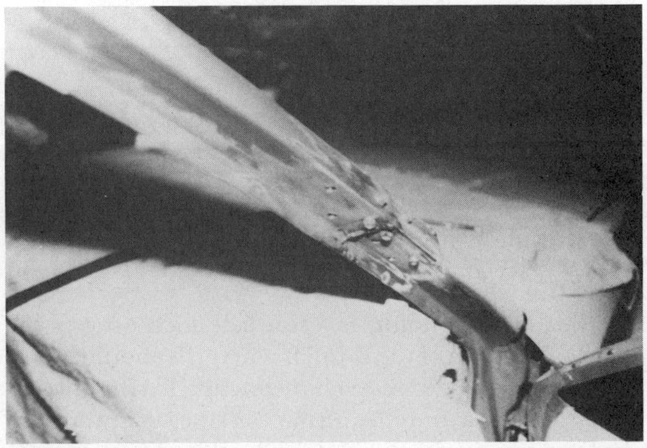

FIGURE 8-60 Finished A-pillar section

To make the offset butt joint, cut the inner piece of the pillar at a different point than the other piece was cut, thus creating the offset (Figure 8-59). Whenever possible, try to make the cuts between the factory spot welds, so that it will not be difficult to drill them out, and make the cuts no closer to each other than 2 to 4 inches. Butt the sections together and continuous weld them all around (Figure 8-60).

SECTIONING B-PILLARS

For sectioning B-pillars, two types of joints can be used: the butt joint with insert and a combination offset and lap joint. The butt joint with insert is usually easier to align and fit up when the B-pillar is a relatively simple two-piece cross section without a lot of internal reinforcing members. The insert provides additional strength.

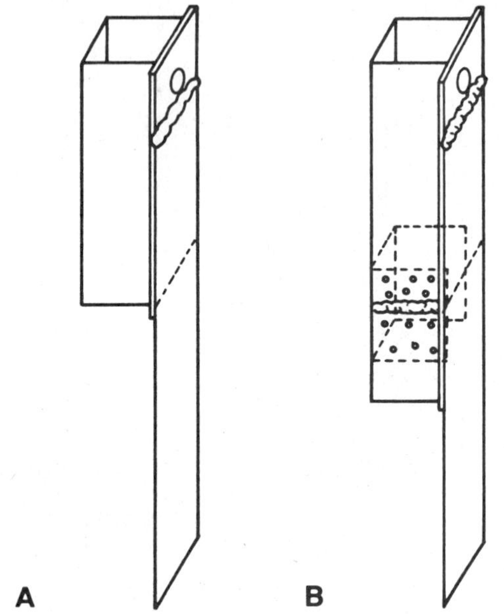

FIGURE 8-61 Sectioning a B-pillar by (A) lap welding the inner panel and (B) plug and butt welding the outer panel

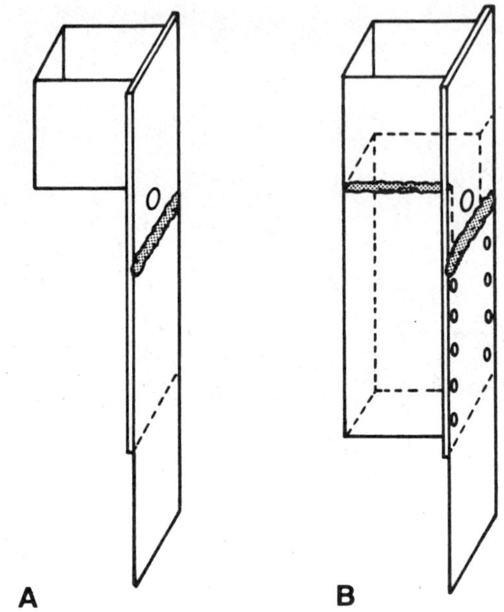

FIGURE 8-62 Creating a combination offset and lap joint by (A) lap welding inside and (B) plug and lap welding outside

The D-ring mount is a seat belt anchoring point located at the top of the B-pillar where the shoulder belt bolts in place. Be sure to cut below the D-ring mount low enough to avoid cutting through the D-ring anchor reinforcement. In the case of the B-pillar, use a channel insert in only the outside piece of the pillar. The D-ring anchor reinforcement welded to the inside piece prevents the installation of an insert there.

Begin by overlapping the new inside piece on the existing one, rather than butting them together, and lap weld the edge (Figure 8-61A). Then, secure the insert in place with plug welds and close the joint with a continuous butt weld around the outer pillar (Figure 8-61B).

On occasion it might be best to obtain a recycled B-pillar and rocker panel assembly and replace them as a unit. This is because any time a B-pillar is hit so hard that it must be replaced, the rocker panel is almost invariably damaged, too. Install the upper end of the B-pillar with either of the two approved types of joints and make a butt joint with insert in the rocker panel in the manner already shown. If the main damage is in the rear door opening, make the butt joint with insert in the front door opening and install the other end of the rocker panel in its entirety. If the main damage is in the front door opening, reverse the procedure.

Generally speaking, the combination offset and lap joint is used more often when installing new parts and when working with separate inside and outside pieces.

1. Cut a butt joint in the outside piece above the D-ring anchor reinforcement.
2. Make an overlap cut in the inside piece below the D-ring anchor reinforcement.
3. Install the inside piece first, with the new piece overlapping the existing one.
4. Lap weld the edge (Figure 8-62A).
5. Place the outside pieces, make plug welds in the flanges, and close the section with a continuous weld at the butt joint (Figure 8-62B).

Usually, it is advantageous to use the offset and lap joint on a B-pillar with three or more pieces in its cross section, where it is difficult to install an insert. In fact, sometimes the offset and lap joint procedure is mandatory, because it is downright impossible to install an insert.

SECTIONING FLOOR PANS

When sectioning a floor pan, be careful not to cut through any reinforcements, such as seat belt anchors. Make sure the rear section overlaps the front section, so the edge of the bottom piece under the vehicle is always pointing rearward. This is so road splash, which moves from the front to the rear, streams past the bottom edge and does not strike it head-on. As with most welding procedures, be sure to apply a weld-through primer (Figure 8-63).

As is the case with trunk floors, when sectioning floor pans it is wise to remember that they are very prone to corrosion from exhaust gases and moisture. For this reason, the finishing process described in step 5 takes on added significance.

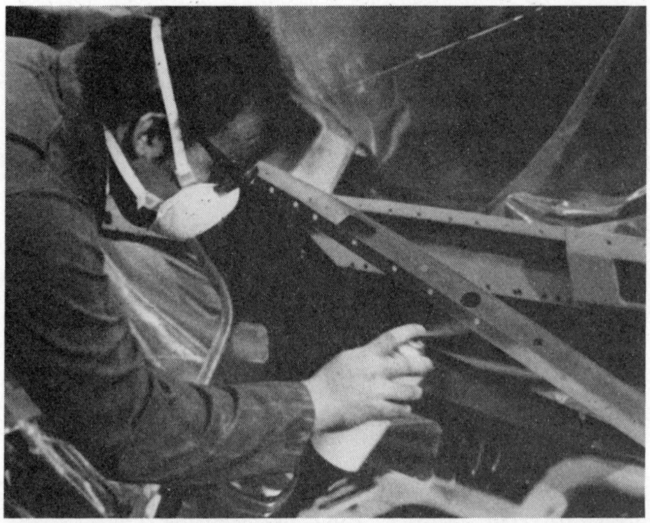

A

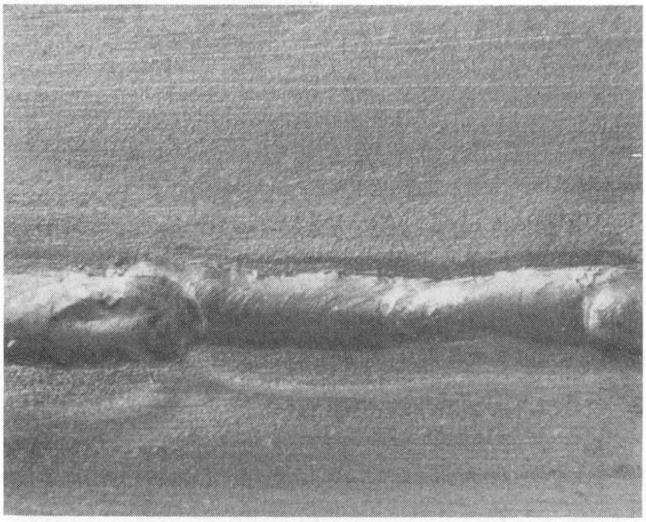

B

FIGURE 8-63 (A) Applying weld-through primer before welding. (B) This weld was made over weld-through primer. Note the absence of skips or porosity.

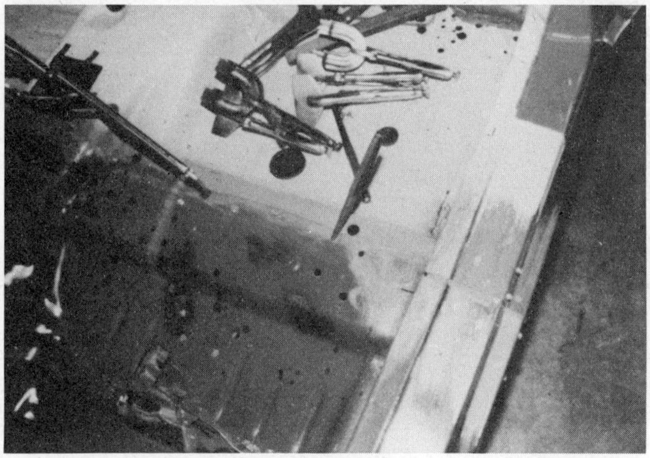

FIGURE 8-64 Plug weld the floor pan lap joint.

1. Join all the floor pan sections using a lap joint.
2. Plug weld the lap joint (Figure 8-64), putting the plugs in from the topside, downward (Figure 8-65A).
3. Caulk the topside, forward edge with a flexible body caulk.
4. On the bottom side, lap weld the underlapping edge with a continuous bead. This technique is prone to some burn-through.
5. Cover the lap weld with a primer, a seam sealer, another coat of primer, and a topcoat (Figure 8-65B). This assures that there will be no carbon monoxide gas entering the passenger compartment. Apply sound-deadening material, if required.

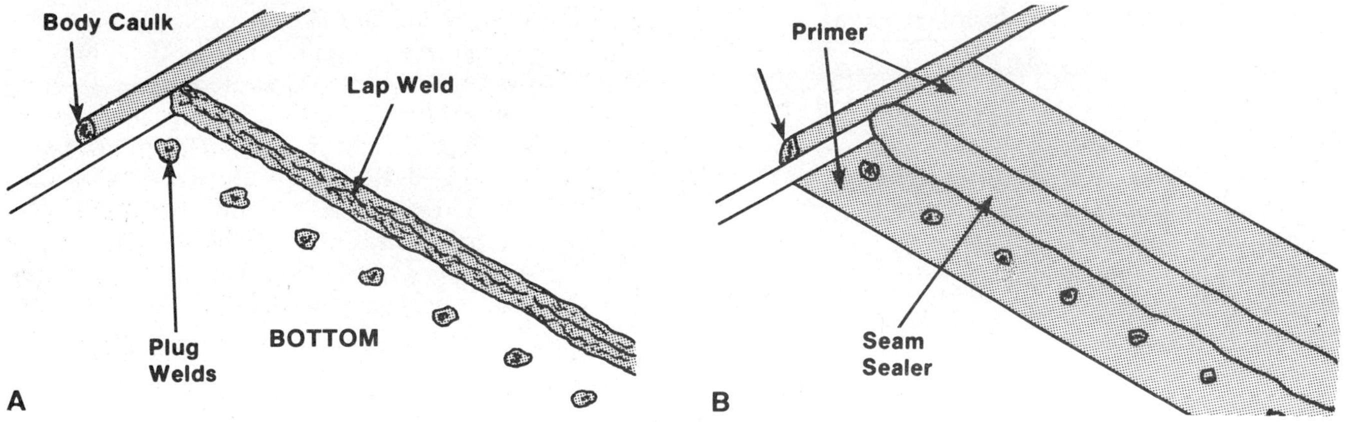

FIGURE 8-65 (A) Plug weld from the top and lap weld the bottom edge, then (B) seal and prime the bottom edge.

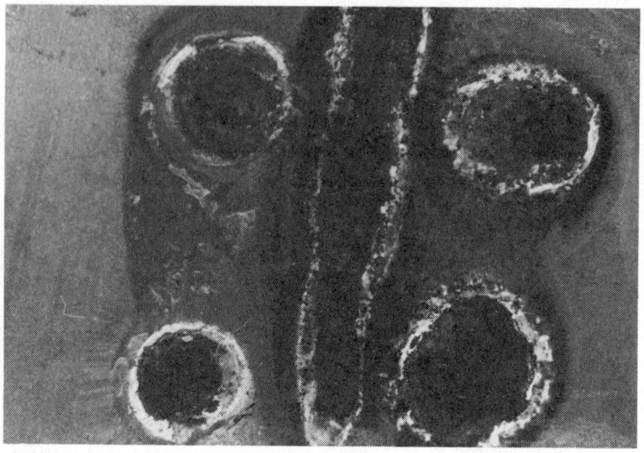

FIGURE 8-66 The left half of this welded joint was tightly fit-up; the right half had gaps.

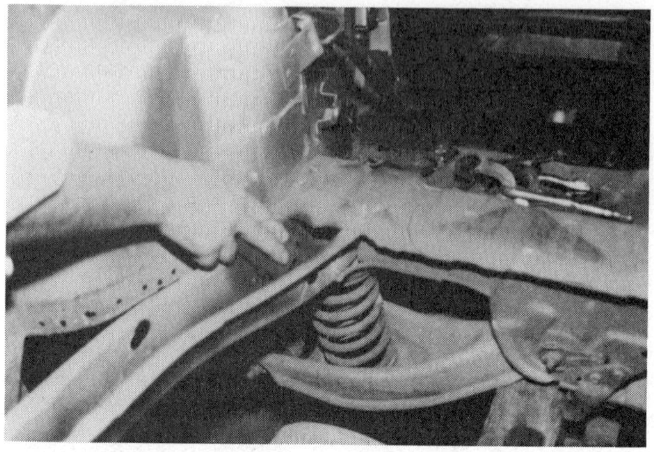

FIGURE 8-67 A trunk floor should be sectioned above the rear flange of the cross member.

SHOP TALK

It is important to remember that a poor fit-up requires a longer welding arc time because it takes longer to fill gaps. In the case of zinc-coated materials, this results in more heat and more melting of the zinc. The more zinc that is melted, the more unstable the arc. Figure 8-66 shows the difference between a tight and loose fit-up. It is easy to see that the right half of this welded joint is corroded more than the left half. The left half was fit-up tightly, while the right half had gaps between the pieces. Make tack welds; use clamps, screws, or hammer blows while welding. In short, do whatever it takes, but keep the fit-up tight.

SECTIONING TRUNK FLOORS

When sectioning a trunk floor, follow the basic procedures just described for the floor pan. However, there are a few variations as described below.

SHOP TALK

With damage that requires sectioning of the trunk floor, the rear rail usually will be sectioned as well.

1. There is generally some kind of a crossmember under the trunk floor in the vicinity of the rear suspension. Whenever possible, section the trunk floor above the crossmember's rear flange (Figure 8-67). Section the rail just rearward of the crossmember.

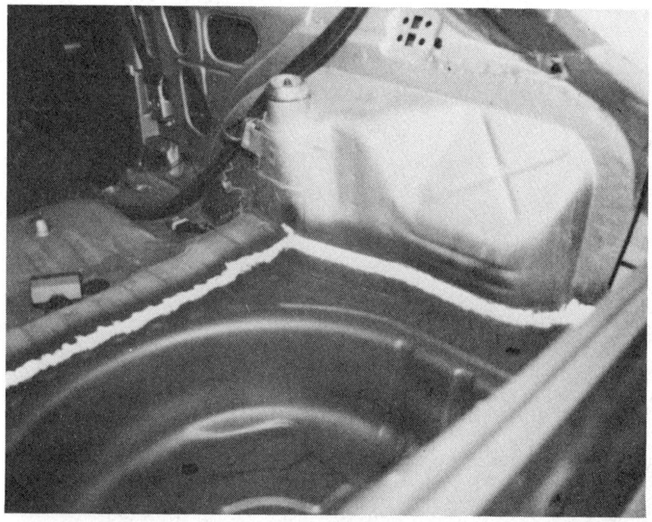

FIGURE 8-68 Caulk the topside, forward edge of the trunk floor.

2. Plug weld the trunk floor lap joint to the crossmember, again putting the plugs in from the topside, as in a floor pan.
3. Caulk the topside, forward edge (Figure 8-68).
4. On the bottom side, a lap weld on the underlapping edge is usually not necessary because of the strength provided by the crossmember. However, on vehicles where the trunk floor section is not above a crossmember, the lower edge must be lap welded.

In both cases, cover the bottom seam with primer, seam sealer, and a topcoat. With a trunk floor, sealing against carbon monoxide gas is critical because of the proximity of the tail pipe.

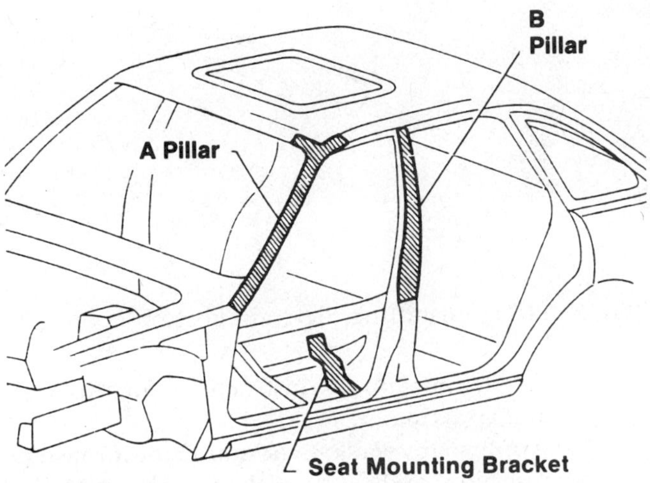

FIGURE 8-69 Urethane foam is used in the A- and B-pillars and under seat mounting brackets.

FOAM-FILLED PANELS

In the past, plastic foam-filled body sections were found only in so-called luxury cars. Now, however, they are being used in more and more unibody vehicles.

Plastic or urethane foam is used as a means of adding rigidity, strength, and sound control to lightweight unibody vehicles. In these uses, foam can be an important tool in the efforts to reduce NVH—noise, vibration, and harshness. NVH has become an important area of engineering work in recent years.

Figure 8-69 shows how the urethane foam is used to fill the A-pillars and the upper portions of the B-pillars. It is also used in the cavity between the front seat mounting brackets and the floor pan. During collision repairs, distortion, straightening, cutting, and heating of the metal in these areas will damage the foam filler. To ensure that the vehicle is fully restored to pre-accident condition, the foam filler must be replaced.

CAUTION

There are several single-part urethane foam products on the market, packaged in aerosol cans. These products are offered to the consumer market for insulation purposes and are not suitable for collision repair work. One-part urethanes cure by absorbing moisture from the air. For this reason, they should never be used for filling vehicle body panels. In an enclosed area they might not cure completely and can cause corrosion.

Automobile engineers recommend that a two-part urethane foam panel filler should have the following characteristics:

- Weight of 0.006-ounce or greater per cubic inch
- No noticeable volume changes due to humidity and temperature when cured
- No adverse effect on electrodeposition primers

Most factory replacement pillars come with the foam in place. The parts are first coated with self-etching primer. They are then filled with urethane foam after the drain tubes and trim mounting clips have been installed. An example of removing a foam-filled A-pillar is as follows:

1. Locate the reference hole on the upper part of the pillar. Measure 4 inches down from this hole and mark the position on the inside of the pillar. Now, measure up 2-3/4 inches from the last mark and mark the outside of the pillar. These are the cutlines (Figure 8-70).
2. Carefully cut the pillar at each marked position. A cutting jig makes this operation much easier and more accurate (Figure 8-71). The jig can be made in the shop from scrap sheet metal.
3. Carefully drill out the spot welds between the two cuts and at the bottom of the pillar and remove the pillar (Figure 8-72).

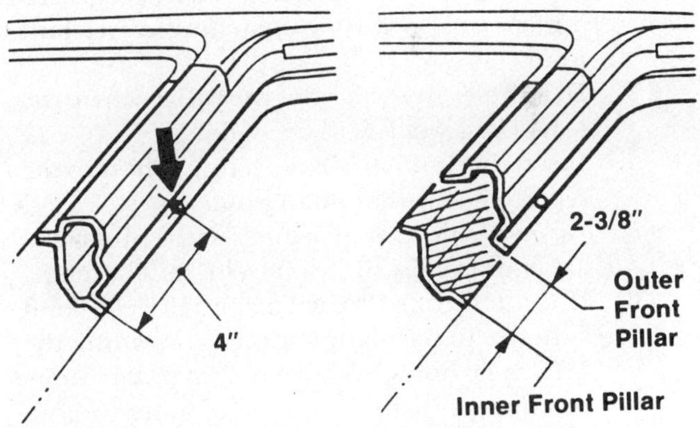

FIGURE 8-70 Marking the A-pillar for cutting

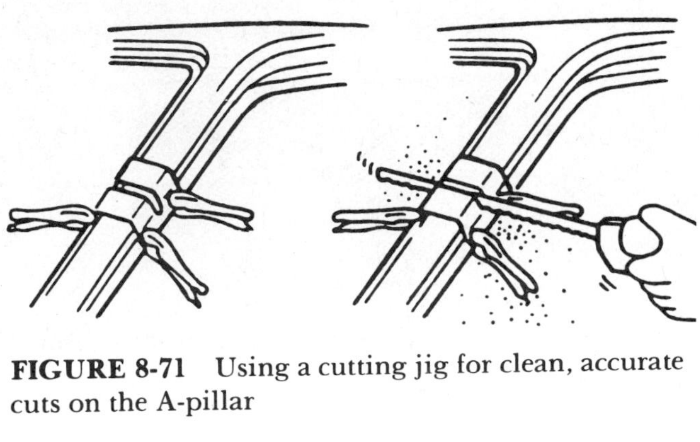

FIGURE 8-71 Using a cutting jig for clean, accurate cuts on the A-pillar

FIGURE 8-72 Drill out the spot welds in the A-pillar as described earlier in this chapter.

To install a foam-filled A-pillar, proceed as follows:

1. Before starting the installation process, remove the old urethane foam filler from inside the pillar for about 3 inches. This will provide space for the new foam and also allow the drain hose to expand as the drain pipe connecting tube is inserted (Figure 8-73).
2. Prepare the replacement pillar for installation. Cut the upper pillar area to the required dimensions to form a matching offset butt joint.
3. Drill holes in the base of the pillar where the MIG plug welds will be made.
4. Insert the drain pipe connecting tube into the replacement pillar drain hose.
5. Apply weld-through primer in the plug weld areas and where the seam welds will be made.
6. If recommended, apply a special weld-bond adhesive to the mating surfaces, avoiding the MIG weld holes. Some vehicle manufacturers recommend using this type of adhesive as a means of increasing the strength and corrosion-resistance of the weld. Be sure to follow the manufacturer's instructions for mixing and applying.
7. Apply soapy water to the drain hose for easier insertion, and position the replacement pillar. Make sure that the drain hose is correctly inserted in the drain hole in the pillar (Figure 8-74).
8. Clamp the pillar in place and check the fit.
9. Remove the pillar and apply weld-through primer to the plug weld areas only.
10. Apply weld-bond adhesive to the rest of the mating surfaces (one side only).
11. Complete the plug and seam welds according to the manufacturer's guidelines. Dress the welds.
12. Apply urethane foam filler into the pillar assembly through the hole located in the upper portion of the inner pillar. Figure 8-75 shows the use of a cartridge-type urethane foam applicator gun.
13. Remove any excess foam from the joint area.
14. Prime the weld joints with a two-part epoxy primer and apply a color coat.
15. Apply an anti-corrosion material to enclosed surfaces not filled with foam.
16. Install the door and front fender, and check for proper alignment (see Chapter 11).

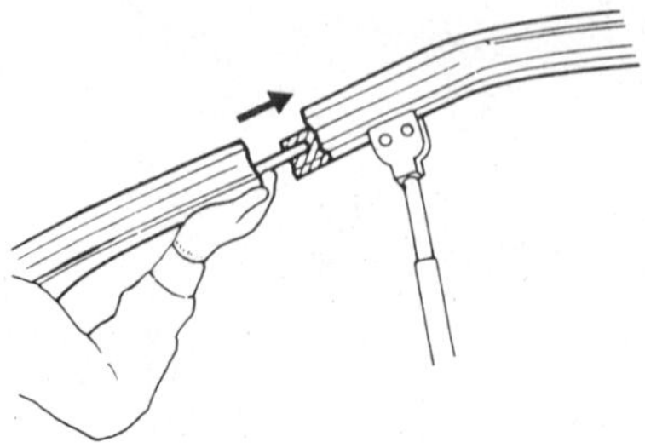

FIGURE 8-74 Installing the drain hose pillar

Follow a procedure similar to the one given here for A-pillars when installing a urethane foam-filled B-pillar.

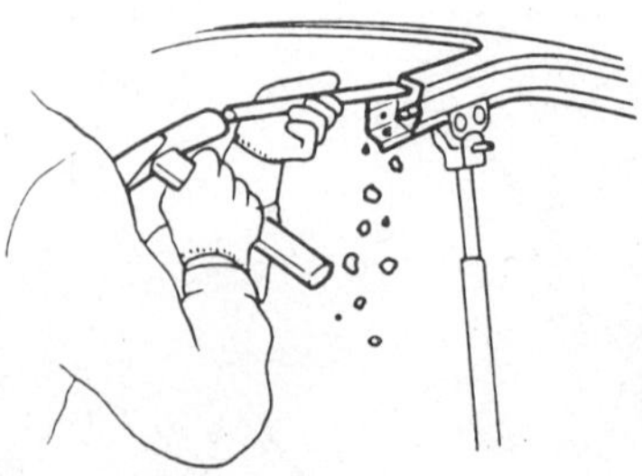

FIGURE 8-73 Removing the old urethane filler from the A-pillar

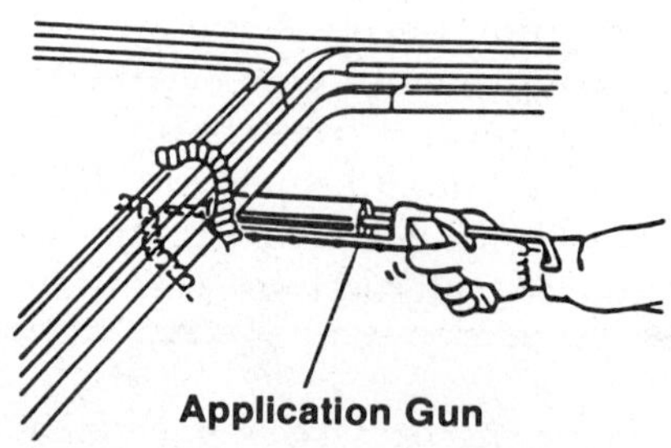

FIGURE 8-75 Applying urethane foam through a hole in the upper A-pillar

FULL BODY SECTIONING

Research has established that full body sectioning can yield satisfactory results, producing a vehicle as strong and serviceable as an undamaged one. Full body sectioning is actually just a logical sequence of sectioning two A-pillars, two rocker panels, and a floor pan. Full disclosure of this procedure to the customer or next owner of the vehicle is an important ethical necessity. A discussion between the insurance representative, vehicle owner, and the collision repair facility is required. All of the repair procedures, including alignment, welding, and corrosion protection, must be fully explained to the owner. It is important that the following points are covered:

- The recycled sections—both body and mechanical—must be of like kind and quality.
- All VIN code identifications and EPA emission control requirements must be met and all suspension, braking, and steering components must be in proper working order. It must be noted that a recycled vehicle identification number would not match the current one.
- Carefully inspect the front and rear sections for proper alignment. If either is out of alignment, proper fit-up of the sectioned joints will be difficult, if not impossible, to achieve.
- The power train warranty, as well as other part warranties, can be affected or invalidated.

Full body sectioning is the most severe or extensive form of collision repair. As with all sectioning, it is absolutely necessary that the work is the highest quality.

For decades, collision repair technicians have used replacement assemblies to make economical repairs on damaged vehicles. One of the most common practices has been to purchase an entire undamaged rear, front, or top section from a salvage yard and then weld it into place on the vehicle being repaired. This repair practice is now most often used in rebuilding late-model vehicles purchased at salvage yards and vehicle auctions. A typical candidate for this type of repair is a vehicle that has sustained severe rear end damage. In such a case, sectioning is more practical and cost effective than conventional methods of repairing severely damaged unitized automobiles. It reduces the extensive time and labor of dismantling the vehicle and disturbs less of the corrosion protection.

The term *clip* is used to describe any number of operations involving replacement assemblies. A clip, as used in this text, means a large assembly of salvage body parts (Figure 8-76). What is included in a clip varies for different makes and models of vehicles. Some manufacturers supply large assemblies, but most clips are cut from wrecks and stocked for recycling to collision repair shops. Clips without rust or other damage have always been a premium salvage item (Figure 8-77).

FIGURE 8-76 Typical clip used in sectioning

FIGURE 8-77 Salvage yards are a prime source of clips.

The techniques and procedures previously described are repeated in a full body sectioning procedure. By sectioning both A-pillars, both rocker panels, and the floor pan, a vehicle can be fully sectioned.

When the individual components are sectioned, aligned, and welded using the proper techniques and procedures, full body sectioning is a suitable and satisfactory procedure. Vehicles repaired by full body sectioning are completely crashworthy. This has been tested and proved time and again. Keep in mind, however, that full body sectioning is not a frequently required procedure. Thus, for ethical reasons, full disclosure must be given to the owner before such repairs are started.

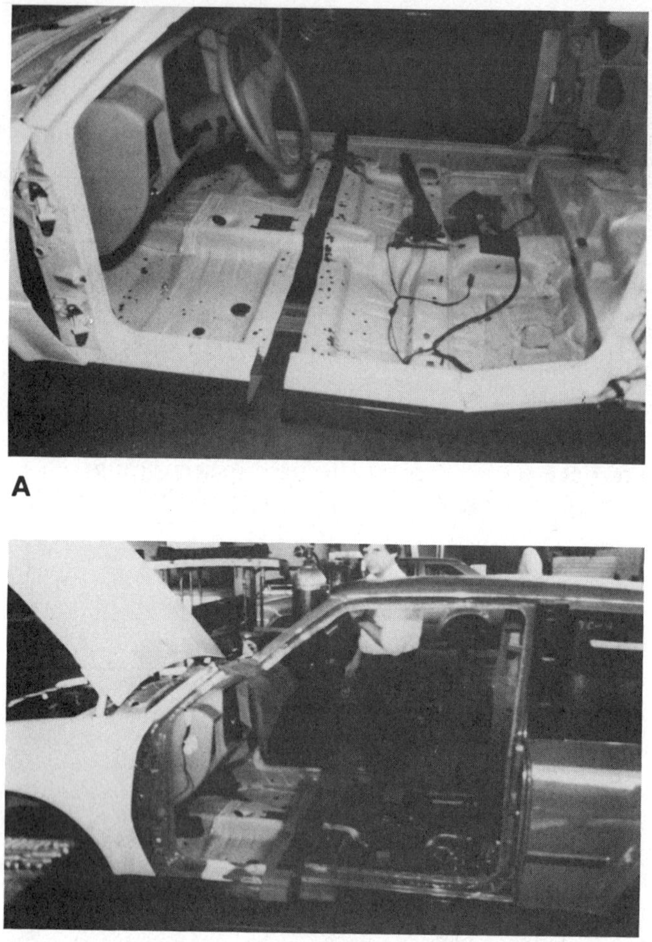

A

B

FIGURE 8-78 (A) Passenger body section and (B) front body section replacement

Figure 8-78 shows a popular compact that was fully body sectioned. The undamaged front half of one vehicle is joined to the undamaged rear half of another. Butt joints and inserts are used in the middle of the A-pillars and in the two rocker panels, and a lap joint is used on the floor pan. The rocker panel and floor pan cuts are made near the middle of the door opening.

Remember that the floor pan might have reinforcements and brackets that need to be removed before sectioning. Reinforcements can be left on the replacement rear half to aid in alignment.

Joining the Full Body Sections

After the front and rear sections have been trimmed to fit, drilled for plug welds, and primed with weld-through primer, follow these steps to join the two sections:

1. Install the rocker panel and pillar inserts. Hold them in place with sheet metal screws.
2. Place the A-pillar inserts in the upper or lower portion of the windshield pillar, depending on the angle and contour of the windshield.
3. Fit the two halves together by first joining the rocker panels and then the A-pillars. Clamp the rocker panel and A-pillar flanges to prevent the sections from pulling apart.
4. Check the windshield and door openings for proper dimensions, using a tram bar or a steel rule. If possible, install the doors and windshield to verify proper alignment.
5. When proper alignment is achieved, secure the floor pan with sheet metal screws. This will pull the two pieces together and hold the sections secure during welding.
6. Using self-centering gauges and a tram bar, double-check the vehicle dimensions and panel alignment before welding.
7. Weld the sections together using the techniques previously described in this chapter for joining rocker panels, A-pillars, and the floor pan.
8. Restore all corrosion protection.

Flat-Rate Times

Flat-rate times to replace an assembly include the time to remove the damaged one and then align and install the replacement. The following items are not included in the flat-rate time:

- Any repair, straightening, or alignment of adjacent parts or assemblies
- Cutting away, pushing, or pulling damaged parts or assemblies for access to attachment points
- Removal of usable parts from the damaged part of the assembly and their installation on another part of the assembly
- Cutting and fitting welded assemblies to be installed
- Filling or drilling attachment holes
- Alterations necessary when parts or assemblies of one model vehicle are installed on a different model
- Removal or replacement of undercoating
- Freeing up parts frozen by rust or corrosion
- Removal/replacement of vinyl roof covers

On the typical full body sectioning repair, with or without the roof panel, additional time would be consumed on at least five of these items. This, plus basic labor and parts, must be taken into consideration to determine whether or not a full body sectioning is justified and wise.

REVIEW QUESTIONS

1. Which of the following statements concerning replacement at factory seams is incorrect?
   a. Replacing panels at factory seams is common in unibody repair.
   b. Damaged rails and panels should not be returned to factory specifications until after they have been removed from the vehicle.
   c. It is easy to destroy more of the factory welds than is necessary when doing the job.
   d. All of the above

2. Which of the following methods is used for separating spot welds?
   a. cutting them out with a plasma torch
   b. cutting them out with a clamp saw bit
   c. grinding them down with a high-speed grinding wheel
   d. all of the above

3. When drilling out spot welds on HSS, Technician A uses the same drill speed used for drilling mild steel. Technician B uses a drill speed that is faster than that used for mild steel. Who is correct?
   a. Technician A
   b. Technician B
   c. Both A and B
   d. Neither A nor B

4. What color is metal that has been arc-brazed?
   a. black
   b. reddish copper
   c. silver
   d. brass-colored

5. When preparing a vehicle for installation of replacement body panels, Technician A grinds the flanges of the structural panels, while Technician B does not. Who is correct?
   a. Technician A
   b. Technician B
   c. Both A and B
   d. Neither A nor B

6. When a replacement body panel overlaps any existing panels, the overlap portion should be ______________.
   a. 3/4 to 1 inch
   b. 1 to 1-1/2 inches
   c. at least 2 inches
   d. at least 3 inches

7. When visually aligning a cosmetic panel, Technician A makes sure that the gaps between panels are tapered. Technician B makes sure that the gaps between panels are even. Who is correct?
   a. Technician A
   b. Technician B
   c. Both A and B
   d. Neither A nor B

8. Which of the following should be avoided when making sectioning cuts?
   a. suspension mounts
   b. crush zones
   c. dimensional reference holes
   d. all of the above

9. A butt joint with insert is used mainly on what type of section?
   a. open surface
   b. crush zone
   c. closed
   d. compound

10. Which type of joint is used on floor pans and trunk floors?
    a. offset butt joint without an insert
    b. overlap joint
    c. staggered butt joint
    d. butt joint with insert

11. What is the preferred tool to remove a recycled part?
    a. metal saw
    b. chisel
    c. cutting torch
    d. none of the above

12. Rocker panels come in which of the following designs?
    a. two-piece
    b. three-piece
    c. four-piece
    d. both a and b

13. Technician A uses an offset butt joint without an insert to repair a rear rail, while Technician B uses the same technique to repair a trunk floor. Who is correct?
    a. Technician A
    b. Technician B
    c. Both A and B
    d. Neither A nor B

14. Where should A-pillars be cut when sectioning?
    a. at the upper end
    b. near the middle
    c. at the lower end
    d. both a and c

CHAPTER

9 Analyzing Mechanical and Electronic Components

OBJECTIVES

- Explain the basics of front, rear, and computer-controlled suspension systems.
- Describe the design and operation of parallelogram, rack-and-pinion, power, and four-wheel steering systems.
- Define the elements of proper wheel alignment, including caster, camber, toe, thrust line, steering axis inclination, and turning radius.
- Differentiate between drivetrain and suspension problems and those caused by a bad CV joint.
- Understand how various brake systems work, and describe the procedures for manual and pressure bleeding.
- Perform key cooling and air-conditioning system repairs and maintenance.
- Inspect an exhaust system, and describe the guidelines for working on an emission control system.
- Use a test light, jumper wire, and various test meters to perform electrical system diagnostic procedures.
- Describe the operation of on-board computers and other automotive electronic system components.

Until a few years ago, mechanical repairs were not generally performed by collision repair shops. Instead, they were sublet to specialty shops. With the introduction of unibody vehicles, it has become important for the collision repair technician to develop some mechanical knowledge and skills. Basically, he or she must be able to remove these components from the repair area and replace them after the body repairs have been completed. It should be noted that some states require collision repair technicians to be certified in order to perform mechanical repairs.

The repair of collision damage to unibody structures often involves service of mechanical parts. In many cases, major driveline parts such as engines, transmissions, and drive axles are mounted directly to structural body panels. In other cases, these parts are mounted to supporting crossmembers, subframe assemblies, or cradle assemblies that are themselves mounted to the panels. The unibody panels provide the structural support for the static load of the mechanical parts mounted to them. Unibody panels must also satisfy dynamic loads developed by the mechanical components. For example, engine and transmission mounts must satisfy severe torsional loads. Drive axle mountings must support the loads of acceleration and braking. Proper alignment of mechanical part mountings can be critical to the operation of the part. The dimensional relationship between engine mountings and other driveline component mountings must be maintained.

Although mechanical components might not be repaired by the collision repair technician, they can give the technician leads as to the type and extent of damage that can be passed on to the mechanic who is going to make the repairs. And in order to do the required analysis and/or repairs, the service manual for the vehicle in question is a must. Never attempt to work with mechanical components without the aid of a service manual.

DRIVETRAIN SERVICE

The vehicle's drivetrain (sometimes called its power train) consists of several important mechanical systems. The drivetrain in a front-wheel-drive unibody vehicle includes the engine, transaxle, drive shafts, and suspension system together, usually in the front end of the vehicle. It is supported by crossmembers that are bolted directly to the frame.

Engine mounts and transaxle supports are through-bolted. These through-bolt mountings must be maintained parallel to each other to allow for the correct movement of the mechanical parts. When they are not in proper alignment, free movement of the parts might be restricted.

Control linkages might not operate correctly unless proper alignment is maintained. For instance, engine performance can be affected by misalignment

FIGURE 9-1 Typical unibody motor mounts

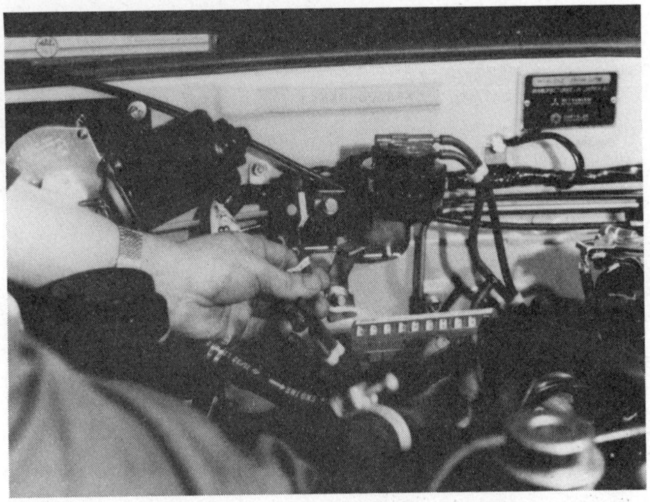

FIGURE 9-2 Label parts as they are disconnected.

of the throttle control linkages, and misalignment in transmission linkages can easily cause erratic transmission performance. Proper drive shaft angles must be maintained to prevent vibration and chatter of the drive shaft and constant velocity joints. While driveline components normally vibrate during vehicle operation, these vibrations are isolated from the passenger compartment through the use of specially designed motor mounts.

Motor mounts prevent metal vibrations at the mounting locations from being transferred to the body of the vehicle (Figure 9-1). Misalignment of the motor mounts can cause vibrations to be transferred directly to the passenger compartment.

To provide the necessary structural support for mechanical parts, special fasteners are frequently used. Cradle assembly mounting biscuits, which are found on some unibody vehicles, use a special self-tapping bolt that cuts the threads in the mounting biscuit when it is installed at the factory. If possible, the same bolt should be reinstalled in the same mounting biscuit when it is reused. When new bolts are required, common replacement fasteners should not be used because they might not supply the structural strength necessary for the mechanical part. Check the individual manufacturer's specifications on the use of all fasteners.

At times it is desirable to completely remove the drivetrain, even if it is not damaged, to make repairs to the body. Because modern unibody vehicles tend to have very crowded engine compartments, removal of the drivetrain allows ready access to structural panels. If more time is needed to work around the drivetrain rather than remove it, the drivetrain should be removed. In some cases, the time to remove the drivetrain is made up with the considerable time savings in the repair or replacement of the body panels. The decision whether to remove the drivetrain or to work around it can be made by the repair technician and estimator. If the mechanical components will interfere with the quality of repairs, by all means remove them as well.

At the factory, the drivetrain is installed from the bottom of the vehicle, and this is the best way to reach it in the shop as well. The following procedure describes the use of a floor jack, jack stands, rocker panel clamps, and crosstube for disconnecting the drivetrain assembly, supporting it from below, and lifting the body away from it.

With the vehicle on the ground, start disconnecting the engine and suspension assembly from the top of the engine compartment. Before disconnecting the vacuum or electrical lines, make sure to attach a piece of masking tape to both sides of the disconnected parts and mark them for easy reassembly (Figure 9-2).

Removal of the mechanical components is usually done before the vehicle is straightened. When removing the drivetrain from a unibody vehicle, the typical procedure is as follows:

1. Disconnect the battery cables from the battery and the body ground strap from the body.
2. Remove the air cleaner to aid visibility and to increase the working area.
3. Drain the cooling system. The radiator hose can be disconnected at either end.
4. Disconnect the vacuum hoses from all components that are not mounted to the drivetrain, such as the power brake booster and fuel vapor canister. To keep out dirt, golf tees are convenient for plugging vacuum lines (Figure 9-3).
5. Disconnect the throttle linkages. Remove only those parts that link the engine to the body. Take care to always separate at clips

FIGURE 9-3 Using golf tees to plug vacuum lines

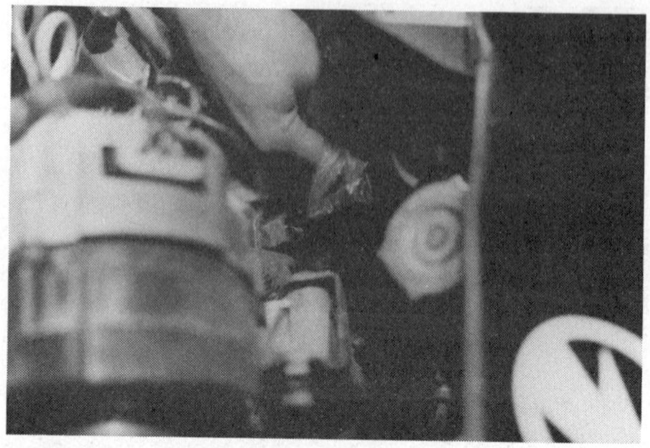

FIGURE 9-5 Plastic and wire ties are ideal for keeping dirt out of disconnected hoses.

FIGURE 9-4 Disconnecting the shift linkage cable(s) from the transaxle

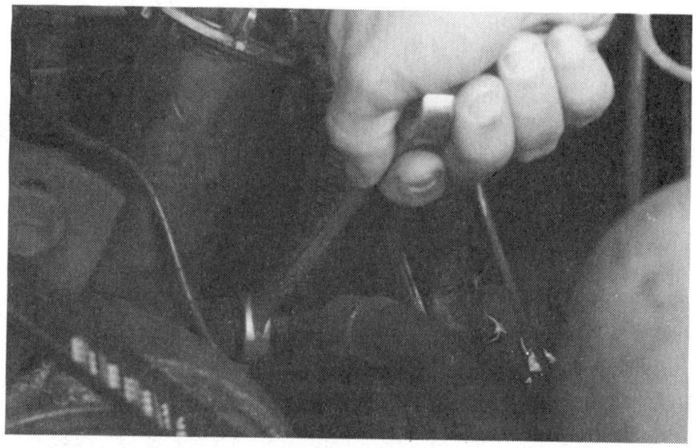

FIGURE 9-6 Disconnecting the power steering pump hoses

and do not modify the length adjustment of any connecting link or cable. This will ensure that the throttle operates properly when reassembled without requiring a great deal of adjustment time.

6. Disconnect the shift linkage cable(s) from the transaxle (Figure 9-4). Be careful to separate the cable from the bracket that mounts to the housing of the automatic or standard transaxle. Disconnect the manual clutch cable (if so equipped).
7. Disconnect the speedometer cable at the union located midway on the cable, or at the transaxle if not equipped with a cable union. Disconnect the transaxle cooling lines at the rubber hoses (if so equipped).
8. Disconnect the heater hoses at the engine (if possible) and use plastic and wire ties to keep out dirt (Figure 9-5). It is usually only necessary to remove hoses from the engine end; this will help prevent damage to the heater core.
9. Disconnect the power steering pump hoses at the in-line connection, rather than at the pump (Figure 9-6).
10. Disconnect the electrical wiring harnesses at the cowl connectors or at the computers.
11. Remove the front engine stabilizer strap.
12. Remove the air-conditioning compressor (Figure 9-7). If the A/C system is not damaged, simply remove the compressor from its mounting bracket and leave it with the body. This will avoid the need for discharging it. If the A/C system is damaged and will need to be repaired, disconnect the lines and the wiring at the compressor. The system must be protected from moisture contamination by plugging all open lines promptly.
13. Remove the radiator fan and shroud. In some vehicles this might not be necessary. Remove the fan and shroud only if additional clearance is needed to get the engine to drop down without damaging anything.

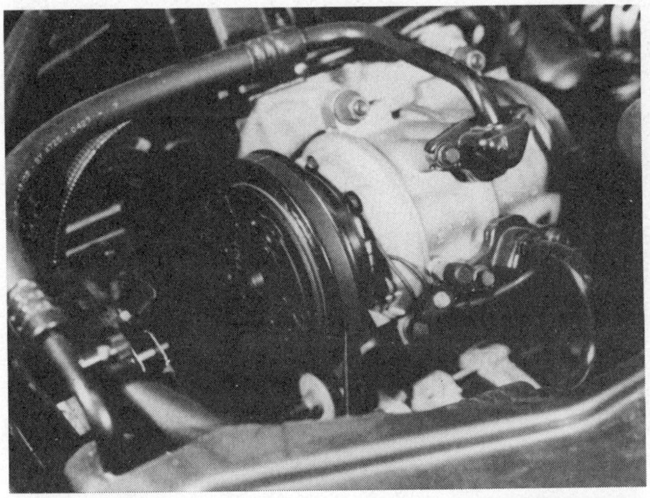

FIGURE 9-7 Typical air-conditioning compressor

FIGURE 9-9 Placing jack stands under the raised vehicle

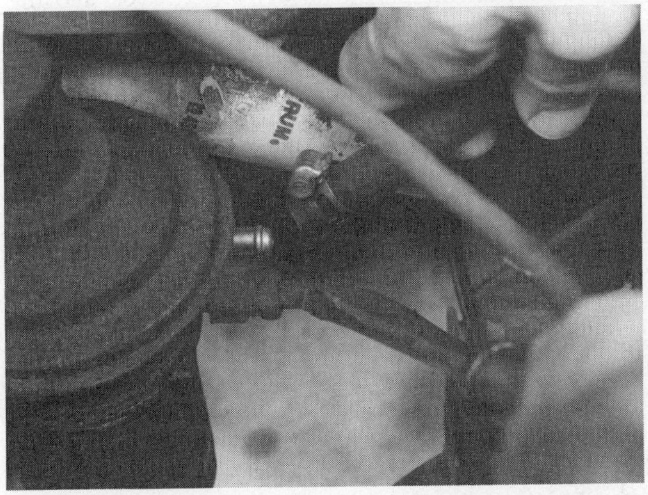

FIGURE 9-8 Disconnecting the fuel lines

FIGURE 9-10 If not using a bench or lift, carefully support the vehicle on 4 × 4s.

14. Look for any other wires attached to the engine and traveling to the body, and disconnect them. Be sure to label as explained previously.
15. Disconnect the fuel lines at the fuel pump (Figure 9-8). This completes the disconnects in the upper engine compartment.
16. To get the vehicle up in the air, install the rocker panel clamps, then install the crosstube as far forward as possible. Place a floor jack under the vehicle and place the jack stands in position on the crosstubes (Figure 9-9). A lift or bench can also be used to raise the vehicle.

    If a bench or lift is not available, place a flatbed cart with 4-inch by 4-inch lumber pieces under the drivetrain and lower the vehicle onto the 4 × 4s (Figure 9-10). Be sure to locate the 4 × 4s and the cart in a position that will allow the removal of the cradle mounting bolts.
17. Once the vehicle is in the air, take the time to remove the brake caliper assemblies from the discs. Doing this will prevent the need to bleed the brake system during reassembly. The caliper assemblies must be fastened to the vehicle instead of being allowed to hang by the brake hoses.
18. Disconnect the exhaust pipe at the coupling behind the engine and remove it (Figure 9-11).
19. It might be necessary to disconnect the right and left tie-rod ends at the steering knuckles.

Remove the three upper strut mounting bolts from each side (Figure 9-12), followed by the four cradle mounting bolts.

FIGURE 9-11 Removing the exhaust pipe

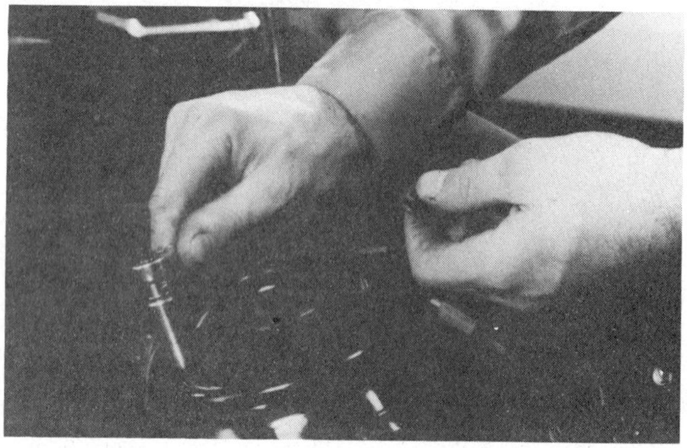

FIGURE 9-12 Removing the upper strut mounting bolts

Carefully lift the vehicle away from the drivetrain assembly by raising the jack that is under the crosstube. As always, support the vehicle properly. Care must be taken when lifting the body away from the drivetrain; as the lifting progresses, keep checking all sides for wires or hoses that might not have been disconnected (Figure 9-13).

Remove the drivetrain and repair the damaged unibody panels using the techniques described in other chapters of this book. While the body is being repaired, a technician can service the drivetrain as required.

SHOP TALK

CV joints are delicate. Special care must be taken to protect them when removing the drivetrain. Wire the struts together inboard after removing the engine. This will prevent damage to the CV joints and rubber boots.

FIGURE 9-13 Lifting the vehicle away from the drivetrain assembly

Reinstallation of the drivetrain can be accomplished by reversing the removal procedure. After the unibody structure has been repaired, the cradle can be positioned by using the line-up holes located at the right front and right rear cradle mounting points. An incorrectly positioned cradle can cause wheel alignment problems.

It must be remembered that the procedures mentioned in this chapter are general; the specifics of drivetrain removal differ slightly for each body style. An example would be a body style without a full cradle, where it is necessary to tie the suspension system to the engine and drivetrain assembly to help prevent the CV joints from separating from their assembly.

On a unibody vehicle with rear-wheel drive, the drivetrain removal procedure will also differ to a degree. The most important thing to remember is to use common sense and always refer to the proper service manuals for exact procedures.

SUSPENSION AND STEERING SYSTEMS

Suspension and steering systems have evolved through many different designs over the years. Today's typical unibody vehicle is equipped with a strut suspension and rack-and-pinion steering. But, before the technician can diagnose steering and suspension problems he/she must understand the design of the various systems and their operating characteristics. First of all, a suspension and steering system performs three basic functions:

1. Acts as the overall connection between the wheels and the vehicle body.

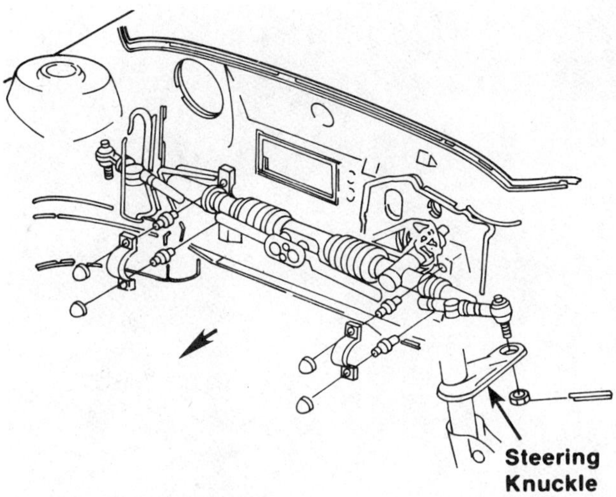

FIGURE 9-14 The steering gear on some unibody vehicles is bolted directly to the dash panel.

2. Damps and controls the ride; that is, it acts to partially absorb road shock and sway.
3. Provides directional control of the vehicle.

In unibody construction, the body provides the critical mounting positions for the suspension and steering. For instance, the inner fender skirts in conventional body-over-frame construction prevent dirt and road splash from entering the engine compartment. In unibody construction, they are called strut towers and fender aprons and have the additional task of providing the upper mounting controls for the strut suspension system.

In some cases, the steering gear of a unibody vehicle is bolted directly to the dash panel (Figure 9-14). In such instances, the panel must hold the steering gear in its correct location. The unibody structure must maintain the proper relationship of the steering and suspension parts to each other. Rear suspension components are also mounted directly to the unibody structural panels. The proper relationship of the rear suspension parts to each other and to the front suspension parts depends on the position of the structural panels.

CAUTION

The manufacturer has designed certain suspension, steering, and alignment features into the unibody vehicle. When any repairs are made, these specifications must be followed. If the specs are changed or tampered with, they might change the handling and safety features built into the vehicle. As an example, do not straighten or bend any of the pieces that make up the suspension system.

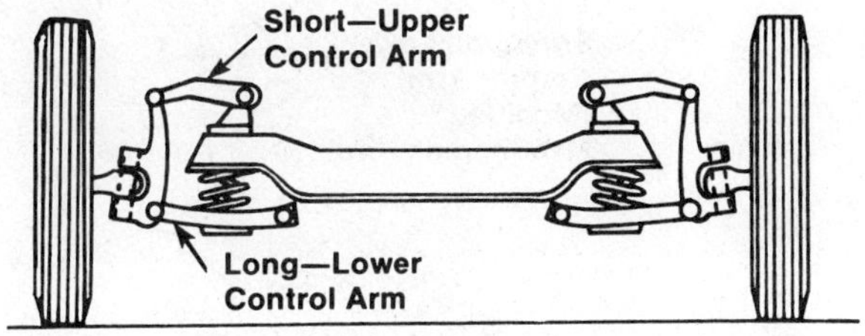

FIGURE 9-15 Short/long arm suspension

FRONT SUSPENSION

There are two main types of unibody front suspension systems: the short/long arm (SLA) independent and the strut. The latter is by far the most commonly used, but independent front systems are making a comeback. The main reason for the return of the modified short/long arm suspension (Figure 9-15) is that it allows a lower hood profile, thus offering improved aerodynamics.

The upper (short) and lower (long) control arms on the traditional independent front suspension function primarily as locators, fixing the position of the system and its components relative to the vehicle. They are attached to the frame with bushings that permit the wheel assemblies to move up and down separately in response to irregularities in the road surface. The outer ends are connected to the wheel assembly with ball joints inserted through each arm into the steering knuckle. The ball joints serve as the pivots for the wheel assembly, permitting the wheels to turn between the upper and lower control arms in whatever direction the vehicle is being steered. The load-carrying ball joint supports the vehicle weight. It is mounted in the arm on which the coil spring is seated or to which the torsion bar is attached.

Strut suspension is dramatically different in appearance from the traditional independent front suspension, but the components operate in the same way to meet suspension demands. Its most distinctive feature is the combination of the main elements into a single assembly. The MacPherson strut assembly typically includes the spring, upper suspension locator, and shock absorber and is mounted vertically between the top arm of the steering knuckle and the inner fender panel. Although there are several strut systems in use (Figure 9-16), all variations operate in basically the same manner.

The design and operation of a strut system is relatively simple compared to the short/long arm suspension system. Like the short/long arm suspension, the strut system has a lower control arm and spring. The strut replaces the shock absorber and the upper control arm and uses only a lower ball joint (Figure 9-17). Strut suspension systems use a coil spring that is part of the strut assembly. In some cases, the coil spring is placed

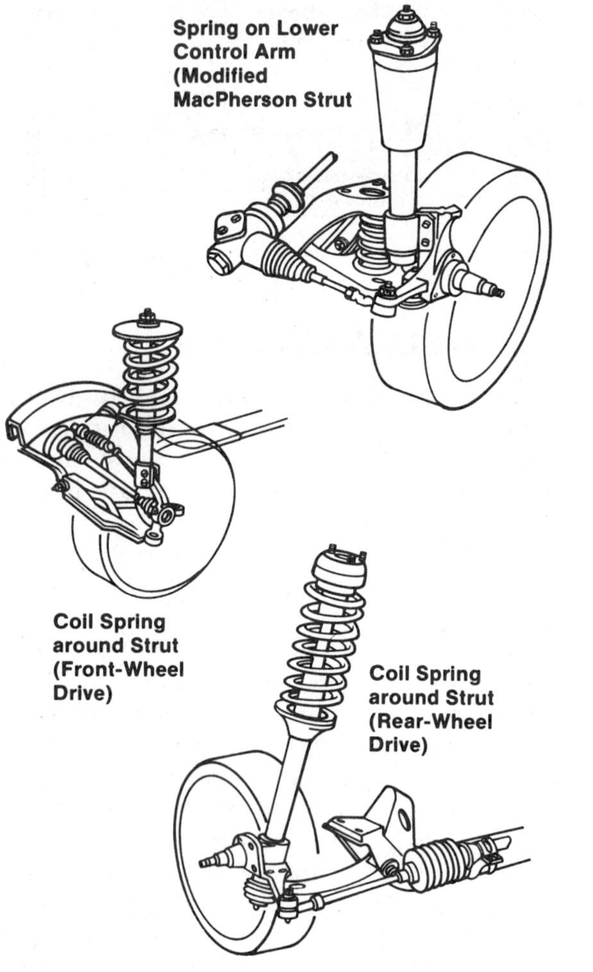

FIGURE 9-16 Several strut type front suspension designs

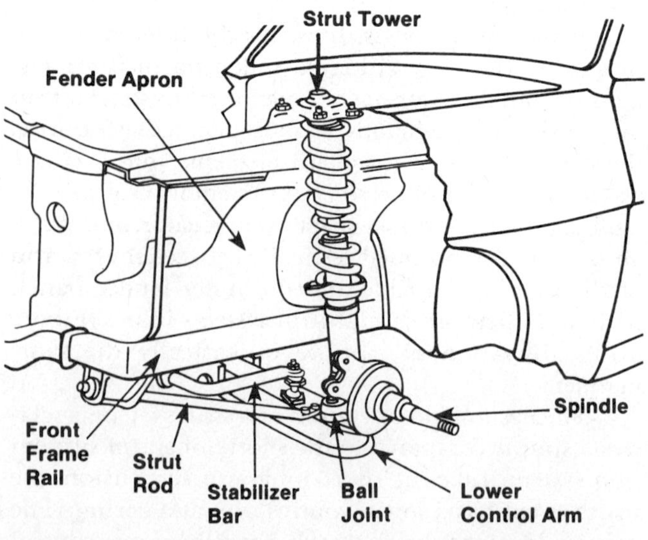

FIGURE 9-17 The major components of a strut suspension system

FIGURE 9-18 The suspension system is mounted to the unibody structural panels.

between the lower control arm and the unibody structure. Without an upper control arm, the upper ball joint is replaced by the bearing plate at the top of the strut. The loads generated by the strut suspension are transferred directly to the unibody structure through the spring mounting.

Many unibody vehicles have lower control arms attached directly to the front rails. Others have lower control arms attached to suspension subframes or crossmembers that are bolted directly to the front rails.

The suspension system mounting locations in the unibody structural panels (Figure 9-18) must be able to withstand the load transferred by vertical movement. In the case of strut suspensions, the vertical load is transferred directly from the upper strut mount to the top panel.

Suspension systems also experience lateral or side-to-side loads during operation; very heavy side loads are transferred through the suspension system during high-speed cornering. The lateral load from the cornering effect is transferred from the wheel through the lower control arm to the supporting structural panels of the unibody.

REAR SUSPENSION

The rear suspension system is an important, though often ignored, area. Generally, rear suspensions require no special service. Broken or worn parts should be replaced. Control arm or leaf spring bushings are constantly flexing. Bushings keep the rear wheels in line with the front wheels, and when worn can upset the settings of the entire suspension and drive shaft systems. Loose, worn, or broken attaching parts will allow the rear wheels to shift, causing premature tire wear as well as short U-joint life. A metallic jingling sound when

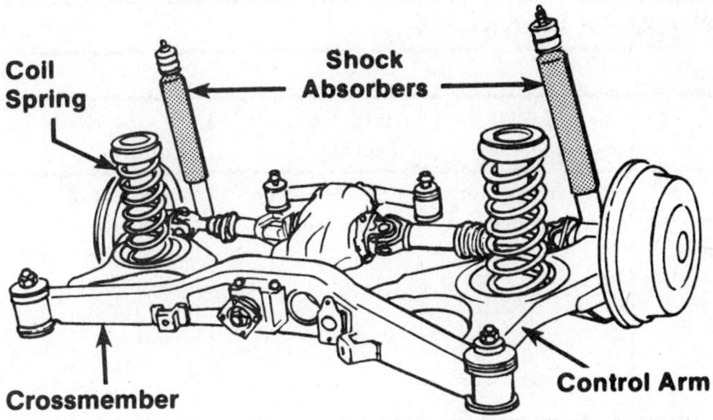

FIGURE 9-19 Typical independent rear suspension system found on many unibody vehicles

driving over small bumps or unusual tracking indicate the need for service. Usually a visual inspection is enough to determine repair requirements.

There are three principal types of rear suspensions: live axle, semi-independent or dead axle, and independent. Unlike live axle and dead axle suspensions, whose uses are restricted exclusively to either rear-wheel-drive or front-wheel-drive models, respectively, independent rear suspensions (Figure 9-19) can be found in large numbers on both FWD and RWD unibody vehicles. The introduction of independent rear suspensions was brought about by concerns for improved traction and ride and prompted the introduction of independent front suspensions. If the wheels can move separately on the road and energy forces can be maintained, traction and ride will be improved.

Table 9-1 diagnoses common suspension problems.

COMPUTER SUSPENSION SYSTEMS

As already described, conventional suspension systems consist of coil springs, struts, and shock absorbers. The new computer suspension systems use sensors to adjust height, leveling of the vehicle, and ride firmness in some models.

Five systems are in current use:

- Electronic air suspension (EAS)
- Electronic level control (ELC)
- Variable damping shock (VDS)
- Springless electronic suspension (SES)
- Computer-controlled suspension (CCS)

Some of these systems are relatively new or in the final experimental stages. Repairing them will become more common as more vehicles come equipped with them. Anyone involved in the collision repair process needs a basic knowledge of how these systems work; the best source of information is the service manual. Keep in mind that computer suspension systems are vulnerable to certain problems, any one of which can upset the ability to maintain the desired distance between chassis and road. A plugged or leaking air hose, a bad solenoid, a defective compressor, or a misadjusted or faulty height sensor, for example, can interfere with the suspension's ability to level itself. Because of this, check the system's operation after a collision to make sure it is functioning correctly.

To understand the basic operation of computer suspensions, look at Figure 9-20. In this EAS system, air

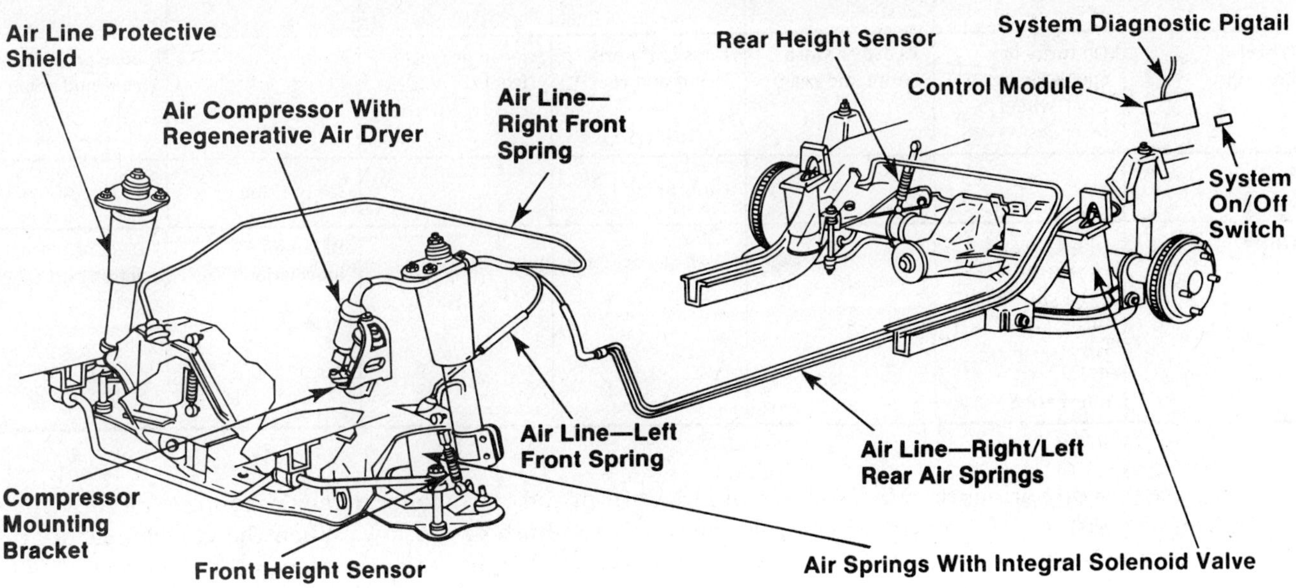

FIGURE 9-20 Typical electronic air suspension system

TABLE 9-1: SUSPENSION PROBLEM DIAGNOSIS

	Problem					
Check	**Noise**	**Instability**	**Pulls to One Side**	**Excessive Steering Play**	**Hard Steering**	**Shimmy**
Tires/Wheels	Road or tire noise	Low or uneven air pressure; radials mixed with belted bias ply tires	Low or uneven air pressure; mismatched tire sizes	Low or uneven air pressure	Low or uneven air pressure	Wheel out of balance or uneven tire wear or overworn tires; radials mixed with belted bias ply tires
Shock Dampers (Struts/ Absorbers)	Loose or worn mounts or bushings	Loose or worn mounts or bushings; worn or damaged struts or shock absorbers	Loose or worn mounts or bushings	—	Loose or worn mounts or bushings on strut assemblies	Worn or damaged struts or shock absorbers
Strut Rods	Loose or worn mounts or bushings	Loose or worn mounts or bushings	Loose or worn mounts or bushings	—	—	Loose or worn mounts or bushings
Springs	Worn or damaged	Worn or damaged	Worn or damaged, especially rear	—	Worn or damaged	—
Control Arms	Steering knuckle control arm stop; worn or damaged mounts or bushings	Worn or damaged mounts or bushings	Worn or damaged mounts or bushings	—	Worn or damaged mounts or bushings	Worn or damaged mounts or bushings
Steering System	Component wear or damage	Component wear or damage	Component wear or damage	Component wear or damage	Component wear or damage	Component wear or damage
Alignment	—	Front and rear, especially caster	Front, camber and caster	Front	Front, especially caster	Front, especially caster
Wheel Bearings	On turns or speed changes: front-wheel bearings	Loose or worn (front and rear)	Loose or worn (front and rear)	Loose or worn (front)	—	Loose or worn (front and rear)
Brake System	—	—	On braking	—	On braking	—
Other	Clunk on speed changes: transaxle; click on turns: CV joints; ball joint lubrication	—	—	—	Ball joint lubrication	Loose or worn friction ball joints

springs replace the coil springs used in a conventional independent suspension. The computer controls the vehicle's height by either telling a battery driven air compressor to pump air into an air spring or telling a valve to let air out. Three height sensors, two in the front and one in the rear, tell the computer if the vehicle is too high or too low. When the vehicle reaches the proper height, the sensors send a trim signal. The pump is shut off and the air spring valve is closed, trapping the air in the spring.

A

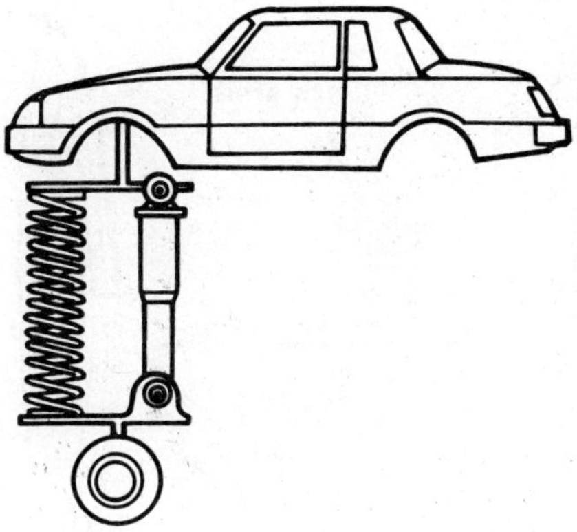

B

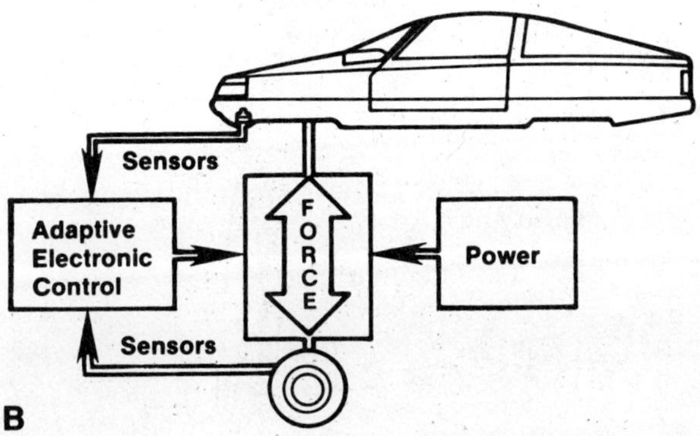

FIGURE 9-21 (A) Passive suspension compromises handling because of static spring rates and shock valving. (B) Active suspension constantly adapts to changing road and driving conditions.

The SES and CCS types are known as *active* suspension systems. They have the ability to adapt themselves to changing road and driving conditions (Figure 9-21). Instead of conventional springs or air springs to support the vehicle, double-acting hydraulic cylinders (called *actuators*) are mounted at each wheel. Each actuator maintains a sort of hydraulic equilibrium with the others to carry the vehicle's weight, while maintaining the desired body attitude. At the same time, each actuator serves as its own shock absorber, eliminating the need for yet another traditional suspension component.

In other words, each hydraulic actuator acts as both a spring (with variable-rate damping characteristics) and a variable-rate shock absorber. This is accomplished in an active suspension system by varying the hydraulic pressure within each cylinder, as well as the rate at which it increases or decreases. By regulating the hydraulic pressure in the individual actuators, each wheel can react independently to changing road conditions.

A steering angle sensor is used to signal the computer when the vehicle is turning. To monitor body motions, a roll sensor, lateral acceleration, and G-sensors are used. The computer also monitors hydraulic pressure within the system and the speed of the pump motor.

Once it has all the necessary inputs, the computer can then regulate the ebb and flow of hydraulic pressure within each individual actuator according to any number of variables and its own built-in program. Another feature of a suspension such as this is that it can be programmed to behave in a variety of unique and previously impossible ways: leaning into turns, for example, or even raising a flat tire on command to change the tire without using a separate jack.

STEERING

The steering linkage connects the output of the steering box to the front wheel steering arms. The two main types of steering linkage now in use are the standard or parallelogram and rack-and-pinion.

Parallelogram Steering

This steering uses two tie-rod assemblies connected to the steering arms to support a long center link. The center link holds the tie-rods in position (Figure 9-22). An idler arm supports the center link on one end; the other end of the center link is attached to the pitman arm. The arrangement forms a parallelogram shape (each end moves equal and parallel to the other end). On some systems a steering damper is attached to the frame and the steering linkage to help absorb road shocks and wheel shimmy.

Rack-and-Pinion Steering

The rack-and-pinion design (Figure 9-23) uses a simplified steering system that weighs less and takes up less space. For these reasons, it is used on most domestic and import unibody vehicles.

The rack-and-pinion type of steering unit takes the place of the idler arm, pitman arm, center link, and gear box used in parallelogram systems. It consists of a pinion gear at the end of the steering shaft, which meshes with a toothed shaft (a bar with a row of teeth cut into one edge) known as the rack. When the steering wheel is turned, the pinion rotates in a circle and moves the rack sideways—left or right—to turn the wheels.

Two inner tie-rod ends are threaded onto the end of the rack and are covered by rubber bellows boots. These boots protect the rack from contamination by dirt, salt, and other road particles. The inner tie-rods are threaded onto the outer tie-rod ends, which connect to the steering arms. Rack-and-pinion has fewer friction points than a standard steering system, so more energy and movement from road forces get through to the steering wheel. This gives the driver a more positive feel of the road. A power-assisted rack-and-pinion system responds faster to input changes, and the boost can be

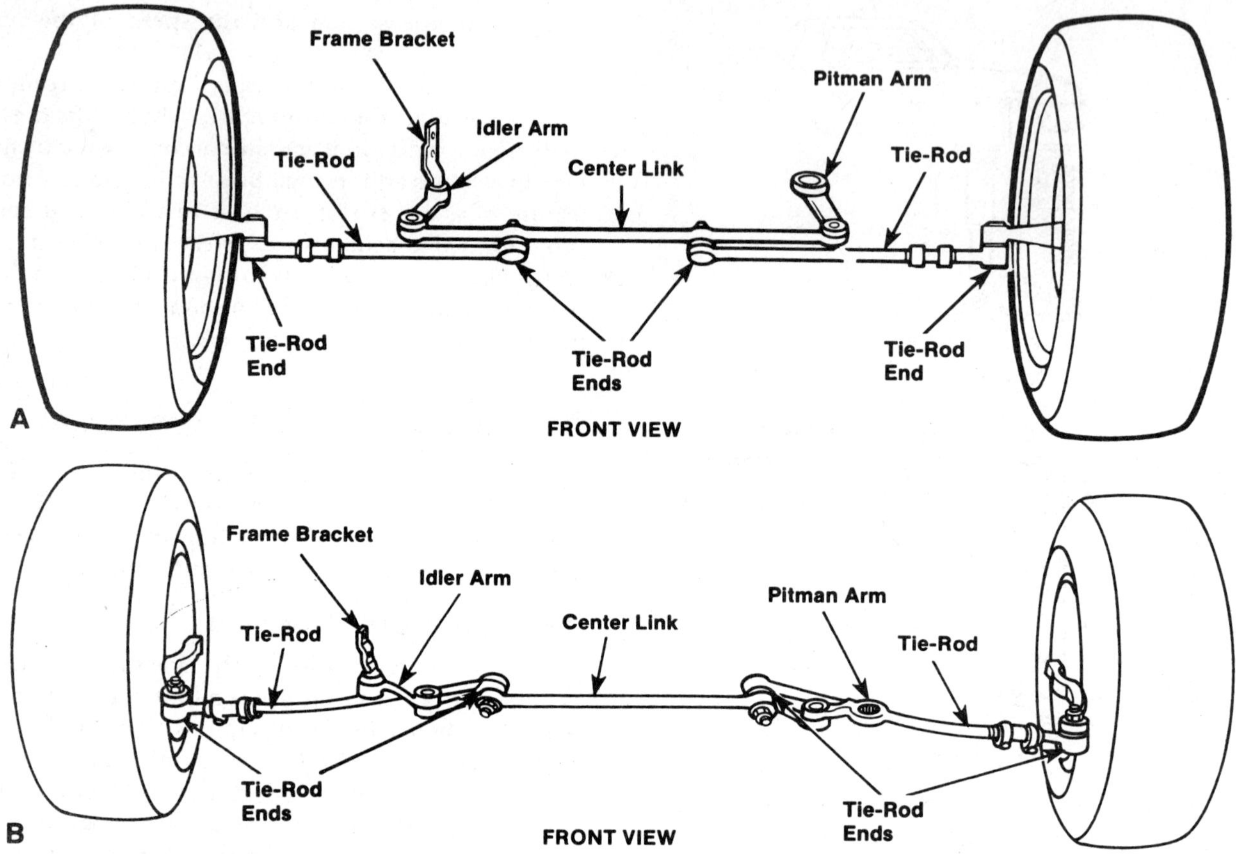

FIGURE 9-22 A typical parallelogram steering system mounts (A) behind the front suspension and (B) ahead of the front suspension

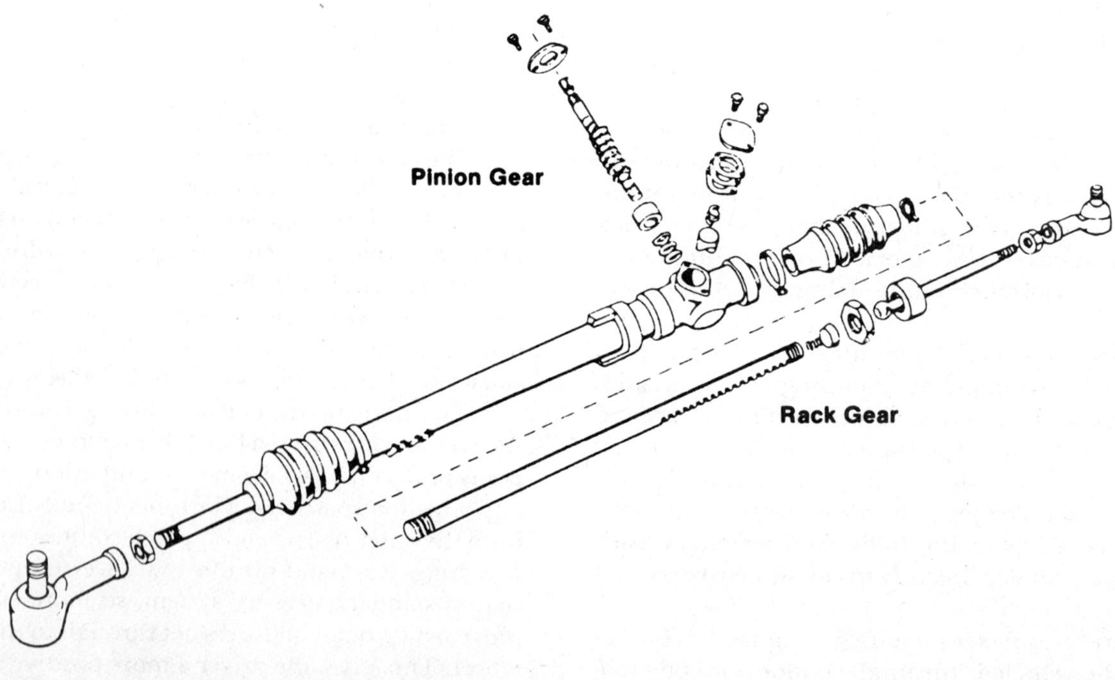

FIGURE 9-23 Rack-and-pinion steering system

adjusted to suit driving conditions or individual drivers. It also steers more easily without a boost than hydraulic systems.

To check a rack-and-pinion system, begin by raising the vehicle and taking the weight off the front suspension. Visually inspect the steering system for any physical damage. Check the boots for leaks, inspect the tie-rods, and examine the mounting points for any distortion. Inspect the tie-rod ends. Grab the tie-rod near the tire and try pushing it up and down. Any vertical looseness indicates damage or wear.

Check the inner tie-rod socket by squeezing the bellows until the socket can be felt (Figure 9-24). With your other hand, push and pull on the tire. Looseness in the socket indicates damage or wear. Take a front tire in each hand, and see if they can be moved back and forth in opposite directions. If excessive movement is noted, wear or damage is likely. Observe the rack-and-pinion at the same time. Any movement might indicate a problem.

SHOP TALK

Remember that rack-and-pinion units must be mounted on a level plane. Being out of alignment as little as 3mm can affect handling. The closer the mounting brackets are to the middle of the rack, the more critical this measurement becomes. Misalignment of the rack-and-pinion will cause changes in the steering geometry during jounce/rebound. This condition cannot be corrected by changing the length of the tie-rods.

Power Steering

The power steering unit is designed to reduce the amount of effort required to turn the steering wheel. It also reduces driver fatigue on long drives and makes it easier to steer the vehicle at slow speeds, particularly during parking.

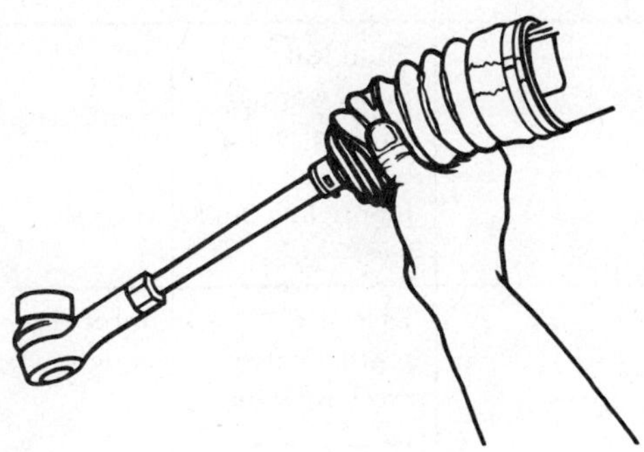

FIGURE 9-24 Check the inner tie-rod socket by squeezing the bellows.

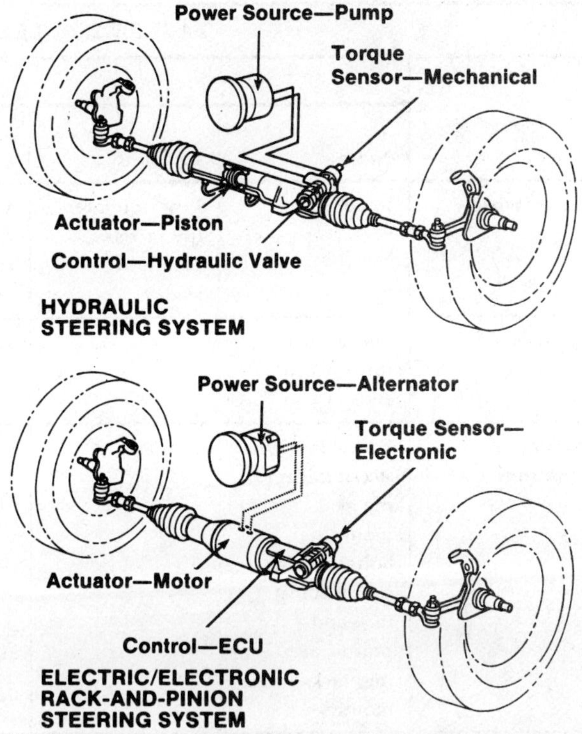

FIGURE 9-25 The alternator in a nonconventional power steering system replaces the hydraulic pump of conventional systems.

Power steering can be broken down into two design arrangements: conventional and nonconventional or electronically controlled. In the conventional arrangement, hydraulic power is used to assist the driver, while in the nonconventional arrangement an electric/electronic rack-and-pinion unit provides power assistance in steering (Figure 9-25).

In an electronically controlled power steering arrangement, the hydraulic pump, hoses, and fluid associated with conventional power steering systems are replaced with electronic controls and an electric motor located concentrically to the rack itself. The design features a DC motor armature with a hollow shaft to allow passage of the rack through it. The outboard housing and rack are designed so that the rotary motion of the armature can be transferred to linear movement of the rack through a ball nut with thrust bearings. The armature is mechanically connected to the ball nut through an internal/external spline arrangement.

Here are some power steering service tips that should be kept in mind:

- *Protect the system.* Protect the system from dirt and moisture at all times. If the system must be open, be sure to plug or tie off all openings with plastic.

TABLE 9-2: STEERING PROBLEM DIAGNOSIS

	Problem					
Check	**Noise**	**Instability**	**Pull to One Side**	**Excessive Steering Play**	**Hard Steering**	**Shimmy**
Tires/Wheels	Road or tire noise	Low or uneven tire pressure; radial tire lead	Low or uneven tire pressure; radial tire lead	Low or uneven tire pressure	Low or uneven tire pressure	Unbalanced wheel; uneven tire wear; over-worn tires
Tie-Rods	Squeal in turns: worn ends	—	Incorrect toe: tie-rod length	Worn ends	Worn ends	Worn ends
Mounts or Bushings	Parallelogram steering: steering gear mounting bolts, linkage connections; rack-and-pinion steering: rack mounts	Idler arm bushing	—	Parallelogram steering: steering gear mounting bolts, linkage connections; rack-and-pinion steering: rack mounts	Parallelogram steering: steering gear mounting bolts, linkage connections; rack-and-pinion steering: rack mounts	Parallelogram steering: steering gear mounting bolts, linkage connections; rack-and-pinion steering: rack mounts
Steering Linkage Components	Bent or damaged steering rack	Incorrect center link or rack height	Incorrect center link or rack height	Worn idler arm, center link, or pitman arm studs; worn or damaged rack	Idler arm binding	Worn idler arm, center link, or pitman arm studs
Steering Gear	Improper yoke adjustment on rack-and-pinion steering	—	—	Improper yoke adjustment on rack-and-pinion steering; worn steering gear or incorrect gear adjustment on parallelogram steering; loose or worn steering shaft coupling	Parallelogram steering: low steering gear lubricant, incorrect adjustment; rack-and-pinion: bent rack, improper yoke adjustment	—
Power Steering	—	—	—	—	Fluid leaks; loose, worn, or glazed steering belt; weak pump; low fluid level	—
Alignment	—	—	Unequal caster or camber	—	Excessive positive caster, excessive scrub radius (incorrect camber and/or SAI)	Incorrect caster

- *Use recommended fluid.* Always replace the fluid lost with the recommended type. Most vehicles require either Dexron or Type F fluid, while some need special fluid specified by the manufacturer. Some fluids claim to meet the specifications for both of the above types. Follow the recommendations to protect the warranty.
- *Bleed the system.* Many systems are self-bleeding, but some require a specific bleeding procedure to eliminate air. Air in the system can cause noise, vibration, and erratic performance. Follow the directions in the service manual.
- *Check the hose routing.* Check the hose routing when reassembling the system. Always route and hang hoses identical to the factory installation. Avoid contact with other parts; especially watch rubbing against moving parts.

To perform a diagnostic check of possible steering problems, follow the procedure given in Table 9-2.

Collapsible Steering Columns

To reduce the chance of injury to the driver, automotive engineers have designed steering columns that collapse in the event of a collision. Lower steering sections are linked by two or more universal joints; these joints allow the sections to fold. The upper column normally incorporates a solid shaft and a tube. The shaft is locked to the tube by one of several methods, each designed to "give" in the event of a collision.

Methods used to lock the shaft to the tube include plastic inserts or steel balls held in a plastic retainer that allow the shaft to "roll" forward inside the tube. There are also collapsing steel mesh or accordion-pleated designs that give way under pressure. During damage analysis, check the steering column for evidence of collapse. Although the vehicle can be steered with a collapsed column that has been pulled back, the collapsed portion must be replaced. All service manuals provide explicit instructions for doing this.

CAUTION

When working on a collapsible steering column, do not hammer on or bump column components. With the column removed from the mounts, it is extremely susceptible to impact damage. A slight impact on the column end can collapse the steering shaft or loosen the plastic inserts that maintain column rigidity. When removing the steering wheel, use a puller. Do not hammer or pound on any components during removal.

Four-Wheel Steering

Four-wheel independent suspensions have been around for years and will continue to be an important part of the automotive industry. Now the industry is offering four-wheel independent steering systems, where the rear wheels also help to turn the vehicle by electrical, hydraulic, or mechanical means. Because these systems are going to be more common in the future, the collision repair industry will have to adapt to them.

To understand the advantages of four-wheel steering, it is wise to review the dynamics of typical steering maneuvers with a conventional front-steered vehicle. The tires are subject to the forces of grip, momentum, and steering input when making a movement other than straight-ahead driving. These forces compete with each other during steering maneuvers. With a front-steered vehicle, the rear end is always trying to catch up to the directional changes of the front wheels. This causes the vehicle to sway or yaw. As a normal part of operating a vehicle, the driver learns to adjust to these forces without thinking about them.

When turning, the driver is putting into motion a complex series of forces. Each of these must be balanced against the others. The tires are subjected to road grip and slip angle. Grip holds the vehicle's wheels to the road, and momentum moves the vehicle straight ahead. Steering input causes the front wheels to turn. The vehicle will momentarily resist the turning motion, causing a tire slip angle to form (Figure 9-26). Once the vehicle begins to respond to the steering input, cornering forces are generated. The vehicle yaws as the rear wheels attempt to keep up with the cornering forces already generated by the front tires. This is referred to as rear-end lag, because there is a time delay between steering input and vehicle reaction. When the front wheels are turned back to a straight-ahead position, the vehicle

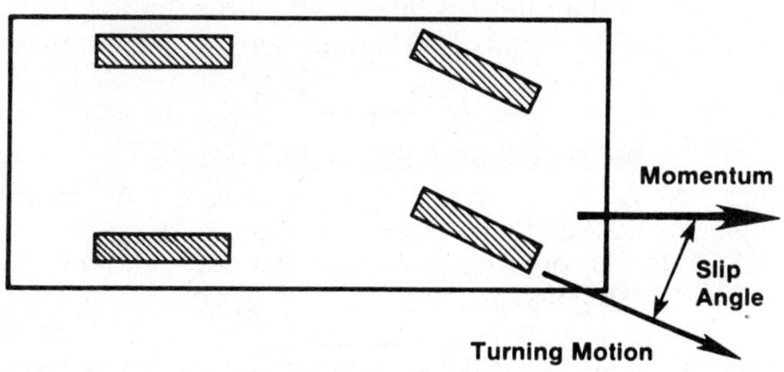

FIGURE 9-26 When a turn is initiated with front steering, the momentum wants to move the vehicle forward while the steering input causes a directional change. This creates a tire slip angle as the vehicle momentarily resists turning.

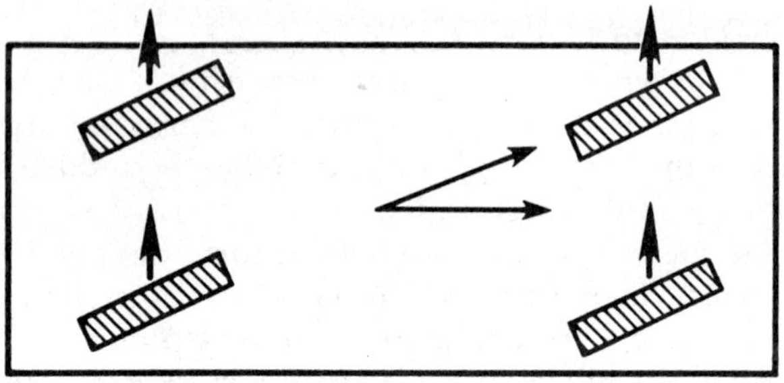

FIGURE 9-27 With four-wheel steering, the entire vehicle moves in one direction.

must again try to adjust by reversing the same forces developed by the turn. As the steering is turned, the vehicle body yaws as the rear wheels again try to keep up with the cornering forces generated by the front wheels.

The idea behind four-wheel steering is that a vehicle requires less driver input for any steering maneuver if all four wheels are steering the vehicle. As with two-wheel steering, grip holds the four wheels on the road. However, when the driver turns the wheel slightly, all four wheels react to the steering input, causing slip angles to form at all four wheels. The entire vehicle moves in one direction rather than the rear half attempting to catch up to the front half (Figure 9-27). There will also be less yaw when the wheels are turned back to a straight-ahead position. The vehicle responds more quickly to steering input because rear wheel lag is eliminated.

Currently there are three types of production four-wheel steering systems: mechanical, hydraulic, and electro-hydraulic designs. Since each system is unique in its construction and repair needs, the service manual for the vehicle must be followed for proper diagnosis, repair, and alignment of a four-wheel system.

WHEEL ALIGNMENT

The best alignment procedure for the modern unibody vehicle is a four-wheel alignment. Approximately 80 percent of today's vehicles not only have front-end alignment specifications, but also require rear-wheel alignment. This is due to the growing number of vehicles with independent rear suspensions. Collision repair of any magnitude usually requires at least a four-wheel alignment check.

To begin alignment work, the body structure must be within specifications. If collision damage has been repaired, the vehicle must be restored to pre-accident tolerances. A four-wheel alignment will not reduce the need for proper body and chassis straightening. Before making adjustments, conduct the following prealignment checks:

1. *Visual Inspection.* This includes checking for uneven tire wear and mismatched tire sizes or types. Remember that on vehicles that have been driven several thousand miles, look for normal wear as well as the results of collision damage. Look for towing damage as well.
2. *Ride Height Measurement.* Each of the front and rear wheels must carry the same amount of weight. The vehicle is designed to ride at a specific height, sometimes referred to as *curb height* (Figure 9-28). Ride height specs are published in the service manuals and some alignment spec books.

   If the vehicle leans to one side or seems to be lower on one side than on the other, something is wrong. To isolate the problem, place a jack in the center of the main crossmember in the front of the vehicle (Figure 9-29). Raise the vehicle several inches, and look at the rear of the vehicle. If the rear looks level, then the problem is in the front suspension, on the side of the vehicle that shows the lean. If the rear is not level, then the problem is with the low side of the rear suspension.
3. *Under Vehicle Checks.* With the vehicle raised, inspect all steering components such as control arm bushings, upper strut mounts,

MEASURING INSTRUCTIONS

Dim. "A" & "B"—with vehicle at curb weight (full tank, fluids and tire pressure), measure from lowest point on bumper to ground, including standard equipment bumper guards.

Dim. "C" & "D"—with vehicle loaded (see passenger and load illustration section for correct load specifications), measure from lowest point on rocker, excluding flanges, to ground.

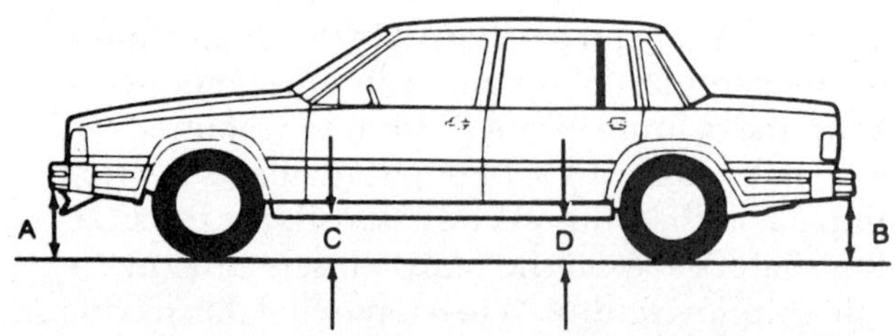

* Measured at curb weight with 2 passengers in front

FIGURE 9-28 Curb height measurement points from the manufacturer's recommendations

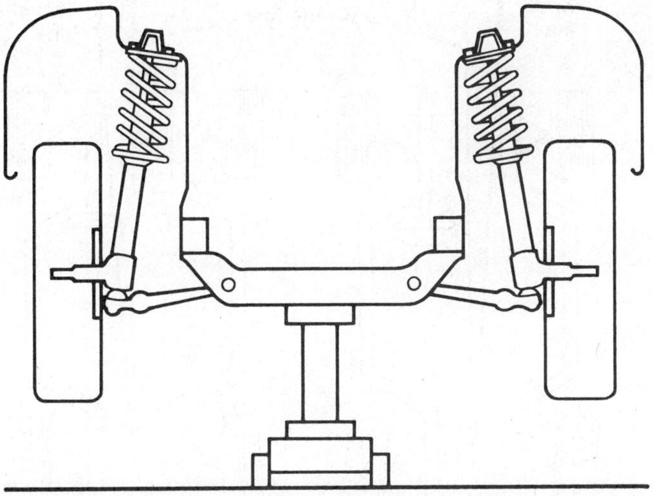

FIGURE 9-29 Raising the vehicle by jacking in the center of the main crossmember

pitman arm, idler arm, center link, tie-rod ends, ball joints, and shock absorbers. Check the CV joints (if equipped) for looseness, popping sounds, binding, and broken boots. Damaged components must be repaired before adjusting alignment angles.

4. *Road Test.* Begin the alignment with a road test. While driving the vehicle, check to see that the steering wheel is straight. Feel for vibration in the steering wheel as well as in the floor or seats. Notice any pulling or abnormal handling problems, such as hard steering, tire squeal while cornering, or mechanical pops or clunks. This helps find problems that must be corrected before proceeding with the alignment.

The proper alignment of a suspension/steering system centers around the accuracy of the following angles:

Caster. This is the angle of the steering axis of the front wheel from true vertical, as viewed from the side of the vehicle. The forward or rearward tilt from the vertical line, as shown in Figure 9-30, illustrates caster. Tilting the steering axis forward is negative caster; tilting backward is positive caster. Caster is the first angle adjusted during an alignment.

Caster is designed to provide steering stability. The caster angle for each wheel should be equal. Unequal caster angles will cause the vehicle to steer toward the side with less caster. Too much negative caster can cause the vehicle to have "sensitive" steering at high speeds. The vehicle might wander as a result of too much negative caster.

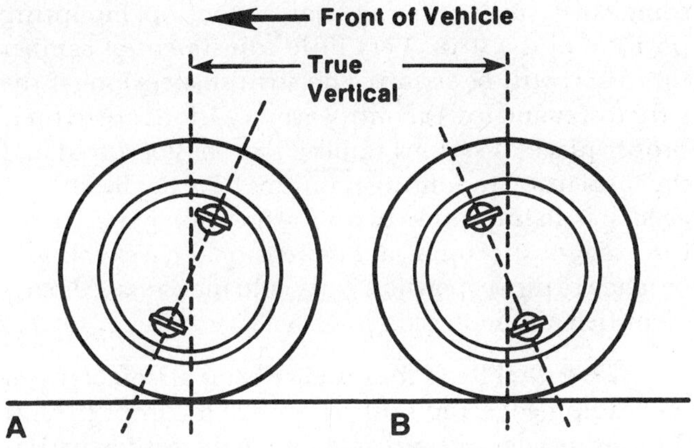

FIGURE 9-30 (A) Positive and (B) negative caster

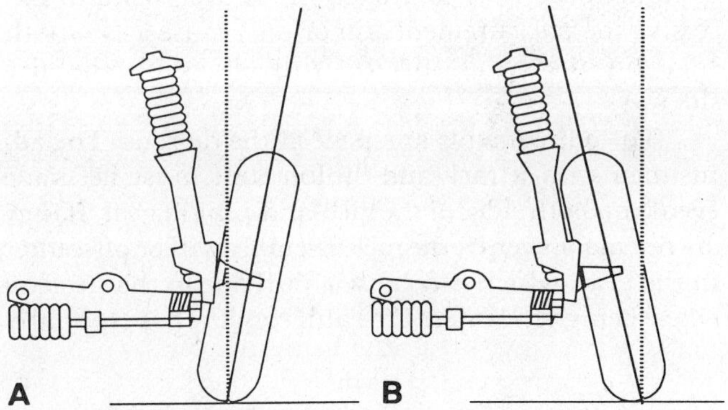

FIGURE 9-31 (A) Positive and (B) negative camber

Several factors can adversely affect caster. The most common problem is worn or loose strut rod bushings and control arm bushings. Caster adjustments are not provided on some strut suspension systems. Where they are provided, they can be made at the top or bottom mount of the strut suspension.

Camber. This is the angle represented by the tilt of either the front or rear wheels inward or outward from the vertical as viewed from the front of the vehicle (Figure 9-31). Camber is designed into the vehicle to compensate for road crown, passenger weight, and vehicle weight. Camber is usually set equally for each wheel. Equal camber means each wheel is tilted outward or inward the same amount. Unequal camber causes tire wear and causes the vehicle to steer toward the side that is more positive.

Camber is controlled by the control arms and their pivots. It is affected by worn or loose ball joints, control arm bushings, and wheel bearings. Anything that changes chassis height will also affect camber. Camber is adjustable on most vehicles; some manufacturers prefer to include a camber adjustment at the spindle assembly. Camber adjustments are also provided on some strut suspension systems at the top mounting position of the strut. Very little adjustment of camber (or caster) will be required on strut suspensions if the strut tower and lower control arm positions are in their proper place. If serious camber error has occurred and the suspension mounting positions have not been damaged, it is an indication of bent suspension parts. In this case, diagnostic angle and dimensional checks should be made to the suspension parts, and the damaged parts identified and replaced.

Toe. This is the distance comparison between the leading edge and trailing edge of the front tires. If the leading edge distance is less, then there is toe-in. If it is greater, it is toe-out (Figure 9-32). Actually, toe is critical as a tire wearing angle. Wheels that do not track straight ahead have to drag as they travel forward. Excessive toe measurements (in or out) cause a sawtooth edge on the tread surface from dragging the tire sideways.

Toe adjustments are made at the tie-rods. Toe adjustments on a rack-and-pinion unit must be made evenly on both sides of the rack-and-pinion gear. If they are not made evenly, the rack assembly will be off-center in the gear, which can cause a pull due to the steering wheel being off-center. This pull condition is especially common with power-assisted rack-and-pinion gears. It can be corrected by making the adjustments equal on both sides of the gear. The steering assembly must be centered before these adjustments are made. Toe is the last adjustment made in an alignment.

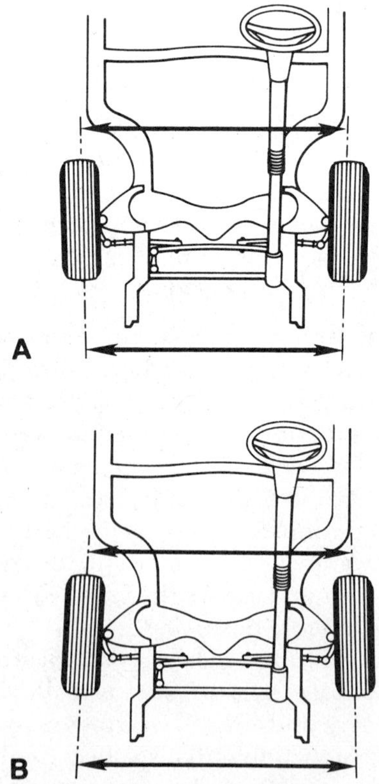

FIGURE 9-32 (A) Toe-in and (B) toe-out

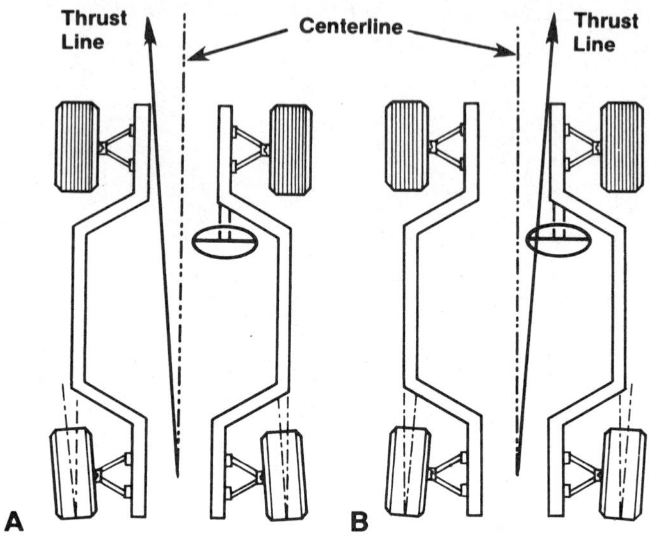

FIGURE 9-33 (A) Left and (B) right thrust direction

Thrust Line Alignment. A main consideration in any alignment is to make sure the vehicle runs straight down the road, with the rear tires tracking directly behind the front tires when the steering wheel is in the straight-ahead position. The geometric centerline of the vehicle should parallel the road direction. This will be the case when rear toe is parallel to the vehicle's geometric centerline in the straight-ahead position. If rear toe does not parallel the vehicle centerline, a "thrust" direction to the left or right will be created (Figure 9-33). This difference of rear toe from the geometric centerline is called the *thrust angle.* The vehicle will tend to travel in the direction of the thrust line, rather than straight ahead.

To correct this problem, begin by setting individual rear toe equal in reference to the geometric centerline. Four-wheel alignment machines check individual toe on each wheel. Once the rear wheels are in alignment with the geometric centerline, set the individual front toe in reference to the thrust angle. Following this procedure assures that the steering wheel will be straight ahead for straight-ahead travel. If you set the front toe to the vehicle geometric centerline, ignoring the rear toe angle, a cocked steering wheel will result.

Steering Axis Inclination (SAI). This angle locates the vehicle weight to the inside or outside of the vertical centerline of the tire. The SAI is the angle

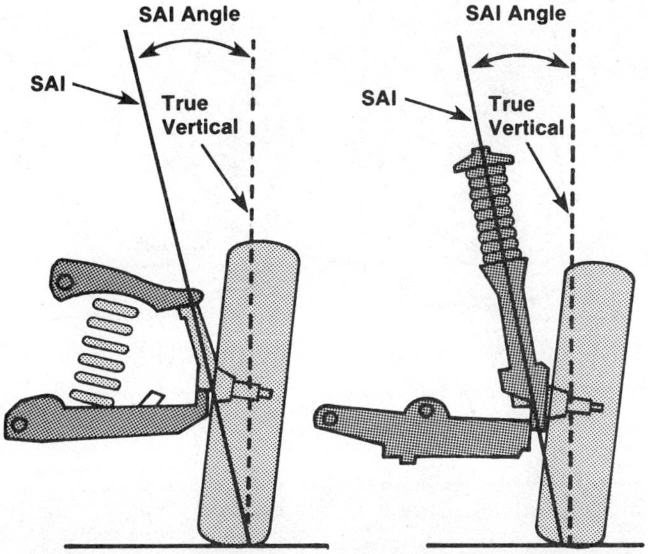

FIGURE 9-34 SAI is a line drawn through the upper and lower pivot points. The SAI angle is how far this line differs from the true vertical.

between true vertical and a line drawn between the steering pivots as viewed from the front of the vehicle (Figure 9-34). It is an engineering angle designed to project the weight of the vehicle to the road surface for stability. SAI also helps the steering system return to straight ahead after a turn. SAI should be checked because it can help locate other problems. Incorrect SAI requires that parts must be replaced to correct SAI angles (Table 9-3).

If the vehicle has 0 (zero) SAI, the upper and lower ball joints (or strut pivot points) would be located directly over one another. Problems associated with this simple relationship include tire scrub in turns, lack of control, and increased effort during turn recovery. If the SAI is tilted, a triangle is formed between ball joints and spindle; an arc is then formed when turning. There is a high point at straight-ahead position and a drop downward turning to each side. This motion travels through the control arms to the springs and finally to the weight of the vehicle. The forces generated in a turn are actually trying to lift the vehicle. The tilting and loading effect of SAI offsets the lifting forces and helps to pull the tires back to straight ahead when the turn is finished.

To check the SAI, measure the angle between true vertical and a line drawn from the upper ball joint to the lower ball joint. With strut systems, the line is drawn down the strut pivot centerline to the lower ball joint. When the SAI angle includes the camber angle, the sum of the two is called the included angle. Problems that occur with this built-in angle can be due to a bent spindle, a bent upper or lower control arm or strut, or a badly worn or damaged ball joint. These damaged components must be replaced before doing the alignment. Doing an alignment without checking all the angles can result in error. For example, if the SAI angle is 7 degrees and the camber angle is 1 degree positive,

TABLE 9-3 SAI DIAGNOSTIC REFERENCE CHART

Suspension System	SAI	Camber	Included Angle	Probable Cause
Short/	Correct	Less	Less	Bent Knuckle
Long Arm	Less	Greater	Correct	Bent Lower Control Arm
Suspension	Greater	Less	Corrrect	Bent Upper Control Arm
System	Less	Greater	Greater	Bent Knuckle
MacPherson	Correct	Less	Less	Bent Knuckle and/or Bent Strut
Strut	Correct	Greater	Greater	Bent Knuckle and/or Bent Strut
Suspension	Less	Greater	Correct	Bent Control Arm or Strut Tower (out at top)
	Greater	Less	Correct	Strut Tower (in at top)
	Greater	Greater	Greater	Strut Tower (in at top) and spindle and/or Bent Strut
	Less	Greater	Greater	Bent Control Arm or Strut Tower (out at top) Plus Bent Knuckle and/or Bent Strut
	Less	Less	Less	Strut Tower (out at top) and Knuckle and/or Strut Bent or Bent Control Arm
Twin I	Correct	Greater	Greater	Bent Knuckle
Beam	Greater	Less	Correct	Bent "I" Beam
Suspension	Less	Greater	Correct	Bent "I" Beam
	Less	Greater	Greater	Bent Knuckle

then the included angle (SAI plus camber) should be 8 degrees. If the camber required a 3 degree adjustment to achieve the included angle specification of 8 degrees, the SAI would be less than specifications and the camber angle would be greater than specifications. This would reveal that something is wrong.

Turning Radius. Turning radius or cornering angle is the amount of toe-out present on turns (Figure 9-35). As a vehicle goes around a corner, the inside tire must travel in a smaller radius circle than the outside tire. This is accomplished by designing the steering geometry to turn the inside wheel sharper than the outside wheel. The result can be seen as toe-out on turns. This eliminates tire scrubbing on the road surface by keeping the tires pointed in the direction they have to move.

DIAGNOSTIC CHECKS

The analysis of ride and handling complaints involves the consideration of diagnostic angles. Suspension system parts must be considered as moving parts in vehicle operation; diagnostic angles evaluate the parts as they move. When a vehicle has a steering problem, the first diagnostic check should be a visual inspection of the entire vehicle for anything obvious: bent wheels, misalignment of the cradle, and so on. If there is nothing obviously wrong with the vehicle, make a series of diagnostic checks without disassembling the vehicle.

One of the most useful checks that can be made with a minimum of equipment is a jounce-rebound check.

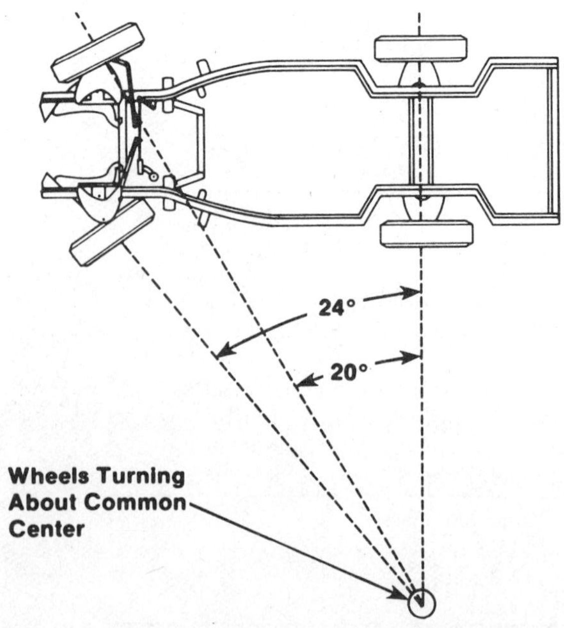

FIGURE 9-35 Turning radius

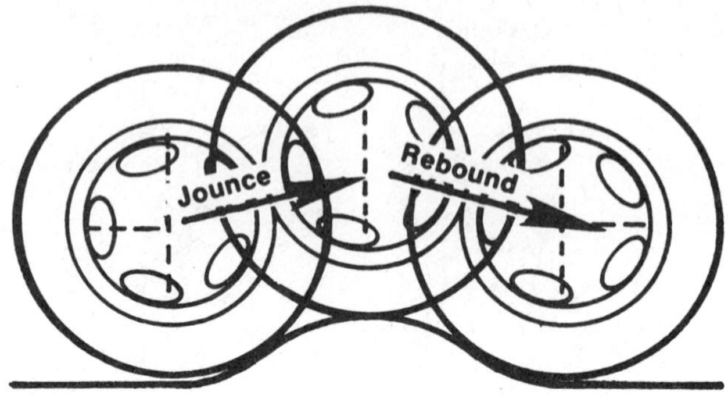

FIGURE 9-36 A wheel moving through jounce and rebound

- Jounce (Figure 9-36) is the motion caused by a wheel going over a bump and compressing the spring. During jounce, the wheel moves up toward the chassis. Jounce can be simulated by sitting on the bumper and pushing down on the vehicle. The vehicle should jounce equally on both sides.
- Rebound is the motion caused by a wheel going into a dip or returning from a jounce and extending the spring. During rebound, the wheel moves down away from the chassis. Rebound can be simulated by lifting up on the bumper. The vehicle should lift equally on both sides.

A jounce-rebound check will determine if there is misalignment in the rack-and-pinion gear. For a quick check, unlock the steering wheel and see if it moves during jounce and/or rebound. For a more careful check, use a pointer and a piece of chalk. Use the chalk to make a reference mark on the tire tread and place the pointer on the same line as the chalk mark. Jounce and rebound the suspension system a few times while someone watches the chalk mark and the pointer. If the mark on the wheel moves unequally in and out on both sides of the vehicle, chances are there is a steering arm or gear out of alignment. If the mark does not move or moves equally in and out on both sides of the vehicle, the steering arm and gear are probably all right.

The next diagnostic check is for turning radius. This check evaluates the proper relationship of the two front wheels as they are turned through a steering arc. To measure turning radius, one wheel is turned on a turn plate or protractor a given amount; the amount of rotation of the opposite wheel is measured with a similar turn plate or protractor. The process is then repeated with the opposite wheel. The results are compared right to left to determine if the two front wheels are rotating through the same arc.

On a typical turning radius check, the left front wheel is turned out 20 degrees and the right wheel rotation is measured. The readings will usually not

TABLE 9-4: RADIAL TIRE LEAD/PULL CORRECTION CHART—FRONT WHEEL DRIVE

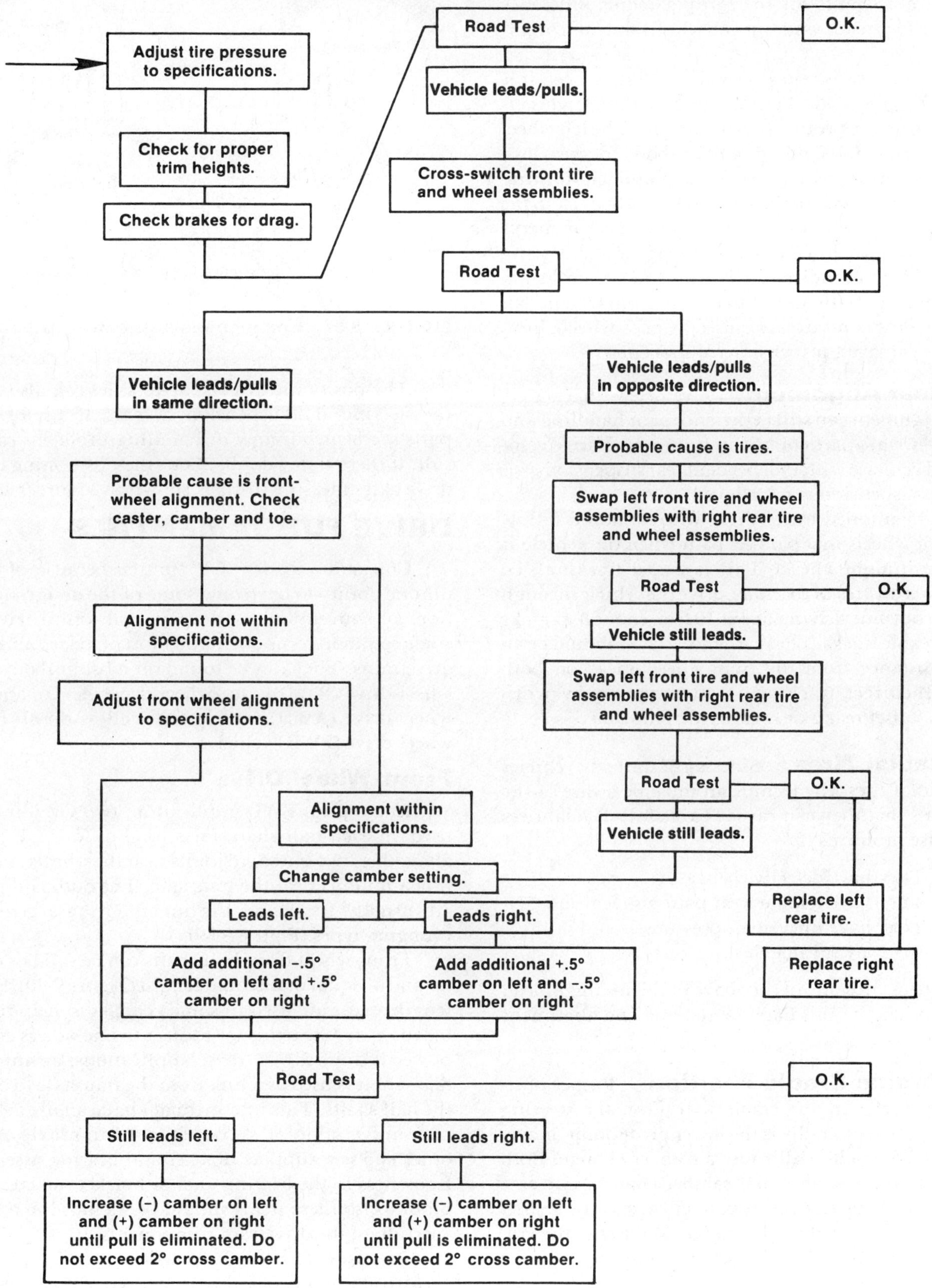

match with the right wheel. The right wheel should turn in the same amount or about 2 degrees less. The difference accounts for the turning radius difference between the inside and outside wheels during cornering.

The process is repeated with the right wheel; it is turned 20 degrees out. The movement of the left wheel is measured on the protractor or turn plate. The left wheel should turn in the same amount or about 2 degrees less.

By design, a vehicle might use a different turning radius from one side to the other. If in doubt, refer to the manufacturer's specifications. If these measurements do not repeat within 2 degrees, damage to the steering arms or gear is indicated. Turning radius measurements are especially useful in determining whether an improper toe condition is caused by poor wheel alignment or damaged suspension components.

Rear Alignment. A vehicle with a perfect front alignment can still experience poor handling and rear tire wear—particularly on front-wheel drive vehicles and vehicles with independent rear suspensions—if the rear suspension is misaligned.

As mentioned earlier, rear wheels should follow the front wheels in a parallel path when the vehicle is moving straight ahead. This is called tracking. To check, wet an area of roadway, drive the vehicle through it, and continue driving about 10 feet, enough to make measureable tracks. The rear-wheel tracks should be an equal distance from the front-wheel tracks on both sides. Incorrect tracking is often caused by worn springs, offset rear axle, or rear toe problems.

Radial Tires. Since most unibody vehicles have radial tires, the technician must be aware of the problems that they can cause (Table 9-4). Radial tires can cause problems if:

- They have defective belts.
- They have telltale wear patterns (Figure 9-37).
- They have uneven air pressure.
- They are mismatched.

Radial tires with any of the above problems might lead the technician to misdiagnose steering and alignment problems.

Engine Cradle Position. Proper positioning of the engine cradle will affect the steering angles. Since the cradle is the lower pivot point, movement of the cradle will cause a camber change. Both wheels will show an equal camber change, one side negative and one side positive. It will also cause an SAI but not an included angle change. Make sure the cradle's position is within the specifications given in the service manual.

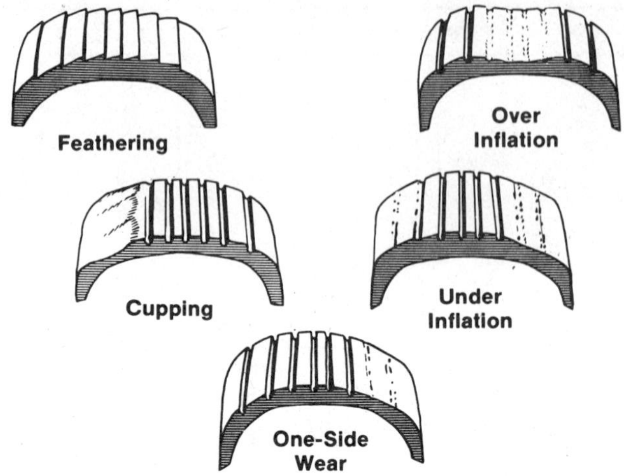

FIGURE 9-37 Common radial tire wear patterns

The positioning of the drive axles can also affect the steering and suspension systems. If any of these parts are bent, shimmy or handling problems can result. If there is any doubt about the positioning of the drive axles measure them as shown in Figure 9-38.

DRIVE FUNDAMENTALS

Constant velocity (CV) joints, because of their unique ability to overcome some of the design limitations of conventional style universal joints, have become common components on most of today's unibody drivetrains. They can be found on most unibody rear-wheel-drive (RWD), four-wheel-drive (4WD), and all-wheel-drive (AWD) vehicles, as well as on all front-wheel-drive (FWD) vehicles.

Front-Wheel Drive

In a typical FWD application, two CV joints are used on each half shaft, for a total of four CV joints altogether: two outboard joints near the wheels and two inboard joints near the transaxle. The outboard joints are usually fixed and the inner ones are generally plunging types (Figure 9-39).

Front-wheel-drive half shafts can be solid or tubular and of equal or unequal length (Figure 9-40). Equal length shafts are used in some vehicles to help reduce torque steer (the tendency to steer to one side as engine power is applied). In these applications, an intermediate shaft is used as a link from the transaxle to one of the half shafts. This intermediate shaft can use an ordinary universal joint to couple to the transaxle. At the outer end is a support bracket and bearing assembly. Looseness in the bearing and/or bracket can create vibrations, so these items should be included in any inspection of the drivetrain components.

Because the half shafts on a front-wheel-drive vehicle turn at roughly one-third the speed of the drive shaft

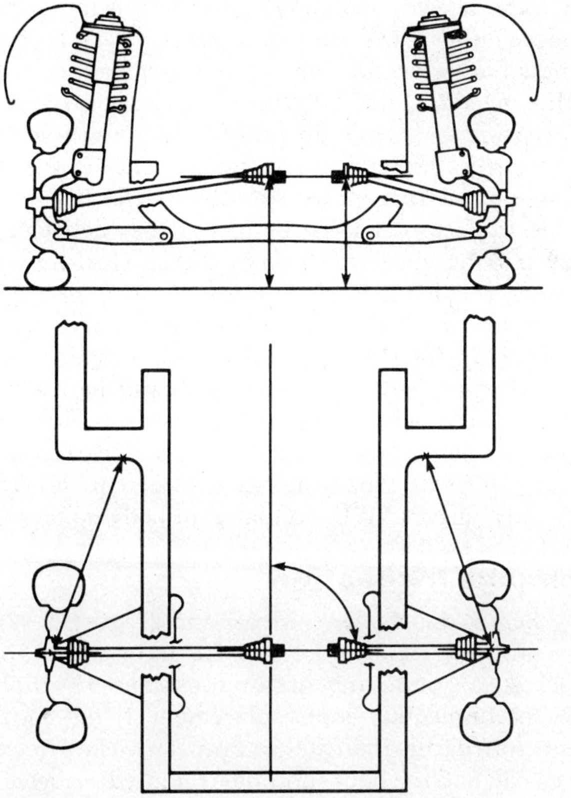

FIGURE 9-38 Measuring drive axle positions

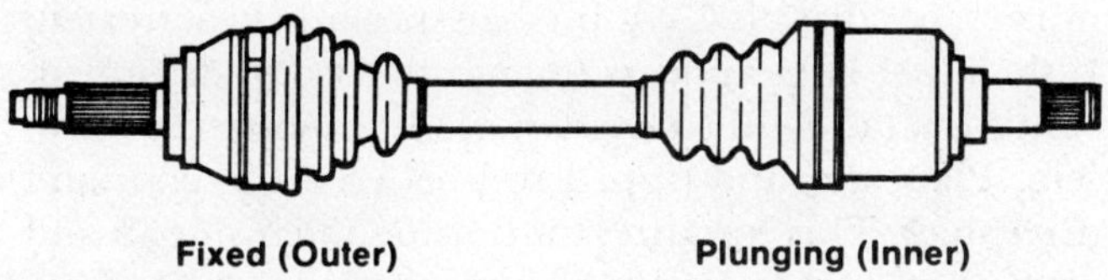

FIGURE 9-39 Inner and outer CV joints on a front-wheel-drive half shaft

in a rear-wheel-drive vehicle, half shaft balance and runout are not very important. The small damper weight that is sometimes attached to one half shaft dampens harmonic vibrations in the drivetrain and stabilizes, not balances, the shaft as it spins.

Rear-Wheel Drive

There are two basic types of CV joint applications found on rear-wheel-drive unibody vehicles: independent rear suspension (Figure 9-41A) and main drive propeller shaft (Figure 9-41B). In RWD with independent rear suspension, CV joints can be found at both ends of the axle shafts (for a total of four). One or both joints on either shaft can be of the plunging variety.

In RWD with a main drive propeller shaft setup, plunging joints are sometimes used at one or both ends of the shaft to eliminate high-speed vibrations. That is, a plunging joint will compensate for changes in shaft length that occur with rear suspension travel.

In main drive propeller shaft applications, the CV joint still needs a protective boot, but it is of a different design than the boots used on front-wheel-drive or rear-wheel-drive axle half shafts. Instead of wrapping around the outside of the joint, it is mounted within the outer housing. This helps to protect the boot but also makes inspection more difficult. CV-jointed propeller shafts revolve at high speeds and require balancing if removed for service.

Four-Wheel Drive

With four-wheel drive there are many possible drivetrain configurations, so the number of CV joints will vary accordingly. CV joints and/or universal joints can be used in various configurations on front and/or rear axle shafts as well as on the front and rear propeller shafts. On the typical 4WD vehicle, a transfer case is

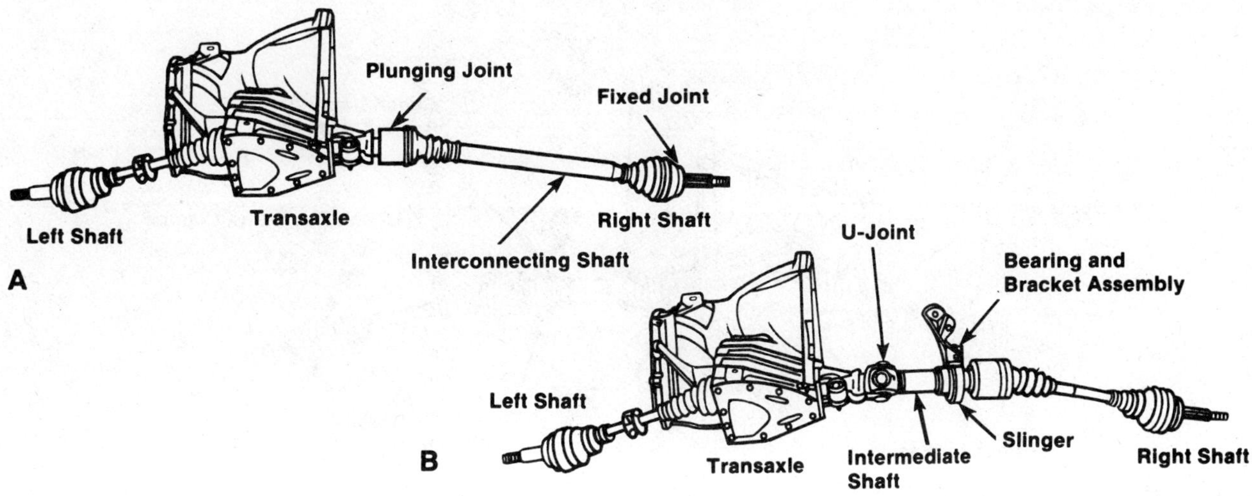

FIGURE 9-40 (A) Unequal and (B) equal length FWD half shafts

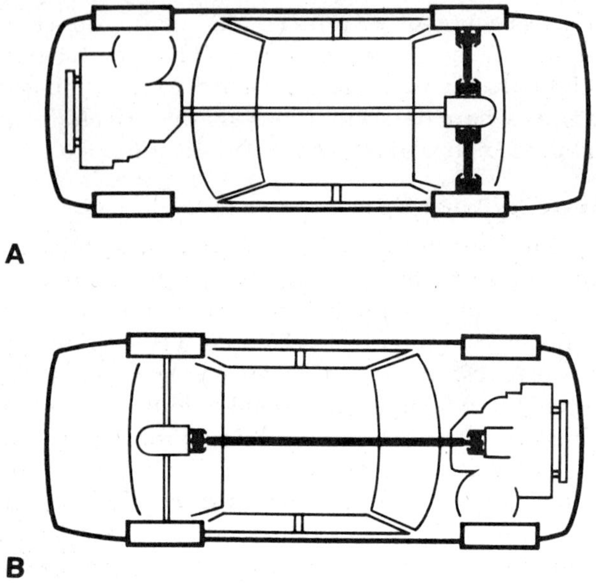

FIGURE 9-41 CV joint positions on RWD with (A) independent rear suspension and (B) a main drive propeller shaft setup

mounted to the side, underside, or back of the transmission. A chain or gear drive within the case receives the power flow from the transmission and transfers it to two separate drive shafts leading to the front and rear axles.

A selector switch or shifter located in the driving compartment controls the transfer case so that power is directed to the axles as the driver desires. Power can be directed to all four wheels, to only two wheels, or to no wheels (neutral). On many vehicles, the driver is also given the option of a "low" 4WD range for extra traction in especially rough conditions, such as deep snow or mud.

The selector switch can also permit the driver to choose either a full-time or a part-time 4WD setting. These names do not mean exactly what they say. In full-time 4WD, all four wheels are constantly open to receiving power from the power train, but they are not necessarily all receiving an equal power flow. Part-time units only deliver power to all the wheels when the driver consciously asks for it, but they deliver equal power to the wheels at all times. But on a full-time 4WD system, the driver is given the option of selecting the equivalent mode of a part-time system. This is done by locking up the differential in the drivetrain transfer case. Part-time units or modes should be used only when poor road conditions require extra traction; otherwise, accelerated and excessive gear and component wear will result. Part-time systems are frequently called all-wheel-drive (AWD) in many unibody models.

Damage Inspection

Begin the damage inspection (Figure 9-42) by checking the condition of the CV joint boots. Splits, cracks, tears, punctures, or thin spots caused by rubbing call for immediate boot replacement. If the boot appears rotted, this indicates improper greasing or excessive heat, and the boot should be replaced. Squeeze-test all boots; if any air escapes, replace the boot. Also replace any boots that are missing. Keep in mind that any discoloring of the housing at the bearing grooves is normal because all CV joints are specially heat treated. If the inner boot appears to be collapsed or deformed, venting it (allowing air to enter) might solve the problem. Place a round-tipped rod between the boot and drive shaft. This equalizes the outside and inside air and allows the boot to return to its normal shape.

Make sure that all boot clamps are tight. Missing or loose clamps should be replaced. If the boot appears loose, slide it back and inspect the grease inside for

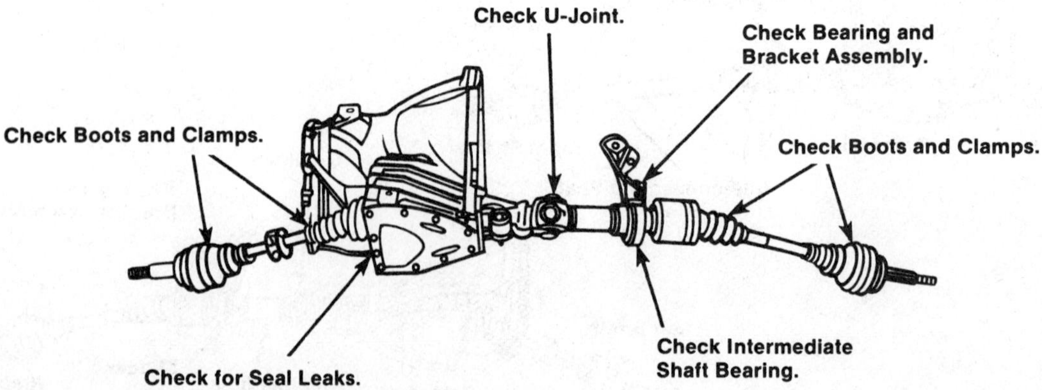

FIGURE 9-42 CV damage inspection points

possible contamination. A milky or foamy appearance indicates water contamination. A gritty feeling when rubbed between the fingers indicates dirt. In either case, as a minimal service before the joint can be dried, the old grease will have to be cleaned out, the joint repacked with fresh grease, and a new boot installed. However, in many cases a water- or dirt-contaminated joint will have to be replaced rather than regreased.

The drive shafts should be checked for signs of contact against the chassis or rubbing. Rubbing can be a symptom of a weak or broken spring or engine mount, or chassis misalignment. On front-wheel-drive transaxles with equal length half shafts, inspect the intermediate shaft U-joint, bearing, and support bracket for looseness by rocking the wheel back and forth and watching for any movement.

Various drivetrain and suspension problems can be confused with symptoms produced by a bad CV joint. The following list of symptoms should help guide the technician to a proper diagnosis:

- *A popping or clicking noise when turning.* This signals a worn or damaged outer joint. The condition can be aggravated by putting the vehicle in reverse and backing in a circle. If the noise gets louder, the outer joint(s) should be replaced.
- *A "clunk" when accelerating, decelerating, or when putting the transaxle into drive.* This kind of noise can come from excessive play in the inner joint of FWD applications, the inner or outer joints in a RWD independent suspension, or from the drive shaft CV joints or U-joint in a RWD or 4WD drivetrain. Be warned, though, that the same kind of noise can also be produced by excessive backlash in the differential gears and transmission.
- *A humming or growling noise.* This is sometimes due to inadequate lubrication in either the inner or outer CV joint. It is more often due to worn or damaged wheel bearings, a bad intermediate shaft bearing on equal length half shaft transaxles, or worn shaft bearings within the transmission.
- *A shudder or vibration when accelerating.* This can come from excessive play in either the inboard or outboard joints, but more likely the inboard plunge joint. This can also be caused by a bad intermediate shaft bearing on transaxles with equal length half shafts. On FWD vehicles with transverse-mounted engines, it can also be caused by loose or deteriorated engine/transaxle mounts. Be sure to inspect the rubber bushings in the upper torque strap on these engines to rule out this possibility. Note, however, that shudder could also be inherent to the vehicle itself.
- *A vibration that increases with speed.* This problem is rarely due to CV joint problems or FWD half shaft imbalance. An out-of-balance tire or wheel, an out-of-round tire or wheel, or a bent rim are the more likely causes. It is possible that a bent half shaft as the result of collision or towing damage could cause a vibration, as could a missing damper weight.

NVH Diagnosis

Diagnosis of noise, vibration, and harshness (NVH in auto engineering shorthand) is becoming more and more important these days, especially in collision damage repair work, as vehicles have become more sensitive to these problems. Even the modern, computer-engineered vehicle engines and drivetrain components generate considerable vibration at certain operating conditions. Vibrations are the worst at the so called "resonant frequencies." These are points at which the vibrations reinforce one another, reaching higher amplitudes (intensity) than any single vibration would create by itself.

Vibration control is a very important science in the design of comfortable, quiet vehicles. Rubber mounts are simple but effective tools in the fight against vibration. Those located between the frame and cradle should be inspected if cradle movement is indicated; any tear or deformation indicates damage. These parts should be replaced in sets.

There are many different types of motor mounts, but all are designed to accomplish the same thing—to control the transmission of engine and drivetrain vibrations to the passenger compartment and the rest of the chassis. Be sure to inspect all mounts, since damage to one indicates likely damage to others. Note that there might be shims or washers in some applications. If so, be sure to position the replacement shims the same as in the OEM installation. Positioning of shims will make a big difference in how the engine is supported. If in doubt, check the service manual.

Before tightening any rubber suspension bushings, let the vehicle down and bounce it a few times. Then the bushings are ready to tighten. If this is not done, the customer might have riding height problems or other difficulties. Tighten the bushings in the normal operating position to prevent premature failure. Watch for other problems with rubber suspension bushings, such as improper load and wrong twist.

The mounting, positioning, and adjustment of both the transmission and throttle linkages affect the

TABLE 9-5: CAUSES OF VEHICLE VIBRATIONS

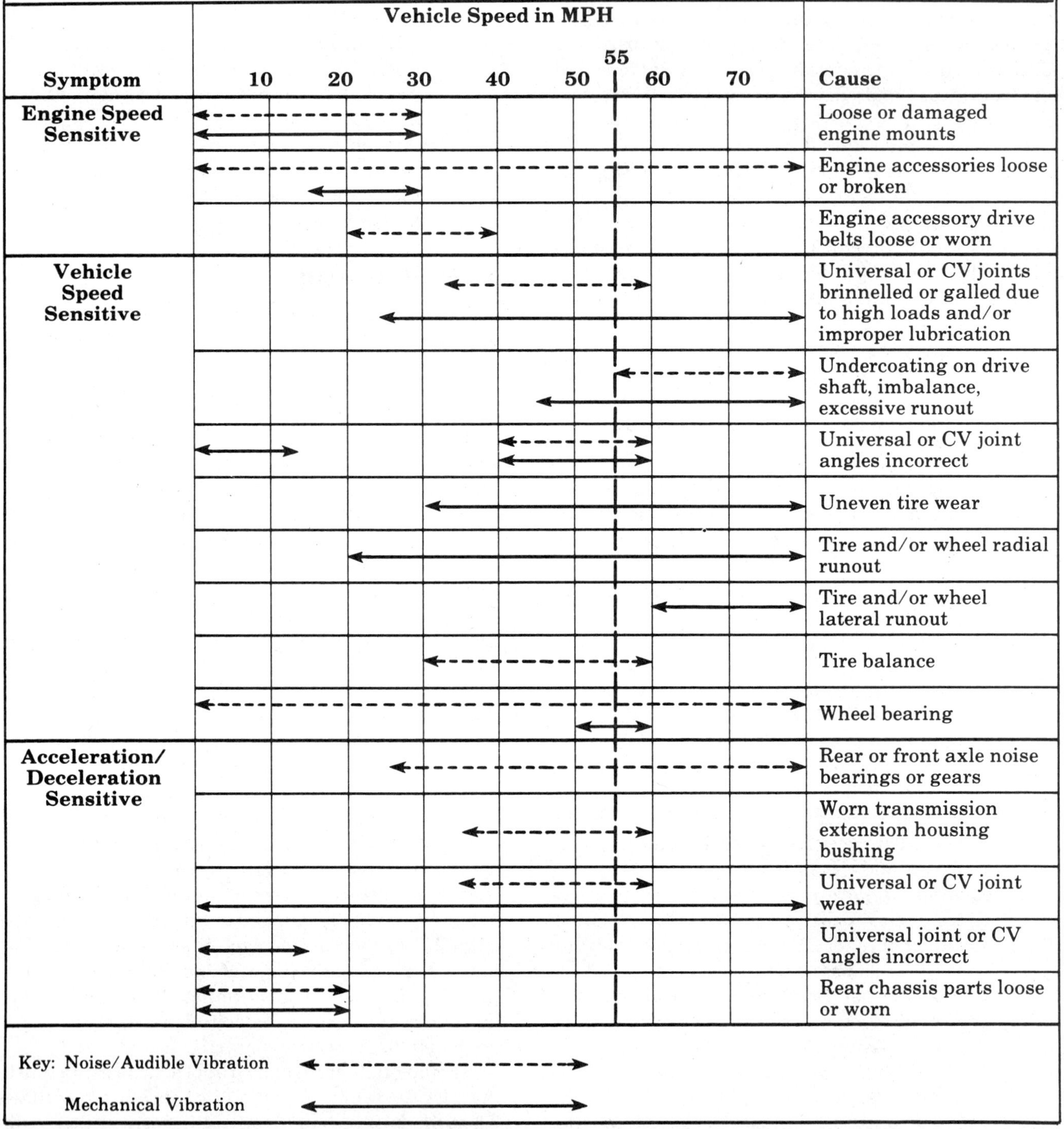

operation of the transmission and throttle. Major system malfunctions can be caused by improper linkage adjustment. Do not disassemble any more than is necessary. Separate at the quick disconnect joints held together by special clips. When replacing damaged components, duplicate adjustment by comparing to the original, if possible. There are special adjusting pins on some models. Check the service manual for proper adjustment procedures.

Table 9-5 is designed to give the collision repair technician a working knowledge of the process of diagnosing noise and vibration situations.

BRAKE SYSTEMS

In the age of unibodies, it is important that the collision repair technician know how the various brake systems operate, how to check for collision-damaged parts, and how to remove the components. Most of the repair itself is still done by brake specialists.

In front-wheel-drive automobiles, the front brakes do as much as 80 percent of the work in stopping the vehicle. Thus, the conventional front/rear split hydraulic system (Figure 9-43A) has given way to a dual diagonal split hydraulic system which combines a front brake with its opposite rear brake (Figure 9-43B). This system allows straight stopping and provides 50 percent braking capacity in case of failure in either of the two hydraulic systems.

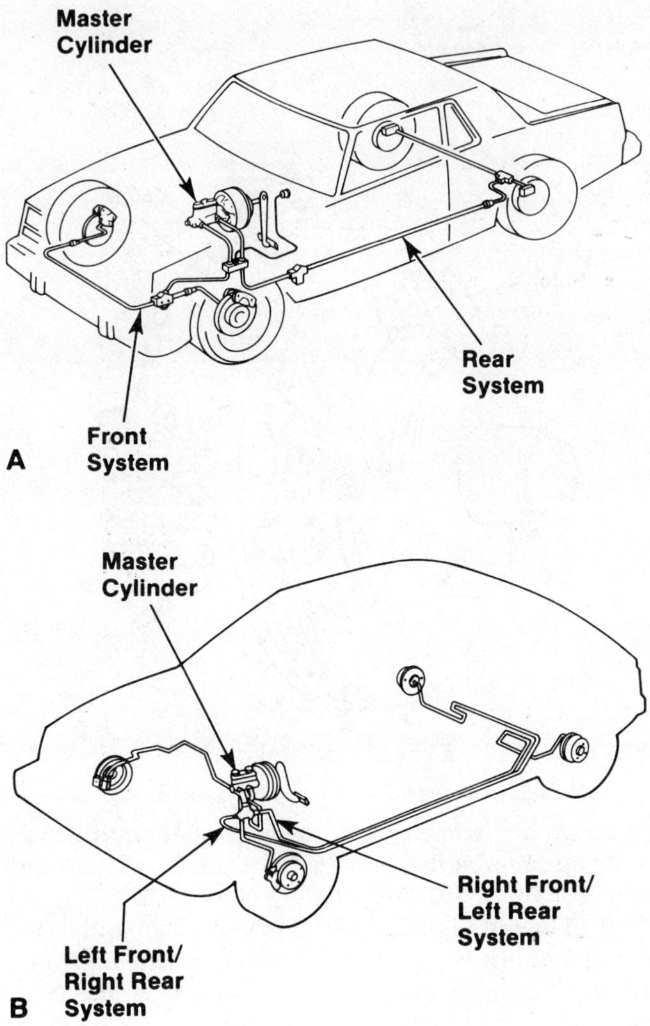

FIGURE 9-43 (A) Front/rear split and (B) dual diagonal split hydraulic braking systems

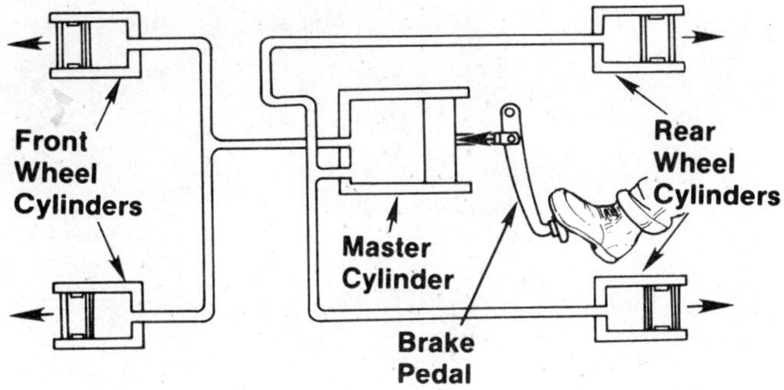

FIGURE 9-44 Hydraulic power actuating system

The brake actuating system employs hydraulic pressure or power-assisted hydraulic pressure. When the driver steps on the brake pedal, the force creates hydraulic pressure in the master cylinder. This pressure is transmitted to the wheel cylinders, where it turns into force again to act upon the drums or discs that stop the vehicle (Figure 9-44).

Servicing wheels and tires has become much more complex in recent years. This is because of the advent of radial tires, directional wheels, directional tires, and on-vehicle dynamic balancing. Here are some tips that might prevent problems:

- Mark the studs and corresponding bolt holes with chalk as the wheel is removed. This will ensure that the dynamic balance of the wheel will be the same when it is reinstalled.
- Mark all wheels with the position they were mounted (that is, front-left, right-rear, and so on).
- Some non-drive wheels have a bearing preload specification. Be sure to follow directions when reinstalling the hubs.
- When replacing brake system parts with salvage parts, be sure to use a matched set of hubs, bearing, and race. These parts have worn together and will not fit properly if not assembled as a matched set.

SHOP TALK

Keep the brake system closed to the atmosphere as much as possible. This will keep moisture from entering the system. Moisture is readily absorbed by the brake fluid and causes sludge and corrosion to form. In time, this will cause partial or complete loss of brake effect. Moisture also lowers the boiling point of the brake fluid. Boiling results in vapor, which has the same effect as air in the system.

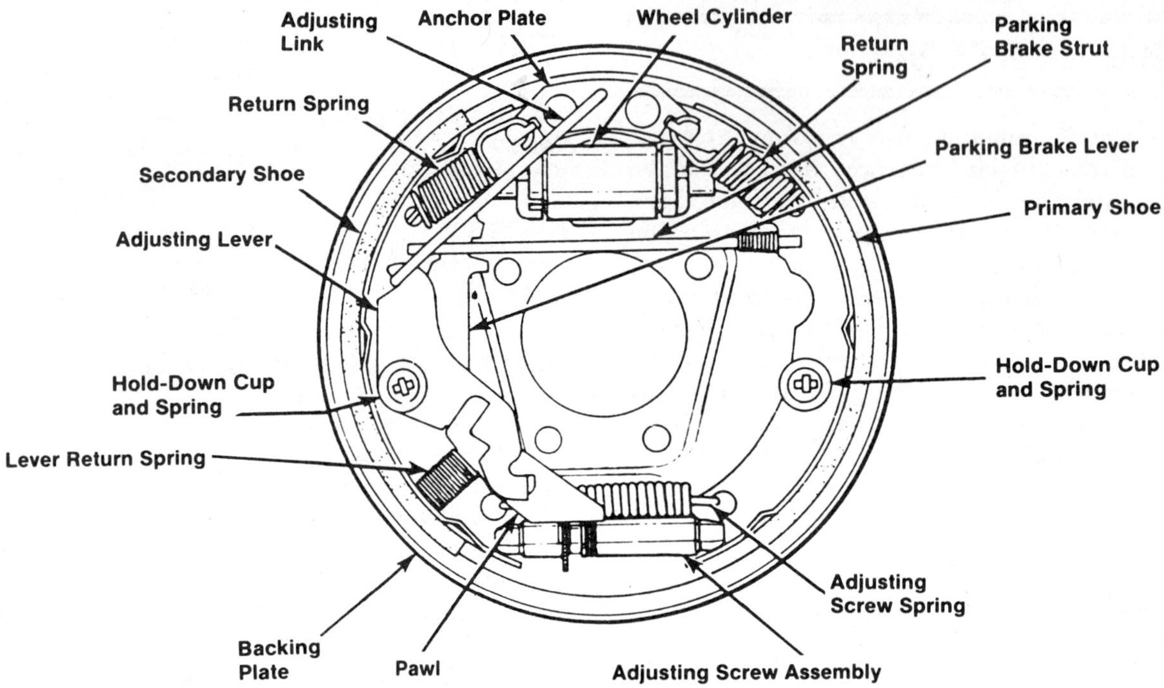

FIGURE 9-45 Typical drum brake

TYPES OF HYDRAULIC BRAKES

Two basic types of hydraulic brakes are used in unibody vehicles. They are drum brakes and disc brakes.

Drum Brakes

A drum brake assembly consists of a cast-iron drum, which is bolted to and rotates with the vehicle wheel, and a fixed backing plate to which are attached the shoes and other components—wheel cylinders, automatic adjusters, linkages, and so on (Figure 9-45). Additionally, there might be some extra hardware for parking brakes. The shoes are surfaced with frictional linings, which contact the inside of the drum when the brakes are applied. The shoes are forced outward, against the action of the return springs, by pistons which are actuated by hydraulic pressure. As the drum rubs against the shoes, the energy of the moving drum is transformed into heat, and this heat energy is passed into the atmosphere.

When the brake shoe is engaged, the frictional drag acting around its circumference tends to rotate it about its hinge point, the brake anchor. If the rotation of the drum corresponds to an outward rotation of the shoe, then the drag will pull the shoe tighter against the inside of the drum, and the shoe will be self-energizing.

Disc Brakes

Disc brakes resemble the brakes on a bicycle: the friction elements are in the form of pads, which are squeezed or clamped about the edge of a rotating disc. With automotive disc brakes, this disc is a separate unit, called a *rotor,* inboard of the vehicle wheel (Figure 9-46). The rotor is made of cast iron, and since the pads clamp against both sides of it, the sides are machined smooth. Usually the two surfaces are separated by a finned center section for better cooling. The pads are attached to metal shoes, which are actuated by pistons, the same as with drum brakes. The pistons are contained within a caliper assembly, a housing that wraps

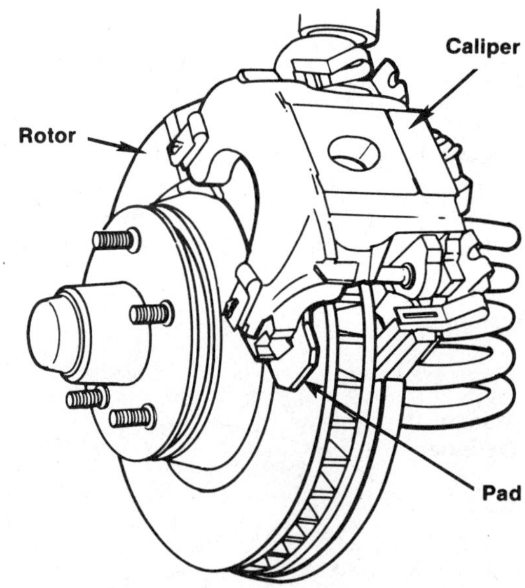

FIGURE 9-46 Typical disc brake

around the edge of the rotor. The caliper is kept from rotating by way of bolts holding it to the vehicle's suspension framework.

The caliper contains the pistons and related seals, springs, and boots, as well as the cylinder(s) and fluid passages necessary to force the friction linings or pads against the rotor. The caliper resembles a hand in the way it wraps around the edge of the rotor and attaches to the steering knuckle. Some models employ light spring pressure to keep the pads close against the rotor; in other caliper designs this is achieved by a unique type of seal that allows the piston to be pushed out the necessary amount, then retracts it just enough to pull the pad off the rotor.

Unlike shoes in a drum brake, the pads act perpendicular to the rotation of the disc when the brakes are applied. This effect is different from that produced in a drum brake, where frictional drag actually pulls the shoe into the drum. Disc brakes are said to be non-energized, and so require more force to achieve the same braking effort. For this reason, they are ordinarily used in conjunction with a power brake unit.

The actual repair work on hydraulic brake system parts such as the master cylinder, combination valve, wheel cylinders, brake shoe and caliper assemblies, and other components, as previously stated, is best left to brake specialists. However, tasks such as disassembly, reassembly, and replacement of some damaged components are well within the scope of a collision repair technician's skills.

MASTER CYLINDER

The master cylinder is the heart of the hydraulic system. It is located in the engine compartment, usually on the driver's side, and is connected to the brake pedal by a special rod. The master cylinder initiates braking when the brake pedal is depressed by pushing out a piston inside the cylinder, exerting pressure that is transferred through the system. To protect against total failure of the system, all vehicles are now required to have two hydraulic systems. The dual master cylinder provides for either a front/rear split or a diagonally split system.

The master cylinder must be checked before the vehicle is put back in service. To check the fluid level, clean all dirt and grease off the unit, then simply pop the wire bracket (or whatever locking device is on the top) and remove the cover. The level should not be more than 1/4 inch below the top of the reservoir. If the level is below this, check the brake line connections and refill the reservoir with fluid. If the system is leaking anywhere but at the brake line connections, the master cylinder should be replaced or rebuilt by a brake mechanic.

Never use the master cylinder and other brake mounts as a pulling attachment. If a pull must be made to correct cowl damage, use a plate and different bolts to anchor the pull.

Handling Brake Fluid

The chemical compounds that make up brake fluid attack and damage most automotive paint finishes. If spilled on paint, a quick flush with water can lessen the damage. The following cautions must be strictly observed when handling brake fluid:

- Use the exact type (DOT 3, DOT 4, or DOT 5) of brake fluid specified in the vehicle service manual or on the master cylinder cover.
- Keep the master cylinder tightly covered. Always recap it immediately after filling.
- Use the smallest can of fluid, and use it all if possible. Brake fluid that has been exposed to the air in a partially filled can should not be reused because it has likely been contaminated with moisture.
- Tightly cap the fluid container after use.
- If any fluid has become contaminated, throw it out.
- Do not reuse old brake fluid.
- Do not reuse an old brake fluid container, since it is not possible to know what else might have been in the can.
- Do not transfer brake fluid from its original container to anything other than a container specifically designed to hold brake fluid, such as a pressure bleeder.

BLEEDING

When it is necessary to remove or replace a brake component, be sure to follow the manufacturer's instructions that appear in the service manual. Remember that anytime the brake system is open, it must be bled. Keep the system open for as short a time as possible to prevent moisture from entering and causing sludge and corrosion.

Bleeding is the process of removing air from the brake system. Air is lighter than liquid and it seeks high points in the hydraulic system. Bleed screws are provided at each of these collecting points: the calipers, wheel cylinders, and on some master cylinders. Bleeding involves opening up these screws in a specific order to let the trapped air escape. Fluid is added to the master cylinder to replace whatever is lost in bleeding.

Bench Bleeding the Master Cylinder

When the master cylinder is removed for rebuilding or replacement, bench bleeding is necessary to ensure that air does not remain in it when it is reinstalled.

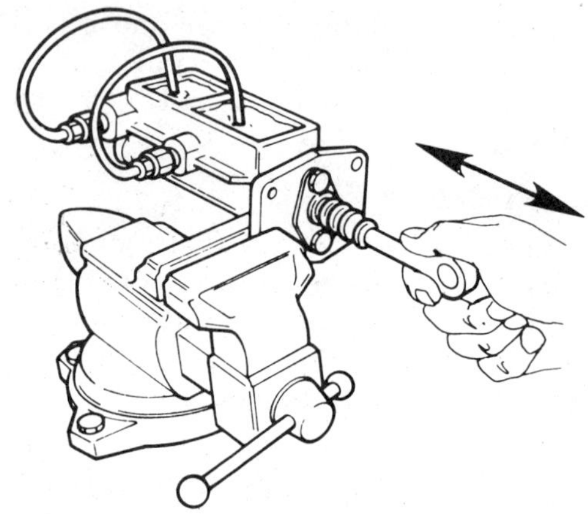

FIGURE 9-47 Pumping air out of the master cylinder

Mount the master cylinder in a vise with the bore angled slightly downward. Attach two short brake lines or purge tubes to the outlet ports so they curl back into the reservoirs with the ends below the fluid level. Stroke the piston back and forth to pump air out of the cylinder and into the reservoir (Figure 9-47). Do this until only clear brake fluid comes out of the tubes. The same method will work using threaded plugs instead of the purge tubes.

Bleeding the System

Whether the pressure or manual bleeding technique is used, both methods follow a common sequence for opening the bleed screws. Begin at the master cylinder if it has bleed screws (Figure 9-48). Move to the combination valve if it has bleed screws. Next, bleed the wheel cylinders or calipers. Start with the wheel located farthest from the master cylinder, and work back from there. The bleeding sequence at the wheels is different for dual front/rear systems than it is for dual diagonal systems. In addition, each manufacturer might have a preferred sequence for any given model design. Check the service manual for specific instructions.

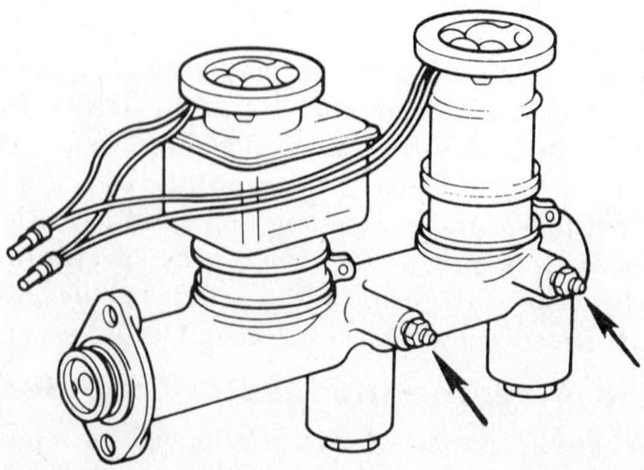

FIGURE 9-48 Master cylinder bleed screws

Some four-piston calipers have two bleed screws. In this case, bleed the lower one first. On diagonal systems, bleed one system at a time. Do one front disc brake first, then the diagonally connected rear drum brake.

Always check the master cylinder first. If the brake fluid falls below the level of the intake ports, air will get into the system. Refill the reservoir and pump the brake pedal slowly a number of times. Often this is all that is needed to purge it of unwanted air. If this does not work, then the system must be bled.

Manual Bleeding

Manual bleeding should be done only if a pressure bleeder is not available. Begin at the master cylinder. Clean the cover before removing it. After removing the cover, clean the diaphragm gasket. Fill the reservoir to 1/4 inch from the top. Apply pressure to the brake pedal slowly and with a smooth action. Open the bleed screw on the first wheel in the sequence. Drain the fluid through the bleeder hose into a jar that has been partially filled with clean brake fluid (Figure 9-49A).

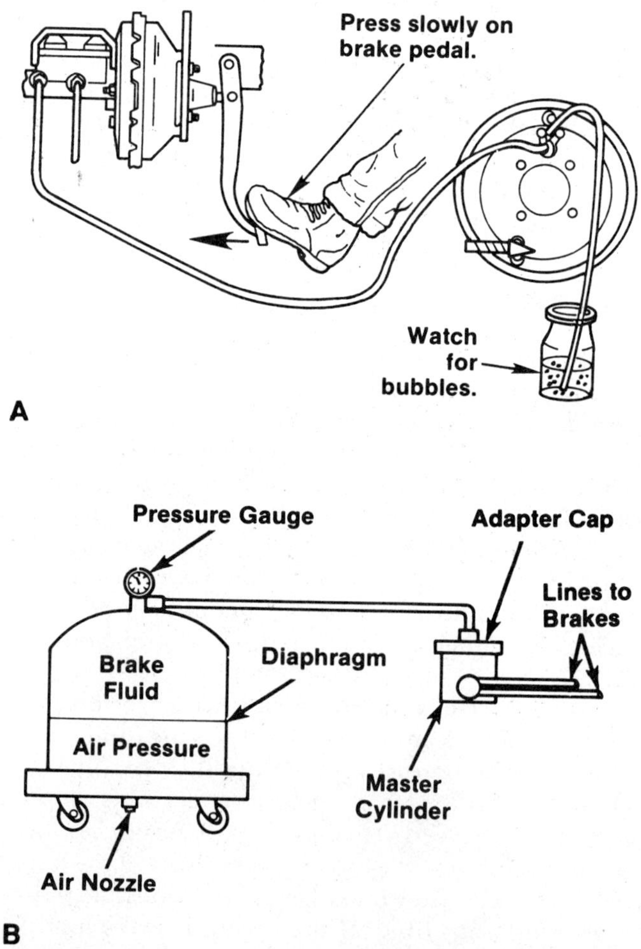

FIGURE 9-49 (A) Manual bleeding operation and (B) pressure bleeding operation

Keep up pedal pressure while the bleed screw is open. When the pedal bottoms out, close the screw and release the pedal. If all the air is not yet purged and air bubbles can be seen in the fluid, repeat the process. When only clear fluid with no bubbles appears, go on to the next wheel in the sequence.

While bleeding the brakes, watch the fluid level in the reservoir. After every six pedal applications more fluid will have to be added so it does not fall below the level of the intake port. If it does, more air will enter the system.

Pressure Bleeding

This procedure is the recommended method of ridding the hydraulic system of air. The pressure unit used in this process is a tank divided into two sections by a flexible diaphragm (Figure 9-49B). Pressurized air comes into the bottom chamber, compresses the fluid in the top, and brings it up to the desired pressure. The fluid then goes into the hydraulic system through a hose attached to a master cylinder adapter cap. When using a pressure bleeder unit, make sure to use the correct adapter for that particular master cylinder.

Bring the pressure up to 15 to 20 psi. Make sure the master cylinder cover is clean so that no loose particles of dirt fall into the reservoir. Remove the gasket, and clean the gasket seat. Fill the reservoir and attach the adapter cap and hose. Check the coupling sleeve and make sure it is fully engaged before opening the fluid supply valve.

Follow the sequence for bleeding as recommended in the service manual. Allow the fluid to flow out of the bleed screws through the hose and into a jar. Once completed, close the supply valve of the pressure bleeder unit. Wrap the coupling sleeve in a rag to prevent brake fluid from dripping onto the vehicle finish. (Flush off any brake fluid immediately with water or it will ruin the paint.) Undo the coupler. Disconnect the unit and check to determine that the brake fluid level in the reservoir is not more than 1/4 inch from the top. Replace the gasket and cover and check the brakes.

The best method of pressure bleeding a brake system is with a vacuum type bleeder. This technique sucks the fluid through the system rather than pumping it and has definite advantages:

- Will not create foaming
- Does not activate the pressure differential valve
- Less chance of fluid contamination

Final Check

Late model vehicles sometimes have air left in the master cylinder bore above the outlet ports. As a final check, raise the rear of the vehicle so the bore angles downward. Take off the reservoir cover and tap the pedal lightly a number of times. Very small bubbles should come up through the fluid to the top of the reservoir. When a spurt of fluid appears and the stream of bubbles ceases, the vehicle is ready to be road tested.

BRAKE LINES

The brake lines are generally the major brake component that a collision repair technician must deal with. When making a collision inspection, check the brake lines for chafing, crimps, loose or missing tube clips, kinks, dents, blockage, and leakage. Leaks are evidenced by fluid seepage at the connections or stains around hose ends. Blockages are not so readily apparent, but are just as detrimental to the function of the braking system, often preventing proper release of the brakes. During a brake application the pressure forces the fluid past the obstruction, but when the pressure is relaxed, the fluid will not readily flow back past the blockage and the brakes drag.

Brake lines are usually steel, except where they have to flex—between the chassis and the front wheels, and the chassis and the rear axle. At these places flexible hoses are used. When replacing damaged brake lines, use the same type of material as the OEM. This includes stainless steel, armor plate tubing, or ribbed hose. Local availability might be limited on special types, but it is important to match the factory materials.

When cutting tubing to length, it is important to duplicate the OEM brake line flare. Modern brake lines use either the double flare or ISO flare styles. Do not use compression fittings in brake line repairs. Replace all supporting clamps removed during the repair. Support springs prevent kinking and serve a very important role; be sure to replace them just as they were, and install new ones if damaged. Always replace brake lines in the original routing to avoid damage to the lines later. Remember, you are repairing the vehicle, not reengineering it. A change in routing may result in rubbing or chafing of brake lines as the suspension moves.

Most brake hoses have a male fitting on one end and a female fitting on the other. Disconnect the female end first, remove the clip or jam nut holding it down, then unscrew the male end. Install the new hose by connecting the male end. If a copper gasket was used, replace it with a new one. When the male end is tight, connect the female end. Tighten it in such a way as to keep the hose from touching any part of the chassis or suspension. Check for interference during turning of the front wheels.

POWER BRAKES

Power brakes are nothing more than a standard hydraulic brake system with a vacuum assist or booster unit added between the pedal and the master cylinder to

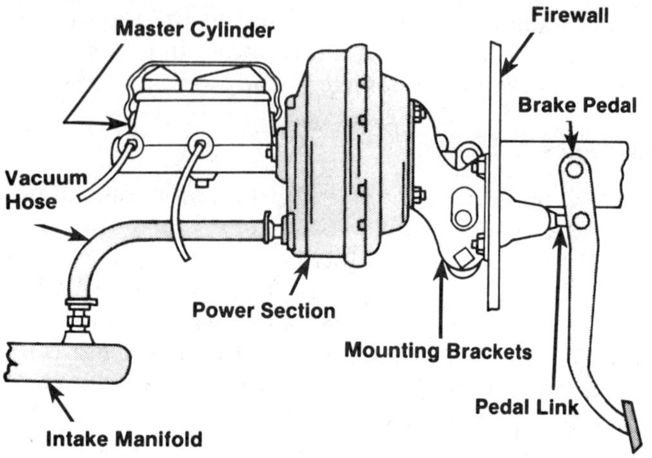

FIGURE 9-50 Typical power brake system

help activate the brakes (Figure 9-50). Most power brake units consist of a piston, control valves, and a vacuum connection from the engine intake manifold. When pressure is off the brake and the vacuum unit is in the released position, the vacuum system intake port is closed. However, a special atmospheric port remains open to allow air to pass from one side of the vacuum piston to the other. This maintains equal pressure on the piston and keeps it in the *off* position. When the brake is applied, the atmospheric port closes and the vacuum port opens. Vacuum from the engine then sucks the unit piston forward against the master cylinder operating rod and actuates the brakes.

When a unibody vehicle is involved in a collision, the power brake booster should be carefully inspected. Pay particular attention to vacuum hoses, check valves, fasteners, and the master cylinder itself. Replace all damaged pieces. This is the system which builds up to 2500 psi and makes the vehicle stop—do not take any chances with it.

SHOP TALK

To check if the booster is working, hold the pedal down and start the engine. Note that when the engine first starts, the brake pedal will tend to go down. This is normal and is related to the buildup of vacuum.

ANTI-LOCK BRAKES

Anti-lock braking systems (ABS) represent an important development in vehicle control. The purpose of ABS is to prevent wheel lockup under heavy braking or slippery conditions, while letting the driver maintain control of the vehicle. This makes it possible to stop the vehicle in the shortest possible distance on all types of surfaces.

Under normal braking conditions, the ABS system works much the same way as a standard brake system. The main difference is that power assist is provided by hydraulic power rather than vacuum. When the system enters the anti-lock mode, hydraulic pressure in the four wheel circuits is controlled to prevent any wheel from locking. Each wheel is designed with a separate electrical valve and hydraulic line. For stability, both rear wheel valves receive the same electrical signal. The system can automatically hold or reduce pressure at each wheel, depending upon the input received from the wheel speed sensors. The input is processed by a diagnostic control module.

To determine when anti-lock is required, the diagnostic control module uses information from the wheel speed sensors. If one wheel slows down too quickly, the diagnostic control module decides that the wheel is beginning to lock and reduces hydraulic pressure to that brake. Although system design can vary from one vehicle to another, most ABS components are similar to those shown in Figure 9-51.

- *Hydraulic Assembly.* The hydraulic assembly performs the function of a master cylinder and hydraulic booster. It also contains the wheel circuit valves used for brake pressure control.
- *Diagnostic Control Module.* The diagnostic control module is a small computer located in the trunk, under the front seats, or in the engine compartment. It handles a variety of tasks, including monitoring the system for proper operation, controlling the fluid pressure to the brakes, detection of wheel lock tendencies, and providing communication with electronic test equipment.
- *Wheel Speed Sensors.* The wheel speed sensors can be mounted on the wheel hubs, in the differential housing, or on the transmission. While the design of the sensors varies slightly from one vehicle to another, the function remains the same: to send wheel speed data to the diagnostic control module. The front wheel sensor is attached to a boss in the steering knuckle.
- *Pump/Motor Assembly.* This is an electric pump that takes low brake fluid from the hydraulic reservoir and pressurizes it for power assist. The pump maintains hydraulic pressure in the hydraulic accumulator; the pump and accumulator provide hydraulic pressure even if the engine stalls. During anti-lock braking, brake pressure is controlled by cycling electric valves.

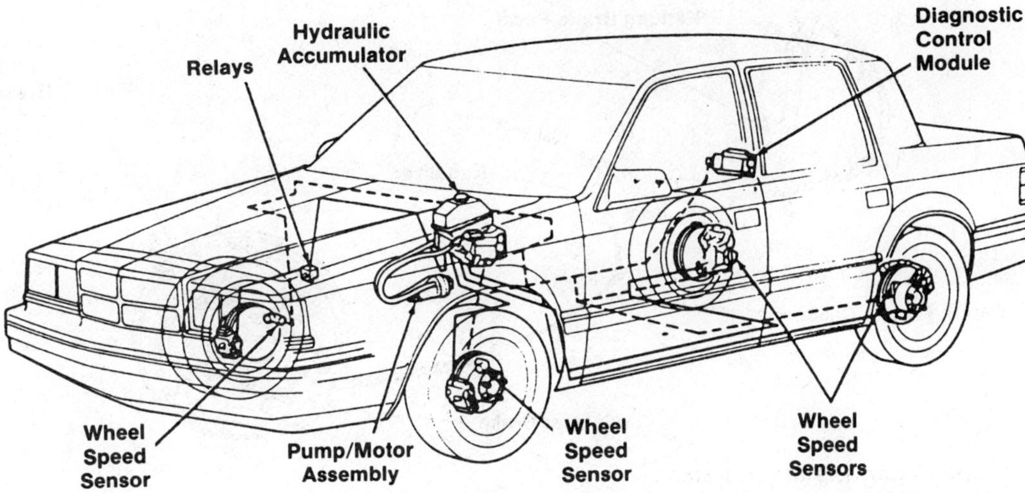

FIGURE 9-51 Typical anti-lock braking system components

- *Hydraulic Accumulator.* The hydraulic accumulator stores extra hydraulic fluid. It also provides power assist on some ABS systems. Because the accumulator stores fluid at pressures as high as 2000 psi, it is very important that caution be used when working on the system to prevent vehicle damage and/or personal injury. Removal or disassembly may result in injury and/or improper ABS system operation, so be sure to follow the service manual instructions.

Integral ABS systems can be easily identified because the master cylinder and brake booster are replaced by a special hydraulic unit. Non-integral ABS systems use a conventional braking system with the ABS components added on.

Service Precautions

There are some general cautions that should be kept in mind when servicing an ABS system. Failure to follow them may cause ABS system damage.

- Disconnect the diagnostic control module before welding.
- The diagnostic control module can be exposed to temperatures up to 185 degrees Fahrenheit for up to 2 hours. It may be exposed to temperatures up to 200 degrees Fahrenheit for no longer than 15 minutes.
- Do not connect or disconnect the diagnostic control module with the ignition on.
- Do not disassemble any component that is noted as nonserviceable.
- Always depressurize the hydraulic accumulator before working on any hydraulic part.

To depressurize the hydraulic accumulator, first turn off the ignition or disconnect the battery. Next pump the brake pedal a minimum of twenty-five times. A noticeable increase in pedal feel will occur when the accumulator is discharged. Then pump it two or three more times.

Because of the sensitivity of parts of the ABS system, a few additional points need to be kept in mind. One is that the cable routing for the front and rear wheel speed sensors is critical. Stretching of the cables due to the movement of the wheels, or contact with moving parts, can easily damage them and result in open circuits. ABS parts are very sensitive and should not be handled roughly. Never grasp the wheel speed sensor with a pliers. Make sure that all clips and retainers have been replaced.

Always refer to the service manual for details on system diagnosis and bleeding procedures.

PARKING BRAKES

Parking brakes can be either hand or foot operated. Parking brakes work mechanically, using a ratio lever assembly connected through a cable system to the rear brakes. As the parking brake is applied, the attached front cable assembly tightens, and tensile force from the front cable is transmitted through the vehicle's brake cable system to the rear brakes (Figure 9-52). The tension pulls the flexible steel cables attached to each of the rear drum brakes, operating an internal lever and strut mechanism and expanding the brake shoes against the drum. Springs return the shoes to the unapplied position when the cable tension is released.

When inspecting the system, check for free movement of parking brake cables in the conduits. Look for

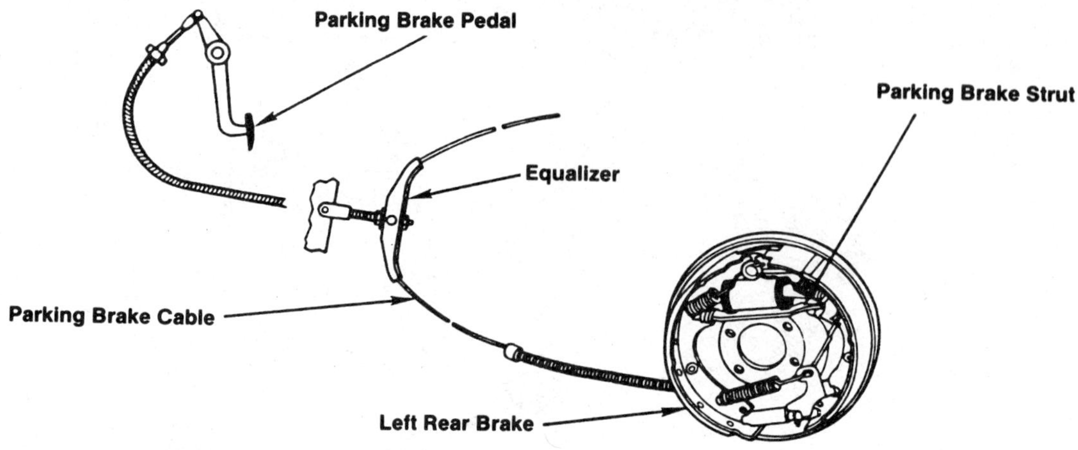

FIGURE 9-52 Typical parking brake system

kinked or broken conduit or cables. Be sure the cable can move freely in the conduit and that all connections are sound. Check all equalizer and linkage parts of the drum brake assembly for damage.

COOLING SYSTEMS

Whenever a vehicle is in a collision, especially when front end damage occurs, there is a good chance that some part of the cooling system has been damaged. To be sure some obvious problems do not slip by before releasing a vehicle, it is a good idea to become familiar with the cooling system (Figure 9-53).

COOLANT

One of the most frequently misunderstood areas of the cooling system is coolant. A common belief is that the stronger the concentration, the better. This is not the case. Pure water can actually transfer heat better than pure coolant, but it does not protect the system from freezing and corrosion. Pure coolant has a higher boiling point (330 degrees Fahrenheit) than water. But, due to its lack of heat transfer ability, pure coolant can cause an engine to overheat. In addition, pure coolant offers no corrosion protection, since the anti-corrosion chemical must be mixed with water to activate.

WARNING

When working on the cooling system, remember that at operating temperatures the coolant is extremely hot. Touching or spilling the coolant can cause serious burns.

The ideal coolant-to-water ratio is 50/50. This ratio provides freezing protection to − 34 degrees Fahrenheit, while acheiving a coolant boiling point of 265 degrees Fahrenheit. Some systems, such as those found on mid-engine vehicles, and those having dual heaters, can require up to three to four gallons of coolant. Always check the vehicle specifications. Measure the level of installed protection with a coolant tester, following the manufacturer's directions.

CAUTION

Never run the engine with the cooling system dry. Lack of coolant and lubricant in the water pump can cause destruction of the pump or the engine.

Coolant Leaks

A low coolant supply reduces cooling capacity. In addition to obvious external leaks resulting in squirting steam and water, coolant leaks may also be internal. Internal leaks can result from a blown cylinder head or intake manifold gasket or a warped cylinder head. Another type of internal leak occurs when the automatic transmission fluid (ATF) cooler tank leaks from the inside into the radiator. A sure sign of this is a thick, pink solution in the radiator caused by the mixing of coolant and transmission fluid. Due to the impact forces in a collision, this area should be a high priority in any post-repair inspection.The loss of ATF due to a fractured radiator or ATF-to-radiator line causes two problems:

- Eventual loss of hydraulic pressure needed to operate the transmission
- Loss of lubrication needed to protect the internal mechanical and friction surfaces

With low pressure and missing lubrication, it does not take long for the clutch plate to literally "burn up" and

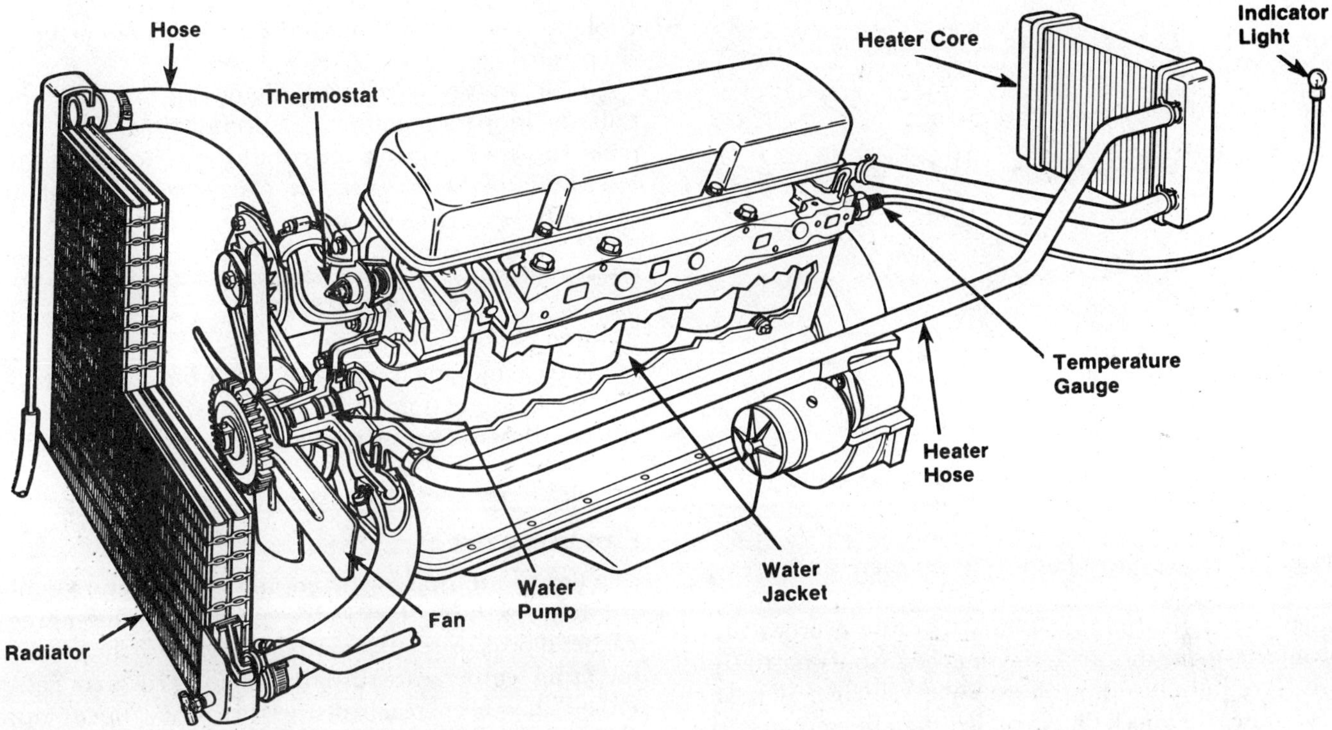

FIGURE 9-53 Cooling system components

the pressure pump and other parts of the transmission to become scored and galled due to metal-to-metal contact. If there is a real transmission problem, most collision repair shops turn the work over to a transmission specialist.

Coolant Recovery Bottle

This bottle is normally plastic and can be easily damaged. Check for cracks or abrasions in the bottle and make sure the hose leading to the radiator is connected and in good shape. Coolant recovery bottles are normally not repaired. If cracked or distorted, replace with a new one.

Refilling a System

Before making any replacements in the cooling system, drain the coolant and dispose of it properly (check local regulations). Never reuse the old coolant. When refilling, make sure that the proper coolant is used in vehicles with aluminum engines or radiators. Some warranties will not be honored if the coolant recommended by the manufacturer is not installed.

After installation of the coolant, the system must be bled of all air to insure proper cooling. Always follow the manufacturer's recommendations. On some vehicles there are bleed valves; on others, the thermostat must be out to bleed the system; and on others, the upper radiator hose must be removed. Check the service manual for specific procedures.

RADIATOR

The coolant flows from the engine to tubes located inside the fins of the radiator where the air flow cools it. If these tubes become plugged, either by being bent or through maintenance neglect, the flow of coolant through the radiator is reduced and overheating can result. This condition is more noticeable at highway speeds and/or heavier loads. If the vehicle is not air conditioned, plugged areas of the radiator can be identified by cold spots felt on the front of the radiator after the vehicle is warmed up. Chances are that a collision-damaged vehicle will have damaged or bent areas in the radiator, even though leaks might not be present. So always be sure to check that area carefully. A radiator pressure tester is the proper way to test the radiator for leaks (Figure 9-54). Pressure testing should always be done if the radiator or any hoses were removed or replaced during collision repair.

Crushed fins can be straightened with a special tool designed for that purpose, and tubes that are not too badly mangled can be soldered. But if large hunks of cooling fins have been pulled loose or if multiple tubes have been crushed or ruptured, a new core is recommended. If the cost to repair core damage begins to approach the cost of a new core, most shops will opt for the new core. Besides, a new core offers greater reliability, especially if the radiator is showing its age.

Due to the aerodynamic designs of some vehicles, the radiator ends up being the low point in the cooling

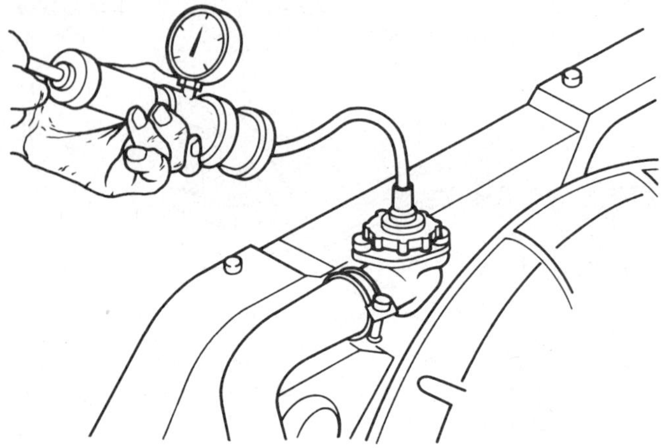

FIGURE 9-54 Using a radiator pressure tester

system. Because the radiator is lower than other areas, the cooling system might not be able to be completely filled after it is drained. This will cause overheating problems due to the cooling system being only partially full. To eliminate this problem, some vehicles will have a separate filler neck that is higher than the radiator. If the vehicle is not so equipped, jacking up the front of the vehicle will place the fill point higher than the rest of the system.

Radiator Cap

Most cooling systems operate under a pressure of about 15 psi. This is because increased pressure on a liquid raises its boiling point. To maintain the correct pressure in the system, the radiator cap must stay sealed up to its rated pressure and release excess pressure (Figure 9-55). A defective radiator cap would allow the boiling point to be lowered, which could cause the coolant to boil even though the engine is not actually overheated.

FIGURE 9-55 The radiator cap must hold a required amount of pressure.

Calcium deposits in a radiator cap can make the radiator inoperative after a period of time. Check to make sure the cap is in good condition. The best way to test a radiator cap is to use a pressure tester designed for that purpose. Replace if necessary.

WARNING

A hot radiator can release hot steam very quickly. When removing a radiator cap from a vehicle, always use extreme caution. Slowly loosen the cap, allowing pressure to escape before removing it fully.

Thermostat

The thermostat is the engine's temperature control. It keeps the coolant in the engine until the engine reaches it peak operating temperature. Once the proper operating temperature (usually 180 to 195 degrees Fahrenheit) has been reached, the thermostat opens up, allowing the coolant to flow to the radiator for cooling. A thermostat that is stuck closed will not allow the coolant to flow through the radiator and engine overheating will result. A quick way to tell if this is happening is to feel the upper radiator hose. If it is cold but the engine is hot, the thermostat is probably stuck closed. On the other hand, if the thermostat is stuck open, the engine will take a long time or possibly fail to reach proper operating temperature. Common thermostat locations are shown in Figure 9-56.

Water Pump

The water pump circulates the coolant throughout the cooling system. It does this through the use of internal blades called impellers that push the coolant through the system. Loose or corroded impellers do not circulate coolant properly and will cause overheating. This is more noticeable at highway speeds. A quick check for proper coolant flow is to look into the radiator neck with the cap removed and the engine warm

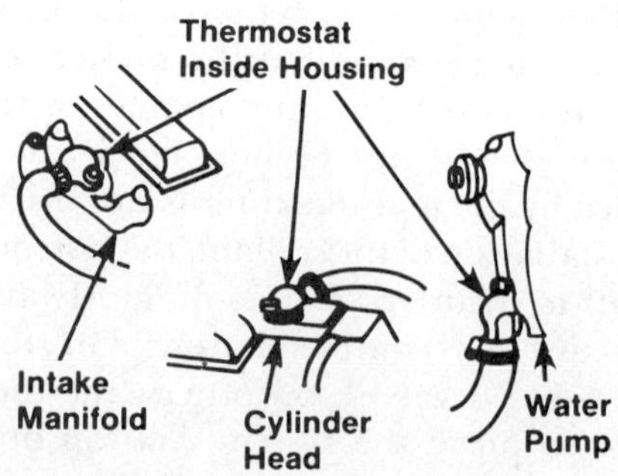

FIGURE 9-56 Thermostat locations

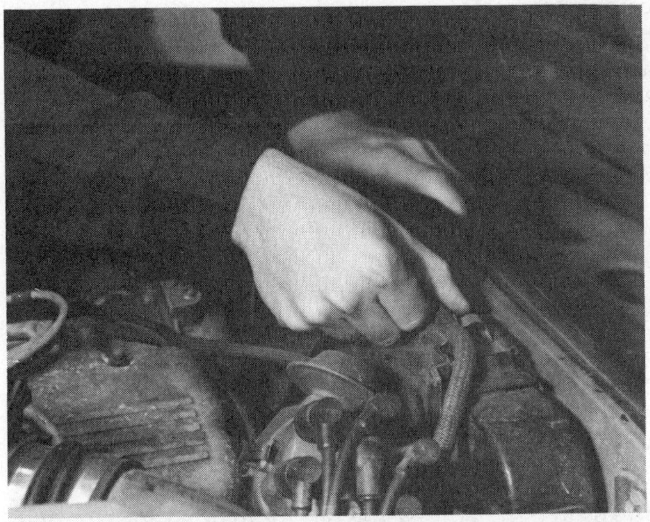

FIGURE 9-57 All radiator hoses must be securely clamped.

enough for the thermostat to be open. Check to be sure that there is coolant movement. If not, the water pump may need to be replaced.

Radiator Hoses

If the cooling system has been worked on during collision repair, check to make sure that the hoses are in good condition and are securely clamped (Figure 9-57). The lower radiator hose is for the coolant coming from the radiator to the water pump. When the water pump is turning, a low pressure area is created in the lower hose as the coolant is drawn into the pump. To keep it from collapsing, there will sometimes be a coiled spring inside the lower hose. Make sure this hose is not showing signs of being collapsed.

Belt Tension

On vehicles where the cooling fan is driven by belts from the engine, proper tension must be maintained or belt slippage will occur. When this happens, the cooling fan does not turn at full speed, resulting in reduced air flow through the radiator at idle. This is a problem only at idle; at highway speeds, the air flow is sufficient to maintain cooling.

Fans and Fan Clutches

Check to make sure the fan blades turn with no interference and that the fan mounts are not rubbing against the radiator or body (Figure 9-58). On some cooling systems there might be two or more fans.

Some vehicles are equipped with a special type of cooling fan with a built-in clutch. Fan clutches can either be filled with fluid or use a thermostatic spring that allows the fan to slip at highway speeds when it is not needed for airflow. This reduces drag on the engine,

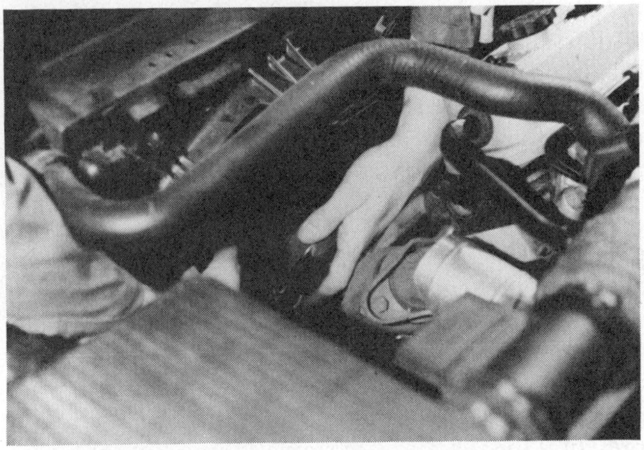

FIGURE 9-58 Checking a fan by spinning it

resulting in increased fuel economy. If a fan clutch does not work due to leaking fluid or a defective spring, fan speed is reduced at idle, causing overheating.

Electric Cooling Fans

The cooling fan may be electric, rather than belt driven. An electric fan is less of a draw on an engine than a belt-driven fan, resulting in better fuel economy. Check for loose electrical connections, and bare, burnt, or cut wires.

Most vehicles have one single-speed cooling fan, but some are equipped with two fans or one two-speed fan. These are designed to increase air flow through the radiator. If engine overheating is present at idle, check to make sure that the fan or fans are operating correctly.

Electric Cooling Fans and Air Conditioners

When the air conditioner is turned on, the electric cooling fan should also come on, regardless of coolant temperature. This prevents the engine from overheating due to the increased load placed on it from the air-conditioning system. If the vehicle has two fans, one is used to cool the A/C condenser.

Cooling Fan Motor

One of the most common problems on vehicles with electric cooling fans is that road film collects on the shaft between the fan motor and the fan blades. This buildup can cause the bearings in the motor to bind and reduce fan speed. This in turn will cause the engine to overheat during idle. This condition will not show up at highway speeds because the airflow is sufficient to avoid overheating. The only way to detect slow fan speed is through the use of a strobe-type tachometer.

Cooling Fan Controls

On most vehicles, the electric fan is controlled by a computer. The coolant sensor reads the coolant temperature and sends an electric signal to the computer. This signal controls the fan relay. The fan relay is turned on when the engine reaches its proper operating temperature. When energized, the relay turns on the fan motor. If an engine overheats at idle and the fan does not operate, the coolant sensor, relay, or computer might be at fault.

Air Dams

With modern fuel-efficient designs dictating how vehicles look, air dams are becoming more common. These are a necessary part of the cooling system because air needs to be directed to the engine in order for proper cooling to be achieved. Make sure there is an air dam present if the vehicle was originally supplied with one.

HEATER OPERATION

The heater may be part of the air-conditioning system or a separate item. Actually, the heater core could be considered a miniature version of the radiator. That is, as hot coolant flows through the heater, a fan blows air over the tubes, warming the air and delivering it to the passenger compartment. This blower fan, located in the heater housing, forces air through the heater core and into the passenger compartment.

The heating distribution system is a duct system (Figure 9-59). Outside air enters the system through a grille, usually located directly in front of the windshield, and goes into a plenum chamber where rain, snow, and some dirt is separated from it. The air from the plenum is directed through the vehicle's heater core and through the air conditioner evaporator, or into a duct that runs across the dash panel of the vehicle. Outlets in the duct direct the airflow into the passenger compartment. Doors inside the system either recirculate the air inside the compartment or circulate outside air, according to the control settings. They also route the air inside the system through or around the heater or air conditioner or to the windshield defroster.

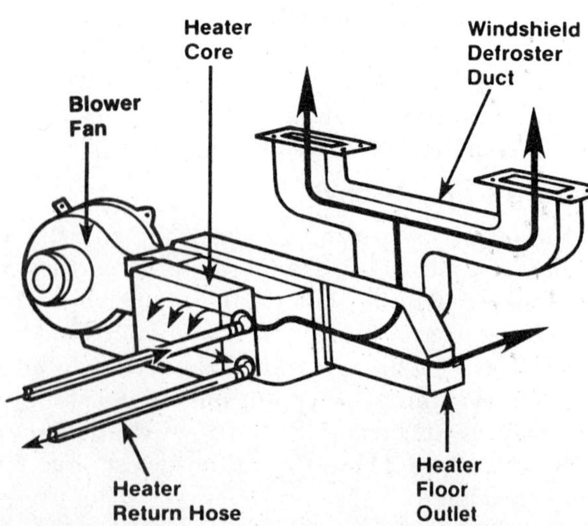

FIGURE 9-59 Typical heat distribution system

Some vehicles have an additional distribution system that brings air into the vehicle through intakes located in the engine compartment. From there it is carried through ducts to side vents located in the cowl side panel of the passenger compartment near the front passenger's feet. The airflow through these vents can be controlled mechanically, by a vacuum system, or electronically.

The heating system is controlled by a number of cables, valves, and switches. Be sure the cables are correctly reconnected if any of them were disconnected during repair. Follow the service manual for troubleshooting. Also keep in mind that the heating system may have collected some dust during repair. Be sure to operate it during cleanup so that the customer is not caught in a cloud of sanding dust after picking up the vehicle and turning on the heater.

AIR-CONDITIONING AND HEATER SYSTEMS

Proper handling of the air-conditioning system during collison repair is both one of the most important and least understood aspects of working with mechanical components. Many needless repairs are caused unknowingly by collision repair technicians who do not understand the importance of following some very strict rules for working with air-conditioning systems. What compounds the problem is that malfunctions often occur several months after the collision repair work is completed, so the customer is unaware of who caused the problem.

A/C OPERATION

An air-conditioning system works on the simple principle that when a liquid is converted to a gas, it absorbs heat from its surroundings. When it is reconverted from this gaseous state back to a liquid, it gives up this heat. The air-conditioning system (Figure 9-60) consists of the following components:

- Compressor/clutch assembly
- Condenser
- Receiver-dryer or accumulator
- Expansion valve
- Evaporator

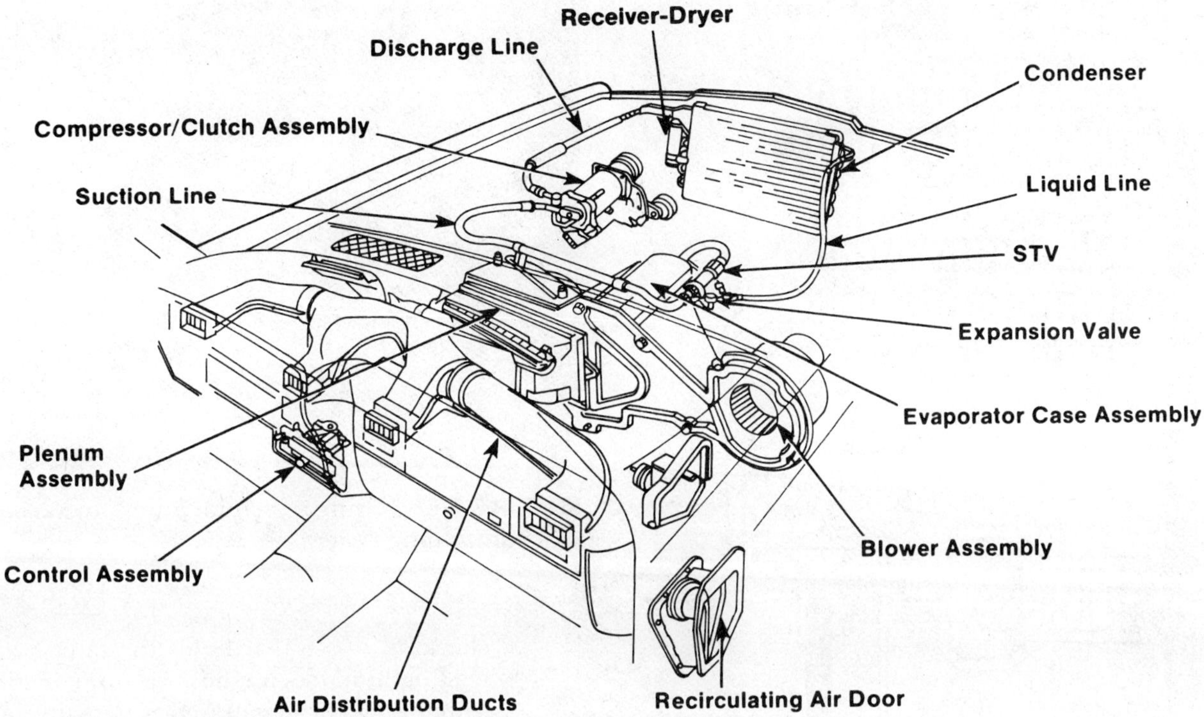

FIGURE 9-60 Typical air-conditioning system

In the basic A/C system, the refrigerant is absorbed and transferred as shown in Figure 9-61 and the following six steps:

1. Refrigerant leaves the compressor/clutch assembly as a high-pressure, high-temperature vapor.
2. By removing heat via the condenser, the vapor becomes a high-pressure, lower temperature liquid.
3. Moisture and contaminants are removed by the receiver-dryer or accumulator, where the cleaned refrigerant is stored until it is needed.
4. The expansion valve converts the high-pressure liquid into a low-pressure liquid by controlling its flow into the evaporator.
5. Heat is absorbed from the air inside the passenger compartment by the low-pressure, low-temperature refrigerant, causing the liquid to vaporize.
6. The refrigerant returns to the compressor/clutch assembly as a low-pressure, higher temperature vapor.

Refrigerants are the chemicals that transfer heat by absorbing it during evaporation and releasing it during condensation. For many years, Refrigerant 12 (R-12) was universally used, but its adverse affect on the ozone layer has led to strict handling controls and the development of other environmentally safe refrigerants.

A/C HANDLING TIPS

Although most actual A/C repair and service work is done in specialty air-conditioning shops, here are pointers that the collision repair technician must keep in mind:

- *Moisture is the Worst Enemy.* Most A/C failures are caused by moisture entering the system. Moisture interacting with refrigerant causes sludge and hydrochloric acid to form. This will attack the delicate parts of the compressor and eat away at the aluminum components of the evaporator and condenser.
- *Refrigerant.* The release of R-12 into the atmosphere is prohibited by current EPA and many state regulations. Never vent the refrigerant into open air; use a recovery/recycling machine.
- *Seal the Openings.* When removing or opening up the A/C unit, seal all openings with appropriate protective barriers. These can be hydraulic caps and plugs, tight-fitting rubber corks, or plastic wraps. Use sturdy rubber bands or wire ties to hold plastic wraps in place securely.

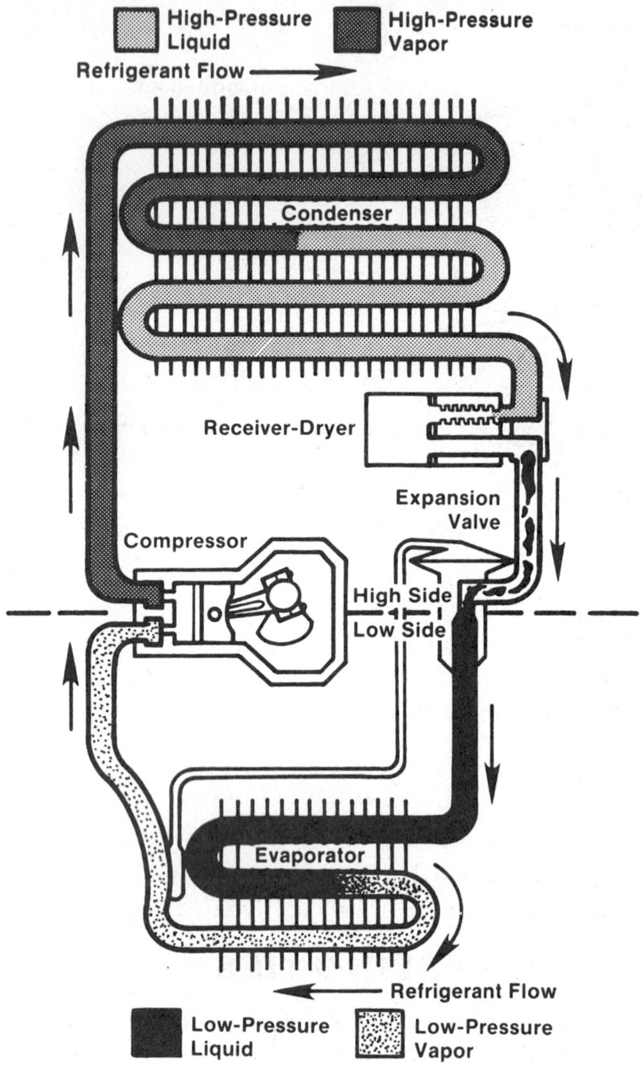

FIGURE 9-61 Refrigerant flow cycle

- *Cautions for Reassembly.* If a system has been open to the atmosphere more than a few hours, do the following:
  1. Change the oil.
  2. Flush each component separately with nitrogen gas before charging.
  3. Replace the receiver-dryer or accumulator.
  4. During evacuation, hold the system at high vacuum for a minimum of 30 minutes.
- *Evacuating Guidelines.* Before recharging, the system must be evacuated as directed in the service manual (Figure 9-62). Evacuating removes all atmospheric gases. It also causes all water to vaporize so the moisture is removed. Incidentally, evacuating does not save the receiver-dryer or accumulator if it has been open to the atmosphere more than a very short time.
- *Pressure Test for Leaks.* Test the system for leaks by closing the system under vacuum and checking to see if it holds the same vacuum reading. If it does not hold vacuum, it will not contain the refrigerant under pressure.

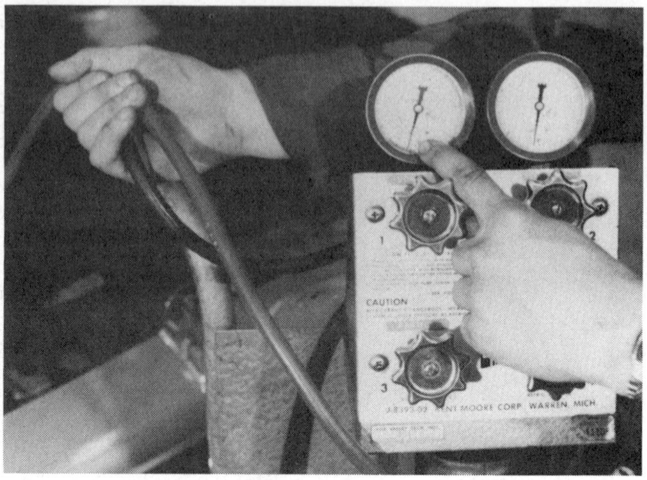

FIGURE 9-62 Using a vacuum pump to evacuate an air-conditioning system

- *Role of the Bridging Diode.* The A/C compressor is driven by the engine through an electric clutch. When this clutch cuts in and out, a spike voltage is developed by the building and collapsing of the magnetic field. This spike voltage could damage the delicate computer equipment in the vehicle. To prevent this, a bridging diode is wired into the air-conditioning circuit. The diode does not allow the current to flow in the direction where it could cause damage. When there is A/C system damage, the bridging diode might be destroyed. If it is not replaced, the air-conditioning clutch might not energize and damage to the computer could result. The diode is normally found on the air-conditioning compressor wiring and can be checked with a test light or ohmmeter.
- *Guidelines for Use of A/C Oil.* Use only oil designed specifically for air-conditioning systems. Do not reuse oil that has been drained from the system. Also do not use oil that has been stored in a partially filled container because it will have absorbed moisture. As stated before, even a small amount of moisture can be very damaging.

EXHAUST SYSTEMS

The exhaust system is used to collect and discharge the exhaust gases caused by the combustion of the air/

fuel mixture within the engine. The exhaust system found on today's unibody vehicles must also aid in the control of exhaust emissions.

WARNING

When inspecting or working on the exhaust system, remember that its components get very hot when the engine is running; contact with them could cause a severe burn.

Engine exhaust consists mainly of nitrogen (N_2). It also contains carbon monoxide (CO), carbon dioxide (CO_2), water vapor (H_2O), oxygen (O_2), nitrous oxides (NO_x), hydrogen (H_2), and unburned hydrocarbons (HC). Three of these exhaust components (CO, NO_x, and HC) are major air pollutants and their emission to the atmosphere must therefore be controlled.

The catalytic converter, mounted in the exhaust system, plays a major role in emission control. The catalytic converter works as a gas reactor, and its catalytic function is to speed up the heat-producing chemical reaction between the exhaust gas components in order to reduce the air pollutants in the exhaust. Figure 9-63 illustrates the three basic types of catalytic converter designs.

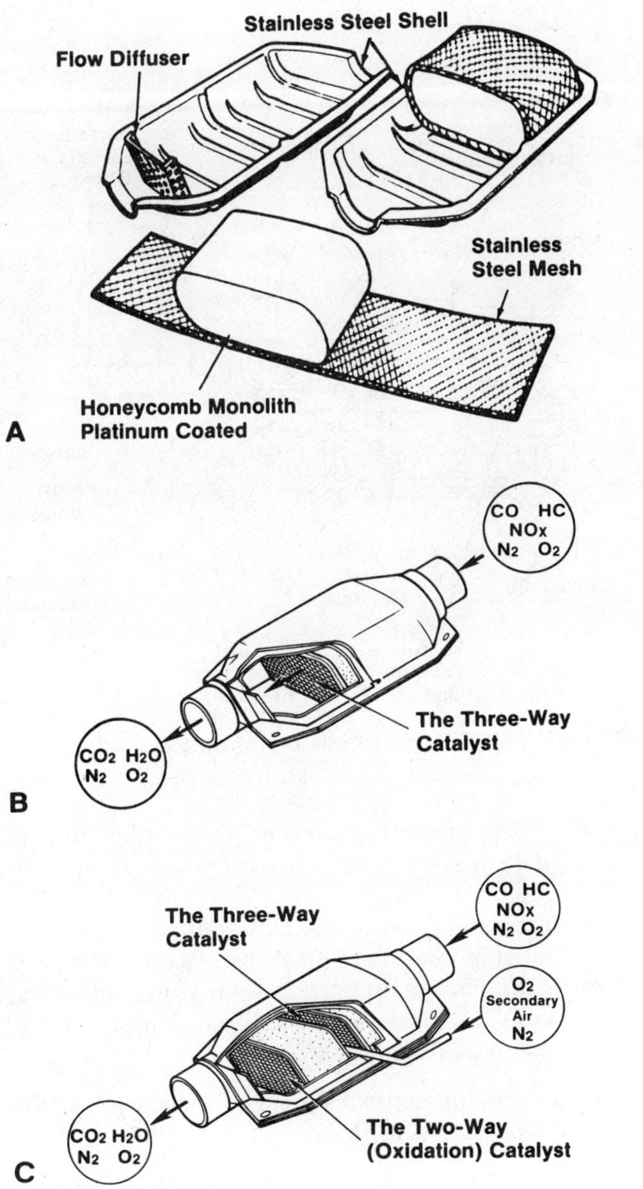

FIGURE 9-63 Types of catalytic converters: (A) single-bed, (B) dual-bed, and (C) three-way

SHOP TALK

Because of constant changes in EPA catalytic converter servicing and installation requirements, check with the vehicle manufacturer or the EPA for the latest data regarding replacement.

In addition to the catalytic converter, the exhaust system includes an exhaust manifold, exhaust pipe, muffler, resonator (optional), and tail pipe (Figure 9-64). All of the components are designed to conform to the available space of the vehicle's undercarriage and yet be high enough from the road to avoid striking obstructions. Most parts of the exhaust system, especially the muffler and tail pipe, are subject to collision damage.

To check the system's condition, grab the tail pipe (when it is cool) and try to move it up and down and side to side. There should be only slight movement in any direction. If the system feels wobbly or loose, raise the rear of the vehicle and check the clamps and hangers that fasten the exhaust system to the underbody. Also, jab at all rusted areas in the system with an old screwdriver. If the blade sinks through the metal at any point, that part needs replacing. Tap on parts of the system with a hammer or mallet (Figure 9-65). A ringing sound indicates that the metal is good. A badly corroded part gives out a dull thud.

To check further, start the engine (never in a closed shop). Stuff a rag in the tail pipe and feel around every joint for leaks. If one is found, try tightening the clamp. If this does not stop the leak, it must be repaired.

If a loud rumble is heard during normal operation, either there is a large leak in one of the components or the muffler is not operating properly. There is only one way to repair faulty exhaust system parts, and that is to replace them. To replace a single component such as the muffler, it might not be necessary to take off any of the parts behind it. However, some systems are welded together and so the complete system must be replaced.

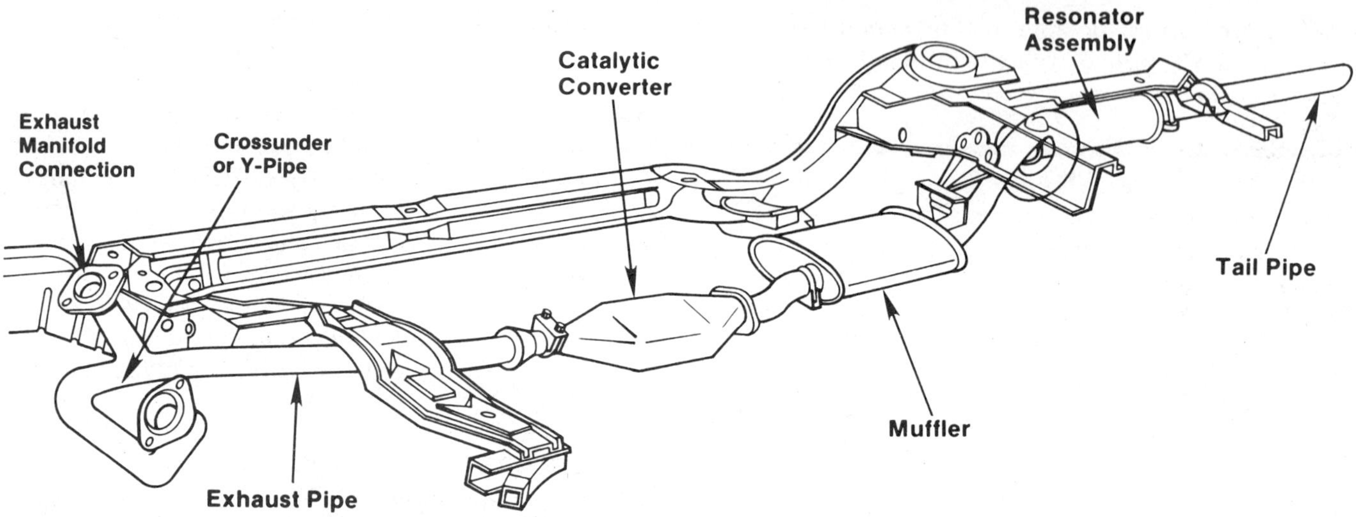

FIGURE 9-64 Typical exhaust system

FIGURE 9-65 Tap on parts of the exhaust system with a hammer to see if the metal is good.

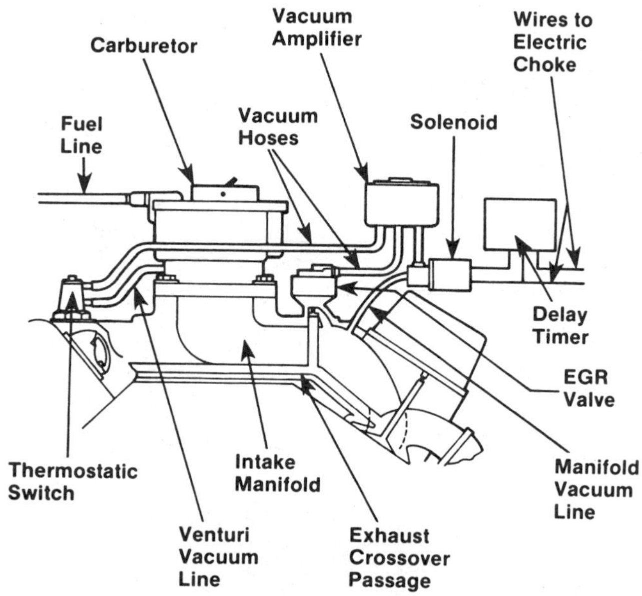

FIGURE 9-66 Exhaust gas recirculation

EMISSION CONTROL SYSTEMS

Many times emission control systems are damaged in a collision and must be serviced as part of the repair. The Clean Air Act, which is a federal law, makes the collision repair technician responsible for the emission control system. The law requires technicians to restore emission control systems to their original design. It also prescribes penalties for shops and technicians who alter emission control systems or fail to restore them to their original design. The following guidelines must be strictly adhered to:

- Damaged parts must be replaced with good parts. Eliminating damaged parts to avoid replacing them is against the law.
- Using parts that prevent proper operation of the emission control system is also against the law.
- Proper repairs to the emission control system must be made to manufacturer's specifications.
- All replacement parts for emission control systems must satisfy the original design requirements of the manufacturer.

The emission control system consists of three subsystems. They are as follows:

1. *Exhaust and Engine Control System.* In addition to the catalytic converter, the other important part of this system is the exhaust gas

recirculation (EGR) valve. This valve opens to allow engine vacuum to siphon exhaust into the intake manifold (Figure 9-66). The EGR valve consists of a poppet and a vacuum actuated diaphragm. When ported vacuum is applied to the diaphragm, it lifts the poppet off its seat. Intake vacuum then siphons exhaust into the engine. On many late-model vehicles, a positive back pressure or negative back pressure EGR valve is often used. This type of valve requires a certain level of back pressure in the exhaust before it will open.

2. *Positive Crankcase Ventilation System.* Crankcase emissions are easily controlled by the positive crankcase ventilation (PCV) system. This system channels blowby gases into the fuel intake area, where they are drawn into the engine and burned (Figure 9-67).
3. *Evaporative System.* This emission control system filters fumes from the gas tank and the carburetor through a charcoal canister which absorbs and stores vaporized fuel (Figure 9-68). When the vehicle is running, these vapors are drawn into the engine and burned, at the same time restoring the capacity of the canister to store fumes.

Some emisson control systems have more controls than mentioned here. The servicing, repair, and replacement procedures for emission control components are given in the service manual. To make it easier for the technician, manufacturers are required by law to install identification labels supplying vacuum routing and connection information, adjustment specifications, and parts numbers when the vehicles are built. These labels are considered part of the emission control system under the law. The labels must be replaced where collision repair service requires their removal or damages the labels.

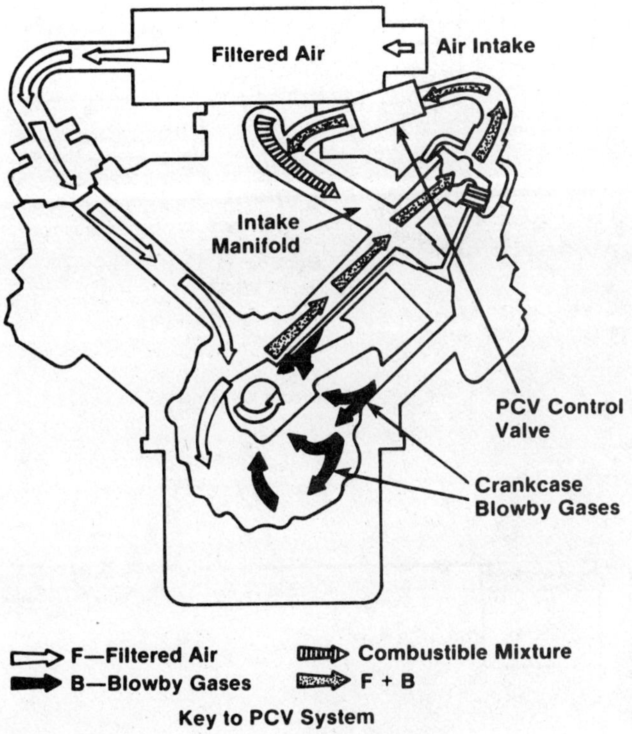

FIGURE 9-67 How the PCV system works

ELECTRICAL SYSTEM SERVICE

Electrical system repair is an essential part of bringing a collision-damaged vehicle back to pre-accident condition. Electrical repairs are becoming

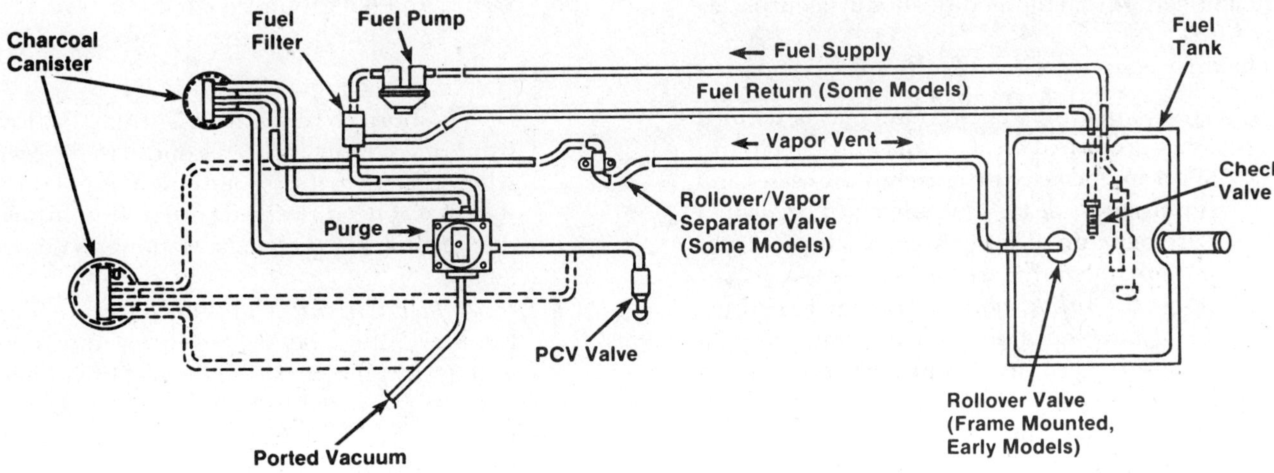

FIGURE 9-68 Typical evaporative system

more and more common during collision repair due to the increased use of computer-controlled systems by auto manufacturers. These repairs can be made complicated by delicate computer circuits that are easily damaged by voltage spikes and static electricity. Yet, following collision repair, all lights, motors, sensors, and switches must operate properly.

ELECTRICAL TERMINOLOGY

Various vehicle systems are controlled by a series of electrical controls and devices. To understand electricity, the technician must become familiar with the following three electrical terms:

1. *Current.* Current is the movement of electricity through a wire or circuit. It is measured in amperes or amps using an ammeter. The common electrical symbol for current is "A."
2. *Voltage.* Voltage is the pressure that pushes the electricity through the wire or circuit. Voltage is measured in volts using a voltmeter. The common electrical symbol for volts is "V."
3. *Resistance.* Resistance is a restriction or obstacle to current flow. It is measured in ohms using an ohmmeter. Resistance may be in the form of a part or load in the system, a poor connection, corrosion, or a purposely designed current-restricting component.

COMMON ELECTRICAL PROBLEMS

Common electrical problems in a collision-damaged vehicle generally mean open, shorted, or grounded circuits; faulty circuit breakers; blown fuses; faulty relays; and bad wiring. Contrary to popular opinion, most electrical problems are not caused by a faulty computer control module. In fact, most experienced technicians believe that the module should be suspected last, not first.

The three common circuit faults are:

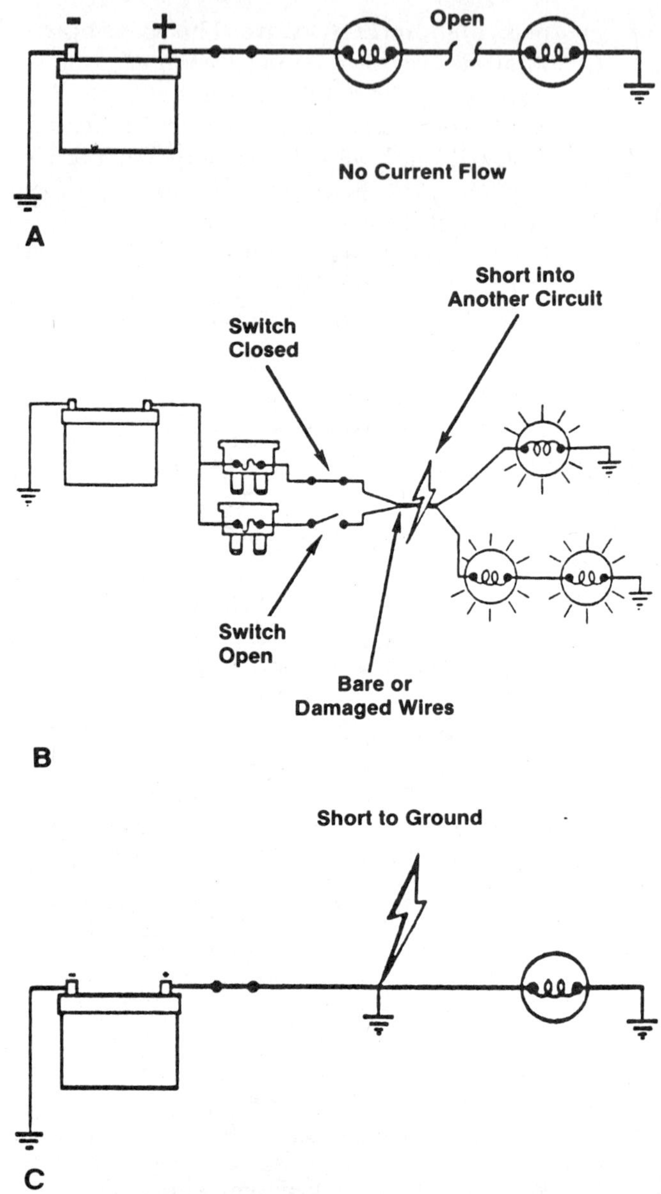

FIGURE 9-69 Typical circuit faults: (A) open circuit; (B) short circuit; and (C) grounded circuit

1. *Open Circuit.* An open circuit may be defined as an unwanted break in an electrical circuit. With an open circuit, current flow ceases and the circuit is dead. This effect can occur in a wire or in an electrical part such as a light bulb filament (Figure 9-69A).
2. *Short Circuit.* A short circuit may be defined as an unwanted wire-to-wire connection in an electrical circuit. A short occurs when the insulation is worn between two adjacent wires and the "metal" in each wire contacts each other, or when wires are damaged or pinched due to collision damage.

   A short circuit allows current to flow from one circuit into a second circuit, even when current to the second circuit is not on at the time. This unwanted current flow allows the second circuit to have power (Figure 9-69B).
3. *Grounded Circuit.* A grounded circuit (Figure 9-69C) may be defined as an unwanted wire-to-metal connection in an electrical circuit. A grounded circuit condition occurs when the insulation on a wire is worn and the "metal" in the wire touches the metal of the vehicle body or frame. When this occurs, the

current is allowed to flow directly to ground without flowing to the electrical parts in the circuit. In other words, the current takes the path of least resistance.

SHOP TALK

Bad grounds are the most common cause of electrical system problems. If just the ground for one part is faulty, it will only affect that one circuit. If a major ground is faulty, many circuits will fail all at once. A major ground is one that connects the battery to the engine block or body. Sometimes a short is intermittent—it shorts momentarily when the vehicle bounces heavily or jars. A flickering dash light is an example of this.

DIAGNOSTIC EQUIPMENT

Locating an electrical fault is possible without major difficulty if the technician uses the following diagnostic tools: test meters, test lights, and jumper wires (Figure 9-70). But, keep in mind that electrical systems can be damaged if the wrong type of test equipment is used.

Test Meters. There are two basic types of test meters: digital volt/ohmmeter (DVOM) and analog volt/ohmmeter (AVOM).

The digital volt/ohmmeter is the meter recommended by auto manufacturers. When diagnosing and servicing today's electrical/electronic systems, a high impedance (10 mega-ohm input) DVOM is recommended in order to avoid damaging sensitive electrical components. It can be used to measure voltage (volts), current (amps), and resistance (ohms) in circuits and circuit components. Digital readouts give the precise measurement needed for proper diagnosis. DVOMs are used in conjunction with the vehicle's service manual.

An analog volt/ohmmeter has a pointer needle that moves across the face of a scale when making electrical measurements. It is difficult to get an exact reading with this type of design. Use of an AVOM can damage sensitive electronic components.

Test Lights. An externally powered test light is often used to determine if there is current flowing through the circuit. One lead of the test lamp is connected to a good ground, the other lead to a point in the circuit. If the lamp lights, current is present at that point. Test lights should not be used randomly, only when specified in the service manual procedures. They are not to be used on solid state circuits.

Jumper Wires. Jumper wires are used to temporarily bypass circuits or circuit components for electrical testing. They consist of a length of wire with an alligator clip at each end. They can be used to test circuit breakers, relays, lights (Figure 9-71), and other components. To protect circuits from damage, the jumper wire should have a replaceable in-line fuse whose rating matches the circuit being tested (usually 5 to 10 amps). Like the test light, jumper wires should only be used when specified in the service manual.

CAUTION

Never permanently bypass a protective device with a jumper wire. Do it only for test purposes.

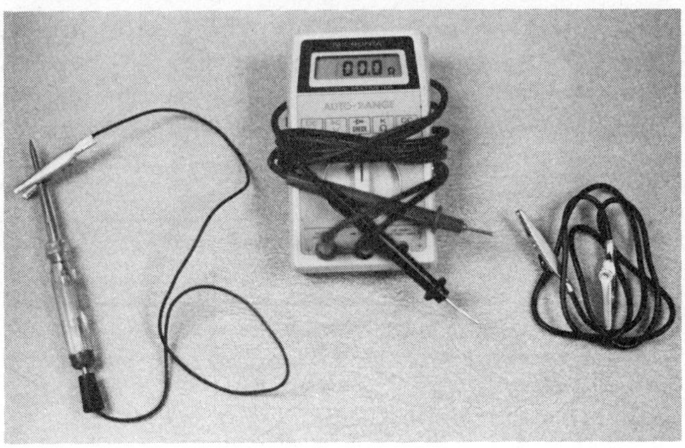

FIGURE 9-70 Typical automotive diagnostic tools (left to right): test light, test meter, and jumper wires

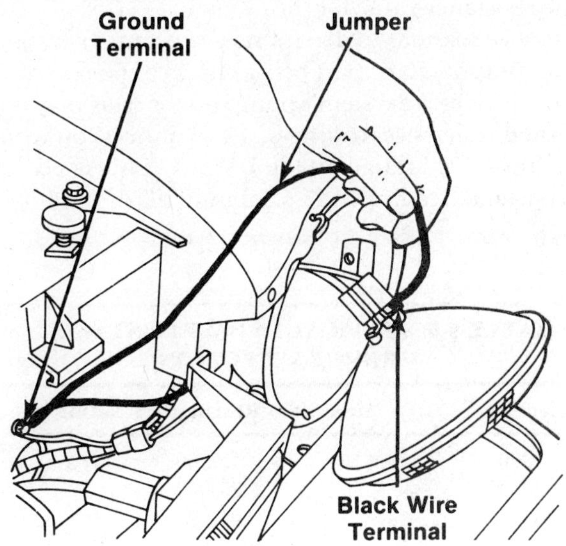

FIGURE 9-71 Using jumper wires to solve a headlight problem

FIGURE 9-72 Measuring open circuit voltage across battery terminals using a DVOM

TESTING A CIRCUIT USING A DVOM

As previously stated, the DVOM can be used to check resistance and voltage, as well as circuit components (Figure 9-72).

Measuring Resistance

When measuring resistance, always disconnect the circuit from the power source. The DVOM must never be connected to a circuit in which current is flowing when making resistance tests. Doing so can damage the meter.

The DVOM offers a selection of ranges for measuring resistance. The ranges are used to measure different resistance values. Many DVOMs offer ranges from 200 to 2,000 K. Use Table 9-6 to select the range scale closest to the resistance value for the test being performed. Use the service manual to determine the normal resistance of the component or part being checked. For example, an anti-lock brake system's pump motor relay may have a normal resistance reading of 45-55 ohms. You would select the "200" range on the DVOM. This is because the resistance reading of 45-55 ohms will be 200 ohms or less.

SHOP TALK

Always refer to the DVOM owner's manual before using the DVOM for diagnostic purposes. You must make sure that you understand how to use the DVOM before performing diagnostics. All DVOMs are fairly similar, but there can be differences between makes.

Checking an Unknown Resistance

1. Connect the DVOM test leads to opposite ends of the circuit or wire being tested.
2. Set the range selector switch on the highest range position. Turn the DVOM on.
3. Reduce the range setting until a "1" or "OL" resistance value is displayed. If a "1" or "OL" reading is displayed during this check, the range selection is lower than the resistance value being checked for.
4. Turn the range selection switch up one setting and read the digital resistance value.

Measuring Voltage. The DVOM allows you to select either alternating current voltage (ACV) or direct current voltage (DCV). The ACV is selected when measuring alternating current voltage. AC current is the type of current that is found in the home. The DCV is selected when measuring direct current voltage. DC current is what is measured in the automobile.

When using the DVOM to measure voltage, the selection of the range scale is very important (Table 9-7). If the voltage reading will be 12-14 volts, select the 20 V range on the DVOM. This range is selected because the voltage reading will be less than 20 volts.

If measuring a 36-volt system, select the 200 V range scale because the 36-volt reading is more than the 20 V range scale will read and is less than the 200 V scale (200 volts).

Checking Continuity

A circuit remains closed only as long as the current can continue from the power source, through the part, and back to the power source. Typically, the service manual will ask for a continuity check of a wire or part.

TABLE 9-6: TYPICAL DVOM RESISTANCE RANGE SELECTION

Selection	Maximum Resistance Reading
200	200
20 K	20,000
200 K	200,000
2,000 K	2,000,000

TABLE 9-7: TYPICAL DVOM VOLTAGE RANGE SELECTION

Selection	Maximum Voltage Reading
2 V	2 volts
20 V	20 volts
200 V	200 volts

The continuity check is done to see if the electrical circuit has a complete path without any opens.

1. Connect the DVOM test leads to the opposite ends of the wire or part being tested.
2. Set the range selector switch on the highest resistance range position.
3. Digital "O" indicates continuity. Digital "1" or "OL" indicates no continuity. Continuity is also present if the DVOM reads a resistance value.

Checking for Shorts

Many times adjoining wires within a wiring harness get pinched together or severed in a collision. A continuity check will detect a severed wire. When checking for a short between two adjoining wires in a harness:

1. Set the DVOM range selector switch the same as for a continuity check.
2. Connect the DVOM test leads to the opposite ends of the adjoining wires.
3. Because the wires are insulated, there should be a reading of "1" or "OL" on the DVOM. If the DVOM reads "0," the wires are shorted.

Testing a Circuit with a Test Light

Whenever a technician has an electrical system problem that does not directly concern the computer—usually a part such as the wipers, horn (Figure 9-73), alternator, or lights—the circuit that supplies current to that part should be tested first.

1. Connect the light to the voltage source and the "hot" side of the part (the side that the current enters). If the part works with the light in place but does not work when the light is removed, the wiring to the part has an open circuit somewhere. If the light is on, but the part does not work, the part is bad.
2. Connect the light to the hot side of the part and then ground the light. Turn on the switch for the part. If the light comes on, there is power going to the part. This means that either the part is bad or it is not properly grounded.
3. Connect the light to the ground side of the part and then to a good ground. If the part works, it is not properly grounded.
4. Connect the light across the part, with one connection on the hot side and one on the ground side. If the light comes on, but the part does not work, the part is defective.

FIGURE 9-73 An externally powered test light can be used to test the horn switch circuit.

DIAGNOSING ELECTRICAL SYSTEMS

Electrical systems can be diagnosed in a variety of ways. Wiring diagrams and diagnostic flow charts are two of the most important tools that can be used by the collision repair technician. In most cases the only tool needed is a DVOM.

Wiring Diagrams

To determine and isolate electrical problems, it is often necessary to trace the electrical circuit. The only way to do this properly is with a wiring diagram found in the service manual. Wiring diagrams look extremely confusing to people who do not know how to read them, but they are really nothing more than road maps. They trace the route that the electrical current takes from the battery to a component and back to the battery.

A wiring diagram of a light circuit is shown in Figure 9-74. Reading a wiring diagram is not difficult if one understands the symbols. Symbol charts can be found at the front of the wiring diagram section of the service manual.

SHOP TALK

The symbol used to identify a part is either a universal symbol or an auto manufacturer's symbol. Since manufacturers use different symbol designs for some parts, it is important that the technician follow the symbol chart for the specific wiring diagram being used.

Most wires noted on wiring diagrams will be identified by their insulation color. Color coding allows the

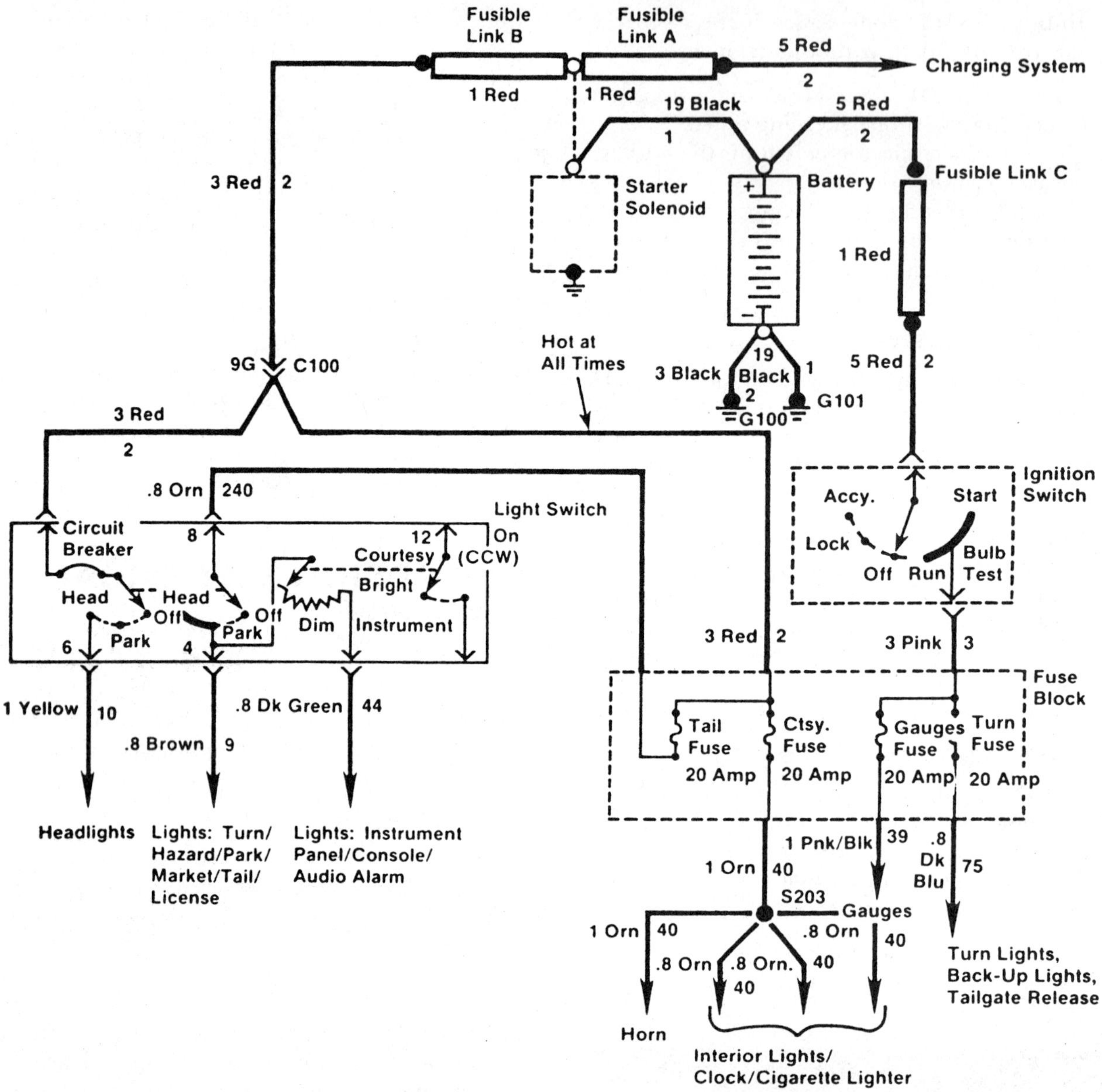

FIGURE 9-74 Typical light circuit wiring diagram

technician to find a specific wire in a harness or in a connector (Figure 9-75). The color code abbreviation chart can also be found at the front of the wiring diagram section of the service manual. Remember, different automakers use different color coding abbreviations on their wiring diagrams.

Sometimes wiring or harness changes are made by the automaker in the middle of a vehicle production year. This means that the color of the wire on the wiring diagram may not match the color of the wire on the vehicle.

Many wiring diagrams found in the service manuals also have circuits numbered. Numbering is used to specify exactly which part of the circuit the service manual is referring to (Figure 9-76).

SHOP TALK

A wiring diagram is more like a book than a picture. That is, it is not possible to understand a wiring diagram just by looking at it. Like a book, it must be read carefully all the way through for a complete understanding. A numbered circuit is simply a "chapter in the book" or a portion of the wiring diagram.

Color	Abbreviations
Aluminum	AL
Black	BLK BK B
Blue (Dark)	BLU DK DB DK BLU
Blue (Light)	BLU LT LB LT BLU
Brown	BRN BR BN
Glazed	GLZ GL
Gray	GRA GR G
Green (Dark)	GRN DKDG DK GRN
Green (Light)	GRN LT LG LT GRN
Maroon	MAR N
Orange	ORN O ORG
Pink	PNK PK P
Purple	PPLPR
Red	RED R RD
Tan	TANT TN
Violet	VLTV
White	WHT W WH
Yellow	YELY YL

FIGURE 9-75 Typical wire color coding chart

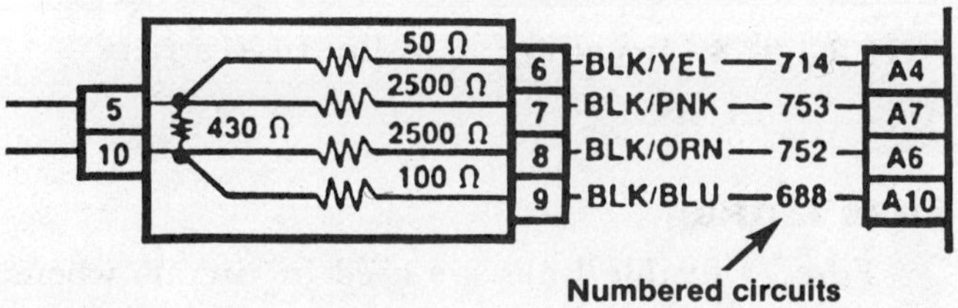

FIGURE 9-76 Numbering each circuit on the wiring diagram helps to locate that specific circuit for testing.

Diagnostic Flow Charts

These are one of the tools of the future for collision repair technicians and insurance appraisers. Most auto manufacturers use some type of diagnostic flow chart to aid technicians in the repair of electrical/electronically controlled systems. Diagnostic flow charts provide a systematic approach to the electrical system and component troubleshooting and repair. They are found in service manuals and are vehicle make and model specific.

During the diagnosis, start at the top of the chart and follow the sequence down (Figure 9-77). The chart will point to the next area to move to after each check is performed. The technician must work through the entire chart, step-by-step, as directed, and perform those repairs or parts replacements as indicated. Many times a diagnostic flow chart will address only a single problem.

CIRCUIT PROTECTIVE DEVICES

When overloads or shorts in a circuit cause too much current to flow, the wiring in the circuit heats up, the insulation melts, and a fire can result, unless the circuit has some kind of protective device. Fuses, fuse links, maxi-fuses, and circuit breakers are designed to provide protection from high current. They may be used singularly or in combination.

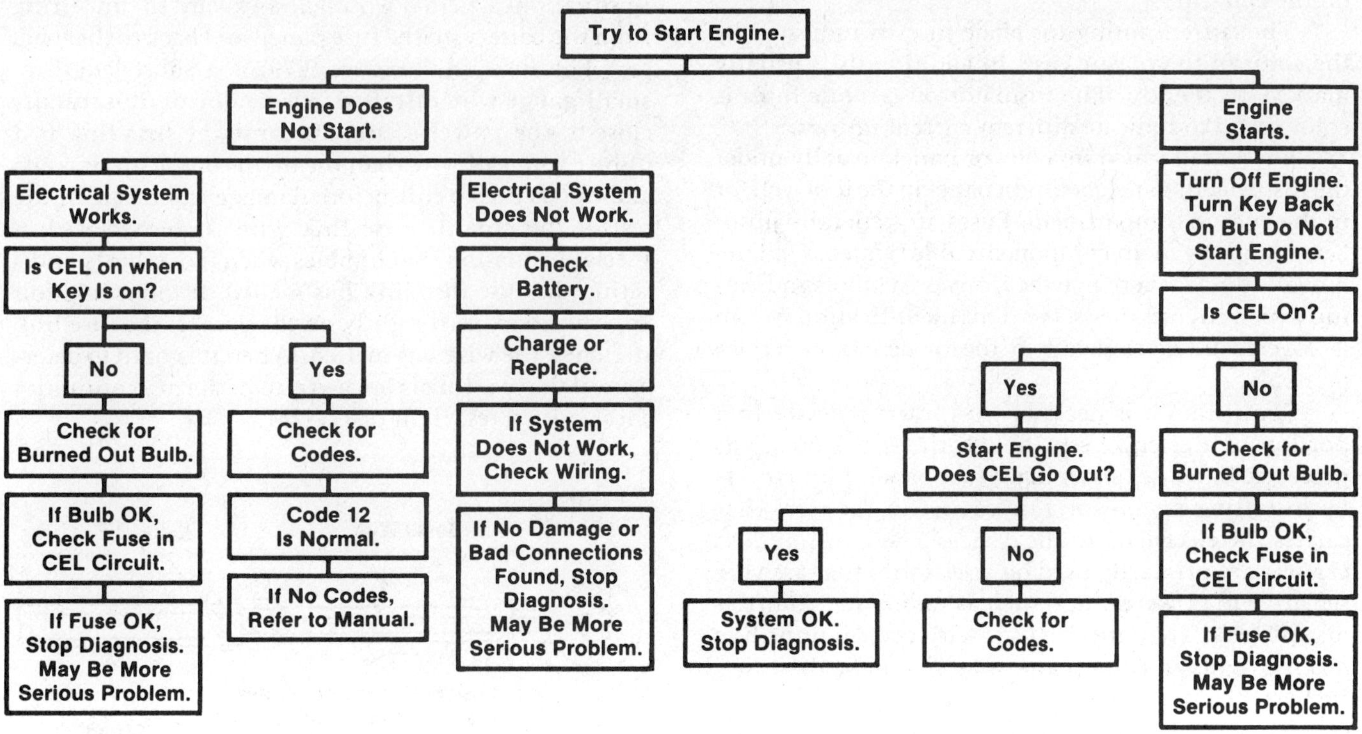

FIGURE 9-77 Typical diagnostic flow chart

Fuses

There are three types of fuses in automotive use: cartridge, blade, and ceramic. The cartridge fuse is found on most older domestic vehicles and a few imports. It is composed of a strip of metal enclosed in a glass or transparent plastic tube. To check the fuse, look for a break in the internal wire or metal strip. Discoloration of the glass cover or glue bubbling around the metal end caps is an indication of overheating.

Late-model domestic vehicles and many imports use blade or spade fuses. To check the fuse, pull it from the fuse panel and look at the element through the transparent plastic housing. Look for internal breaks and discoloration.

The ceramic fuse is used on many European imports. The core is a ceramic insulator with a conductive metal strip along one side. To check this fuse, look for a break in the contact strip on the outside of the fuse. All fuse types can be checked with a circuit tester or multimeter.

Fuses are rated by the current at which they are designed to blow. A three-letter code indicates the type and size of the fuses. Blade fuses have codes ATC or ATO. All glass SFE fuses have the same diameter, but the length varies with the current rating. Ceramic fuses are available in two sizes, code GBF (small) and the more common code GBC (large). The ampere rating is also embossed on the insulator. Codes such as AGA, AGW, and AGC indicate the length and diameter of the fuse. The code and the current rating is usually stamped on the end cap.

The current rating for blade fuses is indicated by the color of the plastic case. In addition, it is usually marked on the top. The insulator on ceramic fuses is color coded to indicate different current ratings.

Fuses are located in a box or panel, usually under the instrument panel, behind a panel in the foot well, or in the engine compartment. Fuses are generally numbered, and the main components abbreviated. On late-model vehicles there may be icons or symbols indicating which circuits they serve. This identification system is covered in more detail in the owner's and service manuals.

Sometimes it is necessary to protect a device in a portion of a circuit even though the entire circuit is protected by a fuse carrier in the fuse panel. This is done by installing an in-line fuse carrier in the wire that carries the electricity to the device. These in-line fuse carriers are primarily used on accessories that are very sensitive to power surges, such as radios and compact disc players. They are also used with add-on units like driving lights and power antennas. Normally these fuse carriers are close to the units they protect.

There are several types of in-line fuse carriers. One of the most popular and easiest to install is shown in Figure 9-78.

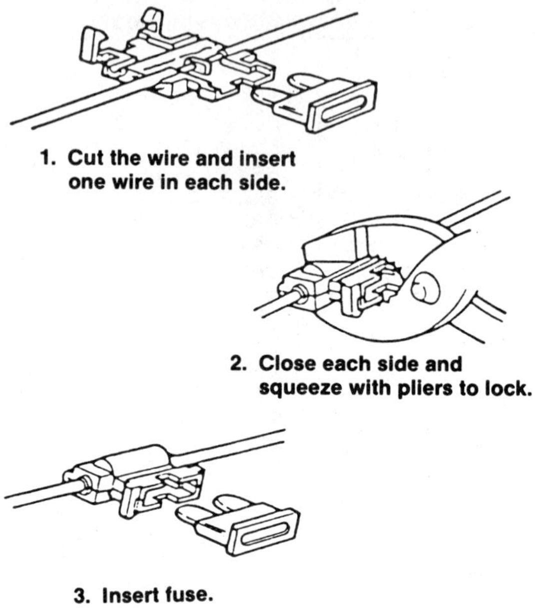

FIGURE 9-78 Installing an in-line fuse carrier

Fuse Links

Fuse or fusible links are used in circuits where maximum current controls are not so critical. They are often installed in the positive battery lead that powers the ignition switch and other circuits that are live with the key off. Fuse links are normally found in the engine compartment near the battery. Fuse links are also used in situations where it would be awkward to run wiring from the battery, to the fuse panel, and back to the load.

The fuse link (Figure 9-79) is a short length of small-gauge wire installed in a conductor. It is usually close to the power source. Because the fuse link is a lighter gauge of wire than the main conductor, it melts and opens the circuit before damage can occur in the rest of the circuit. Fuse link wire is covered with a special insulation that bubbles when it overheats, indicating that the fuse link has melted. If the insulation appears good, pull lightly on the wire. If the fuse link stretches, the wire has melted. When it is hard to determine if the fuse link is burned out, perform a continuity check with a test light or DVOM.

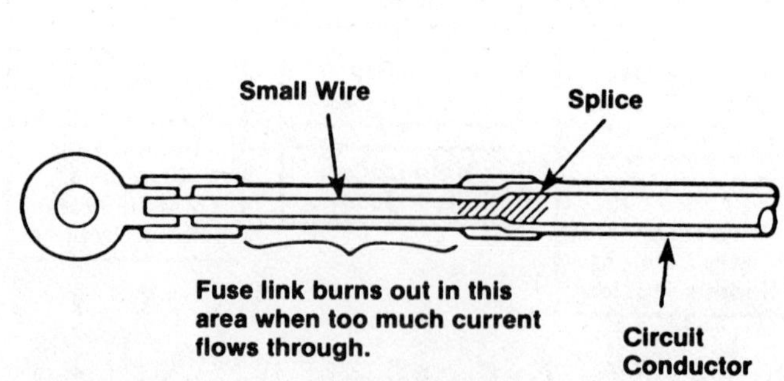

FIGURE 9-79 Typical fuse link

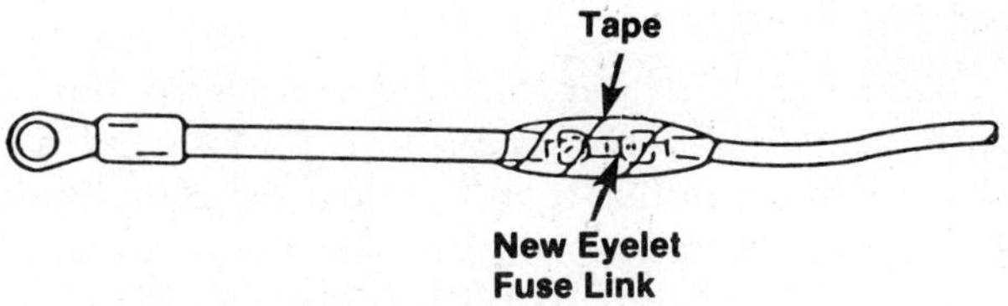

FIGURE 9-80 Replacing a fuse link

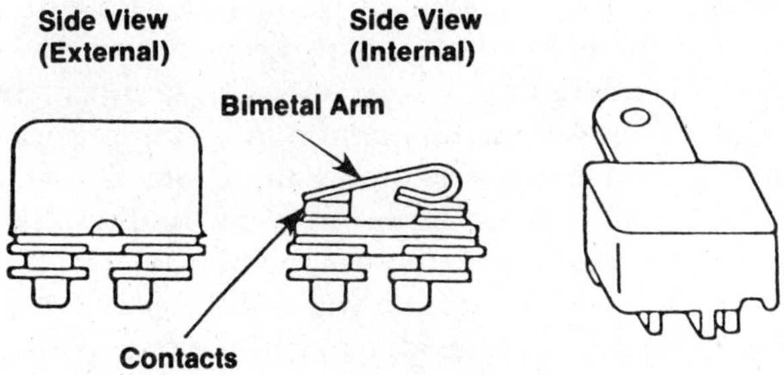

FIGURE 9-81 Typical cycling circuit breaker

CAUTION

Do not mistake a resistor wire for a fuse link. The resistor wire is generally longer and is clearly marked "Resistor—do not cut or splice."

When replacing fuse links, first cut the protected wire where it is connected to the fuse link. Then, tightly tape or solder a new eyelet terminal fuse link of the same rating as the link it is replacing (Figure 9-80). Since the insulation on manufacturers' fuse links is flameproof, never fabricate a fuse link from ordinary wire because the insulation may not be flameproof.

Maxi-Fuses

A number of new electrical systems use maxi-fuses in place of traditional fuse links. Maxi-fuses look and operate like two-prong blade fuses, except they are much larger and can handle more current. (Typically, a maxi-fuse is four to five times larger.) Maxi-fuses are located in their own underhood fuse block.

Maxi-fuses are easier to inspect and replace than fuse links. To check a maxi-fuse, look at the fuse element through the transparent colored plastic side housing. If there is a break in the element, the maxi-fuse has blown. To replace it, pull it from its fuse box or panel. Always replace a blown maxi-fuse with a new one having the same ampere rating.

Another advantage of maxi-fuses is that they allow the vehicle's electrical system to be broken down into smaller circuits that are easy to diagnose and repair. For example, in some vehicles a single fuse link controls one-half or more of all circuitry. If it burns out, many electrical systems are lost. By replacing this single fuse link with several maxi-fuses, the number of systems lost due to a single fuse failure is drastically reduced. This makes it easy to pinpoint the source of trouble.

Circuit Breakers

Some circuits are protected by circuit breakers (abbreviated c.b. in service manuals). They can be fuse panel mounted or in-line. Like fuses, they are rated in amperes.

Each circuit breaker conducts current through an arm made of two types of metal bonded together (known as a bimetal arm). If the arm starts to carry too much current, it heats up. As one metal expands faster than the other, the arm bends, opening the contacts and breaking the current flow. A circuit breaker can be cycling (Figure 9-81) or noncycling.

In the cycling type, the bimetal arm will begin to cool once the current to it is stopped. Once it returns to its original shape, the contacts are closed and power is restored. If the current is still too high, this cycle of breaking the circuit will be repeated. In this case the problem should be found and repaired.

Cycling circuit breakers are generally used in circuits that are prone to occasional overloads. These include devices such as power windows, in which a jammed or sticking window can overwork the motor. In this situation a circuit breaker prevents the motor from burning out.

In automotive work, two types of noncycling circuit breakers are used. One is reset by removing the power from the circuit. There is a coil wrapped around the bimetal arm (Figure 9-82A). When an overcurrent exists and the contacts open, a small current passes through the coil. This current through the coil is not

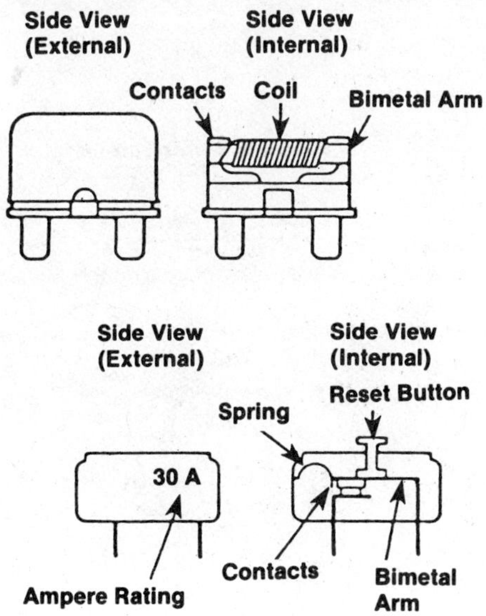

FIGURE 9-82 Resetting noncycling circuit breakers (A) by removing power from the circuit and (B) by depressing a reset button

enough to operate a load, but it does heat up both the coil and the bimetal arm. This keeps the arm in the open position until power is removed. The other type is reset by depressing a reset button. A spring pushes the bimetal arm down and holds the contacts together (Figure 9-82B). When an overcurrent condition exists and the bimetal arm heats up, the bimetal arm bends enough to overcome the spring and the contacts snap open. The contacts stay open until the reset button is pushed and the contacts snap together again.

REPAIRING WIRING AND CONNECTIONS

Some damaged wiring can be repaired if correct procedures and careful workmanship are applied when handling wiring harnesses (Figure 9-83). Using the correct technique is very important, since a repair can affect the electrical and electronic performance of the circuit. Be sure to follow the manufacturer's recommendations for wire repair. These recommendations can be found in the vehicle service manual.

When servicing electrical wiring never tug on the connectors. Use the special tools designed to separate the connectors. This will minimize the possibility of having electrical problems caused by intermittent contact. Be sure to route the wiring in the same location as the OEM. In addition, it is necessary to protect the electrical connectors from moisture and corrosion by using dielectric grease.

Damaged wiring can usually be repaired if proper procedures and materials are used. While there are many methods that can be used to repair damaged wiring, the two most popular ways are either solder repair or splice sleeves.

Solder Repair

All copper wire joints should be soldered, if possible. A pencil type soldering iron is recommended for this procedure rather than a soldering gun (Figure 9-84). This is because guns have heavy coils that can create induced current in nearby components and damage them. Be sure to use rosin-core solder, not acid-core, when soldering electrical connections. Acid fluxes create corrosion and can damage electronic components. In addition, dielectric grease is frequently used to protect connections from corrosion. When using this product, which has no acid content, follow the manufacturer's instructions and be sure not to cut into the

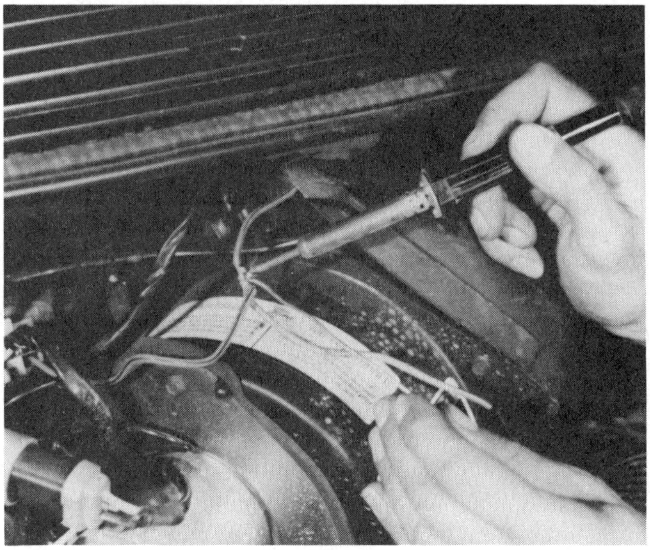

FIGURE 9-84 Pencil soldering copper wire

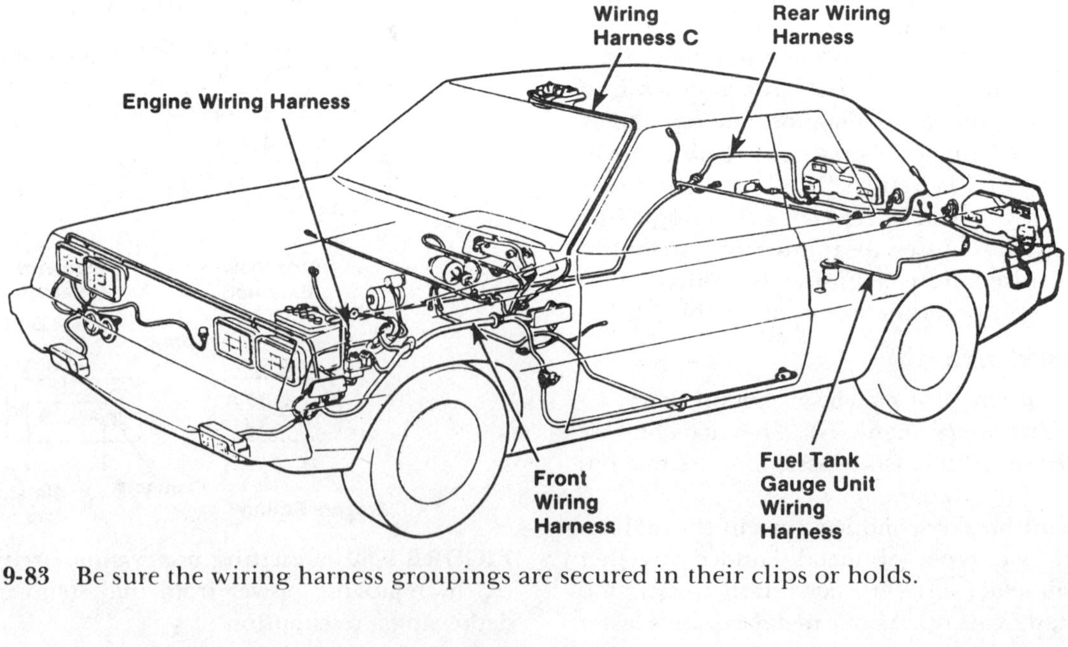

FIGURE 9-83 Be sure the wiring harness groupings are secured in their clips or holds.

moisture-sealed connector joints with test equipment probes.

A shrink tube can also be used over the joint to further protect the connection from corrosion. When heated, the tube shrinks in size and provides a coating over the repair much like the original insulation.

Use the following procedure to make a solder connection:

1. Remove all traces of dirt and corrosion.
2. Strip the insulation back about 1/2 inch on each wire.
3. Wind the wires tightly together.
4. Heat the wire joint, not the solder. Touch the solder to the joint and allow it to flow into the joint. Do not use too much solder.
5. Allow to cool.
6. Gently pull on both ends of the wire to make sure the repair is secure.

SHOP TALK

When soldering under the instrument panel, hold the soldering iron above the joint. The solder will always flow toward the heat source.

Splice Sleeves

Some auto manufacturers are using aluminum in their wiring harnesses. Since aluminum cannot be soldered, special cautions must be taken to ensure a good electrical contact. Splice sleeves (Figure 9-85) are good for this task since they do not require solder and are available for use on aluminum wire to prevent galvanic corrosion. A special tool, however, is required to crimp them onto the wire.

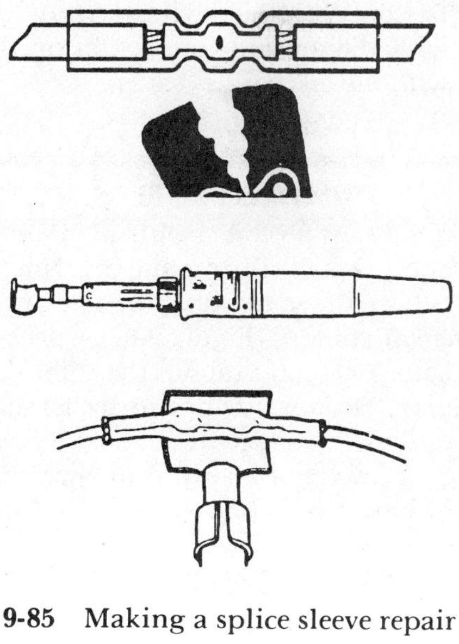

FIGURE 9-85 Making a splice sleeve repair

To install a splice sleeve, proceed as follows:

1. Remove all traces of dirt and corrosion.
2. Strip the insulation back about 3/8 inch on each wire.
3. Match the splice sleeve to the wire diameter.
4. Insert the wire into the splice sleeve until the wire meets the stop in the center of the sleeve.
5. Select the proper sized jaw on the crimp tool for the splice sleeve being used.
6. Crimp both ends of the splice sleeve.
7. Heat the splice sleeve with a heat gun until the sealer oozes out of the ends.
8. Allow to cool.
9. Gently pull on both ends of the wire to make sure the repair is secure.

WORKING WITH BATTERIES

Computers are always using a small amount of a vehicle's electrical power. This creates a steady "draw" on the battery which can result in complete discharge of the battery in about two weeks. The battery should be disconnected if the vehicle will be at rest for any length of time.

Because of the potential electrical problems caused by a collision, always disconnect both battery cables as soon as a vehicle is received at the collision repair shop (Figure 9-86). When removing or disconnecting the battery, the ignition must be off to prevent voltage spikes. Voltage spikes are voltage surges that can destroy many microcircuits in today's electronic systems.

Remove and isolate the negative (-) ground battery cable before removing the positive cable. To isolate the battery cable connector, wrap it completely using electrical tape. This isolation will keep the connector from

FIGURE 9-86 Disconnect both battery cables before doing any collision repair work.

accidentally grounding against the vehicle's metal structure and creating a voltage spike. Once the negative battery cable connector has been removed and isolated, the positive battery cable connector can be removed from the battery. It is a good idea to remove the battery completely from the vehicle before doing any collision repair work. Once the battery has been placed on a bench, it should be checked for a cracked case and similar damage. Use a hydrometer to check the state of the charge (specific gravity). In many cases, the battery will have to be recharged or replaced.

SHOP TALK

Disconnecting the battery can erase stored electrical/electronic system fault codes on some vehicles. Always reference the service manual for specific information concerning disconnecting and connecting the battery.

When installing a battery, make sure that the ignition is off. Reconnect the positive battery cable and then the negative battery cable.

Special care is necessary when charging or jump starting the battery to avoid damaging computer circuits. Use the following safety measures when charging or jump starting:

- Hydrogen gas is formed during charging or jump starting and can cause an explosion, so charge only in a well-ventilated area.
- Keep batteries away from welding operations, open flames, or other heat sources.
- Do not charge batteries with cracked casings.
- Do not smoke near batteries.
- Batteries will charge faster if they are warm.
- Do not charge a frozen battery.

Charging a Battery

The procedure for charging a battery with a battery charger is as follows:

1. Disconnect the negative battery cable.
2. Check the battery casing for damage.
3. Check the water level and add water if necessary.
4. Loosen the vent caps, if the battery is so equipped.
5. Attach the charger clamps to the battery.
6. Set the charger to the recommended settings.
7. Turn on the charger.
8. When charging is completed, turn off the charger.
9. Remove the charger clamps.
10. Replace the vent caps (if applicable).
11. Attach the negative battery cable.

Jump Starting

Although it is a common practice in some collision repair shops, avoid jump starting whenever possible. The discharged battery can explode or create voltage spikes, which can damage electronic components. This is true of both the vehicle being started and the vehicle providing the jump. When jump starting cannot be avoided, use the following procedure:

1. Inspect the battery casing for damage.
2. Position the two vehicles so the jumper cables will reach. Make sure the vehicles are not touching.
3. On the vehicle to be started, turn off all switches and lamps.
4. Connect one end of the jumper cable to the positive terminal of the discharged battery.
5. Connect the other end of the jumper cable to the positive terminal of the booster battery.
6. Connect one end of the other cable to the negative terminal of the booster battery.
7. Connect the other end to an engine bolt head, A/C compressor, alternator mounting bracket, or other ground. Do not connect it to the negative battery terminal of the discharged battery.
8. Make sure that the jumper cables are not in the way of moving engine parts.
9. Start the engine in the vehicle with the booster battery. Let it run a minute, then start the other vehicle.
10. Leave all switches off. Reduce engine speed to idle on both vehicles to prevent damage to the electrical systems.
11. Disconnect all cables in reverse order, starting with the ground connection on the vehicle with the discharged battery.

CAUTION

Never jump start a vehicle with a "maintenance-free" battery that is fully discharged. Not all such batteries have the same color "eyes" or hydrometers when discharged (Figure 9-87). Check the top of the battery case to find out the color codes for the battery. Do not attempt to recharge a fully discharged maintenance-free battery. Jumping or recharging can create an explosion hazard. These batteries have no way to vent gas buildup and therefore should be replaced.

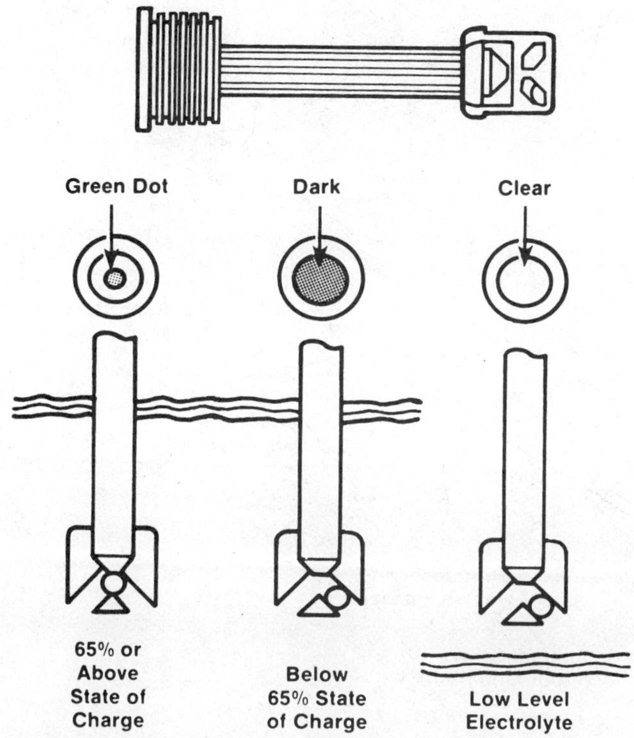

FIGURE 9-87 Typical design and operation of built-in hydrometers on maintenance-free batteries

FIGURE 9-88 Always wear eye protection when charging a battery.

When handling batteries, keep in mind that the electrolyte in the battery is a mixture of water and sulfuric acid. Be sure to wear eye protection whenever handling or charging the battery (Figure 9-88). Do not tilt the battery when carrying it. Be careful not to let electrolyte get on clothing, skin, or painted surfaces. If the battery acid does spill, neutralize with baking soda. Immediately wash the skin or eyes with water and seek medical attention.

LIGHTING AND OTHER ELECTRIC CIRCUITS

Automotive lighting systems have become increasingly more sophisticated. Headlights and tail lights have grown into multiple-light systems, and turn signal indicators have changed from optional accessories into standard equipment and evolved further into emergency and hazard warning systems (Figure 9-89). Indicator lights on the instrument panel commonly warn of failure or improper operation of the charging system, seat belts, brake system, parking brakes, door latches, and other items on the vehicle.

Headlights have both a high and a low beam. The two-light systems have one headlight on each side, each of which has a high-beam and a low-beam filament. The four-light systems have two lights on each side. Two of these should be lit for low beam and all four for high beam. The headlights used in the two-light system and the four-light system are different in size. Until recently, all headlights were round, but recent styling developments created a need for a rectangular light, which is also supposed to be more efficient.

There are other electrical circuits in modern vehicles that operate such accessories as:

- Power and heated seats
- Power windows
- Power door and window locks
- Power and heated outside mirrors
- Automatic headlight dimmers
- Cruise control

Other electrical devices include radios, cassette players, speaker systems, chimes, buzzers, graphic displays, analog instruments, and computer commands. Speedometers, odometers, and various gauges are operated either mechanically or electrically.

Windshield Wipers and Washers

A typical windshield wiper operates on a small single- or multi-speed electric motor. A switch on the steering wheel assembly or instrument panel activates the motor. The windshield washer pump generally has its own motor and reservoir. The pump forces the washer fluid through tubing to the nozzles (Figure 9-90). The nozzles spray the washer fluid on the windshield.

Horn

Most horn systems (Figure 9-91) are controlled by relays. When the horn button, ring, or padded unit is

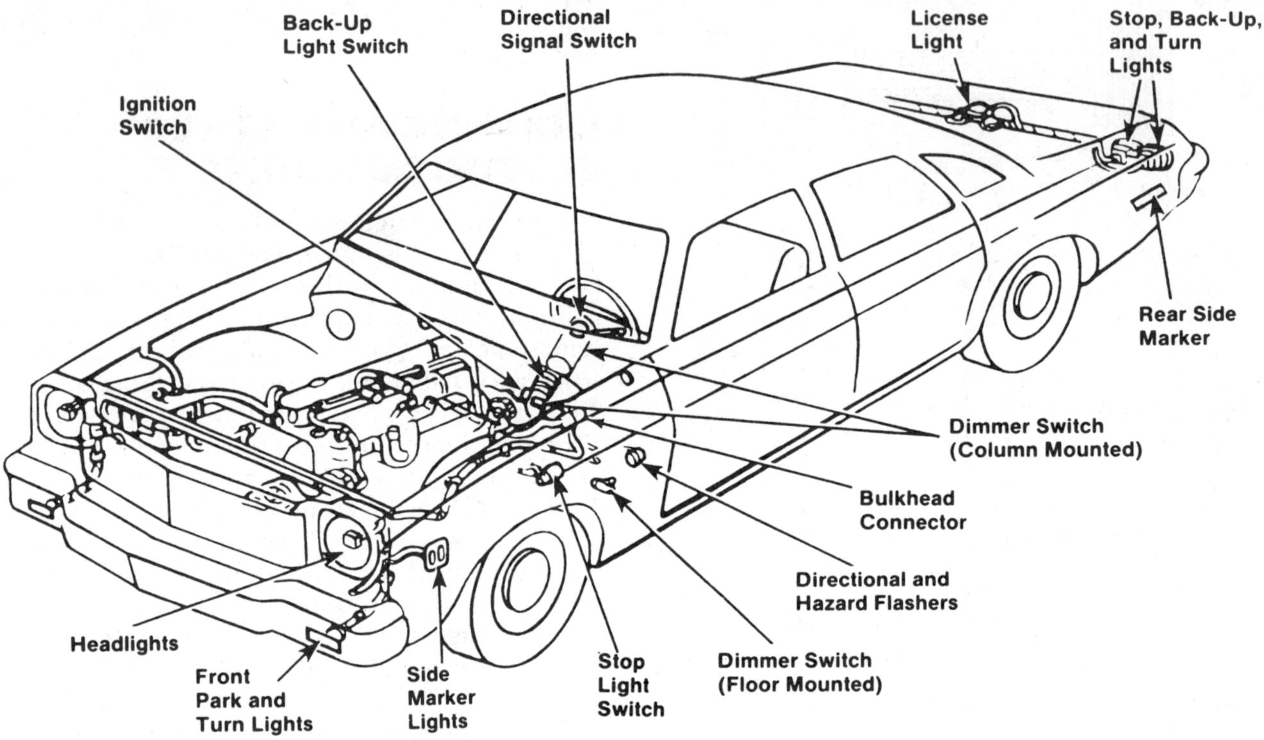

FIGURE 9-89 Typical automotive lighting systems

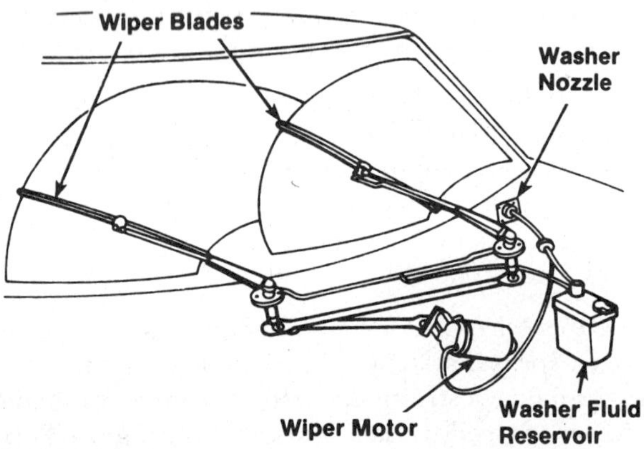

FIGURE 9-90 Windshield wipers and washers

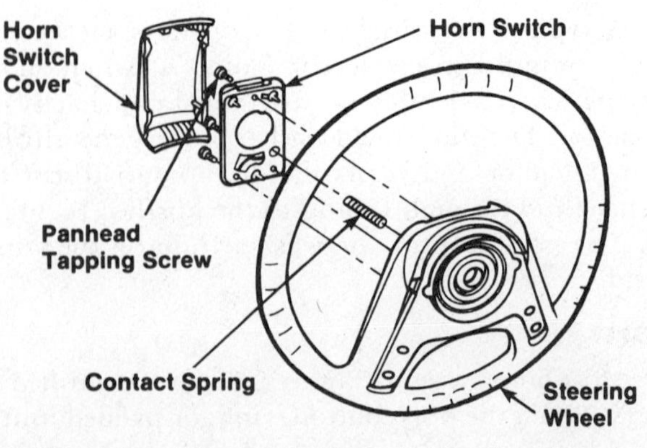

FIGURE 9-91 Typical horn switch exploded view

depressed, electricity flows from the battery through a horn lead, into an electromagnetic coil in the horn relay to the ground. A small flow of current through the coil energizes the electromagnet, pulling a movable arm. Electrical contacts on the arm touch, closing the primary circuit and causing the horn to sound.

ELECTRONIC SYSTEM SERVICE

Many electronic diagnostic and repair procedures are not that difficult. If the technician goes back to the basics of electrical circuits and can follow a few simple rules, he/she can find and repair most electronic problems.

SHOP TALK

The use of the word electronic in this book refers to on-board computers and other "black-box" type items, while the electrical systems just described refer to wiring and electrical components such as alternators, lights, motors, and the like.

UNDERSTANDING ELECTRONIC DISPLAYS AND ON-BOARD COMPUTERS

Electronic instrument panel displays are becoming more and more popular. The reason for this is that technology has become less expensive and more reliable. In addition, many customers like the high-tech, state-of-the-art image of electronic displays. But, in collision repair, electronic displays call for some special cautions. These complex and expensive parts must be handled carefully to avoid damaging them.

Today's modern vehicles have an average of five programmable computers. The memories on these computers can be erased and reprogrammed. There are up to 20 additional computers with memory that cannot be reprogrammed. Computers can be found under the instrument panel, under the seats, under the hood, behind kick panels, or in the trunk.

Most instrument panels are driven by some type of electrical signal. These gauges can be either analog or digital (Figure 9-92). A digital display uses numbers instead of a needle or graphic symbol. In an analog display, an indicator moves in front of a fixed scale to give variable readouts. The indicator is often a needle, but it can also be a liquid crystal or graphic display. An example is a speedometer in which the speed is shown by a set of vertical bars that light up or dim as the speed changes.

The advantage of analog displays is that they show relative change better than digital displays. They are useful when the driver must see something quickly and the exact amount of change is not important. For example, an analog tachometer shows the rise and fall of the engine speed better for shifting than a digital display. Here the driver does not have to know exactly how many rpm the engine is running. The most important thing is how fast the engine is reaching the red line on the gauge.

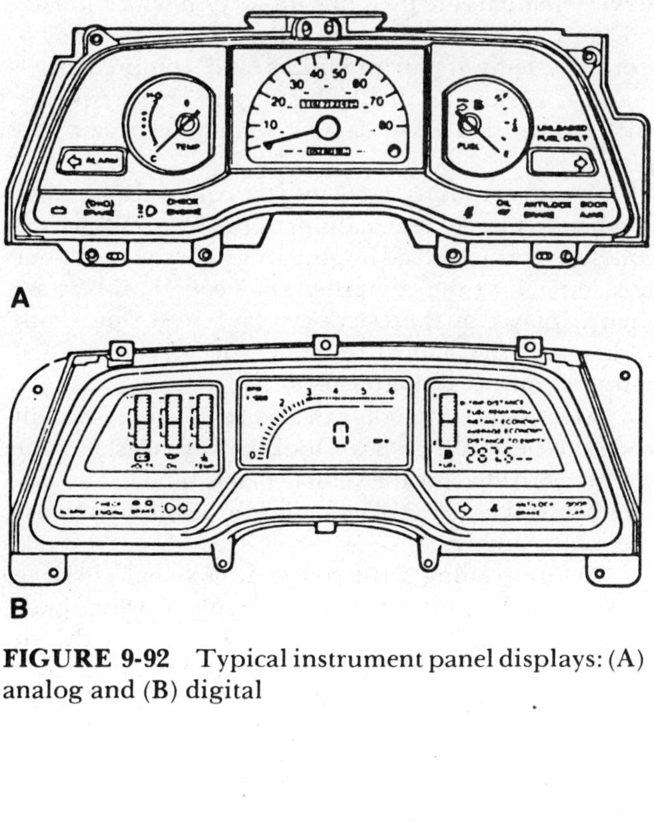

FIGURE 9-92 Typical instrument panel displays: (A) analog and (B) digital

A digital display is better for showing exact data such as miles or operating hours. Many speedometer/odometer combinations are examples of both analog and digital. The choice of display types is a matter of designer and buyer preferences.

An analog electrical signal is continuously variable. An analog current is like the water flowing from a faucet, which is gradually turned up and down. Sometimes it flows a lot, sometimes only a little, and sometimes not at all. As an example, a temperature sensor causes the current to change as the temperature changes. As the temperature rises, the resistance decreases. This causes an increase in the circuit current. As the sensor cools, the current likewise decreases.

The changing current is used to drive a gauge. The higher the temperature or pressure, the more current flows in the gauge circuit. The current creates a magnetic field that moves the pointer. In a temperature gauge, the higher the current (temperature), the greater the magnetic field and the more the pointer moves. These are called magnetic gauges and are used widely.

A digital electrical signal has only two states. It is either on or off. If a switch is turned on and off many times, the number of pulses can be counted. For example, a sensor can be made to turn on and off each time a wheel moves a certain distance. The number of pulses that are counted in a given period of time allows the computer to display the speed. The pulses can also be used by the computer to change the odometer reading.

Types of Electronic Displays

There are three types of electronic displays used today. They are:

- *Light-Emitting Diode (LED).* These are used as either single indicator lights, or they can be grouped to show a set of letters or numbers. LED displays are commonly red, yellow, or green, and use more power than other displays. They can also be hard to see in bright light.
- *Liquid Crystal Diode (LCD).* These have become very popular for many uses, including watches, calculators, and instrument panel gauges. They are made of sandwiches of special glass and liquid. That is where the term "liquid" comes from. A separate light source is required to make the display work. The display has wires on the glass. When there is no voltage, light cannot pass through the liquid. When

voltage is applied, the light passes through the segment. LCDs do not like cold temperatures, and the action of the display slows down in cold weather. These displays are also very delicate, and must be handled with care. Any rough handling of the display can damage it.

- *Vacuum Fluorescent Display (VFD).* These displays use glass tubes filled with argon or neon gas. When current is passed through the tubes they glow very brightly. These displays are both durable and bright.

USE OF SENSORS

All gauges and electronic displays require input from a sensor before they can register a reading. On vehicles without electronic control, the sensor input signal passes directly to the gauge. For example, on older vehicles the temperature gauge is connected directly to a sensor that monitors engine temperature. A rise in temperature results in the sensor generating increased current in the temperature gauge circuit. This causes the pointer of the magnetic temperature gauge to move upward on an analog display scale.

In newer electronic engine control systems, the input from the temperature sensor is used in two ways. The sensor input signal is first directed to the control computer. The computer uses the sensor input to manage a variety of systems, including air/fuel ratio, fuel injection timing, spark timing, and switching of emission control system components. The computer also passes the temperature sensor signal onto the instrument panel temperature gauge. This gauge can be an analog gauge, electronic display, or a simple temperature warning light.

SELF-DIAGNOSTICS

Most electronic systems have built-in self-diagnostic capability. Examples would be ABS braking systems and air bags. In many vehicles, the display lights on the instrument panel are used as part of this self-check procedure. In vehicles of this type, the electronic control system performs a self-check each time the ignition key is turned to the ON position.

The self-check makes sure that all of the bulbs, fuses, sensors, and electronic control modules are working. If the self-check finds a problem, it might store a fault code for later servicing. It may also instruct the computer to turn on a trouble light to show that service is needed. For instance, some systems run an instrument cluster self-test on start-up. It goes through a series of nine tests, called prove out, checking for function of the display, connectors to the sensors, and the condition of the electronic displays themselves. During prove out, all illuminated parts of the display are briefly lighted. Passing all parts of this test tells that the display is working properly. If portions of the display do not light, or if they remain lit after the test, there is a problem. The technician should check whether he/she has properly reinstalled all of the connectors to the back of the replacement display.

ON-BOARD COMPUTERS

On-board computers are now used to control numerous vehicle systems, such as the engine, brakes, transmission/transaxle, suspension, and ignition. These computers may share data and functions or they may act independently of one another based on the demands of the driver and prevailing road conditions.

As mentioned earlier, input devices called sensors act as the eyes and ears of the computer. Specifications for ideal sensor input are stored in the computer memory. The computer reads sensor input, compares it to the data in memory, and makes a decision based on this comparison. The computer then sends electrical signals to controlling output devices such as solenoids, switches, and relays. These output devices convert the signals into mechanical actions to adjust system operation.

The engine control module is normally the most powerful and complex of all on-board computers. It maintains a constant check on engine performance, making thousands of split-second decisions in response to input from a dozen or more sensors.

The computer itself is a highly reliable electronic device. Trouble within an electronically controlled system is more often caused by damaged wiring, connections, or sensors. The diagnostic flow charts in vehicle service manuals are used in conjunction with numerical fault codes to help technicians troubleshoot electronic systems in an organized and comprehensive manner. Following the fault code and chart sequence will ensure that only damaged components are replaced.

A fault code is recorded into the computer's memory whenever the system malfunctions. These numerical codes are normally read by plugging a special break-out box tester into the computer. The code numbers are then displayed on the tester's display screen. Since fault codes vary from model to model, the service manual must be consulted (Figure 9-93).

Electronic scan tools can also be used to read fault codes. When using a scan tool, be sure to select the proper cartridge for the vehicle make, model, and system being checked. Take the time to learn how to use the tool properly.

Before reading fault codes, do a visual check to make sure the problem is not a result of wear, loose connections, or faulty vacuum hoses. Inspect the air cleaner and throttle body or fuel injection system. Do not forget the PCV system and hoses and make sure the vapor canister is not flooded. Look at the wiring harnesses and connectors and the charging and alternator system; the connectors are especially important. The low level electrical signals in today's electronic circuits cannot tolerate the increased resistance caused by corrosion in connector contacts.

Fault codes are determined by counting the number of flash codes that appear. As an example, without starting the engine, turn the ignition key on and off three times within 5 seconds, ending with it on. The "Check Engine" or "Power Loss" light will glow a short time to test the bulb, then start flashing. Count the first set of flashes as tens. There will be a 1/2-second pause before the light starts flashing again. This time count by ones. Add the two sets of flash codes together to obtain the fault code. For example, 3 flashes, a 1/2-second pause, followed by 5 flashes is read as code 35. Watch carefully, because each fault code will be displayed only once.

Look up the fault code in the service manual. This will tell which circuit to check. Once the fault codes are flashed, the computer will signal a code 55. If the light does not flash at all, there are no fault codes stored. The problem could be a mechanical one, or the computer could be defective. Keep in mind that replacing the computer should be one of the last options. Check everything else first.

Fault codes are also used in diagnosing malfunctions in other automotive systems, such as anti-lock brakes. For example, during normal operation, the ABS warning lamp either blinks 7-10 times and then turns off or lights for 5-10 seconds and then turns off. When a

Code 12 No tach reference to ECM
Code 13 Oxygen sensor circuit
Code 14 Shorted coolant sensor circuit
Code 15 Open coolant sensor circuit
Code 21 Throttle position sensor circuit (1981-83)
Code 21 Throttle position sensor voltage high (1984-86)
Code 22 Throttle position sensor voltage low (1984-86)
Code 23 Open or ground MC solenoid (carburetted)
Codes 23 & 25 Manifold air temperature circuit voltage high (FI)
Code 24 Vehicle speed sensor circuit
Code 31 Wastegate solenoid
Code 32 Barometric or altitude sensor
Codes 33 &34 MAP or vacuum sensor circuit
Code 35 Idle speed or idle air control circuit shorted
Code 41 No distributor reference pulses to ECM
Code 42 EST or EST bypass circuit grounded or open
Code 43 ESC retard signal too long
Code 44 Air/fuel mixture lean
Codes 44 & 45 Faulty 02 sensor
Code 45 Air/fuel mixture rich
Code 51 Faulty PROM or installation
Code 52 Calpak missing
Code 53 EGR vacuum sensor vacuum incorrect

FIGURE 9-93 These are examples of fault codes, but be sure to always check the service manual for the specific vehicle being worked on.

fault has been detected, the lamp lights and remains on. To enter the ABS fault code diagnostic mode on GM vehicles, it is necessary to jump ALDL terminals H and A (ground). To enter the ABS fault code diagnostic mode on Ford vehicles, turn the ignition to the ON position. Consult the appropriate service manual for other vehicle makes.

PRINTED CIRCUITS

Many modern electronic components use printed circuit boards. A printed circuit board is made of thin sheets of nonconductive plastic onto which conductive metal such as copper has been deposited. Parts of the metal are then etched or eaten away by acid. The remaining metal lines form the conductors for the various circuits on the board. A printed circuit board can hold many complex circuits in a very small area. The circuit board can be connected to a power supply or ground wiring through the use of plug-in connectors mounted on the circuit board. Printed circuits also are very reliable.

There are several precautions that should be observed when working on a printed circuit.

1. Never touch the surface of the board. Dirt, salt, and acids on your fingers can etch the surface and set up a resistive condition. It is possible to knock out an entire circuit in the instrument panel with a fingerprint.
2. The copper conductors can be cleaned with a commercial cleaner or by lightly rubbing an eraser across the surface.
3. The printed circuit board is damaged very easily because it is very thin. Be careful when plugging in bulbs not to tear the surface. Handle printed circuits with care.

ELECTRONIC THEFT DETERRENT SYSTEMS

Automotive electronic theft deterrent systems (Figure 9-94) may go by many names according to each manufacturer's preference. They may be called a keyless entry system, alarm system, theft alarm system, anti-theft protection system, and so on. But they all have the same purpose: prevent theft of the vehicle and its contents, while at the same time possibly lowering the insurance premium.

Once a popular aftermarket item, theft deterrent systems are fast becoming standard equipment on many automobiles, and options on others. As is the case with other automotive electronic systems, computers control theft deterrent system activities. As would be expected from a computer-controlled system, the theft deterrent system uses sensors and relays to function. In many

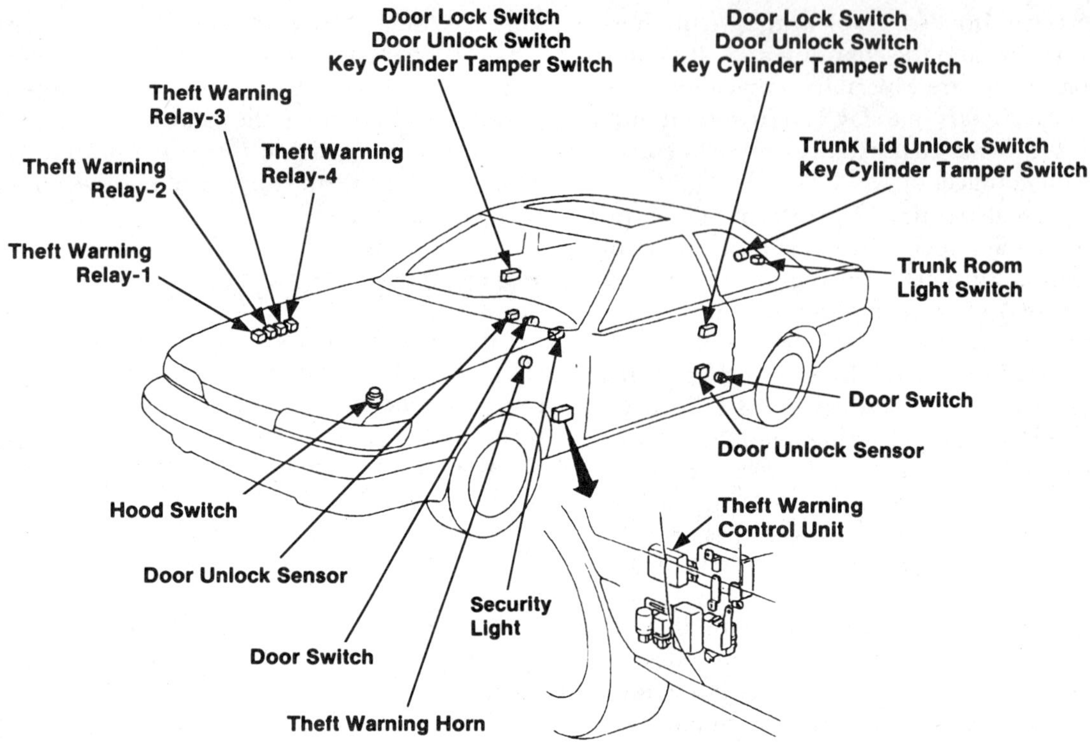

FIGURE 9-94 Nissan's electronic theft deterrent system blinks the headlights, blows the horn, and cuts off the starter motor if someone tries entering without a key or tampers with the lock cylinders.

cases, the sensors are tied in with door locks, trunk locks, and the starter circuit. The theft deterrent system requires some care when handling the components.

Activating the theft deterrent system may disable the ignition and/or sound a horn and flash the courtesy lights to scare the would-be burglar. Since the various locks are located in vulnerable positions, chances are good that they will be damaged in an accident. Once the components are restored to their pre-accident condition, system operation can be checked with a DVOM and jumper wires. These two tools will also aid in diagnosing problems in the system.

PROTECTING ELECTRONIC SYSTEMS

The last thing a technician wants to do when a vehicle comes into the shop for collision repair is create problems. This is especially true when it comes to electrical systems and electronic components. It is a must to be aware of the proper ways to protect automotive electrical systems and electronic components during storage and repair. They are as follows:

- Disconnect the battery cables (negative cable first) before doing any kind of welding. To avoid the possibility of explosion, completely remove the battery when welding under the hood or on the front end. Also, make sure the ground connection is clean and tight. Position the ground clamp as close as possible to the work area to avoid current seeking its own ground.

SHOP TALK

Whenever disconnecting or removing the battery on a computer-controlled vehicle, remember that the memory for radio station selection, seat position, climate control setting, and any other "driver programmable" options is erased. When delivering the vehicle, advise the customer that he or she must reprogram these settings. Or better yet, record them before beginning and reprogram them yourself before delivery.

- Static electricity can cause problems. Avoid it by grounding yourself before handling new displays. One way is to touch a good ground with one hand before handling the display with the other hand. Another way is to use a grounding strap that attaches to the wrist and then to the vehicle.
- Avoid touching bare metal contacts. Oils from the skin can cause corrosion and poor contacts.

- Be careful about the placement of welding cables. Keep the electrical path as short as possible by placing the ground clamp near the point of welding. Also, do not let the welding cables run close to electronic displays or computers.
- Take care when handling electronic displays and gauges. Never press on the gauge face because this could damage it.
- If a fault code indicates a problem with the oxygen sensor, extra caution is required. The oxygen sensor wire carries a very low voltage and must be isolated from other wires. If not, nearby wires could add more induced voltage. This gives false data to the computer and can result in a driveability problem. Some manufacturers use a foam sleeve around the oxygen sensor wire to keep it separate and insulated from other wires.
- The sensor wires that connect to the computer should never be rerouted. The resulting problem might be impossible to find. When replacing sensor wiring, always check the service manual and follow the routing instructions.
- Remove any computer which could be affected by welding, hammering, grinding, sanding, or metal straightening. Be sure to protect the removed computer by wrapping it in a plastic bag to shield it from moisture and dust.
- Be careful not to damage wiring when welding, hammering, or grinding.
- Be careful not to damage connectors and terminals when removing electronic components. Some may require special tools to remove them.
- Always route wiring in the same place it was originally. If not, electronic crossover from the current-carrying wires can affect the sensing and control circuits. Reuse or replace all electrical shielding for the same reason.

REVIEW QUESTIONS

1. Who has the authority to decide whether to remove the drivetrain or work around it?
   a. estimator
   b. collision repair technician
   c. both a and b
   d. neither a nor b

2. In unibody construction, what provides the critical mounting positions for the suspension and steering systems?
   a. body panels
   b. drivetrain
   c. upper and lower control arms
   d. cradle assembly mounting biscuits

3. The strut replaces the shock absorber and the ______________.
   a. lower ball joint
   b. lower control arm
   c. upper control arm
   d. all of the above

4. Upon examining a collapsed steering column, Technician A decides to attempt a repair of the collapsed portion. Technician B says the collapsed portion must be replaced. Who is correct?
   a. Technician A
   b. Technician B
   c. Both A and B
   d. Neither A nor B

5. What is the definition of camber?
   a. The forward or backward tilt of the steering axis
   b. The distance between the centerline of the ball joints and the centerline of the tire at the point where the tire contacts the road surface
   c. The amount of toe-out present on turns
   d. The tilt of either the front or rear wheels inward or outward from the vertical as viewed from the front of the vehicle.

6. To determine whether an improper toe condition is caused by poor wheel alignment or damaged suspension components, Technician A makes a turning radius check; Technician B makes a jounce-rebound check. Who is correct?
   a. Technician A
   b. Technician B
   c. Both A and B
   d. Neither A nor B

7. Which of the following computer suspension systems is an active system?
   a. SES
   b. CCS
   c. both a and b
   d. neither a nor b

8. Which of the following statements concerning brake systems is incorrect?
   a. Front-wheel drive vehicles now have a dual diagonal split hydraulic brake system.
   b. The major brake component that a collision repair technician must deal with are the brake lines.
   c. Brake lines need not be replaced in their original routing.
   d. The master cylinder initiates braking when the brake pedal is depressed.

9. When bleeding a four-piston caliper with two bleed screws, Technician A bleeds the lower one first, while Technician B starts with the upper. Who is correct?
   a. Technician A
   b. Technician B
   c. Both A and B
   d. Neither A nor B

10. Which emission control subsystem is responsible for channeling blowby gases into the fuel intake area?
    a. engine control
    b. positive crankcase ventilation
    c. evaporative
    d. exhaust gas recirculation

11. What is the engine's temperature control?
    a. thermostat
    b. radiator cap
    c. cooling fan
    d. coolant

12. When evacuating an A/C system that has been open to the atmosphere, Technician A holds the system at high vacuum for 10 minutes. Technician B holds it at high vacuum for at least 30 minutes. Who is correct?
    a. Technician A
    b. Technician B
    c. Both A and B
    d. Neither A nor B

13. Which of the following tools are used to locate electrical faults?
    a. jumper wires
    b. test light
    c. DVOM
    d. all of the above

14. A ____________ is a short length of small gauge wire installed in a conductor.
    a. fuse
    b. fuse link
    c. maxi-fuse
    d. printed circuit

15. Which type of electronic display is both durable and bright?
    a. vacuum fluorescent
    b. light-emitting diode
    c. liquid crystal diode
    d. liquid fluorescent

CHAPTER

10 Restoring Corrosion Protection

OBJECTIVES

- Define corrosion and describe the common factors in rust formation.
- Identify the principal methods of corrosion protection used on unibody vehicles.
- Describe the anti-corrosion materials used to prevent and retard rust formation.
- Choose the correct anti-corrosion materials and equipment for specific applications.
- Outline the correct corrosion treatment procedures for each of the four general corrosion treatment areas.
- List the four types of seam sealers and explain where each should be used.
- Describe the problems of corrosion caused by dissimilar metal body trim and accessories, and explain how to deal with them.

The automobile industry has changed dramatically in the last few years. The coming of the unibody vehicle has been responsible for a great deal of this change. For instance, in unibody construction, the vehicle's body panels are no longer just cosmetic sheet metal. They now constitute the structural integrity of the vehicle. This means that rust is not just an eyesore. The unibody vehicle has more welded joints in critical structural areas where corrosion can do serious damage (Figure 10-1). Corrosion is an ever present danger to the unibody vehicle since rusting of the thin-gauge structural sheet metal panels can affect the driveability of the vehicle and the safety of its passengers.

This has resulted in a change in the meaning of the words *corrosion protection* to the automotive industry. Examples of this change include the following:

- Corrosion protection is the phrase that is replacing rustproofing, undercoating, and sound deadening. Corrosion protection implies a vehicle lifetime maintenance responsibility for the consumer. Rustproofing, undercoating, and sound deadening had suggested a one-time application to the new vehicle by automobile dealerships and rustproofing franchises.
- Auto manufacturers are including in their collision repair manuals recommendations for sheet metal repair or replacement. They suggest that the collision repair shop should apply an anti-corrosion material to the part repaired or replaced so that corrosion protection is restored.
- In affiliation with I-CAR, the insurance companies and the collision repair industry are promoting corrosion protection restoration.
- The increased usage of replacement panels in the collision repair shop requires widespread corrosion protection treatment.
- The advent of the unibody vehicle has made corrosion protection a must.

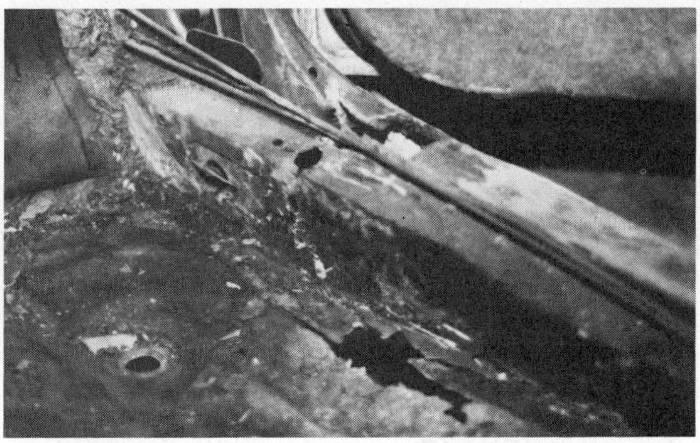

FIGURE 10-1 This rocker panel and floor pan have been damaged beyond repair by widespread corrosion.

Manufacturers have responded to this challenge with corrosion protection measures of dramatic effectiveness. Collision repair shops must similarly use proper materials and techniques to ensure that the repair area resists corrosion as effectively as the original part.

WHAT IS CORROSION?

Corrosion of metal involves the combining of the metal with oxygen to form a metal oxide, or rust (Figure 10-2). This helps explain why another name for corrosion is oxidation. Examples of oxidation are:

- Iron + Oxygen = Iron Oxide
- Zinc + Oxygen = Zinc Oxide
- Aluminum + Oxygen = Aluminum Oxide

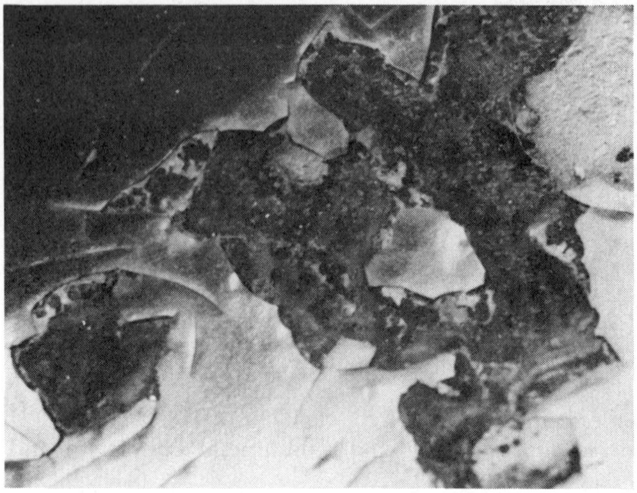

FIGURE 10-2 Rust never sleeps.

Oxidation occurs in two steps. First, the stable metal and oxygen atoms break down into unstable ions with positive and negative charges. Then these unstable ions seek stability by combining with each other to form metal oxides (Figure 10-3).

There is one more essential ingredient, since oxidation is an electrochemical reaction. That ingredient is an electrolyte, which helps the formation of ions and the flow of the positively and negatively charged particles. In the typical environment of vehicles, moisture provides this third essential ingredient, even more so with the addition of salt, acids, and other pollutants. When iron is subjected to an electrolyte, the stable iron atoms break down into unstable iron ions with a positive charge (Fe++). At the same time, hydrogen and oxygen in water (H_2O) and oxygen (O_2) break down into hydroxyl ions (OH−). Since positives and negatives attract each other, these unstable ions combine to form iron red oxide (Fe_2O_3)—rust (Figure 10-4).

There are only two practical ways to protect auto body steel against corrosion. One is to seal off the atmosphere—the oxygen and moisture—from its surfaces. The other is to put it close to a metal, such as zinc, that is more chemically active. This provides protection through a process called *sacrificial corrosion*. The sacrificial metal corrodes instead of the steel. Another way

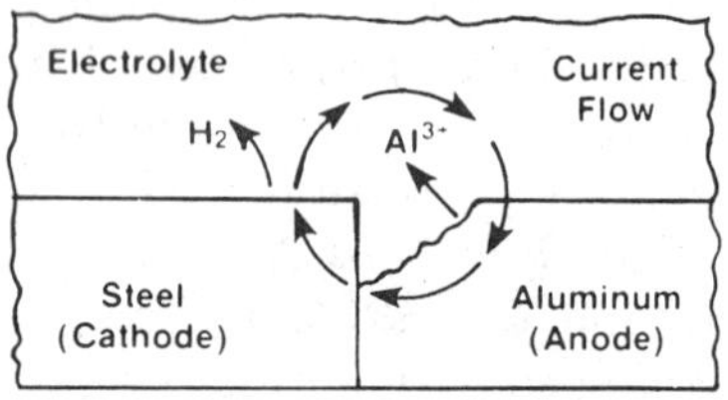

Galvanic or Bimetallic Corrosion
Galvanic corrosion occurs when two metals of dissimilar activity are placed in contact with one another in the presence of an electrolyte. The more active metal will anodize and corrode in preference to the less active metal which, as the cathode, is protected.

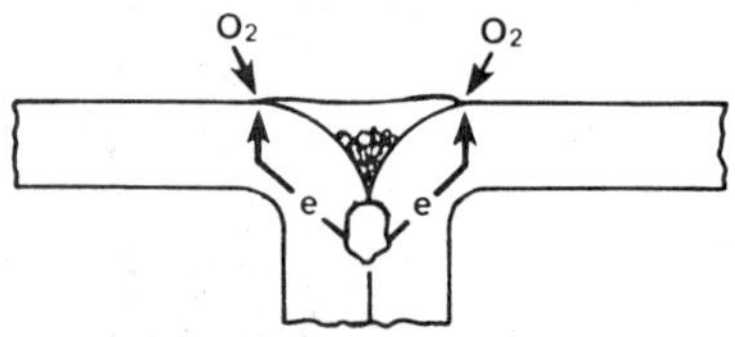

Crevice Corrosion at Joint
Crevice corrosion is caused by a gradient between the oxygen at the surface of the electrolyte and oxygen-starved electrolyte at the bottom of the crevice. Typical of weldments, sheet-metal joints, and rough surfaces where water might be trapped, the oxygen gradient also causes a rough microfinished surface to corrode faster than a smooth surface.

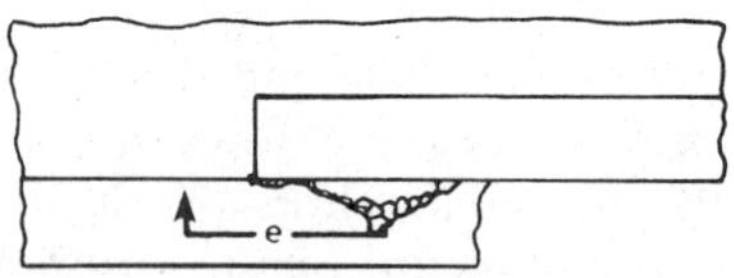

Crevice Corrosion at Lap Joint
Crevice corrosion also can occur between tightly sealed joints where the concealed metal surface is oxygen starved and electrolyte might seep between irregularities in the mating surfaces. This condition also occurs where moisture-bearing materials (such as felt) are in contact with the steel.

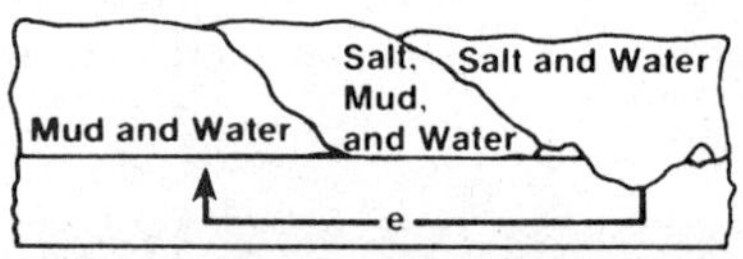

Poultice Corrosion
Electrolyte composition gradients are probably the most common cause of corrosion. Clumps of mud frequently collect under car fenders. The varying concentrations of salt and water encourage corrosion.

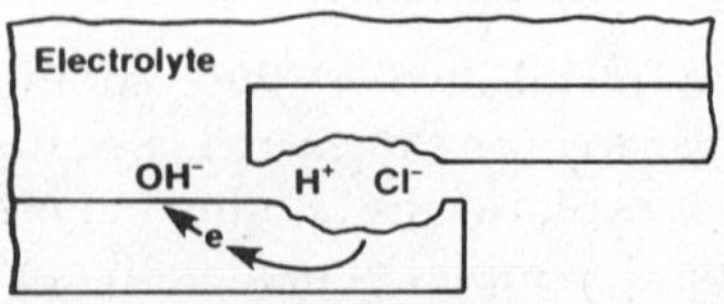

Crevice Corrosion at Lap Joint
Crevice corrosion showing acid formation and increased chloride ion concentration within the crevice.

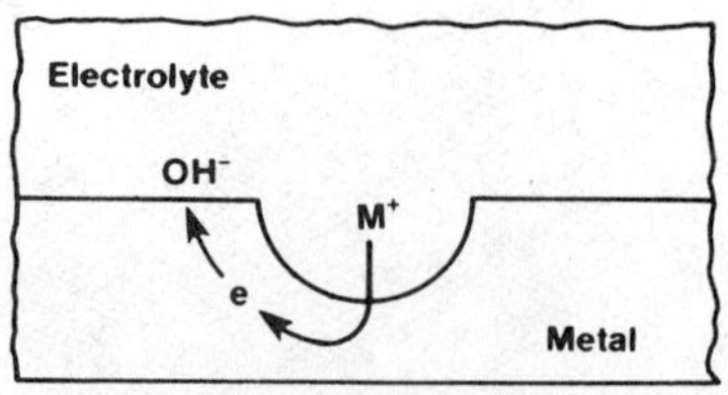

Pitting Corrosion
Similar to crevice corrosion, pitting corrosion occurs at localized areas where oxygen has been depleted, pH has become lowered and chloride has become enriched.

FIGURE 10-3 The many different kinds of corrosion

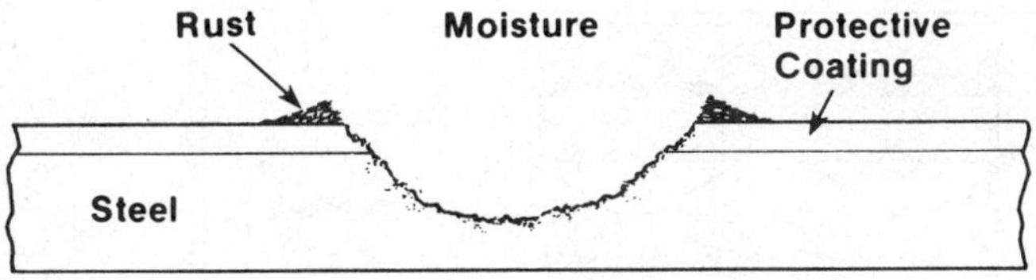

FIGURE 10-4 How rust is formed

to say this is that zinc provides galvanic protection to steel. *Galvanic* refers to the electrochemical process involved. All measures taken at the factory to prevent corrosion involve one of these two methods. These same principles must be applied during collision repair.

CORROSION PROTECTION BY AUTOMAKERS

The fight against vehicle corrosion really began in earnest in the early 1960s with the introduction of the anodic bath electrodeposition rust-inhibiting primer. This process involved three steps. First, a phosphate bath was used to remove surface rust, machine oils, and the like. Then the vehicle body was submerged in electrically charged primer to draw the primer deep into cracks and crevices. Finally, a topcoat of additional primer was applied. Some of these earlier vehicles, however, were only "slipper dipped" in primer baths up to their door pillars. Anodic dip processes ultimately proved less effective than needed because of complex chemical reasons.

In 1986, cathodic electrodeposition or "E-coat" became the automotive industry's first line of defense against corrosion. At about the same time, the development of small crystal phosphate occurred. This provided much better subcoat adhesion. The E-coat made possible the electrodeposition of tough epoxy primers, known to be aggressive corrosion fighters. Like the earlier anodic process, cathodic electrodeposition submerges auto bodies in paint baths, but the results are much more satisfactory (Figure 10-5). The epoxy primer sticks to sheet metal far better than previously used primer materials. It also has a much better ability to climb into remote body cavities such as window pillars. The E-coat has proven to be so tough, it is virtually immune to attack by rust during aggressive salt spray tests.

Another big advantage of cathodic electrodeposition is that it lays the primer on thickly. The thickness of E-coats is about 1.2 mils, compared to 0.6 mil for the anodic method. This added thickness helps cover minor surface imperfections in the metal.

FIGURE 10-5 Cathodic electrodeposition is an effective corrosion fighter.

During recent years, manufacturers have increasingly begun to use galvanized or zinc-coated sheet steel, further contributing to the corrosion resistance of automobiles (Figure 10-6). This zinc coating can be applied in a variety of ways, including hot dipping and electroplating. Additionally, a variety of techniques has been developed for applying a zinc-rich compound during the mill rolling process initially used to produce the raw sheet metal stock. Some of the more popular zinc coatings used by auto manufacturers are:

- *One-Sided Galvanized.* Low carbon steel with a zinc galvanized coating on one side applied by the hot dipping process.
- *One-and-One-Half-Sided Galvanized.* Steel in which the unexposed side has a zinc coating and the exposed side has a lighter alloy-coated surface.
- *Two-Sided Galvanized.* Steel coated evenly with zinc on both sides.
- *Two-Sided Galvannealed.* Steel coated on both sides with zinc and heat treated so the coating is fully alloyed (mixed) with the base metal.
- *Zincrometal.* Low carbon steel coated with a zinc-rich primer coating.
- *Electrogalvanized.* Steel which is zinc coated by electroplating.

The heavy use of zinc in late-model vehicles is perhaps the single most important change in the corrosion protection of unibody vehicles. Many vehicles today have every part coated with zinc in some way. Regardless of how the zinc is applied—by hot dipping, electroplating, or painting—it offers three kinds of protection. The first, of course, is the same as any coating: it

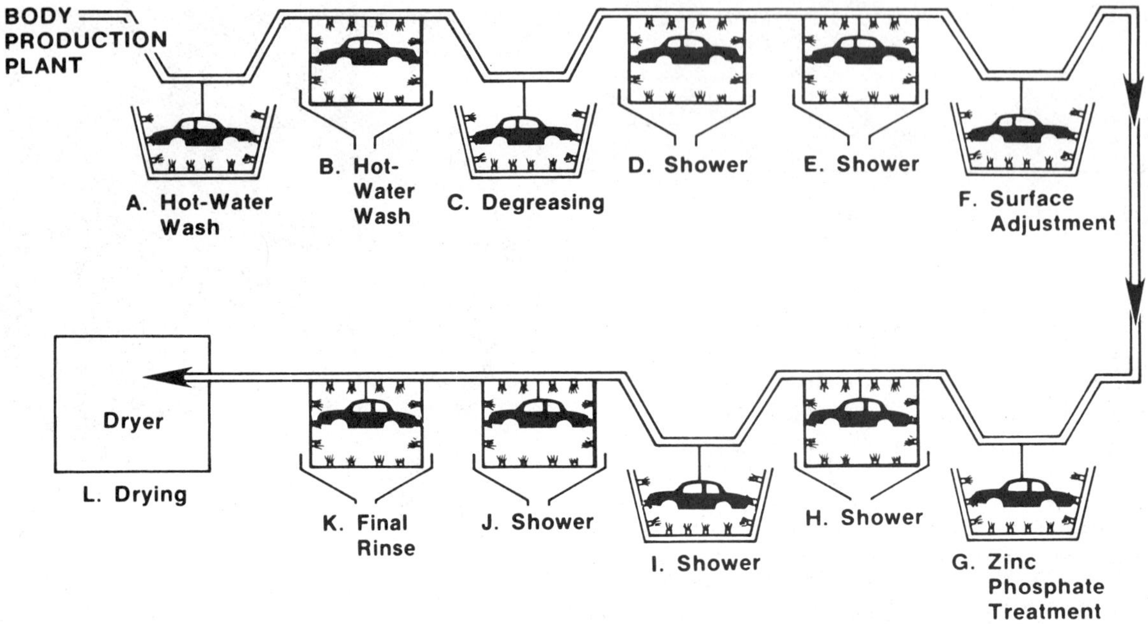

FIGURE 10-6 The galvanizing process: (A) and (B) metal chips, dirt, and other foreign matter are removed with hot water; (C) oil is removed with a weak alkali degreasing agent; (D) and (E) the degreasing agent is washed off with water; (F) nucleus of zinc phosphate film is applied to panel surfaces; (G) body is dipped into zinc phosphate tank for crystallization; (H), (I), and (J) zinc phosphate liquid is washed off with water; (K) final rinse to prevent blistering; and (L) drying

seals off the atmosphere. The second is galvanic. Because it is chemically more active than steel, it sacrifices itself to protect the steel. The third is the coating effect of the zinc oxide formed during galvanic protection. When the original coating is damaged, such as in a nick or chip, the zinc undergoes sacrificial corrosion. As it does, it forms zinc oxide. Zinc oxide is a dense white coating that creeps over the surface of the steel exposed by the nick and seals over it. In a very real sense, the zinc coating repairs itself.

When dealing with zinc coating during collision repair, there are two important things to remember:

1. If at all possible, do not remove it.
2. If it must be removed, replace it.

Preserving and restoring the zinc protection as described later in this chapter is one of the most important parts of a successful unibody collision repair.

OTHER FACTORY PROTECTION

There are two other basic types of corrosion protection that are used by automakers to protect their vehicles:

- Paint
- Anti-corrosion compounds

A paint system such as those described in later chapters will provide a barrier between the atmosphere and the steel surface (Figure 10-7). When this barrier is in place, the moisture and impurities in the air cannot interact with the steel surface and the steel is protected from corrosion. If the paint surface or barrier is broken by a stone chip or scratch, the steel

FIGURE 10-7 A properly applied paint system is an effective barrier against corrosion.

in this area is no longer isolated from the moisture and impurities in the air. Corrosion will then take place in this region. Corrosion will spread between the paint and steel surface. If the adhesion of the paint to the steel is poor, large sections of the paint can be separated from the steel. This will result in a large area of the steel left unprotected, and severe rust will quickly follow. If impurities are present between the paint and the steel, then oxygen in the air can pass through the paint, reacting with the impurities and the steel to form rust. In this case, corrosion will take place on the steel surface and the protective paint barrier will be destroyed. Paint, by itself, is only effective as long as the paint film remains intact.

Anti-corrosion compounds are additional coatings applied over the paint film. These protective coatings can be applied either by the manufacturer or as an aftermarket process. The two most popular types of anti-corrosion compounds are:

- Petroleum-based compounds
- Wax-based compounds

Anti-corrosion compounds are primarily used in enclosed body sections (Figure 10-8) and other rust-prone areas.

All of these recent developments have made five-year and longer corrosion protection warranties standard in the industry. Of course, new vehicle corrosion warranties are changing. They are getting better and better as manufacturers implement more and more corrosion protection measures at the factory. Maintaining these warranties depends upon repair work performed according to strict guidelines and professional standards. Factory parts and recommended repair procedures must be used to protect the warranty.

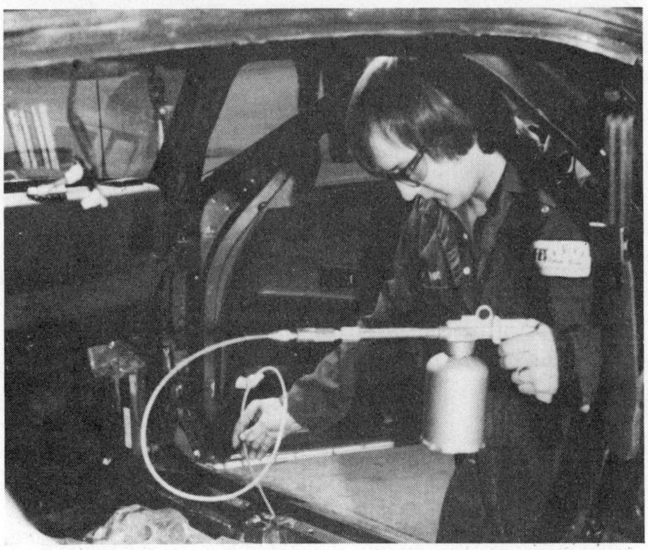

FIGURE 10-8 Enclosed body sections require the use of anti-corrosion compounds.

CAUSES FOR LOSS OF FACTORY PROTECTION

Even with all of the care taken to protect vehicles, corrosion protection breakdown still occurs. The breakdown falls into three general categories:

- Paint film failure
- Collision damage
- Repair process

The paint film is the result of the entire process of coatings, primers, and color coats that the manufacturer applies. When the paint film fails, corrosion begins. Stone chips, moisture, and improper surface preparation can all lead to paint film failure.

During a collision, the protective coatings on a vehicle are damaged (Figure 10-9). This occurs not just in the areas of direct impact, but also in the indirect damage zones. Seams pull apart, caulking breaks loose, and paint chips and flakes. Locating the damage and restoring the protection to all affected areas remains a key challenge for the collision repair technician.

Vehicle repair is often one of the major causes of protective coating damage. For example, repair procedures often require cutting body panels and seams either mechanically or with a plasma arc cutter. Even minor straightening and stress relieving procedures can damage these protective coatings and start corrosion. Welding temperatures cause zinc to vaporize and be lost from the weld area. Abrasive operations during repair and refinishing can also leave areas unprotected. After all welding and repair work has been completed, these damage points need careful attention to eliminate con-

FIGURE 10-9 Corrosion protection is usually affected during a collision.

taminants. Then steps must be taken to shield the metal from the atmosphere by sealing all surfaces thoroughly.

Other cautions that should be taken to preserve the factory corrosion protection are:

- Remove only the minimum amount of paint film from damaged areas.
- Be extremely careful not to scratch any part except that to be repaired. If there is an accidental scratch, take necessary corrective measures.
- When clamping or holding the affected panels during repairs, clamping tools can cause scratches on the panel. These areas must be treated to avoid rusting.
- While grinding (Figure 10-10), cutting, or welding panels, place protective covers over adjacent surfaces to protect them from the flame, sparks, or metal chips.
- Cover any opening of the body sills and similar areas with masking tape to prevent metal chips from entering during the grinding, cutting, or welding operation.
- Completely remove any metal chips from inside the vehicle body. Use a vacuum cleaner, not dry compressed air, to remove metal chips. If dry compressed air is used, metal chips can accumulate in corner areas.

Certain environmental and atmospheric conditions help to influence the rate of corrosion. They are:

- *Moisture.* As the amount of sand, dirt, and water on the underside of the vehicle body increases, so will the chances of corrosion accelerate. Floor sections that have snow and ice trapped under the floor matting will not dry. Likewise, if holes at the bottom of the doors and side sills are not kept open, water will accumulate. Remember, moisture is one of the requirements for rust.
- *Relative Humidity.* Corrosion will be accelerated in areas of high relative humidity, especially those areas where temperatures stay above freezing and where atmospheric pollution exists.
- *Temperature.* A temperature increase will accelerate the rate of corrosion to those parts that are not well ventilated.
- *Air Pollution.* Industrial pollution (acid rain), the salt air in coastal areas, or the increased use of corrosive road salt will accelerate the corrosion process. Road salt will also accelerate the disintegration of paint surfaces.

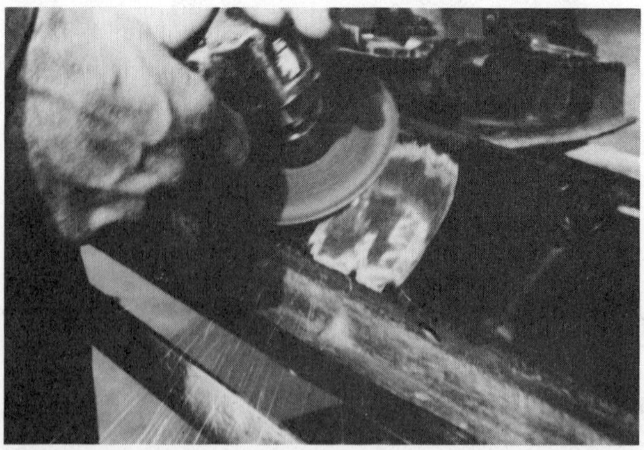

FIGURE 10-10 Be sure surrounding areas are protected when grinding panels.

SHOP TALK

Old rules of protecting a vehicle in the winter, like weekly washes and parking in a warm garage, may do more harm than good, according to many auto body experts. In most automatic car washes, wash water is recirculated from a reservoir beneath the line. That means the vehicle is washed with the same water that rinsed off the previous vehicle. In the winter, the wash water forms a strong salt solution that is sprayed on each vehicle that goes through the wash. Not only do you get salty water over the paint, but a salt solution is sprayed underneath to settle in all the nooks and crannies where the rust takes hold. As just stated, temperature increase is a rust accelerator. So a warm garage, instead of protecting the vehicle, actually speeds up the rusting process. In fact, it is estimated that regular—daily or nightly—parking of a snow-encrusted vehicle in a warm garage can accelerate the rusting process by a factor of three. Warm in this case is any temperature above freezing.

- *Acid Rain.* Acid rain is the term given to rain containing effluents from manufacturing and chemical industries. It causes discoloration and even destruction of the paint film that could lead to corrosion damage. As shown in Figure 10-11, the higher the pH number the less chance of acid rain problems. Any condition below a 6.0 pH reading is considered acid. Acid rain repairs are covered in Chapter 14.

Another type of corrosion that must be considered when working on automobiles is known as galvanic corrosion. As explained earlier, this occurs when two dissimilar metals are placed in contact with each other.

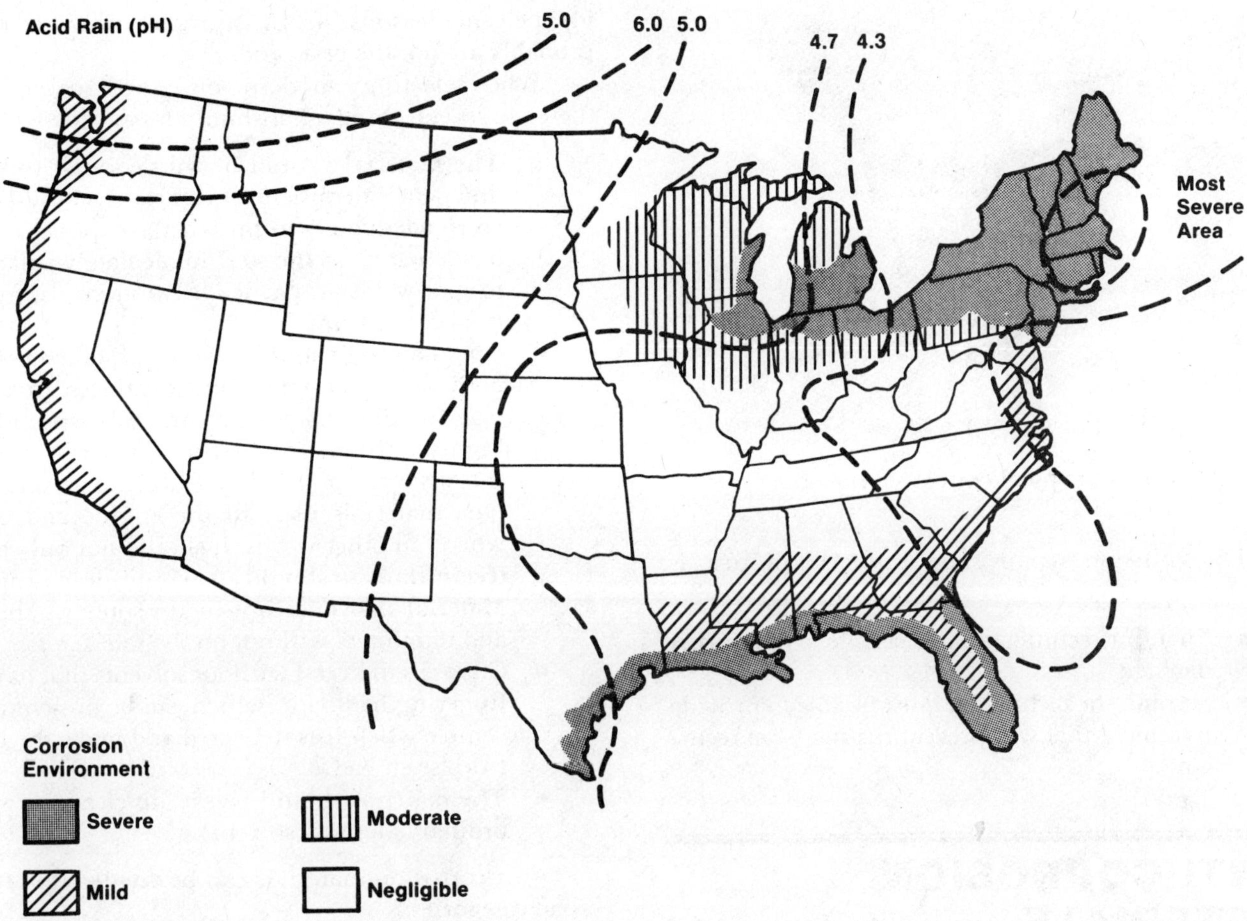

FIGURE 10-11 Acid rain has become a factor in much of the United States.

The more chemically active of the two will corrode, protecting the other metal in the process. As shown in Table 10-1, this is why zinc (a metal of high galvanic activity) will sacrifice itself to protect steel. In the case of other metals, as mentioned later in the chapter, galvanic corrosion can cause problems.

TABLE 10-1: RELATIVE GALVANIC ACTIVITY OF METAL

Magnesium	Most active
Aluminum	↑
Zinc	
Chromium	
Iron	
Cadmium	
Cobalt	
Nickel	
Tin	
Lead	↓
Copper	Least active

Regardless of the cause, if corrosion prevention is not practiced, the cost to the collision repair shop and insurance company is comebacks and lost customers. Inadequate preparation that leaves dirt, grease, or acids on the metal will cause the loss of adhesion. Rust will start, a little at first, creating corrosive "hot spots" at the points of failure (Figure 10-12). That is, a piece of unprotected steel corrodes at a rate that is approximately uniform throughout its surface. If part of the piece is protected, either by a paint film or a galvanizing layer of zinc, something different happens. The corrosive action becomes concentrated in the areas of the surface that are not protected. These areas then rust much faster than if the whole piece was uniformly unprotected.

When repairing a vehicle that has the original factory finish and corrosion protection intact, the technician is working in a potential area of a corrosive hot spot. If the rust protection is not restored, the corrosive reactions will be focused in this area like a spotlight focuses on its target. The result will be corrosive activity so severe that complete rust-through can result in a matter of a few months' time.

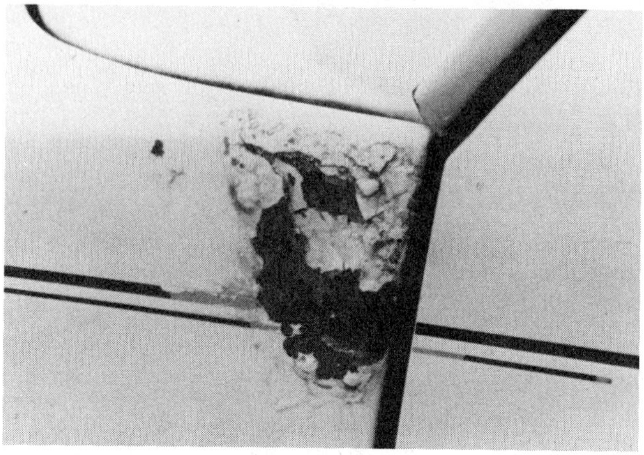

FIGURE 10-12 Typical corrosive "hot spot"

The collision repair shop's interest in rust is twofold:

- First, the technician must be able to repair rust damage.
- Second, the technician must be able to provide treatment that will prevent rusting from recurring.

ANTI-CORROSION MATERIALS

The collision repair shop's efforts in protecting vehicle bodies from rusting should focus on creating a clean, chemically neutral surface on the sheet metal, then sealing the surface under layers of paint. Under certain conditions, as mentioned earlier in the chapter, a wax- or petroleum-based anti-corrosion compound is used to exclude air and moisture from the metal surface. More and more new vehicles come off the assembly line today with anti-corrosion materials that are available to the collision repair shop. Being able to replace or install these materials is very important.

Corrosion prevention has not always been a popular collision repair shop operation. The original rustproofing was called undercoating, and it was an asphalt-based product that was sheer agony to apply, because it got not only on the underside of the vehicle, but also on everything else within 20 feet of the application bay. But, worst of all, it did not work very well. In time, the solvents would evaporate, the asphalt would harden and crack, and the moisture that causes oxidation would actually become trapped under the undercoating.

Asphalt-based undercoats do have benefits in terms of sound deadening and preventing stone marks under fenders. And they are useful today on fiberglass panels for the same reasons. As a rustproofer, however, they probably are not the best product.

When selecting a modern anti-corrosion material, there are several things that should be considered:

- The material should be thin enough to flow and penetrate pinchweld cracks. It should also creep adequately to protect the exposed metal of such areas as the steel immediately adjacent to spot welds—a particularly tough rustproofing proposition.
- The material should have good adherence to both bare metal and painted surfaces. In addition to adhering to the surface, it should be highly resistant to water, cutting from stones thrown up from the road, and ordinary solvent-type materials used in the engine and elsewhere. In other words, it should not only protect initially, it should also continue to protect. Material that does not retain some pliability and toughness will not do the job.
- Choose a material without solvents that have a lingering bad odor, which can be present in a vehicle when it is delivered and make the best work seem bad.
- The material should be easy to clean up with ordinary and safe solvents.

Anti-corrosion materials can be divided into four broad categories:

- *Anti-corrosion Compounds.* As already mentioned, either wax- or petroleum-based anti-corrosion compounds are resistant to chipping and abrasion; they can undercoat, deaden sound, and completely seal the surface of a vehicle from the destructive results of rust and corrosion. They should be applied to the undercarriage and inside body panels so that they can penetrate into joints and body crevices to form a pliable, protective film.

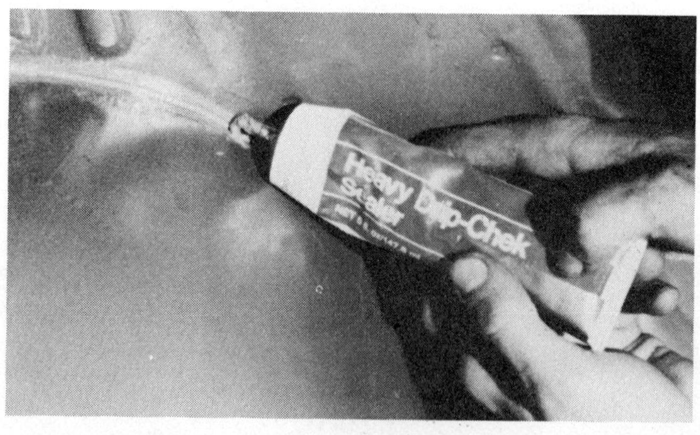

FIGURE 10-13 Applying a seam sealer

FIGURE 10-14 Applying weld-through primer

- *Seam Sealers (Figure 10-13).* These prevent the penetration of water or mud into panel joints and serve the important role of preventing rust from forming between adjoining surfaces.
- *Weld-through Primers (Figure 10-14).* Weld-through primers are used where it is difficult to obtain coverage with anti-corrosion compounds. An example would be the area between the two pieces of base metal at a weld joint.
- *Rust Converters.* These chemicals convert ferrous (red) iron oxide to ferric (black/ blue) iron oxide. Rust converters also contain some type of latex emulsion which seals the surface after the conversion is complete. These products offer an interesting alternative for areas that cannot be completely cleaned of rust, but should not be considered a replacement for the products and procedures recommended in this course. Rust converters are only effective on tight, well-bonded rust, not loose, scaled rust.

SHOP TALK

Be sure to carefully read and follow the manufacturer's instructions on the container. Several of these anti-corrosion materials can be used on the same part or section.

New and better anti-corrosion materials are constantly appearing on the automotive market. But it must be remembered that the backbone of any serious effort to improve products and services is careful research. Research helps to separate ideas that seem to work from those that really do. Research helps to predict the likelihood of success for a specific approach.

Manufacturers, as already mentioned, have always been interested in corrosion protection of their new vehicles. Typically auto manufacturers test their products with a daily cycle as follows:

- The vehicle is driven several miles on a gravel road to promote stone chips.
- Then the vehicle is subjected to a salt spray.
- Next, it is again driven on the gravel road.
- The vehicle is then exposed to 100 degrees Fahrenheit at 100 percent relative humidity for several hours.
- Then the vehicle is allowed to "rest" for 12 hours at ambient temperatures.

To pass the corrosion test, new vehicles should withstand 50 to 100 cycles of this punishment before rust perforation occurs.

During the past few years, I-CAR has assumed a leadership position in conducting research on unibody collision repair. Through its research, I-CAR has become a source of "state-of-the-art" information on corrosion protection for the collision repair and insurance industries. This kind of testing and proven information can give the collision repair technician confidence that, if the recommendations given in this chapter are followed, the repair work will meet the exacting requirements of modern unibody repair.

In the I-CAR corrosion test, collision damaged vehicles, after repair, are tested in a specially made corrosion chamber (Figure 10-15) as follows:

1. Salt spray is applied four times each day at hourly intervals.
2. This is followed by drying and rest at room temperature.

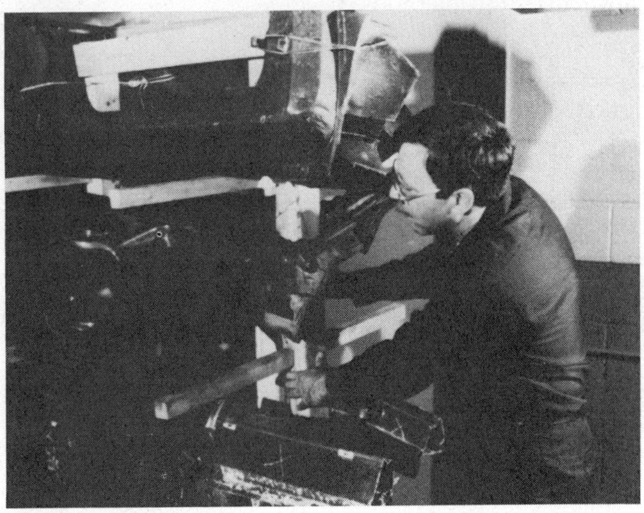

FIGURE 10-15 Use of the corrosion chamber allows I-CAR research technicians to approximate five years of on-the-road use on a vehicle in just eight weeks.

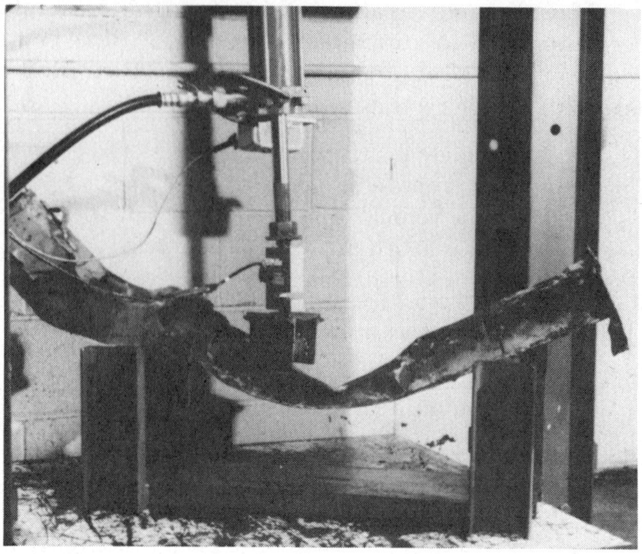

FIGURE 10-16 Strength testing is accomplished through both bending and compression.

3. Each day the vehicle is exposed to 105 degrees Fahrenheit at 100 percent humidity for eight hours.
4. This is followed by a 12-hour rest at room temperature.

This torturous sequence is kept up for eight weeks—approximating five years of on-the-road use. The corrosion chamber test is followed by strength testing by bending and compression (Figure 10-16). This test sequence demonstrates that the repair methods and materials used in collision repair play a key role in corrosion protection. Consult I-CAR *Advantage* bulletins and other trade publications for updated information on anti-corrosion methods.

BASIC SURFACE PREPARATION

Surface preparation is one of the most important steps in assuring long-term corrosion protection of body panels and other metal parts. Without the proper surface preparation (especially on bare metal), the rest of the repair procedure and refinishing efforts will be wasted. The conventional system generally consists of the following three-step process called metal treating.

- *Cleaning to Remove Contaminants.* Use a wax and grease remover to dissolve and float off oil, grease, and other surface contaminants. Apply the wax and grease remover with a clean, white cloth (Figure 10-17). Work small areas, 2 to 3 square feet, wetting the surface liberally and keeping it wet. While the surface is still wet, fold a second cloth and wipe the surface to remove the contaminants. Turn the cloth frequently while drying the surface.
- *Cleaning with Metal Conditioner.* The purpose of the metal conditioner is to deep clean and etch the metal. Dilute the metal conditioner with water in a plastic bucket according to the label instructions. Apply it to the metal—a spray bottle is recommended (Figure 10-18). Then rinse with clear water and wipe dry with a clean cloth.
- *Applying Conversion Coating.* The conversion coating forms a zinc phosphate coating that is chemically bonded to the metal. This layer makes an ideal surface for the primer and pre-

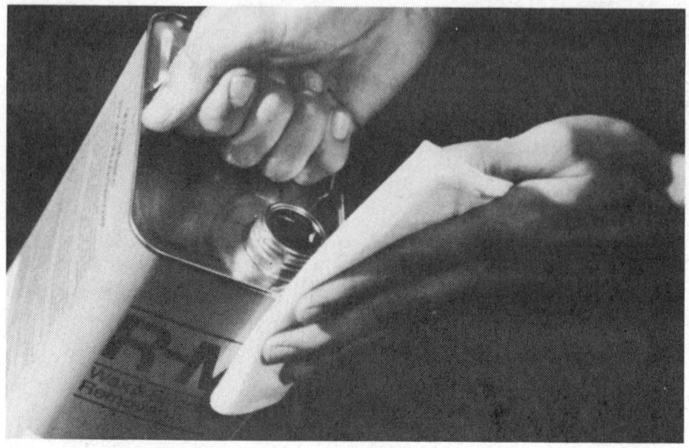

FIGURE 10-17 Saturate a clean, white cloth with wax and grease remover to remove surface contaminants.

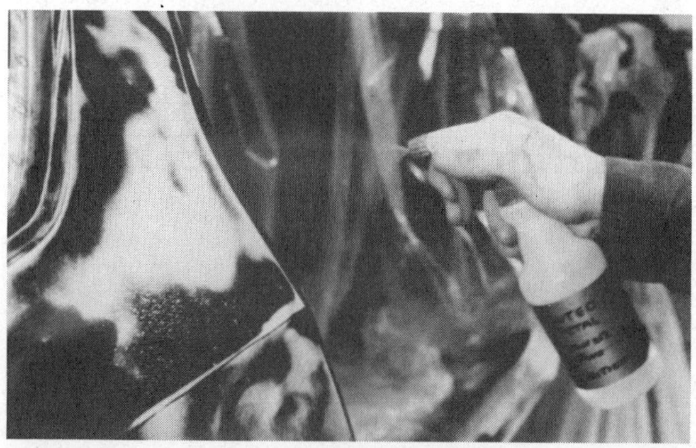

FIGURE 10-18 Use a spray bottle to apply metal conditioner and conversion coating.

vents rust from creeping under the paint film. Use conversion coating on galvanized and uncoated steels and aluminum. Be sure to use the correct product for each type of surface. The conversion coating is mixed with water according to the instructions on the container. Using a spray bottle, apply the coating to the metal surface. Then leave the coating on the surface 2 to 5 minutes. Work only as much area as can be coated and rinsed before the coating dries. Reapply if the surface dries before the rinsing. Rinse the coating from the surface with clean water, or with a damp sponge or cloth that is rinsed occasionally in clean water. Wipe dry with a clean cloth and allow the surface to air dry.

CORROSION TREATMENT AREAS

The corrosion treatment areas that must be considered during collision repairs can be grouped in four categories. They are:

- *Enclosed Interior Surfaces.* Includes body rails and rocker panel assemblies.
- *Exposed Joints.* Includes quarter panel-to-wheelhousing and quarter panel-to-trunk floor joints.
- *Exposed Interior Surfaces.* Includes floor pan, apron, and hood sections.
- *Exposed Exterior Surfaces.* Includes fenders, quarter panels, and door skins.

The term *exposed* as used in this chapter refers to a panel surface that is accessible without having to remove a welded component.

ENCLOSED INTERIOR SURFACES

Of all the areas to be protected during a repair, the enclosed interior surfaces are the most important. These include underbody structures such as front rails, rear rails, and rocker panels. The reason for the importance of these enclosed surfaces is that they represent the principal load-carrying parts of the unibody vehicle. Corrosion of these components can have a severe effect on the crashworthiness and durability of the vehicle.

Metal conditioners and conversion coatings described earlier in this chapter are not recommended for use inside enclosed surfaces. The reason is that the chemicals and moisture might be difficult, if not impossible, to fully remove from the inside seams. Since stone chipping is not a hazard here, the primer should develop adequate adhesion in these areas without conversion coating or metal conditioner.

With this in mind, begin the process on enclosed surfaces with a thorough cleaning and degreasing. Because of the closed construction of these components, the cleaning must be done before the part is welded into the repair area.

After cleaning and degreasing, the enclosed surfaces must be protected with a primer. As described in Chapter 15, there are many different primers used by the auto trade. However, for corrosion protection of enclosed interior surfaces, the ones used most are:

- *Self-Etching and Two-Part Epoxy Primers.* Not all of these relatively new materials are recommended by the manufacturers for use on structural components; be sure to check the compatibility of these primers before choosing them. When using a self-etching or two-part epoxy primer, always follow the manufacturer's instructions.
- *Weld-through Primers.* These primers, which are rich in zinc, protect weld joints better. Remember, the heat from welding can destroy factory corrosion protection (see Chapter 7).

SHOP TALK

Do not use lacquer-based primers on bare metal, because they do not provide adequate corrosion protection.

The application of both primer and anti-corrosion compound to the inside of enclosed surfaces must be done only with the coating manufacturer's recommended equipment. This is normally the airless or pressure-feed type of spray gun (see Chapter 15), although siphon equipment is sometimes recommended. Aerosol cans or conventional spray guns will not do the job in enclosed surfaces because it is not possible to spray the material directly on the surface. This work requires special spray wands to reach all the way inside cavities and joints where the material must go (Figure 10-19).

WARNING

When working with spray equipment, be sure to use a respirator and wear proper protective clothing and rubber gloves.

This spray equipment uses compressed air behind the fluid to force the fluid through the wand and nozzle. The fluid is broken up into a very fine atomized state,

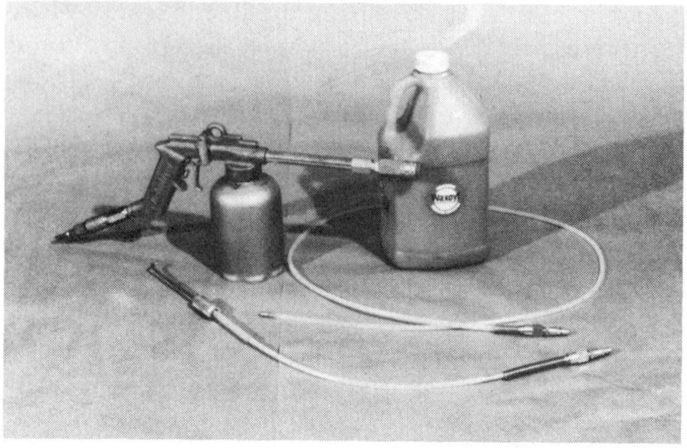

FIGURE 10-19 Special spray wands are required for spraying enclosed surfaces.

sometimes called a *fog*. When the material is introduced inside the enclosed surface in this atomized state, it spreads rapidly and evenly into all areas, including tiny crevices.

To use a spray wand, insert it into the cavity to the farthest point that it is needed. There are several wand styles available. Before spraying with a wand, be sure that the:

- Spray pattern is checked.
- Spray equipment is filled with the desired primer or anti-corrosion compound.

Begin the spraying, and pull the wand out at an even rate, coating the inside of the cavity evenly as it moves along.

To ensure perfect cavity sealing it is necessary for the technician to know and pay attention to the following basic rules:

- Before starting, the spray pattern must be checked and corrected by adjusting the air pressure.
- Make especially certain that the airless spraying device is always filled when spraying.
- When spraying areas where anti-corrosion compound could get into the passenger area, spray with reduced pressure.
- Long-term corrosion protection is ensured when all enclosed surfaces have been coated with a continuous film. This is achieved by a slow and even passing of the wand through the cavity. The effect is improved by simultaneously moving the wand back and forth.
- After use, the equipment must be cleaned with kerosene, mineral spirits, and cleaning solvent. Spray solvent through all wands and hang them to drain overnight. Failure to clean the gun and wands on a regular basis can result in clogging or total blockage of the wands, which will prevent proper function.

The general corrosion protection process for an enclosed interior surface is as follows (if welding is not used during the repair, steps 2, 3, and 4 can be omitted):

1. Clean the enclosed interior surface with wax and grease remover.
2. Apply weld-through primer to bare metal areas only. Do not apply over paint, primer, or galvanized surfaces.
3. Apply weld-through primer only in the immediate weld area since this product has poor adhesion characteristics. After welding, thoroughly remove all welding residue and surplus primer from the joint area.
4. Research has shown that wire brushing is not the best way to clean a weld area. Wire brushing can leave scratches in the original primer, which are not always filled by the new primer. The primer tends to "float" over the scratches, creating very small voids in which corrosion can start. A better way to clean the weld area is to use a plastic abrasive. Another way is to sandblast with a captive sandblaster, if available, or with regular sandblasting equipment. Thorough cleaning must be stressed here as well.
5. After the area is thoroughly cleaned, apply a primer—two-part epoxy or self-etching is usually recommended. Self-etching primers must be top-coated. Be sure to allow sufficient drying time according to the primer manufacturer's recommendations.
6. Apply anti-corrosion compound according to the manufacturer's directions. The compound is applied using service or access holes (Figure 10-20) and drain holes. When the compound is dry (approximately 1 hour) the holes must be cleared.

To spray specific enclosed interior surfaces, special techniques might be required. These area considerations include:

- *Trunk*. Remove the spare tire, tools, floor mat, floor extensions, and padding on each side of the trunk to spray quarter panels. The rear quarter panel behind the wheel is coated from inside the trunk using a spray wand, spraying downward in the recess between the trunk and the quarter panel. Do not spray any exposed surfaces of the trunk.

  Spray the deck lid by inserting the wand into existing holes, making sure the material reaches the edges. Use a pressure of approx-

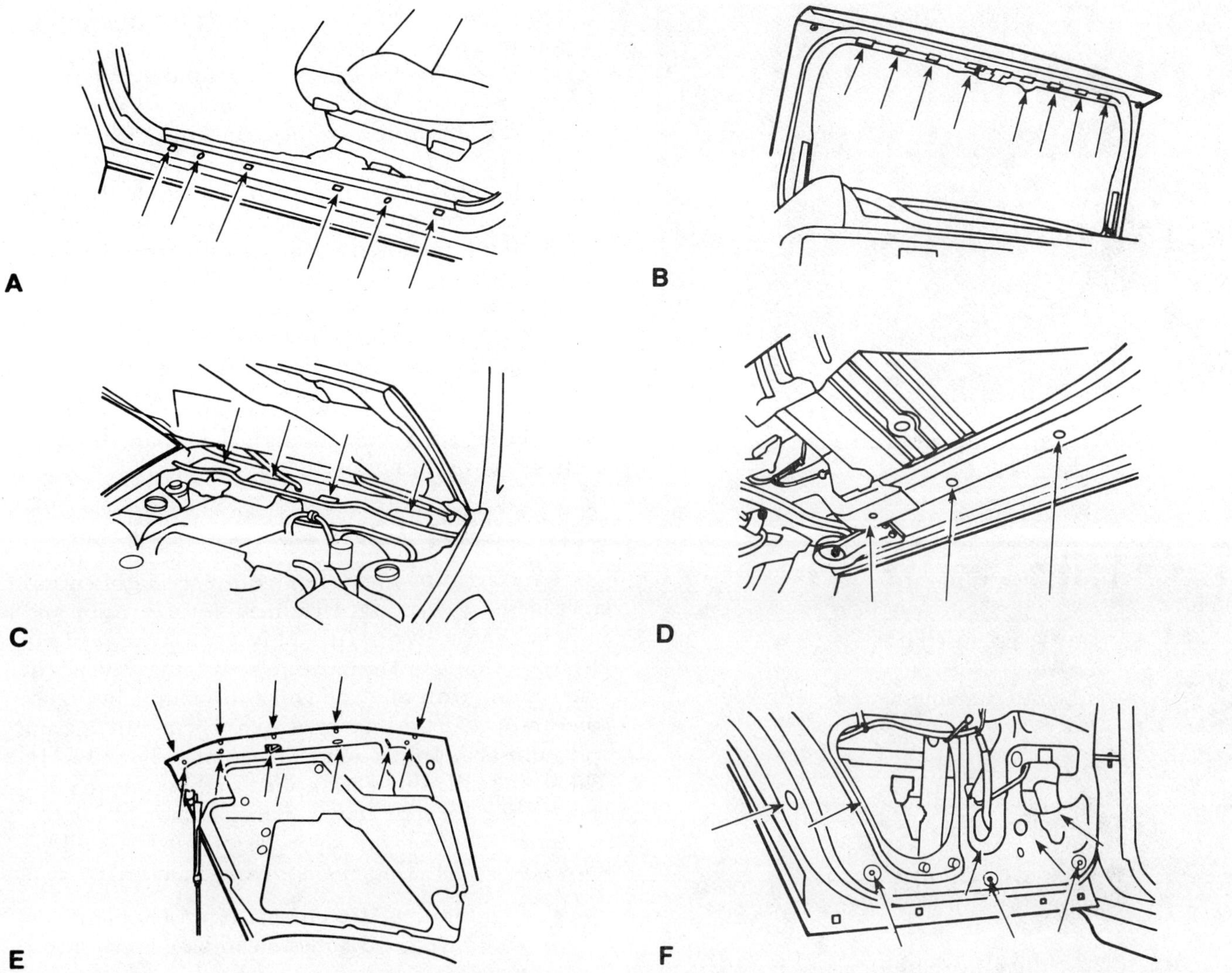

FIGURE 10-20 Service or access hole locations: (A) interior of body sill; (B) rear end of deck lid; (C) inside cowl top; (D) inside rail; (E) front edge of hood; and (F) lower part of door

imately 60 psi. When replacing a trunk floor, use a weld-through primer in all weld zones. When the spraying is completed, replace the padding, floor extensions, floor mat, tools, and spare tire. Wipe off the overspray with a cloth dampened lightly with enamel reducer, solvent, or kerosene.

- *Doors.* Doors can be treated through their access holes after the interior panel has been removed, or a 1/2-inch hole can be drilled in each door, approximately 6 to 9 inches above the bottom. Center punch and drill the hole, taking care to avoid power window and door lock motors.

With the windows up, insert the wand into the hole as far as possible. Slowly retract the wand while spraying the length of the bottom third of the door (Figure 10-21). Just before the wand is withdrawn from the hole, point it downward in order to assure direct coverage into the inside corner of the open door. Plug the hole with a 1/2-inch plastic body plug (Figure 10-22).

- *Rear Lock Pillar and Quarter Panel.* The reverse side of the rear lock pillar and quarter panel can sometimes be sprayed from the trunk area. To insure coverage, the front edge of the wheel well and the quarter panel area should be coated via one or more 1/2-inch holes drilled in the rear lock pillar or by removing the ventilator cover. Insert the wand in the drilled hole in the rear lock pillar, and spray while gradually

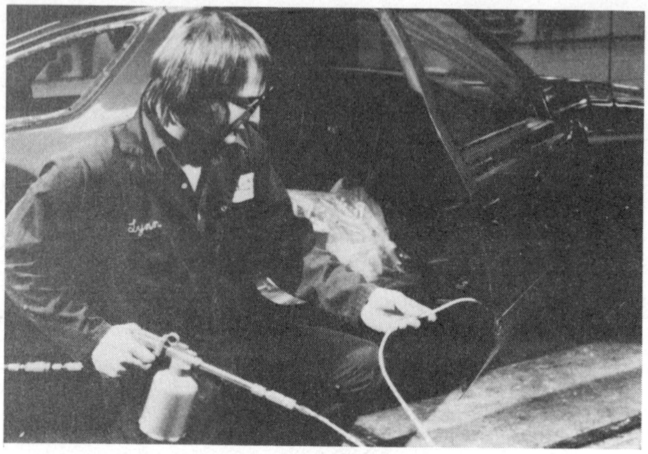

FIGURE 10-21 Applying anti-corrosion compound through a door access hole

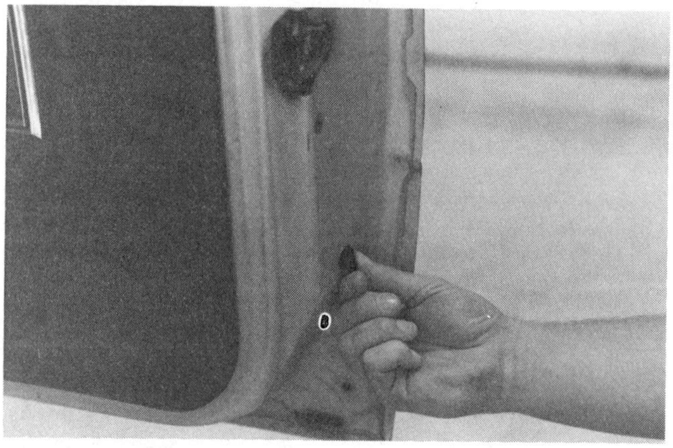

FIGURE 10-22 Plugging the door access hole with a plastic body plug

withdrawing the wand using a spray pressure of 80 psi. Plug the drilled hole with a plastic body plug.

- *A-Pillar or Front Hinge Pillar (and Four-Door B-Pillar).* Drill a hole in the front step plate at the center of the curve where it meets the front supports. Insert the wand and spray thoroughly. Also spray the lower edge of the front fender and any boxed-in areas in the vicinity. Drill a hole in the center of the curve formed by the step plate and rear support. Spray with the wand. If the vehicle is a four-door, the wand must reach the rear of the B-pillar. If it does not, drill a hole from the rear of the B-pillar.
- *Behind Front Fender.* Some vehicles might have a boxed cavity behind the front fender into which the wand can be inserted either from under the hood or alongside the front hinge pillar.
- *Rocker Panel.* Check to see if the rocker panels are boxed. If not, work from both ends. Before drilling any holes, check underneath both ends of the rocker panels for existing holes. If satisfactorily located, these can be used to spray the entire length of the rocker panel with the wand. If it is inconvenient or undesirable to work the rocker panel from above, drill a hole from below into the rocker panel at about the center and spray in both directions. Be sure to spray on both sides of reinforcements, if present.

It is very important when sectioning a rocker panel assembly to use a weld-through primer in the weld areas, as shown in Figure 10-23.

EXPOSED JOINTS

Exposed joints and seams require special attention since these areas are highly vulnerable to corrosion and must be protected correctly. This is true because of the effect of welding on the metal, as well as the tendency of water, snow, dirt, mud, and other contaminants to become trapped in the joint area. As a general rule, a seam sealer must be applied over all the joints (Figure 10-24). The seam sealer must be applied so there are no gaps between the material and the panel surface.

The following are some factors that should be known when selecting and applying seam sealers:

- *Paintability.* All seam sealers must be paintable and have good adhesion to bare and primed metal.

FIGURE 10-23 Applying weld-through primer to a rocker panel assembly

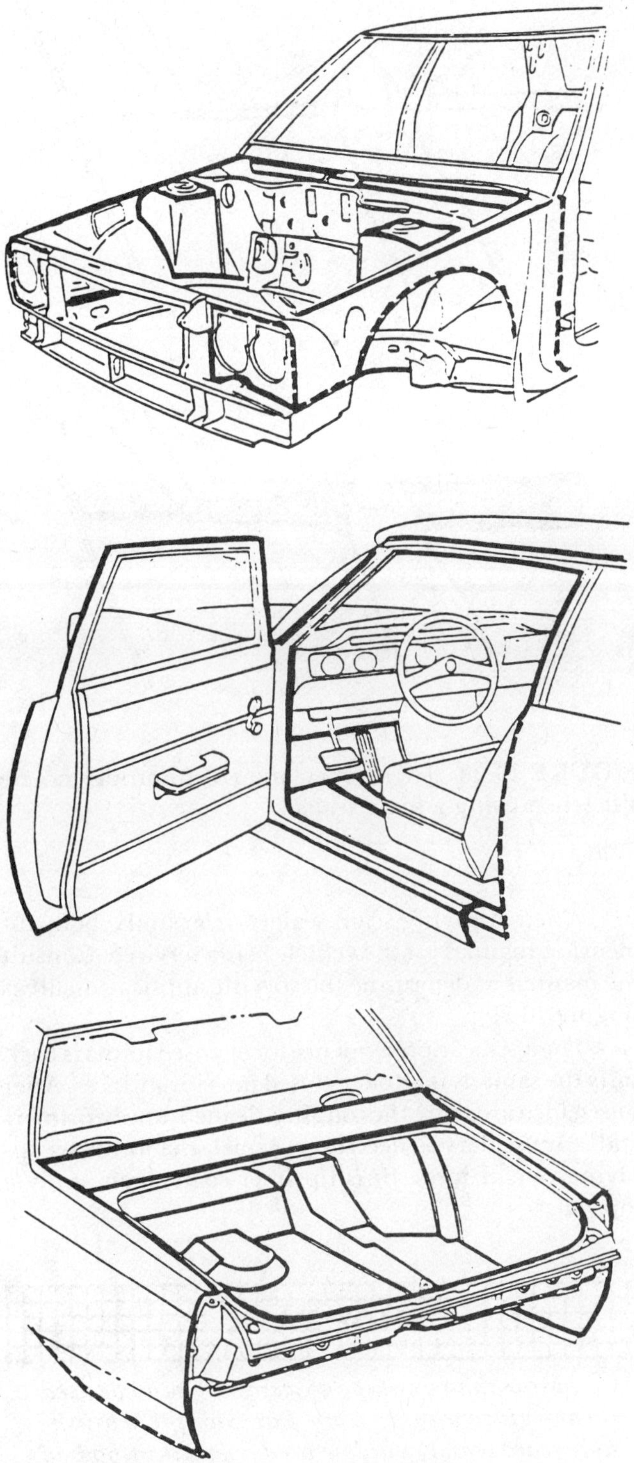

FIGURE 10-24 Seam sealer locations

- *Flexibility.* This is a critical issue with today's unibody vehicle. The seam sealer must be able to withstand the motion associated with the automobile.

- *Silicone Sealants.* These should not be used as a seam sealer. They typically are not paintable, attract dust and dirt over time, and do not offer the adhesion of other types of seam sealers.

There are four types of seam sealers that are commonly used in unibody corrosion protection:

- *Thin-bodied Sealers.* These are designed to fill seams under 1/8-inch wide. Thin-bodied sealers will shrink slightly to provide definition to the joint, while remaining flexible to resist vibration. Adhesion is good to both primed metal and bare metal surfaces. Since many of the seams are on a vertical surface, sag control is important so the sealer does not run out of the seam. Typically thin-bodied sealers carry names such as drip-chek or flow grade.
- *Heavy-bodied Sealers.* These are used to fill seams from 1/8- to 1/4-inch wide (Figure 10-25). These sealers can be tooled to hide the seam or can be left in bead form. Shrinkage should be minimal, with good resistance to sagging and high flexibility to resist cracking. Heavy-bodied sealers are used on both pinchweld and overlap seams. They are typically dispensed from cartridges, but some are available in squeeze tubes.
- *Brushable Seam Sealers.* These are used on body seams where appearance is not important. These seams are normally hidden and not seen by the customer, unless under close examination. Brushable seam sealers are designed to resist salt and automotive fluids such as gasoline, transmission fluid, and brake fluid. Any

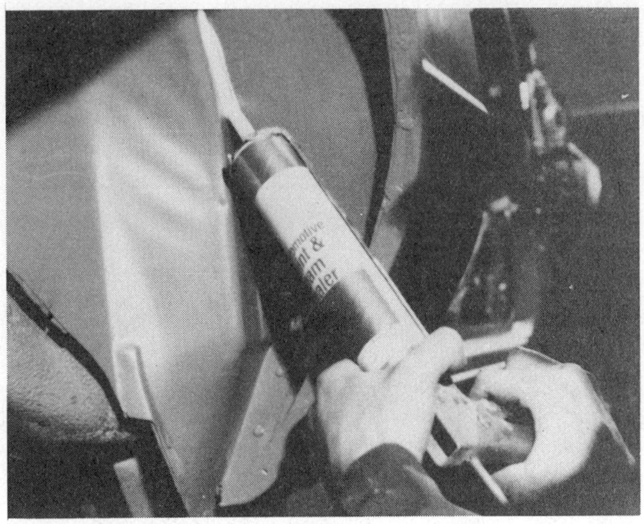

FIGURE 10-25 Applying a heavy-bodied seam sealer

FIGURE 10-26 Application of a brushable seam sealer

seams such as those under the hood that might be exposed to automotive fluids should have a brushable seam sealer (Figure 10-26).

- *Solid Seam Sealers.* These sealers, containing 100 percent solids, are used to fill larger voids at panel joints or seams. They come in strip caulking form, and are designed to be pressed into place (Figure 10-27).

Be sure to follow the manufacturer's instructions carefully for the use of seam sealers.

When applying seam sealer, it is important to use a nozzle with a small hole in the end, and then spread out the bead with a fingertip. If the nozzle hole is small, the finish can be kept neat (Figure 10-28A). If the nozzle hole is too large, the sealer will spread too wide and might cause a poor looking finish (Figure 10-28B).

A neat seam sealer application can also be obtained if automotive masking tape is used to make a parting line for the sealer application area.

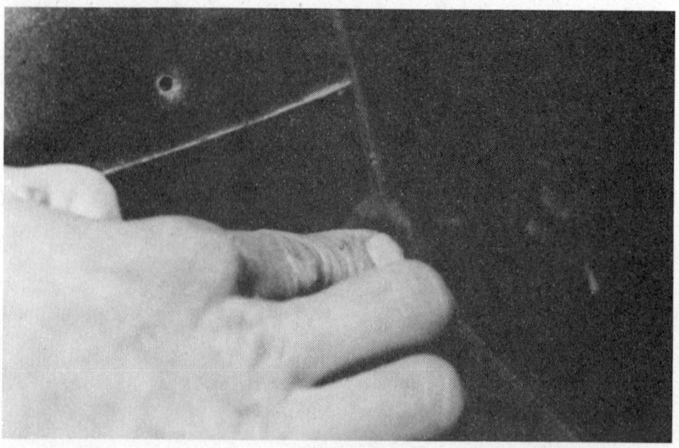

FIGURE 10-27 Application of a solid seam sealer

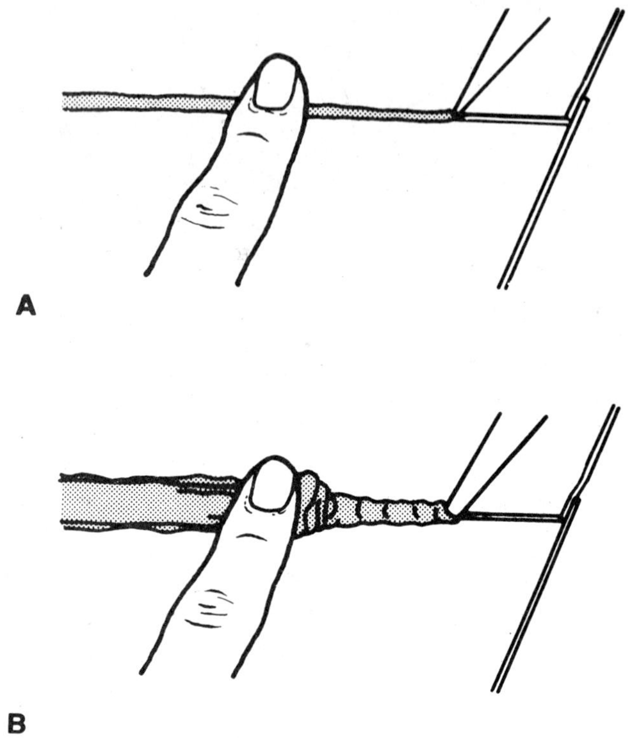

FIGURE 10-28 (A) Proper and (B) improper nozzle size when using a seam sealer

When applying seam sealers, refer to the body dimension manual for the vehicle being serviced. Consult the manual to determine the specific application areas (Figure 10-29).

The application sequence for exposed joints is basically the same as it is for enclosed interior surfaces. After the welded areas are thoroughly cleaned, apply primers and seam sealers as necessary. Finish the joints by applying one or more finish primer coats, then apply a topcoat.

SHOP TALK

More than one type of seam sealer can be used on any given joint or seam. For example, a brushable seam sealer could be used over a thin-bodied sealer.

The corrosion protection process for exposed joints and seams can be summed up as follows:

1. Thoroughly clean the joint or seam.
2. Apply primers and seam sealers.
3. Apply final primer coat(s).
4. Apply a topcoat.

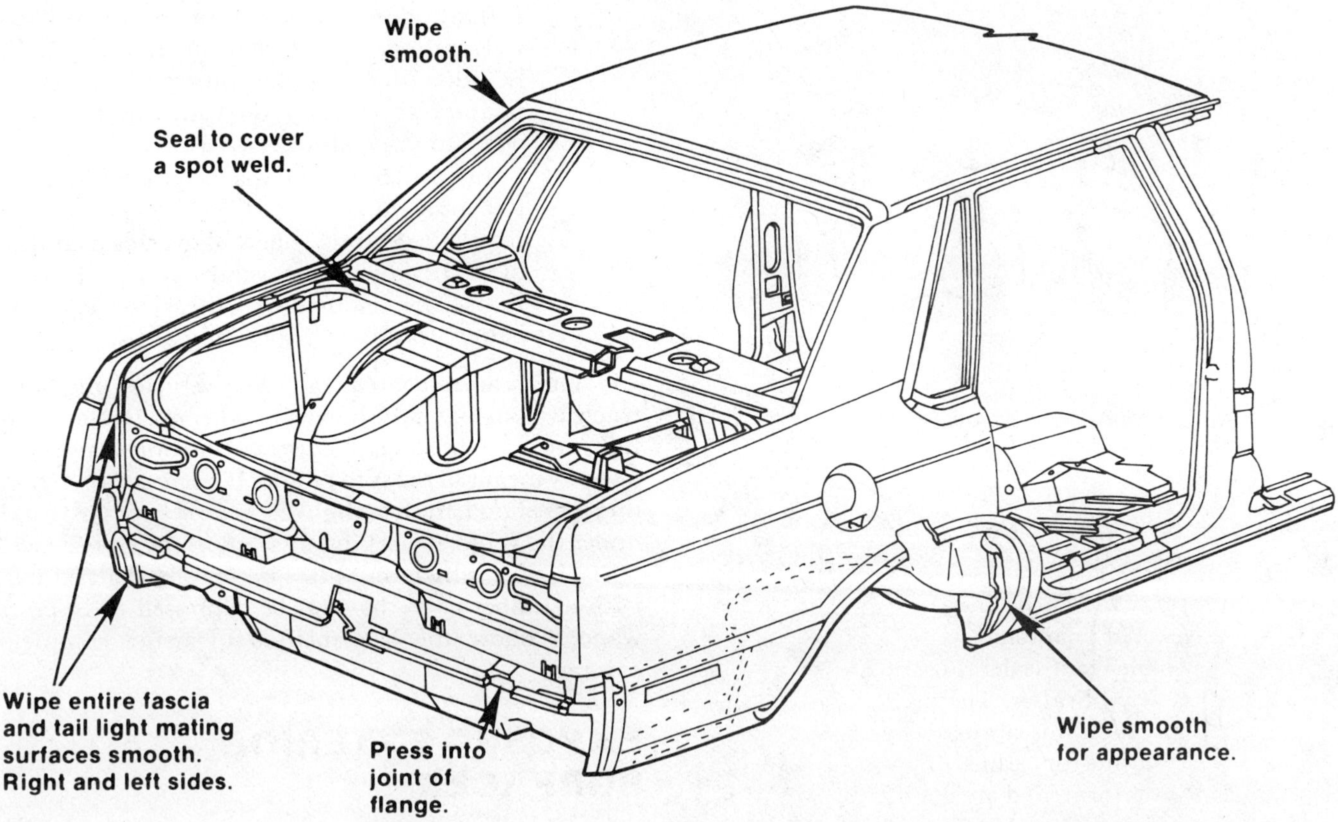

FIGURE 10-29 The recommendations given in the body dimension manual should be followed for seam sealer applications.

EXPOSED INTERIOR SURFACES

The bottom surfaces of the underbody and inside of the wheelhousing can be damaged by flying stones, causing rust to develop. These areas are given an anti-corrosion treatment with a material such as shock-absorbing wax-based anti-corrosion compound. As a general rule do not spray anti-corrosion compound into the passenger compartment; apply it from below the vehicle. Corrosion damage can be detected by removing the floor matting (Figure 10-30).

Metal conditioners and conversion coatings are not recommended on interior surfaces. There are two reasons for this:

- These surfaces are not exposed to physical damage the way exterior surfaces are.
- These surfaces contain joints and seams that should not be contaminated with etching and conversion coating chemicals. Also, they are generally difficult to rinse clean.

The corrosion protection process for exposed interior surfaces begins with a thorough cleaning with a wax and grease remover. Once the surface is completely air dried, spray the first coating of wax- or petroleum-based anti-corrosion compound to all welded areas and panel joints (Figure 10-31A). Then apply a second coat over the entire area (Figure 10-31B). Protect surround-

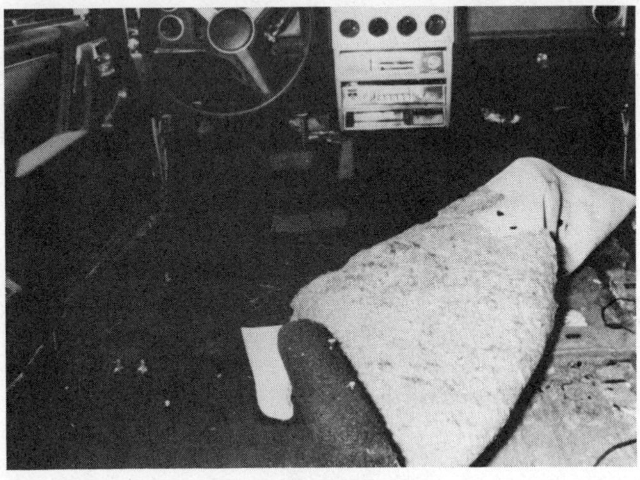

FIGURE 10-30 Remove the floor matting to look for corrosion damage.

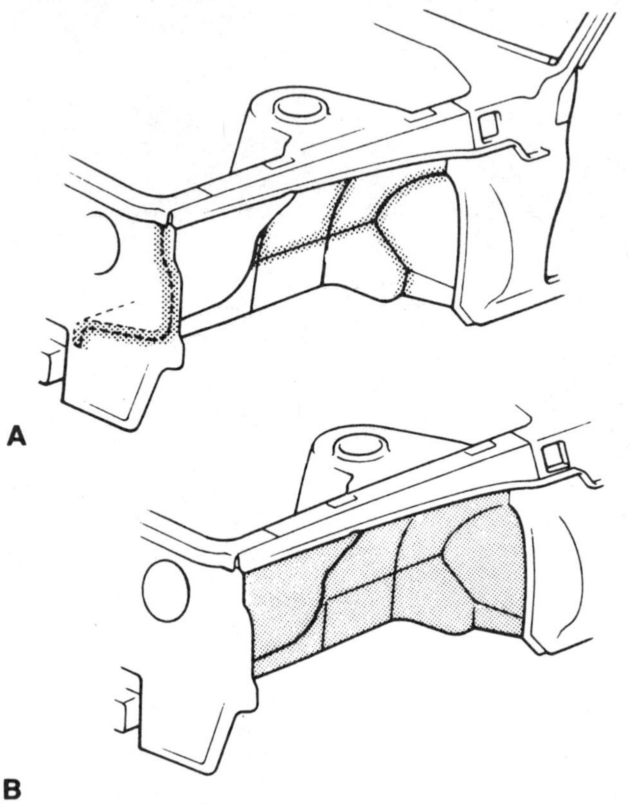

FIGURE 10-31 (A) Apply the first coat of anti-corrosion compound to the welded areas and panel joints, and (B) the second coat to the entire area.

ing areas with masking paper and/or tape to prevent the anti-corrosion compound from sticking to undesirable areas.

In areas where there is a moderate corrosion environment, the proper primer might be all that is needed for exposed interior surface corrosion protection. Although there are several different types available, two-part epoxy primers most closely duplicate the baked-on electrodeposition coating, or E-coat, used by vehicle manufacturers. Nearly any topcoat can be applied over the two-part epoxy primer.

Self-etching primers can also be used to some extent for exposed interior surfaces. Lacquer-based primers will not provide proper adhesion when used on bare metal, even if the surface has been properly cleaned and a conversion coating was used. This point cannot be overemphasized: lacquer-based primers should never be used directly on the bare metal of modern unibody vehicles.

To restore corrosion protection to a specific exposed interior surface such as under the hood, proceed as follows:

1. Lift the hood; spray the front quarter panel as well as the area between it and the wheel well. Be sure to apply the anti-corrosion compound right down to the fender beads. Use a wand to reach all recessed areas.
2. Spray the large open spaces with a 45-degree flat spray nozzle.
3. Spray the leading edge and the side channels of the hood with the wand.
4. Loosen or remove the battery and coat the battery tray and surrounding areas.

The headlight areas on some vehicles might be reached from under the hood. On other makes the headlight areas can be reached from existing holes under the hood by means of the spray wand. If not accessible from under the hood, the headlight areas can be sprayed from under the vehicle, working forward in the front wheel well with the vehicle on a lift. This may be a baffled area with a rubber edge that can be depressed to insert a wand. Choose the best method to assure complete coverage.

EXPOSED EXTERIOR SURFACES

Exposed exterior surfaces are subjected to many more chips and nicks than interior surfaces. This is why the use of conversion coating is of critical importance on these surfaces. Conversion coating provides the kind of superior paint film adhesion that prevents creeping rust from working its way under the paint when chips and nicks do occur.

Exposed exterior surfaces are of two types:

- Cosmetic
- Underbody, on which an airless spray unit is used (Figure 10-32)

Corrosion protection procedures for exposed exterior cosmetic surfaces are as follows:

1. Clean with a wax and grease remover.
2. Apply a metal conditioner.
3. Rinse with water.
4. Apply a conversion coating.
5. Rinse with water and allow to dry.
6. Apply a primer—two-part epoxy primer is recommended.
7. Apply a primer-surfacer or primer-sealer.
8. Apply the topcoat.

If a lift is available, it makes underbody corrosion protection work easier (Figure 10-33). However, the job can be done on the floor. When applying anti-corrosion material to the underbody, start by spraying the fenders

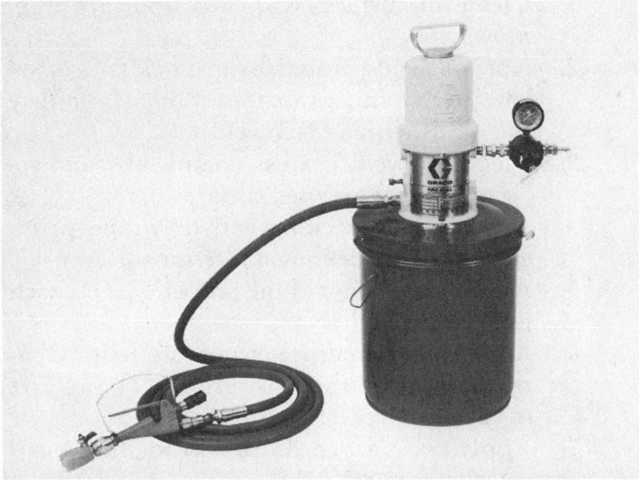

FIGURE 10-32 Typical airless underbody spray unit

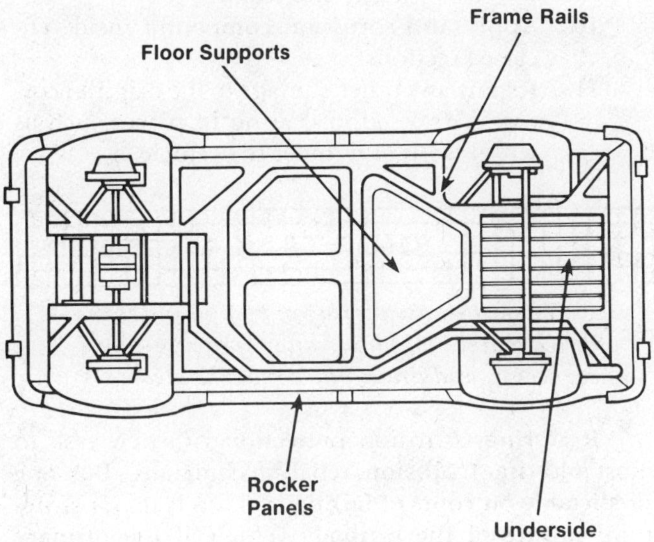

FIGURE 10-33 Important underbody corrosion protection areas

and wheel wells, paying particular attention to the fender beads. On some vehicles it will be necessary to remove the wheels to do an adequate spraying job. Plastic fender skirts should be removed.

Spray the remaining underbody and splash pans adjacent to the front and rear bumpers. Spray the underside of the floor pan, weld joints, frame, tank straps, and seams. Coat the exterior of the gas tank, inside of the gas door, and gas filler area. Remove any loose debris or sound deadener, particularly around joints, before spraying. Loose sound-deadening material or dirty surfaces at critical locations will only create pockets for rust to form and will prevent the anti-corrosion material from reaching the metal.

CAUTION

Do not spray the brakes, brake linings, or catalytic converter. In addition, do not spray the radiator core, battery, air-conditioning condenser, fans, belts, hoses, or pulleys.

In general, anti-corrosion procedures for exposed exterior underbody surfaces are as follows:

1. Clean with a wax and grease remover.
2. Apply a metal conditioner.
3. Rinse with water.
4. Apply a conversion coating.
5. Rinse with water and allow to dry.
6. Apply a primer—self-etching primer is recommended.
7. Apply anti-corrosion material and sound-deadening materials to restore the vehicle to pre-accident condition.
8. Remove overspray with enamel reducer, solvent, or kerosene.

EXTERIOR TRIM AND ACCESSORIES

To prevent corrosion, it is very important to install a barrier between dissimilar metal components such as aluminum bumpers and stainless steel body trim. Plastic or rubber isolating pads accomplish this effectively. Installing body trim must be done correctly to avoid galvanic corrosion. For example, when installing trim requires drilling holes in a new or repaired panel, drill all holes before applying the primers and topcoats, then coat the inside edges of all holes completely. When using a kit for replacement trim, be sure to use all parts supplied with the kit. If parts are not purchased as a kit, duplicate the original assembly. Clearly, there is a great variety of body trim and accessories requiring many different application techniques. In all cases, be sure to follow the manufacturer's recommendations to avoid problems (Figure 10-34).

Fasteners are an important part of any unibody vehicle. But, as the number of structural parts multiplies, so too does a corrosion problem—galvanic (or dissimilar metal) corrosion. This added susceptibility to galvanic corrosion comes at a time when heavy use of road salts and the growing acidity of rain and other condensation are giving rise to increasingly corrosive conditions. Moreover, galvanic corrosion of fasteners in unibody structures is much more than a mere cosmetic

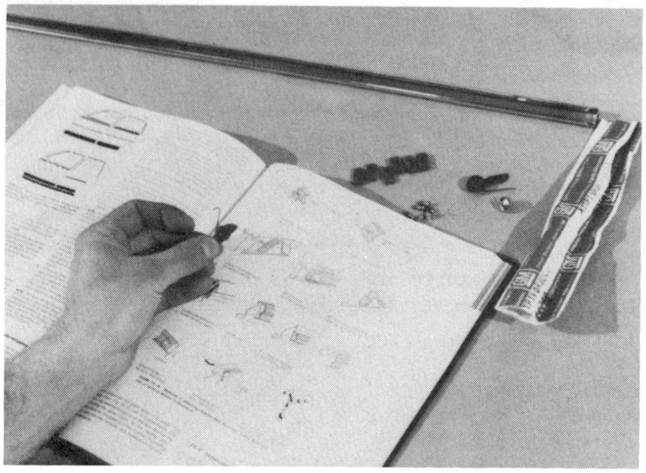

FIGURE 10-34 Follow the manufacturer's instructions when using an exterior body trim kit.

problem. Breakdown of fasteners and the surface around them can lead to premature and catastrophic failure of the unibody structure.

Galvanic corrosion could be totally eliminated if it were possible to isolate the vehicle from electrolytes such as rain and condensation. Since this is not feasible, one or more of the following methods must be used instead:

1. Select materials of construction close together on the galvanic series (see Table 10-1).
2. Insulate dissimilar metals from one another.
3. Avoid threaded joints for materials far apart in the galvanic series.
4. Apply all coatings with caution.

In modern automotive assemblies, examples exist where most, if not all, of these basic rules for controlling galvanic corrosion are broken or forgotten of necessity.

SUMMARY OF ANTI-CORROSION PROCEDURES

The correct corrosion treatment procedures for protecting each type of vehicle surface have been covered in this chapter. However, a typical repair seldom involves work on only one type of surface. The following procedure combines all of the individual procedures into an efficient sequence that makes practical sense in any collision repair shop:

1. Clean all surfaces with wax and grease remover.
2. Apply a weld-through primer to all uncoated bare steel joints prior to welding. Complete the welds with a MIG welder.
3. Clean all welded areas completely, and reclean interior exposed surfaces.
4. Treat enclosed sections with the proper primer. Also, prime exposed interior surfaces.
5. Apply seam sealer of the proper type for each joint, and reprime.
6. Apply metal conditioner and conversion coating to exposed exterior surfaces, taking care to exclude these chemicals from seams.
7. Apply two-part epoxy primer to all exposed exterior surfaces.
8. Apply primer-surfacer to any exterior surfaces that might need additional filling.
9. Topcoat all exterior and interior surfaces, as well as all joints and seams.
10. Apply anti-corrosion compound inside enclosed sections.
11. Repair any other damage to the original corrosion protection coating in places such as wheel wells and under the vehicle.

SHOP TALK

When the anti-corrosion compound is dry—approximately one hour after spraying—all service, access, and drain holes must be cleared.

Restoring corrosion protection is a new task to most old-time collision repair technicians. But one predictable outcome of failing to do so is major structural failure of the unibody vehicle if the primary structural components are affected. Corrosion of these structural components can severely impair the handling and crashworthiness of the vehicle. This damage might not be evident until an accident, but then it will be too late.

Serious problems can also arise from the improper application of anti-corrosion materials in certain areas of late-model vehicles. Areas that are particularly sensitive are:

- Parts that generate or conduct heat
- Electrical accessories
- Labels and identification numbers
- Moving parts

Here is a list of specific areas where spraying anti-corrosion material must be done with extra care or not at all:

1. In seat belt retractor areas and passive restraint guide rails and related parts, it might cause the seat belt to operate improperly. It might also get on the belt material, causing appearance problems.
2. Hidden headlight covers, motors, and linkages located in the header and fascia panels in front of the radiator core support are vulnerable areas. Coatings applied to the radiator will result in higher operating temperatures and shorter engine life.
3. Exhaust systems and heat shields that are sprayed, including turbochargers or superchargers, can result in a foul odor or, under certain conditions, can be a fire hazard.
4. Bumper energy-absorber performance might be altered by anti-corrosion material on the sliding member.
5. Power window motors, channels, guides, and tracks as well as window regulators and cable mechanisms, can be made inoperative.
6. When cruise control transducers become plugged with anti-corrosion material, the result can be erratic speed control or an inoperative unit.
7. The engine, along with its components and accessories, must not be sprayed.
8. The transmission/transaxle, U-joints, universal joints, drive shafts, seals, and differentials generally do not require corrosion protection. Coating on the drive shaft can cause imbalance. This imbalance might result in vibration while the vehicle is in motion.
9. Brake drums, rotors, and backing plates should be kept free of anti-corrosion material to prevent any erratic braking action.
10. Transmission vents or rear axle vents must not be sprayed. Accumulated material can cause a blockage, which will result in a heat-related pressure rise that will force fluids or lubricants past the seals.
11. Exterior surfaces of the fuel tank grommet and the surface of the fuel tank filler tube within 2 inches of the fuel tank grommet must not be sprayed.
12. Antenna masts and power antenna drain holes that are clogged with anti-corrosion material will result in water collecting inside the mast or motor. Plenum (cowl) areas, especially around windshield wiper components, must not become clogged. The trapped water can cause damage by freezing or corroding the parts.
13. All locks, key cylinders, power lock motors/solenoids, latches, linkages and hinges should not be sprayed with anti-corrosion material. Also keep drain holes open at the bottom of doors, rocker panel sills, and so forth to avoid water accumulation.
14. Air filter service life can be greatly shortened by coating with anti-corrosion material.
15. The life of air lift shock absorbers will also be shortened by spraying on the boots. The anti-corrosion material collects dust and grit and will result in the chafing of the boots.
16. Shift linkages, speedometer cables, and other moving cables should not be sprayed.
17. All electrical components such as lighting, ignition, and charging systems should not be sprayed.
18. Theft prevention label locations on fenders, quarter panels, hood, deck lid/hatch, doors, and fascia should never be sprayed with anti-corrosion material.

REVIEW QUESTIONS

1. To protect auto body steel from corrosion, Technician A seals off oxygen and moisture from the surface. Technician B protects the steel through "sacrificial corrosion." Who is correct?
   a. Technician A
   b. Technician B
   c. Both A and B
   d. Neither A nor B

2. Which of the following statements is incorrect?
   a. Unlike the anodic bath process, cathodic electrodeposition does not involve submerging the entire auto body.
   b. E-coats are thicker than those produced by the anodic method.
   c. Epoxy primers stick well to sheet metal.
   d. All of the above

3. What is the name given to steel whose unexposed side has a zinc coating and whose exposed side has a lighter alloy-coated surface?
   a. one-sided galvanized
   b. one-and-one-half-sided galvanized
   c. two-sided galvanized
   d. two-sided galvannealed

4. What is used to change red iron oxide to black/blue iron oxide?

a. any anti-corrosion material
b. anti-corrosion compound
c. rust converter
d. body sealer

5. Which of the following can lead to paint film failure?
a. improper surface preparation
b. stone chips
c. moisture
d. all of the above

6. Technician A uses a vacuum cleaner to remove metal chips from inside a vehicle body, while Technician B uses dry compressed air. Who is correct?
a. Technician A
b. Technician B
c. Both A and B
d. Neither A nor B

7. What can result when two dissimilar metals are placed in contact with each other?
a. higher pH number
b. crystal phosphate
c. cathodic electrodeposition
d. galvanic corrosion

8. What is the recommended method of application for metal conditioner?
a. clean, white cloth
b. sponge
c. paintbrush
d. spray bottle

9. Which of the following statements is incorrect?
a. Conversion coatings can be used on aluminum.
b. Of all the areas to be protected during a collision repair, exposed interior surfaces are the most important.
c. When applying conversion coating, work only as much area as can be coated and rinsed before the coating dries.
d. Rails and rocker panels are examples of enclosed interior surfaces.

10. What type of primer does not provide adequate corrosion protection under enclosed interior surface conditions?
a. weld-through
b. lacquer-based
c. two-part epoxy
d. both a and b

11. Technician A cleans welded areas with a plastic abrasive. Technician B uses a captive sandblaster. Who is correct?
a. Technician A
b. Technician B
c. Both A and B
d. Neither A nor B

12. Which of the following conditions has a tendency to accelerate corrosion?
a. temperature decreases
b. high relative humidity
c. low relative humidity
d. both a and b

13. Which type of seam sealer is used for filling seams less than 1/8 inch wide?
a. thin-bodied
b. brushable
c. solid
d. heavy-bodied

14. Technician A sprays anti-corrosion compound directly into the passenger compartment, while Technician B applies the treatment from below the vehicle underbody. Who is correct?
a. Technician A
b. Technician B
c. Both A and B
d. Neither A nor B

15. Which of the following areas requires spraying anti-corrosion material very carefully or not at all?
a. seat belt retractors
b. engine
c. electrical components
d. all of the above

CHAPTER 11 Replacing Doors, Hoods, Deck Lids, and Body Panels

OBJECTIVES

- Replace both welded and adhesive-bonded door skins.
- Replace an SMC door skin.
- Discuss general custom body panel installation procedures.
- Name the ways vehicle manufacturers provide for adjustments of mechanically fastened panels.
- Perform hood-to-hinge, hood height, and hood latch adjustments.
- Perform deck lid, fender, and door adjustments.

A collision-damaged vehicle can require a variety of different repair operations, depending on the nature and location of the damage sustained. Damaged panels can be pulled back into alignment, and bulges, dents, and creases can be made to look like new using the techniques discussed earlier. A professional approach, combined with attention to detail, can sometimes salvage metal that is seemingly beyond repair. However, there will be instances where the damage is simply too great and replacement is the only answer. This chapter covers replacement procedures for doors, hoods, and deck lids, as well as the installation of custom body panels. Keep in mind as well that panel replacement is often the only satisfactory remedy for extensive corrosion damage (Figure 11-1).

DOOR SKIN REPLACEMENT

Like other damaged panels, a door skin can be bumped back into shape, pulled into shape, or replaced. The decision is based on the amount of work hardening in the buckles, the accessibility to the backside of the door skin, and the amount of door frame damage. Figure 11-2 shows a damaged door skin that can be repaired, while the door in Figure 11-3 requires complete replacement. Another option is replacement of just the door skin.

The door skin wraps around—and is flanged around—the door frame. The skin is secured to the frame either by welding or with adhesives. Typical replacement procedures for both types of skins are given here.

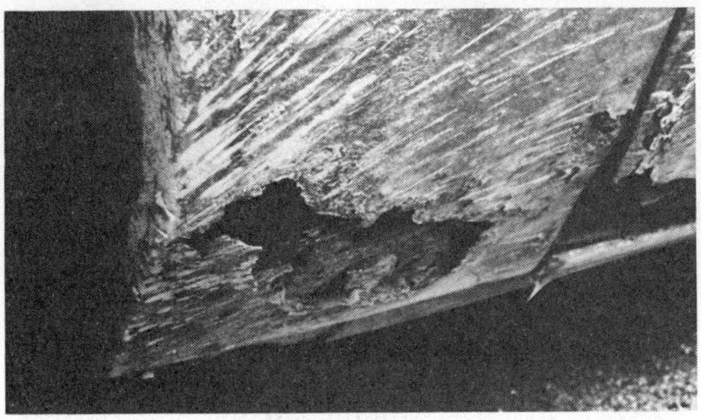

FIGURE 11-1 Extensive corrosion damage calls for replacement of this panel.

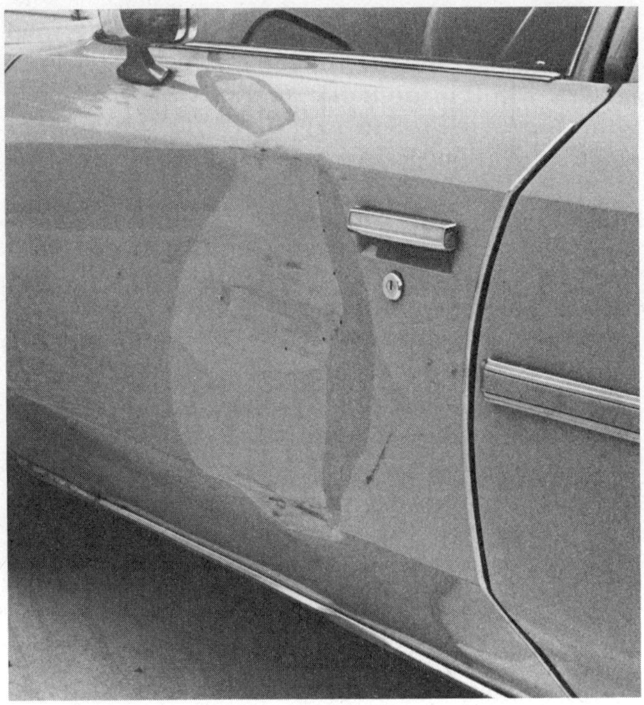

FIGURE 11-2 This door skin can be repaired.

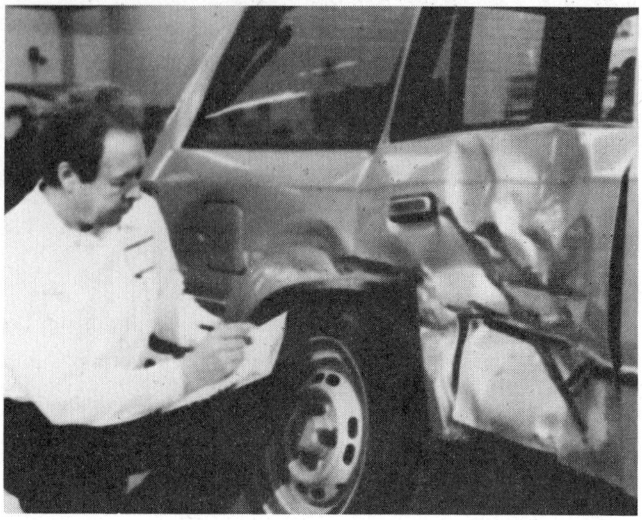

FIGURE 11-3 This door is a candidate for replacement.

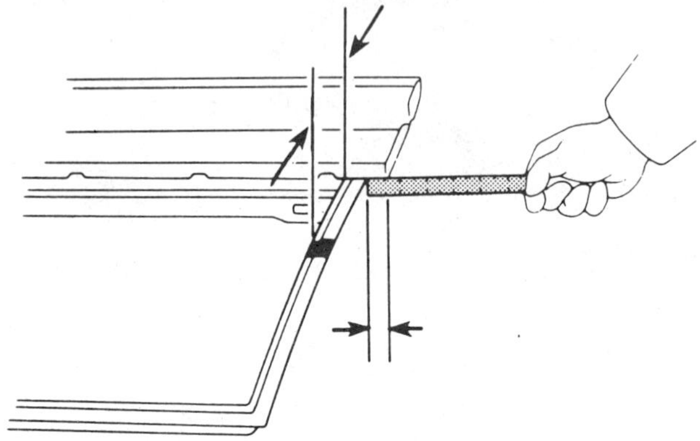

FIGURE 11-4 Measuring door skin location and alignment

REPLACING WELDED DOOR SKINS

1. Before removing the door, check to see if the hinges are sprung and observe the alignment of the door with respect to its opening.
2. Observe how the skin is fastened to the door to determine how much interior hardware must be removed.
3. Remove the trim panel and disconnect the negative side of the battery to isolate all power door accessories. Remove all hardware from the door.
4. To prevent loss, place all removed parts and hardware inside the vehicle.
5. Straighten and/or align the inner door frame.
6. Remove the door glass to prevent breakage while repairing the door. Place it inside the vehicle for protection.
7. Remove the door and move it to a suitable work area.
8. Using an oxyacetylene torch and wire brush, remove the paint from the spot welds in the skin hem. Using a drill and spot weld cutter or grinder, remove the spot welds.
9. Apply tape to the door frame and measure the distance between the lower line of the tape and the edge of the skin. Also measure the distance between the edge of the skin and the door frame (Figure 11-4).
10. Use a plasma arc cutter or grinder to remove the brazed portion of the skin from the door frame.
11. Grind off the edge of the door skin hem flange (Figure 11-5). Only grind off enough metal so that the skin can be separated from the flange; do not grind into the frame. Also, do not use a cutting torch or power chisel to separate the panels. The frame can become distorted or be accidentally cut.
12. Separate the reinforcing strip on the top of the skin (if applicable).
13. Using a hammer and chisel, start to separate the skin from the frame. Use a pair of tin snips to cut around any spot welds that could not be cut or ground out (Figure 11-6).
14. When the skin moves freely, remove it. Use clamping pliers to remove what remains of the hem flange. Any remaining spot welds,

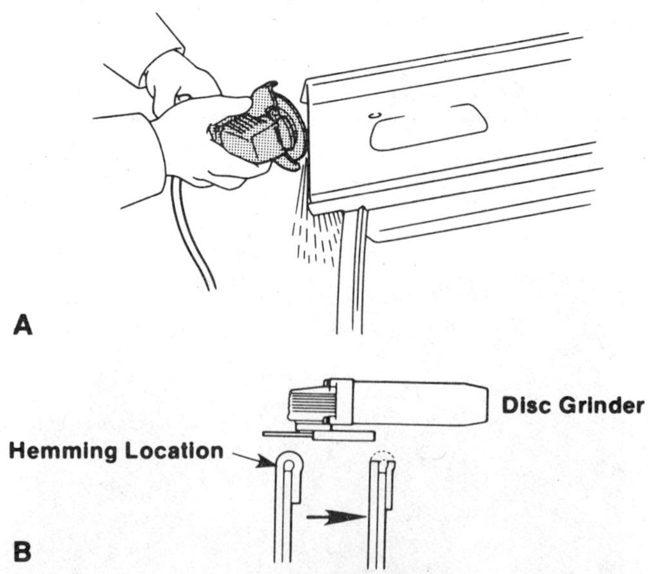

FIGURE 11-5 (A) Grinding off the door skin hem flange; (B) hem flange cross-section

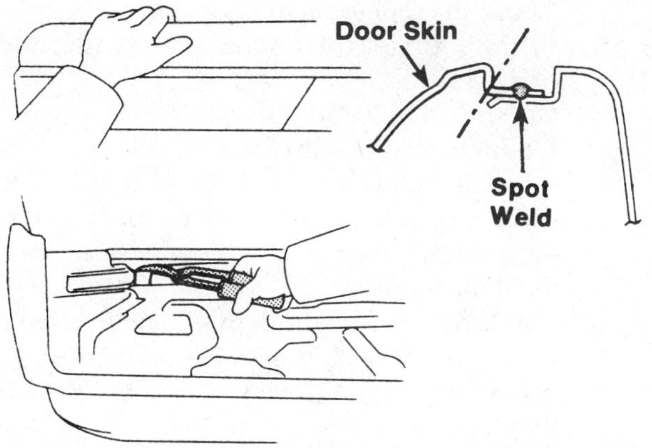

FIGURE 11-6 Cutting around spot welds

braze, and rust should be ground off using a disc grinder.

15. With the skin removed, examine the frame for damage. Repair any remaining inner door damage at this time. If necessary, remove damage to the inner flange with a hammer and dolly.
16. Apply weld-through primer to the areas to be welded. Cover other bare metal with an anti-corrosion primer or other corrosion protection material.
17. Prepare the new skin for installation. Using a drill or hole punch, make holes for plug welds. Remove the paint from the weld and braze locations with a sander. Apply weld-through primer to the bare metal.
18. Some door skins are accompanied by a silencer pad that must be attached to the skin. To do this, clean the skin with alcohol, then heat it and the silencer pad with a heat lamp. Finally, glue the pad to the skin.
19. Apply body sealer to the backside of the new skin. Apply the sealer evenly, 3/8 inch from the flange, in a 1/8-inch-thick bead.
20. Using clamping pliers, attach the new skin to the door and align it properly. Braze where required.
21. Use a hammer and dolly to flange the hem (Figure 11-7). Cover the dolly face with masking tape to avoid marring the skin. Bend the hem flange gradually in three steps, being careful not to throw the skin out of alignment. Do not create bulges or creases in the skin.
22. After working the flange within 30 degrees of the frame, use a flanging tool to finish the hem flange (Figure 11-8). Again, finish the hem in three steps, being careful not to deform the skin.

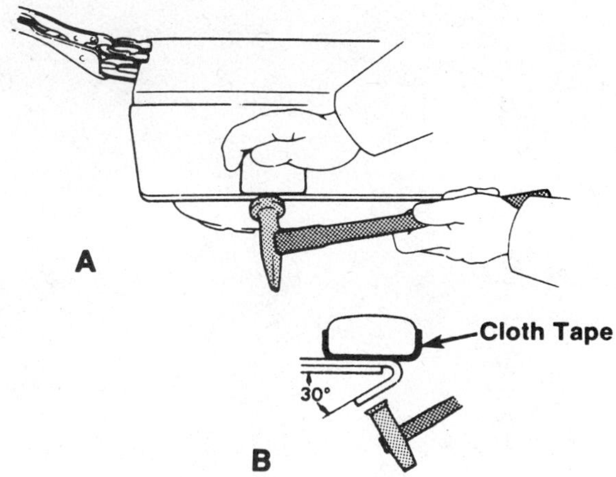

FIGURE 11-7 (A) Dollying the door skin hem flange; (B) door skin cross-section

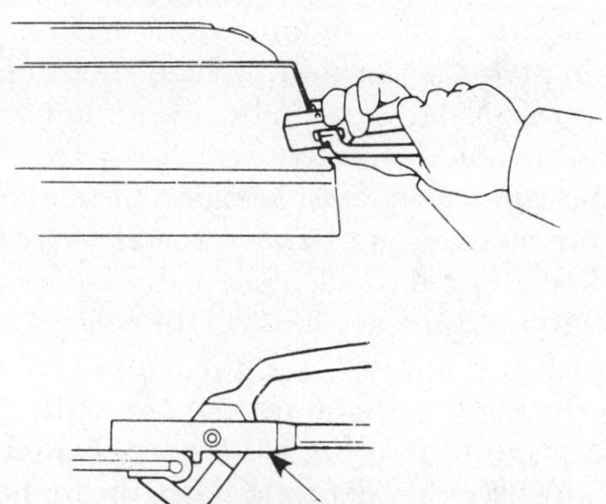

FIGURE 11-8 Forming the door skin hem flange with a flanging tool

23. Weld the plug or spot weld locations of the glass opening (Figure 11-9), then tack weld the hem flange.
24. Apply seam sealer to the hem flange. Apply anti-corrosion compound to the inside of the welded and brazed areas.
25. Drill holes into the new door skin to accommodate moldings, trim, and so forth. Make sure all edges are cut-in prior to installing any parts.
26. Place the door on the vehicle to check its alignment, prepare it for refinishing, and then reinstall it.
27. Align the door with the adjacent panels. Check for proper operation.

As discussed in the previous chapter, all joints and seams require special attention because of their vulnerability to corrosion. This is also the case when replac-

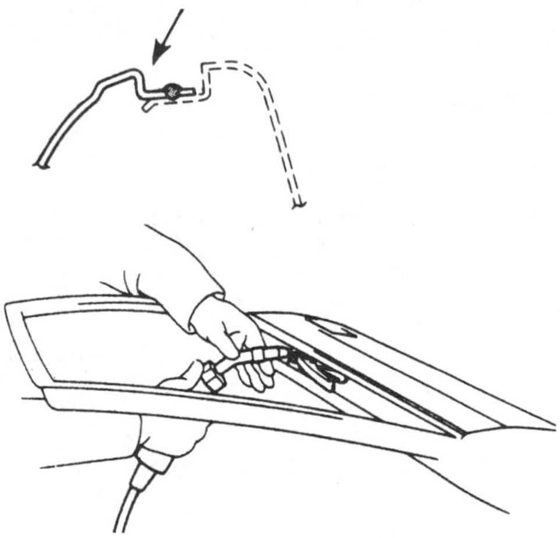

FIGURE 11-9 Welding the door glass opening

ing door skins; use of an appropriate seam sealer will keep water and contaminants out of the door and provide long-term protection. When choosing a seam sealer, probably the two most important factors are paintability and flexibility.

All seam sealers must be paintable and have good adhesion to bare and primed metal. A seam sealer should be allowed to adequately dry before painting. Actual drying time depends on the seam sealer itself, how thick it is applied, and the temperature and humidity during the drying period. Normally, the lower the temperature is below 70 degrees Fahrenheit, the longer the necessary dry time. Also, the higher the humidity is above 50 percent, the longer the necessary dry time. Never attempt to use a seam sealer without first reading the instructions provided by the manufacturer.

REPLACING ADHESIVE-BONDED DOOR SKINS

1. After removing all hardware, trim material, and the glass, remove the door from the vehicle.
2. Use a grinder and 3-inch, No. 24-grit roll-type disc to grind the edge of the door. This will make it possible to safely separate and remove the damaged skin.
3. Peel off the remaining hem flange using a metal chisel.
4. Use sandpaper to remove the rust from those areas which are too tight for a grinder. Then use the grinder and No. 24-grit disc to clean all areas where the door skin meets the door frame.
5. If necessary, straighten the door frame.
6. Thoroughly clean the edge of the door frame, inside and out, with a good adhesive cleaner.
7. Check the replacement door skin for fit.
8. Prime both sides of the door frame with adhesive primer.
9. Cut the tip of the adhesive nozzle to provide a 1/8-inch bead of adhesive.
10. Apply the adhesive in a continuous bead. The adhesive can be applied either to the inside of the door skin creases or to the door skin side of the door frame.
11. Carefully position the door skin on the frame.
12. Flange the door skin in the usual manner. Be sure to wipe away any excess adhesive along the flange.
13. Seal the crimped seam with seam sealer.
14. To protect against rust and corrosion, apply an anti-corrosion compound to the inside of the door.
15. Bond or tack weld the tabs at the top edge of the door skin. Use the method that best suits the design of the door. Keep in mind that this step will not be necessary on certain vehicle makes.
16. Paint and reassemble the door.

REPLACING SMC DOOR SKINS

As described in Chapter 2, SMC panels are becoming popular on unibody vehicles. Many doors are made completely of SMC, except for ultra-high-strength steel door intrusion beams, and the steel used for lock and hinge reinforcements. To replace an SMC door skin, proceed as follows:

1. If not already done, cut away the center of the skin. Air shears or a pneumatic saw work well because the cut depth is easily controlled.
2. Remove the remaining door skin by heating the bonding areas and applying pressure with a pry bar or chisel. Be careful not to destroy the door frame flange.
3. Sand the door frame flange to remove all remaining adhesive. Clean the bonding areas of the replacement skin with soap and water, and allow to dry. Sand the bonding areas to expose the glass fibers. Wipe dry with a clean cloth.
4. Apply a bead of two-part adhesive to the door frame flange (Figure 11-10).
5. Set the skin on the door frame and lightly clamp. Do not squeeze too tight; it is desirable to leave adhesive between the skin and frame.

To complete the job, allow the adhesive to cure, paint the door, and then reassemble and mount it on the vehicle.

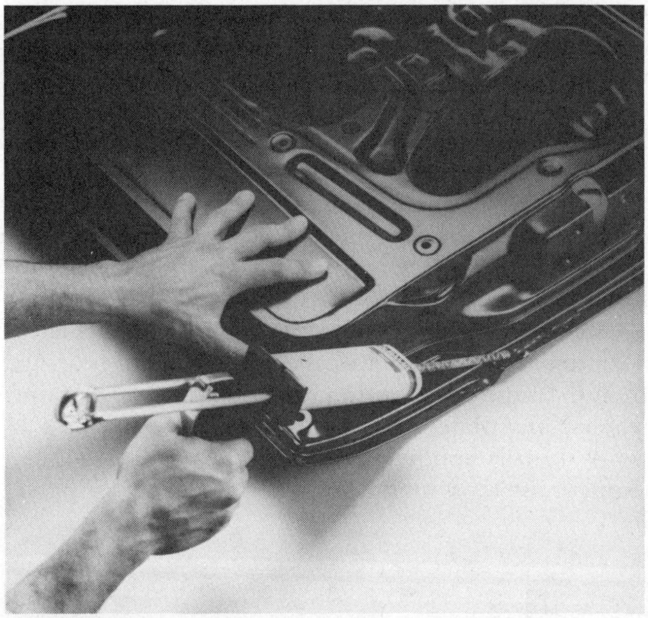

FIGURE 11-10 Applying a two-part adhesive to the door frame flange

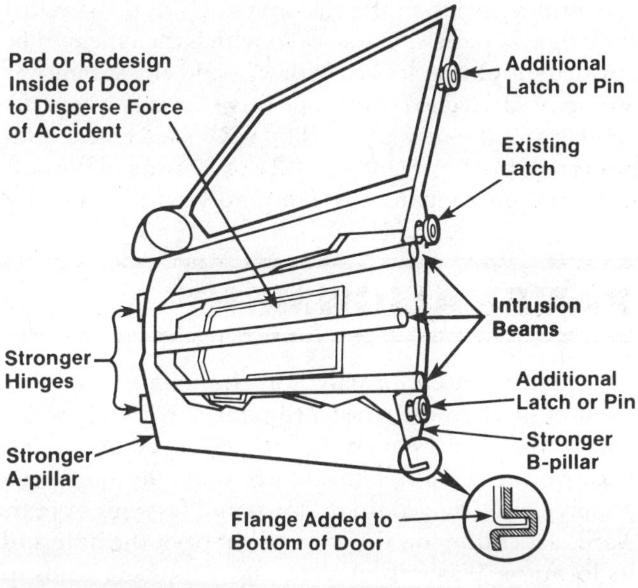

FIGURE 11-11 Inner components of a typical door with intrusion beams

Further information on the repairs of SMC panels can be found in Chapter 13.

Door Frame Metal Reinforcements

All the doors of unibody vehicles have inner reinforcements at various locations. There are several attached to the door frame at the handle and window crank mounting points.

There is also the door intrusion beam. This beam is welded or bolted to two metal support brackets, which are attached to the door frame. Due to new standards set by the National Highway Traffic Safety Administration (NHTSA), many automobile designs will have to change to withstand harsher side impacts. Manufacturers have until 1994 to comply with the new standard. One-third of the vehicles now on the road already comply to the new standards, as shown in Figure 11-11.

There are some other door frame reinforcements, such as at the hinge locations and the door lock plate.

CUSTOM BODY PANELS

Customizing is a growing trend. Many vehicle owners, when faced with the need to replace damaged panels, spend the extra money to give their vehicles a performance facelift. A number of companies are marketing spoiler and air dam kits that not only enhance the appearance of the vehicles, but also improve the

FIGURE 11-12 Vehicles equipped with air dams, side skirts, and spoilers achieve improved aerodynamics.

aerodynamic performance during high-speed driving (Figure 11-12). Air dams restrict air passing under the vehicle body, thus reducing turbulence and resistance to airflow. The rear spoiler alters the airstream at the rear end to increase traction.

Spoilers, side skirts, air dams, and so on are commonly made from molded polyurethane plastic that usually comes with a silicone release which must be removed with a wax and grease remover. Then, the panel must be scuff sanded with No. 240-grit sandpaper followed by No. 320-grit paper to improve adhesion.

Most custom body panels utilize original fasteners on the vehicle as well as additional fasteners provided by the kit manufacturer. Double-sided adhesive tape is sometimes used to fit the custom panels tightly against the original panels. The area to which the tape will be applied should be cleaned with wax and grease remover and scuff sanded to ensure a positive bond between the vehicle and the new panel. The key to successful add-ons is proper alignment, careful positioning of the new fasteners, and preparation of the panel mating surfaces.

PANEL ALIGNMENT

Manufacturers provide for alignment or adjustment of mechanically fastened panels in several ways. One of the most frequent is slotted or oversized holes. Because the openings are larger than the bolts, the panels can be moved up or down and forward or rearward, depending on the size and shape of the hole and its location. Washers are used under the bolt heads to provide a sufficient bearing surface on which the bolt can be tightened. The hood hinge connection shown in Figure 11-13 is an example of an adjustable fastener with slotted holes.

FIGURE 11-13 Slotted holes permit hood adjustments.

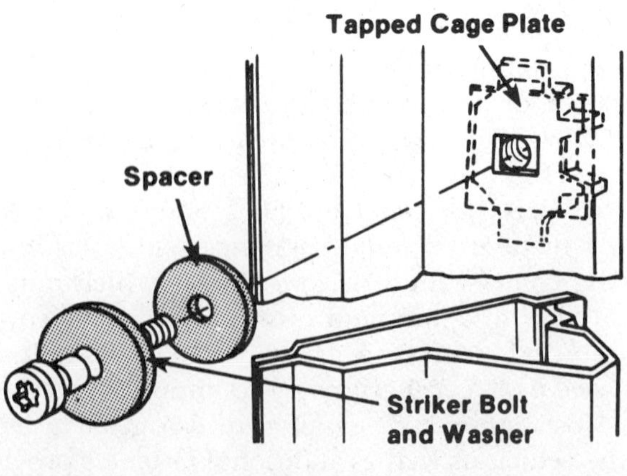

FIGURE 11-14 A caged plate permits striker adjustments.

Another common fastener that allows for panel adjustments is a caged plate (Figure 11-14). The caged plate is like a large nut threaded to accept two or more bolts, depending on the number of bolts in the connection. The plate is housed in a "cage" of thin sheet metal spot welded to the supporting panel (Figure 11-15). The cage is larger than the plate, so that the plate can be moved around; the cage prevents the plate from falling away from the panel. Oversized holes in the panel allow the panel to be adjusted in any direction. Caged plates are often used in doors and door pillars where a cluster of slotted holes would weaken the weight bearing strength of the panel without the reinforcement of the steel backing plate.

A shim is another means of making a panel adjustment. By loosening a bolt, the shim can be slipped

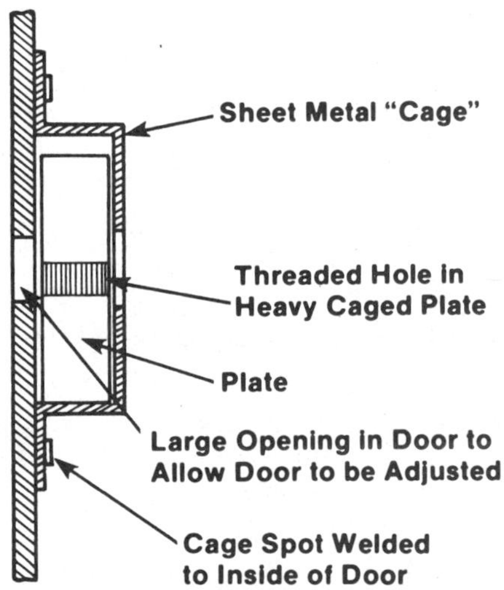

FIGURE 11-15 Caged plate cross-section

FIGURE 11-16 The welding of unibody panels greatly reduces the use of shims.

under the bolt head and around the threads. When the bolt is retightened, the position of the attached panel is changed. Shimming body panels was once a very common operation. But with the welded panels of unibody construction, there are fewer body panels that can be shimmed (Figure 11-16).

Another means of adjusting panels is the adjustable stop (Figure 11-17); this is a threaded bolt with a rubber cap that is attached to the panel by a fixed nut in a clip. A locknut secures the stop in position. Adjustable stops are usually found under the hood to prevent it from vibrating and rattling on its hinges. The rubber cap prevents the stop from chipping the paint and denting the hood.

REPAIRING ADHESIVE-BONDED WELD SEAMS

Adhesives, in addition to fastening door skins, are used in the welded seams around the doors, outer rocker panels, and roof side rails, as shown in Figure 11-18. There are the areas where the vehicle is most prone to flexing with changing loads and road conditions. The addition of weld-bond adhesives in these joints helps to reduce this flexing. It also helps to reduce vibration and noise in these areas. When areas involving weld-bond joints are repaired, it is important for the adhesive to be properly applied before welding. This will ensure that the structure matches its pre-accident performance.

Most weld-bond adhesives are two-part epoxies. They must meet requirements for strength, low and high temperature performance, and moisture resistance. Several product brands meet these requirements; always follow the manufacturer's instructions.

Repair Procedure

The following description of a weld-bonded adhesive joint repair is based on replacing the outer rocker panel. Other weld-bonded adhesive repair procedures are similar.

FIGURE 11-17 Adjustable stops are often found under the hood.

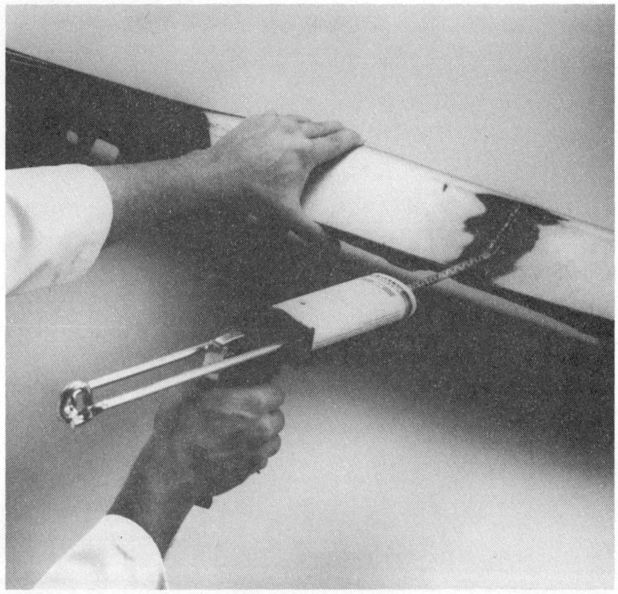

FIGURE 11-18 Weld-bond adhesive is used in the welded seams around the doors, outer rocker panels, and roof side rails.

CAUTION

Before cutting, be sure to support the roof.

To remove the damaged part(s), proceed as follows:

1. Cut off the damaged portions as shown in Figure 11-19 to make removal easier.
2. Using an air grinder, cut the welds labeled "N" in Figure 11-20.
3. Using a mini belt sander, spot cut the welded portions from the inside as shown in Figure 11-21. A spot weld cutter can also be used.

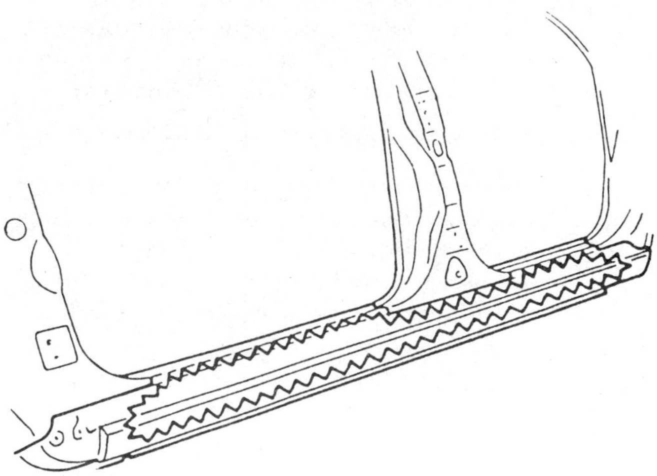

FIGURE 11-19 Cut off damaged portions of the outer rocker panel to make removal easier.

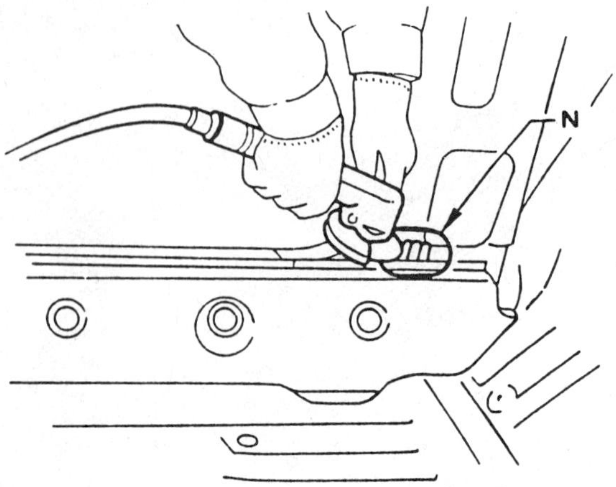

FIGURE 11-20 Using an air grinder, cut the welds labeled "N."

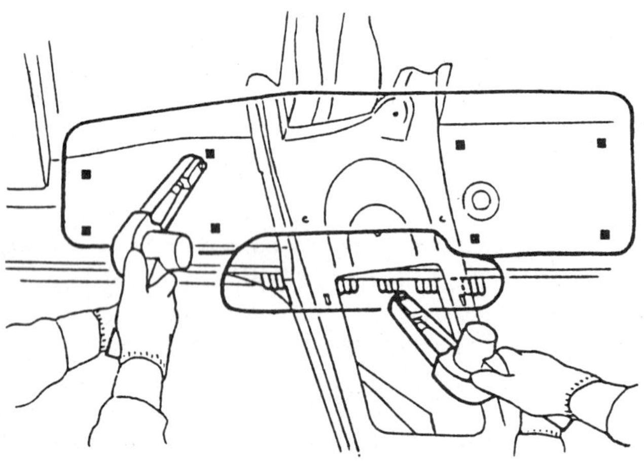

FIGURE 11-21 Using a mini belt sander, spot cut the welded portions from the inside.

4. Using a drill, spot cut completely through the welded parts labeled "B" in Figure 11-22. Use these holes as MIG plug weld holes when installing the replacement part. The outer rocker panel is now ready to be removed.

To install the replacement part, proceed as follows:

1. Prepare the new rocker panel by applying a coat of weld-through primer to the MIG plug weld hole locations (Figure 11-23).
2. As shown in Figure 11-24, apply sealant to the joint area between the rear wheel arch and outer rocker panel.
3. Apply an approved epoxy weld-bond adhesive to the mating surfaces of the rocker panel. Make sure adhesive does not get in the immediate area of the plug welds (Figure 11-25).

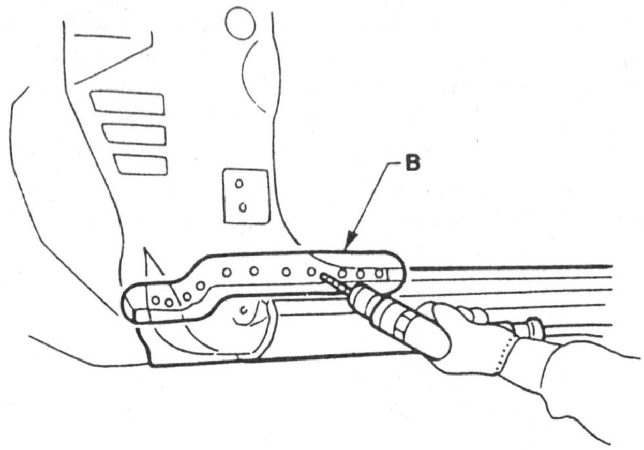

FIGURE 11-22 Using a drill, spot cut completely through the welded parts labeled "B."

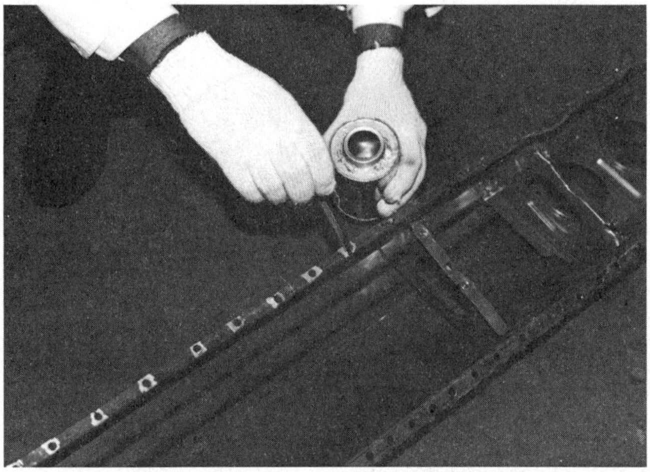

FIGURE 11-23 Prepare the new rocker panel by applying a coat of weld-through primer to the MIG plug weld hole locations.

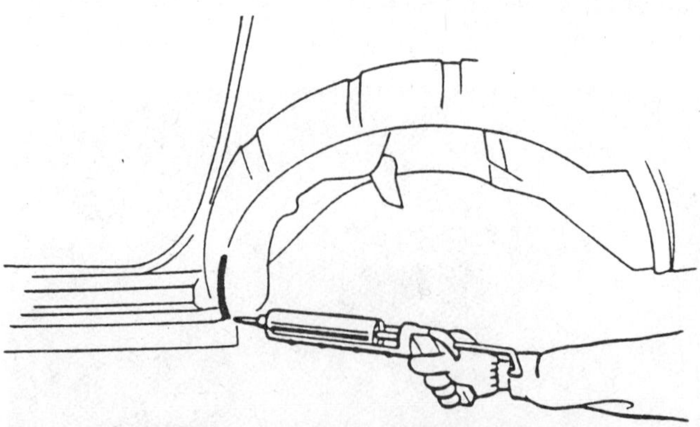

FIGURE 11-24 Apply sealant to the joint area between the rear wheel arch and outer rocker panel.

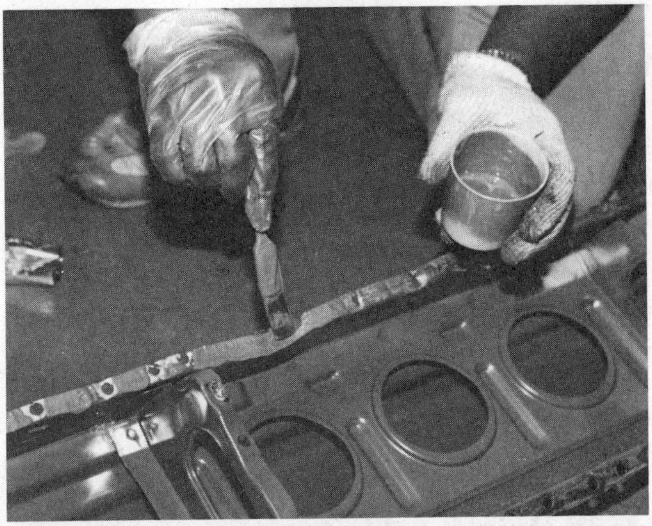

FIGURE 11-25 Applying weld-bond adhesive to the rocker panel mating surfaces

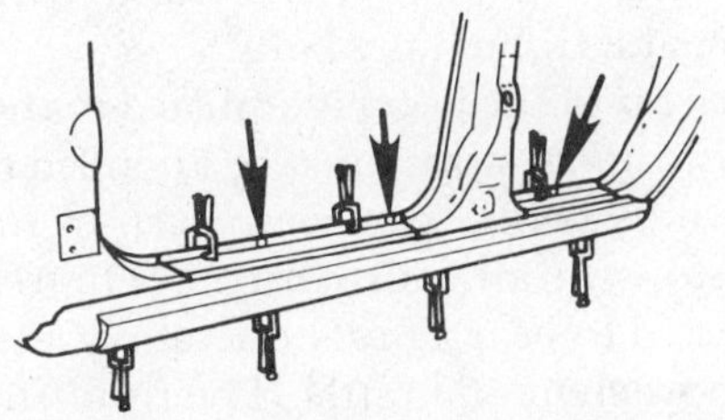

FIGURE 11-26 Installing the new rocker panel

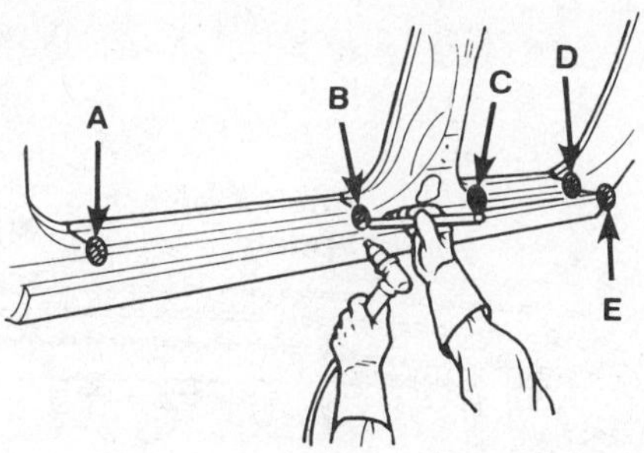

FIGURE 11-27 MIG plug welding the locating holes

4. Align the locating holes, place the new panel in position, and clamp as illustrated in Figure 11-26.
5. Braze at the areas indicated A through E in Figure 11-27 and MIG plug weld the holes.
6. Dress the MIG plug welds and brazed joints with an air grinder to smooth the surfaces, then finish with a sander. Follow with an application of seam sealer.
7. To finish the installation, apply an anticorrosion compound to the inside surfaces of the rocker panel, as described in Chapter 10.

As with other materials used in collision repair, the use of weld-bond adhesives requires a number of personal safety cautions that must be observed. They are:

- Epoxy systems can create skin irritation, so wear gloves and avoid skin contact.
- If skin contact has occurred, wash hands with soap and hot water. Then apply a skin cream.
- If adhesive accidentally contacts the eyes, wash immediately with clean water for 15 minutes. Then consult a physician.
- Spot welding in weld-bond joints can generate gases that can be harmful if inhaled. Be sure to work in a well-ventilated area and wear a respirator. A cartridge respirator is sufficient for occasional brief welding. For longer welding jobs, a fresh air supply respirator is necessary.

HOOD ADJUSTMENTS

The hood is the largest adjustable panel on most vehicles, and adjusting it is usually a two-person job. The hood can be adjusted at the hinges, at the adjustable stops, and at the hood latch. The adjustments allow it to be moved up and down and forward and rearward to align it with the fenders and cowl. The hood should align with the fender, side to side, with a

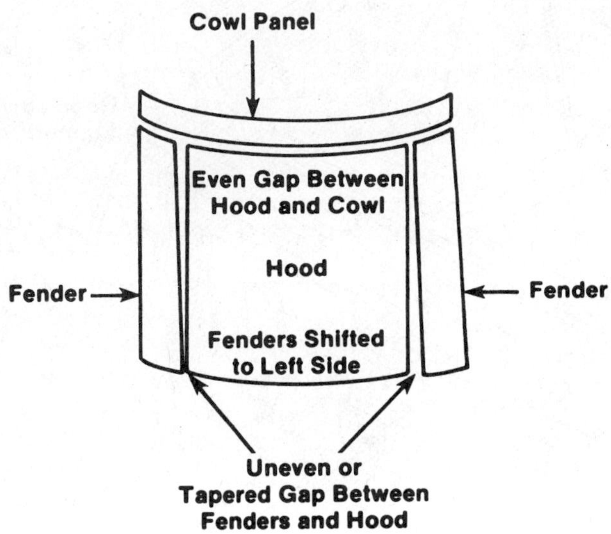

FIGURE 11-28 Checking the gaps around the hood

gap of approximately 5/32 inch in between. The front edge of the hood should be even with the front edge of the fender and there should be enough of a gap at the back edge to clear the cowl panel. A misaligned hood is shown in Figure 11-28.

The rear of the hood is attached to a pair of hinges that are bolted to either the cowl or the inner fender (Figure 11-29). The holes in the hinges are slotted to allow the hinges to be raised or lowered at the cowl or fender and to allow the hood to be moved forward or rearward at the hinges.

The front of the hood is held in place by a hood latch (Figure 11-30). This latch is used to align the front of the hood so that it closes tightly and aligns with the fenders and fascia. Slotted holes are usually provided on the latch for alignment purposes. Most vehicles have adjustable stops mounted on the radiator core support or along the inner edges of the fenders to provide up-and-down adjustment of the corners and prevent hood flutter at the front of the hood. In addition, the hood is equipped with a safety catch to prevent it from flying off should it accidentally open when the vehicle is in motion.

FIGURE 11-29 Typical hood hinge

Hood-to-Hinge Adjustment

Adjusting the hood forward and rearward can be accomplished by slightly loosening the bolts attaching it to the hinges. Then, close the hood and line it up properly, carefully raising it far enough for another technician to tighten the bolts. The front of the hood must align with the front of the fenders, and sufficient clearance must exist between the hood and cowl to allow the hood to be raised without rubbing the cowl.

Hood Height Adjustments

To correct the alignment of the hood up and down, slightly loosen the bolts holding the hinges to the fenders or cowl. Then, slowly close the hood and raise or lower its back side as necessary. When the back of the hood is level with the adjacent fenders and cowl, slowly raise the hood and tighten the bolts.

Once the rear of the hood is adjusted to the correct height, the adjustable stops must be attended to. Some vehicles have only two stops, one in each front corner; others have stops in all four corners. The rear stops must be adjusted to bear against the hood; this eliminates hood movement and rattle. The front stops control the height of the front corners of the hood. Turn the stops in or out until the front is even with the top of the fenders. Be sure to retighten the locknut on the stops after adjusting them.

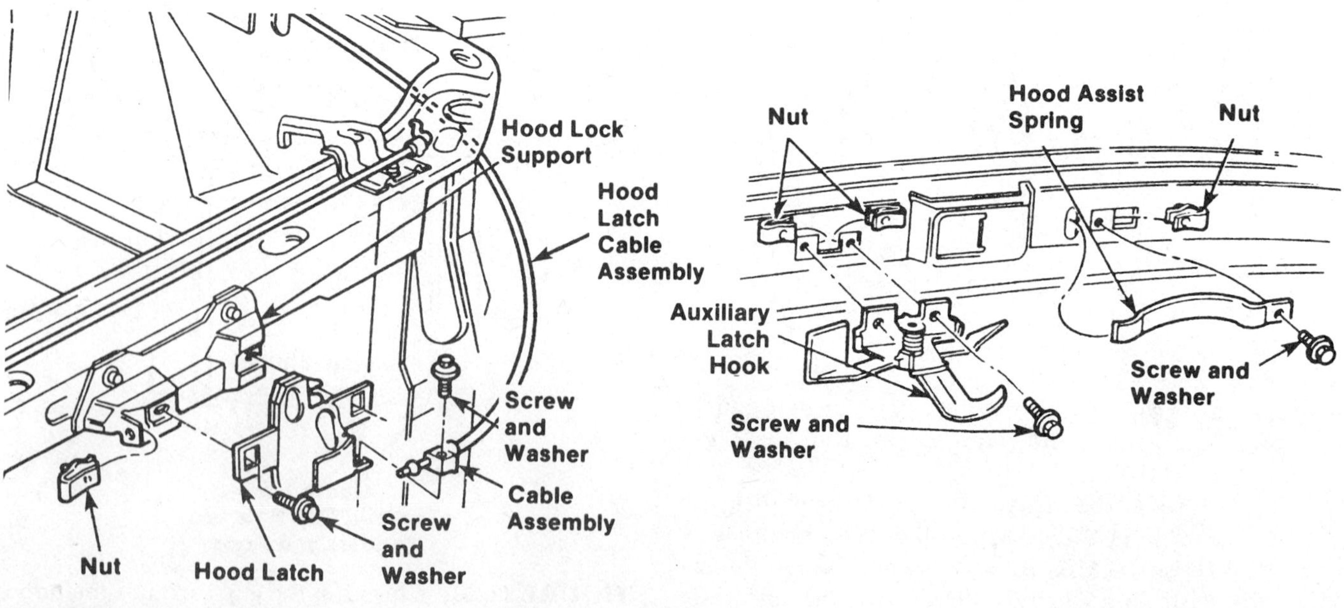

FIGURE 11-30 Components of a typical hood latch assembly

SHOP TALK

In order to lower the hood, it might be necessary to lower the adjustable stops under the corners of the hood. If the hood is lowered without lowering the stops, it might bend and crease.

Hood Latch Adjustment

After making the hood-to-hinge and height adjustments, test the hood for proper latching. If it must be slammed excessively hard to engage the latch, the latch should be raised. The latch should also be raised if the hood does not contact the front stops when closed. A general hood latch adjustment is as follows:

1. Remove the hood latch from the radiator core support and lower the hood.
2. Check that all the gaps around the hood are properly aligned.
3. Reinstall the hood latch and lower the hood until it engages or contacts the auxiliary latch or safety catch.
4. Attempt to raise the hood. If it does open, adjust the auxiliary latch or safety catch so that it engages. Sometimes the hook can be bent until the auxiliary latch "catches."
5. Lower the hood slowly to see if it shifts to one side or the other when it is locked. The sticker bar bolted to the hood should be centered in the "U" of the latch. When the hood is latched, it should be even with the surrounding sheet metal and fit tightly.
6. Loosen the hood latch just enough to be able to move the latch, yet still maintain a tight fit.
7. Move the latch from side to side to align it with the auxiliary latch hook. Move the latch up or down as required to obtain a flush fit between the top of the hood and the fenders when upward pressure is applied to the front of the hood.
8. Tighten the hood latch and open the hood.
9. Loosen the rubber adjustable stops located on the radiator core support.
10. Close the hood and observe the left and right front corners. If necessary, adjust the stops to eliminate any looseness at the front of the hood and ensure a tight fit.
11. Tighten the stops.
12. Check to see that the side stops (if any) are in place and in good condition.
13. Always check the vehicle's service manual for specific hood latch adjustment procedures.

DECK LID ADJUSTMENTS

The deck lid (Figure 11-31) is very similar to the hood in construction. Two hinges connect the deck lid to the rear body panel; the trailing edge is secured by a locking latch. Weatherstripping seals the deck from dust and water. For the seal to be effective, the deck lid must be in continuous contact with the weatherstripping when the deck lid is closed.

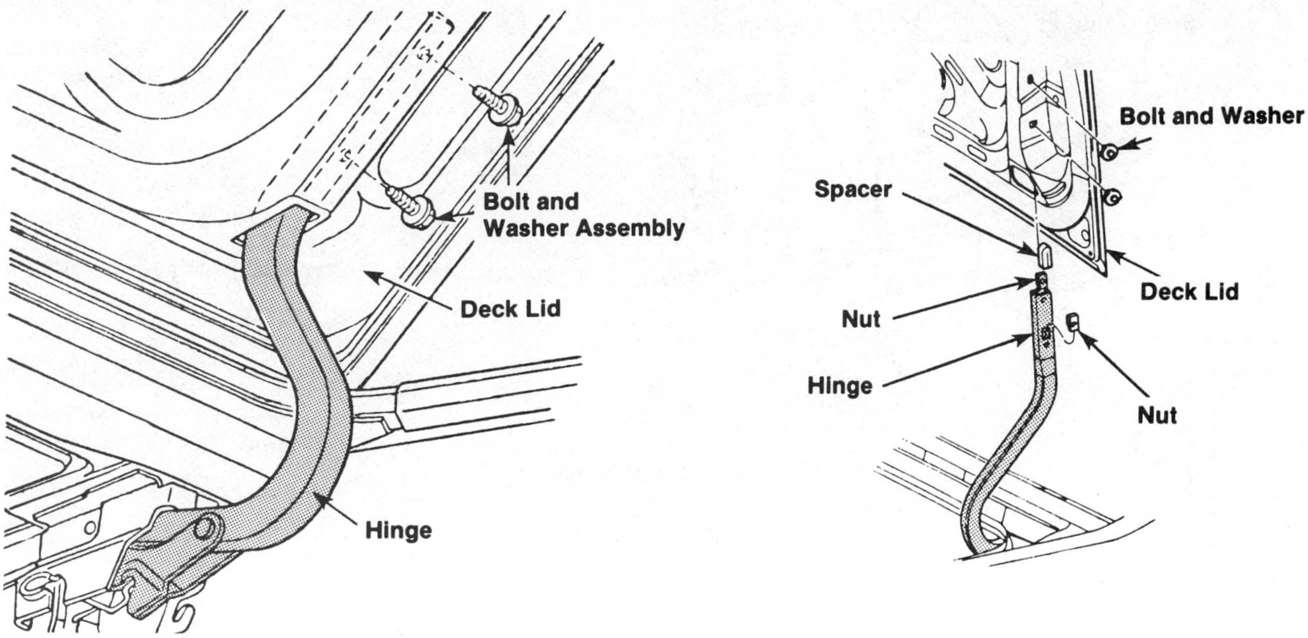

FIGURE 11-31 Major parts of a typical deck lid

The deck lid must be evenly spaced between the adjacent panels. Slotted holes in the hinges and/or caged plates in the deck lid allow it to be moved forward, rearward, and side to side. To adjust the deck lid forward or rearward, slightly loosen the bolts on both hinges. Close and adjust the deck lid as required, then raise it and tighten the bolts.

In some cases, it might be necessary to use shims between the bolts and the deck lid to raise or lower the front. If the front edge must be raised, place the shim(s) between the hinge and the lid in the front bolt area. To lower the front edge, place the shim(s) at the back of the hinge.

Figure 11-32 shows a typical deck lid lock assembly. The lock assembly is usually located in the deck lid, with the striker plate bolted to the rear body panel. On some vehicles, these positions are reversed. The deck lid striker plate can usually be adjusted up, down, and sideways to properly engage, align, and tightly hold the deck lid. The necessary adjustments are similar to those given for the hood.

Hatchbacks are usually difficult to align because of their size. For the most part, many lids of this type are nearly horizontal in design, which makes them more prone to water and dust leaks. Some models use adjustable hinges, while others use welded hinges. Hatchbacks also use gas-filled door lift assemblies, or springs, one at each upper and lower corner of the lid. Some play might be available in the door lift support brackets to allow adjustment of the hatchbacks. For specific adjustment procedures, check the service manual.

FENDER ADJUSTMENTS

Fenders are usually bolted (Figure 11-33) to the radiator core support or to the inner fender panel, and to the cowl behind the door and under the vehicle. When these bolts are loosened the fender can be moved forward and rearward.

SHOP TALK

The screws holding the splash guard to the fender might also have to be loosened to adjust the fender.

The fender can also be moved in and out so that it is flush with the door and parallel with the hood. This fender-to-door adjustment can be made by shimming the two large bolts attaching the fender to the cowl. The top bolt is usually in the door hinge pillar; the bottom bolt is either in the hinge pillar or under the vehicle in the rocker panel area. By shimming the top bolt, the upper fender can be moved out. Shimming the bottom bolt will move the lower portion of the fender out. If the fender is in too far and not flush with the door, the protruding door edge will cause noise and wind turbulence when the vehicle is in motion.

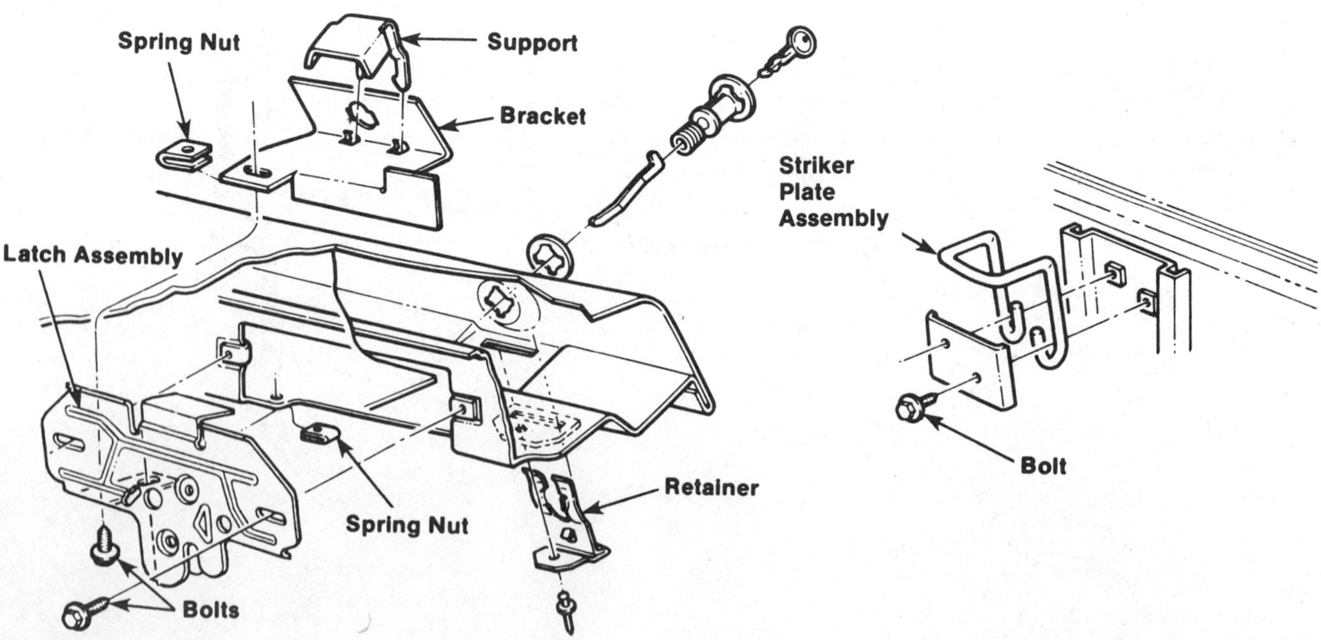

FIGURE 11-32 Typical deck lid lock assembly

FIGURE 11-33 Fender connections

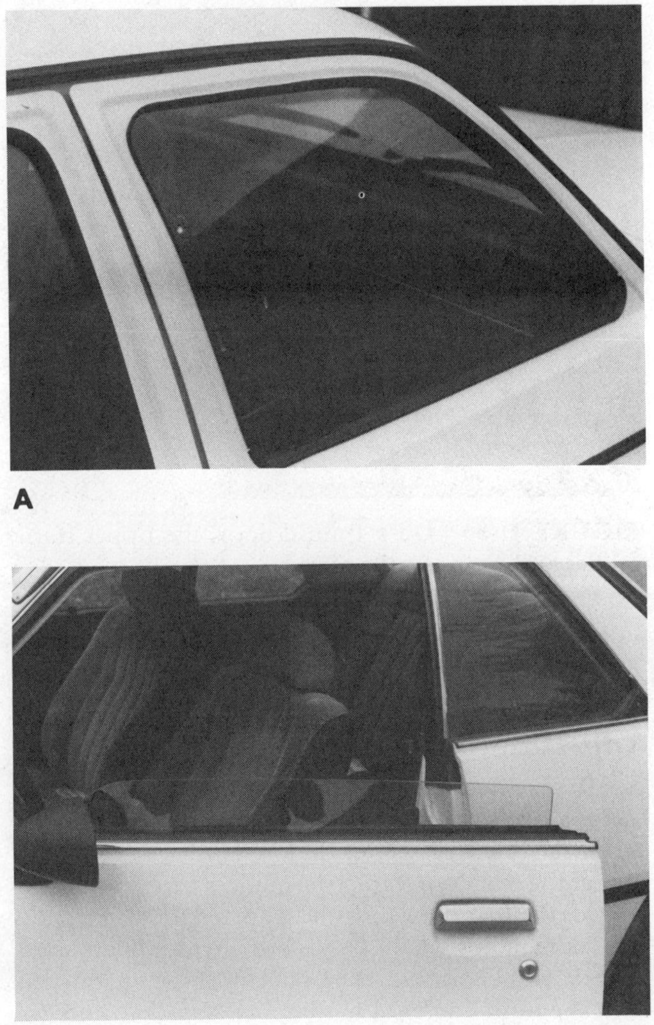

A

B

FIGURE 11-34 (A) Framed door and (B) hardtop door

Often the fender and hood adjustments must be made simultaneously to achieve a satisfactory result. The gap between fender and hood should be no more than 5/32 inch. The door-to-fender gap should also be no more than 5/32 inch. And, of course, the front of the fender and the hood should be aligned as well. The result will be even spacing all around the fender.

DOOR ADJUSTMENTS

Doors must be accurately adjusted so that they close easily, do not rattle, and do not leak water or dust. This section will describe various door adjustments. Complete information on servicing door and vent window glass can be found in Chapter 12.

Types of Doors

Two types of doors are used on today's unibody vehicles: framed and hardtop (Figure 11-34). The framed door uses an upper frame structure that surrounds the glass; the hardtop type does not. Both types use one-piece window glass. Some vehicles use a vent glass assembly and a one-piece glass section.

The glass mechanism on framed doors is rather simple: the upper frame serves as a guide and support for the glass (Figure 11-35). Hardtop doors do not have this advantage, so the glass generally rests against the top door opening. A soft rubber gasket is used in the door opening to protect the glass when it is fully raised. The glass must have some means of support and height control when it is partially lowered. Hardtop door glass is prone to damage if it is allowed to come in contact with the roof drip rail when the door is closed. The glass must also be properly tilted to make contact with the upper gasket; if it does not tilt in far enough, the door will leak water or dust. If it is tilted in too far, the door

FIGURE 11-35 On a framed door, the upper frame guides and supports the glass.

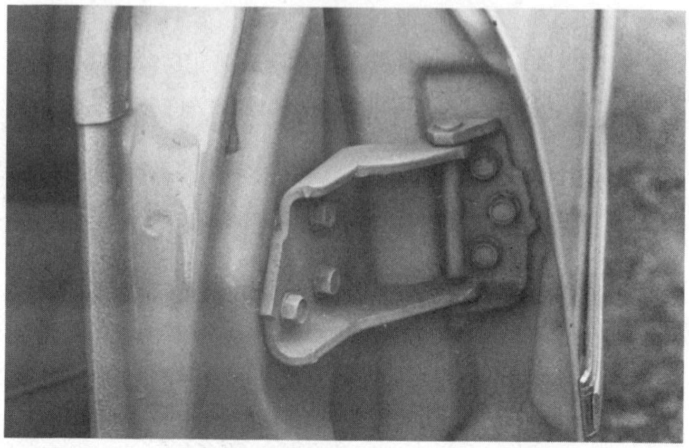

A

B

FIGURE 11-36 (A) Bolted and (B) welded door hinges

will be hard to close and the gasket can be damaged. The framed door requires the least amount of adjustments for the glass.

Adjustment Procedures

Doors must fit the openings in which they are hung, as well as align with the adjacent body panels. When the doors on a four-door sedan need adjusting, start at the rear door. Since the quarter panel cannot be moved, the rear door must be adjusted to it. Once the rear door is adjusted, the front door can then be adjusted to fit the rear door; next, the front fender can be adjusted to fit the front door. On hardtop models, the windows can then be adjusted to fit the weatherstripping. The windows are usually adjusted starting with the front vent assembly and working toward the back. The vent assembly is adjusted to fit the front door pillar, and the front window is then adjusted to it. The rear door window is adjusted to the front window rear edge and the opening for the rear door assembly.

Doors are attached to the body with hinges. The hinges can be bolted or welded to both the door and body. Figure 11-36A shows a hinge bolted to both the door and body. Figure 11-36B shows a hinge welded to the door and bolted to the body. Obviously, no adjustments can be made to the welded door hinge. The bolted hinge, however, can be adjusted forward, rearward, up, and down. The use of shims also allows the hinge to be moved in or out as desired.

To adjust a door, follow these steps:

1. Determine which hinge bolts must be loosened to move the door in the desired direction.
2. With a padded pry bar or scissor jack and wooden block (Figure 11-37), loosen the hinge bolts just enough to permit movement of the door.
3. Move the door the necessary distance. Tighten the hinge bolts and check the fit to be sure there is no binding or interference with the adjacent panel.
4. Repeat the operation until the desired fit is obtained, then check the striker pin alignment for proper door closing.
5. Remove the striker pin and check the door position in relation to the opening.
6. The door and glass must be checked to assure proper alignment to the roof rail and vertical weatherstripping.

On some vehicles, a special wrench must be used to loosen and tighten the hinge bolts. If the hinges are to be removed, a line should be scribed around each one to mark the position; this makes reinstallation easier. Be careful not to cut the paint or corrosion will result. It might be necessary to loosen the fender at the rear bottom edge in order to reach the bolts. If the hinge pins

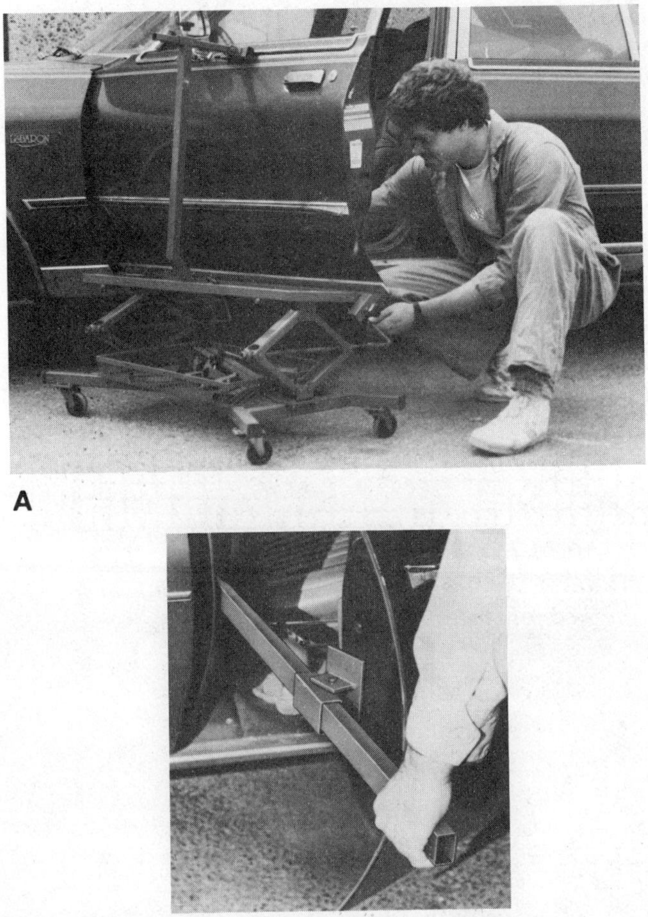

A

B

FIGURE 11-37 Use of a (A) scissor jack and (B) pry bar make door adjustments easier.

are worn out, the hinges might have to be changed. Some hinges use bushings around the pins. When these bushings are worn out, replace them. This will serve to retighten the pin in the hinges and also readjust the door to a degree.

In-and-out adjustments are also very important. Not only must the door fit the opening, it must also be reasonably aligned in and out to fit the body panels. The door must provide a good seal between the weather-stripping and the body opening. The weatherstripping should be compressed sufficiently in the opening to prevent water, dust, and drafts from entering the vehicle.

Care must be taken when adjusting the in-and-out movement of the door. If the door is moved out on the top hinge, it will not only affect the top of the door, it will also move the opposite bottom corner in. If the bottom of the door is moved in on the hinge, it will move the opposite corner out. The key is to move the door in or out equally on both hinges; this way, it will only affect the front of the door because the amount of adjustment decreases toward the back. The center door post, striker pin, and lock will determine the position of the rear of the door. The front edge of the door should always be slightly in to stop wind noises at the leading edge of the door panel.

When door adjustments are necessary, it is sometimes advantageous to remove the striker plate. Doing so allows the door to be centered in the opening. The striker plate has not been adjusted properly if the door rises or is forced down when it is closed. The striker plate should merely provide a slight rise to the door when it is closed. The striker plate can be moved up and down, in and out, and back and forth.

When removing or replacing door hinge pins, the technician must remove the hinge spring first. To remove the spring, cover it with a heavy cloth to prevent it from flying loose. Use a spring compressor tool (Figure 11-38) to remove the spring, then remove the hinge pins. The door can then be removed from the vehicle. When the door is reinstalled, it is usually necessary to again use the compressing tool. The spring must be seated properly in the tool before compressing it to prevent it from slipping and causing damage or injury.

To replace the door side hinge strap, follow the procedure given in the service manual.

Station Wagon Rear Door Adjustments

Most unibody station wagons have a three-way tailgate. (Compact wagons are the exception; most of these have a hatchback style rear gate.) The three-way tailgate has a unique hinge and locking arrangement that allows it to be operated as a tailgate with the glass fully down or as a door with the glass up or down.

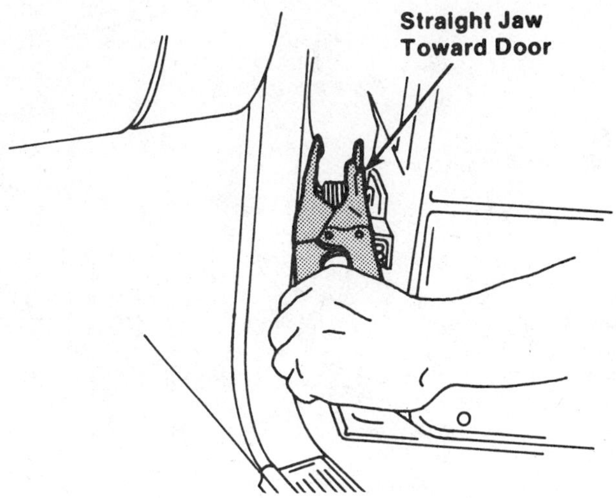

FIGURE 11-38 Using a spring compressor tool

The lock system on the three-way tailgate is designed to perform the following functions:

- Allow the tailgate to be opened and closed as a door with the glass up or down, while at the same time preventing accidental operation of the upper left lock, which allows the tailgate to be opened as a gate.
- Allow the tailgate to be opened and closed as a gate with the glass down, while at the same time preventing accidental operation of the lower right lock, which allows the gate to be opened as a door.

Before doing any station wagon door adjustments, closely examine the door area to determine where the misalignment exists. It might be necessary to remove the right upper striker pin and adjust the tailgate as a regular door. If the right lower striker pin forces the door up or down and out of alignment, readjust the pin to provide a smooth open/close action.

The left hinges can be adjusted to position the tailgate in the opening and to align it flush with the adjacent body panels. They are also adjustable for smooth lock operation. Closely examine the hinges to determine what adjustments are available. On some vehicles the lower left hinge provides up-and-down as well as in-and-out adjustment for the tailgate. Some vehicles provide for certain types of adjustments on the body side of the hinge; others allow adjustment to the door side of the hinge. After adjusting the left hinges, it is necessary to reinstall and adjust the strikers. If all adjustments have been done correctly, the tailgate should operate smoothly and fit properly.

SHOP TALK

Prior to performing any adjustments, the position of the hinge, latch, and striker should be marked for easier realignment.

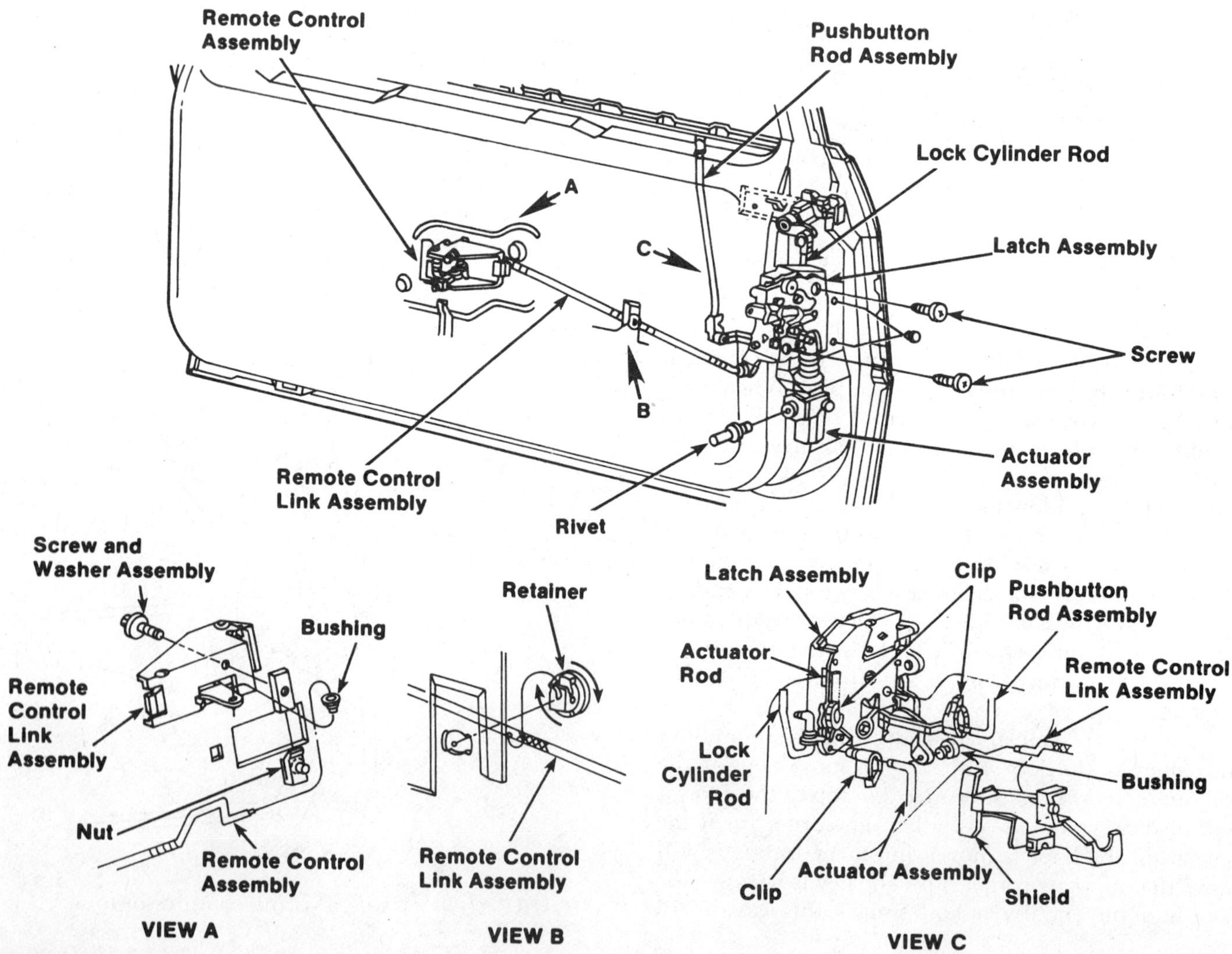

FIGURE 11-39 Typical door lock assembly

DOOR AND DECK LID LOCKS

Door lock assemblies usually consist of the outside door handle and linkage, the inside door lock mechanism, and the inside locking rod (Figure 11-39). Various types of exterior door handles are available and usually depend on the vehicle model. Among those in use are the pushbutton and the lift handle type. On the pushbutton type, the button directly contacts the lock lever that releases the mechanism to open the door. Most exterior door handles operate the lock mechanism by using one or more rods. The handles can be replaced by raising the window and removing the interior trim, panel, and water shield to gain access to the attaching hardware and linkage.

Exterior door handles are attached by screws, bolts, or rivets to the door panel. Some causes of malfunctioning exterior door handles are worn bushings; bent or incorrectly adjusted lock cylinder rods; lack of lubrication on the handle, linkage, or latch; and worn or damaged latches.

Inside door lock mechanisms are generally the pull handle type. The mechanisms are connected to the lock by one or more lock cylinder rods and are accessible by removing the interior trim panel and water shield. The mechanism is attached to the door handle with screws, bolts, or rivets, depending on the vehicle make. The rods are usually long and must be supported to prevent them from engaging or getting caught in other components within the door panel. Clips or bushings are used to secure the rods in place.

Deck lid or hatch doors usually do not have exterior or interior door handle mechanisms. They operate with a key (or instrument panel switch on powered units) and lock mechanism. Some are adjustable; others are riveted in place.

The lock cylinder on door and deck lids is usually held on the panel with a retainer. A sealing gasket is also used on the cylinder to seal out water and dust. The retainer must be disengaged to remove the lock cylinder. A pointed tool or screwdriver can be used to accomplish this. On some vehicles the cylinder can be removed once the linkage is disconnected. The cylinder is either directly connected to the latch or connected by a lock cylinder rod. The lock is usually lubricated with dry powdered graphite.

An optional feature on some vehicles is that all the latches can be locked at one time from the driver's door control panel. The control panel electrically operates a power cylinder to control the latches. The power unit is bolted or riveted to the inner door frame. Failure of this system to operate could be due to faulty wiring or connections or a defective power unit.

Some vehicles have an interior pushrod or button to lock the doors. On this type, the doors cannot be opened from the inside once they are locked. This is a safety feature that is especially useful when small children are placed in the rear seat area.

Keyless door entry systems are becoming a more common feature on late-model vehicles. The system consists of two main components:

- Five button keypad on the outside panel of the driver's door

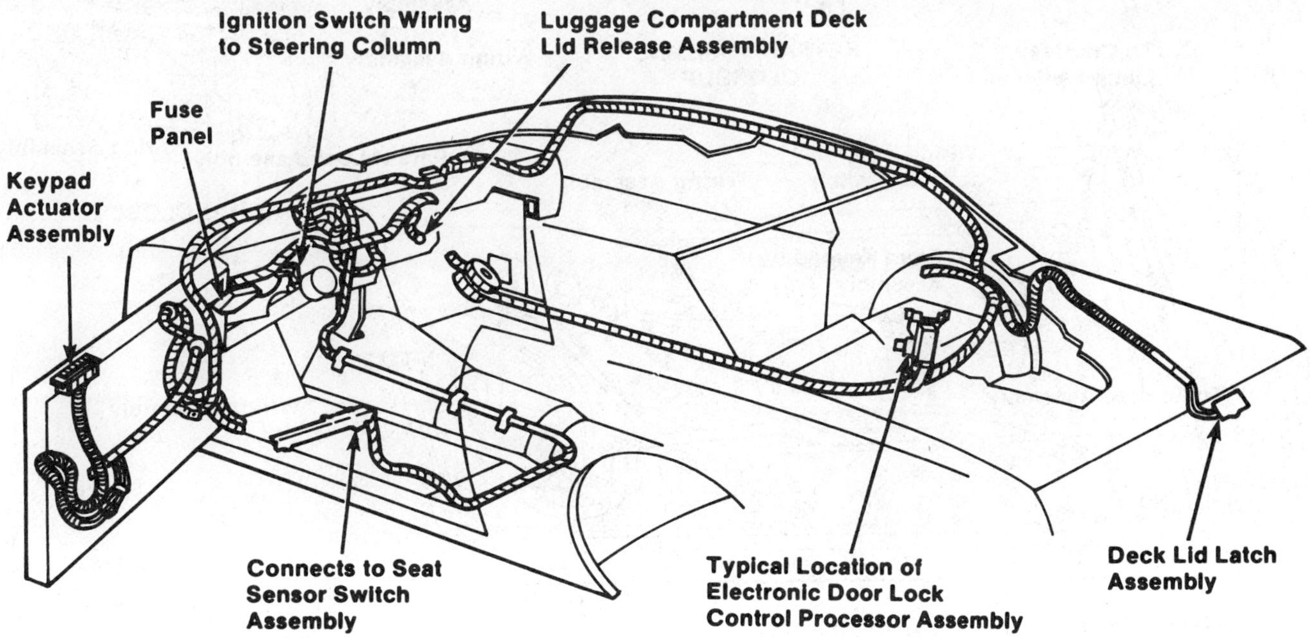

FIGURE 11-40 Typical keyless entry system

- Electronic microprocessor/relay module

A typical keyless entry system (Figure 11-40) performs the following functions:

- It unlocks the driver's door. A keypad code is programmed into the system at the factory. The factory-programmed code is permanently recorded on the owner's warranty card (usually located inside the luggage compartment deck lid) and on a separate code card.

  Some systems also allow the owner to select and program their own personal code (a birthdate or part of a social security number, for example) by pressing a specified sequence of keypad buttons. When either the factory-programmed code or the owner's code is entered, the driver's door unlocks.
- It unlocks the other doors of the vehicle if the keypad button is pressed within a certain time limit of the driver's door unlocking.
- It turns on the interior lamps and the illuminated keyhole on the driver's door. The lamps are turned on by pressing any keypad button or lifting the door handle.
- It unlocks the luggage compartment deck lid when the keypad button is pressed within a certain time limit after the driver's door is unlocked.
- It locks all the doors automatically when:
  —The driver's seat is occupied.
  —All the doors are fully closed.
  —The ignition switch is turned on.
  —The transmission selector passes through the reverse position.
- It locks all the doors from outside the vehicle when the keypad buttons are pressed at the same time.

Collision damage or a malfunction in the keyless entry system might require the removal of related components from the vehicle. Figure 11-41 shows a keypad and wiring harness. The keypad can be secured by clips or screws. The microprocessor/relay module is located either under the instrument panel or on the package tray in the luggage compartment. It is secured with screws or bolts. After reassembly of the keyless entry system components, test the system to ensure that it is fully operational.

INSTALLING SOUND-DEADENING PADS

Reducing noise, vibration, and harshness (NVH) as already mentioned has become an important focus of

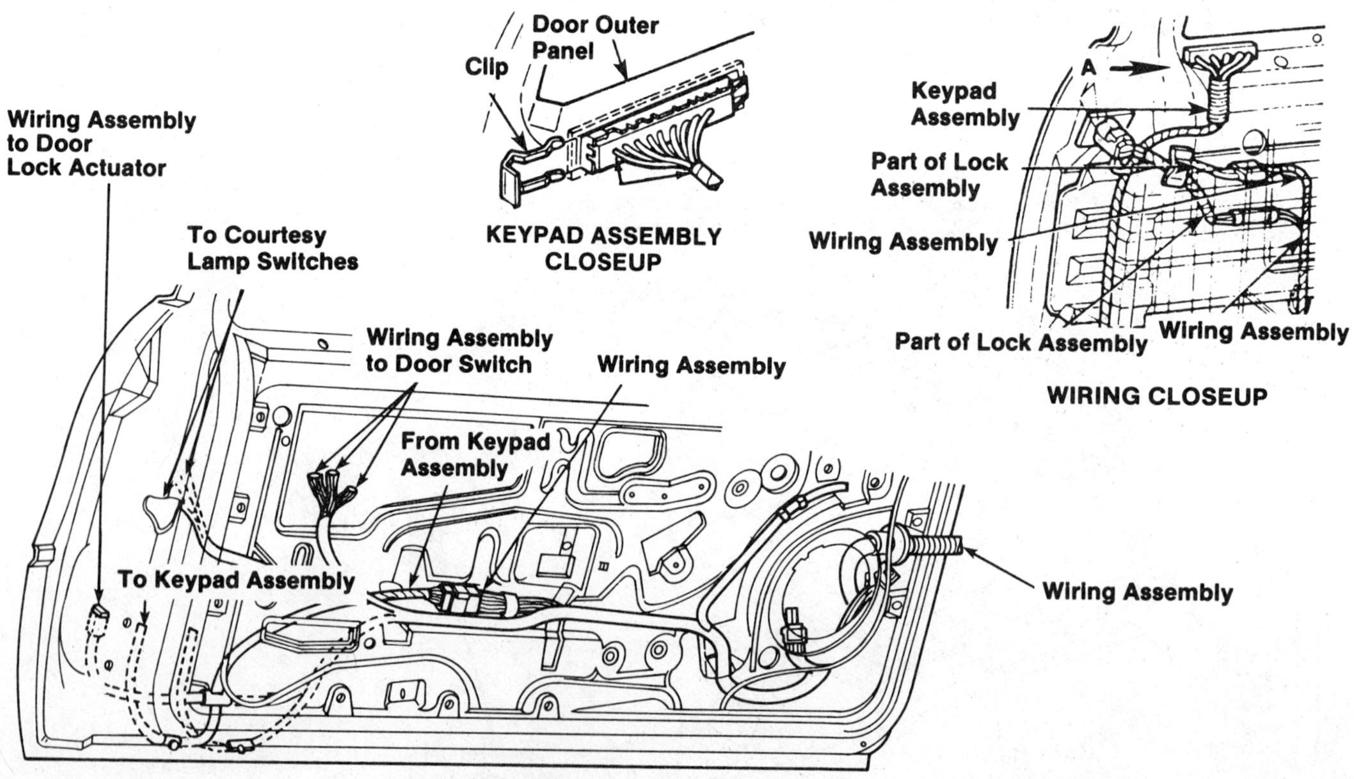

FIGURE 11-41 Door keypad and wiring harness

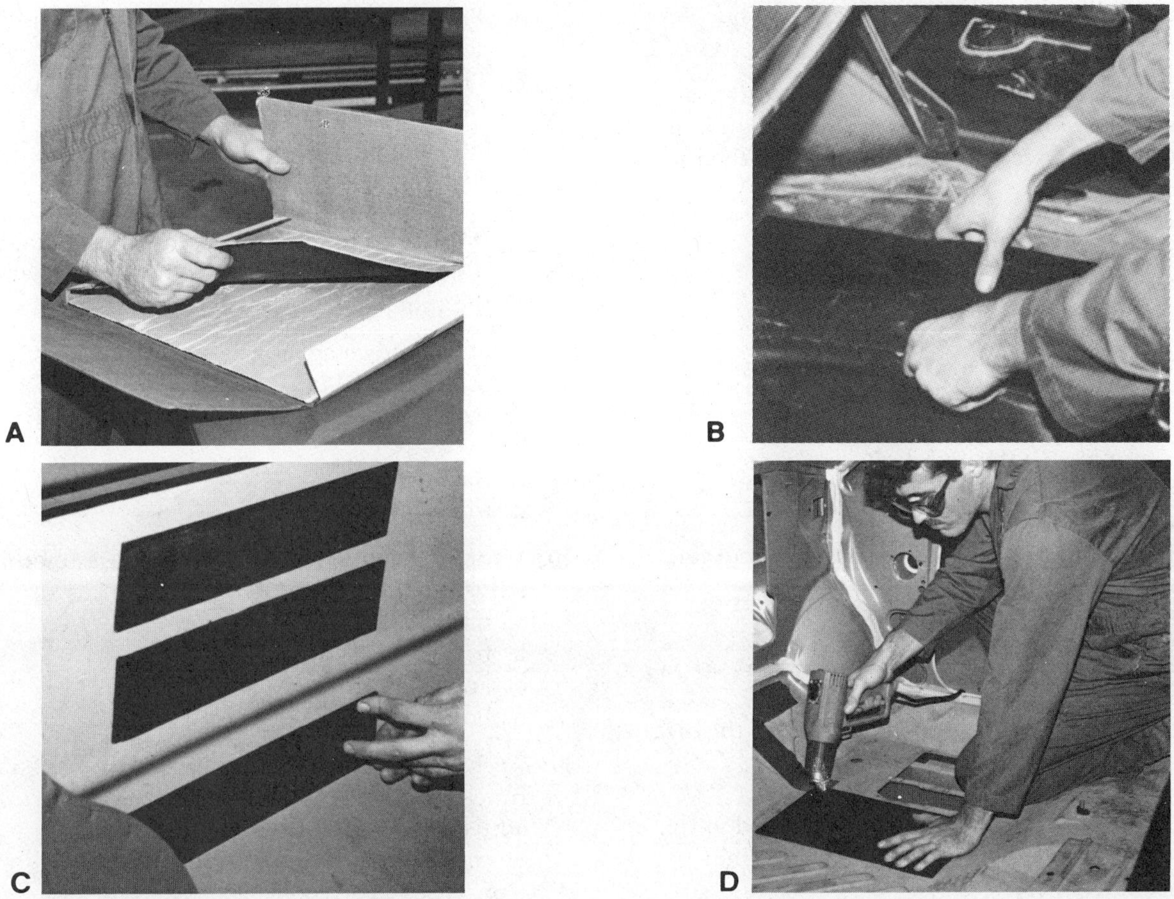

FIGURE 11-42 (A) One type of product is available in 12-inch by 12-inch and 16-inch by 16-inch squares. (B) Sound-deadening pads come with an adhesive back. Just cut to shape, peel, and apply. (C) Another product comes in a 3-inch by 50-foot roll. (D) Sheets can be heated with a heat gun to make them flexible to conform to the sheet metal.

automotive engineering in recent years. Sound-deadening pads are bonded to the inside surfaces of trunk cavities and doors. This sound-deadening material is made of a plastic or asphalt-based material about 1/8 to 3/16 inches thick.

During collision repair, these pads must be replaced to ensure that the sound-deadening qualities of the body assembly match its pre-accident performance. Guidelines to aid in replacing sound-deadening pads are as follows (Figure 11-42):

- The original factory material is bonded and, in some cases, heat formed to the surface contours. This is a combustible material that must be removed from any area to be welded.
- After repair, the area must be properly refinished to provide corrosion protection before the sound-deadening material is applied.
- Some pads are available as factory service parts. Be sure to check the factory body repair manual for details. If a factory pad is not available, the sound-deadening qualities must be restored with a suitable replacement product. These should be available from several sources. One product is available in 12-inch by 12-inch and 16-inch by 16-inch squares with a pressure-sensitive adhesive on the back. Just cut to size and shape, peel off the backing, and apply. The sheets can be heated with a heat gun to make them flexible to conform to the contour of the sheet metal. Another product comes in a 3-inch wide by 50-foot long roll. The material bonds aggressively to plastic, metal, and other automotive surfaces.

REVIEW QUESTIONS

1. The decision to repair or replace a door is based on ______________.
   a. the amount of damage to the door frame
   b. the accessibility to the backside of the door skin

   c. the amount of work hardening in the buckles
   d. all of the above

2. Technician A uses a power chisel to separate a welded door skin from the door skin hem flange. Technician B uses a welding torch to do the job. Who is correct?
   a. Technician A
   b. Technician B
   c. Both A and B
   d. Neither A nor B

3. When replacing an adhesive-bonded door skin, what is the alternative to bonding the tabs at the top edge of the skin?
   a. fastening with through-bolts
   b. fastening with rivets
   c. tack-welding
   d. There is no alternative to adhesive bonding.

4. Which of the following statements concerning custom body panels is incorrect?
   a. One of the keys to successful add-ons is preparation of the panel mating surfaces.
   b. Seldom do custom body panels utilize the original fasteners on the vehicle.
   c. Rear spoilers increase vehicle traction.
   d. Double-sided adhesive tape can be used to fit custom panels tightly against the original sheet metal.

5. Which of the following fasteners is used on unibody vehicles?
   a. caged plate
   b. adjustable stop
   c. adjustable fastener with slotted holes
   d. all of the above

6. The hood should align with the fender, side to side, with a gap of approximately ____________ in between.
   a. 1/32 inch
   b. 5/16 inch
   c. 1/16 inch
   d. 5/32 inch

7. Technician A occasionally uses shims to raise or lower the front of a deck lid. Technician B maintains that shims should never be used on deck lids. Who is correct?
   a. Technician A
   b. Technician B
   c. Both A and B
   d. Neither A nor B

8. Which of the following statements concerning fender adjustments is incorrect?
   a. The gap between fender and hood should be no more than 5/32 inch.
   b. The screws holding the splash guard to the fender might have to be loosened to make the fender adjustment.
   c. The fender and hood adjustments should never be made simultaneously.
   d. If the fender is in too far, the protruding door edge can cause noisy wind turbulence when the vehicle is moving.

9. Most unibody station wagons feature a ____________ tailgate.
   a. hatchback style
   b. two-way
   c. three-way
   d. four-way

10. Which type of door requires the least amount of adjustment for the glass?
    a. hardtop
    b. framed
    c. any GM vehicle door
    d. both a and b

11. Technician A says that a possible cause of a malfunctioning exterior door handle is an incorrectly adjusted lock cylinder rod. Technician B says that worn bushings are a possible cause of the problem. Who is correct?
    a. Technician A
    b. Technician B
    c. Both A and B
    d. Neither A nor B

12. A keyless entry system automatically locks all the doors when ____________.
    a. the ignition switch is turned on
    b. the driver's seat is occupied
    c. the transmission selector passes through the reverse position
    d. all of the above

13. Which of the following statements is incorrect?
    a. Sound-deadening material is combustible.
    b. Some sound-deadening pads are available as factory service parts.
    c. After a repair, apply sound-deadening material before dealing with corrosion protection.
    d. None of the above

14. Where are caged plates often used?
    a. doors
    b. door pillars
    c. both a and b
    d. neither a nor b

CHAPTER

12 Other Collision Repairs

OBJECTIVES

- Describe gasket, full cutout, and partial cutout glass replacement procedures.
- Describe the basics of vent window and tailgate door glass service.
- Perform a headlight adjustment.
- Explain how to replace a bumper and test the energy absorbers.
- Install adhesive and rivet-on body side moldings.
- Explain how to repair and replace damaged vinyl tops.
- Conduct detailed seat belt and air bag inspection and service.

After all sheet metal damage has been repaired, there is still much work to be done to restore a damaged vehicle to its pre-accident condition. The front or rear bumper, windshield, lights, and windows often require replacement. Damage to the roof panel might necessitate replacing the vinyl top material. Moldings and weatherstripping often have to be replaced, as well as interior trim, seats, and instrument panel. Such a wide diversity exists among manufacturers in these areas that only typical examples of each type of repair can be noted here. Always refer to a manufacturer's service and/or body repair manual when in doubt as to the best procedure for replacing body accessories.

Today's vehicles are built with a lot of glass that affords greater visibility for driving. Most of the glass built into the vehicle is centered in the windshield, doors, and rear window. Frequently, this glass is broken as a result of a collision, flying gravel, or vandalism.

Glass also plays an important role in aerodynamic styling (Figure 12-1). Air drag is reduced dramatically by fitting the glass accurately to the sheet metal and eliminating trim. This approach reduces wind noise and provides a more sleek, uncluttered appearance. Stationary glass panels in modern unibody vehicles also play an important role in the strength of the body structure. That is, the stationary glass in most vehicles is an integral part of the upper body structure. It provides lateral bracing to help prevent roof collapse during a rollover. For this reason, when stationary glass is replaced, it must be installed in the same manner as it was by the manufacturer to maintain the strength of the upper body. If the glass is installed incorrectly, the collision repair technician could be putting a vehicle back on the road that is not structurally sound. It is, therefore, important for the collision repair technician to be familiar with the various techniques used to remove and install glass. Glass must also be removed from areas of major damage before the damage can be straightened.

When replacing glass, be aware of the following:

- If a collision repair shop fails to restore the original structural integrity of a vehicle during

FIGURE 12-1 Glass plays an important role in aerodynamic styling. Air drag is reduced dramatically by fitting the glass accurately to the sheet metal and eliminating metal trim.

repair, it could be violating federal law. The Federal Traffic and Motor Vehicle Safety Act says: "No motor vehicle repair business shall knowingly render inoperative, in whole or in part, any device or element of design installed on or in a motor vehicle . . . ."

- Butyl tape alone *cannot* be used to completely replace urethane because it has no structural bonding properties. It only holds the windshield in place and prevents leaks. *Never* use just butyl tape alone to install a windshield.

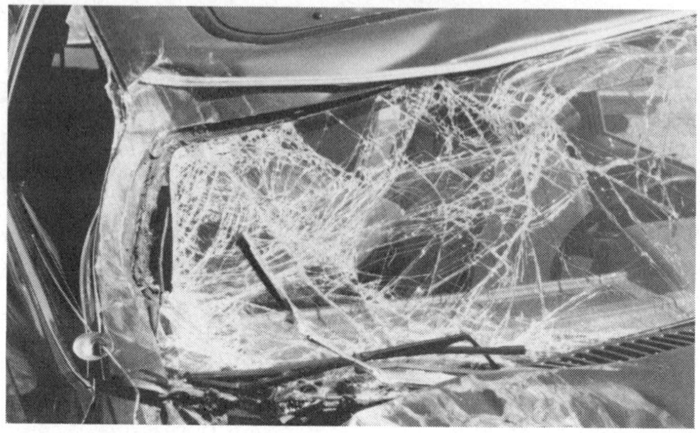

FIGURE 12-3 Lamination kept this windshield intact upon impact.

TYPES OF GLASS

For years, the glass used in most vehicles was either laminated or tempered (Figure 12-2). Both of these are considered safety glass because of their construction. Today, other types of glass are also found in modern unibody vehicles.

Laminated plate glass is used for windshields. It consists of two thin sheets of glass with a layer of clear plastic between them. Some glass manufacturers have increased the thickness of the plastic material for greater strength. When this type of glass is broken, the plastic material will tend to hold the shattered pieces in place and prevent them from causing injury (Figure 12-3). The plastic material is usually clear to provide an unobstructed view from all angles.

Tempered glass is used for side and rear window glass but never for windshields. It is a single piece of heat-treated glass and has more resistance to impact than regular glass of the same thickness. The strength of tempered glass results from the high compression of its surfaces, which comes about by rapidly heating the glass to approximately 1100 degrees Fahrenheit. The high temperature softens the glass, and it is then cooled rapidly by blowing air on both flat surfaces. The resulting rapid contraction adds compressive stress to the surface of the glass, which strengthens it. Because of this compression, the glass cannot be cut, drilled, or ground after the tempering process.

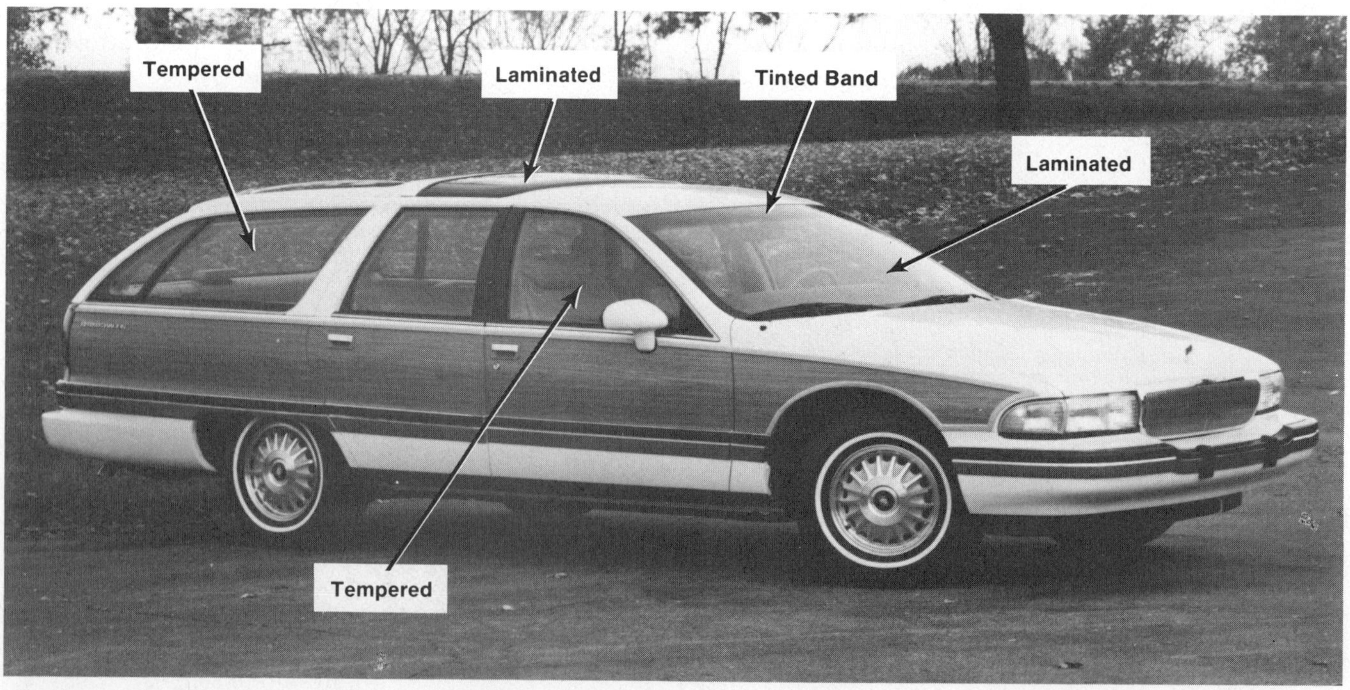

FIGURE 12-2 Laminated and tempered glass have been used in vehicles for years.

When tempered glass is broken, the pieces of glass are small and have a granular texture. The shattered glass has an interlocked structure to it that obstructs visibility; this is one reason why tempered glass is not used in windshields. Another reason it is not used is that this type of glass does not readily give and would, therefore, cause more severe head injuries during a collision. Tempered glass will also shatter if previously damaged or stressed. For example, a chipped edge or a stone striking it can weaken the glass so that it suddenly shatters some time after the initial defect occurred.

Both laminated and tempered glass can be tinted. Tinted laminated glass contains a shaded vinyl material to filter out most of the sun's glare. This type of windshield is helpful in reducing eye strain and fatigue, and prevents fading of the interior. Some windshields are shaded to reduce the sun's glare by means of a dark tinted band or section across the top part of the windshield, as seen in Figure 12-2. Tinted glass is usually recommended if the vehicle is to be equipped with air conditioning.

Glass can also be tinted by adding small quantities of metal powder to the other normal ingredients of glass to give it a particular color. The addition of cobalt gives the glass a blue tint. Iron gives the glass a reddish tint.

Glass can also be fitted with a defrosting circuit or antenna. Self-defrosting glass is common in rear windows (Figure 12-4A). Before heat treatment, metal powder, which conducts electricity, is printed on the glass surface in the form of heating wires. The metal powder is baked on the surface during the tempering process.

An antenna wire for radio reception is placed either between the layers of laminated glass (windshield) or printed on the surface of the glass (rear window) (Figure 12-4B). Some glass has antenna wires and heating wires side by side.

Antilacerative glass is similar to conventional multilayered glass, but it has one or more additional layers of plastic affixed to the passenger compartment side of the glass. This glass is used in the windshield only and is added protection against shattering and cuts during impact.

The newest glass used in unibody applications is called *modular* or *encapsulated* (Figure 12-5). This fits the contours of a vehicle more closely than glass used in the past. It also differs from ordinary auto glass in that it comes with a plastic trim molding attached to its edge. There are variations in the molding design. Some are one piece; others are two piece, with one glued to the glass and the other snapped into place after the glass is installed.

When the molding is part of the glass unit, it saves time and money on the assembly line by reducing the number of separate parts to be inventoried and handled.

A

B

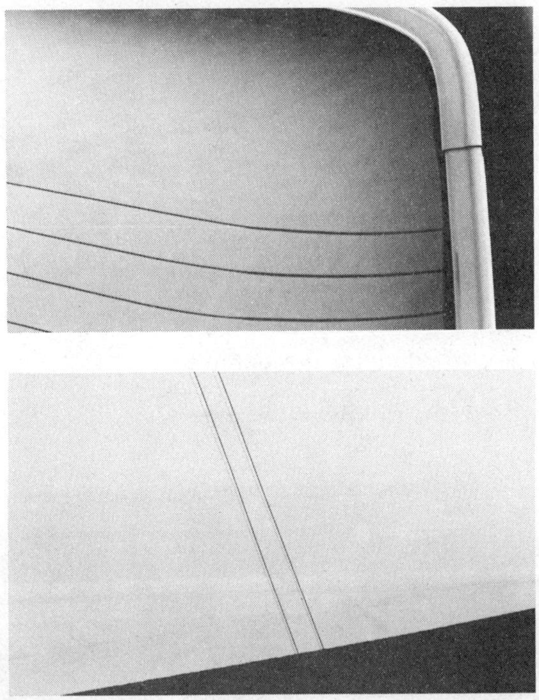

FIGURE 12-4 Glass may contain (A) defrosting elements or (B) antenna wires.

FIGURE 12-5 Modular glass in a unibody vehicle

This helps the manufacturers, but creates some new problems for collision repair shops and glass installers. Modular glass is not easy to work with. Its close fit makes removal of undamaged glass more difficult. In fact, special tools for removing modular glass are recommended by the manufacturers (Figure 12-6). One such tool is the vibrating power knife. It comes with several blades to fit the various joint areas between the glass and sheet metal. The vibrating blades gently cut the urethane adhesive all around to separate the glass from the body (Figure 12-7).

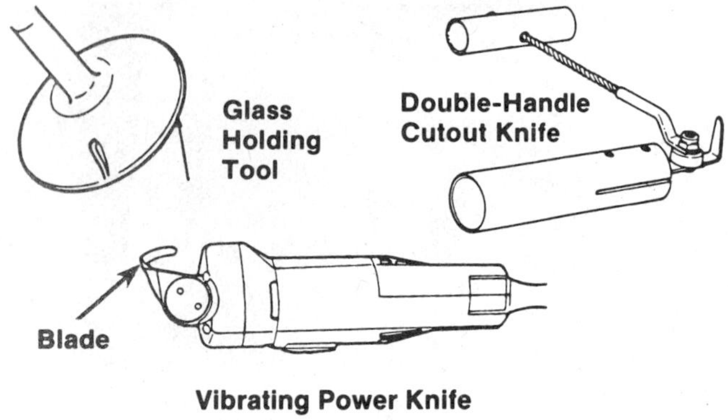

FIGURE 12-6 Special tools for removing modular glass

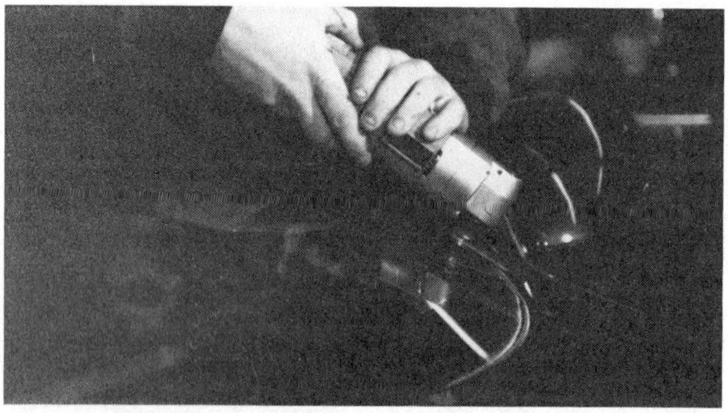

FIGURE 12-7 Vibrating power knife in use

FIGURE 12-8 Windshield installed with a gasket

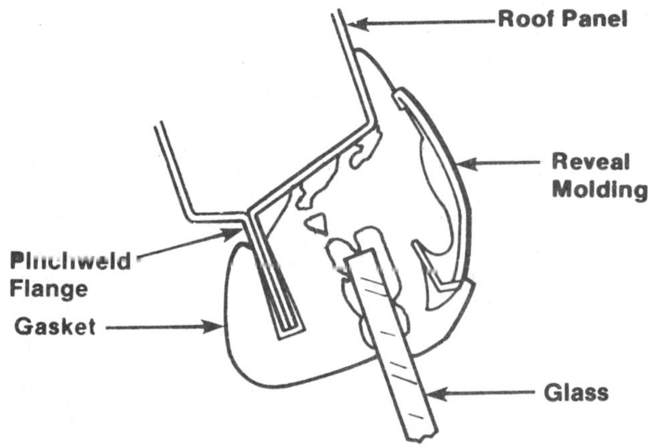

FIGURE 12-9 Cross section of a typical gasket installation

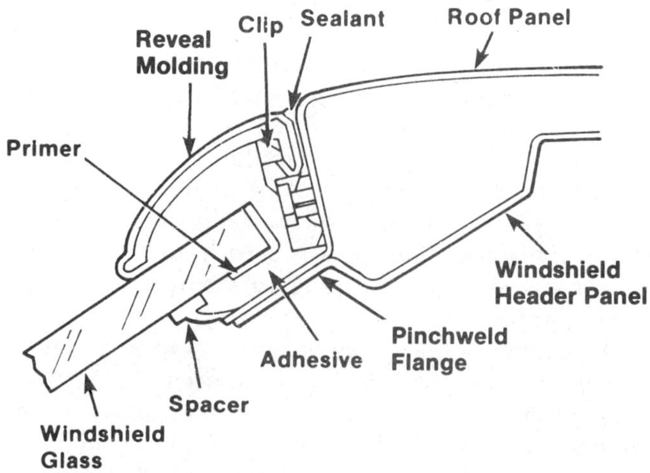

FIGURE 12-10 Cross section of a typical adhesive-bonded windshield

REMOVAL AND REPLACEMENT OF THE WINDSHIELD

Replacement of the windshield involves two different methods based on the materials used: gasket installations or adhesive installations. The adhesive installation is further divided into two methods: the full cutout and the partial cutout method.

The gasket installation method is more common in older vehicles but still finds use in present day vehicles (Figure 12-8). The gasket is grooved to accept the glass, the sheet metal pinchweld flange, and sometimes the exterior reveal molding (Figure 12-9). The gasket locking strip must be removed before the glass can be removed from the opening.

The adhesive installation, as the name implies, uses an adhesive material to secure the glass in place of a gasket (Figure 12-10). The adhesive method has virtually replaced the gasket method for installing windshields. The use of adhesives permits the windshield to be mounted flush with the roof panel, decreasing wind drag and noise (Figure 12-11). Adhesive-bonded windshields also increase the overall rigidity of the vehicle, minimizing body twist and helping to keep the glass in

FIGURE 12-11 Typical adhesive-bonded windshield

place during a collision. Rubber stops and spacers separate the glass from the metal. Reveal moldings are held in place by adhesive grooves in the body or by clips. The adhesive installation is more advantageous than the gasket installation because the pinchwelds do not have to be as exact since no pressure is exerted on the glass outside of the reveal moldings. The adhesive can be dispensed from a cartridge or it can be applied in strip or tape form. Typical adhesives in use include urethane and butyl rubber.

As previously mentioned, adhesive installations are further divided into two methods. The partial cutout method takes advantage of the fact that most of the original adhesive is in good condition and of sufficient thickness to allow it to remain and be utilized as a base for the application of new adhesive. When the original adhesive is defective or requires complete removal, the full cutout method is utilized.

GASKET METHOD

Before removing the windshield, the interior and exterior moldings must be removed. In most cases, the garnish moldings used on the interior face of the windshield consist of several strips that are secured in place by screws or retaining clips. All of the garnish moldings (Figure 12-12) should be removed first using a special tool. If unavailable, a suitable spring clip removal tool can be made from banding strap steel as shown in Figure 12-13. Slip the end of the tool over the clip and pull to remove the molding (Figure 12-14). If possible, the rearview mirror should be removed as well.

On the exterior of the vehicle, remove the reveal moldings and other trim or hardware (such as windshield wiper arms) in the area adjacent to the glass being removed. The reveal molding is usually secured in place by retaining clips that are attached to the body

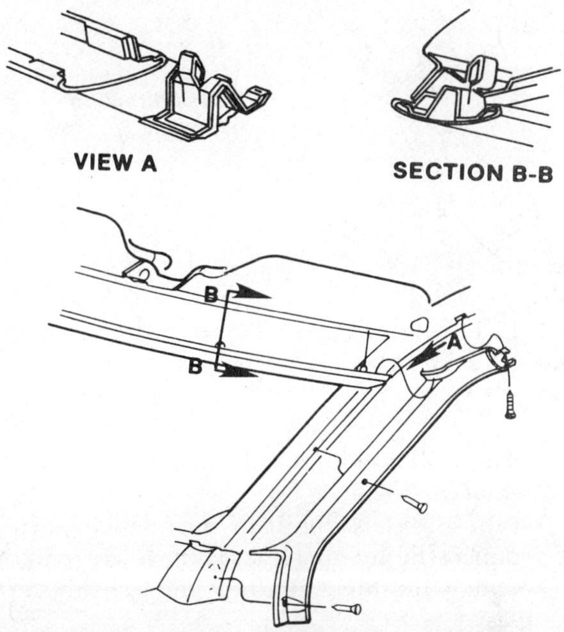

FIGURE 12-12 Interior garnish molding

opening. A projection on the clip engages the flange on the reveal molding, retaining the molding between the clip and body metal.

To replace a windshield using a gasket, proceed as follows:

1. Place protective covers over adjacent panels.

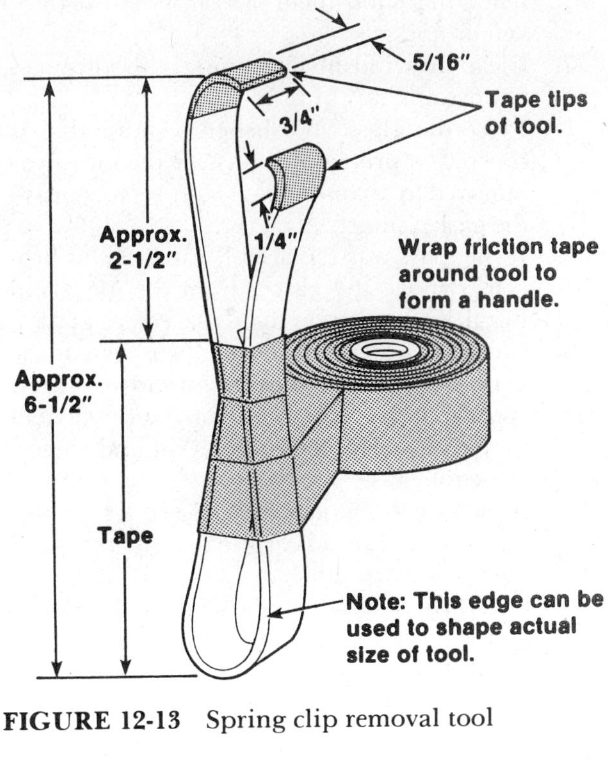

FIGURE 12-13 Spring clip removal tool

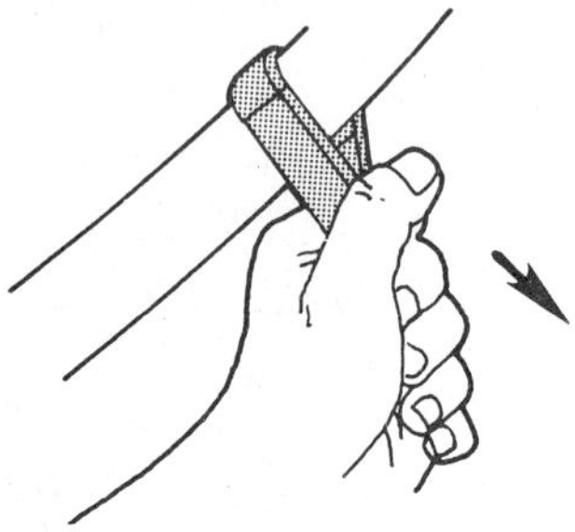

FIGURE 12-14 Pull on the tool to remove the garnish molding.

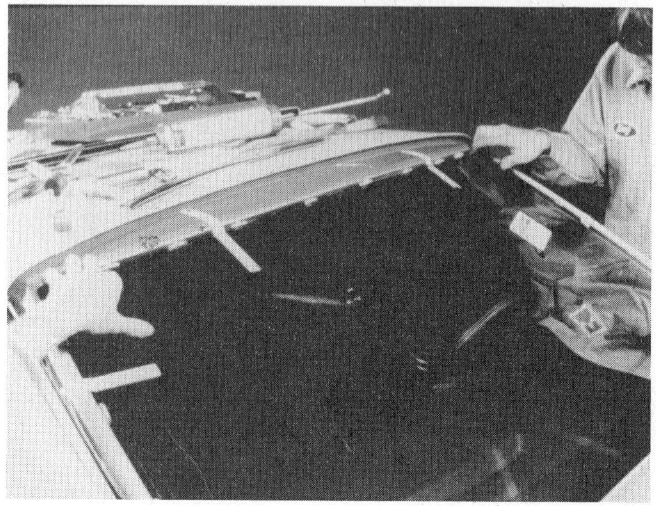

FIGURE 12-15 Marking the replacement glass with masking tape for alignment purposes

2. Be sure all moldings, trim, and hardware are removed.
3. If the glass has a built-in antenna wire, disconnect the antenna leads at the lower center of the windshield and tape the leads to the glass.
4. Locate the locking strip on the outside of the gasket. Pry up the tab and remove the locking strip to open the gasket all the way around the windshield.
5. Use a putty knife to pry the gasket away from the pinchweld inside and outside of the vehicle.
6. With an assistant, push the windshield and gasket from the windshield body opening. If the glass was not cracked and is to be reused, do not exert uneven pressure on the glass. Always wear safety goggles and gloves when removing windshield or rear window—broken or not.
7. Clean the windshield opening with solvent to remove any dirt or residual sealer.
8. Place the glass on a bench or table that is covered to protect the glass. If the glass was removed to accommodate body repairs, leave the gasket intact.
9. If the glass was replaced because it was broken, remove the gasket from the glass and install the gasket on the replacement glass.
10. Cracks that develop in the outer edge of the glass are sometimes caused by low or high spots or poor spot welds in the pinchweld flange. Examine the pinchweld and correct any problems.
11. Install setting blocks and spacers. If the original blocks are not available, cut pieces of used gasket for blocks.
12. Carefully install the glass on the blocks to verify the fit. Center it and check the gap between the glass and the pinchweld. The gap should be uniform around the entire pinchweld. After the windshield is lined up, apply several strips of masking tape, bridging from the glass to the vehicle body (Figure 12-15).
13. Slit each piece of tape at the end of the glass, then lift the windshield out and set it aside. When it is time to permanently install the windshield, line up the tape on the vehicle with the tape on the glass to get the proper positioning.
14. Insert a venetian-type cord in the pinchweld groove of the gasket. Starting at the top of the glass, the cord ends should meet in the lower center of the glass. Tape the ends of the cord to the inside surface of the glass (Figure 12-16). Squirt a soapy solution in the pinchweld groove for easier installation.
15. Apply waterproof sealer to the base of the gasket.
16. With an assistant, install the glass and gasket assembly in the body opening and use the masking tape to align it. Slip the bottom groove over the pinchweld.
17. Very slowly pull the cord ends so that the gasket slips over the pinchweld flange. Work the bottom section of the glass in first, then the sides, and finally the top section; the glass might crack if the cord is pulled from one end only.
18. Dispense a small bead of waterproof sealer around the body side of the gasket.
19. Remove excess sealer with solvent.

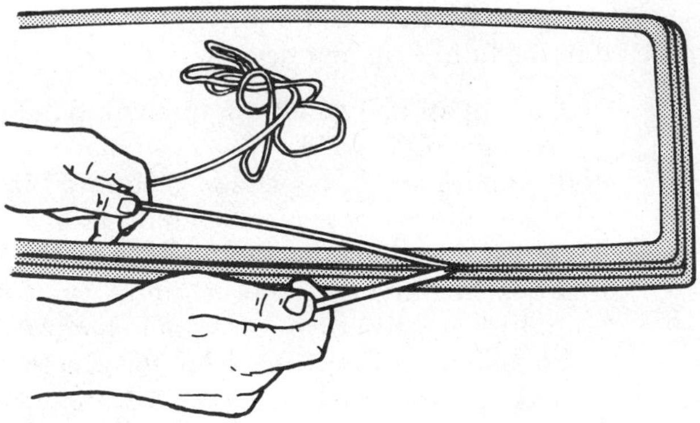

FIGURE 12-16 Taping the venetian-type cord to the inside of the glass

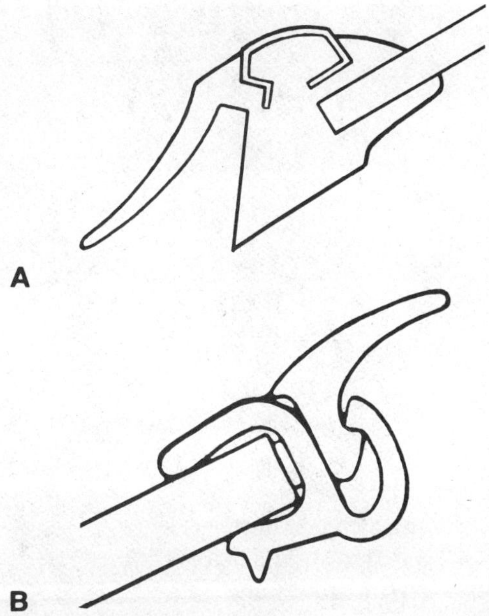

FIGURE 12-17 (A) One type of modular glass unit comes with a one-piece plastic molding applied. No other molding is applied after installation. (B) The two-piece modular glass molding design has one molding applied to the glass in the factory, and the other snapped into place after installation.

20. Install the reveal and garnish moldings.
21. Check the windshield for water leaks using a low pressure stream of water.
22. Place a soapy solution in the locking strip groove. Replace the locking strip by spreading the groove and feeding the strip into the opening. The soapy solution lubricates the groove and makes it easier to slide the strip through.

FULL CUTOUT METHOD

Urethane adhesive and butyl ribbon sealer are used together in the full cutout adhesive installation. But only the high-strength urethane adhesive provides the structural strength required with unibody vehicles. The butyl sealer is used only to hold the windshield in place while the urethane cures.

CAUTION

Be sure to wear safety glasses, work gloves, and coveralls when performing any glass removal or replacement procedure.

When using the full cutout method the first step is to remove the windshield according to the following procedure:

1. Protect both the interior and the exterior of the vehicle from damage. Cover the front seat and instrument panel. Also cover the fender surfaces next to the windshield area. Remove the rearview mirror. Take off the wipers, antenna, moldings, and any other parts that might get in the way. Remove the cowl vent panel if it makes it easier to reach all around the edge of the windshield.
2. Using a hook tool, lift the vinyl molding at the bottom of the windshield. Once it can be securely grasped, pull it out from around the windshield. Do this carefully so the molding (Figure 12-17) is not damaged.
3. With a utility knife, score the exposed urethane all the way around the outside of the windshield. Cut as close to the glass as you can. This step makes it easier to insert the vibrating power knife in the next step.
4. Using a vibrating power knife, cut through the adhesive next to the glass (Figure 12-18). Cut as close to the glass as possible. Be very careful. Make sure safety gloves and safety glasses are worn because the glass might splinter. Piano wire can also be used to cut the adhesive. Use a 3- to 4-foot length of the lightest gauge wire that can be purchased. The cutting procedure with piano wire is a two-person job (Figure 12-19). If removing a piece of modular or encapsulated glass—especially if it is undamaged and can be saved—most automakers recommend using the vibrating power knife to remove the glass instead of piano wire.
5. Carefully lift out the glass.

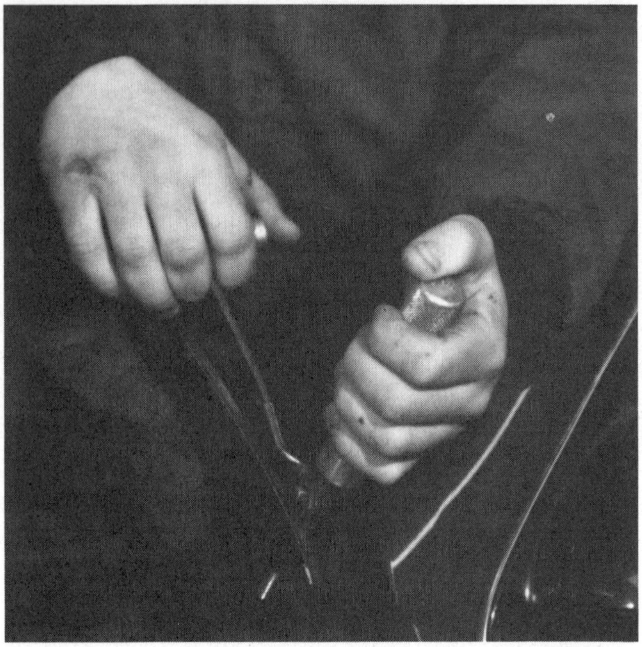

FIGURE 12-18 A vibrating power knife works well in areas where it can be held close to the edge of the glass.

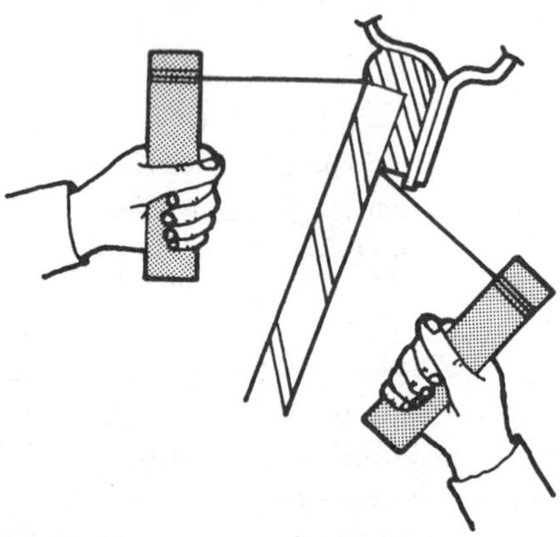

FIGURE 12-19 Cutting through windshield adhesive with piano wire

To prepare for the installation of the windshield, proceed in the following manner:

1. Cut out all the remaining urethane adhesive from the pinchweld with a razor blade knife. Be careful not to cut or damage the headliner. Make sure no loose pieces of old adhesive are left. If the windshield had been replaced before with butyl tape or other unknown materials, it is especially important to remove the old adhesive completely from the pinchweld area. This will ensure the desired strength in the new joint.
2. Next, check for rust in the pinchweld area. If there is rust, remove it with a wire brush. Then sand if needed. Use a grit recommended by the adhesive manufacturer.
3. Before doing anything else, dry set the new windshield. Make sure there will be at least 1/4 inch of bonding surface around the entire edge of the windshield. Adjust setting blocks as needed to get proper alignment. After the windshield is lined up, apply several strips of masking tape, bridging from the windshield to the vehicle body.
4. Slit the pieces of tape at the edge of the glass (Figure 12-20). Lift the windshield out and set it aside. When it is time to set the new windshield on the bead of urethane adhesive, line up the tape on the vehicle with the tape on the windshield to get the proper positioning.
5. Use a urethane adhesive cleaner on a lint-free dry cloth to wipe the pinchweld area. This cleans it and removes any loose urethane. Let it dry.
6. Use a urethane primer in the pinchweld area. Do not use a liquid butyl primer or a ribbon sealer primer. Apply the primer to all bare metal in the pinchweld area. It is not necessary to prime over any urethane left on the pinchweld surface.

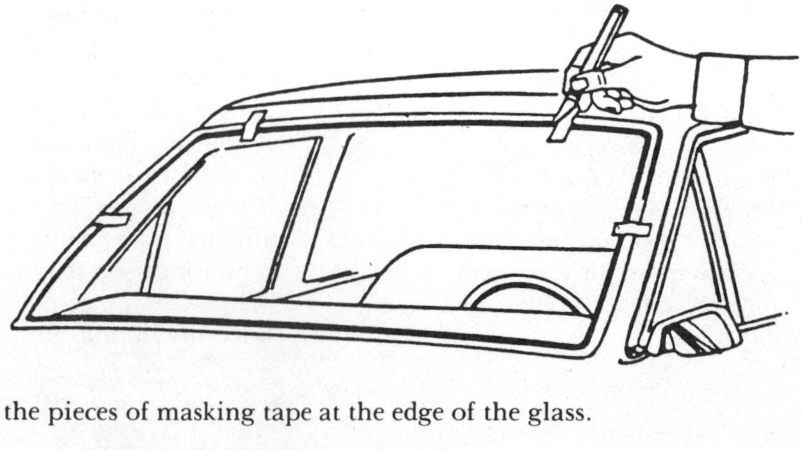

FIGURE 12-20 Slit the pieces of masking tape at the edge of the glass.

7. Clean the inside of the new windshield with a cleaner that leaves no film. Check to see what is required by the adhesive manufacturer for use with the product. Wipe dry with a clean, lint-free cloth.
8. Put a 1/2-inch coat of urethane primer on the inside contact surface of the windshield. Be sure to apply it all the way around. Let it dry for the required time.
9. Install square ribbon sealer to the inside edge of the pinchweld. For shallow pinchwelds, use 5/16- by 5/16-inch. For deeper pinchwelds, use 3/8 by 3/8-inch. Ribbon sealer is made square because it provides better support than round sealer. It also does not squeeze into the space that is needed for the urethane.
10. Apply a fast-curing, high-strength urethane adhesive following the manufacturer's instructions. Apply it directly behind the pinchweld as shown in Figure 12-21. Cut the opening on the cartridge nozzle at a 45-degree angle that will allow a bead size just a little bit larger than the ribbon sealer (Figure 12-22). Apply the urethane adhesive by holding the dispenser at a 45-degree angle as shown in Figure 12-23. If too much adhesive is applied, excessive squeeze-out occurs.

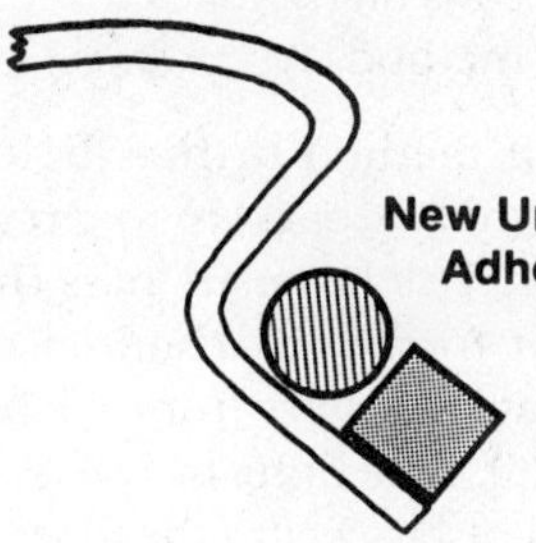

FIGURE 12-21 Cross-sectional view showing application of ribbon sealer and urethane adhesive in the full cutout method

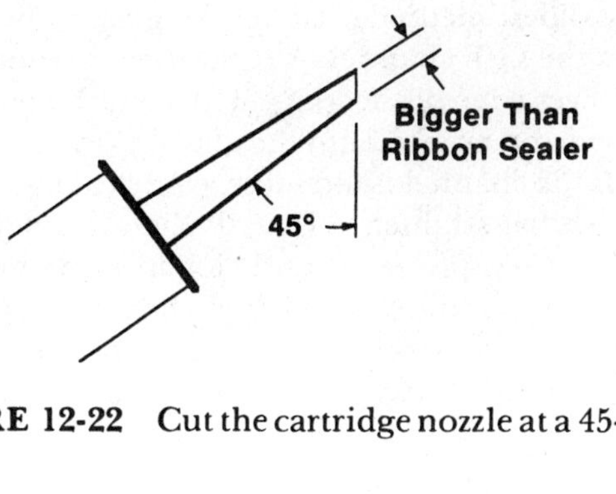

FIGURE 12-22 Cut the cartridge nozzle at a 45-degree angle.

FIGURE 12-23 Applying the urethane adhesive

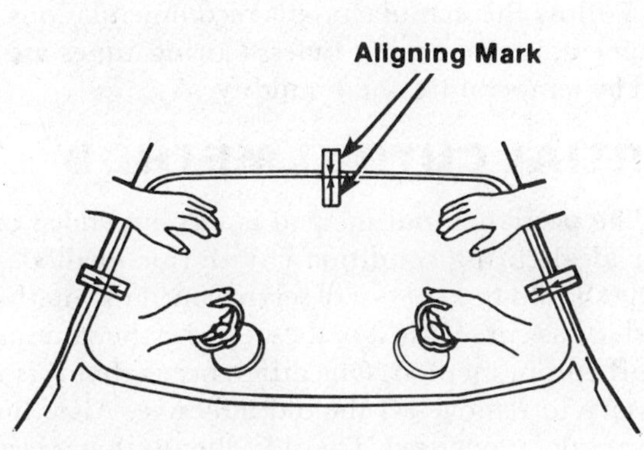

FIGURE 12-24 Line up the windshield glass with the masking tape.

The final step in the full cutout operation is the installation of the windshield. This is accomplished as follows:

1. Set the windshield into place. Line up the pieces of masking tape on the glass and body (Figure 12-24). Press firmly so the glass makes contact with the ribbon sealer and the urethane adhesive.
2. Look for squeeze-out along the edge of the windshield. Using a putty knife, paddle the squeezed adhesive and spread it evenly. Smooth the surface and remove any excess adhesive (Figure 12-25). Fill all spaces between the glass and the metal. Remember, at least 1/4 inch of the windshield surface must be bonded to the vehicle body.

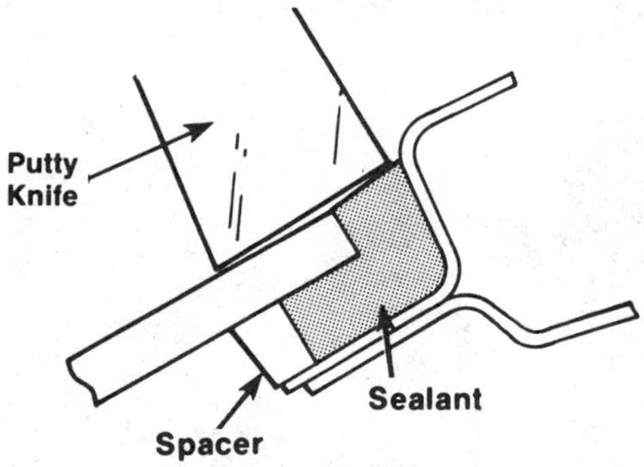

FIGURE 12-25 Smoothing the adhesive with a putty knife

3. Install the moldings. Clean up the area with a liquid adhesive cleaner.
4. Attach the cowl vent panel, wiper, antenna, and other parts that were removed at the beginning of the repair.

The ribbon sealer will hold the windshield in place while the urethane adhesive cures. The urethane adhesive will reach its maximum strength in four or five days. Follow the manufacturer's recommendations for urethane adhesive curing times. Curing times are affected by temperature and humidity.

PARTIAL CUTOUT METHOD

The partial cutout method is recommended only under ideal curing conditions. With this method, the vehicle should be kept out of service until the urethane adhesive has cured. Most of the steps are the same as in the full cutout method. One difference is that it is not necessary to remove all the old adhesive. Also, butyl ribbon sealer is not used. The old adhesive that is left on the pinchweld serves as the base for the new adhesive.

SHOP TALK

If the old urethane adhesive is not bonded securely to the vehicle body, remove all of it and use the full cutout method. The same is true if butyl ribbon sealer or some other unknown material is discovered in the pinchweld. Do not put urethane over it in this case—use the full cutout method instead.

Because there is no butyl ribbon sealer to hold the glass in position, allow enough time for the urethane to cure before driving the vehicle. The buildup of strength depends on temperature and humidity. Follow the adhesive manufacturer's guidelines for curing time under various conditions.

CAUTION

Where ideal conditions of high temperature and high relative humidity are not present or when the customer is not willing to leave the vehicle in the shop for the additional day (or more) for curing, the full cutout method should be used.

REMOVING AND REPLACING REAR WINDOWS

The removal and replacement procedures for rear and many stationary side windows follow the above methods rather closely. Methods vary slightly for different vehicle makes. However, many of the operations that are applied to one make of vehicle can readily be applied to others.

On some vehicles, the rear windows are removed toward the inside of the vehicle rather than to the outside. This procedure will require the removal of certain interior trim parts and accessories and the use of a special knife. For this type of removal, consult the manufacturer's service manual for specific procedures.

DOOR GLASS SERVICE

Various methods are used to attach the door glass to the lifting mechanism:

- Bolt-through method
- Adhesive bonding method
- Sash channel method

The bolt-through method utilizes bolts or rivets that have plastic or rubber gaskets to prevent direct contact with the glass. The fasteners pass through the window and secure it to the lift channel or bracket. Rubber spacers separate the glass from the bracket and fasteners (Figure 12-26). The bolts or rivets are used to attach the lift channel or bracket to the glass. The bolts are inserted through the glass.

The use of adhesives is another method of securing the lift bracket to the glass. Usually a U-channel, with insulator stays, is used to prevent the glass from contacting the metal channel or lift bracket (Figure 12-27).

The oldest method of attaching glass to lift channels uses the sash channel. A rubber seal or tape is put on the lower edge of the glass. Then a sash channel is positioned and tapped onto the glass by using a rubber mallet. If the channel is too loose, electrical tape can be used as a shim to tighten it. Usually the edges of the sash channel can be squeezed slightly for an even tighter fit.

Door glass requires servicing when it is broken or must be removed for other repairs. It might also have to be removed to replace a broken channel assembly.

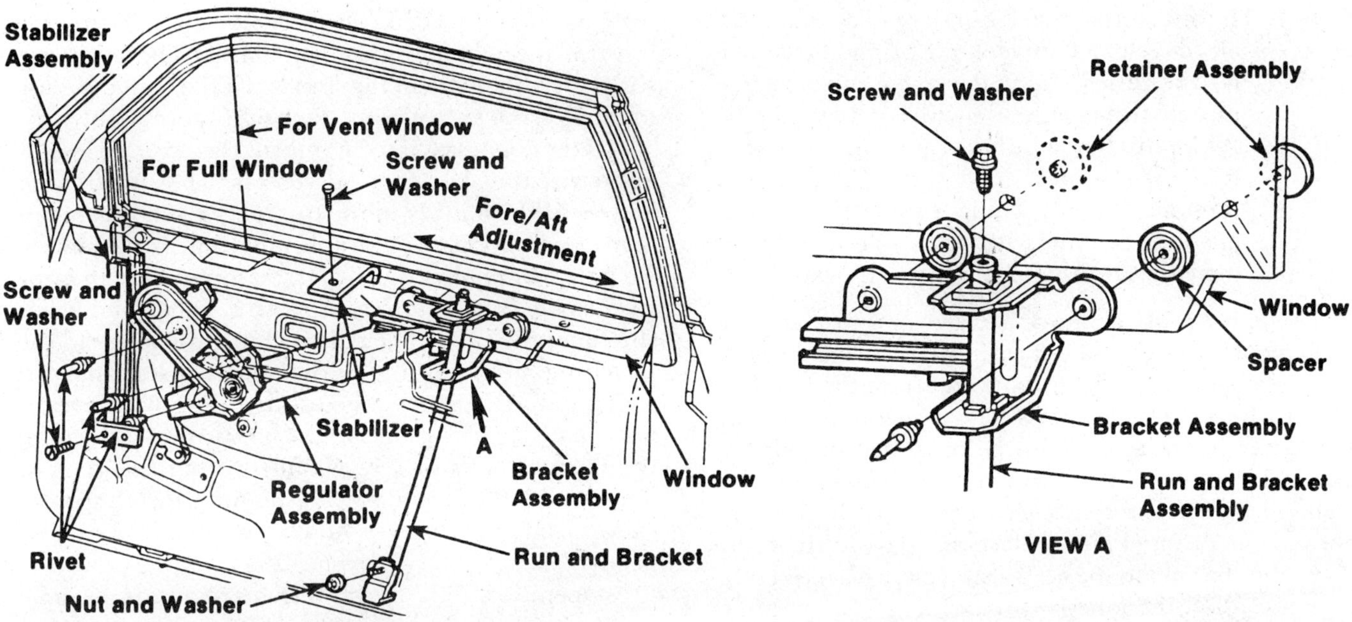

FIGURE 12-26 Window glass secured with rivets and spacers

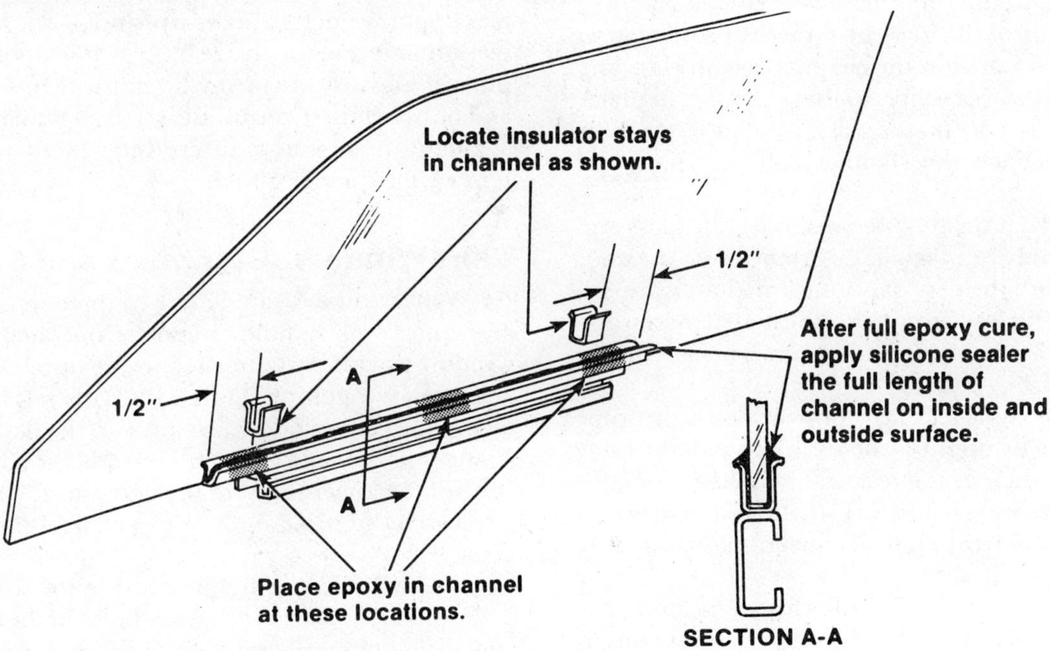

FIGURE 12-27 Adhesive-bonded glass in bracket

Removing the door glass also requires removing the door trim panel, water shield, lower window stop, and the hardware securing the channel bracket to the lift channel. The forward edge of the glass assembly must be tilted up to remove it from the door. At this point the channel assembly can be removed from the glass.

Mark the position of the lift channel on the glass. The channel can then be unbolted or the rivets drilled out. If the glass is glued into the lift channel, use this procedure:

1. Remove the channel from the glass by applying heat from a welding torch along the full bottom length of the channel. Slowly pass the tip back and forth for 60 to 90 seconds, then grip the channel with pliers and pull it loose. Repeat the heating operation if necessary.

2. If the original glass is to be reused, scrape off all traces of adhesive. If the original channel is to be reused, clamp it in a vise and burn out the remaining adhesive with a welding torch. While still hot, wire brush any adhesive traces from the channel.
3. Remove the remaining adhesive from the glass and channel with lacquer thinner. Complete the cleaning operation with water.

When the lift channel is clean and dry, the replacement glass can be glued in place. If using new glass, transfer the position markings from the original glass to the replacement glass. The vehicle manufacturer's service manual can also be consulted for correct channel position. Then, apply an epoxy adhesive to the channel. Position the spacers about 1/2 inch from the ends of the channel and replace the glass in the sash. Tape the channel to the glass using cloth-backed body tape, and allow the adhesive to cure for at lease one hour prior to reinstallation. After the adhesive cures, apply a thin bead of silicon adhesive along the length of the channel on both sides of the glass.

To reinstall glass in a bolted or riveted channel, rest the top part of the glass on a piece of softwood or carpeting, then position the channel on the glass and use light blows with a rubber mallet to force it in place. If the channel is loose on the glass, use tape to close the gap. Then reattach the channel bolts or rivets and spacers.

After the channel has been installed, the glass and channel assembly can then be positioned into the door and secured with the necessary attaching hardware. Install the lower glass stop and adjust it if necessary. Finally, all the hardware and trim panels can be reinstalled.

To remove quarter glass (on rear doors on some vehicle makes), it might be necessary to remove some interior items such as the rear seat, window regulator handle, trim panel(s), and water shield. While supporting the glass, remove the lower frame-to-glass attaching screws and then lift out the glass.

If the window binds or is stiff, check the channel. It might be necessary to add lubricant to the glass runs or guides. A door glass that tips forward (or rearward) and binds can be caused by improper adjustment of the lower sash brackets, a loose channel or cam roller, or a stabilized channel that is out of adjustment. If the channels are too tight or lack proper lubricant, the glass or rubber will bind. To free them up, dry silicone spray should be applied. Oil is not recommended for use on the rubber channels. It can cause the rubber to swell and deteriorate.

Vehicle doors that use a full trim panel sometimes have a set of brackets at the top of the door to which the trim panel is attached. If the brackets are set too far inward, the window glass will bind. The antirattle slides are another item that can cause the window glass to bind. If not set correctly, they will cause problems.

If the door glass is to be adjusted because it does not align with the edge of the quarter glass, be sure to check and see if the fault is in the quarter glass adjustment. Some quarter glass is movable and some is stationary. To adjust or remove quarter glass, some of the interior such as the seats, armrest, inner trim panel, and water shield might have to be removed to gain access to the attaching mechanism. On some types of stationary or swing-out glass, it is not necessary to remove the seats. Stationary quarter glass can be adjusted by loosening the retainer bolts or screws and shifting the glass to align it with the door glass. The manufacturer's specifications and procedures should be consulted for specific details.

Operation of the movable quarter glass is the same as that for door glass; stops are used to control the up and down movements. The up-stop controls the up and forward movement of the glass. This stop must be set to obtain the proper spacing between the door and quarter glass. Some vehicles use two up-stops—one for the front section and one for the rear. Just like door glass, the quarter glass is provided with adjustments for inward and outward tilt. Consult the service manual and make a careful inspection to determine what is necessary to remove the quarter glass.

Vent Window Alignment and Service

Vent windows can be one of three types: friction, gear and crank handle, or power operated. The vent window assembly is attached to the upper door frame and window opening (Figure 12-28). Screws are used to secure the vent window assembly to the door frame. A channel at the bottom of the assembly is adjustable to control the lower travel of the door glass. The channel also can be adjusted in or out to control the tilt of the glass.

To replace the vent window (Figure 12-29), an awl or dowel can be used to align the holes in the door frame and vent window. Then, reinstall the mounting screws and torque them to the manufacturer's specifications.

Tailgate Glass Service

Tailgate glass is generally secured to a metal channel with screws or bolts. Removal or replacement of tailgate glass is similar to the procedure for door glass. To remove tailgate glass, run the glass to the down position and remove the inner trim panel cover. Then run it to the up position and out of the tailgate until the glass channel screws are visible and accessible. Remove

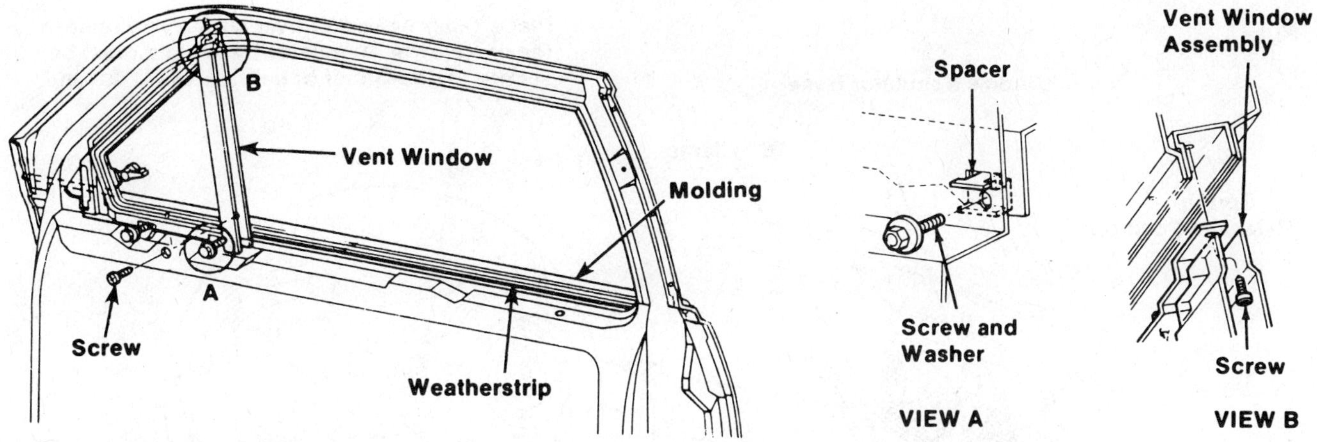

FIGURE 12-28 Front vent window assembly

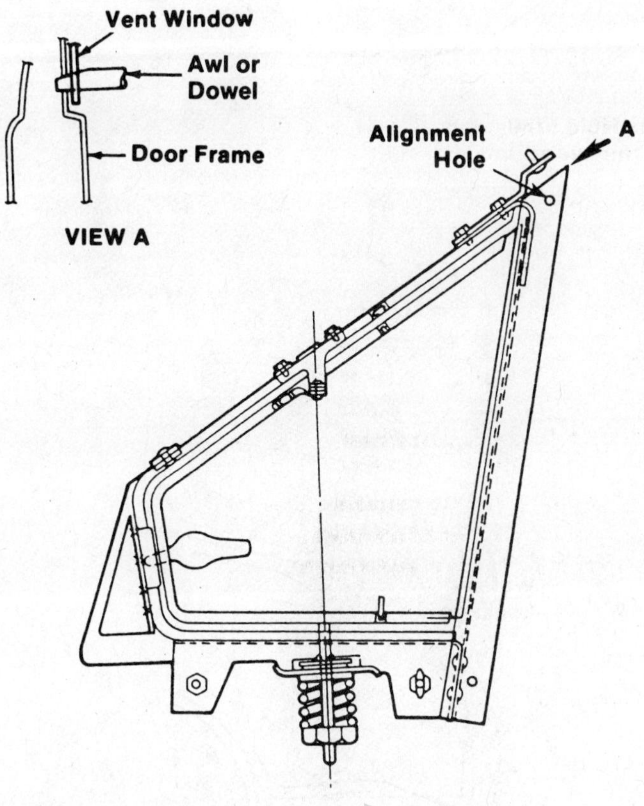

FIGURE 12-29 Vent window installation

the glass channel screws to disconnect the glass from the channel assembly. Slide the glass out of the opening, then remove the attaching hardware from inside the tailgate. Finally, remove the slides from the glass in four places. The installation procedure is a reverse of the removal procedure. Always check the service manual for specific instructions.

Manual and Power Regulators

Regulators to raise or lower glass can be manual or electric. On older vehicles, regulators were also used on vent windows to move them in and out. Both types of regulators are very similar, the only difference being the use of the handle crank mechanism on manual regulators as opposed to the electromotor-driven gear mechanism on electric regulators. The lift arms are the same for both types. The regulator and its associated parts are sometimes riveted to the door structure instead of being bolted. In this case, drill out the rivets when servicing is required and reinstall the necessary parts using a rivet gun.

For those window regulators that are spot welded to the inner door panel, use a spot weld cutter to remove visible welds. If necessary, use a chisel between the regulator and inner panel to separate the two structures. Generally, the replacement regulator is installed with U-clips or screws.

If the regulator is to be removed without removing the glass, be sure to secure the glass in the up position to prevent it from dropping inside the door panel. Heavy cloth tape or a wedge can be used for this purpose. Always consult the service manual for the proper removal and installation of the regulator.

Electric regulators require additional care because of the use of counterbalance springs in their design (Figure 12-30). On some models, the counterbalance spring must be released before servicing the regulator motor or other associated parts. This can be accomplished by one of two methods:

- Position the regulator near an access hole in the inner door panel that is large enough to allow the regulator to be clamped to the panel. Using a special shop-fabricated tool (Figure 12-31), release the spring tension. After servicing the

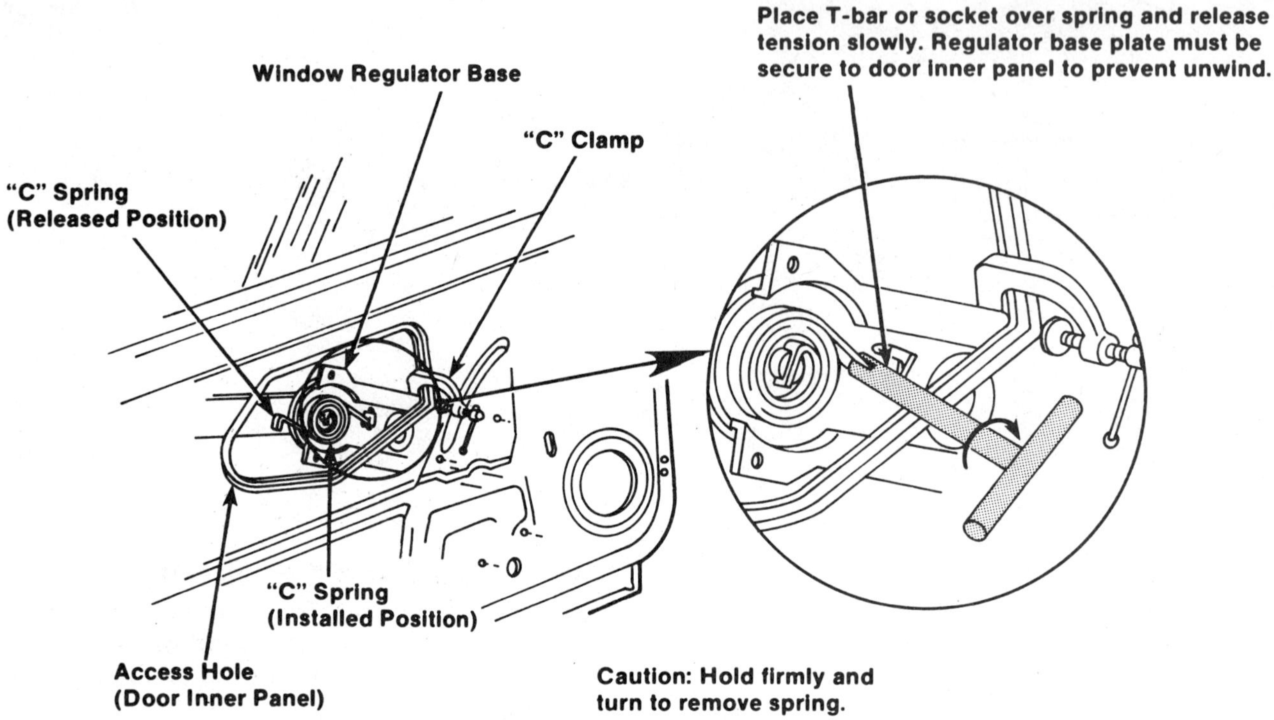

FIGURE 12-30 Counterbalance spring used in electric regulators

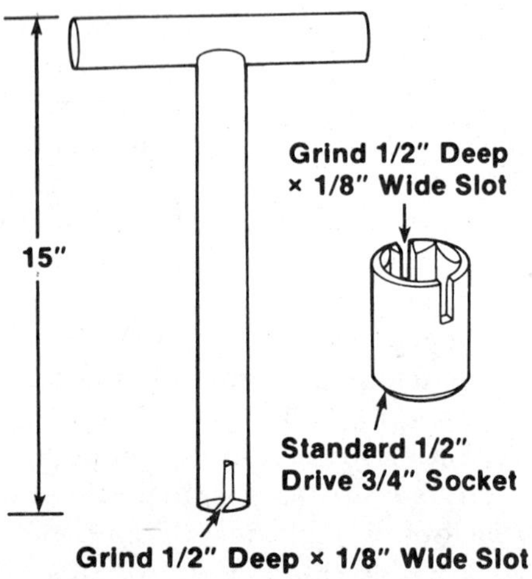

FIGURE 12-31 Shop-fabricated spring removal tool

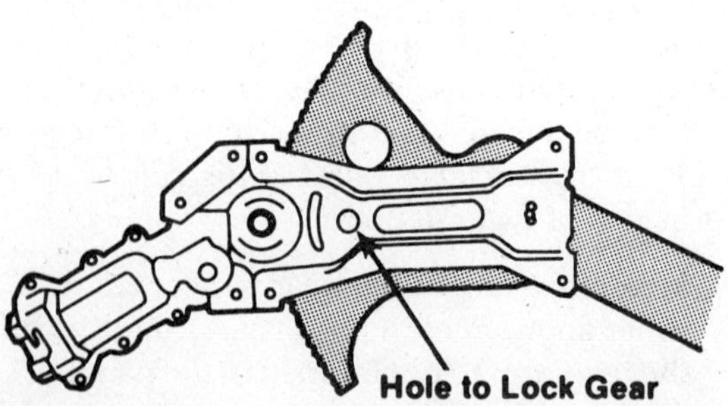

FIGURE 12-32 Secure the regulator with a bolt or screw.

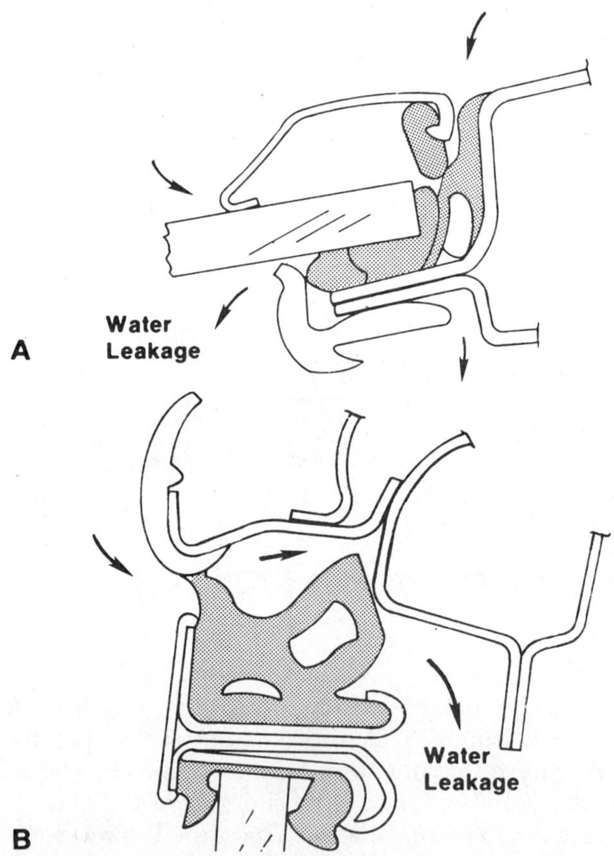

FIGURE 12-33 Water leakage around (A) a windshield and (B) door weatherstripping

motor or part, the spring can be reinstalled in its original position. Removing the spring avoids possible injury and damage to the door.
- Secure the regulator to the inner door panel by drilling a hole through the regulator gear and back plate and inserting a screw or bolt through the hole (Figure 12-32). Use a nut to lock the assemblies together. Be sure to remove the screw or bolt after servicing the regulator.

Door adjustment and replacement/repair are covered in Chapter 11.

LOCATING LEAKS

Water leaks and wind noises are frequent customer complaints. Such problems are often difficult to locate. Water leaks frequently occur at panel joints and glass-to-metal joints due to cracked or insufficient sealer (Figure 12-33A). Dust and water leaks also occur at doors, windows, trunk lids, and windshields whenever the weatherstripping becomes damaged or loose or when the doors or window glass are improperly adjusted (Figure 12-33B).

Wind noises are high-frequency sounds heard while driving. They are heard mainly when the windows are closed. This is generally due to loose, worn, or improperly applied weatherstripping, which allows air to leak into the passenger compartment (Figure 12-34A). Wind noise is also produced when the wind hits a projection such as molding (Figure 12-34B). This disturbance produces an eddy or swirl behind the object, thus creating a noise (the principle of flute and bugle sounds).

A loose body molding, a poorly aligned front fender, or an improperly adjusted hood are just some examples of causes of wind noises. A troubleshooting chart on how to identify and solve wind noises is given in Table 12-1.

Common methods used to locate air and water leaks are:

- Spraying water on the vehicle
- Applying soap and water solution to the vehicle
- Driving the vehicle over very dusty terrain
- Directing a strong beam of light on the vehicle and checking for light between the panels

Before making a leak test, remove all applicable interior trim from the general area of the reported leak. The spot where dust, water, or light enters the vehicle might be some distance from the actual leak. Therefore, remove all trim, seats, or floor mats from areas that are suspected as possible sources of the leak.

Leak Checks Using Water

Place one person inside the vehicle with all the doors and windows closed. Then starting at the bottom

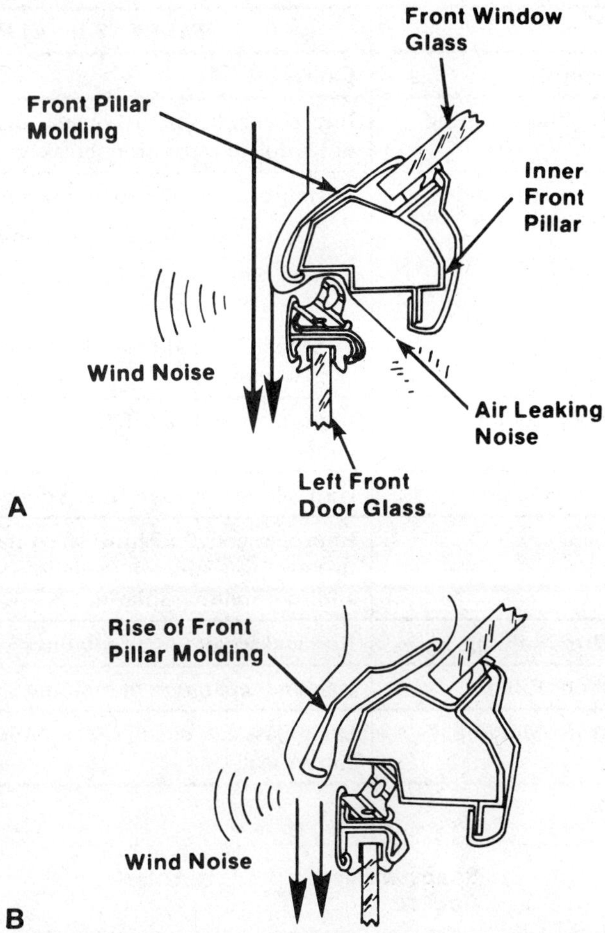

FIGURE 12-34 Wind noise caused by (A) loose weatherstripping and (B) loose molding

and working up, spray the vehicle with a forceful stream of water in the suspected area of the leak. The person inside the vehicle should act as an observer to locate exactly where the water enters.

Another way to discover water leaks around a windshield or rear window is to apply a soapy solution around the outside edge of the glass. Then, from inside the vehicle, apply compressed air to the window-to-panel joint. Any gap in the sealer will result in bubbling of the soap solution (Figure 12-35).

Leak Checks Using Light

Leaks can often be located by moving a strong light source around the vehicle, while an observer remains inside the vehicle. This method is useful only if the leakage course is in a straight path. If the path is curved, the light beam will not pass through the turns and curves.

Making the Repair

Plugs and grommets are used in floor pans, instrument panels, and trunk floors to keep dust and

TABLE 12-1: ELIMINATING WIND NOISE

Sources	Causes	Corrections
Weatherstripping	Imperfect adhesion to contact surface and improper contact of lip due to separation, breakage, crush, and hardening	Repair or replace weatherstripping.
Door Sash and Related Parts	1. Improper weatherstripping contact due to a bent door sash 2. Gap caused by corner piece improperly installed. 3. Gap caused by corner sash badly finished. 4. Separation and breakage of the rubber on the door glass run	1. Repair. 2. Install properly. 3. Repair with body sealer and masking tape. 4. Repair.
Door Assembly	Improper weatherstripping contact due to improper fitting door.	Correct door fit.
Door Glass	Gap caused due to ill fitting door glass	Align door glass.
Body	Improper body finishing on contact surface for door weatherstripping (uneven panel joint, sealer installed improperly, and spot welding splash)	Repair contact surface.
Drip Molding	Rise and separation of molding	Repair or replace.
Front Pillar	Rise and separation of molding	Repair or replace.
Waist Molding	Door glass gap due to rise of molding and deformation of rubber seal.	Repair.

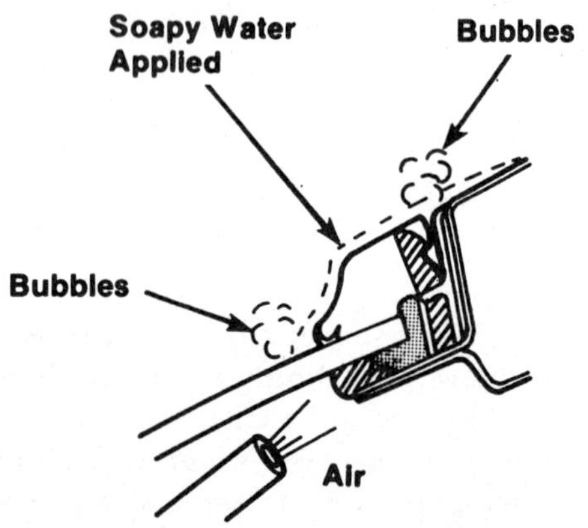

FIGURE 12-35 Testing for leaks using soapy water and compressed air

water out of the interior. These items should be carefully checked to ensure they are in good condition.

Windshields and rear windows usually develop water leaks that can be repaired without removing the glass. A majority of leaks occur at the top of the windshield or rear window. If several leaks are detected, it is better to seal the entire area rather than at each leak point.

To repair a water leak on an adhesive-bonded windshield, first clean the leak area and blow it dry. Then, trim off the surplus adhesive that extends beyond the edge of the glass (Figure 12-36). After removing the adhesive, dry the area using compressed air. Use a solvent to clean off any oil or grease that might be present.

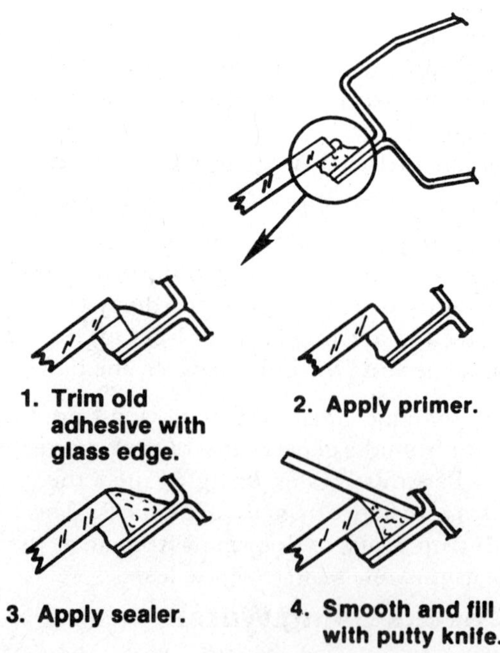

FIGURE 12-36 Steps in sealing a leak in an adhesive-bonded windshield

Prime the repair area with a urethane primer-sealer and let it dry according to the manufacturer's instructions. Then, apply the sealer along the cleaned area and use a putty knife to smooth it out. The sealer should be spread so that it is even with the top edge of the glass and tapered back to the molding clip area. Be sure to work it into any existing crevices. While the sealer is still soft, water check the area again. Use a very soft stream of water so as not to disturb the sealer. If no leaks are detected, reinstall the trim and remove any surplus sealer from the glass or vehicle.

Glass that uses the gasket installation method commonly develops leaks between the gasket and the glass or between the gasket and the vehicle body. Mark the exact leak area and dry it using compressed air. Apply beading compound or joint sealer to the marked area. This type of caulking material never hardens; it remains soft and flexible. It might be necessary to use a flat, pointed piece of wood to make an opening between the gasket and glass (or body) so that the cartridge nozzle can be inserted. Place the cartridge nozzle adjacent to the wood and move it forward, filling the opening. Remove surplus sealer with a solvent or enamel reducer. When the sealing is complete, recheck the area using the water spray method.

HEADLIGHTS

The majority of the headlights on today's vehicles are either the two or four sealed beam systems. If the vehicle was involved in a front end collision, the headlights should always be adjusted after they have been replaced.

The two sealed beam system has the low and high beams built into the same bulb. In the four sealed beam system, two beams have both low and high; the other two beams are used only on the high beam. High or low beam is selected by a switch on the floorboard or steering column. Quick disconnects are used for the left and right sealed beams. A typical halogen headlight system consists of adjusting screws for horizontal and vertical movement (Figure 12-37).

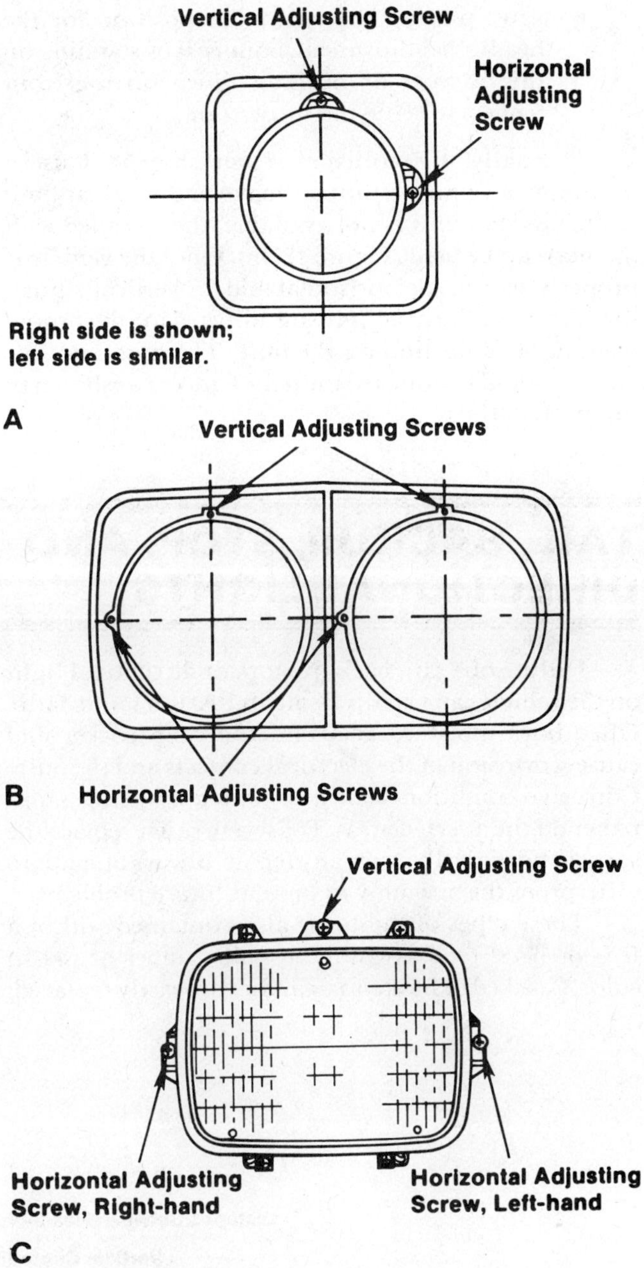

FIGURE 12-37 Headlight adjusting screws

Before making headlight adjustments, make the following inspections to ensure that the vehicle is level. Doing so will go a long way toward preventing an incorrect headlight setting.

- If the vehicle is heavily coated with snow, ice, or mud, clean the backside with a high-pressure stream of water. The additional weight can alter the ride height.
- Ensure that the gas tank is half full. Half a tank of gas is the only load that should be present on the vehicle. The exception to this is a vehicle that is constantly driven with a loaded trunk, such as a salesperson's vehicle. Such vehicles should have the adjustments made with the normal load present.
- Check the condition of the springs and shock absorbers. Worn or broken suspension components will affect the setting.
- Inflate all tires to the recommended air pressure. (Take into consideration cold or hot tire conditions.)
- If collision damage requires straightening of the frame, make sure that the wheel alignment and rear axle tracking path are correct.

- After placing the vehicle in position for the headlight adjustment, bounce it by standing on the bumper or pushing down on the front fenders to settle the suspension.

Normally, the collision repair shop will use a headlight alignment unit to make the necessary adjustments. If a unit is not available, the so-called wall method can be used (Figure 12-38). Once the vehicle is properly set up, the horizontal and/or vertical adjusting screws on the headlight are adjusted for the proper reading or indication on the unit. The high and low beams should be adjusted until they appear as shown in Figure 12-39.

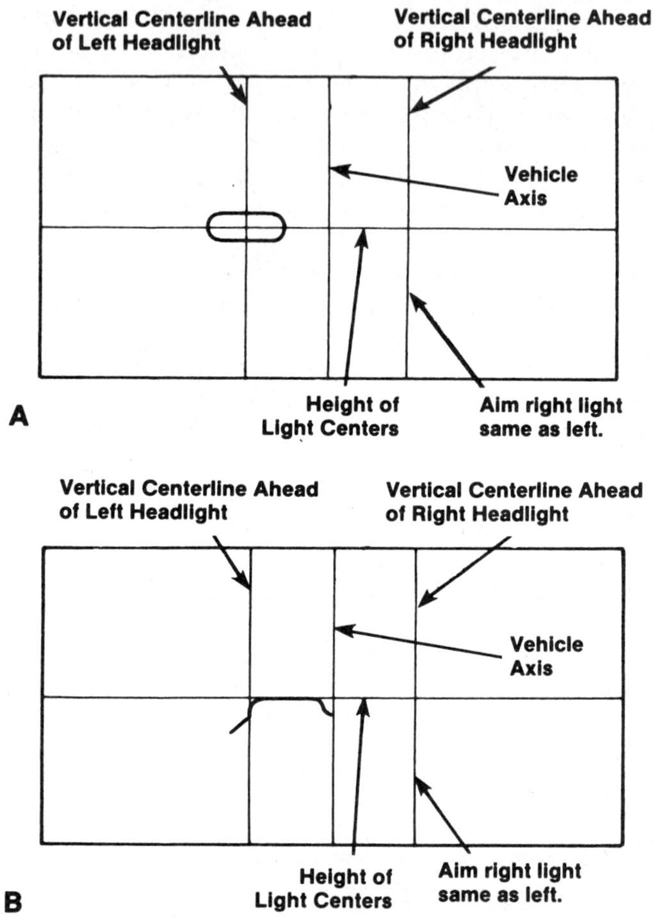

FIGURE 12-39 (A) Correct high beam and (B) correct low beam headlight adjustment

TAIL, BACKUP, STOP, AND DIRECTIONAL LIGHTS

Failure of a tail, backup, stop, or directional light on the vehicle can usually be attributed to a faulty bulb. Oftentimes moisture gets into the bulb socket and causes corrosion of the electrical contacts and the bulb. Corrosive conditions can be repaired by using sandpaper on the affected areas. For severe cases, replace the socket and/or bulb. After any repair, always attempt to waterproof the assembly to prevent future problems.

These types of lights are also contained within a lens or bezel type assembly, usually amber or red in color. Cracked or broken assemblies are easily replaced;

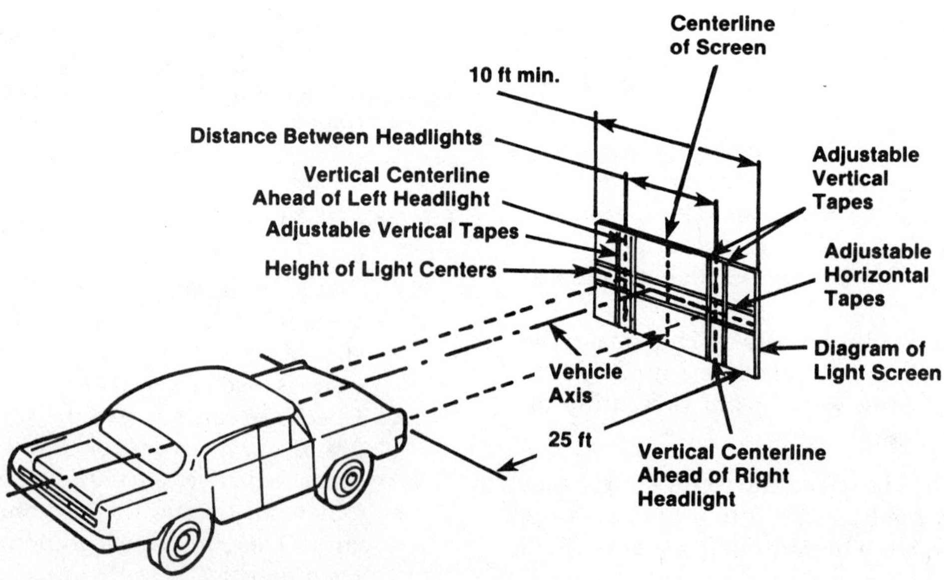

FIGURE 12-38 Aligning headlights on a wall

they are secured by attaching hardware readily accessible to the collision repair technician.

RATTLE ELIMINATION

Rattles and squeaks are normally caused by sheet metal that is loose or rubbing adjacent parts, loose bolts and screws, and loose or improperly adjusted doors, hood, or body panels. Other rather simple things such as a broken or loose exhaust mount or an improperly secured jack, tire, or articles in the trunk can also be causes of rattle.

Oftentimes a noise will be pinpointed by the customer to be in a certain area of the vehicle when in fact it might be caused by something in another area of the vehicle. This effect is caused by the sound traveling through the body. Usually a thorough investigation and a test drive of the vehicle is recommended so that the source of the noise can be located.

Most noise repairs involve the readjustment of parts, the replacement of parts, tightening loose attaching hardware, and welding broken parts. Check the hood for proper alignment at the front and the back. If the paint is knocked off or scratched on one end, it is probably hitting or rubbing against another part. Check the hood latch for looseness and proper fit as well as the adjustable stops. If the back of the hood flutters, readjust the hood so that it properly fits at the back edge. Also check the grill, wheelhousing, trim moldings, and bumper brackets for tightness.

Many areas on the body of the vehicle can also cause rattles and squeaks. The most susceptible areas are the instrument panel, doors, steering column, and seat tracks. It is also possible for weatherstripping to squeak, especially when it becomes very dry. Lubrication should be applied to applicable moving parts such as door hinges. All attaching hardware should be checked for tightness, especially in the area of the suspected noise source.

BUMPER REPLACEMENT

Prior to 1972, bumpers were rigidly bolted to the vehicle's frame. At best, these bumpers only resisted the bending forces of an impact; they transferred the energy shock directly to the frame and, thus, to the occupants of the vehicle. On the contrary, modern bumpers are designed to absorb the energy of a low-speed impact, minimizing the shock to both the frame and the occupants. Between 1973 and 1983, auto manufacturers were

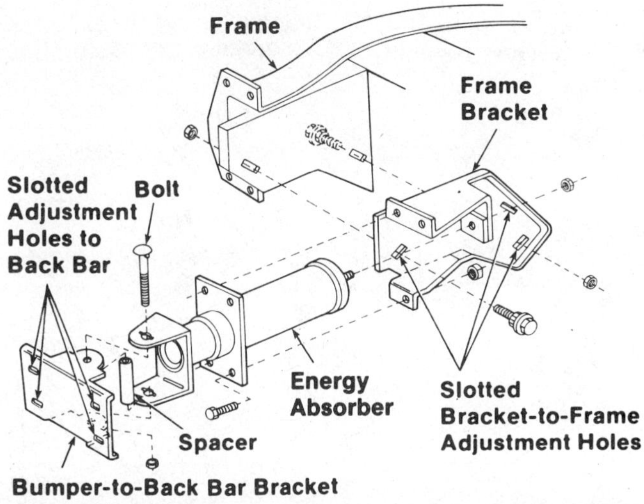

FIGURE 12-40 Typical energy absorber mounting arrangement

required to equip vehicles with bumpers that could withstand 5 mph collisions without incurring damage to the vehicle. In 1983, the standards were relaxed to 2.5 mph. To comply with these regulations, manufacturers fitted bumpers with energy absorbers. Most energy absorbers are mounted between the bumper face bar or bumper reinforcement and the frame (Figure 12-40).

Removing a bumper is basically a matter of removing the correct bolts (Figure 12-41). This job is made easier if the bumper is supported by a jack. On some vehicles, stone deflectors, parking lights, and other items must be disconnected before the bumper can be removed from the vehicle. Figure 12-42 shows a typical bumper with a urethane cover and related hardware. Bumpers with energy absorbers should be unbolted from the end brackets. Fixed bumpers should be removed by unbolting the brackets from the frame.

Several cautions must be observed when removing bumpers with energy absorbers.

- The shock type absorber is actually a small pressure vessel and should never be subjected to heat or bending. If cutting or welding near a shock type absorber, remove it.
- If the energy absorber is bound due to the impact, relieve the pressure before attempting to remove the bumper. Secure the bumper with a chain to prevent its sudden release and drill a hole into the front end of the piston tube to vent the pressure. Then, remove the bumper and energy absorber.
- Wear approved safety glasses when handling, drilling into, or removing a bound energy absorber.

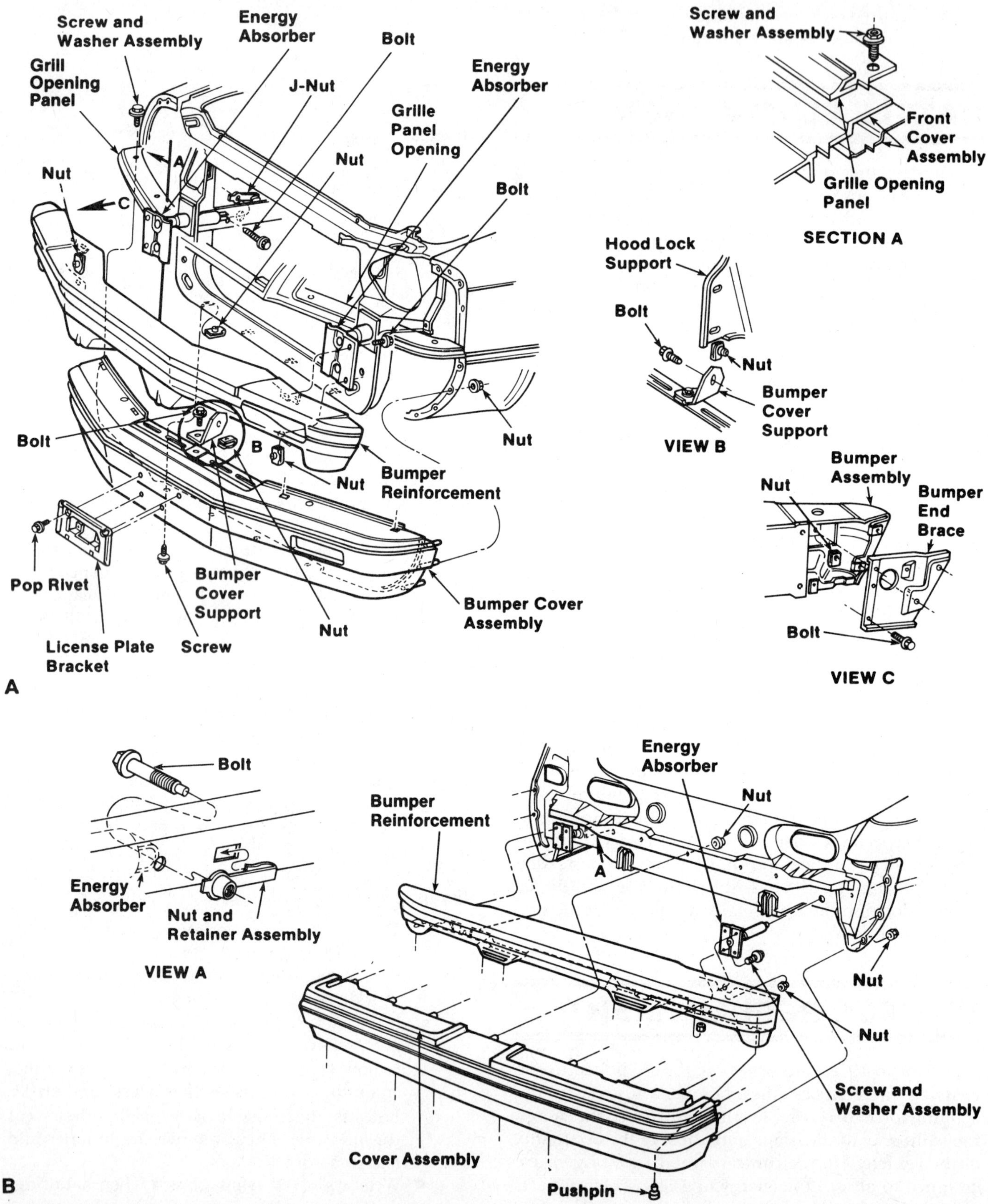

FIGURE 12-41 (A) Typical front bumper and (B) typical rear bumper attachment

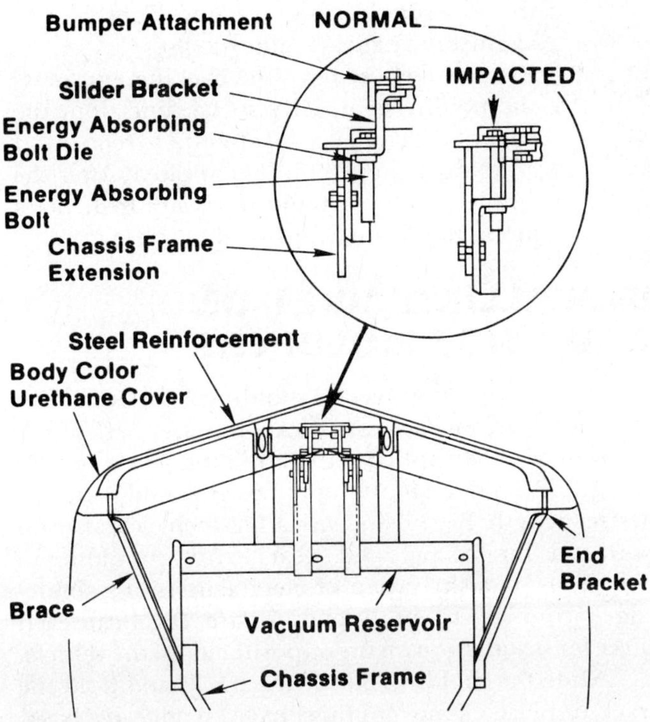

FIGURE 12-42 Typical bumper with urethane cover and energy absorbers

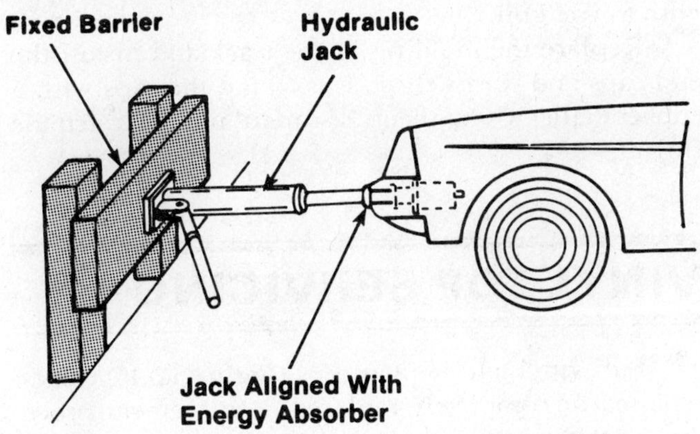

FIGURE 12-43 Testing energy absorber operation

When replacing a bumper, always test the energy absorbers as follows:

1. Position the vehicle facing a solid, fixed barrier (Figure 12-43).
2. Shift the transmission into park and fully engage the parking brake.
3. Place a hydraulic jack between the barrier and the bumper. Position the jack squarely over the energy absorber.
4. Compress the absorber 3/8 inch.
5. Release the jack. If the bumper does not return to its normal position, replace the absorber.

Be sure to torque all bolts to the manufacturer's specifications. After bolting the bumper in place, adjust it so that it is an equal distance from the fenders and front grille. The clearance across the top must be even. Adjustments are made at the mounting bolts. The mounting brackets allow the bumper to be moved up or down, side to side, and in and out. If necessary, shims can be added between the bumper and the mounting bracket to adjust the alignment.

INSTALLING BODY SIDE MOLDINGS

A variety of fasteners can be used to secure moldings to a vehicle. These include clips, bolts, studs welded to the body, adhesives, and rivets. The latter two methods are used to install add-on body side moldings by dealers, trim shops, and collision repair shops. Instructions for both the adhesive and rivet-on methods follow.

INSTALLING ADHESIVE BODY SIDE MOLDINGS

1. Park the vehicle on level ground. The surface to which the molding will be applied should be cleaned and at least 70 degrees Fahrenheit to ensure proper adhesion.
2. For greater protection, the molding should be applied to the outermost surface of the vehicle. If the vehicle has an outermost body line, install the molding 1/8 inch above or 1/8 inch below the body line, but not on the body line itself. If the vehicle does not have a prominent body line, select the outermost surface of the body contour.
3. Thoroughly clean the area with water and detergent. Use a clean rag wetted with wax and grease remover to remove waxes and silicones. Use a clean cloth for each side of the vehicle.
4. If the body molding will not be aligned above or below a body line that can be used as a guide, mark the correct height of the molding with a piece of chalk and a steel rule. Mark the height at the rear and front of the vehicle and at each door gap.
5. Stretch a piece of masking tape from front to rear, connecting the marks (Figure 12-44).

FIGURE 12-44 Applying masking tape as a straightedge

Keep the tape taut and sight along its length to ensure a straight line.

6. To cut the molding to length, start 3 inches from the front edge of the fender and measure to the edge of the door. Allow for 1/8-inch clearance between the molding and the edge of the fender. Cut the molding to size, then repeat this procedure for the rear quarter panel piece of molding.
7. When measuring for door pieces, leave 1/8-inch clearance at both ends of the molding. To prevent binding when the door is opened, cut the front ends of the molding at a 45-degree angle, using a single-edge razor blade.
8. Peel 6 inches of the protective backing paper from the cut end of the molding. Do not touch or dirty the exposed adhesive after removing the backing. Begin installing the molding 1/8 inch from the rear fender edge. Align it with the top edge of the tape and lightly press against the panel (Figure 12-45).

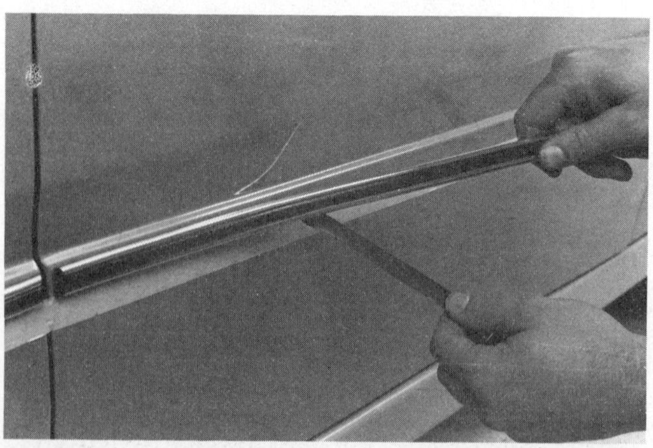

FIGURE 12-45 Aligning the molding with the tape

9. Progressively remove the backing and press the molding against the surface and along the edge of the tape. Do not attempt to reposition a molding piece after it is applied. After the whole length of molding is applied, press along the entire length with a roller.

INSTALLING RIVET-ON BODY SIDE MOLDINGS

Before applying rivet-on moldings, park the vehicle on a level surface and clean the area where the molding will be applied. After applying masking tape as a guide, hold the molding in position and mark the desired length. Be sure to leave a 1/2-inch clearance on both ends for end caps and spears. Cut the molding to length, using a sharp pair of metal cutters or a single-edge razor blade to cut the ends square. Duplicate each piece for installation on the opposite side of the vehicle.

Slide the molding out of the track, and hold the track in place. Some moldings have an adhesive backing to assist in positioning the track. With the track in place, use a drill bit that is 1/64 inch larger than the rivet diameter to drill holes in the track. Also, drill holes 3/4 inch from each end of the molding. Install the rivets with a rivet gun.

Replace the molding in the track and install the end caps and spears. If necessary, tap the caps with a rubber mallet to seat them. Be careful not to scratch the finish.

VINYL TOP SERVICING

Full vinyl and landau tops (Figure 12-46) can be repaired or completely replaced. Replacement procedures vary depending on the roof design, attachment method, and trim molding design. Most vinyl tops are glued directly to the roof panel. Others are padded with foam rubber.

Vinyl tops are attached to the roof panels with adhesive. A number of mechanical fasteners are also used to secure the vinyl top to the edges of the roof: clips over weld-on studs, reveal moldings, finishing lace, drive nails, screws, drip scallop moldings, weatherstripping retainers, or finishing plugs. While some vinyl tops come with a preapplied adhesive backing, this chapter deals only with the adhesive application of an unpadded vinyl top. Foam rubber pads are applied following the basic procedure described for vinyl top installation.

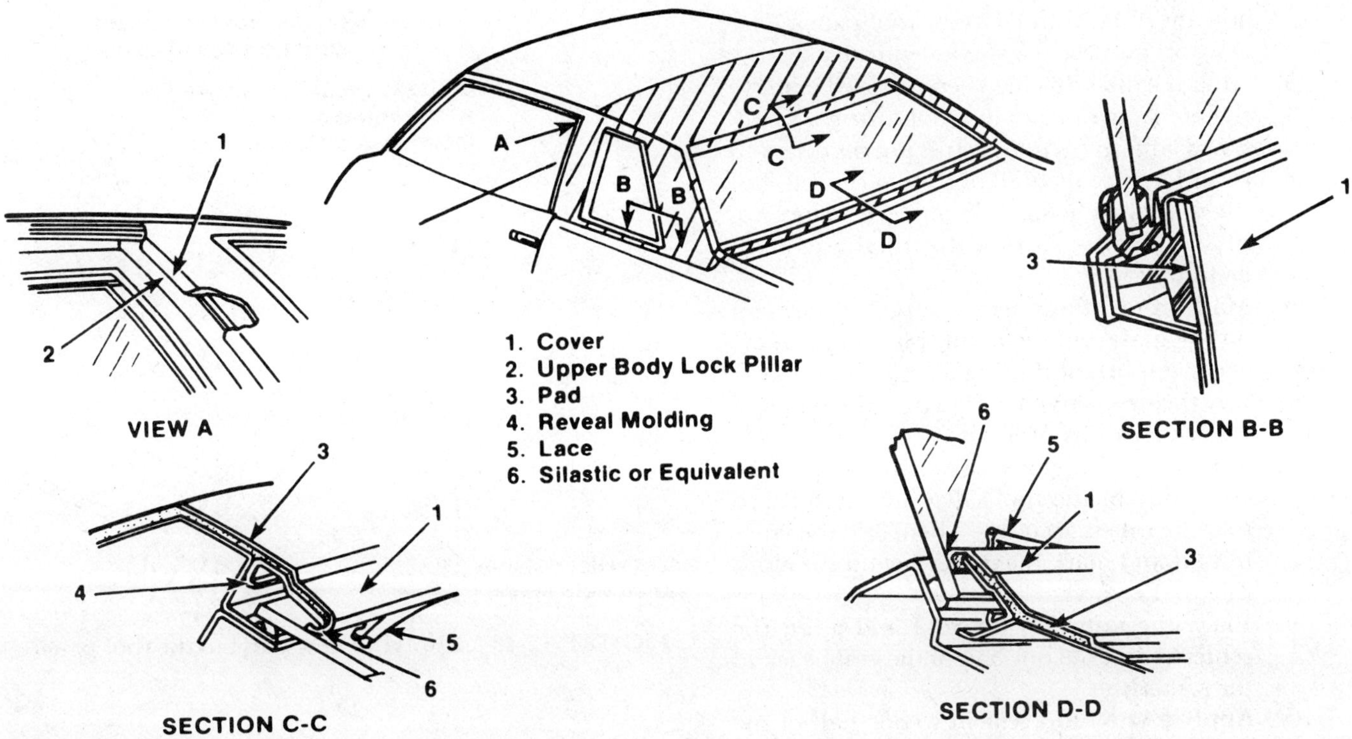

FIGURE 12-46 A landau vinyl top installation

Vinyl top repairs fall into three categories. One is complete replacement of the vinyl material. This can be necessary due to direct damage to the roof or to weathering of the material. If not properly maintained, the vinyl material will discolor, blister, and peel. Partial removal of a vinyl top is sometimes necessary to repair damage to the rear quarter panel or windshield A-pillar. Minor defects—such as cuts and abrasions—can be repaired with a patching compound or rebonded with a soldering iron.

COMPLETE REPLACEMENT

When a vinyl top has been damaged beyond repair, it can be replaced by following this procedure.

1. Unpack the new vinyl material and spread it out to remove the wrinkles.
2. Remove the trim moldings that cover the edges of the vinyl—windshield and rear window moldings, drip rail moldings, upper quarter panel moldings, and any emblems, ornaments, or nameplates attached through the vinyl material. If necessary, remove the quarter panel inner trim to gain access to the ornament retaining nuts. Some vehicles also require removal of interior garnish moldings to gain access to weatherstripping retaining screws. Be very careful when removing drive nails. Avoid stretching the holes so that the nails can be reinstalled in the same holes.
3. Clean excess cement and sealer from the roof edges and the windshield and rear window. If necessary, use a heat gun to soften the cement and ease removal.
4. Using a heat gun, soften the edges of the vinyl material so that it can be pulled from the roof. Use a pair of pliers to grasp the vinyl edge and peel the vinyl material from the roof. If necessary, use the heat gun as shown in Figure 12-47 to soften the cement. Carefully observe

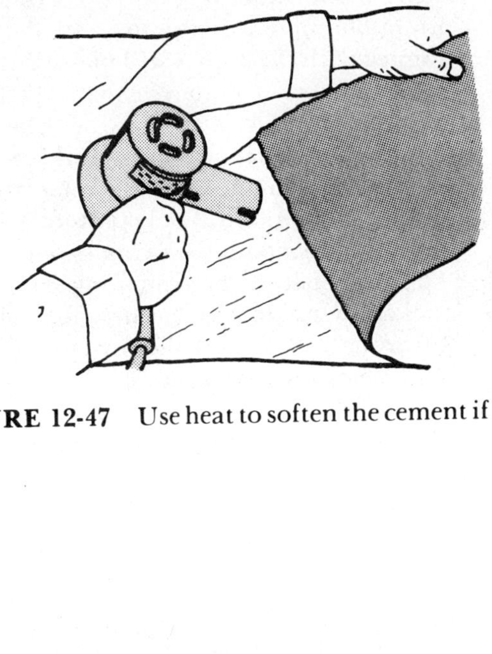

FIGURE 12-47 Use heat to soften the cement if necessary.

how the vinyl is fitted to the roof panel. The new top must be installed identically.

5. After the old vinyl has been removed, remove the old cement from the roof, using a xylol-based adhesive remover. It is not necessary to completely remove all of the old cement, but the roof must be smooth and level before applying the new vinyl or the installation will be lumpy.
6. Mask around the roof to avoid getting cement on the glass and the paint. Masking is especially important if using a spray adhesive.
7. Lay the new vinyl material over the roof and verify that it fits properly.
8. Remove the vinyl and use chalk to mark the centerline of the roof. Measure carefully so that the line is accurate. Then, fold the vinyl in half and mark the centerline on the cloth side of the vinyl.
9. Place the vinyl on the roof and align the centerlines. Fold one half of the vinyl back to the centerline.
10. Apply a vinyl top cement to one half of the vinyl. Do not apply cement to areas that cover the upper quarter panel, roof sides, or the area around the door, windshield, and rear windows.
11. Starting at the centerline, also apply vinyl top cement to the exposed half of the roof panel. Do not apply cement as yet to the sides or around windows. Be sure to carefully follow the adhesive manufacturer's instructions. Some vinyl top cements must be brushed or rolled on. Others are designed to be sprayed on. However the cement is applied, it must be applied evenly. There should be no skips, voids, globs, or pools of cement.
12. After allowing for the manufacturer's suggested tack time, carefully roll the vinyl over the cement, smoothing out any wrinkles or air bubbles. Be careful to keep the vinyl aligned with the roof. A stiff plastic squeegee can be used to work out air bubbles (Figure 12-48). To work out wrinkles, use a heat gun to soften the vinyl and cement. Hold the gun 1 to 2 inches from the vinyl and rotate the gun in a circular motion. Apply pressure with the squeegee to smooth the wrinkles out.
13. Apply cement to the remaining half of the roof and vinyl material as described in steps 10 through 12. Smooth the vinyl over the roof and stretch out all wrinkles.
14. Apply cement to the body and backside of the vinyl material in the area below the rear window (if applicable). Position the material and remove any wrinkles.

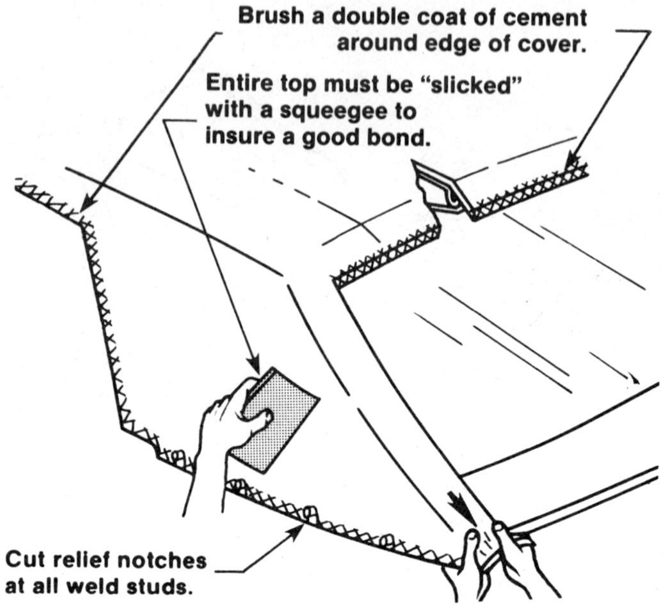

FIGURE 12-48 Applying new vinyl to the roof panel

15. Apply cement to the roof sides and door openings and to the backside of the vinyl material. Make angle cuts at the corners as necessary. Then, stretch the vinyl material and cement it in place. Pull the material as taut as possible.
16. On vehicles with frenched rear windows, carefully apply cement around the rear window on the padding and plastic panel. Use a spatula type tool to fold the vinyl material back under the rear window opening panel.
17. Trim the vinyl material at the windshield and rear window openings. Leave 1/2 inch of material at the openings for cover attachment in the window opening recess.
18. Apply cement to the rear window opening and the backside of the vinyl material. Make angle cuts as necessary at the corners and cement the material in the rear window opening recess. Trim the vinyl material around the molding retainer weld studs.
19. Apply adhesive to the windshield opening and the backside of the vinyl material. Make angle cuts as necessary at the corners and cement the material in the windshield opening recess. Trim the vinyl material around the molding retainer weld-on studs.
20. At each roof rear quarter panel, locate and punch two holes for the roof side ornament on models so equipped. Install each side or-

nament with the studs entering the punched holes. Then, from inside the vehicle, install the retaining nuts on the studs and apply a silicone sealer around the nuts.

21. Reinstall the interior trim panels, garnish moldings, weatherstripping, and exterior moldings.

SHOP TALK

Any exterior moldings using positive bolt-on fasteners must be installed prior to installation of interior trim parts.

22. Clean the entire work area.

PARTIAL REMOVAL

Remove all trim from the affected area. Pull up the vinyl material and peel it away. Place rags in the fold of the vinyl material to prevent it from creasing. If the repair area requires welding or heat shrinking, make sure that the vinyl material is pulled far enough away so that it will be unaffected by the heat. If necessary, place a piece of nonflammable material between the repair area and the vinyl material to absorb the heat. After the repair and refinishing have been completed, apply cement to the area and smooth the vinyl material back into place. Pull it tight and work out any wrinkles. Clean up any excess cement, and reapply the trim.

After the repair has cured for several days, wrinkles can appear in the vinyl material. Use a household iron pressed over a damp cloth to soften the vinyl material and smooth the wrinkles out (Figure 12-49). A heat gun and a stiff plastic squeegee can also be used. If the wrinkles cannot be removed, pull the vinyl material up and reglue the area.

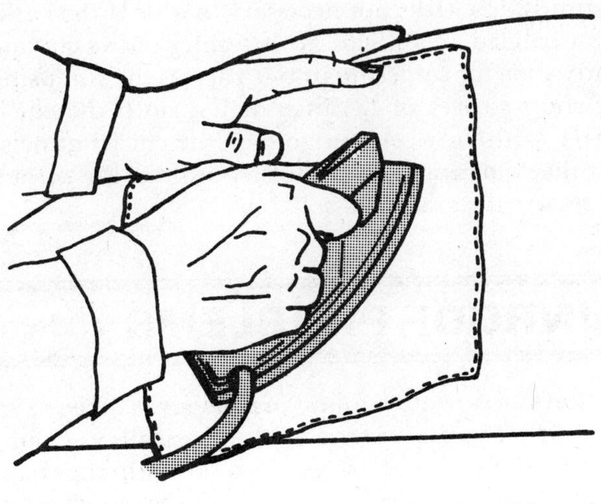

FIGURE 12-49 Smoothing out wrinkles with an iron

PATCHING A VINYL TOP

A torn vinyl top can sometimes be repaired with a vinyl patching compound and a grain mold die. The grain mold die is made from a special die material or from body filler. To make the die, clean a 6- to 8-inch square portion of the undamaged vinyl top with vinyl cleaner. Spray the area with a silicone mold release, then apply a properly catalyzed amount of die material or body filler to the area. Cover the material with a scrap piece of vinyl and place a block of wood on top of the vinyl to force the material into the grain pattern. Allow the die to harden; then peel it up, trim the edges, and spray with a silicone mold release.

After the die has been made, patch the torn vinyl following this procedure:

1. Trim off any loose fabric and frayed edges. If the damaged area is large, cement the loose vinyl material back into place and allow it to dry. Cut out at least 1/16 inch of material on both sides of the tear line to allow for the vinyl patching compound.
2. Clean the area with a vinyl cleaner and clean cloth. Do not spray the cleaner directly onto the vinyl material.
3. Apply the vinyl patching compound to the repair area with a trowel. Use the trowel or a putty knife to work the compound under any loose edges of the vinyl material. Apply the first layer to the metal-vinyl edge to make a root fillet (Figure 12-50). Wipe off any excess compound from the vinyl material.
4. Because most vinyl patching compounds have a thin viscosity, several coats will be necessary to completely fill the void. Follow the manufacturer's instructions for curing time between coats. A heat gun can be used to speed the curing time. Hold the gun 1 inch above the compound and move it in a circular motion for about 20 seconds. Press down on the material around the repair to keep it flat.
5. Build up the repair area until the compound is slightly above the level of the vinyl material (Figure 12-51). Usually three layers are adequate to complete the job, but larger holes might require four or five layers.
6. Before the last application hardens, draw the side edge of the trowel over the face of the compound to strike off the excess buildup and produce a level surface. Also, remove any

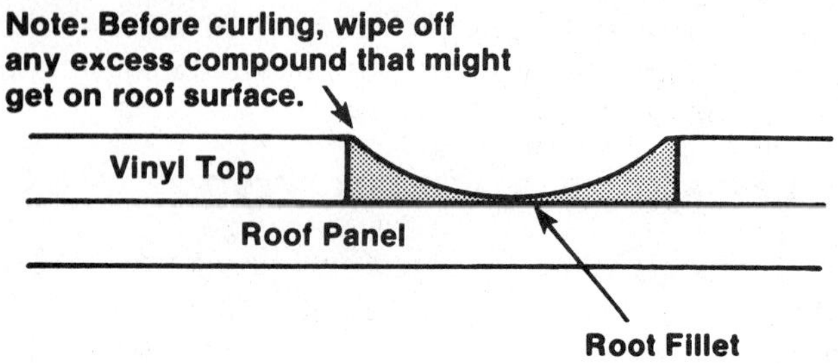

FIGURE 12-50 Filling the repair area to the metal-vinyl edge

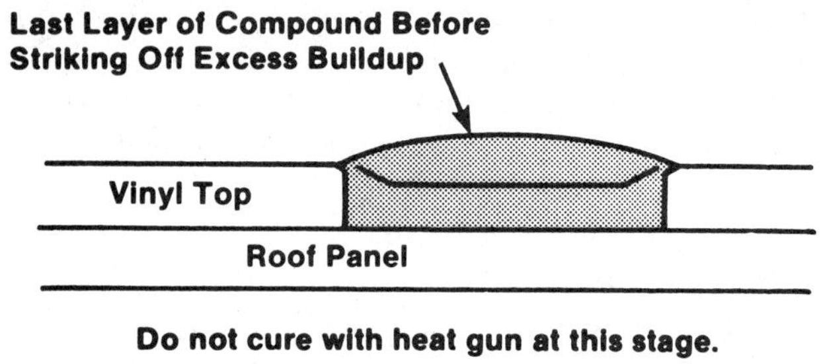

FIGURE 12-51 Build up the last layer above the vinyl material.

excess compound from the vinyl material adjacent to the repair area; otherwise, it will cure to the material and cause unwanted buildup on the surrounding area.

7. Cure the leveled compound by heating for about 30 seconds. The patched area should develop a shiny glazed appearance. Have the grain mold die ready in your other hand.

WARNING

Be careful not to blister or scorch the vinyl with the heat gun.

---

8. Remove the heat and immediately press down forcefully with the die. Hold the die on the area for at least 10 seconds to obtain a good impression.
9. Cool the area with a moist sponge.
10. Spray vinyl prep conditioner on a clean cloth and lightly dab the area. Do not rub.
11. Spray the repair area with a matching color vinyl paint or dye.

MINOR SURFACE CUTS

The edges of a surface cut that has not penetrated through the cloth backing can be welded with a soldering iron. Set the transformer heat range on the iron to approximately 225 degrees Fahrenheit.

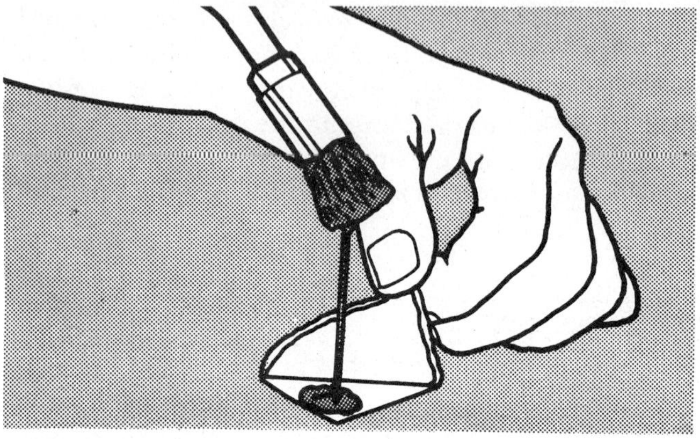

FIGURE 12-52 Gluing a cut in a vinyl top

Clean the soldering iron tip thoroughly with an abrasive pad to avoid vinyl buildup on the tip. Lightly slide the tip across the cut, using very short strokes of 1/4 inch or less until the cut is covered with vinyl. Go over the cut lengthwise with the soldering tip to smooth out the surface. The grain effect can be restored by carefully etching the grain pattern into the vinyl with the sharp edge of the tip.

If required, the gloss or shine on the surface can be removed with a dulling agent and liquid vinyl. Several color coats of liquid vinyl may be required; the last one should be a fog coat to minimize gloss.

MAJOR CUTS

A major cut—one that has penetrated through both the vinyl and the cloth backing—can also be repaired with the soldering iron. Set the transformer heat range to approximately 225 degrees Fahrenheit and clean the tip of the iron with an abrasive pad.

If the damage is more than 1 inch long, apply a light coat of cement to secure the loose vinyl material (Figure 12-52). (It is not necessary to glue if the cut is over a padded area.) Start at the center of the cut and lightly slide the soldering iron tip across the cut, using very short strokes of 1/4 inch or less until the cut is covered with vinyl. Again go over the cut lengthwise with the soldering iron tip to smooth out the surface and restore the grain.

SUNROOF PROBLEMS

Sunroofs are often found in unibody vehicles (Figure 12-53). The glass is replaced in basically the same manner as movable glass described earlier in the chapter. As in any glass service work, check the service manual for the exact methods of solving any sunroof prob-

FIGURE 12-53 Sunroofs are popular in unibody vehicles.

lems. Remember that specific directions vary from one manufacturer to another.

If the sunroof is motorized, it can be checked in the same manner as any other motorized electrical circuit described in Chapter 9. If the motor is faulty, it should be replaced.

PASSIVE RESTRAINT SYSTEMS

All vehicles built or sold in the United States must now have one or both of these passive restraints: automatic seat belts and/or air bags. Both have made a tremendous impact on driver safety over the last few years. It is very important that collision repair technicians are able to service these systems; a poor job can place the driver at risk.

SEAT BELTS

There are two types of seat belts in common use:

- *Active Restraint.* This system is one that a vehicle's occupant must make a manual effort to use. The driver and passenger(s) must fasten the seat belts for crash protection.
- *Passive Restraint.* This system is one that operates automatically. No action is required of the occupant to make it functional. Federal law requires that all vehicles sold in the United States after the 1990 model year be equipped with a passive restraint. In some vehicles, this means that air bags are standard equipment. In others, passive seat belts are employed. Regardless of the type of passive restraint system used, the manual lap belt must always be engaged for optimum safety.

The passive seat belt system automatically puts shoulder belts on the driver and front seat passenger. It operates by means of electric motors which draw the belts from a "parking" area in front, near the A-pillar, to an "engaged" area at the B-pillar (Figure 12-54). The movement is accomplished with carriers and track assemblies at the top of the door frame. One end of each belt attaches to the carrier; the other end is secured to the inertia lock retractor. The retractors are mounted in the center console. When the doors are closed and the ignition turned on, the belts move rearward and secure the occupants. Manual lap belts must always be used with passive shoulder belts.

When servicing or replacing lap and shoulder belts, refer to the following precautionary items:

1. Keep sharp edges and damaging objects away from belts.
2. Avoid bending or damaging any portion of the buckle or latch plate.
3. Do not attempt repairs on lap or shoulder belt retractor mechanisms. Replace with new parts.
4. Tighten all seat and shoulder belt anchor bolts as specified in the service manual.

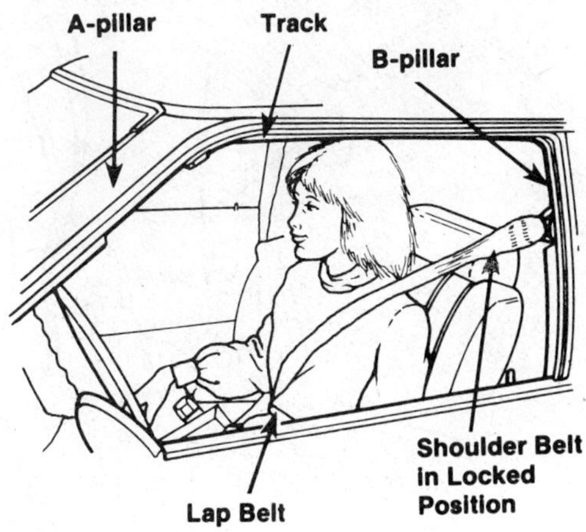

FIGURE 12-54 A typical passive seat belt in operation

Unrepaired damage to the seat belt system could lead to the technician, appraiser, collision repair shop, and insurance company being held liable if the system fails in the event of another collision. Manufacturer's recommendations for seat belt replacement and system testing following a collision vary, if they exist at all. However, in any case, a visual functional inspection of the belts themselves is very important. The following inspections provide typical, detailed procedures for seat belt inspection.

Drive Track Assembly Inspection

The drive track assembly (Figure 12-55) is very often damaged in a collision. Since body dimensions will affect the assembly, they must be correct to avoid any twisting or bending in the track, which can cause binding of the lock box.

Collision repair of the drive track assembly requires care to maintain proper body mounting surfaces. The use of shims on the drive track assembly is not recommended. Upper body damage must be properly repaired, returning the vehicle to exact dimensional specifications, before any repairs are performed on the seat belt system.

Some vehicle manufacturers recommend the replacement of the drive track assembly that was in use during a collision. If the collision was minor, a qualified technician might determine that the belts show no damage and continue to operate properly. When replacing any drive track, follow the manufacturer's recommendations for bolt tightening and torquing.

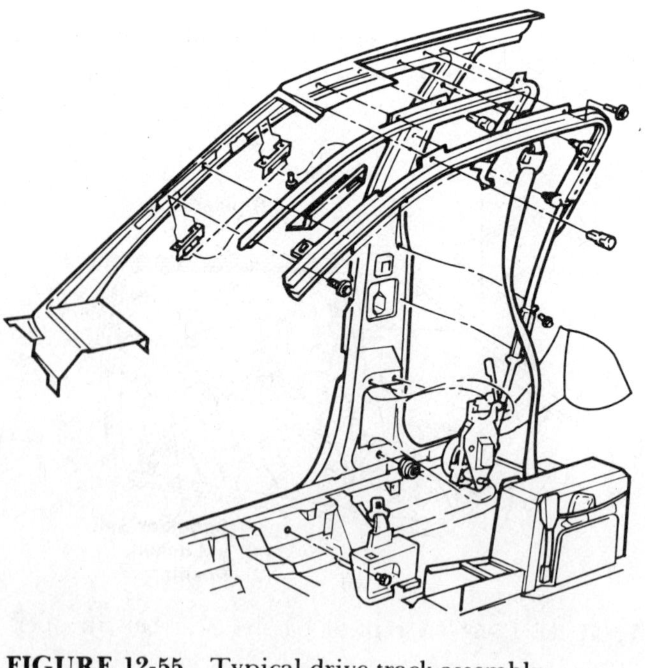

FIGURE 12-55 Typical drive track assembly

The drive motor is usually located at the base of the drive track assembly, behind the front seat side trim panel. Its purpose is to pull the tape that positions the belt. If a check of the drive motor reveals that it is faulty, replace it. Like any motorized system, the seat belt parts need periodic lubrication. Remember that:

- Noise complaints could mean the system needs lubrication.
- The system must be kept free of dirt to keep it from binding.
- Lubricant should not get on trim or belt webbing.

Webbing Inspection

1. Pay special attention to where the webbing contacts maximum stress points, such as the buckle, D-ring mount, and retractor. Collision forces center on these locations and can weaken the belt. Signs of damage at these points require belt replacement.
2. Check for twisted webbing due to improper alignment when buckling.
3. Fully extend the webbing from the retractor. Inspect the webbing and replace with a new assembly if the following conditions are noted:
   - Cut or damaged webbing
   - Broken or pulled threads
   - Cut loops at the belt edge
   - Color fading as a result of exposure to sun or chemical agents
   - Bowed webbing
4. If the webbing cannot be pulled out of the retractor or will not retract to the stowed position, check for the following conditions and clean or correct as necessary.
   - Dirty webbing coated with gum, syrup, grease, or other foreign material
   - Twisted webbing
   - Retractor or loop on B-pillar out of position

Buckle Inspection

The technician must find out whether or not the buckle works and if the buckle housing cover has been damaged.

1. Insert the tongue of the seat belt into the buckle until a click is heard. Pull back on the webbing quickly to assure that the buckle is latched properly.
2. Replace the seat belt assembly if the buckle will not latch.

3. Depress the button on the buckle to release the belt. The belt should release with a pressure of approximately 2 pounds.
4. Replace the seat belt assembly if the buckle cover is cracked, the push button is loose, or the pressure required to release the button is too great.

Retractor Inspection

Retractors for lap belts should lock automatically once the belt is out fully. They either work or they do not; so, they are simple to check. But it is a different story for combination lap/shoulder belt retractors. These operate from sensors that detect sudden changes in motion.

The two types of seat belt retractors used with passive seat belt systems are either webbing sensitive or vehicle sensitive. Webbing sensitive seat belt retractors can be tested by grasping the seat belt and jerking it. The retractor should lock up. If it does not, replace the seat belt retractor.

Vehicle sensitive seat belt retractors will not lock up using the same procedure as the webbing sensitive retractors. To test these belts, a brake test is required. Perform this test in a safe place. A helper is required to check the retractors on the passenger side and in the rear seats if the vehicle is equipped with rear lap/ shoulder belts.

Do the brake test for each belt by driving the vehicle at 5 to 8 mph and quickly applying the brakes. If the belt does not lock up, replace the seat belt assembly. During this test, it is important for the driver and helper to brace themselves in the event the retractor does not lock up. Replace any seat belt assembly that does not lock up. If the vehicle is not driveable when it first comes into the shop, check the retractors after the vehicle has been repaired. Make note of their condition on the damage report. Most retractors are not interchangeable. That is, an "R" marked on the retractor tab indicates that it is for the right side only, and an "L" should be used on the left side only.

On some vehicles both the webbing sensitive and vehicle sensitive retractor tests are required. Check the service manual.

Seat Belt Tensioner Inspection

Some vehicles are equipped with seat belt tensioners. These contain a small explosive charge that is detonated during a collision. This causes the seat belts to tighten during severe frontal impacts. The tensioners, located in the B-pillar, can be used on vehicles with or without air bags. Tensioners may not deploy if the seat belt is not buckled. To determine if the tensioners have been deployed or to replace deployed tensioners, follow the procedures in the service manual.

SHOP TALK

The passive seat belt system is a part of the inertia switch circuit. During an impact, the inertia switch cuts off the power to the restraint system and the fuel pump. Reset the switch to restore operation to both. It is also important that shoulder belts be replaced whenever a belt is stressed in a collision.

Anchorage Inspection

The seat belt anchors are found where the retractors attach to the vehicle body. High impact forces may occur here during a collision; therefore, be sure to carefully inspect the anchor areas by looking for:

1. Cracks and distortion in the metal where the retractors and D-ring mounts anchor.
2. Cracks, distortion, and looseness in the bolts that attach the retractors to the body. These bolts are usually large, prevailing torque fasteners.
3. Dirt and corrosion in the anchorage area.

Loose bolts should be replaced. If there is damage to the metal in the mounting area, proper repairs must be completed before reattaching the anchor. Clean the area prior to welding. Be sure to apply correct MIG welding procedures and restore corrosion protection to the area. When spraying anti-corrosion compounds, make sure they do not enter the retractor. This can keep it from operating properly.

CENTER BELTS

Most vehicles have center belts in the rear seats, and many have them in the front as well. Usually these belts do not have retractors. Instead they have adjustable slide locks. These locks should be tested to make sure they work. If not, the seat belt assembly must be replaced. Check the webbing and buckles just as in the other belt assemblies, paying special attention to the buckles. Chances are they are not used much and thus may get dirty and not operate properly.

WARNING LIGHT AND SOUND SYSTEMS

When the ignition is turned to the on or run position, the FASTEN SEAT BELT light should come on. There should also be a buzzer or chime. In some vehicles there is a computerized voice that directs the occupants to fasten their seat belts. If these warning light and sound systems do not come on, perform a visual inspection, checking for the following:

1. First, look for a blown fuse or circuit breaker.
2. If that checks out fine, and there is sound but no light, check for a damaged or burned out bulb.
3. If the bulb lights but there is no sound, check for damaged or loose wiring, switches, or buzzer.

Running some of the basic electrical checks described in Chapter 9 should locate the problem. Having the proper wiring diagram and service manual on hand can help. Dismantling and inspecting seat belt systems takes a systematic approach and a focus on detail. Their important function in passenger protection calls for careful attention.

SHOP TALK

Emergency releases are often used to let occupants out of the vehicle if the seat belt system jams or is inoperative (Figure 12-56). If the emergency release was used, it should be reconnected before servicing.

AIR BAGS

The air bag will only deploy during frontal impacts (Figure 12-57). The air bag will not deploy in side or rear impacts, in rollover situations, or in multiple impact collisions. The frontal impact must fall within 30 degrees of the vehicle's centerline for the air bag to activate (Figure 12-58).

The vehicle speed necessary for deployment will depend upon the object being struck:

- If the vehicle hits an immovable object, such as a concrete wall, the required speed to cause deployment is 12 mph.

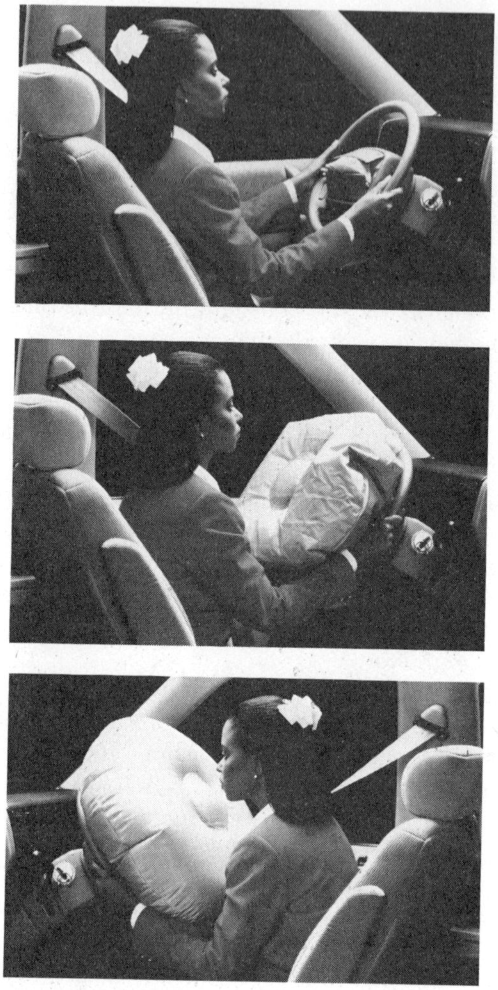

FIGURE 12-57 Deployment sequence of a typical air bag system

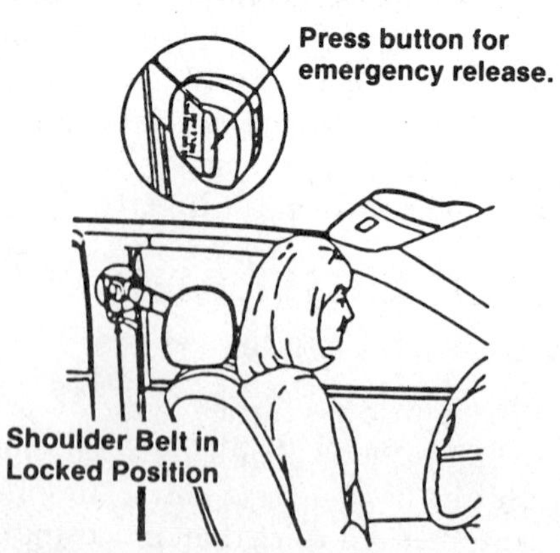

FIGURE 12-56 The emergency release lets occupants out of the vehicle when the seat belts are not functioning.

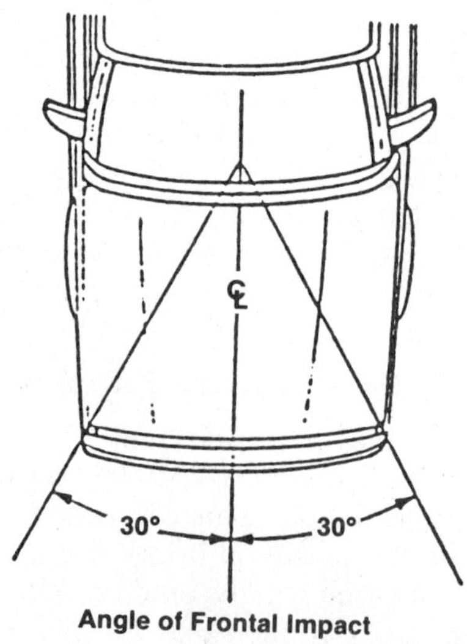

FIGURE 12-58 Air bags will deploy only during frontal impacts with 30 degrees of the vehicle centerline.

- If the vehicle hits a movable object, such as another vehicle which would give with the impact, the speed required for deployment is 24 mph.
- Impacts that occur from striking a parking block at low speeds, or other types of minor bumping, will not cause deployment.

If a vehicle strikes an immovable object, the energy is absorbed only by the vehicle. However, if two vehicles collide, both vehicles have energy absorbing designs that will soften the impact and slow their deceleration. Therefore, a higher speed is required to deploy the air bag when two moving vehicles collide.

It is virtually impossible for the air bag to accidentally deploy while the vehicle is being driven, and there have been no documented cases of this happening. Manufacturers have conducted tests in which air bags were intentionally deployed while someone was driving the vehicle. During these tests, the driver was able to maintain control of the vehicle after the air bag deployed. It is, however, possible for the collision repair technician to cause an unwanted deployment while repairing the air bag system. This is very easy to guard against, and can save the shop the cost of the air bag unit. To avoid an unwanted deployment:

- Disarm the system prior to repairs, following the procedures given in the service manual.
- Disarm the system prior to performing welding on the vehicle.
- Determine if there is a backup energy reserve. Check the service manual for the time needed for the reserve module to discharge.

It is important that the technician refer to the service manual for specific recommendations and procedures for disarming the air bag system before making repairs. The manual may recommend that the battery cable be taped, or isolated, to prevent contact during repairs. The manual may also direct the technician to remove the air bag module.

Air Bag System Parts

While the location and design of the air bag system varies from manufacturer to manufacturer, all air bag systems have the following similar major parts and systems:

1. *Electrical System (including the impact sensors and diagnostic control module).* Its main functions are to conduct a system self-check to let the driver know that it is operative, to detect an impact, and to send a signal that inflates the air bag.
2. *Air Bag Module.* The air bag module is the most visible part in the system. It is located in the steering wheel assembly, behind the center cover, on all vehicles. The assembly consists of the bag assembly and the inflator module (Figure 12-59). On vehicles equipped with a passenger side air bag, the module is located in the glove box area of the instrument panel.

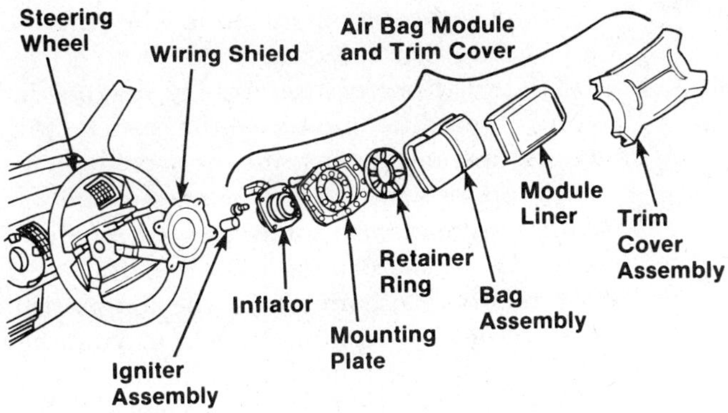

FIGURE 12-59 Components of a typical air bag module.

3. *Inflator Module.* This module receives a signal (current) from the diagnostic control module. The current heats the igniter and starts a thermal reaction with the chemicals in the inflator module. The thermal reaction creates nitrogen gas, which inflates the air bag.
4. *Bag Assembly.* The air bag itself is made of nylon material and the inside is coated with neoprene. The outside of the air bag is room temperature during a deployment. The heat generated during deployment burns the labels that can be found on the underside of the inflator module. Both bags will deploy at the same time on vehicles with both driver's side and passenger side air bags, even if no one is in the passenger seat. The passenger side air bag is about three times larger than the driver's side air bag. This is because the driver will always be positioned behind the steering wheel, but front seat passenger location can vary.

   The air bag is designed to deflate almost immediately after it inflates by using:
   - Relief holes in the back of the bag.
   - An uncoated (no neoprene) strip around the circumference of the bag.
5. *Coil or Clock Spring.* The coil, or clock spring, is mounted directly behind the steering wheel. Inside the clock spring are coiled electrical conductors. These conductors will roll and unroll as the steering wheel is turned. The clock spring has two electrical connectors. One is attached to the wiring harness at

the base of the steering column. The other is attached to the inflator module.

6. *Diagnostic Control Module.* The diagnostic control module monitors the system for faults, operates the instrument panel indicator lamp, provides an energy reserve, and records crash data. (In some cases, the module must be replaced after recording four deployments.) Disconnecting the diagnostic control module from the system may not be enough to prevent the air bag from deploying during service; check the service manual.
7. *Energy Reserve Module.* Most air bag systems have an energy reserve built into the system. The energy reserve module maintains enough voltage to deploy the air bag if the battery cables are cut during a collision. It is located in the diagnostic control module on some systems or can be a separate part.
8. *Impact Sensors.* The impact sensors are switches that close when the vehicle slows down at a specific rate. They verify the direction and severity of the impact. After a collision, visually inspect the sensors for damage or mislocation. To cause a deployment, at least one forward sensor and one rear sensor must close at the same time. Depending on the system used, there will be 3 to 5 sensors on the vehicle.
9. *Knee Bolster or Diverter.* The knee bolster or diverter is an extra cushioned area that protects the driver's knees and prevents the driver from sliding under the air bag during collision. It is located underneath the steering column trim panel, or is designed into the lower trim panel.

Disarming the Air Bag System

To disarm the air bag system prior to diagnostics, follow the procedures found in the service manual. The procedures vary from one manufacturer to another. Therefore, follow the service manual instructions to prevent an unwanted deployment of the air bag, and to determine if any special tools or procedures are required. The manufacturer may specify a special tool that will fool the diagnostic control module into thinking that an air bag is still attached. This allows the diagnostic control module to go through its normal self-check sequence. Some manufacturers use a jumper wire across the upper clock spring connector, while some use a special load tool that plugs into the upper clock spring connector in place of the air bag module.

The following general safety cautions must be taken into consideration when working on or around an air bag system:

- Before beginning any kind of work on the system, disconnect the negative side of the battery to prevent accidental deployment of the air bag. On some vehicles, it is necessary to wait 10 minutes for the reserve power supply to discharge.
- The air bag can be deployed at anytime. This is because the bag assembly is electrically coupled directly into the battery through the inflator module and various impact sensors.
- When carrying a live (undeployed) air bag module, be sure the trim cover is pointed away from your body when walking. This will help reduce the chances of serious injury if the air bag suddenly deploys.
- When laying an air bag module down, make sure the trim cover is faceup to minimize a "launch effect" of the module if the air bag suddenly deploys.
- Defective or used air bag modules must be replaced as a whole unit. Do not attempt to service individual parts of these modules.
- Wear safety glasses and protective gloves when handling deployed air bags, and wash hands with soap and water when finished. Inflator modules contain irritating and toxic chemicals.
- Never probe the electrical connectors on an air bag system. This could deploy the air bag.

Servicing the Air Bag System

Always begin the servicing with a visual inspection of the air bag system, checking the basic parts that are easily visible for obvious problems. Always inspect the air bag module, instrument panel indicator lamp, all impact sensors, all wiring harnesses, and the electrical connectors. Look for obvious impact damage, parts mislocation, stress to parts, or signs of pulled or broken wires.

Always use the service manual to locate hidden parts. Damage from a collision or mishandling during a nonrelated repair can set up a fault, which will disarm the air bag system. If the vehicle is accessible with a fault code procedure or display parameter, troubleshooting can be very straightforward.

Most air bag systems have self-diagnostic capabilities (see Chapter 9). The service manual is needed to make full use of on-board diagnostics. Diagnosis begins with the instrument panel indicator lamp. Usually it will light during system start-up. It will stay lit or blink 6-10 seconds depending on the system, and then turn off. If the light stays on there is a fault present in the system.

Fault Code Diagnostics. It is important that the collision repair technician refer to the vehicle's service manual to determine how to use the instrument panel indicator lamp to make a diagnosis of the air bag system. If there is a fault in the system, the service manual may direct the technician to enter the fault code diagnostic mode by jumping electrical connectors and cycling the ignition. Once the fault code diagnostic mode has been entered, the instrument panel indicator lamp will display the fault code. Each flash code is on for about 1/2 second and off for about 1/2 second. The number of flash codes in a series determines the digits of a fault code. These digits can then be compared to the diagnostic flow chart found in the service manual to determine the exact air bag system problem. The procedures used to obtain fault code(s) using the instrument panel indicator lamp will vary slightly from one vehicle to another.

There are two types of fault codes that can be accessed using the service manual. The diagnostic control module stores both of these codes in its memory. A fault code that is current means that a problem still exists in the air bag system. The current code will remain in the system until the fault is repaired and the code is cleared from the diagnostic control module's memory. History codes represent system faults that have been corrected and are no longer active. The history code may represent damage that has been repaired or perhaps an intermittent problem. By accessing the history file in the diagnostic control module, the technician can read the fault codes that have been set.

After the air bag system has been diagnosed and the faults within the system repaired, the fault codes should be removed from the diagnostic control module's memory. The way to clear fault codes differ from one vehicle manufacturer to another. On some vehicles, the fault code is erased when the fault is repaired. Some air bag systems require the use of a scan tool to clear or erase fault codes. The technician must refer to the service manual to determine specifically how to clear fault codes.

A test drive should be performed following the code clearing process. This ensures that the fault codes do not reappear or that no other faults exist in the system.

Replacing Air Bag Parts

Air bag parts are serviced by replacement only. At no time should any repairs be made to the air bag module, impact sensors, diagnostic control module, or the clock spring. Other service guidelines follow.

- Replacement parts should be new parts only, no salvage parts.
- Parts must be matched identically to the vehicle and the location.
- Follow proper torque specifications given in the service manual.
- Use all fasteners that come with the replacement part. Do not reuse old fasteners or substitute other fasteners.

Methods of replacing the impact sensors following a deployment vary from manufacturer to manufacturer. They must be properly positioned in order to perform properly. When replacing impact sensors, all arrows must point forward, all mounting brackets and angles must be correct, all fasteners must be properly torqued, and all wire harnesses and pigtails must be routed properly. Any deviation may affect the operation of the system in a subsequent collision.

The clock spring should be maintained in its correct index position at all times. Failure to keep the clock spring centered relative to the steering wheel's straight ahead, neutral position can cause the air bag system to default into a nonoperative mode.

Pulling the steering wheel to compensate for alignment variations or to service the couplings or gear box seals can cause damage to the clock spring. It could also alter the deflection attitude of the air bag for deployment. Even though the procedure for removal of the steering wheel is essentially similar to that used on vehicles not equipped with air bags, caution should be used during removal.

After an air bag deployment, check if the heat generated during deployment caused scorching or melting of the clock spring housing or the electrical connectors. In cases where scorching is evident, replace the clock spring. When scorching is not evident, the replacement policy will vary from one manufacturer to another. Refer to the service manual to see if the clock spring is to be replaced after following deployment.

Following an air bag deployment, there will be some dust left inside the vehicle. The dust is only to keep the air bag material from sticking to itself during deployment. To clean the vehicle following a deployment, vacuum the seats and carpet, and wipe down all other interior areas with a damp cloth.

Proper disposal of the air bag depends upon whether it has been deployed or not. If it has been deployed, dispose of it like any other scrap part. If an air bag module is determined to be faulty, but undeployed, it must be replaced. Since it is considered a live module, the air bag must either be manually deployed following the directions given in the service manual or returned to the manufacturer for disposal. If the air bag will be returned to the manufacturer, wait for the replacement part to be sent. Remove the faulty air bag and pack it in the shipping container the replacement air bag was shipped in. (There should be instructions for this procedure with the replacement module.) By using the shipping container from the replacement module, all of the labeling and packaging will meet the United States Department of Transportation requirements.

After servicing the vehicle, a final inspection of the system should include checking to make sure the instrument panel lamp turns off after system prove-out. Carefully recheck the wiring and harness routing.

REVIEW QUESTIONS

1. What type of glass consists of two thin sheets of glass with a layer of clear plastic between them?
   a. tinted
   b. tempered
   c. laminated
   d. none of the above
2. What type of glass fits the contours of a vehicle very closely?
   a. modular
   b. tempered
   c. laminated
   d. channel
3. Glass installation using a gasket is more common in ______________.
   a. older vehicles
   b. newer vehicles
   c. vehicles with modular glass
   d. none of the above
4. Which of the following items is used in the full cutout windshield replacement method?
   a. butyl ribbon sealer
   b. setting blocks
   c. utility knife
   d. all of the above
5. In the partial cutout windshield replacement method, what serves as the base for the new adhesive?
   a. butyl ribbon sealer
   b. butyl tape
   c. masking tape
   d. the old adhesive
6. Technician A repairs a minor abrasion in a vinyl top with patching compound. Technician B repairs a minor abrasion in a vinyl top by rebonding it with a soldering iron. Who is correct?
   a. Technician A
   b. Technician B
   c. Both A and B
   d. Neither A nor B
7. Which of the following should not be done before adjusting vehicle headlights?
   a. Inflate the tires to the recommended pressures.
   b. Clean off any snow, ice, or mud from the backside.
   c. Fill the gas tank.
   d. Check the condition of the suspension components.
8. What is required to test an energy absorber?
   a. hydraulic jack
   b. chain
   c. drill
   d. all of the above
9. Which fastening method is used to install add-on side moldings?
   a. rivets
   b. adhesives
   c. both a and b
   d. neither a nor b
10. When measuring for door molding pieces, Technician A leaves 1/8 inch of clearance at both ends of the molding. Technician B makes no allowance for clearance. Who is correct?
    a. Technician A
    b. Technician B
    c. Both A and B
    d. Neither A nor B
11. What can be used to remove wrinkles from a newly repaired vinyl top that has cured for several days?
    a. heat gun and squeegee
    b. iron and damp cloth
    c. both a and b
    d. neither a nor b
12. Prior to patching a vinyl top, Technician A cleans the repair area by spraying vinyl cleaner directly on the vinyl material. Technician B sprays the cleaner on a cloth and uses it to clean the vinyl material. Who is correct?
    a. Technician A
    b. Technician B
    c. Both A and B
    d. Neither A nor B
13. When repairing a major cut in a vinyl top, when should adhesive be used to secure the loose vinyl material?
    a. never
    b. if the cut is more than 1 inch deep
    c. if the cut is more than 1 inch long
    d. always
14. What type of seat belt operates automatically, with no action required by the vehicle's occupant to make it functional?
    a. passive restraint
    b. retractor
    c. active restraint
    d. anchorage

CHAPTER

13 Repairing Auto Plastics

OBJECTIVES

- List some of the growing number of plastic applications in unibody vehicles.
- Identify automotive plastics through the use of international symbols (ISO codes) and by making a trial-and-error weld.
- Describe the basic differences between welding metal and welding plastic.
- Outline the basics of hot-air and airless welding.
- Repair interior and unreinforced hard plastics, and instrument panels.
- Perform two-part adhesive repairs.
- Repair RRIM and other reinforced plastics.

The term *plastics* refers to a wide range of engineering materials that are synthetically compounded. This means that, unlike metals, they do not occur in nature and must be manufactured. Plastics are made from a variety of materials, including crude oil, petroleum, coal, natural gas, and other natural materials.

Because plastic is much lighter in weight than sheet metal, it has become an increasingly important part of the unibody vehicle. In 1975, it represented only about 3 percent of the total weight of the average vehicle. Today, more and more plastic is being used in automobile manufacturing, including bumper and fender extensions, fascias, fender aprons, grille opening panels, fuel lines and tanks, stone shields, instrument panels, and trim panels (Figure 13-1). Many of the new reinforced plastics are nearly as strong and rigid as steel, and some are even more stable dimensionally; they are also extremely corrosion resistant. Tests are being made on plastic engine blocks and plastic frame parts. Plastic suppliers are projecting increased use of plastic in floor pans, windows, steering shafts, springs, wheels, bearings, and other mechanical components.

This increasing use of plastic has resulted in new approaches to unibody collision repair. Many damaged plastic parts can be repaired more economically than they can be replaced, especially if the part does not have to be removed from the vehicle. Cuts, cracks, gouges, tears, and punctures are all repairable; and when necessary, some plastics can also be re-formed after distortion back to their original shape. Repair is also quicker than replacement, since parts are not always available; this means less downtime for the vehicle.

TYPES OF AUTO PLASTICS

Two types of plastics are used today in automotive construction:

1. *Thermoplastics.* These plastics are capable of being repeatedly softened by heating, with no change in their appearance or chemical makeup. They soften or melt when heat is applied and harden when cooled. Thermoplastics are weldable with a plastic welder.
2. *Thermosetting Plastics.* These plastics undergo a chemical change by the action of heating, a catalyst, or ultraviolet light. They are hardened into a permanent shape that cannot be altered either by reapplying heat or catalysts. Thermosets are not weldable, although they can be "glued" using an airless welder.

The use of plastic types can change from vehicle to vehicle, even within the same model year. This can happen if the manufacturer switches suppliers or as a result of design or production changes.

Table 13-1 shows some of the more common plastics with their full chemical name, common name, and where on the vehicle they might be found. Their designation as to thermosetting or thermoplastic is also included.

There is another group of plastics that the collision repair technician should be aware of. These are called

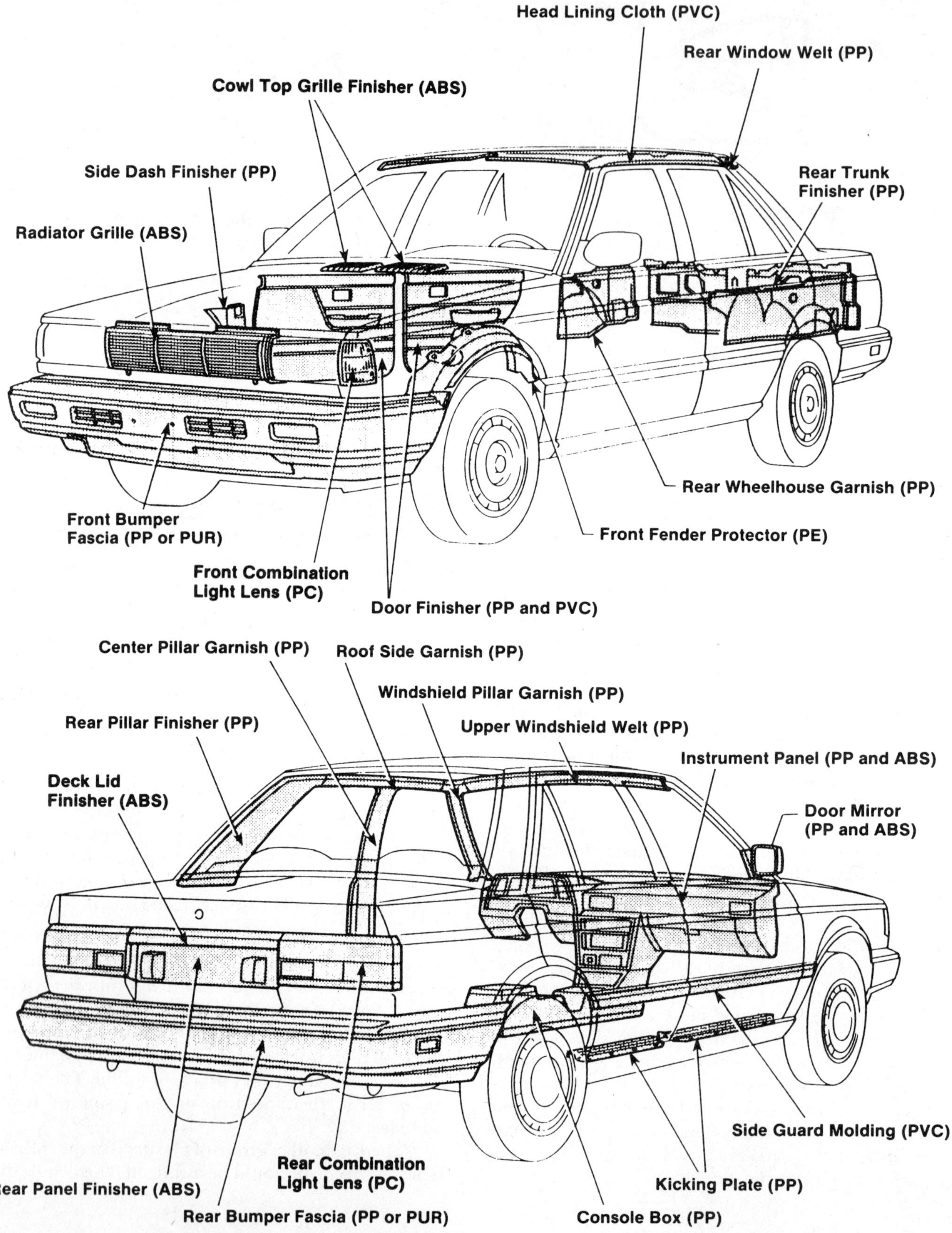

FIGURE 13-1 Plastic locations in today's unibody vehicles

TABLE 13-1: SYMBOL, CHEMICAL NAME, TRADE NAME, AND DESIGN APPLICATIONS OF COMMONLY USED AUTOMOTIVE PLASTICS

Symbol	Chemical Name	Trade Name	Design Applications	Thermosetting or Thermoplastic
AAS	Acrylonitrile-styrene	Acrylic Rubber	—	Thermoplastic
ABS	Acrylonitrile-butadiene-styrene	ABS, Cycolac, Abson, Kralastic, Lustran, Absafil, Dylel	Body panels, dash panels, grilles, headlamp doors	Thermoplastic
ABS/MAT	Hard ABS reinforced with fiberglass	—	Body panels	Thermosetting
ABS/PVC	ABS/Polyvinyl chloride	ABS Vinyl	—	Thermoplastic
EP	Epoxy	Epon, EPO, Epotuf, Araldite	Fiberglass body panels	Thermosetting
EPDM	Ethylene-propylene-diene-monomer	EPDM, Nordel	Bumper impact strips, body panels	Thermosetting
PA	Polyamide	Nylon, Capron, Zytel, Rilsan, Minlon, Vydyne	Exterior finish trim panels	Thermosetting
PC	Polycarbonate	Lexan, Merlon	Grilles, instrument panels, lenses	Thermoplastic
PPO	Polyphenylene oxide	Noryl, Olefo	Chromed plastic parts, grilles, headlamp doors, bezels, ornaments	Thermosetting
PE	Polyethylene	Dylan, Fortiflex, Marlex, Alathon, Hi-fax, Hosalen, Paxon	Inner fender panels, interior trim panels, valances, spoilers	Thermoplastic
PP	Polypropylene	Profax, Olefo, Marlex, Olemer, Aydel, Dypro	Interior moldings, interior trim panels, inner fenders, radiator shrouds, dash panels, bumper covers	Thermoplastic
PS	Polystyrene	Lustrex, Dylene, Styron, Fostacryl, Duraton	—	Thermoplastic
PUR	Polyurethane	Castethane, Bayflex	Bumper covers, front and rear body panels, filler panels	Thermosetting
TPUR	Polyurethane	Pellethane, Estane, Roylar, Texin	Bumper covers, gravel deflectors, filler panels, soft bezels	Thermoplastic
PVC	Polyvinyl chloride	Geon, Vinylete, Pliovic	Interior trim, soft filler panels	Thermoplastic
RIM	"Reaction injection molded" polyurethane	—	Bumper covers	Thermosetting

TABLE 13-1: SYMBOL, CHEMICAL NAME, TRADE NAME, AND DESIGN APPLICATIONS OF COMMONLY USED AUTOMOTIVE PLASTICS (CONTINUED)

Symbol	Chemical Name	Trade Name	Design Applications	Thermosetting or Thermoplastic
RRIM	Reinforced RIM-polyurethane	—	Exterior body panels	Thermosetting
SAN	Styrene-acrylonitrite	Lustran, Tyril, Fostacryl	Interior trim panels	Thermosetting
TPR	Thermoplastic rubber	—	Valance panels	Thermosetting
UP	Polyester	SMC, Premi-glas, BMC, Selection Vibrin-mat	Fiberglass body panels	Thermosetting

composite plastics, or "hybrids," and are blends of different plastics and other ingredients designed to achieve specific performance characteristics.

PLASTIC IDENTIFICATION

There are several ways to identify an unknown plastic. One possibility is the use of international symbols, or ISO codes, which can often be found molded into plastic parts. Manufacturers are using these symbols more and more; unfortunately, there are still some who do not. Keep in mind that one problem with this method is that it is usually necessary to remove the part to read the symbol.

In the event that the part in question is not identified by a symbol, the body repair manual may be of help. While these manuals often name the types of plastic used in a particular application, keep in mind that they are updated frequently, often as much as twice a year. For this reason, it is crucial that the very latest edition be consulted when working on a late-model vehicle.

The burn test has fallen out of favor as a technique for identifying plastics for several reasons. For one, having an open flame in a collision repair shop environment creates a potential fire hazard and is environmentally unsound. Another reason the burn test is not as popular as it once was is because it is not always reliable; many parts are now being manufactured from composite plastics that use more than one ingredient. A burn test is of no help in such cases.

A reliable means of identifying an unknown plastic (assuming it is probably a thermoplastic that is weldable) is to make a trial-and-error weld on a hidden or damaged area of the part (Figure 13-2). Try several different filler rods until one sticks. Most suppliers of plastic welding equipment offer only half a dozen or so different types of plastic filler rods, so the range of possibilities is not that great. The rods are color coded, so once the rod that works is found, the base material is identified.

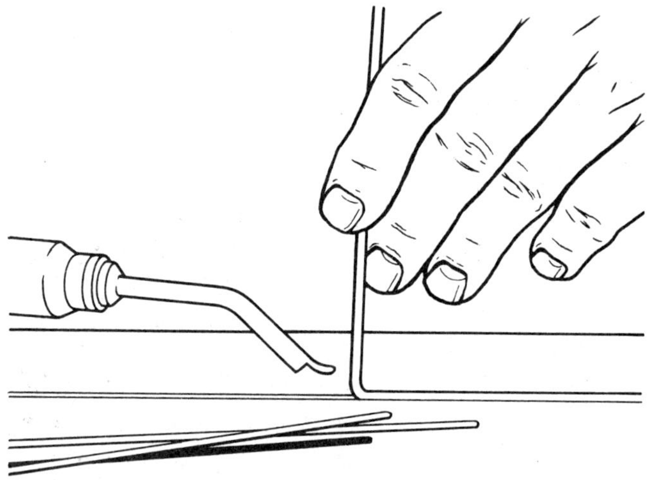

FIGURE 13-2 Making a trial-and-error weld

Table 13-2 contains additional information on types of automotive plastics.

PRINCIPLES OF PLASTIC WELDING

The welding of plastics is not unlike the welding of metals. Both methods use a heat source, welding rod, and similar techniques (butt joints, lap joints, and the like). Joints are prepared in much the same manner, and similarly evaluated for strength. Due to differences in the physical characteristics of each material, howev-

TABLE 13-2: HANDLING PRECAUTIONS FOR PLASTICS

Code	Material Name	Heat Resisting Temperature* °F	Resistance To Alcohol or Gasoline	Notes
AAS	Acrylonitrile Acrylic Rubber Styrene Resin	176	Alcohol is harmless if applied only for short time in small amounts (example, quick wiping to remove grease).	Avoid gasoline and organic or aromatic solvents.
ABS	Acrylonitrile Butadiene Styrene Resin	176	Alcohol is harmless if applied only for short time in small amounts (example, quick wiping to remove grease).	Avoid gasoline and organic or aromatic solvents.
AES	Acrylonitrile Ethylene Rubber Styrene Resin	176	Alcohol is harmless if applied only for short time in small amounts (example, quick wiping to remove grease).	Avoid gasoline and organic or aromatic solvents.
EPDM	Ethylene Propylene Rubber	212	Alcohol is harmless. Gasoline is harmless if applied only for short time in small amounts.	Most solvents are harmless, but avoid dipping in gasoline, solvents, etc.
PA	Polyamide (Nylon)	176	Alcohol and gasoline are harmless.	Avoid battery acid.
PC	Polycarbonate	248	Alcohol is harmless.	Avoid gasoline, brake fluid, wax, wax removers, and organic solvents.
PE	Polyethylene	176	Alcohol and gasoline are harmless.	Most solvents are harmless.
POM	Polyoxymethylene (Polyacetal)	212	Alcohol and gasoline are harmless.	Most solvents are harmless.
PP	Polypropylene	176	Alcohol and gasoline are harmless.	Most solvents are harmless.
PPO	Modified Polyphenylene Oxide	212	Alcohol is harmless.	Gasoline is harmless if applied only for quick wiping to remove grease.
PS	Polystyrene	140	Alcohol and gasoline are harmless if applied only for short time in small amounts.	Avoid dipping or immersing in alcohol, gasoline, solvents, etc.

er, there are notable differences between welding metal and welding plastics.

When welding metal, the rod and base material are made molten and are puddled into a joint. And while metals have a sharply defined melting point, plastics have a wide melting range between the temperature at which they soften and the temperature at which they char or burn. Also unlike metals, plastics are poor conductors of heat and thus difficult to heat uniformly. Because of this, the plastic filler rod and the surface of the plastic will char or burn before the material below the surface becomes fully softened. The decomposition time at welding temperature is shorter than the time required to completely soften many plastics for fusion welding. The result is that a plastic welder must work within a much smaller temperature range than a metal welder.

Because a plastic welding rod does not become completely molten and appears much the same before and after welding, a plastic weld might appear incomplete to the technician who is used to welding only metal. The explanation is simple: since only the outer surface of the rod has become molten while the inner core has remained hard, it is possible to exert pressure on the rod to force it into the joint and create a permanent bond. When heat is taken away, the rod reverts to its original form. Thus, even though a strong and per-

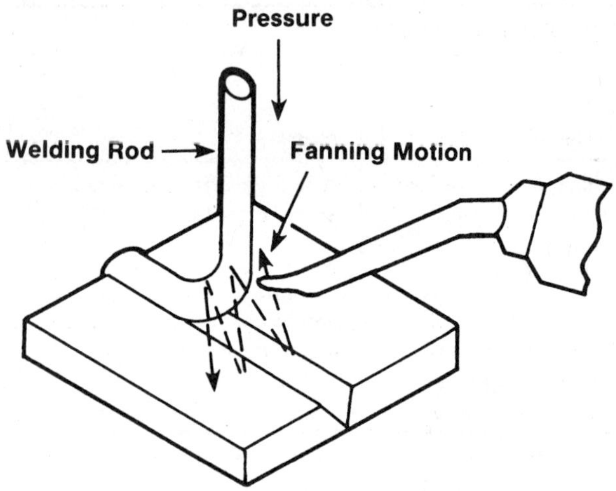

FIGURE 13-3 Successful plastic welding requires the proper combination of heat and pressure.

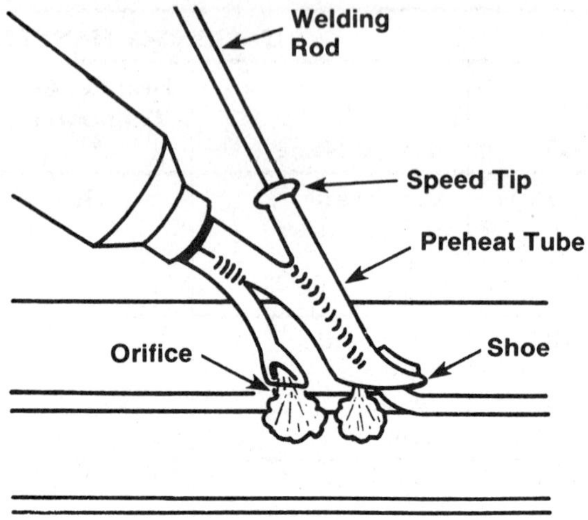

FIGURE 13-4 Plastic welder utilizing a high-speed tip

manent bond has been obtained between the rod and base material, the appearance of the rod is much the same as before the weld was made, except for molten flow patterns on either side of the bead.

When welding plastics, the materials are fused together by the proper combination of heat and pressure. With the conventional hand welding method, this combination is achieved by applying pressure on the welding rod with one hand, while at the same time applying heat and a constant fanning motion to the rod and base material with hot air from the welding torch (Figure 13-3). Successful plastic welds require that both pressure and heat be kept constant and in proper balance. Too much pressure on the rod tends to stretch the bead; too much heat will char, melt, or distort the plastic. With practice, plastic welding can be mastered as completely as metal welding.

SPEED WELDING

Speed welding incorporates the basic methods utilized in hand welding. Its primary difference lies in the use of a specially designed and patented high-speed tip (Figure 13-4), which produces more uniform welds and works at a much higher rate of speed. As with hand welding, constant heat and pressure must be maintained.

The increased efficiency of speed welding is made possible through preheating of both the rod and base material before the point of fusion. The rod is preheated as it passes through a tube in the speed tip; the base material is preheated by a stream of hot air passing through a vent in the tip ahead of the fusion point. A pointed shoe on the end of the tip applies pressure on the rod, thus eliminating the need for the operator to apply pressure. At the same time the shoe smooths out the rod, creating a more uniform appearance in the finished weld.

In speed plastic welding, the conventional two-hand method is replaced by a faster and more uniform one-hand operation. Once started, the rod is fed automatically into the preheat tube as the welding torch is pulled along the joint. Speed tips are designed to provide the constant balance of heat and pressure necessary for a satisfactory weld. The average welding speed is about 40 inches per minute.

Speed welding does have its disadvantages. Because increased speeds must be maintained to achieve the best possible weld, the speed welding torch is not suited for small, intricate work. Also, when the operator is new to this technique, the position in which the torch is held might seem clumsy and difficult. However, experience will enable the operator to successfully make all welds that can be made with a standard welder, including butt welds, V-welds, corner welds, and lap joint welds. Speed welds can be made on circular as well as flat work. In addition, inside welds on tanks can be speed welded, provided the working space is large enough to manipulate the torch.

HOT-AIR PLASTIC WELDING

There are a number of manufacturers who make hot-air plastic welding equipment for the collision repair industry, and all use the same basic technology. A ceramic or stainless steel electric heating element is used

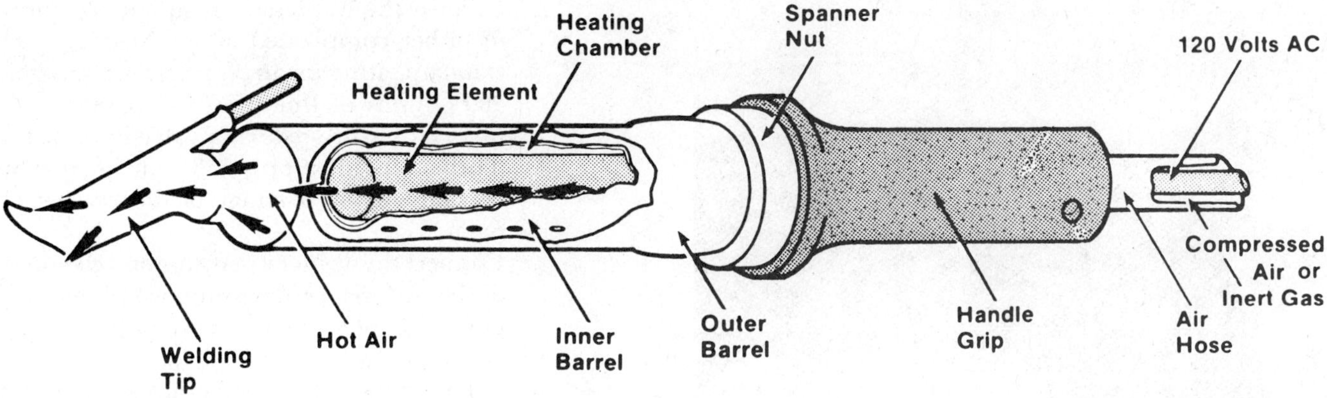

FIGURE 13-5 Typical hot-air welder

to produce hot air (450 to 650 degrees Fahrenheit), which in turn blows through a nozzle and onto the plastic. The air supply comes from either the shop's air compressor or a self-contained portable compressor that comes with the welding unit. Never use oxygen or other flammable gases with any type of plastic welder. Most hot-air welders use a working pressure at the tip of around 3 psi. A pair of pressure regulators is required to reduce the air pressure first to around 50 psi, and then finally to the working pressure of 2-1/2 to 3-1/2 psi. A typical hot-air welder is illustrated in Figure 13-5.

The barrel of the torch gets sufficiently hot so that skin contact could cause a burn if the hot air is directed against the skin long enough. The torch is used in conjunction with the welding rod, which is normally 3/16 inch in diameter and made from the same material as the plastic being repaired. This will ensure that the strength, hardness, and flexibility of the repair is the same as the damaged part. Use of the proper welding rod is very important; even a merely adequate weld is impossible if the wrong rod is used.

One of the problems with hot-air welding is that the 3/16-inch-diameter rod is often thicker than the panel to be welded. This can cause the panel to overheat before the rod has melted. Using a 1/8-inch rod with the hot-air welder can often correct such warpage problems. (This works on the same principle as using a 0.023 wire instead of a 0.035 wire for MIG welding.)

Three types of welding tips are available for use with most hot-air plastic welding torches. They are:

- *Tacking Tips.* These are used to tack broken sections together before welding (Figure 13-6). If necessary, tack welds can be easily pulled apart for realigning.
- *Round Tips.* These tips are used to fill small holes and make short welds, welds in hard-to-reach places, and on tight or particularly sharp corners (Figure 13-7).

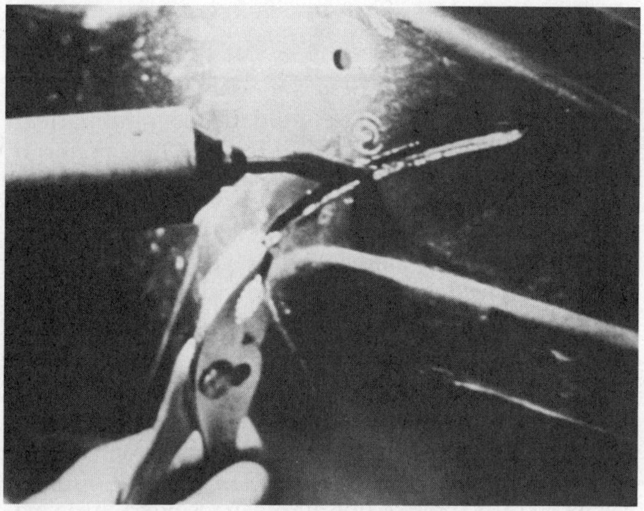

FIGURE 13-6 Tacking welding tip

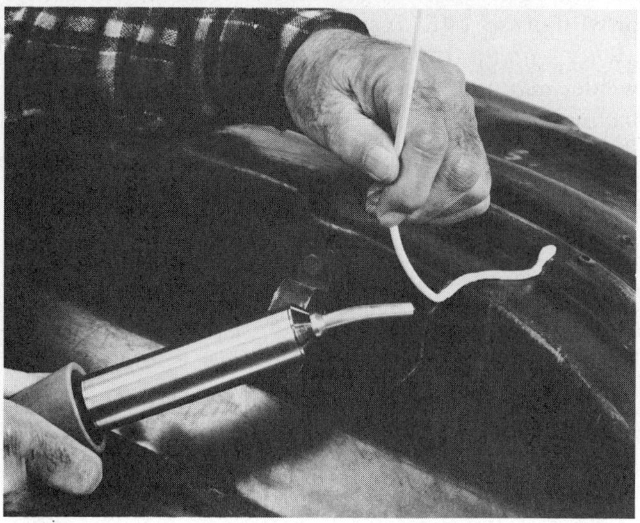

FIGURE 13-7 Round welding tip

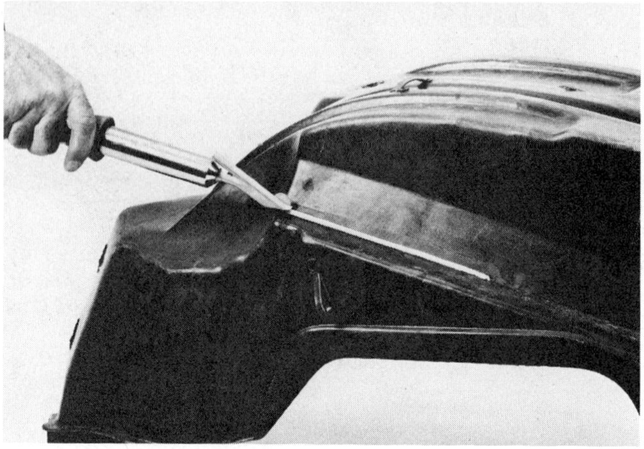

FIGURE 13-8 Speed welding tip

- *Speed Tips.* These are used for long, fairly straight welds. They hold the rod, automatically preheat it, and feed it into the base material, thus allowing for faster welding speeds (Figure 13-8).

Some hot-air welder manufacturers have developed specialized welding tips and rods to meet specific needs. Check the product catalog for information on these materials.

SETUP, SHUTDOWN, AND SERVICING

Naturally, no two hot-air plastic welders are exactly alike; their design varies from one manufacturer to another. The setup, shutdown, and service procedures that follow, although typical of all hot-air welders, should nonetheless be regarded as general guidelines only. For specific instructions, always refer to the owner's manual and other material provided by the welder manufacturer. Keep in mind that some manufacturers advise against using their welder on plastic that is any thinner than 1/8 inch because of the likelihood of distortion. In other cases, it is acceptable to weld plastics as thin as 1/16 inch, provided they are supported from underneath during the operation. Again, it is very important to read and follow the specific directions for the welder being used.

Setup and Shutdown

To set up a typical hot-air welder, proceed as follows:

1. Close the air pressure regulator valve by turning the control handle counterclockwise until it is loose. This will prevent possible damage to the gauge from a sudden surge of excess air pressure.
2. Connect the air pressure regulator to a supply of either compressed air or inert gas. The standard rating for an air pressure regulator is 200 pounds of line pressure. If inert gas is used, a pressure-reducing valve is needed.
3. Turn on the air supply. The starting pressure depends on the wattage of the heating element.
4. Connect the welder to a common 120-volt AC outlet. A three-prong grounded plug or temporary adapter must be used with the welder at all times.
5. Allow the welder to warm up at the recommended air pressure. It is essential that either air or inert gas flows through the welder at all times, from warm-up to cool-down, to prevent burnout of the heating element and damage to the gun.
6. Select the proper tip and insert it into the torch with pliers to avoid touching the barrel while hot.
7. After the tip has been installed, the temperature will increase slightly due to back pressure. Allow 2 to 3 minutes for the tip to reach the required operating temperature.
8. Check the temperature by holding a thermometer 1/4 inch from the hot air end of the torch. For most thermoplastics, the temperature should be in the 450- to 650-degree-Fahrenheit range. Information supplied with the welder usually includes a chart of welding temperatures.
9. If the temperature is too high to weld the material, increase the air pressure slightly until the temperature goes down. If the temperature is too low for the particular application, decrease the air pressure slightly until the temperature rises. When increasing and decreasing the air pressure, allow at least 1 to 3 minutes for the temperature to stabilize at the new setting.
10. Damage to the welder or heating element will not occur from too much air pressure; however, the element can become overheated by too little air pressure. When decreasing the air pressure, never allow the round nut that holds the barrel to the handle of the welder to become too hot to the touch. This is an indication of overheating.
11. A partially clogged dirt screen in the regulator or a fluctuation in the line voltage can also cause over- or underheating. Watch for these symptoms.
12. If the threads at the end of the barrel become too tight, clean them with a good, high-temperature grease to prevent seizing.

13. When the welding is finished, disconnect the electrical supply and let the air flow through the welder for a few minutes or until the barrel is cool to the touch. Then disconnect the air supply.

Maintenance

If it becomes necessary to change the heating element, proceed as follows:

1. While pushing the end of the barrel against a solid object, hold the handle tightly and push in. The pressure on the barrel will compress the heating element spring.
2. Use a spanner wrench to loosen the spanner nut. Keep the pressure on the handle and back the nut off all the way by hand.
3. Hold the barrel and place the complete welder on a bench. Remove the barrel.
4. Gently pull the heating element out of the handle. At the same time, unwind the spring, which is spiraled into the handle, until it is completely out of the handle.
5. Grasp the socket at the end of the spring tightly. Rock the element while pulling until the element is disconnected.
6. To install the new element, reverse the above procedure. Turn the element clockwise (about 1-1/2 turns) while pushing the spring gently back into the handle. This prevents kinking of the spring.

Other, more involved service procedures are best left to qualified repair technicians. Many manufacturers make it clear that disassembling the welder automatically invalidates the warranty.

AIRLESS PLASTIC WELDING

Although only in existence for a relatively short period of time, airless plastic welding has become very popular with the collision repair industry (Figure 13-9). Compared to the hot-air method, it is less expensive, easier to learn, simpler to use, and more versatile. The hot-air method usually calls for 3/16-inch-diameter welding rods, while 1/8-inch-diameter rods are used with the airless method. Using thinner rods not only provides a quicker rod melt, it also helps eliminate two troublesome problems: panel warpage and excess rod buildup.

When setting up an airless welder, the first and most important step is to put the temperature dial at the appropriate setting, depending on the specific plastic

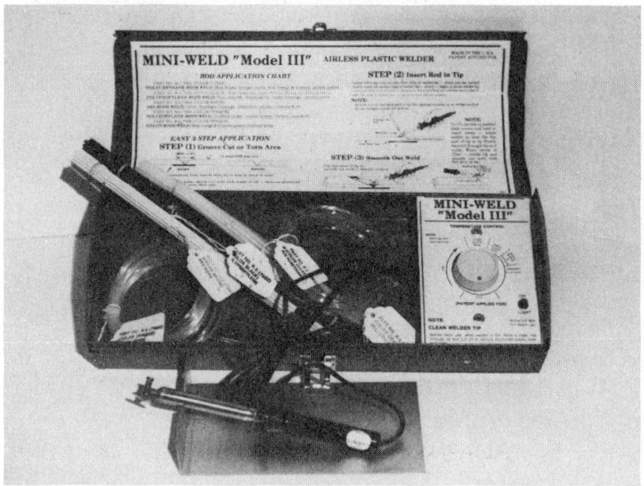

FIGURE 13-9 Typical airless welder

being worked on. It is crucial that the temperature setting is correct; otherwise, the entire welding operation will be jeopardized. It will normally take about 3 minutes for the welder to fully warm up. As for the selection of the welding rod, there is another factor to consider besides size; namely, compatibility. Make sure the rod is the same material as the damaged plastic or the weld will more than likely be unsuccessful. To this end, many airless welder manufacturers provide rod application charts. When the correct rod has been chosen, it is good practice to run a small piece through the welder to clean out the tip before beginning.

ULTRASONIC PLASTIC WELDING

The technology of ultrasonic plastic welding was originally developed solely as a means of assembly, but it is also suitable for making repairs. It relies on high-frequency vibratory energy to produce bonding without melting the base metal. Hand-held systems (Figure 13-10) are available in 20 and 40 kHz; they are equally adept at welding large parts and tight, hard-to-reach areas. Welding time is controlled by the power supply. And although the use of ultrasonics in the collision repair shop is still in its infancy, the strong, clean, precise welds the process produces can only increase its popularity in the years to come.

Most commonly used injection-molded plastics can be ultrasonically welded without the use of solvents, heat, or adhesives. Ultrasonic weldability depends on such factors as the plastic's melting temperature, elasticity, impact resistance, coefficient of friction,

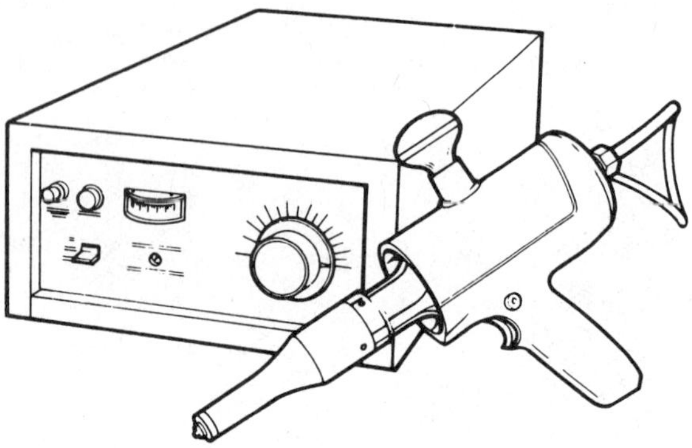

FIGURE 13-10 Typical hand-held ultrasonic welder and power supply

and thermal conductivity. Generally, the more rigid the plastic, the easier it is to weld ultrasonically. Thermoplastics such as polyethylene and polypropylene are ideal for ultrasonic welding, provided the welder can be positioned close to the joint area.

In the typical ultrasonic system, the vibration is generated in the transducer and then transmitted through the sonotrode, which is the equivalent of an electrode. The sonotrode tip directly contacts the workpiece, and an anvil supports the assembly. The best results are achieved when the sonotrode tip and anvil are contoured to accommodate the specific geometry of the parts being joined. An automatic ultrasonic welding cycle consists of lowering the sonotrode or raising the anvil, applying a clamping force, introducing the ultrasonic pulse or vibration, and retracting the sonotrode or anvil.

For spot welding of flat sheets, the sonotrode tip is contoured to a radius of about 50 to 100 times the thickness of the sheet adjacent to the tip. The anvil face is usually flat. This provides a friction-type drive in which slippage can occur between the tip and the top sheet or between the anvil and the bottom sheet.

As in resistance spot welding, wear of the sonotrode tip and anvil depends upon the properties and geometry of the parts being welded. Tips made of high-speed tool steel are generally satisfactory for welding relatively soft materials, such as aluminum, copper, iron, and low carbon steel. Tips of hardenable nickel-base alloys usually provide good service life with hard, high-strength metals and alloys. The material used for the sonotrode tip is also satisfactory for the anvil face.

Longer tip life and more effective welding are possible using tips and anvils with rough faces because they tend to prevent slippage between them and the workpieces. The roughening can be done by sandblasting.

When sonotrode tips begin to show wear, they may be reconditioned by cleaning and burnishing. Light sanding with 400 grit paper is usually sufficient. If the wear is excessive, the tips should be replaced. Most ultrasonic welding machines have mechanically attached tips to simplify replacement.

Occasionally, the tip might tend to stick to the weld surface, particularly if improper machine settings are used. The sticking can be alleviated by increasing the clamping force or decreasing the welding time. With some materials, it is effective to apply a lubricant, such as a diluted soap solution, to the surfaces being joined. If these measures are not adequate, tip sticking can usually be eliminated by welding with a tip containing an insert of special grade tungsten carbide.

PLASTIC WELDING PROCEDURES

The basic procedures for hot-air and airless welding are very similar. To make a good plastic weld with either procedure, keep the following factors in mind:

- *Welding Rod Material.* Rods are frequently color coded to indicate the material they are made of. Unfortunately, the coding is not uniform among manufacturers, so it is important to use the reference information provided. If the rod is not compatible with the base material, the weld will not hold.
- *Temperature.* Too much heat will char, melt, or distort the plastic; too little heat will not provide positive weld penetration between the base material and the rod.
- *Pressure.* Too much pressure stretches and distorts the weld.
- *Angle between Rod and Base Material.* If too shallow, a proper weld will not be achieved.
- *Speed.* If the torch movement is too fast, it will not permit a good weld; too slow a speed can char the plastic.

The basic repair sequence is generally the same for both processes. That is:

1. Prepare the damaged area.
2. Align the damaged area.
3. Make the weld.
4. Allow it to cool.
5. Sand. If pinholes, voids, and the like exist, bevel the edges of the defective area and add another weld bead. Resand.
6. Apply topcoat.

GENERAL WELDING TECHNIQUES

Welding plastic is not difficult when done in a careful and thorough manner. The following guidelines cannot be stressed enough:

- The welding rod must be compatible with the base material in order for the strength, hardness, and flexibility of the repair to be the same as the part.
- Always test a welding rod for compatibility with the base material. To do this, melt the rod onto a hidden side of the damaged part, let the rod cool, then try to pull it from the part. If the rod is compatible, it will adhere.
- Pay close attention to the temperature setting of the welder; it must be correct for the type of plastic being welded.
- Never use oxygen or other flammable gases with a plastic welder.
- Never use a plastic welder, heat gun, or similar tool in wet or damp areas. Remember: electric shock can kill.
- Become proficient at horizontal welds before attempting the more difficult vertical and overhead types.
- Make welds as large as they have to be. The greater the surface area of a weld, the stronger the bond.
- Before beginning a speed weld, run a small piece of the welding rod through the welder to clean out the torch tip.
- Consult a supplier for the brands of tools and materials that best fit the shop's needs. Always read and follow the manufacturer's instructions carefully.

BASICS OF HOT-AIR WELDING

The typical hot-air plastic welding procedure is as follows:

1. Set the welder to the proper temperature (if a temperature adjustment is provided).
2. Wash and clean the part with soap and water first. Allow to dry and then clean with a plastic cleaner; do not use conventional prep solvents or wax and grease removers. To remove silicone materials, use a conventional cleaner first, making sure to completely remove all residue.
3. V-groove the damaged area.

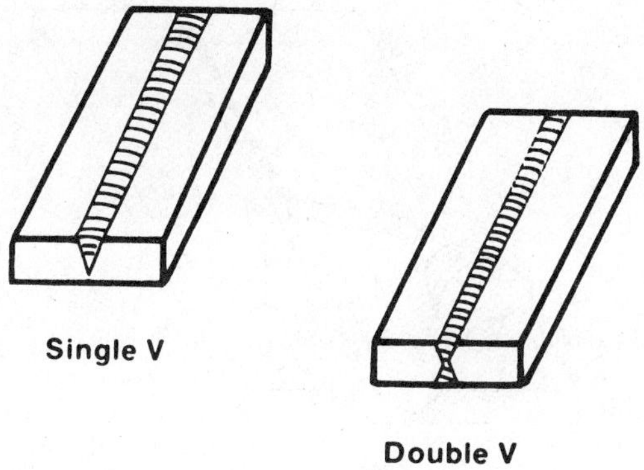

FIGURE 13-11 Single- and double-V butt welds produce strong joints.

4. Bevel the part 1/4 inch beyond the damaged area.
5. Tack weld or tape the break line with aluminum body tape.
6. Select the welding rod and welding tip best suited to the type of plastic and the damage.
7. Make the weld. Allow it to cool and cure for about 30 minutes.
8. Grind or sand the weld to the proper contour and shape.

When welding plastic, single- or double-V butt welds (Figure 13-11) produce the strongest joints; lap fillet welds are also good. When using a round or V-shaped welding rod, the damaged area is prepared by slowly grinding, sanding, or shaving the adjoining surfaces with a sharp knife to produce a single- or double-V. For flat ribbon welding rods, V-grooving is not necessary. Wipe any dust or shavings from the joint with a clean, dry rag. The use of cleaning solvents is not generally recommended because they can soften the edges and cause poor welds.

TACK WELDING

For this process, the edges to be joined must be aligned. On long tears where backup is difficult, small tack welds can be made along the length of the tear to hold the two sides in place while doing the permanent weld (Figure 13-12). For larger areas, a patch can be made from a piece of plastic and tacked in place. To tack weld, proceed as follows:

1. Hold the damaged area in alignment with clamps or aluminum body tape.
2. Using a tacking welding tip, fuse the two sides to form a thin hinge weld along the root of the crack. This is especially useful for long

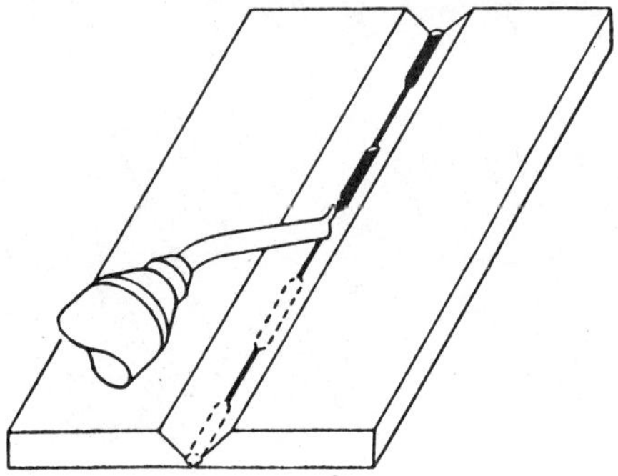

FIGURE 13-12 Tack welds are useful on long tears.

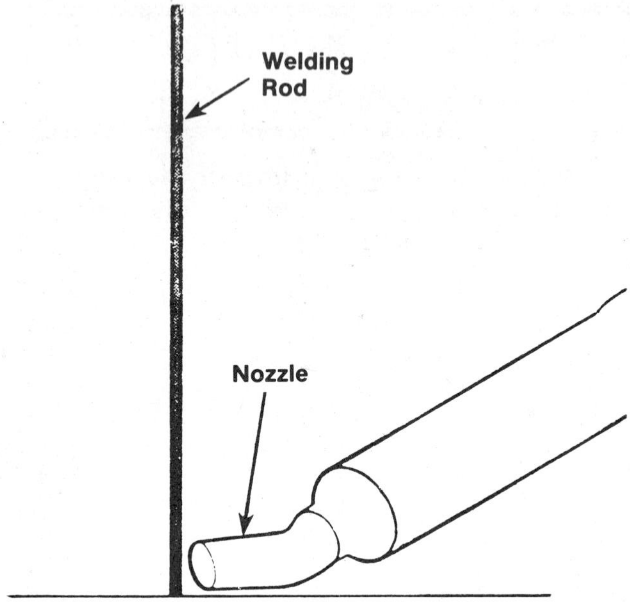

FIGURE 13-13 Keep the nozzle parallel to, and the rod at a right angle to, the base material.

cracks because it allows for easy adjustment and alignment of the edges.

3. Start the tacking by drawing the point of the welding tip along the joint. Press the tip in firmly, making sure to contact both sides of the crack; draw the tip smoothly and evenly along the line of the crack. No welding rod is used when tacking.
4. The point of the tip will fuse both sides in a thin line at the root of the crack. The fused parts will hold the sides in alignment, though they can be separated and retacked if adjustment is necessary. Fuse the entire length of the crack.

WELDING THE V-GROOVE

V-groove butt welding involves several different stages, each of which must be mastered for a quality job to be performed.

Starting the Weld

Prepare the rod for welding by cutting the end at approximately a 60-degree angle. When starting a weld, the nozzle should be held about 1/4 to 1/2 inch above and parallel to the base material. The rod is held at a right angle to the work as shown in Figure 13-13, with the cut end of the rod positioned at the beginning of the weld.

Direct the hot air from the tip alternately at the rod and the base material, but concentrating more on the rod. Always keep the rod in line with the V while pressing it into the seam. Light pressure is sufficient for achieving a good bond. Once the rod begins to stick to the plastic, start to move the torch and use the heat to control the flow. Be careful not to melt or char the base material or to overheat the rod. As the welding continues, a small bead should form along the entire weld joint. A good start is essential because this is where most weld failures begin. For this reason, starting points on multiple-bead welds should be staggered whenever possible.

Continuing the Weld

Once the weld has been started, the torch should continue to fan from rod to base material. But because the rod now has less bulk, a greater amount of heat must be directed at the base material. Experience will help develop the proper technique.

Feeding the Rod

Throughout the welding process the rod is gradually being used up, making it necessary for the welder to renew his or her grip on the rod. Unless this is done carefully, the release of pressure might cause the rod to lift away from the weld and allow air to become trapped under the weld and weaken it. To prevent this, the welder must develop the skill of continuously applying pressure on the rod while repositioning the fingers. This can be done by applying pressure with the third and fourth fingers while moving the thumb and first finger up the rod (Figure 13-14). The rod is cool enough to do this because only the bottom of it is heated. However, care should be observed in touching new welds or aiming the torch near the fingers.

Finishing the Weld

As the end of a weld is approached, maintain pressure on the rod as the heat is removed. Hold the rod still

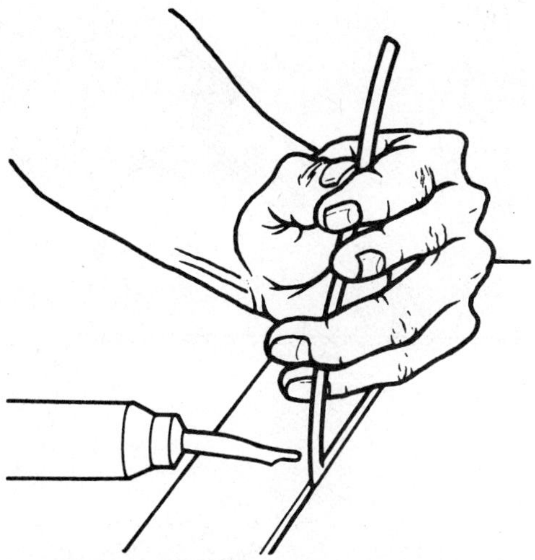

FIGURE 13-14 Repositioning the grip on the welding rod

for a few seconds to make sure it has cooled enough so it will not pull loose, then carefully cut the rod with a sharp knife or clippers. Do not attempt to pull the rod from the joint. About 15 minutes cooling time is needed for rigid plastic and 30 minutes for thermoplastics such as polyurethane.

Rough Grinding the Weld

The weld area can be smoothed by grinding with No. 36 grit sandpaper. A 9- or 10-inch disc on a low-speed grinder will smooth large weld beads. Excess plastic can be removed with a sharp knife before grinding. Care must be taken not to overheat the weld area because it will soften. To speed up the work without damaging the weld, periodic cooling can be done using water.

Checking the Weld

After rough grinding, the weld should be checked visually for defects. Any voids or cracks will make it unacceptable. Bending should not produce any cracks because a good weld is as strong as the part itself.

Finish Sanding

The weld area can be finish sanded using 220 grit sandpaper followed by a 320 grit. Either a belt or orbital sander may be used, plus hand sanding as required. If refinishing is to be done, follow the procedure designed specifically for plastics.

BASICS OF SPEED WELDING

On panel work, speed welding is commonly used. Following are some techniques that are essential for quality speed welding.

Starting the Weld

With the speed torch held like a dagger and the hose on the outside of the wrist, bring the tip over the starting point a full 3 inches from the base material so the hot air will not affect it (Figure 13-15). Cut the welding rod at a 60-degree angle, insert it into the preheat tube, and immediately place the pointed shoe end of the tip on the base material at the starting point. Hold the torch perpendicular to the base material and push the rod through until it stops against the base material at the starting point. If necessary, lift the torch slightly to allow the rod to pass under the shoe. Keeping a slight pressure on the rod with the left hand and only the weight of the torch on the shoe, pull the torch slowly toward you. The weld is now started.

CAUTION

Once the speed weld is started, do not stop. If the forward movement must pause for any reason, pull the speed tip off the rod immediately; if this is not done, the rod will melt into the preheat tube.

Continuing the Weld

In the first inch or two of travel, the rod should be helped along by pushing it into the preheat tube with slight pressure. Once the weld has been properly started, the torch is brought to a 45-degree angle; the rod will now feed without a need for pressure. As the torch moves along, visual inspection is needed to monitor the quality of the weld being produced.

The angle between the torch and base material determines the welding rate. Since the preheat hole in the speed tip precedes the shoe, the angle of the torch to the base material determines how close the hole is to the plastic and how much preheating is being done. It is for this reason that the torch is held at a 90-degree angle

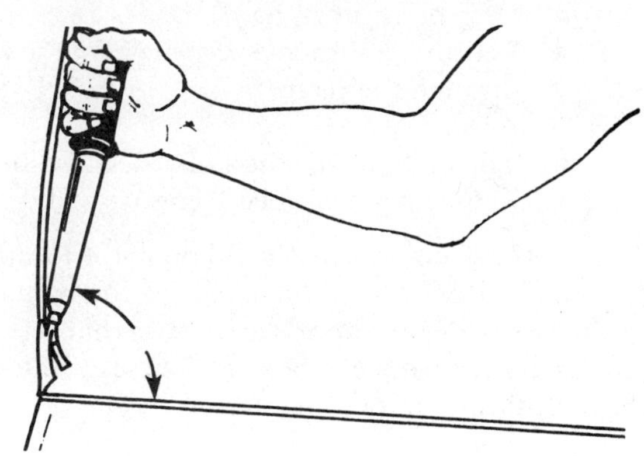

FIGURE 13-15 Starting a speed weld

when starting the weld and at 45 degrees thereafter (Figure 13-16A). When a visual inspection of the weld indicates a welding rate that is too fast, the torch should be brought back to the 90-degree angle temporarily to slow it down, then gradually moved to the desired angle for proper welding speed. It is important that the torch is held in such a way that the preheat hole and the shoe are always in line with the direction of the weld, so that only the base material in front of the shoe is preheated. A heat pattern on the base material indicates the area being preheated. The rod should always be centered in that pattern.

Finishing the Weld

It is important to remember that, once started, speed welding must be maintained at a fairly constant rate. The torch cannot be held still. To stop welding before the rod is used up, bring the torch back past the 90-degree angle and cut off the rod with the end of the shoe (Figure 13-16B). This can also be accomplished by pulling the speed tip off the remaining rod. When cutting the rod with the shoe, the remaining rod must be removed promptly from the preheat tube or it will char or melt, clog the tube, and make it necessary for the tube to be cleaned out by inserting a new rod in it.

A good speed weld in a V-groove will have a slightly higher crown and more uniformity than the normal hand weld. It should appear smooth and shiny, with a slight bead on each side. For best results and faster welding speed, the shoe on the speed tip should be cleaned occasionally with a wire brush to remove any residue that might cling to it and create drag on the rod.

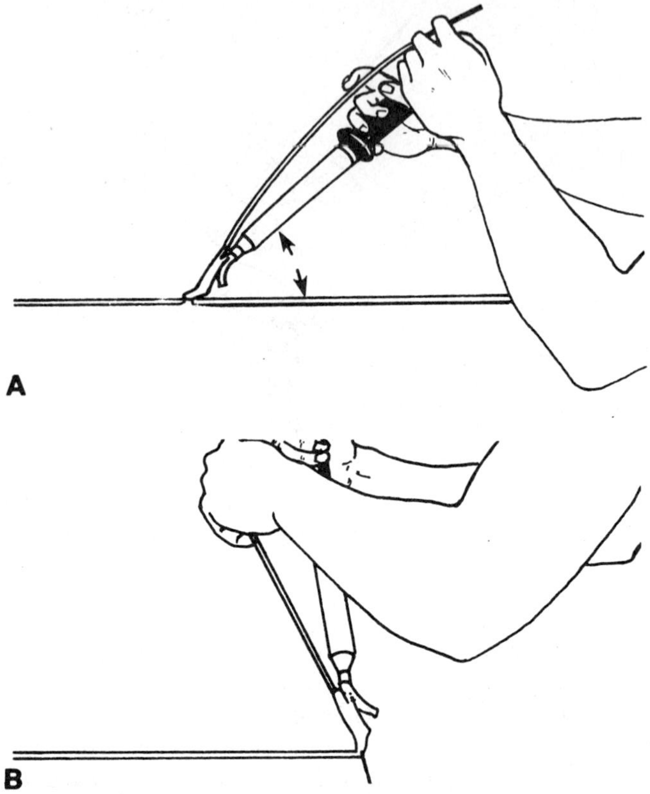

FIGURE 13-16 (A) Continuing and (B) finishing a speed weld

BASICS OF AIRLESS WELDING

There are two methods of airless welding (Figure 13-17):

1. *Single-sided*
   - Used when the piece is not removed from the vehicle
   - Used on thinner, hard plastics
   - Requires a V-groove 75 percent of the way through the piece
2. *Two-sided*
   - The preferred method of airless welding
   - Requires a two-sided V-groove

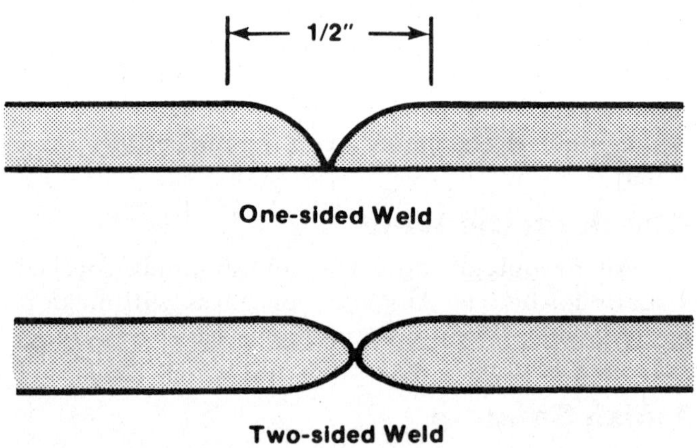

FIGURE 13-17 Proper V-grooves are essential to making airless welds.

The method chosen will depend upon the location of the damaged piece and the extent of the damage. There are two airless welding techniques that refer to how the tip is moved during the weld. These techniques are the melt-flow and the stitch-tamp.

Melt-Flow Technique

The melt-flow technique is the most commonly used airless welding method. It can be utilized for both single-sided and two-sided repairs. A typical melt-flow procedure is as follows:

1. With the welding rod in the preheat tube, place the flat shoe part of the tip in the V-groove.
2. Hold it in place until the rod begins to melt and flow out around the shoe.

3. A small amount of force is needed to feed the rod through the preheat tube. The rod will not feed itself. Care should be used not to feed it too fast.
4. Move the shoe slowly, crisscrossing the groove until it is filled with melted plastic.
5. Work the melted plastic well into the base material, especially toward the top of the V-groove.
6. Complete a weld length of about 1 inch at a time. This will allow smoothing of the weld before the plastic cools.

Stitch-Tamp Technique

This method is used primarily on hard plastics, like ABS and nylon, to insure a good base/rod mix. It is also ideal for use with TPUR and similar plastics. After completing the weld using the melt-flow procedure, remove the rod. Turn the shoe over and slowly move the pointed end of the tip into the weld area to bond the rod and base material together. Stitch-tamp the entire length of the weld. After stitch-tamping, use the flat shoe part of the tip to smooth out the weld area.

Single-Sided Welds

These welds are usually performed when the piece cannot be removed from the vehicle. A cosmetic body filler is not needed since the welding rod itself serves as the filler.

To make a single-sided weld, proceed as follows:

1. Set the temperature dial on the welder for the plastic being welded. Allow it to warm up to the proper temperature.
2. Clean the part by washing with soap and water, followed by a good plastic cleaner.
3. Align the break using aluminum body tape.
4. V-groove the damaged area 75 percent of the way through the base material. Angle or bevel back the torn edges of the damage at least 1/4 inch around the damaged area, using a die grinder or similar tool.
5. Clean the preheat tube, and insert the rod. Begin the weld by placing the shoe over the V-groove and feeding the rod through. Move the tip slowly for good melt-in and heat penetration.
6. When the entire V-groove has been filled, turn the shoe over and use the tip to stitch-tamp the rod and base material together into a good mix along the length of the weld.
7. Resmooth the weld area using the flat shoe part of the tip, again working slowly. Then cool with a damp sponge or cloth.
8. Shape the excess weld buildup to a smooth contour, using a razor blade and/or abrasive paper.

Two-Sided Welds

When making a two-sided weld, be sure to do the following:

- Proper cleaning
- Support of the backside during the weld
- Use of adhesion promoter

Then proceed as follows:

1. Allow the welder to heat up, and clean the preheat tube.
2. Clean the part with soap and water and plastic cleaner.
3. Align the front of the break with aluminum body tape, smoothing it out with a stiff squeegee or spreader.
4. V-groove 50 percent of the way through the backside of the panel.
5. Weld the backside of the panel using the melt-flow method. Move slowly enough to achieve good melt-in.
6. When finished, smooth the weld with the shoe.
7. Quick-cool the weld with a damp sponge or cloth.
8. Remove the tape from the front of the piece. V-groove deep enough that the first weld is penetrated by the V-groove.
9. Weld the seam, filling the groove completely.
10. Use a razor blade or slow speed grinder to reshape the contour.

REPAIRING INTERIOR AND UNREINFORCED HARD PLASTICS

Dash pads or padded instrument panels are expensive to replace and are therefore perfect candidates for repair (Figure 13-18). Most dash pads are made of vinyl-clad urethane foam. Surface dents in foam dash pads, armrests, and other padded interior parts are common in collision repair. These dents can often be repaired by applying heat as follows:

1. Soak the dent with a damp sponge or cloth for about half a minute. Do not wipe dry. Keep the dented area moist.
2. Using a heat gun, heat the area around the dent, holding the gun 10 to 12 inches from the surface. Keep it moving in a circular motion at all times, working from the outside in.

3. Heat the area to around 130 degrees Fahrenheit. Do not overheat the vinyl because it might blister, but keep heating until the area is too uncomfortable to touch.
4. Using gloves, massage the pad, forcing the material toward the center of the dent. The area might have to be reheated and massaged more than once. In some cases, heat alone might repair the damage, without the need for massage.
5. When the dent has been removed, cool the area quickly with a damp sponge or cloth.

Repairing Cuts

To repair a cut in the dash pad, proceed in the following manner:

1. Set up the welder for the proper temperature for welding urethane.
2. Wash the pad with soap and water, dry, and clean with plastic cleaner.
3. If the damaged area is brittle, warm with a heat gun. If there are curled or jagged edges, cut them away.
4. V-groove at least 1/4 inch into the foam padding. Bevel the edges as much as possible. Rough up the area for about 1/4 inch around the V-groove.
5. Turn the welder so the shoe is facing up. Feed the rod slowly through the welder. Start the weld at the bottom of the groove and completely fill it with melted plastic until it is flush with the surface.
6. Smooth out the excess rod buildup. Feather it out over the beveled edges for at least 1/4 inch on each side of the groove.
7. Cool the weld and grind away any remaining excess rod buildup. Rough up the vinyl for about 2 inches beyond the weld on each side for good filler adhesion.
8. Use a flexible filler material designed for use on vinyl to get the desired contour. Allow the filler to cure.
9. Contour sand to remove the tacky glaze. If any filler is accidentally sanded through, apply a skim coat and re-sand.

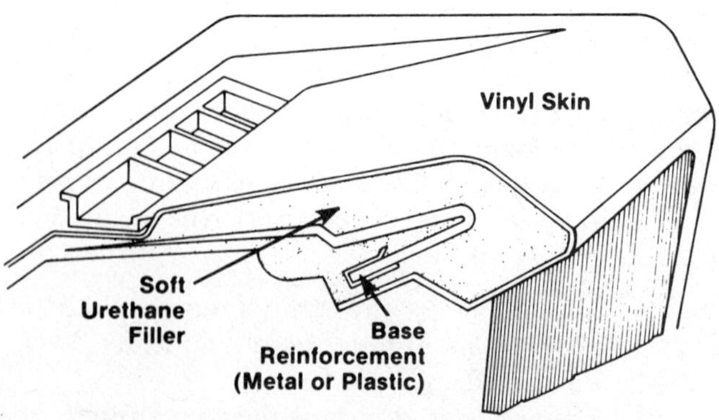

FIGURE 13-18 Dash pad cross section

Using Heat to Reshape Plastics

Thermoplastic parts such as urethane bumper covers that are bent, stretched, or deformed in a collision can often be straightened with heat. This is because plastic has a "memory," which means that the piece wants to keep the shape in which it was molded. If it is bent or scratched, it will return to its original shape if heat is applied. Use a concentrated heat source, such as a heat lamp or high-temperature heat gun. To reshape a distorted bumper cover, use the following procedure:

1. Thoroughly wash the cover with soap and water.
2. Clean with plastic cleaner. Make sure to remove all road tar, oil, grease, and undercoating.
3. Blow or wipe dry.
4. Apply heat directly to the distorted area. When the opposite side of the cover becomes uncomfortable to the touch, it has been heated enough.
5. Use a paint paddle, squeegee, or wood block to help reshape the piece if necessary.
6. Quick-cool the area by applying cold water with a sponge or rag.

Tab Replacement for Exterior Plastic Surfaces

When a bumper cover has been torn away from the mounting screws, the likely result will be broken or torn mounting tabs. Any damage to these areas of the bumper cover must be repaired with a two-sided weld because of the added strength and durability this type of weld provides.

If the material is a thermoplastic, either the hot-air or airless method may be used. If the piece is a thermosetting plastic, airless welding must be used. The repaired piece will be as strong as the original, undamaged part.

The following is the procedure for rebuilding a mounting tab that has been torn off:

1. Begin by cleaning the piece as described before.

2. Bevel back the torn edges of the mounting tab at least 1/4 inch on both sides.
3. Rough up the plastic and wipe dust free.
4. Use aluminum body tape to build a form in the shape of the missing tab. The edges are turned up to form the thickness of the new tab.
5. Set the temperature dial on the welder for the type of plastic being repaired. Allow the unit to warm up.
6. Begin the weld, pushing the rod slowly through the preheat tube. Slightly overfill the form, working the melted plastic into the base material.
7. Smooth and shape the weld. Quick-cool the weld area.
8. Remove the tape and V-groove along the tear line on the other side about halfway through the piece.
9. Fill in the groove and quick-cool. Finish the weld to the desired contour using a slow-speed grinder with a No. 60- or 80- grit disc.

ULTRASONIC STUD WELDING

Ultrasonic stud welding, a variation of the shear joint, is a reliable technique that can be used to join plastic parts at a single point or at numerous locations. In many applications requiring permanent assembly, a continuous weld is not required. The power requirement is low because of the small weld area, and the welding cycle is short, almost always less than half a second.

Figure 13-19 shows the basic ultrasonic stud weld joint before, during, and after welding. The weld is made along the circumference of the stud; its strength is a function of the stud diameter and the depth of the weld. Maximum tensile strength is achieved when the depth of the weld equals half the diameter of the stud. The radial interference (dimension A) must be uniform and should generally be 0.008 to 0.012 inch for studs having a diameter of 0.5 inch or less. The hole should be a sufficient distance from the edge to prevent breakout; a minimum of 0.125 inch is recommended.

In the joint, the recess can be on the end of the stud or in the mouth of the hole, as shown in the examples. With the latter, a small chamfer can be used for rapid alignment. To reduce stress concentration, a large fillet radius should be used at the base of the stud. Recessing

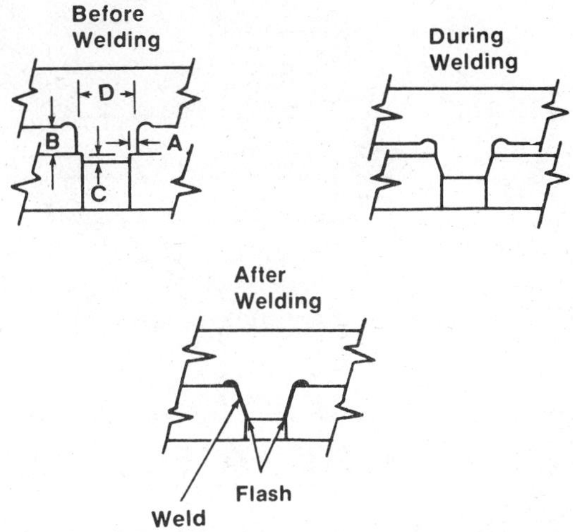

FIGURE 13-19 Basic ultrasonic stud welding joint

the fillet below the surface allows flush contact of the parts.

Other ways in which the ultrasonic stud weld can be used are illustrated in Figure 13-20. A third dissimilar material can be locked in place, as in view A. View B shows separate molded rivets in place of metal self-tapping screws. Unlike metal fasteners, this produces a relatively stress-free assembly.

Figure 13-21 shows a variation that can be used where appearance is important. The stud is welded into a boss, whose outside diameter can be no less than twice the stud diameter. When welding into a blind hole, it might be necessary to provide an outlet for air. Two possibilities are shown: a center hole through the stud (view A), or a small, narrow slot in the interior wall of the boss (view B).

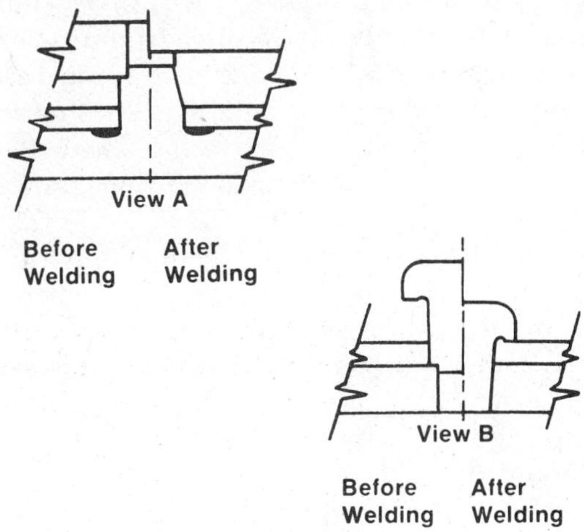

FIGURE 13-20 Variations of the ultrasonic stud welding joint

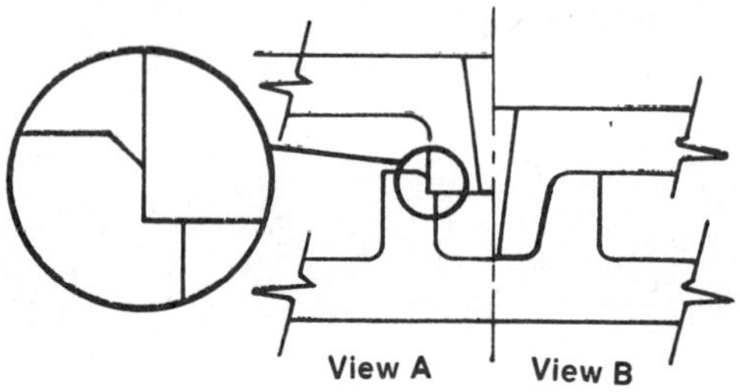

FIGURE 13-21 Welding a stud in a blind hole

ADHESIVE REPAIR SYSTEMS

Adhesive repair systems are of two types: two-part and cyanoacrylate.

Two-part adhesive systems are an acceptable alternative to welding for many plastic repairs. These systems consist of a base resin of polyester, epoxy, or urethane, and a hardener or catalyst. A number of different manufacturers supply two-part adhesive systems to the automotive aftermarket.

Cyanoacrylates, or CAs, have changed significantly in the last few years, with many new formulations becoming available. CAs are sometimes known as super glues and can be a valuable tool for the repair of plastic parts. CAs set up very quickly, and compare well to other repair systems for strength and flexibility. Identification of the plastic being repaired is less important than with some other repair products.

A cyanoacrylate system may contain one or more viscosities of CA and an activating agent, as well as a structural filler product. The activating agent is used to accelerate the bonding process of the CA. Depending upon the specific use, the activating agent may be applied to the piece first, with the adhesive applied after, or it may be sprayed over the adhesive. Care must be used not to apply too much activating agent. If too much is used, the product foams causing a weaker bond.

Structural filler products are available in a powdered form. Filler is placed into voids formed where the plastic is broken or missing. Once the void is filled, adhesive is added to the filler, and a chemical reaction occurs. The activating agent is then applied to complete the process.

It should be mentioned that CAs do not work equally well on all plastics. There is no hard and fast rule. When CAs are used, be sure to use products from reliable suppliers and follow the manufacturer's guidelines for using them.

Before attempting a repair with either type of adhesive system, make sure to read and understand the product literature and follow all recommendations. The main advantage of adhesives is their versatility. Not all plastics can be welded, but adhesives can be used in all but a few instances. If adhesive repair is chosen, the best way to identify the plastic type is by using the flexibility test. In this test, the base material's flexibility is compared to a sample disk of the repair material. Use the repair material that most closely matches the base material. When working with an adhesive system, use the manufacturer's categories to decide on a product. There might be only two categories or as many as five.

It is important to keep in mind that there are differences between manufacturers' product lines, so mixing product lines is not acceptable. It is very important to choose a product line and use it for the entire repair. Remember that:

- Most product lines have two or more adhesives designed for different types of plastic.
- The product line usually includes an adhesion promoter, a filler product, and a flexible coating agent.
- Some product lines are formulated for a specific base material. For instance, one manufacturer offers individual products for use on such plastics as TPO, urethanes, or Xenoy regardless of plastic flexibility.
- A product line might use a single flexible filler for all plastics, or there might be two or more flexible fillers designed for different types of plastic.

TWO-PART REPAIRS

Regardless of which manufacturer's products are used, two-part adhesive repairs share the following common preparation steps:

- Clean the part, first with soap and water and then with a plastic cleaner (Figure 13-22).
- Make sure both the part and the repair material are within the specified temperature range for proper curing and adhesion.
- Mix the two parts of the adhesive thoroughly and in the proper proportions.
- Apply the material within the time guidelines given in the product literature. Use heat if indicated by the manufacturer.
- Follow the cure time guidelines given in the product literature.
- Support the part adequately during the cure time to ensure that the damaged area does not move before the adhesive cures.

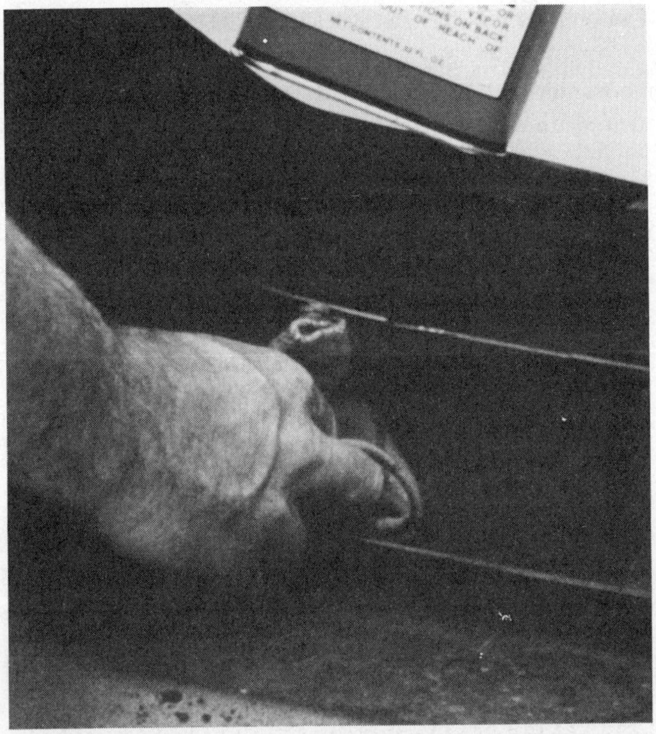

FIGURE 13-22 Cleaning the part with plastic cleaner

FIGURE 13-23 Applying a sprayable adhesion promoter

- Follow the product literature for guidelines on when to reinforce a repair.

Some plastics require the use of an adhesion promoter (Figure 13-23). These include TPO, PP, and E/P plastics. There is a simple test to perform that indicates whether or not the plastic will require an adhesion promoter. Lightly sand a hidden spot on the piece, using a high-speed grinder and No. 36 grit paper. If the material gives off dust, it can be repaired with a standard structural adhesive system. If the material melts and smears or has a greasy or waxy look, then an adhesion promoter will be required. Many plastic fillers and adhesives contain an adhesion promoter; check the label.

The following is a typical two-part epoxy adhesive repair of a polyolefin bumper cover:

1. Clean the entire cover with soap and water, wipe or blow-dry, then clean with a good plastic cleaner.
2. V-groove and taper the damaged area using a slow speed grinder and 3-inch No. 36 grit disc. Grind about a 1-1/2-inch taper around the damage for good adhesion.
3. Use a sander with No. 180 grit paper to featheredge the paint around the damaged area, then blow dust free. Depending on the extent of the damage, the backside might need reinforcement. To do this, use steps 4 through 6.
4. Clean the backside of the cover with plastic cleaner, then apply a coat of adhesion promoter according to the manufacturer's recommendations.
5. Dispense equal amounts of both parts of the flexible epoxy adhesive and mix to a uniform color. Apply the material to a piece of fiberglass cloth using a plastic squeegee (Figure 13-24).
6. Attach the saturated cloth to the backside of the bumper cover and fill in the weave with additional adhesive material.
7. With the backside reinforcement in place as shown in Figure 13-25, apply a coat of adhesion promoter to the sanded repair area on the front side. Let the adhesion promoter dry completely.
8. Fill in the groove with adhesive material. Allow it to cure according to the manufacturer's recommendations.
9. Rough grind the repair area with No. 80 grit paper, then sand with 180 grit, followed by 240 grit. If additional adhesive material is needed to fill in a low spot or pinhole, be sure to apply a coat of adhesion promoter first, followed by a skim coat of adhesive material.

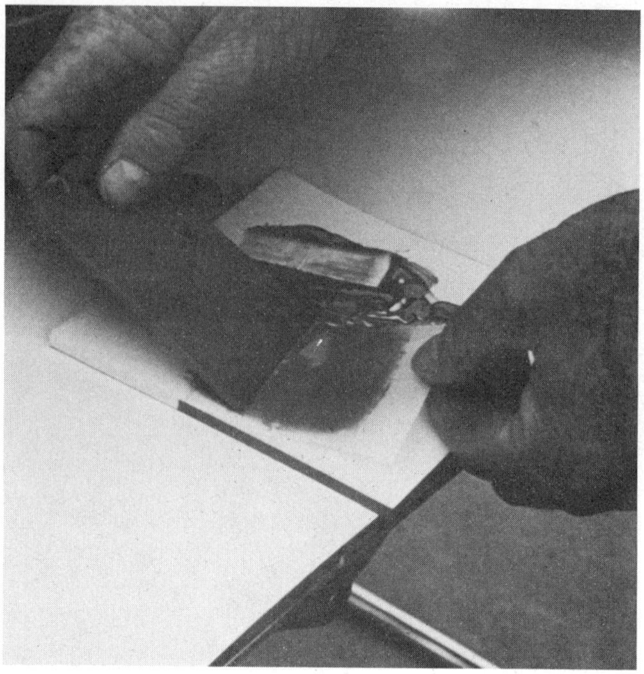

FIGURE 13-24 Applying flexible epoxy adhesive to a piece of fiberglass cloth

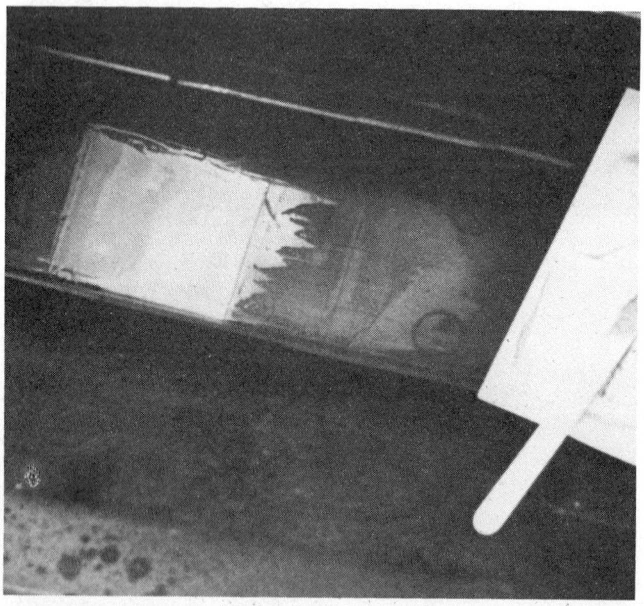

FIGURE 13-25 The repaired reinforcement

SHOP TALK

Never attempt to use flexible putty on polyolefins—it will not work.

Two-Part Adhesive Dash Repair

There are some two-part adhesive products that are recommended for dash pad repair. A typical repair procedure is as follows:

1. Thoroughly clean the part.
2. Sand or grind away the broken or loose vinyl covering to get to the foam beneath. A V-groove need not be used.
3. Apply the adhesive material according to the manufacturer's directions. Make sure the recommended cure times are followed.
4. After curing, sand or grind to contour. Apply a skim coat of adhesive if necessary. Allow the adhesive to cure, then block sand with medium grit paper, followed by fine paper.
5. Apply a sealer, then retexture.
6. Refinish as per paint system recommendations (see Chapter 16).

SHOP TALK

Some dash pad damage might be the result of age and exposure to sunlight. In these cases, repair will not work. To check for this condition, press on the pad to flex the vinyl covering near the damage. If more cracks appear, it is probably beyond repair and should be replaced.

Adhesive Repair of Other Flexible Parts

When repairing other flexible parts with adhesive materials, the procedure to be followed is similar to the one presented previously, with these exceptions:

1. Skip the adhesion promoter.
2. Use either a skim coat of adhesive material or a flexible filler to fill in imperfections—but do not use both.

Minor Cut and Crack Repairs. Adhesives can be used for repairing minor cuts and cracks. First, the repair area must be cleaned with water and a plastic cleaner. It is very important for the mating surfaces to be clean and free of wax, dust, grease, or any other substance. It is not necessary to use solvents (other than a plastic cleaner) for cleaning. Allow the part to warm to 70 degrees Fahrenheit before applying any adhesives.

After cleaning, the next step is to prepare the damage using an adhesive kit. The kit should have two elements: an accelerator and an adhesive. Spray one side of the crack with the accelerator, then apply the adhesive to the same side of the crack.

Carefully position the two sides of the crack in their original position and press them firmly together. Hold for a full minute to achieve good bond strength. Cure for 3 to 12 hours for maximum strength, following the instructions and precautions on the label.

If the original paint was not damaged and the repair was properly positioned, painting might not be required. Where painting is required, be sure to read the manufacturer's instructions that appear on the paint container.

Gouge, Tear, and Puncture Repair.

The procedure for repairing gouges, tears, and punctures is a chemical bonding process; it requires no special tools or skills. As always, the first step is to clean around the damaged area thoroughly with a wax, grease, and silicone-removing solvent. Apply it with a water-dampened cloth, then wipe dry.

To prepare for the structural adhesive, bevel the edges of the damaged area back about 1/4 to 3/8 inch using a 3-inch, medium grit grinding disc, as shown in Figure 13-26. This will produce a coarse surface for good adhesion; in any repair, the mating surfaces should be scuffed. Use a slow speed when grinding (2000 rpm or less). If the sanded area has a "greasy" appearance, it would be a good idea to apply a coat of adhesion promoter. Also, apply additional adhesion promoter after each sanding step.

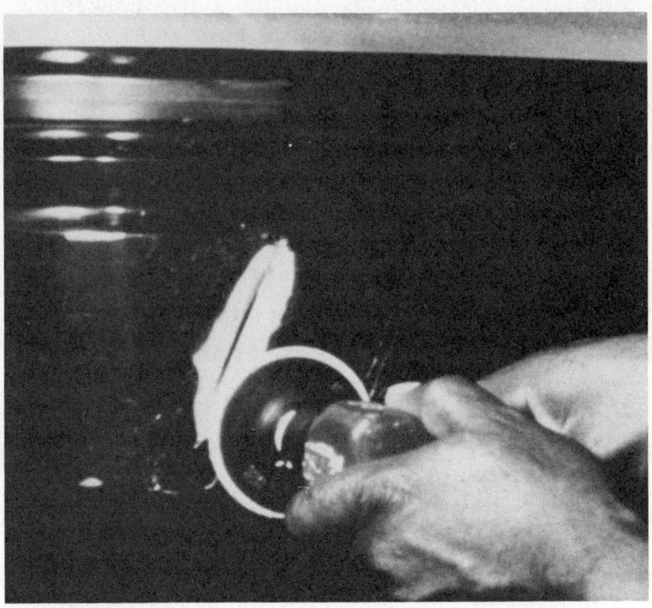

FIGURE 13-26 Beveling the damaged area with a 3-inch grinding disc

The next step is to featheredge the paint around the damaged area with a finer grit disc. Remove the paint, but as little of the plastic as possible, and blend the paint edge into the plastic. Continue removing paint until there is a paint-free band about 1 to 1-1/2 inches wide around the damaged area. The adhesive should not overlap the painted surface. Carefully wipe off all paint and dust in preparation for the next step. The repair area must be absolutely clean before proceeding.

Next, flame treat the area to improve adhesion. Use an oxyacetylene torch with a controlled flame, and a 1-inch cone. Direct the flame onto the beveled area carefully and keep it moving until the area is slightly brown. Be extremely careful not to warp the surface or burn the paint.

The next step is to apply aluminum tape to the repair area. Clean the inner surface with solvent, then install the tape. Cover the damage completely.

Before applying the structural adhesive, the backside of the part should be thoroughly cleaned. Tape it with aluminum body tape to provide support for the repair. For best results, make the tape concave like a dish so that the adhesive overlaps the repair area on the backside. It might be possible to install the tape without loosening or removing any parts from the vehicle, but on some makes partial disassembly is necessary.

Prepare the adhesive material as directed by the manufacturer. Most adhesive compounds come in two tubes. On a clean, flat, nonporous surface such as metal or glass, squeeze out equal amounts of each tube. Then, using a squeegee and an even paddling motion to reduce air bubbles, completely mix the two parts until a uniform color and consistency are achieved.

Paddle the adhesive into the hole, using the squeegee or a plastic spreader (Figure 13-27). This must be done quickly and carefully; the adhesive material will begin to set in about 2 to 3 minutes. Two applications are usually required. The first application simply fills the bottom of the damaged area, so it is not necessary to worry about contour at this time. However, an attempt should be made to fill most of the area's volume. Cure for 1 hour at room temperature or 20 minutes with a heat lamp or gun at 200 degrees Fahrenheit.

Before the final application of adhesive, use a fine grit disc to grind down the high spots. Wipe the dust from the repair area. After the first application has been ground and wiped clean, mix the second application of the adhesive as before, and paddle it into an overfill contour of the repair area. A flexible squeegee or spatula is useful in forming the panel contours.

When the adhesive material has dried, establish a rough contour to the surrounding area with No. 80 grit paper on a sanding block. Then feathersand with a disc sander and No. 180 paper, followed by No. 240 paper to

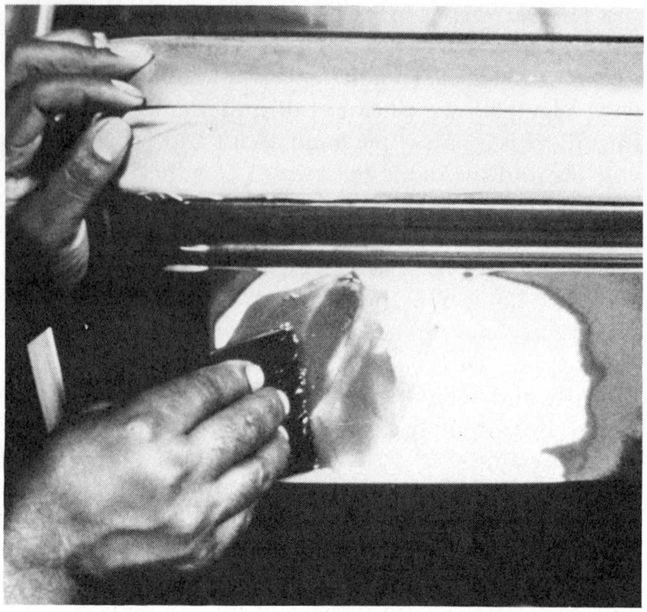

FIGURE 13-27 Applying structural adhesive with a squeegee

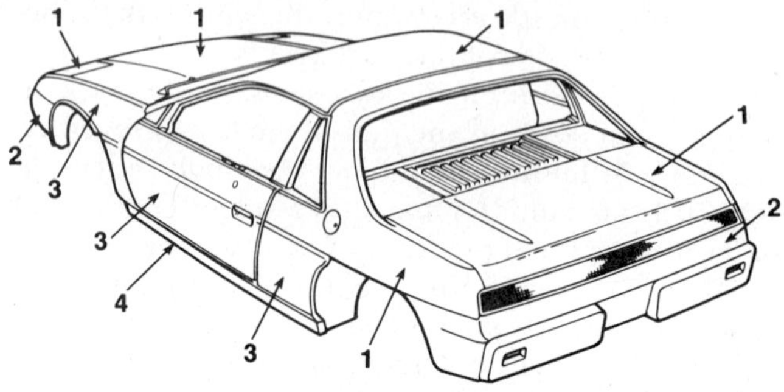

FIGURE 13-28 Typical reinforced plastic body panel identification

achieve levelness with the surface of the part. Check the repair for low areas, pits, or pinholes. Spread additional material over any defects.

Final feathering and finish sanding can be done with a disc sander and No. 320 grit disc. When final sanding is completed, remove all dust and loose material. The surface is now ready for priming and painting (see Chapter 16).

REPAIR OF REINFORCED PLASTIC

Reinforced plastic—including sheet molded compound (SMC), fiber-reinforced plastic (FRP), and reinforced reaction injection molded polyurethane (RRIM)—parts are being used in many unibody vehicles. For example, Figure 13-28 shows the outer unibody panels that are commonly made of reinforced plastics. Table 13-3 provides an overview of reinforced plastic repair materials.

TYPES OF DAMAGE

The damage that generally occurs in reinforced plastic panels includes:

- One-sided damage, such as a scratch or gouge.
- Punctures and fractures.
- Panel separation, where the panel pulls away from the metal space frame.
- Severe damage, which requires full or partial panel replacement.
- Minor bends and distortions of the space frame, which can be repaired by pulling and straightening.
- Severe kinks and bends to the space frame, which require replacement of that piece along factory seams or by sectioning.

Remember, it is likely that combinations of these types of damage will occur on a single vehicle. As with conventional unibody designs, damage to the space frame can include sidesway, sag, collapse, and twist.

Depending on the location and amount of damage, there are four different types of repairs. These are:

- Single-sided repair
- Two-sided repair
- Panel sectioning
- Full panel replacement

In most cases, the reinforced plastic panels will begin to separate from the space frame about the time that the metal begins to deform. It is important to realize that these panels may have a bracing effect on the steel components during a collision and during subsequent pulling and straightening. This may make it necessary to remove the panels from damaged areas before performing the pulling operations. These panels would only be damaged anyway, and would have to be replaced.

A thorough examination of the vehicle is required. Make certain to examine all affected reinforced plastic panels.

TABLE 13-3: REINFORCED PLASTIC REPAIR MATERIAL SELECTION CHART

Type of Repair	Applicable Repair Product				
	Panel Adhesive	Patching Adhesive	Structural Filler	Cosmetic Filler	Glass Fiber Reinforcement
Panel Replacement	X				
Panel Sectioning	X		X[1]	X[1]	X
One-Sided Repairs				X[1]	
Two-Sided Repairs	X[2]	X[2]	X	X	X

Notes: 1. Some panel adhesives can also be used as structural and cosmetic fillers, depending on sanding characteristics.
2. Panel adhesives can also be used as patching adhesives, but not vice versa.

- Check the entire panel for signs of damage.
- Check all panel seams for adhesive bond failure.
- Examine the back of the panel to determine the extent of the damage.

Keep in mind that while the external surface of the panel will exhibit signs of damage, the full extent of the damage may not be easily judged from the external appearance.

TOOLS AND MATERIALS

Most of the tools used for repairs of reinforced plastics should already be available in a well-equipped collision repair shop. The only special tools will be the adhesive applicators, which allow two-part adhesives to be dispensed at a constant rate. The two types of applicators, pneumatic and hand-operated, are described in Chapter 2.

Either type of applicator can be used by following these simple rules:

- Follow the manufacturer's instructions.
- Check for proper adhesive flow.
- Check for consistent mix of two-part adhesives.
- When changing cartridges, run a new test bead.
- If saving part of a cartridge, leave the static mixing nozzle in place.

TYPES OF ADHESIVES

Many of the materials that are used for reinforced plastic repair are two-part adhesive products. That means they have a base material and a hardener. They must be mixed together in the proper ratio as called for by the product manufacturer, and both parts must be thoroughly and adequately mixed together before use.

After mixing, remember that each product has a work life or open time. This is the time in which it is possible to realign or change the position of the panels being replaced and still have the adhesive set up for a good bond. This work life will be provided by the manufacturer.

Two-part adhesives also have a cure time during which the piece should not be moved. This allows the adhesive to set up properly. The cure time of some of the products used in reinforced plastic repair can be shortened with the application of heat.

Temperature and humidity can affect work life and cure time. The product manufacturer will provide a temperature range for the adhesive, as well as guidelines for heat curing. Excessive humidity can slow the curing process.

FILLERS AND TYPES OF GLASS CLOTH

Two filler products are specifically formulated for use on reinforced plastic. They are:

- *Cosmetic Filler.* This is typically a two-part epoxy or polyester filler used to cover up minor imperfections. Do not use fillers designed for sheet metal on reinforced plastic.
- *Structural Filler.* Used to fill the larger gaps in the panel structure, structural fillers add to the structural rigidity.

All two-part products will shrink to some degree. The use of heat will help to speed the drying time and will eliminate some of the shrinkage. Be cautious about using heat around reinforced plastic, but at the same time, heat must be employed to cure the filler.

Check with the product manufacturer for temperatures and dry times. If the product is not properly heated to a full cure, shrinkage will occur as the product cures with time. The rule of thumb is to heat the material to a surface temperature that is higher than any temperatures that the vehicle will be subject to when it is on the road. If it is a black vehicle sitting in the sun in midsummer in Florida, this could be about 170 degrees Fahrenheit or more.

Check with the product manufacturer for recommendations for heat curing, but generally 200 to 250 degrees Fahrenheit for 20 to 40 minutes should do it. The conventional type of heat applicator found in the shop will rarely be able to heat a surface to beyond 180 degrees Fahrenheit. Also remember that at lower temperatures, the product will have to be heated longer. Also, if there is high humidity, the cure process will take longer.

There are several different types of glass cloth available. Rovings and mattings are not appropriate for reinforced plastic repair. Choose unidirectional cloth, woven glass cloth, or nylon screening. The main thing to remember is that the cloth weave must be loose enough to allow the adhesive to fully saturate the cloth, leaving no air space around the weave.

SINGLE-SIDED REPAIRS

Single-sided damage is surface damage or damage that does not penetrate or fracture the rear of the panel. In some cases damage might pass all the way through a panel, but no pieces of panel have been broken away. If the break is clean and all of the reinforcing fibers have stayed in place, then a single-sided repair would be adequate. In this case, the technician must make the bevel very deep to penetrate the fibers in the panel (Figure 13-29A). The broken fibers must come into contact with the adhesive.

The following is a typical single-sided repair procedure:

1. Clean the repair area with soap and water.
2. Clean the repair area using mild wax and grease remover.
3. Remove any paint from the surrounding area by sanding with No. 80 grit sandpaper.
4. Scuff sand the area surrounding the damage.
5. Bevel the damage to provide an adequate area for bonding.
6. Mix two-part filler according to the manufacturer's instructions.
7. Apply the filler and cure as recommended.

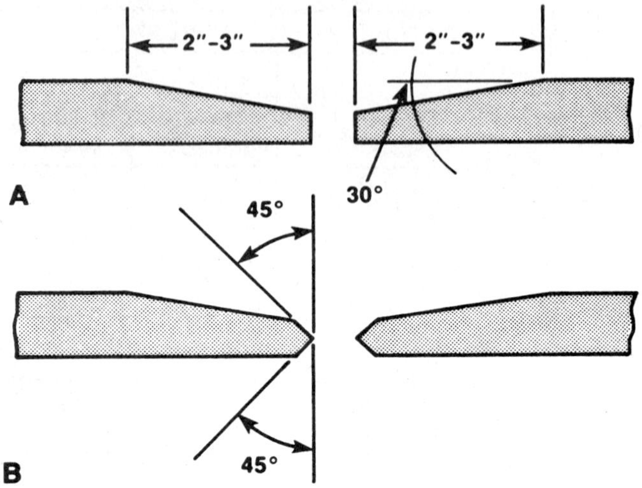

FIGURE 13-29 Beveling the inside and outside of the repair area permits better adhesion for (A) single-sided and (B) two-sided repairs.

Once the filler has been sanded, apply additional coats as required and resand. The product manufacturer will provide grit recommendations.

TWO-SIDED REPAIRS

Damage that passes all of the way through the panel and involves the reinforcing fibers, such as cracks and holes, usually requires a two-sided repair (Figure 13-29B). A backing strip or patch is used to restore the required strength. The patch also forms a foundation on which to build and form the exterior surface to match the original contour of the panel.

To make a two-sided repair in a reinforced plastic panel, proceed as follows:

1. Clean the surface surrounding the damage with a good wax and grease remover. Use a No. 36 grinding disc to remove all paint and primer at least 3 inches beyond the repair area.
2. Grind, file, or use a hacksaw to remove all cracked or splintered material away from the hole on both the inside and outside of the repair area.
3. Remove any dirt, sound deadener, and the like from the inner surface of the repair area. Clean with reducer, lacquer thinner, or a similar solvent.
4. Scuff around the hole with No. 80 grit paper to provide a good bonding surface.
5. Bevel the inside and outside edge of the repair area about 30 degrees to permit better patch adhesion.

FIGURE 13-30 Applying resin to the fiberglass strips

6. Clean the repair area thoroughly with reducer or thinner.
7. Cut several pieces of fiberglass cloth or mat large enough to cover the hole and the scuffed area. The exact number of pieces will vary depending on the thickness of the original panel, but five is usually a good number to start with.
8. Prepare a mixture of resin and hardener following the label recommendations. Using a small paintbrush, saturate at least two layers of the fiberglass cloth with the activated resin mix (Figure 13-30) and apply it to the inside or back surface of the repair area. Make sure the cloth fully contacts the scuffed area surrounding the hole.
9. Saturate three more layers of cloth with the mix. Apply it to the outside surface, making certain that these layers fully contact the inner layers and the scuffed outside repair area (Figure 13-31).
10. With all of the layers of cloth in place, form a saucer-like depression in them. This is necessary to increase the depth of the repair material. Use a squeegee to work out any air bubbles.

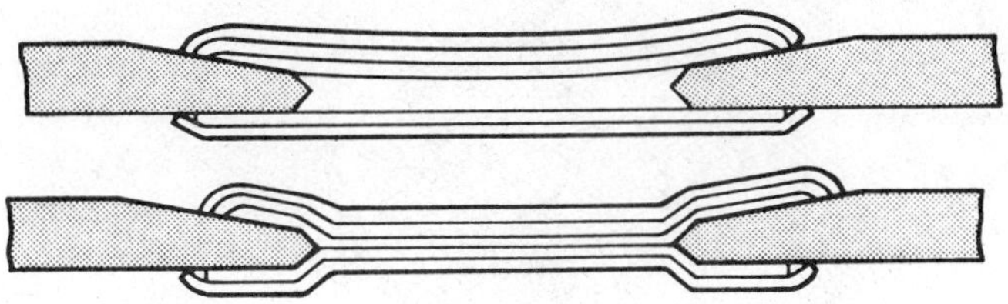

FIGURE 13-31 The additional layers of saturated fiberglass cloth should contact the outside repair area.

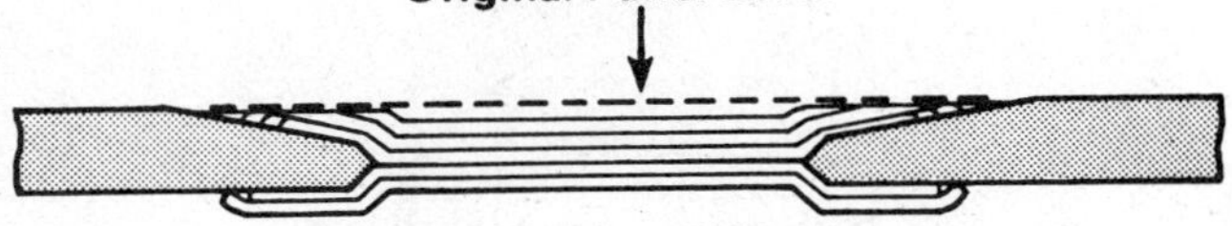

FIGURE 13-32 Sand the patch slightly below the contour of the panel.

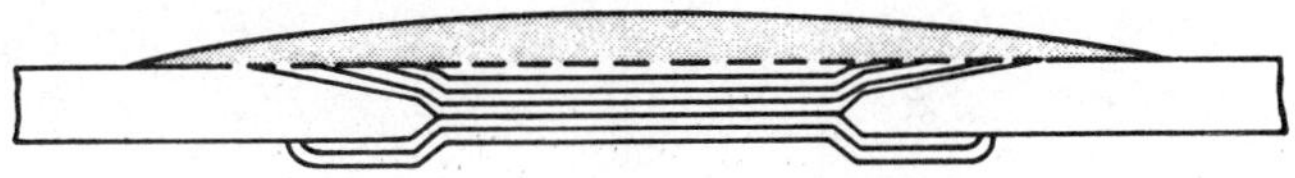

FIGURE 13-33 The hardened patch is ready for sanding.

11. Clean all tools with lacquer thinner immediately after use.
12. Let the saturated cloth patch become tacky. An infrared heat lamp can be used to speed up the process; if one is used, keep it 12 to 15 inches away from the surface. Do not heat the repair area above the recommended temperature, because too much heat will cause distortion.
13. With No. 50 grit paper, disc sand the patch slightly below the contour of the panel (Figure 13-32).
14. Prepare more resin and hardener mix. Use a plastic spreader to fill the depression in the repair area, leaving a sufficient mound of material to grind down smooth and flush.
15. Allow the patch to harden (Figure 13-33). Again, a heat lamp can be used to speed the curing process.
16. When the patch is fully hardened, sand the excess material down to the basic contour, using No. 80 grit paper and a sanding block. Finish sand with No. 120 or finer grit paper.

There are several ways to hold the patch in place from the front side of the panel. One method is to use a pull rod (Figure 13-34). Drill a hole in the middle of the patch, and insert the end of the rod. Position the patch, and pull it snug with the pull rod. Apply heat using a heat gun on the front, or have someone else apply the heat to the back of the patch.

Another method of holding a patch in place is to drill two small holes, roughly an inch apart, in the middle of the patch (Figure 13-35). A wire is run through both holes so that the ends stick out the front of the patch. Pull the patch tight to the panel with the wires, and use a paint paddle or stick to tie the wires off.

This same two-sided repair can be made by attaching sheet metal to the backside of the panel with sheet

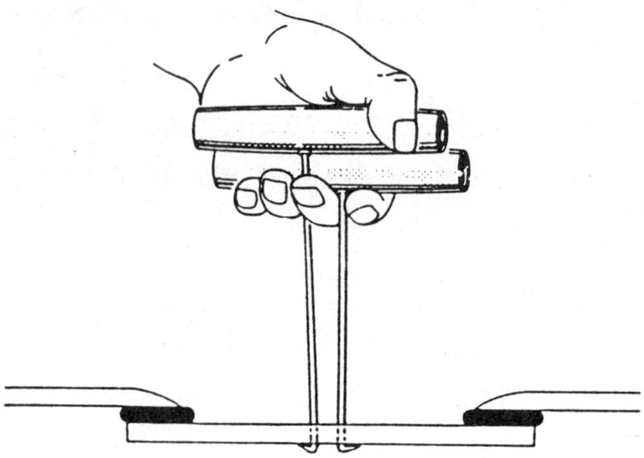

FIGURE 13-34 Holding the patch in place with a pull rod

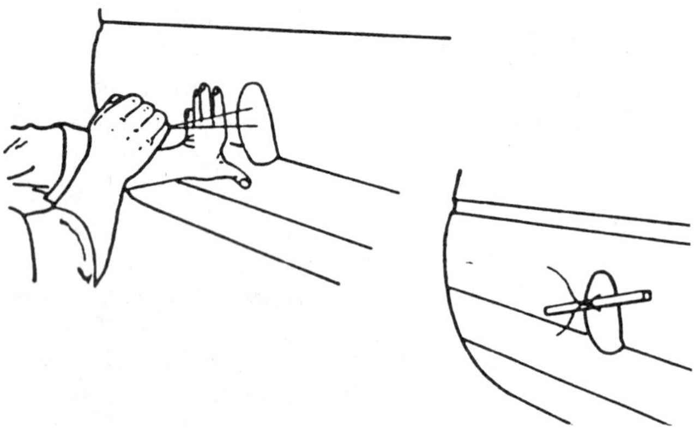

FIGURE 13-35 Holding the patch in place using wire and a stick

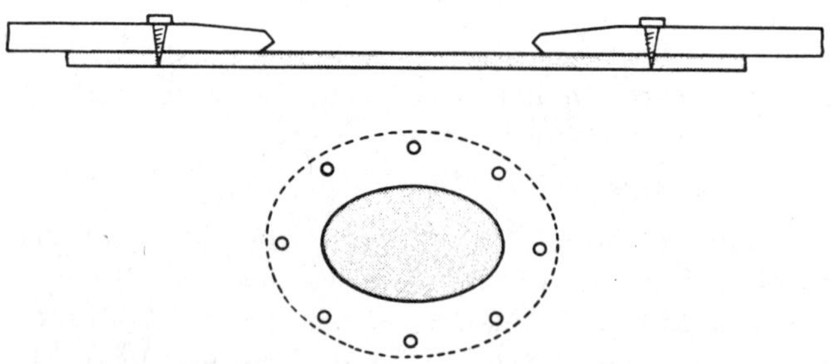

FIGURE 13-36 Sheet metal attached to the backside of a damaged panel

metal screws (Figure 13-36). Sand the sheet metal and both the inner and outer sides of the hole to provide good adherence for the repair material. Before fastening the sheet metal, apply resin and hardener mix to both sides of the rim of the hole. Follow the above procedure for the remainder of the repair.

When the inner side of the hole is not accessible, apply a fiberglass patch to the outer side only. After the usual cleaning and sanding operations, apply several additional layers of fiberglass cloth to the outer side of the hole. Before it dries, make a saucer-like depression in the cloth to provide greater depth for the repair material.

PANEL REPLACEMENT AND SECTIONING

Proper sectioning requires that the collision repair technician understands the areas of the vehicle that are most appropriate for sectioning and how to avoid problems with horizontal bracing, rivets, and concealed parts (Figure 13-37). The replacement panel used will depend on the amount and location of the damage. Using the left rear quarter panel as an example, there are three possibilities. As shown in Figure 13-38, the entire quarter panel can be ordered, or just a front or rear half. It must be remembered that reinforced plastic is a very forgiving and workable material. Just because

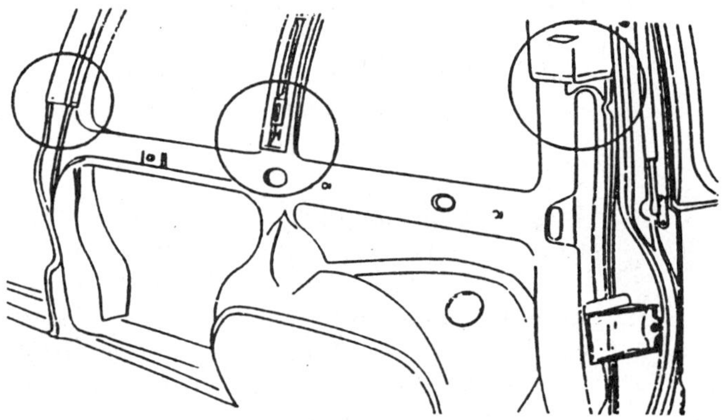

FIGURE 13-37 Common sectioning locations

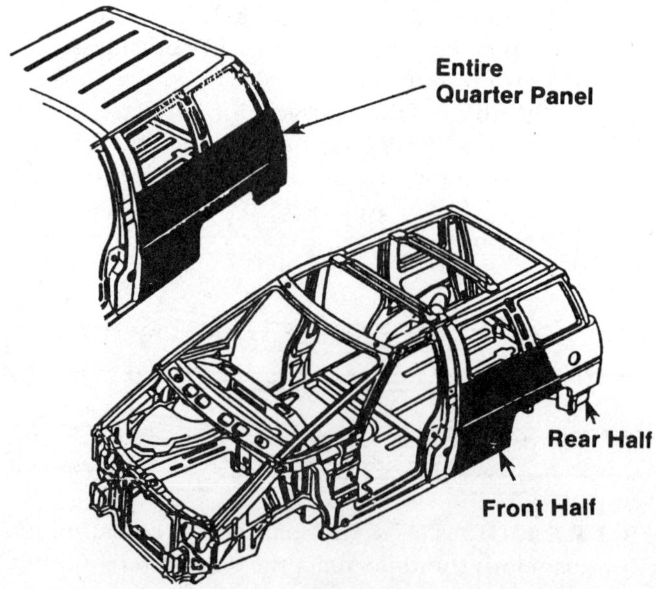

FIGURE 13-38 Replacement panel options

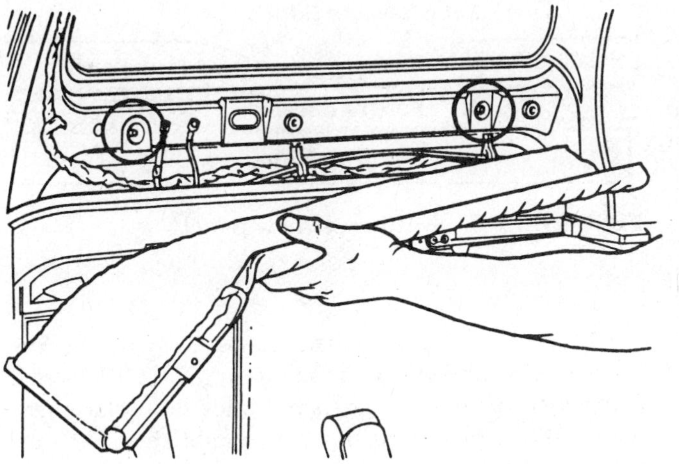

FIGURE 13-39 Removal of interior trim

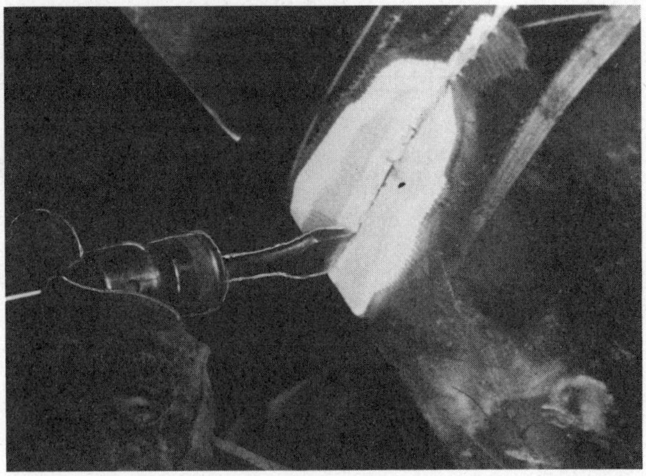

FIGURE 13-40 Using an air chisel to separate the panel from the space frame

quarter panels come split at the wheel well, the sectioning point does not have to be located there. Using the recommended backing strips to reinforce the joints, sectioning can be done almost anywhere.

The mill and drill pads are used to help the factory hold panels in place while the adhesive cures. These mill and drill pads will also help the technician to hold, align, and level replacement panels. If a panel is to be sectioned, it should be done between mill and drill pad locations.

Preparing for Sectioning

The interior trim must be removed to locate the horizontal bracing and mill and drill pads (Figure 13-39). Examination of the back of the panel is the only way to positively gauge the extent of panel damage, as well as to determine the location of the horizontal bracing, mill and drill pads, and electrical and mechanical components.

Once the interior trim is removed, a "window" can be cut. Controlling the depth of the cut is very important. Space frame components, as well as electrical lines and heating/cooling elements, may be located behind the panels. When cutting the window, know what is behind the panel being cut into, or limit the depth of the cut to 1/4 inch to avoid doing damage.

Now that the window has been cut, the rest of the panel can be removed from the space frame. This can be done using heat and a putty knife, or by carefully using an air chisel (Figure 13-40). Choose a flat chisel, beveled on one side only. Be careful not to damage the space frame.

If the door surround panels are to be left attached to the vehicle, the air chisel method may not be the best choice for separating the seam between the panel to be saved and the panel to be scrapped. Use the heat and putty knife method to separate the seams to avoid doing damage to the door surround pieces.

For vehicles in which the rear of the quarter panel is covered by the liftgate surround panel, there are three options for freeing the rear edge of the quarter panel:

- If the liftgate surround panel is damaged, cut away the damaged portion and section in the new part of the surround panel after the quarter panel section is removed.
- If the end of the quarter panel is damaged but the liftgate surround panel is unharmed, cut down the side of the surround panel. Then remove the remaining quarter panel from under the surround panel by heating and prying, or drilling.
- If damage is confined to the side panel, look closely at the rear inner edge of the quarter panel facing the liftgate surround panel; a bonded seam should be apparent. This seam can be separated by heating and prying.

The adhesive bond must be broken to remove the remaining quarter panel. Using a heat gun at approximately 400 degrees Fahrenheit, heat the outboard section of the panel. Concentrate the heat in the area of the bond. Heat a span of about 5 to 7 inches. After a short time the adhesive will soften and the panel can be removed using a panel cutter, chisel, or pry bar. Often more than one person may be needed to assist in the heating and prying operations. Apply the heat and pry at the same time. Remember that if damage is done to the space frame, corrosion protection must be restored.

As shown in Figure 13-41, there is a panel reinforcement that runs the length of the panel horizontally. It is made of reinforced plastic and is bonded to the back of the panel. This reinforcement contains the mill and drill pad bolts and acts as a spacer to hold the panel in its correct place.

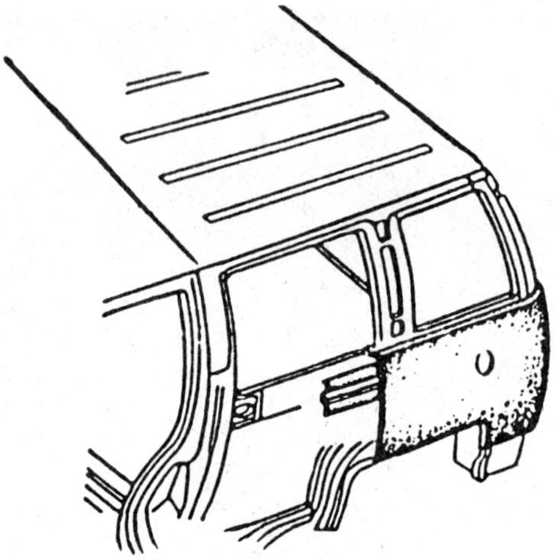

FIGURE 13-41 Typical horizontal panel reinforcement

When replacing a panel use this reinforcement as a sectioning point, as well as an anchoring point for the new panel, by leaving several inches of the reinforcement bonded to the space frame. Determine the sectioning location, and make the cut, using extra care to avoid cutting through the reinforcement. Cut through the outer panel only at the sectioning location.

Decide how much of the reinforcement will be kept, and make a cut on the scrap side of the panel. Cut through both panel and reinforcement. Remove the scrap panel in the usual way, then remove the strip of panel left attached to the reinforcement. On the replacement panel, mark off the corresponding piece of panel reinforcement that must be removed to fit the panel. Be cautious when estimating how much reinforcement must be removed from the replacement panel. Leave enough reinforcement to enable the part to be trimmed to fit.

Fitting the New Panel

After removing the scrap panel, prepare the space frame for the new panel. First remove the old adhesive from the space frame. In some cases the adhesive can just be peeled away from the frame. Usually other means must be used, such as heat and a putty knife or a sander. If heat and a putty knife are used to remove the adhesive, then all of it must be removed, since the heat will break down the adhesive.

Bevel the outside edges of the existing panel to a 20-degree taper. Sand and clean the backsides of the panel where the backing strips will be attached. Backing strips are made using scrap material which duplicates the original panel contour as closely as possible. They should extend about 2 inches beyond either side of the sectioning location. Clean the backing strips and remove the paint from those places where adhesive will be applied.

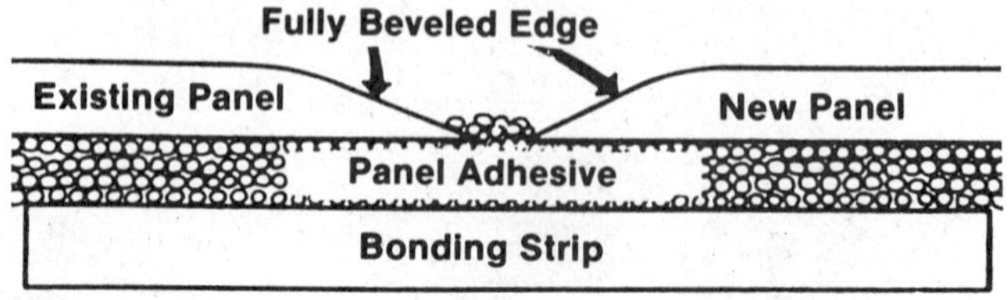

FIGURE 13-42 Fitting the new panel

Measure the replacement panel for fit, trim the panel to size, and check the fit again. Leave a 1/2-inch gap between the existing panel and the replacement panel. When proper fit has been established, the new panel can be prepared for the adhesive. Sand or grind bevels into the panel where they mate to the existing panel (Figure 13-42). Bevel the mating edges of the new panel to a shallow taper, just like on the existing panel. Make sure to bevel all of the way through the panel; do not leave a shoulder.

Apply the adhesive material in a continuous bead all of the way around the panel. Make sure the bead is 3/8 to 1/2 inch in diameter. Check for horizontal bracing, and apply a bead of adhesive to correspond with it. Then fit the panel onto the vehicle and clamp it into place. Install the mill and drill pad nuts and tighten securely.

The work life will be recommended by the adhesive manufacturer. This time should be followed to ensure that the proper fit has been achieved, and that the panel surface. While the procedure illustrated in Figure 13-43 and described here relates to a rear quarter panel section, the principle can be applied to any type of curved reinforced plastic panel.

USING MOLDED CORES

Naturally, holes are much more difficult to repair in a curved portion of a reinforced plastic panel than those on a flat surface. Basically, the only solution in such a case (short of purchasing a new panel section) is to use the molded core method of replacement. This is often the quickest and cheapest way to repair a curved surface. While the procedure illustrated in Figure 13-43 and described here relates to a rear fender section, the principle can be applied to any type of curved reinforced plastic panel.

1. First off, locate an undamaged panel on another vehicle that matches the damaged one; this will be used as a model. A new or used vehicle can be employed, since the model vehicle will not be harmed if care is taken.
2. On the model vehicle, mask off an area slightly larger than the damaged area. Apply additional masking paper and tape to the sur-

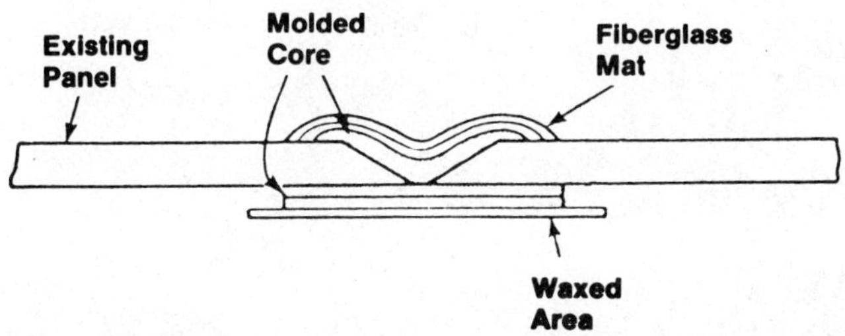

FIGURE 13-43 The repair area covered with fiberglass matting

rounding area, especially on the low side of the panel. This will prevent any resin from getting on the finish.

3. Coat the area with paste floor wax. Leave a wet coat of wax all over the surface. A piece of waxed paper can be substituted for the coat of wax; be sure that it is taped firmly in place (Figure 13-44A).
4. Cut several pieces of thin fiberglass mat material in sizes ranging from 2 by 4 inches to 4 by 6 inches.
5. Mix the fiberglass resin and hardener following the label instructions on both products.
6. Starting from one corner of the mold area, place pieces of fiberglass mat on the waxed area so each edge overlaps the next one; use just one layer of matting (Figure 13-44B).
7. Apply the resin/hardener to the matting with a paintbrush (Figure 13-44C). Force the mixture into the curved surfaces and around corners with the tips of the bristles.
8. Use the smaller pieces of matting along the edges and on difficult curves. Additional resin/hardener can be applied if needed, brushing in one direction only to force the material into the indentations. In all cases, use only one layer of matting.

CAUTION

Be sure that the resin/hardener does not get on any part of the model vehicle that is not coated with wax.

9. After matting has been applied to the entire waxed area, allow the molded core to cure a minimum of 1 hour.
10. Once the molded core has hardened, gently work the piece loose from the model vehicle (Figure 13-44D). The core should be an exact reproduction of this section of the panel.
11. Remove the wax or wax paper from the model vehicle. Then polish this section of the panel.
12. Since the molded core is generally a little larger than the original panel, place it under the damaged panel and align. If necessary, trim down the edges of the core and the damaged panel slightly where needed for better alignment. The edges of the damaged panel and core must also be cleaned.
13. Using fiberglass adhesive, cement the molded core in place on the inside of the panel (Figure 13-44E). Allow the core and panel to cure.
14. Grind back the original damaged edges to a taper or bevel, maintaining the desired contour.
15. Lay a fiberglass mat, soaked in resin/hardener, on the taper or bevel and over the entire core. Once the mat has hardened, level it with a coat of fiberglass filler. Then prepare it for painting (Figure 13-44F).

In some instances, it might not be possible to place the core on the inside of the damaged panel. In this case, the damaged portion must be cut out to the exact size of the core. After the panel has been trimmed and its edges beveled, tabs must be installed to support the core from the inside. These tabs can be made from pieces of the panel or from fiberglass strips saturated in resin/hardener.

After cleaning and sanding the inside sections, attach the tabs to the inside edge of the panel and bond with fiberglass adhesive. Clamping pliers can be used to hold the tabs in place while bonding. Taper the edge of the opening and place the core on the tabs. Fasten the core to the tabs with fiberglass adhesive. Grind down any high spots so that layers of fiberglass mat can be added.

Place the saturated mats over the core, extending about 1-1/2 to 2 inches beyond the damaged area in all directions. Work each layer with a spatula or squeegee to remove all air pockets. Additional resin/hardener can be added with a paintbrush to secure the layers. Allow the resin/hardener time to cure sufficiently, then sand the surface level. For a smooth surface, use fiberglass filler to finish the job.

REPAIRING RRIM

Reinforced reaction injection molded (RRIM) parts are becoming more common. RRIM is usually found in fenders and bumper covers.

RRIM is a two-part polyurethane composite plastic. Part A is the isocyanate. Part B contains the reinforced fibers, resins, and a catalyst. The two parts are first mixed in a special mixing chamber, then injected into a mold. Since RRIM is a thermosetting plastic, heat (100 to 140 degrees Fahrenheit) is applied to the mold to cure the material. The molded product is made to be stiff yet flexible. This allows RRIM to absorb

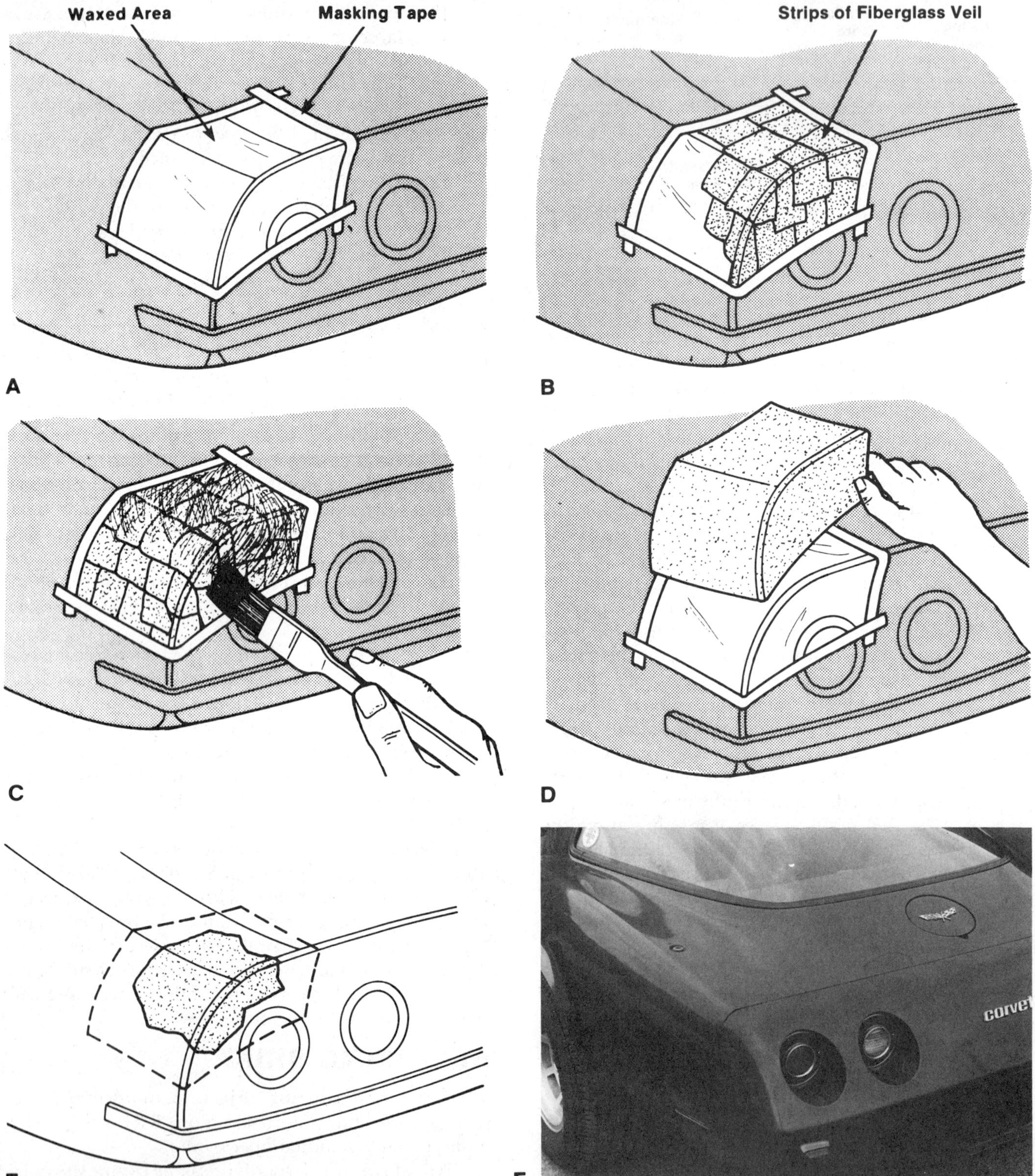

FIGURE 13-44 Steps in making a molded core: (A) coat the area being used as a mold model with paste floor wax or a piece of waxed paper; (B) place pieces of fiberglass mat over the waxed or waxed paper surface; (C) apply resin/hardener to the matting; (D) remove the molded core from the model vehicle; (E) cement the core in place; (F) the completed job.

minor impacts without damage and this makes it an ideal material to use for exposed areas such as front fenders, bumpers, and front and rear fascias. Although minor impacts might not dent RRIM, damage will likely occur in a heavier collision.

Gouges and punctures can be repaired using a structural adhesive. If the damage is a puncture that extends through the panel, a backing patch is required. To make a typical backing patch repair, proceed as follows:

1. Clean the damaged area thoroughly using the plastic cleaner recommended by the manufacturer and a clean cloth. Wipe dry.
2. Remove any paint film in and around the damage with an orbital sander and No. 180 grit disc.
3. Using a No. 50 grit disc, enlarge the damaged area, tapering out the damage for about 1 inch. Wipe or blow away any loose particles.
4. Clean the backside of the damaged area with the plastic cleaner.
5. Use a No. 50 grit disc to scuff sand the area. Extend the area to about 1-1/2 inches beyond the damage. Align the front of the panel with body tape, if necessary.
6. Cut a piece of fiberglass cloth to cover the damaged area and the part of the panel that has been scuff sanded.
7. Mix the adhesive according to the manufacturer's recommendations. Apply a layer of adhesive to the backside of the panel about 1/8 inch thick.
8. Place the fiberglass patch into position on the adhesive. Cover it with a sheet of waxed paper. Use a roller to force the adhesive into the fibers of the patch.
9. Remove the wax paper and add another layer of adhesive. Work out the adhesive to just beyond the edges of the patch. Allow the adhesive to cure following the manufacturer's recommendation.
10. Now move to the front side of the panel. Apply a layer of adhesive, completely covering the damaged area. Build it up to slightly higher than the surrounding contour. Allow it to cure.
11. Apply heat to help speed the cure of the patch.
12. Contour the adhesive to the adjoining surface by block sanding using No. 220 grit paper.
13. Finish by feathering with an orbital sander and a No. 320 disc.
14. Follow the recommendations given in Chapter 16 for priming and painting.

REVIEW QUESTIONS

1. When tack welding, Technician A does not use a welding rod. Technician B says a welding rod must always be used. Who is correct?
   a. Technician A
   b. Technician B
   c. Both A and B
   d. Neither A nor B

2. Technician A says the burn test is a reliable means of identifying plastics. Technician B says it is not. Who is correct?
   a. Technician A
   b. Technician B
   c. Both A and B
   d. Neither A nor B

3. In speed welding, the torch is held at a ________ -degree angle to the base material to start the weld.
   a. 45
   b. 90
   c. 30
   d. 120

4. Which type of welding tips are ideal for working in hard-to-reach places?
   a. round
   b. speed
   c. both a and b
   d. neither a nor b

5. The recommended welding rod size for airless welding is ________.
   a. 3/16-inch diameter
   b. 1/32-inch diameter
   c. 1/16-inch diameter
   d. 1/8-inch diameter

6. Ultrasonic weldability depends on the plastic's ________.
   a. melting temperature
   b. elasticity
   c. impact resistance
   d. all of the above

7. A good speed weld in a V-groove will have ________.
   a. a smooth and shiny appearance
   b. more uniformity than a normal hand weld
   c. a slightly higher crown than a normal hand weld
   d. all of the above

8. Which of the following statements is incorrect?
   a. CAs work equally well on all automotive plastics.
   b. PP and TPO are examples of plastics that require an adhesion promoter as part of the repair process.
   c. The best way to identify a plastic for adhesive repair is by using the flexibility test.
   d. Both a and b

9. Which is more versatile?
   a. adhesive bonding
   b. hot-air welding
   c. airless welding
   d. none of the above

10. When Technician A grinds the base material, it melts and smears, so he or she uses an adhesion promoter to make the repair. Under the same circumstances, Technician B says that an adhesion promoter is not needed. Who is correct?
   a. Technician A
   b. Technician B
   c. Both A and B
   d. Neither A nor B

11. At what angle should the welding rod be cut before speed welding begins?
   a. 45 degrees
   b. 50 degrees
   c. 60 degrees
   d. 90 degrees

12. Technician A uses a pneumatic adhesive applicator when repairing a reinforced plastic, while Technician B uses the hand-operated type. Who is correct?
   a. Technician A
   b. Technician B
   c. Both A and B
   d. Neither A nor B

13. Which of the following statements concerning minor cut and crack adhesive repairs is incorrect?
   a. Allow the parts to warm to 70 degrees Fahrenheit before applying the adhesive.
   b. The adhesive is applied after the accelerator.
   c. Both sides of the cut or crack must be sprayed with accelerator.
   d. For maximum strength, allow 3 to 12 hours of curing time.

14. After completing an airless plastic weld, what is used to smooth it out?
   a. disc sander and No. 320 grit (or finer) disc
   b. cheese grater file
   c. slow-speed grinder and No. 36 grit disc
   d. flat "shoe" part of the torch tip

15. Which of the following is the correct repair method for RRIM?
   a. adhesive
   b. welding
   c. both a and b
   d. neither a nor b

CHAPTER

14 Preliminary Steps of Vehicle Refinishing

OBJECTIVES

- Identify the six basic areas of the I-CAR finish matching system.
- Define *value, hue, chroma,* and *metamerism,* and explain their significance to refinishing.
- Explain the methods used to determine the type of finish on a vehicle.
- Outline the basics of the tinting process, including analyzing color and matching metallic finishes.
- Explain flip-flop and discuss ways of correcting it.
- Make a letdown panel for use with a tri-coat finish.

The first thing the customer will check when a damaged vehicle is delivered is how the repair area looks. The appearance of the vehicle is the single factor that most often causes complaints. It is the painter's responsibility to produce an acceptable match between the repair area and the original finish. The three main factors that affect the appearance of the vehicle are the color of the vehicle, the texture of the finish, and the condition of the finish. These must be taken into account when planning a refinishing operation. A number of potential problems exist that can cause a color variation, and many of them are outside of the painter's control.

Because of the variety of things that can cause mismatch problems, it is very important to follow a step-by-step process on every job. If one does not, it is difficult to find the cause of the problem. The I-CAR system was developed to provide a reliable method of eliminating finish matching problems. If you follow it very carefully through each step of the procedure, it will lead you to where you want to be. It is like following a road map (Figure 14-1).

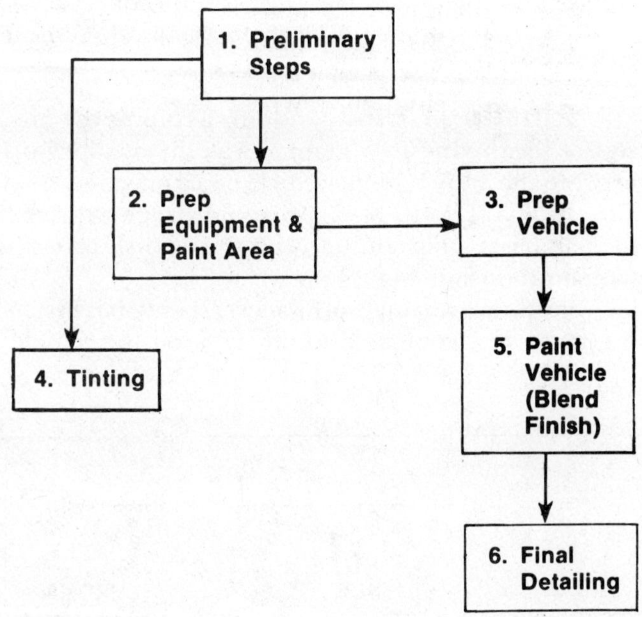

FIGURE 14-1 The six basic areas of the I-CAR finish matching system

Before starting the six-step refinishing procedure, it is important to understand the theory of color, the basic use of the various types of spray guns and systems, and the importance of preparation.

COLOR THEORY

There are many types of artificial light; incandescent light in the home, fluorescent light in schools, shops, factories, and offices, mercury vapor light in parking lots, and so forth. Each of these can affect the way colors are seen. Differences in eyesight can also affect the colors that are seen. It is important to define color and to have a systematic way to describe it. Three qualities relative to describing color are value, hue, and chroma.

Value (Lightness/Darkness). Value refers to the degree of darkness or lightness of the color. The value scale runs up and down the vertical axis of the color tree. White is at the top of the vertical axis, with gradually darkening shades of gray in the middle, and the blackest black at the bottom.

Hue (Color, Cast, Tint). Hue is the color that is seen, moving around the outer edge of the color tree. It moves from blue to red to yellow to green. Colors move around the hue scale (that is, a blue can be moved toward the green side and become aqua). Red can be either more blue (purple or maroon) or more yellow (orange). Yellow can be made more red (orange) or more green (chartreuse).

Chroma (Intensity, Richness, Saturation, Grayness). Chroma refers to a color's level of intensity and richness. It moves along the spokes that radiate outward from the central gray axis of the color tree. Weak, washed-out colors with the least chroma are at the core of the color tree, and highly chromatic colors that are rich, vibrant, and intense are at the outer edge.

When the eye sees a colored object, that object is absorbing all of the light except for the color that it appears to be. A red ball appears red because the ball absorbs all of the colors in the light shining on it except for the reds. In contrast, a black object absorbs almost all light, while chrome absorbs none.

Plotting Color. Plotting color is the process of identifying in a graphic way the relationship between the paint and the vehicle based on value, hue, and chroma (Figure 14-2). This knowledge will help the painter become familiar with the refinishing evaluation process.

Plotting is not an exact science. It only has meaning when it is compared to the color on the vehicle.

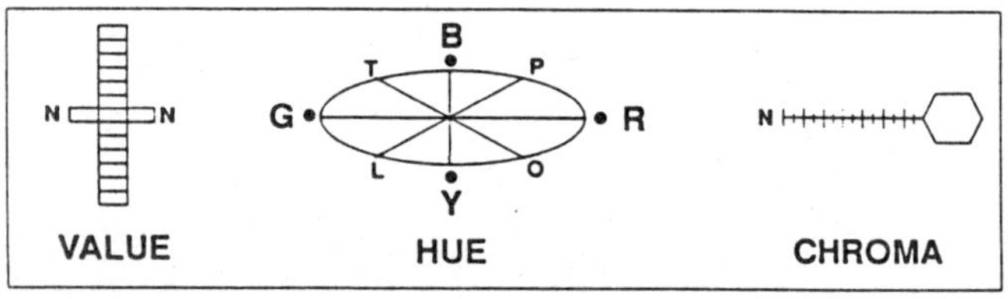

METALLICS ONLY

SIDE TONE VALUE VEHICLE IS	☐ LIGHTER	☐ DARKER
SIDE TONE HUE VEHICLE IS MORE	☐ BLUE ☐ GREEN	☐ RED ☐ YELLOW

COLOR: Light Yellow

CODE	MOVEMENT	SCALE	TOTAL
402	WHITE	336.0	336.0
452	YELLOW	49.0	385.0
451	LT. YELLOW	13.5	398.5
457	YELLOW/RED	9.0	407.5
405	BLACK	11.5	419.0
465	BINDER	3.71	790.5
485	BALANCER	184.5	975.0

FIGURE 14-2 Chart for plotting color

What is important is that the painter understands and applies the plotting system, and that he/she recognizes the differences between changes in value, hue, and chroma.

Metamerism. Different light sources have different effects on paint, causing it to appear to change colors. Known as *metamerism*, this phenomenon exists because a light source is made up of varying levels of colored light. A given paint formula can contain a blue that is not noticeable in daylight but becomes evident when viewed under mercury vapor lights. This is because the amounts of blue light differ between daylight and mercury vapor light. This is the reason it is possible for a refinishing job to match the vehicle color under shop lighting conditions, and then show up as a less than perfect match when the vehicle is viewed in natural daylight. The color match seen under shop lighting conditions can vary depending on the type of lights used. For example, incandescent lights tend to give paint a red cast; fluorescent lights could give paint a yellow or blue cast, depending on the type of fluorescent in the light; and both cool white light and soft white light can alter the appearance of paint color.

Sometimes the refinish paint formula is not made of the same pigments as the OEM finish. Even though the refinish paint might match the OEM finish under daylight or simulated daylight, it could look totally different under another light source. Paint manufacturers try to choose refinish formulas that minimize this problem and make matching easier. That is why it is important to use only the tints called for in the formula.

There are acceptable variances in color that occur at both the refinishing and OEM levels. In other words, paint can fall to either side of the color standard and still be acceptable, as evidenced by a vehicle finished in standard blue; the color applied at the factory could be a bit on the green side. Thus, there can be a perceptible difference in the two finishes even though they are officially the same color. But no matter what the reason for color variance, the painter must match the only standard that really matters—the vehicle itself. For better or worse, the vehicle is the standard.

PRELIMINARY STEPS

The importance of the preliminary steps (Figure 14-3) cannot be overstated. All of the preparation done now will affect what is to come. Basically, the preliminary steps begin with a mental plan of the entire refinishing job. Thinking any task through before rushing

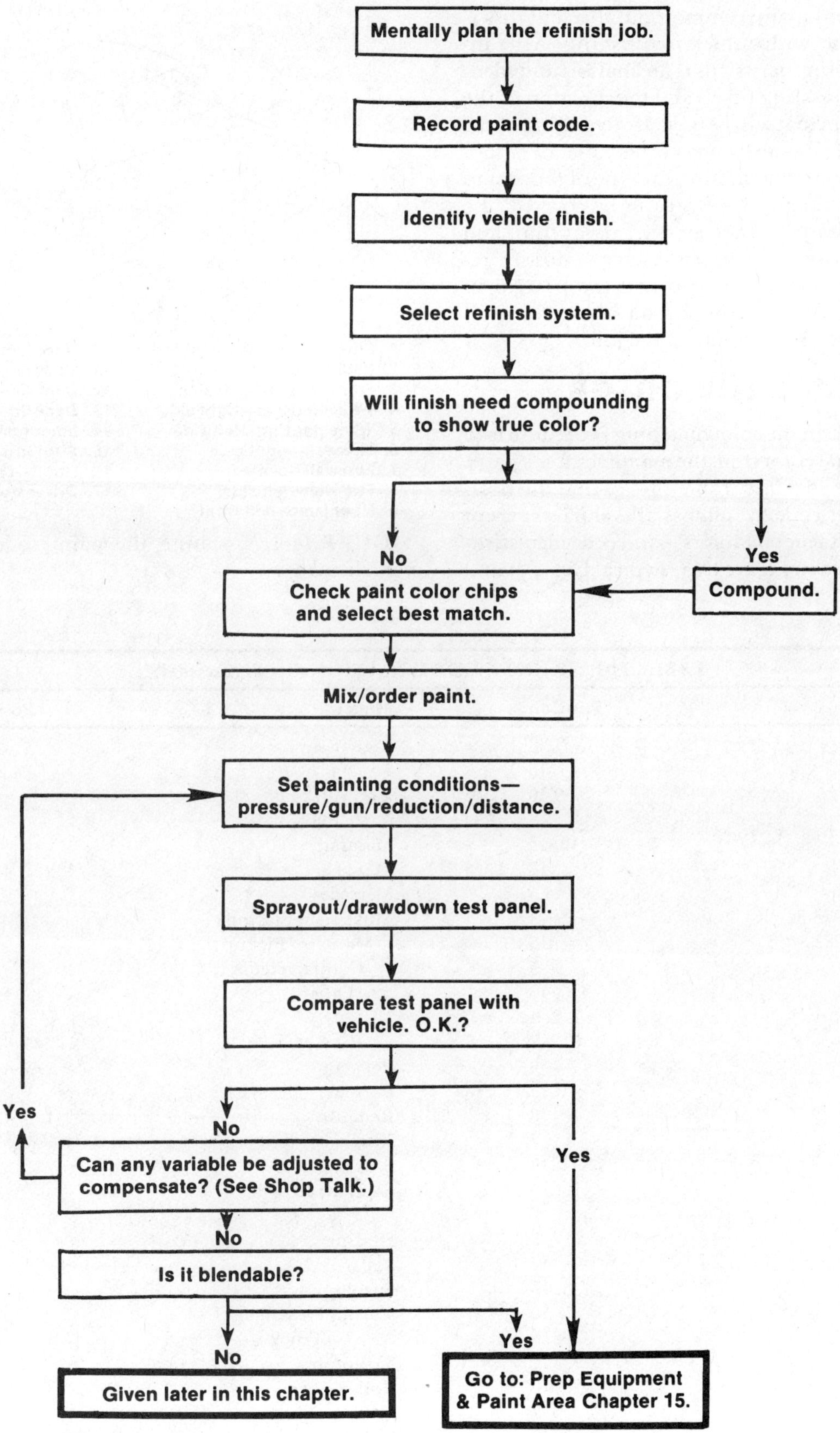

FIGURE 14-3 Preliminary steps of a refinishing job

into it helps avoid many common pitfalls, and this is certainly the case with vehicle refinishing. Also included in the preliminary steps is an analysis and identification of the existing finish and the selection of the refinishing system that will be used. If any tinting of the paint is required, it is also done at this time.

The conditions determining the type of topcoat to be used are: the extent of the area to be covered, the extent of deterioration of the existing paint film, and whether or not the vehicle has been previously repainted. The type of paint used and the process in which it is applied are very important factors that will determine the success or failure of the job.

IDENTIFYING THE COLOR

The first step in the color matching procedure is to obtain the original color from the manufacturer's paint code. Use Figure 14-4 and Table 14-1 to find the position of the paint code on almost all vehicles, except General Motors. General Motors' paint code identification label is shown in Figure 14-5. Typical Chrysler and

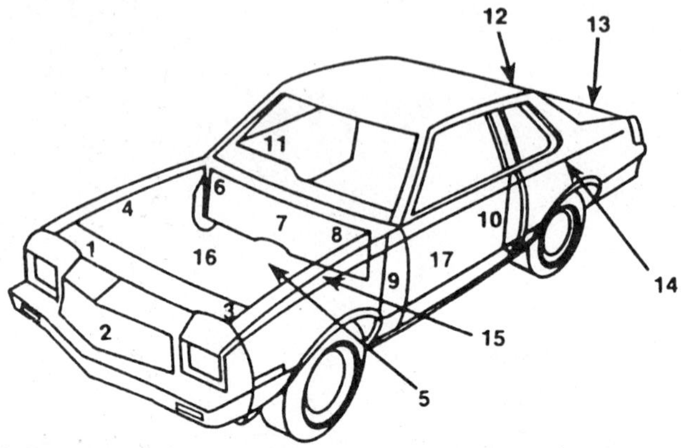

1. Front support—right side
2. Radiator support—center
3. Front support—left side
4. Wheelhousing—right side
5. Wheelhousing—left side
6. Fire wall—right side
7. Fire wall—center
8. Fire wall—left side
9. Door jamb—left front
10. Door jamb—left rear
11. Door jamb—right front
12. Deck lid—right side
13. Deck lid—left side
14. Inner rear compartment
15. Strut tower—left side
16. Hood—underside
17. Driver door panel pocket

FIGURE 14-4 Locating the paint code on various vehicle makes

TABLE 14-1: PAINT CODE LOCATION (See Figure 14-4)

Model	Position	Model	Position
Acura	9	Jaguar	1, 9
Alfa Romeo	4, 13	Lancia	4, 8, 13
AMC	9,10	LUV 1972-80	7
Arrow	3	1981-82	2
Audi	12, 13	Masarati	2, 4
Austin Rover	17	Mazda	1, 2, 3, 4, 6, 8
BMW	4,5	Mercedes	2, 7, 9
Challenger 1978-82	3	Mitsubishi Starion	7
1983	16	Montero/Pickup	3
Champ	5	Cordia/Tredia	4
Chrysler	3, 5, 16	Others	1, 2, 3
Chrysler Imports	1, 2, 4	Nissan	1, 3, 4, 6, 8, 15,*
Citroen	4, 6, 7, 8	Opel—GM	1, 2, 3, 4, 5, 8
Colt 1974-82	5	Peugeot	2, 3, 4, 5, 8
1983-84	3	Porsche	9
Colt Vista	16	Renault	1, 3, 4, 5, 8
Conquest	7	Rover	1, 3, 4, 5
Courier	3, 4, 10	SAAB	5, 6, 8
Daihatsu	1, 6, 7	Sapporo 1978-82	3
Datsun	2	1983	16
Dodge D50	3	Sterling	9
Ferrari	12	Subaru	2
Fiat	4, 8, 13	Suzuki	7, 11
Ford	10	Toyota Passenger	7, 8, 14
Ford Europe	2, 6, 8	Truck	4
GM Imports	2, 12, 13, 14	Triumph	5, 6, 9
Honda	8, 10	Volkswagen	2, 11
Hyundai	6, 7	Volvo	6, 7, 8
Isuzu	2, 10	Yugo	12

*Under right front passenger seat

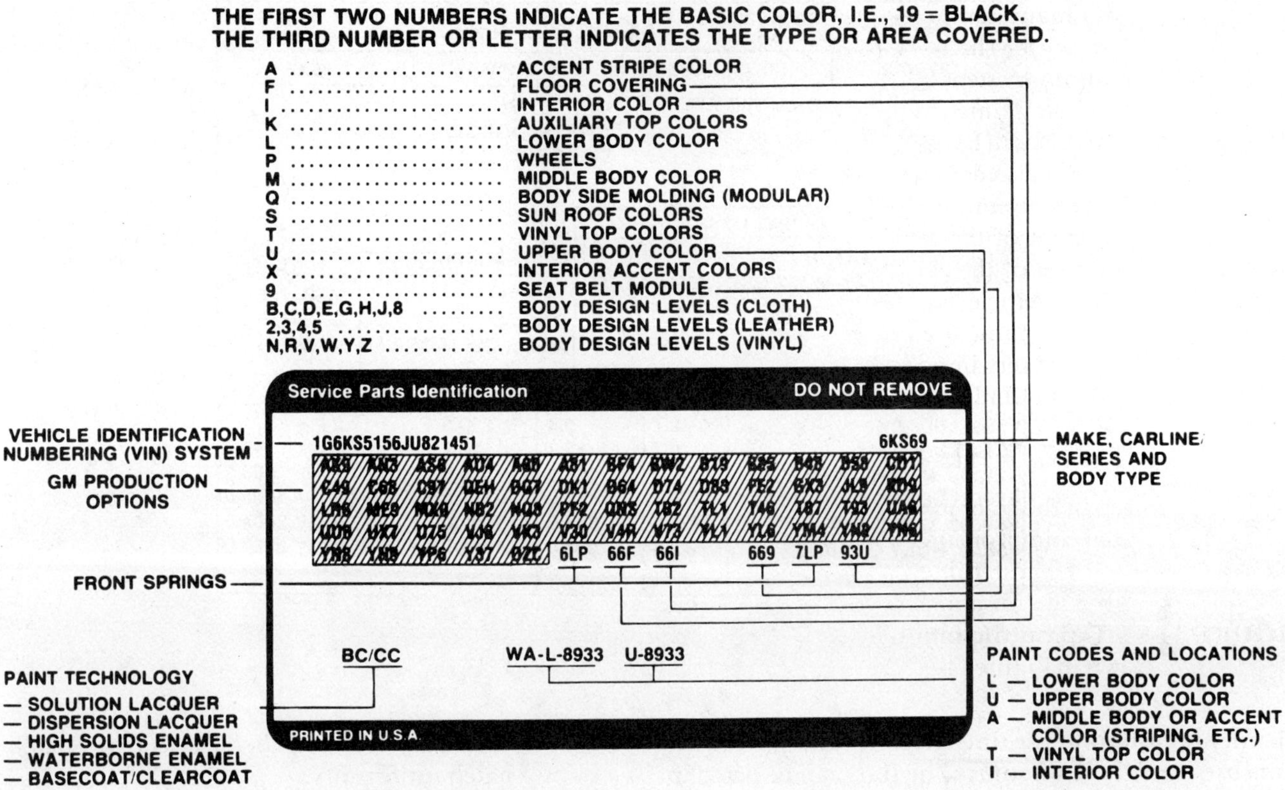

FIGURE 14-5 GM paint code identification label

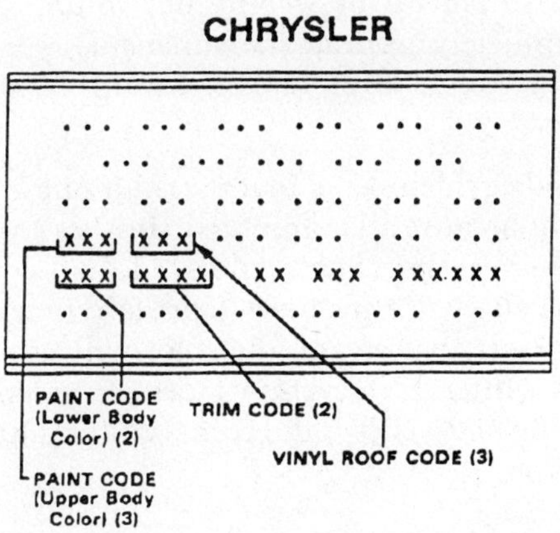

FIGURE 14-6 Color information found on Chrysler's body identification plate

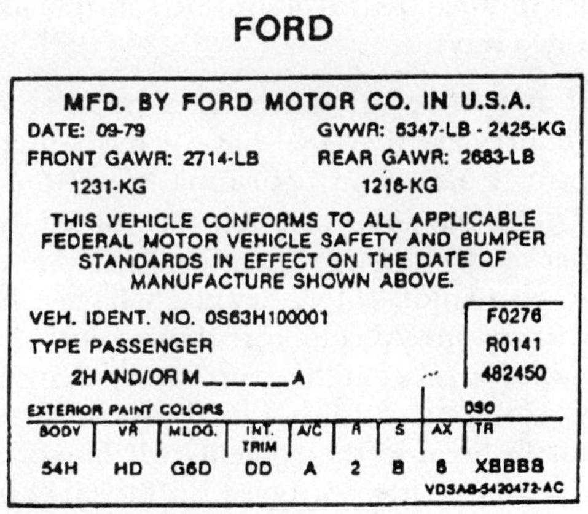

FIGURE 14-7 Color information found on Ford's body identification plate

Ford body identification plates, with color information, are shown in Figures 14-6 and 14-7.

Most collision repair and refinishing shops have an aftermarket refinishing color directory. This book contains color chips and color information for almost all makes and models of vehicles. First locate the vehicle manufacturer's paint code. This permits the painter to identify the color chip next to it. As a double check, it is wise to compare the color chip with the vehicle color, for there is always the chance that the vehicle has been repainted with a different color.

When tinting, always try to find the formula for the paint and understand what it means. The formula will be available from the mixing system or from the local paint jobber. Take some time to understand what tinting bases the formula contains. Generally, if a tint is present in a very small quantity, a little will go a long way. Also use the paint manufacturer's literature to

35 -34755-E	VERSION No. 1	ACRYLIC ENAMEL	EFF. 08-JAN- 89 (REVISED)
COLOR CODE	**COLOR NAME**	**COLOR USAGE**	**YEAR**
DR5	GRAPHIC RED	CHRYS. PLYM.	89-91
PR5		CHRYS DODGE	89-92
		DODGE TRUCK	89-91

SPECIAL INFORMATION
BLENDABLE MATCH
NEW BASECOAT-MUST USE NEW LABEL

REGULATORY	MATCHES
LEAD	34 -34755
PHOTO	37 -34755
MSDS CD: ACR-LL	33 -34755

COLOR NUMBER	1 PINT	1 QUART	2 QUART	3 QUART	GALLON
F5B81 BLACK	2.7	5.4	10.8	16.2	21.6
F5W80 WHITE	6.2	12.4	24.7	37.1	49.4
V2V109 MIXING CLEAR	83.5	167.0	334.0	501.0	668.0
F5P92 MONASTRAL VIOLET	232.0	464.0	928.0	1392.0	1856.0
V6V300 DRIER	250.5	501.0	1002.0	1503.0	2004.0
F5E86 MOLY ORANGE	470.5	941.0	1882.0	2823.0	3764.0

FIGURE 14-8 Typical tint formula

identify the effects of the tints in the formula. Note the tints used and their amounts. The tints can be listed in units, grams, or ounces, depending on the system (Figure 14-8).

If the color match is correct, order the topcoat by color stock number. Refinish suppliers supply topcoat colors in two ways:

- If it is a recent model or a popular color, chances are they will have it ready-mixed in pint, quart, and occasionally gallon cans. These ready-mixed colors are called *factory packaged.*
- If it is an older color, they might have to mix it. *Custom-mixed* colors are those colors that are mixed to order at the paint supply distributor. Custom-mixed colors can always be identified easily because the contents of the container must be written or typed on the label by the person who mixed the paint.

In recent years, most of the major automobile paint manufacturers have made a color mixing system available to painters. With such an "intermix" system, it is possible to mix thousands of colors at a savings of up to 35 percent of the cost of factory-packaged colors.

With this service, inventory can be reduced, keeping only the fast-moving current colors on hand. Yet, the service takes up little space. It is fast and easy to mix a color that matches the manufacturers' standards.

- Formulas are scientifically developed by the laboratory from base tints, which are rigidly controlled by plant chemists to assure batch-to-batch uniformity.
- Formulas come completely indexed on microfiche, along with directions for mixing exact quantities.
- Measurements by weight, not volume, ensure completely accurate measurements.
- Pour-spouts allow precise control of material flow.

It must be remembered that the color might not be exactly right because all automotive finishes gradually change color over time. Some colors fade lighter, others go darker. Yellow, for instance, fades fairly rapidly. If the yellow fades to a cream, the color will usually go lighter and whiter. If the yellow fades to a green, composed of blue and yellow, the color will go bluer and usually darker.

SHOP TALK

With metallic finishes, air pressure, spray gun adjustment, distance, speed, agitation, and reduction can affect finish appearance. Be sure to check and adjust these before deciding to tint.

DETERMINING THE OLD FINISH

Before planning any refinishing job, it must be determined what type of paint was used originally. Methods of doing this include the following:

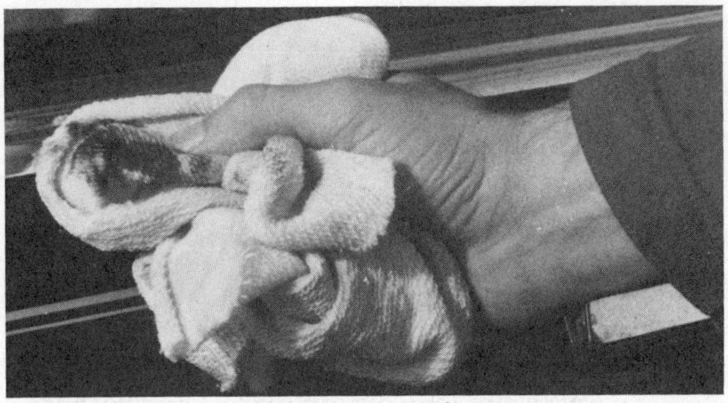

FIGURE 14-9 The paint film left a mark on the rag, indicating an air-dry type paint.

1. *Visual Inspection Method.* If the skin texture near a body line is rough or if what is known as a "polished texture" appears when the paint is rubbed, it indicates that a polish-type paint was used.
2. *Solvent Application Method.* The paint film is rubbed with a white cloth soaked in lacquer thinner to determine the extent to which it will dissolve. If the paint film dissolves and leaves a mark on the rag (Figure 14-9), it is some type of air dried paint. If it does not dissolve, it is either an oven-dried or a two-part reaction type paint. Acrylic urethane paint film will not dissolve as easily as an air-dried paint, but sometimes the thinner will penetrate sufficiently to blur the paint gloss.
3. *Heat Application Method.* First, the area is wet sanded with No. 1000 grit paper to dull the paint film. The area is then heated with an infrared lamp. If a gloss returns to the dulled appearance, the paint is acrylic lacquer.
4. *Hardness Method.* All paints do not dry to the same hardness. Generally two-part reaction and oven-dried paints dry to a harder film than air-dried paint. The most common method of determining paint film hardness and thickness is with either an electromagnetic thickness gauge (Figure 14-10A) or a mechanical thickness gauge (Figure 14-10B). Follow the manufacturer's instructions for the proper use of these gauges.

SELECTING THE REFINISHING SYSTEM

There have been great changes in automobile finishes in the last few years. The successful painter must keep up with new developments. The painter should be familiar with the uses and properties of current OEM and repair finishes. Each refinishing system has specific requirements for use and application.

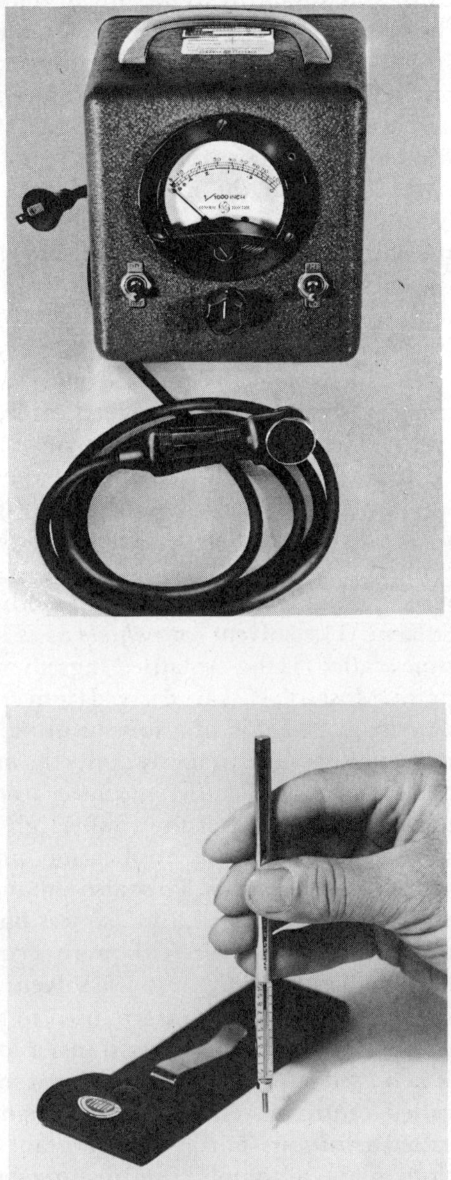

A

B

FIGURE 14-10 (A) A meter for measuring paint thickness electromagnetically; (B) Tinsley mechanical thickness gauge

Contents of Paint

Basically, automotive paints are composed of four components: pigment, binder, solvent, and additives.

- *Pigment.* This is one of two nonvolatile film forming ingredients (that part which remains in the dried film) found in paint. It provides color and durability to the finish. It also gives the paint the ability to hide what is underneath. In addition, pigment can also improve strength and adhesion, change gloss, and modify flow and application properties.

- *Binder.* The other nonvolatile film-forming paint ingredient, the binder holds the pigment in liquid form, makes it durable, and gives it the ability to stick to the surface. The binder is the backbone of paint. It is generally made of a natural resin (such as rosin), drying oils like linseed or cottonseed, or a man-made synthetic resin (such as methyl methacrylate, polyurethane, polystyrene, polyvinyl chloride, and so forth). The binder dictates the type of paint to be produced because it contains the drying mechanism. Binder is usually modified with plasticizers and catalysts. They improve such properties as durability, adhesion, corrosion resistance, mar-resistance, and flexibility.
- *Solvent.* The solvent (or *vehicle* as it is sometimes called) is the "volatile" ingredient of the paint. Most solvents are derived from crude oil. The main function of a solvent in paint is to make it possible to properly apply the material, and it must be of sufficient power to dissolve the binder portion of the film. High-quality solvents improve the application and film properties of topcoats. They also enhance gloss and minimize paint texture, so less buffing is needed. They also help with more accurate color matches. In addition to the solvent already in paint, diluted solvents are used to give the paint a viscosity that makes it easier to apply. When used with lacquer, the diluted solvent is called a *thinner.* When used with enamel, it is called a *reducer.* This is an important distinction in the automobile refinishing business. Remember that a lacquer is *thinned,* while an enamel is *reduced.* The ratio of pigment/binder/solvent that makes up paint is illustrated in Figure 14-11. A recent development in auto refinishing is the use of volatile organic compound (VOC) coatings. These are coatings that have a lower than normal solvent content and higher pigment/binder ratios. They are commonly used with high-volume, low-pressure spraying systems.
- *Additives.* With the extensive changes made in paint technology during the last decade, additives have become more common in many paint materials. While comprising no more than 5 percent of the paint at most (and usually much less), additives perform a variety of vital functions. Some speed up drying and heighten gloss; others slow drying; and still others lower gloss. And some perform a combination of functions—they eliminate wrinkling, provide faster final drying, prevent blushing (a milky, misty look), and improve chemical resistance. Those additives that speed up cure and heighten gloss are often referred to as *hardeners.* Those that slow down drying are called *retarders.* And those that lower gloss are called *flatteners.*

Types of Body Finishes

The exterior of an automobile is painted with a basic finish of either lacquer or enamel. Other terms such as nitrocellulose, acrylic, alkyd, polyester urethane (polyurethane), or acrylic urethane sometimes can be used to describe an automobile paint. However, these products are still either lacquer or enamel. The difference between the two is how they dry and cure. Lacquers dry by the evaporation of solvents. Enamels and urethanes dry by solvent evaporation and by cross linking paint molecules.

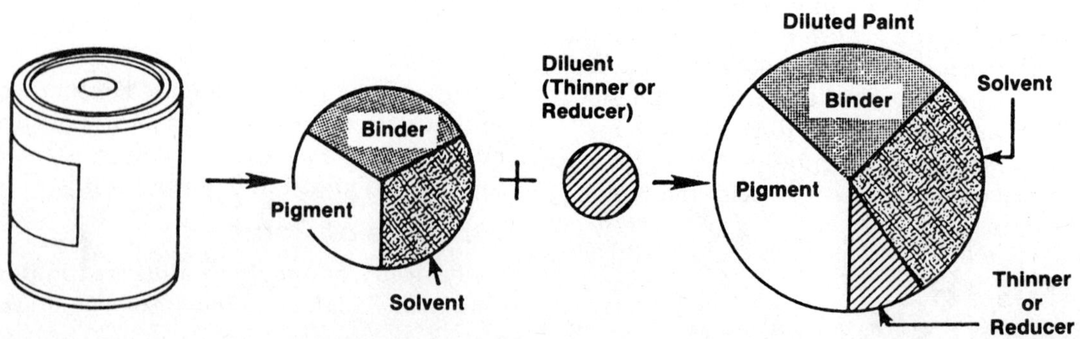

FIGURE 14-11 Ratio of pigment/binder/solvent that makes up paint

WARNING

Polyurethane and acrylic urethane enamels, as well as the two-part or catalyzed type finishes (acrylic and alkyd), use isocyanates as a hardener. Isocyanates are highly reactive and can become health hazards if not handled carefully. Because isocyanates react with many common substances, they must be used, stored, and disposed of with care. Many paints and painting products contain isocyanates and, therefore, share the same potential health hazards regardless of brand.

The OEM coating processes commonly in use today are:

- *Single Stage:* High solids lacquer or baked enamel
- *Base Coat/Clear Coat (two stage):* Baked polyester or acrylic base coat; baked acrylic enamel or acrylic urethane clear coat
- *Tri-coat (three stage):* Baked polyester or acrylic base coat, baked mica intermediate coat, and baked acrylic urethane clear coat

These finishes are baked in huge ovens to shorten the drying times and cure the paint. The oven is important to reflow the paint; this assures leveling and a mirrorlike gloss without need for compounding. Refinish lacquers, on the other hand, are made to air dry and often require compounding.

There is a difference between the lacquer used by auto manufacturers and that used in the collision repair and refinishing shop. Both are fast drying, but refinish lacquers are made to air dry and often require compounding to achieve the proper gloss (Figure 14-12).

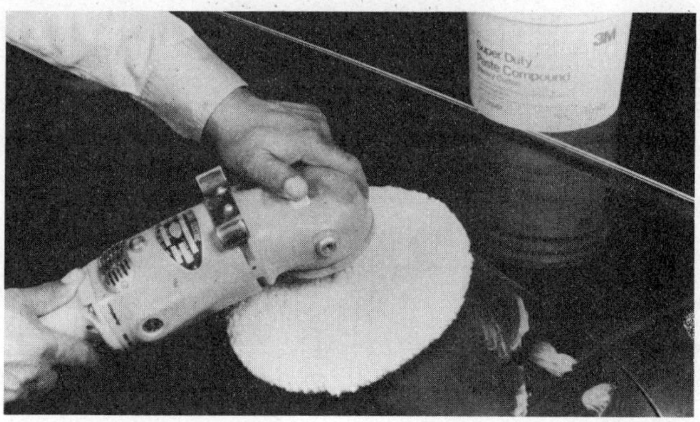

FIGURE 14-12 Refinish lacquers require compounding by hand or machine to bring out the gloss.

Since high-temperature ovens are generally not used in the shop, painters need slow drying thinners to achieve maximum leveling and gloss.

The thermosetting acrylic enamel finishes used by the vehicle manufacturers must also be baked in ovens in order to dry and cure properly. Very few, if any, collision repair and refinishing shops have these large bake ovens, but even if they did, the high temperature might damage a vehicle's upholstery, glass, or wiring. Therefore, shops require an easy-to-apply, yet durable enamel that will:

- Air dry without baking.
- Dry when baked at low temperatures.

More information on the various lacquers and enamels used as topcoats in the automobile industry is given in Chapter 16. As shown in Figure 14-13, in addition to the topcoat, there are other materials that play an important role in the refinishing operation.

It is very important for the painter to choose one system and stick with it for the entire repair. Paint chemistry is very different from manufacturer to manufacturer. The old practice of using one brand of paint, another brand of thinner, and so on, no longer works. Refinishing systems should not be mixed. Each manufacturer will include specific directions for the use of their product, such as amount of reduction. The manufacturer also gives directions for working with different types of substrates. Some might require an adhesion promoter or sealer. In any case, follow the manufacturer's directions for the use of each product line.

Types of Refinishing

With a wide number of refinishing materials available, the painter uses them to do only one of three types of jobs:

- Spot repairs
- Panel repairs
- Overall refinishing

Spot Repair. This type of repair is often called "ding and dent" work because the damaged area is usually small. Scratches or a breakdown of the substrate due to rust or corrosion are also candidates for spot repair.

Spot repair generally involves:

- Minor body repair.
- Metal conditioning.
- Application of primer-surfacers and primer-sealers.
- Topcoat application that is blended into the old finish surrounding the repair.

Functional Nomenclature	Resin/Pigment Ratio	Primary Objective	Use and Features
Primer	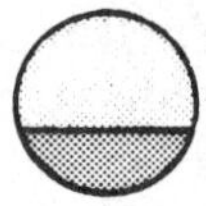	Adhesion and anticorrosion	Applied directly to panel surface
Primer-Surfacer	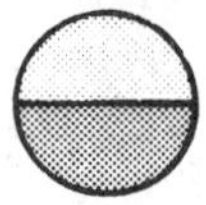	Adhesion, anticorrosion, and smoothness	Intermediate between primer and surfacer. Applied to metal surface or over primer.
Putty		Filler	To smooth out rough spots
Primer-Sealer	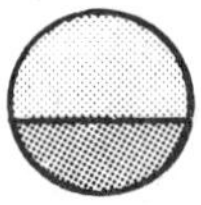	Prevent absorption of topcoat	Intermediate between surfacer and topcoat. Prevents absorption of topcoat and helps avoid troubles arising from old (previous) paint.
Topcoat		Upgrades external appearance	Gives color, gloss, and body to help upgrade merchandising value

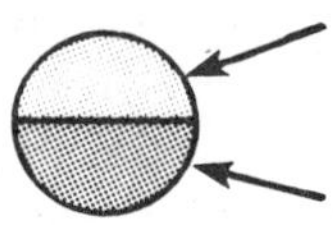

Resin Portion (The more resin, the better gloss and less absorption)

Pigment Portion (The more pigment, the better covering of rough spots and smoothness, but less gloss and more absorption.)

FIGURE 14-13 Functions of automobile paint materials

Panel Repair. Basically, panel repair work uses the same technique as spot repair. The exception is that the area covers an entire panel or panels of the vehicle (door, hood, fender, and so on). Also, the surrounding areas are masked off and the match is made at the panel joints (Figure 14-14). As in spot repair, any imperfections within the panel must be featheredged and the primer(s) applied before the topcoats go on. Sometimes when the color match is especially difficult, it is recommended that the new color be blended into the adjoining panels.

Both spot and panel repairs could be considered small area finishing. Oftentimes the painter must decide whether to do a spot or a panel repair. If the area to be refinished is large, panel repair can be easier.

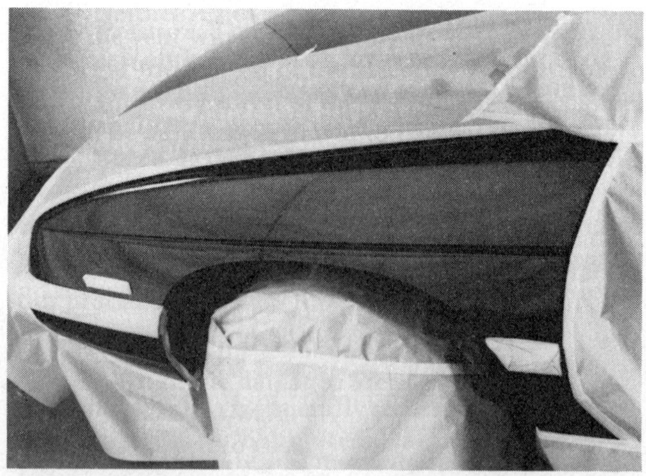

FIGURE 14-14 The areas surrounding the panel to be refinished are masked off with masking paper and tape.

Overall Refinishing. Overall refinishing is just what it says: the whole vehicle is painted. Some reasons for refinishing the entire vehicle include:

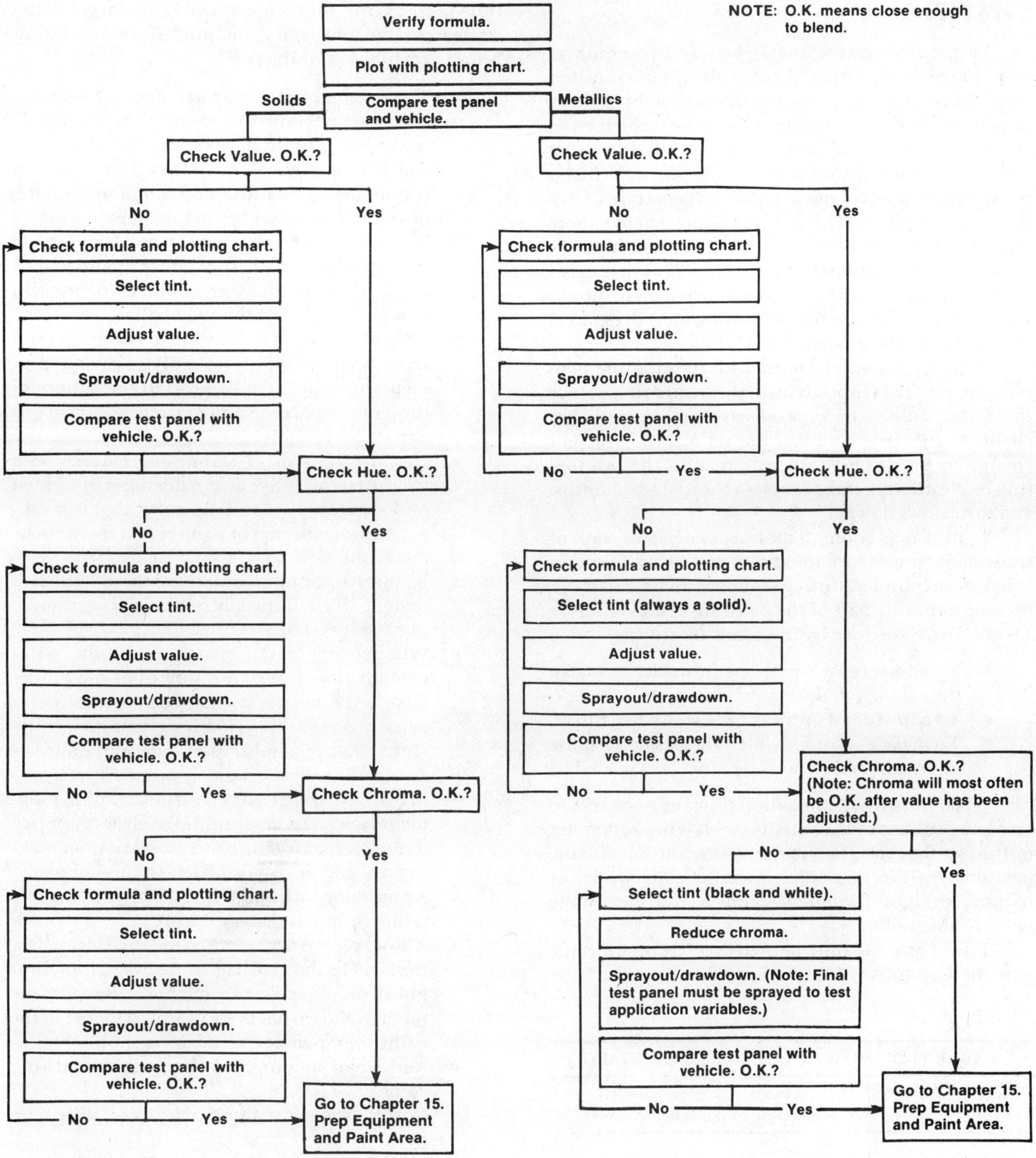

FIGURE 14-15 The I-CAR tinting system

- Size and/or number of spots to be refinished
- Dull, cracked, or worn finish
- Vehicle owner wishes to change color

There are also specialty paint shops that do mostly overall painting, and custom shops that develop glamour finishes for custom, antique, and classic vehicles.

TINTING

Every color weathers in its own way, depending on its pigment composition. A color also weathers differently depending on its care and the part of the country in which the vehicle is driven. In general, vehicles that are in a garage most of the time change less; those that are rubbed and polished a lot change more. Vehicles parked under trees change depending on the type of tree spray or drippings. Those in the South change more rapidly due to increased ultraviolet radiation from the sun. Meanwhile, those in industrial areas or in areas of the country where there are natural chemicals in the air, such as alkali flats, change depending on the chemical to which they are exposed.

If a refinish material made with the same pigmentation as the OEM finish is used, the weathering will be the same, unless the type of care or exposure has changed. In a few months, the refinished area will change to the same weathered color as the original finish. Weathering is the fastest in the first few months, and then slows down.

Tinting may be one of the least understood tools of finish matching. Once the theory of tinting is understood and a complete tinting sequence is laid out, it can become a routine part of the painting process (Figure 14-15). There are three basic reasons for tinting:

- To adjust color variations in shades to match the color from the manufacturer.
- To adjust color on an aged or weathered finish.
- To make a color for which there is no formula or for which there are no paint codes available.

To do finish matching, the painter must be able to recognize colors as they actually are. It is important not only to see the color that is to be worked on, but also the overtones within that color, including the shades of darkness or lightness and the richness or fullness of the color (Table 14-2).

Tinting should only be used as a last resort. If the color of the refinish job varies from the original vehicle finish, check the following possible reasons for the mismatch and understand the guidelines for tinting before deciding to tint the paint:

- The original finish may have faded. Check the paint on unexposed areas such as doorjambs or under the deck lid or hood to determine if the finish has faded. If this is the case, you can restore the paint's luster by compounding the original finish well beyond the repair area.
- Was the wrong color used? Check the vehicle manufacturer's paint code and the paint manufacturer's stock number of the color used to make sure that it is the right color.
- The pigment and/or metallic flakes might not have been mixed thoroughly. Leaving pigment, metallic flake, or pearl in the bottom of the can could cause a mismatch, so be sure to agitate thoroughly.
- Has the amount of thinner or reducer been measured carefully? Overreducing will lighten or desaturate a color. Remember that it is easy to add more thinner or reducer, but it cannot be taken out.
- Be sure to clean and compound the old finish to remove all chalking and oxidation before making a color comparison.
- When using a test panel, allow the paint enough time to dry. Be sure to allow proper flash and dry times for each coat, because paint usually gets darker as it dries. If using a lacquer clear coat, remember that compounding the clear will make the paint appear darker. If testing a base coat/clear coat finish, color judgment cannot be made until the clear is applied to the base coat. Further information on making a test panel is given later in this chapter.
- An agitator cup is preferred when spraying metallics or pearls (Figure 14-16).
- Allow color to dry before trying to adjust it. Dry time can be shortened by using heat lamps, heat guns, or other drying methods. Be sure the method chosen has been approved for use in the collision repair and refinishing shop.
- Add small amounts of tints when adjusting color.
- Correct the most noticeable color difference first.
- Vary the spraying technique. The three shades shown in Figure 14-17 were sprayed out of the same gun. The section on the top was sprayed dry with the gun quite far away. The section in the middle was sprayed with the gun at the normal distance from the panel. The bottom

TABLE 14-2: HOW COLORS ARE DESCRIBED

Lighter—Darker (Value)	1. Direct look (panel to panel) 2. Side angle look (panel to panel)
Cast differences (Hue)	1. Redder 2. Bluer 3. Greener 4. Yellower
Cleanliness (Chroma)	1. Grayer (dirtier or more muddy) 2. Brighter (cleaner appearance)

FIGURE 14-16 Agitator paint cup.

FIGURE 14-17 Matching colors with a spray gun: (A) lighter than normal—dry spray; (B) standard color—normal spray; and (C) darker than normal—wet spray.

section was sprayed wet with the gun held close.

In other words, the application technique could cause the color to vary. Painters who spray wet end up with a darker color than those who spray drier, especially with metallics. Paint applied in heavy, wet coats dries slowly, giving metallic flakes more time to sink deep into the paint film. Because these "buried" flakes reflect less light, the paint looks darker. The reverse happens when paint is applied in lighter coats that dry faster. Table 14-3 lists the application factors that must be considered when applying color.

TABLE 14-3: APPLICATION FACTORS THAT AFFECT THE FINISH COLOR

Make Adjustments to:	To Make the Application: Darker	Lighter
Application (spraying technique)	1. Hold gun closer to the job 2. Decrease gun speed 3. Don't wait as long between coats	1. Hold gun farther from the job 2. Increase gun speed 3. Wait longer between coats
Equipment	1. Use larger fluid tip 2. Decrease fan width 3. Decrease fan pressure 4. Increase volume of material flow	1. Use smaller fluid tip 2. Increase fan width 3. Increase fan pressure 4. Decrease volume of material flow
Shop conditions	1. Decrease temperature	1. Increase temperature

Oddly enough, a mismatch in a panel repair will usually show up more than a mismatch in a spot repair—even though the spot repair is smaller. That is because a panel—such as a vehicle door—has a distinct edge. And the repair, obviously, cuts off at that edge. Mismatched panels—as in the case of front and rear doors—will be right next to each other and will show a sharp contrast.

A spot repair, on the other hand, is performed by blending the repair into the surrounding area. In spot repairs, the first coat is applied to the immediate area being repaired. Subsequent coats extend beyond this area gradually. Finally, a blend coat extends beyond the color coats. Thus, if there is a slight mismatch, the blend coat and the last color coat will allow enough show through of the old finish to make the color difference a gradual one.

Analyzing Color

Through various application techniques, value, hue, and chroma can be adjusted so that the painter can achieve a good match. To determine what must be done, the color must first be analyzed to determine whether it is too light or too dark, looking at the finish from an angle and head-on. Then the painter must check the hue to see if the color sprayed is redder, bluer, greener, or yellower than the original finish. Before adjusting begins, the color must be checked to see if the finish just sprayed has more or less chroma than the original. The sprayed portion must always be allowed to dry before any analysis and adjustments are made.

Adjustments for value rely primarily on shop conditions, spraying techniques, and solvent usage. Other variables include the amount of paint applied, the air pressure at the spray gun, and the amount of color added to the mix.

Once the value has been adjusted, tinting might be required to get the right hue. Each color can only vary in hue in two directions.

- Colors that are either greener or redder in hue include:
  Blues — Purples
  Yellows — Beiges
  Golds — Browns
- Colors yellower or bluer in hue are:
  Greens — Blacks
  Maroons — Grays or Silvers
  Whites
- Colors yellower or redder in hue are:
  Bronzes — Reds
  Oranges
- Colors bluer or greener in hue include:
  Aqua — Turquoise

Charts and manuals available from paint manufacturers can help make the decision on what tint color to use for the appropriate system. Once the color necessary to correctly adjust the hue is determined, the amount must then be calculated, utilizing the least amount necessary to effectively change the color. The color must be thoroughly mixed; the gun triggered to clear the chamber; and then a small panel can be sprayed, allowed to dry, and checked against the original color.

After the color is correct in value and hue, the chroma is adjusted. Attempting to make a color brighter at this point will throw off the previous two corrections. To gray the finish, a wet coat must be sprayed followed by a coat sprayed at half trigger at a slightly greater distance using a small amount of white mixed with a very small amount of black.

There are three angles the painter should use to look at a vehicle's finish when determining whether a color adjustment is necessary (Figure 14-18):

- *Face or Head-On.* Viewing the finish from an angle that is perpendicular to the vehicle.
- *Near-Spec.* Viewing the finish from an angle just past the reflection of the light source.

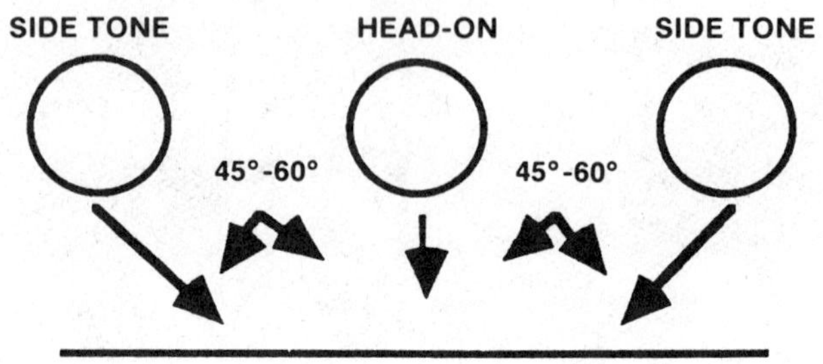

FIGURE 14-18 Angles the painter should use to determine whether or not a color adjustment is necessary.

- *Side Tone.* Viewing the finish at an angle of less than 45 degrees.

The color of the refinished area should be the same as the rest of the vehicle. If there is any difference, the color should be corrected until it is the same when viewed from all three angles.

Tinting can be divided into two categories: major color tinting and minor color tinting. Major color tinting consists of making up a color for which there is no mixing formula available. Find a color chip as close as possible to the desired color and look at that formula. Break the formula down into percentages. For example, gold metallic is 45 percent coarse metallic, 20 percent sparkle metallic, 15 percent gold toner, 15 percent yellow gold, 3 percent soft white, 2 percent soft black. Using these percentages, make only half of a can; stir and tint to match the desired color.

Minor color tinting is used to adjust a color in a given repair situation to achieve an acceptable color match. To achieve this, each major paint supplier has a basic color tinting kit, a set of instructions, and a tinting guide with color chips that is available to the painter (Figure 14-19). Any painter, once familiar with the tinting information that is available, should be able to do minor color tinting to achieve top quality color matches. For more information on tinting colors and/or for a copy of a company's tinting guide, contact your local paint jobber or a paint manufacturer's sales and service representative.

Matching Metallic Finishes

In most cases, solid color finishes—when properly prepared, thinned, and sprayed—will provide a good color match. The matching of metallic (polychrome) finishes, however, is probably the most skillful operation the painter has to perform. There are more vehicles on the road with metallic finishes, which means there are more metallic repairs to be made.

The reason that difficulty is experienced in matching metallic colors is that they are made with a pigment

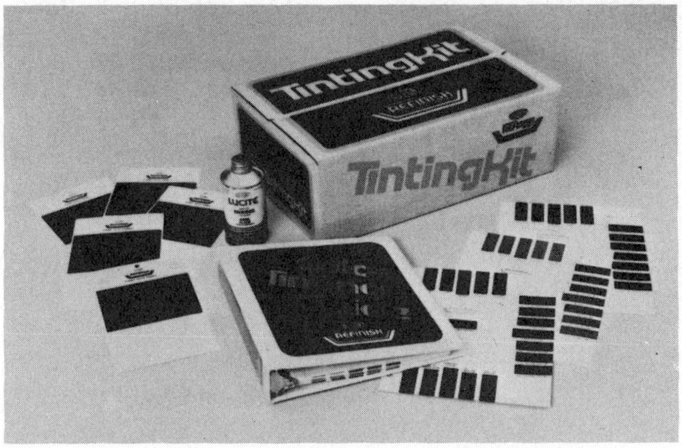

FIGURE 14-19 Typical tinting kit

and aluminum flake in the binder that allows light to penetrate beyond the surface of the paint film. When viewing a metallic color at right angles or perpendicular to the surface, this is the face of the color (Figure 14-20A). When viewing it at a 45 degree angle or less, it is the pitch or side tone of a color (Figure 14-20B).

As the position of metallic and/or pigment particles changes in the paint film, the color shade of the metallic finish changes accordingly. Each metallic particle is like a tiny mirror. That is what changes the

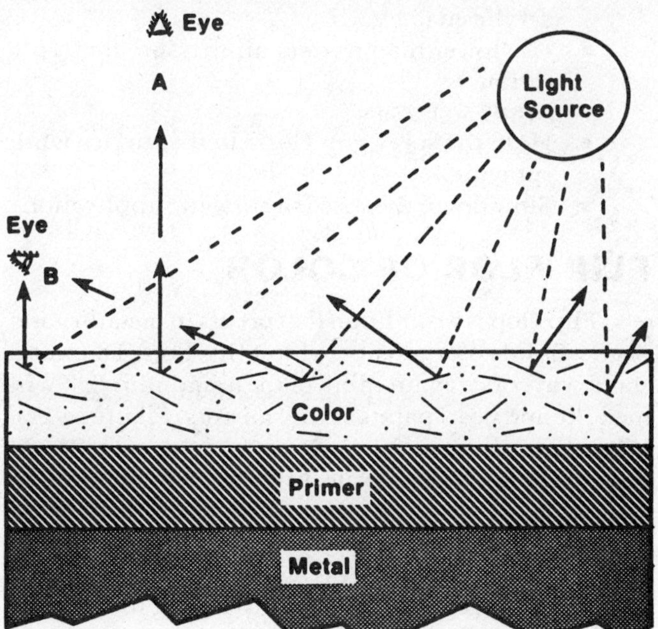

FIGURE 14-20 Metallic color construction showing (A) face and (B) side tone appearance of color

appearance of metallic colors when viewed from different angles. Metallic color also appears to be different when viewed under different kinds of light, such as daylight, shade, sun, or artificial light.

Hints on Color Tinting

Here are some additional tips that might prove helpful when tinting:

- Check the color in daylight as well as artificial light; it might not look the same in both. When a refinish color matches in one light but not in another, it often indicates that the same pigments were not used in the refinish material as in the original finish. The paint manufacturer is usually very careful to use the same pigments in the refinish material that were used in the original finish. To control the uniformity of colors, every batch of paint is checked for exact color match in three different lights—yellow, blue, and daylight.
- Determine what the color problem is and select the proper tinting colors. Do not use mixed colors, because they probably have overcasts of the wrong shades. Adjust the color to make the hue redder, greener, bluer, or yellower.
- To understand the strength of a tinting color, put a few drops on a quart lid with a few drops of white, then mix the two and make a finger smear on the lid.
- Do all tinting systematically.
  —Use a measuring device such as those described in Chapter 15.
  —Keep a list of tinting colors used.
  —Keep a record of the amounts used.
- The color formula is a help in tinting because it shows the original toners used.
- Be sure to mix all tinting colors thoroughly before using; also thoroughly mix a tinting color every time any color is added.
- Add tinting colors in small amounts because it is very easy to overtint. Keep in mind that more color can be added but it cannot be taken out.
- Do not tint the whole can of paint at one time. Do it gradually with small samples until a color match is achieved.
- Be conservative when tinting near the limits of the color range. Correct the most noticeable color differences first.
- Use caution when adding white to metallics or pearls and always use low-strength whites.
- Stay with the same pearls and metallic flakes used in the formula.

- To check the color, spray out a small amount and allow it to dry. Compare it to the panel to be matched. When it is possible, an old panel from the vehicle is good to use because it can be masked in the center for an excellent comparison.
- Keep the tint on the light side until the final match is determined. Do not make a final judgment of the color match while the color is still damp because it will change until it is completely dry.
- Once the color is tinted close enough, complete the repair. Many times that last "just a little bit closer" is the thing that ruins a blendable match.

Adjusting Value

The value of a metallic color is increased when the aluminum flakes lie nearly horizontally at the top of the paint film and hide most of the pigment particles underneath them (Figure 14-21A). The flat positioning of the tiny metallic mirrors reflects a higher than normal amount of light. This causes lighter than normal color shades of the metallic finish. A match to these shades can be achieved by one or a combination of the following procedures:

- Use more solvent than is called for on the label directions.
- Use medium or fast drying thinners to thin acrylic lacquer. Use only fast drying solvents to reduce acrylic enamel.
- Use a higher than recommended air pressure at the gun.
- Apply light to medium coats with adequate flash time between coats.
- Hold the spray gun farther from the surface while spraying.
- Increase the speed of spraying applications.

The value of a color is decreased when the aluminum flakes are positioned nearly perpendicular to the surface (Figure 14-21B). This effect can be achieved on the vertical sides of a vehicle the same as on the horizontal panels and is not simply the settling of the flakes.

Value can be decreased by the use of very slow solvents and extra wet coats, or by one or more of the following:

- Reduce the color with 10 to 25 percent less solvent than the label indicates.
- Use 9 parts slowest drying solvent to 1 part retarder for color reduction in temperatures of 65 to 85 degrees. Usually a 100 percent reduction works well for dark shades of lacquer metallic colors; 25 percent reduction works well for acrylic enamels.
- Use lower than normal air pressure for application.
- Apply wet coats.
- Hold the spray gun closer to the surface while spraying.
- Slow down the speed of spraying applications.

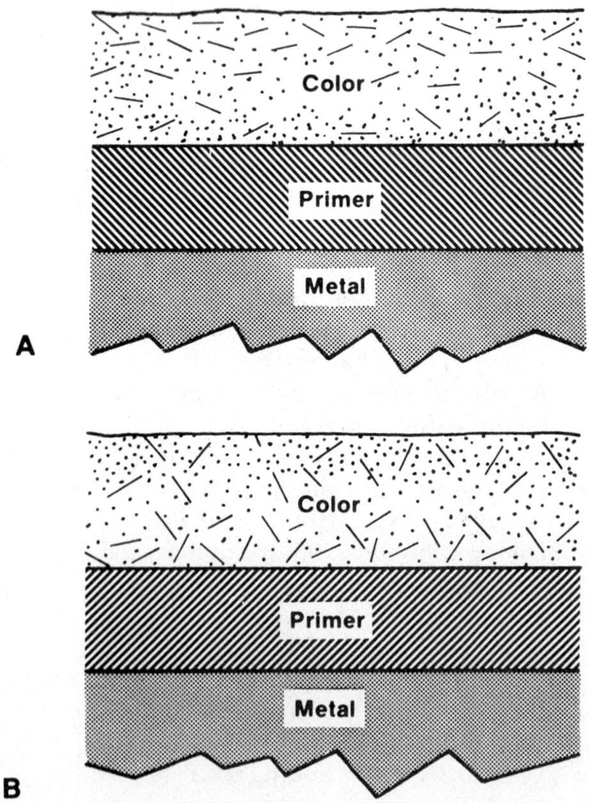

FIGURE 14-21 (A) Increased and (B) decreased value of metallic color. Note the position of the aluminum flakes in each case.

FLIP-FLOP OF COLOR

Flip-flop is a condition that occurs in metallics and pearls. Solid colors do not exhibit flip-flop. To understand why, once again think of the aluminum flakes in metallic and pearl paints as tiny mirrors (Figure 14-22). When light shines onto a mirror's surface, it is reflected back in exactly the opposite direction. Similarly, when light shines onto the aluminum flakes in a vehicle's metallic finish, it is reflected back in a straight line in the opposite direction. The brightness caused by this reflection is greatest when you are standing in the angle from which the light is reflecting. The brightness could drop off rapidly as one moves away from this angle, depending on the pigmentation of the paint. The direction and intensity of the light being reflected back through the paint film is the flip-flop phenomenon.

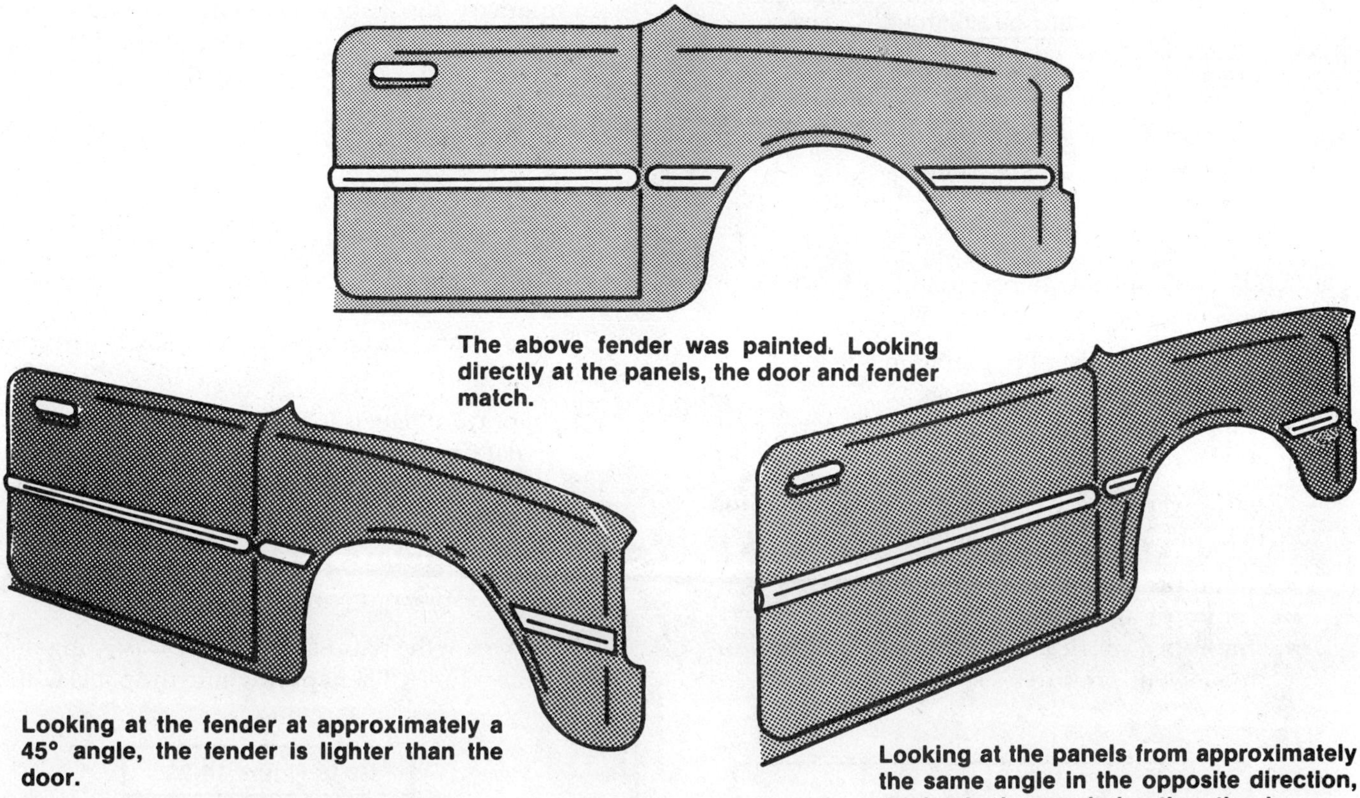

FIGURE 14-22 Evidence of flip-flop

The first approach to correcting the problem is to adjust the spraying technique to compensate for this effect. Spraying the fender a little wetter will slightly darken the appearance when looking directly into the panel. When viewed from an angle, the resulting appearance is lighter. This occurs because the aluminum flakes are positioned flatter and deeper in the paint film.

Spraying the panel slightly dryer reverses the effect, giving a lighter appearance when looking directly at the panel. This is because the aluminum flakes are closer to the surface. The result is a darker appearance viewed at an angle, as light becomes trapped. Adjusting the spraying technique should be used to correct minor conditions of flip-flop.

If spray techniques cannot correct the condition, the addition of a small amount of white will eliminate the sharp contrast from light to dark when the surface is viewed at various angles. The white acts to dull the transparency, giving a more uniform, subdued reflection through the paint film. Care should be taken when adding white since the change occurs quickly. Once too much white is added, recovering the color match becomes virtually impossible.

When confronted with an extremely difficult flip-flop condition, the best method involves adding white, plus blending the color into the adjacent panels. When blending, extend the color in stages. With acrylic lacquer, for example, use a recommended clear blending material, and thin the color when spraying farther into the adjacent panel. The blending agent protects the metallic edges, eliminating the halo or bright edge where the blend ends.

To increase the value, it is only necessary to increase the amount of metallic in relation to the solid color tints (Figure 14-23). Decreasing value involves a little more work. When the vehicle is darker than the test panel, decrease the value as illustrated in Figure 14-24.

A good painter must know how to handle metallic colors. They are very sensitive to the solvents with which they are reduced and the air pressure with which they are applied. Metallic colors are also affected by a number of variables. A variable is one of the spraying conditions such as temperature, humidity, and ventilation, or a part of the spraying process such as amount of reduction, evaporation, speed of solvents, air pressure, and type of equipment. If a painter is to get good color matches, it is important to understand how certain variables affect the shades of metallic colors.

Variables are divided into two categories: positive and negative. Positive variables are those things that a

	ON FACE	ON SIDE TONE	ACTION
Vehicle is ____ compared to test panel.	Lighter	Lighter	Add finest metallic in formula.
	Lighter	Darker	Add coarsest metallic in formula.

FIGURE 14-23 Methods of increasing value

painter does to duplicate the original finish, which in turn results in a good color match. They are:

- Slowness of solvent evaporation.
- Wetness of color application.
- Proper spraying technique and the correct air pressure (see Chapter 15).

Negative variables are those that cause the color shades to be off. Most common are:

- Improper reduction.
- Improper agitation.
- Improper application; primarily too high or too low air pressure.

In summary, the shades of metallic colors are controlled by:

- Choice of solvents.
- Color reduction.
- Air pressure.
- Wetness of application.
- Spraying techniques.

TEST PANELS

There are three basic types of test panels used by painters:

1. Sprayout panels for solid and base coat/clear coat finishes.
2. Letdown panels for tri-coat finishes.
3. Drawdown bars.

Sprayout Panels

To make a sprayout panel, proceed as follows:

1. Prepare the paint as though it was going on the vehicle. Then spray it onto the panel with

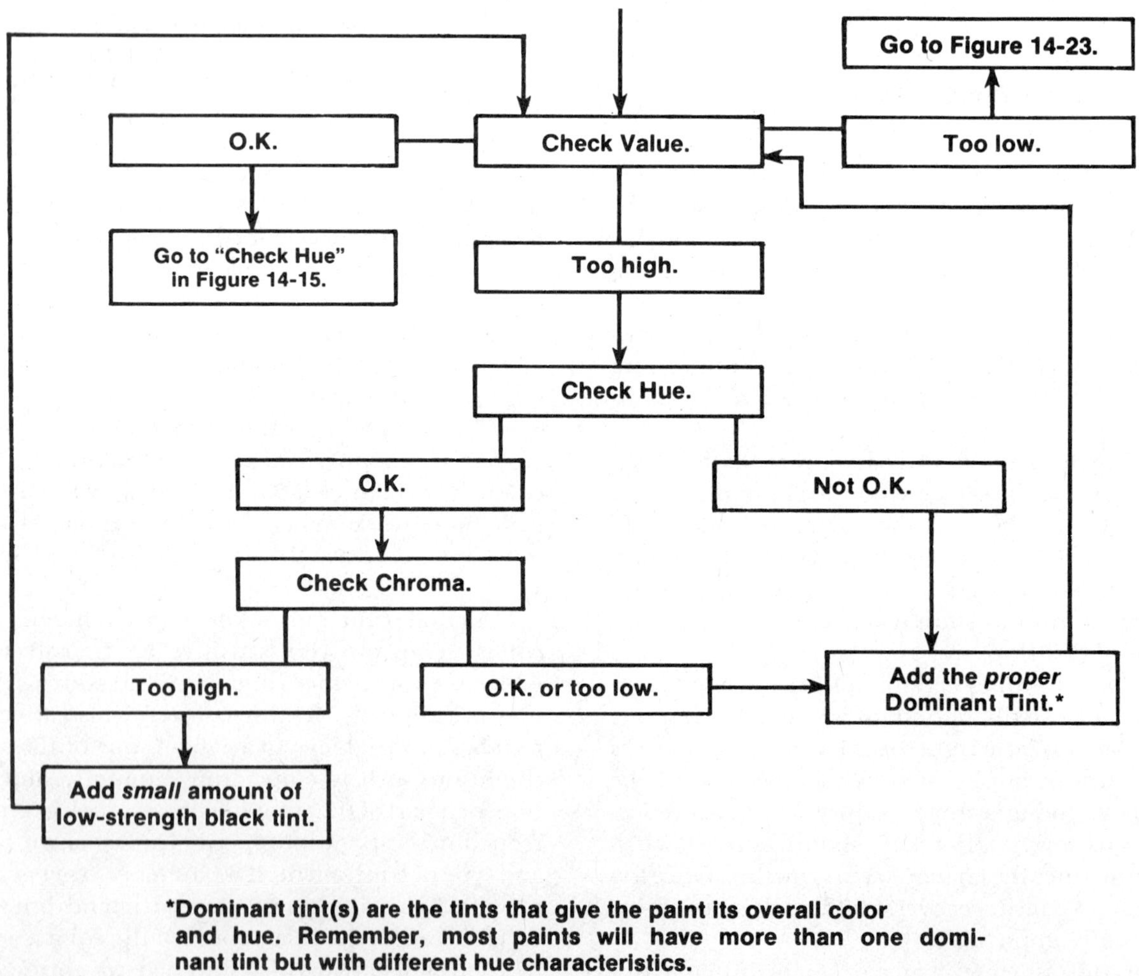

FIGURE 14-24 Methods of decreasing value

the same technique that is to be used when painting the vehicle.

2. Apply as many coats as necessary to achieve full hiding. Allow correct flash time between coats.
3. Clear coat the panel if it is a base coat/clear coat finish.
4. Wait until the panel is dry before using it to check the color match.

Letdown Panels

Letdown panels for tri-coat finishes are essential to determine the correct amount of midcoat color that must be applied to achieve an exact color match. The midcoat color is the most critical portion of the tri-coat finish. Because gun pressure, reduction, and spray technique can affect the finished job, the extra time spent spraying one or more letdown panels will be repaid many times over with a job that satisfies the customer and does not come back later.

Here is how to make a letdown panel for a tri-coat finish.

1. Prepare the letdown panel with the same primers being used on the vehicle. If a primer-sealer is going to be used, apply the primer-sealer to the letdown panel also. Generally, a light color primer-surfacer (or primer-sealer) is preferred for tri-coat finishes.
2. Apply the base coat color to hiding, using the same air pressure and spray pattern that will be used on the vehicle. Duplicating the actual spray techniques when preparing the letdown panel is an important point. Make sure not to vary the procedures.
3. After the panel has dried, divide it into four equal sections (Figure 14-25). Next, mask off the lower three quarters of the panel, exposing the top quarter.
4. Apply one coat of mica midcoat color over the top quarter of the panel.
5. After the mica coat has flashed, remove the masking paper and move it down to the middle of the panel, exposing the top half.
6. Apply another coat of mica midcoat color over the exposed top half of the panel.
7. After this second coat has flashed, remove the masking paper and move it down to expose three quarters of the panel.
8. Apply another coat of mica midcoat color over the exposed three quarters of the panel.
9. After flashing, remove the masking paper entirely.
10. Apply a fourth coat of mica midcoat color. As always, spray the coating in the same way as would be done on the vehicle.
11. After the entire letdown panel has dried, mask off the panel lengthwise this time.
12. Apply the manufacturer's recommended number of clear coats to the exposed half.

If the refinishing is being done over a flexible surface, products and approaches vary. Some refinishing products require additives in the clear coat, others call for flex agents to be used in the base coat, midcoat, and clear coat. Still others do not require an additive or flex agent in any of the three coats. It may be necessary to spray a separate letdown panel with the flex agent if additives are used in any of the three layers because there could be slight color variations caused by the additives. Pay attention to the manufacturer's requirements when tri-coating flexible surfaces.

Drawdown Bar

The drawdown bar is a precision tool with a machined blade designed to give an even paint film distribution. To use it, a black and white sprayout panel is taped to a perfectly flat surface, such as an aluminum

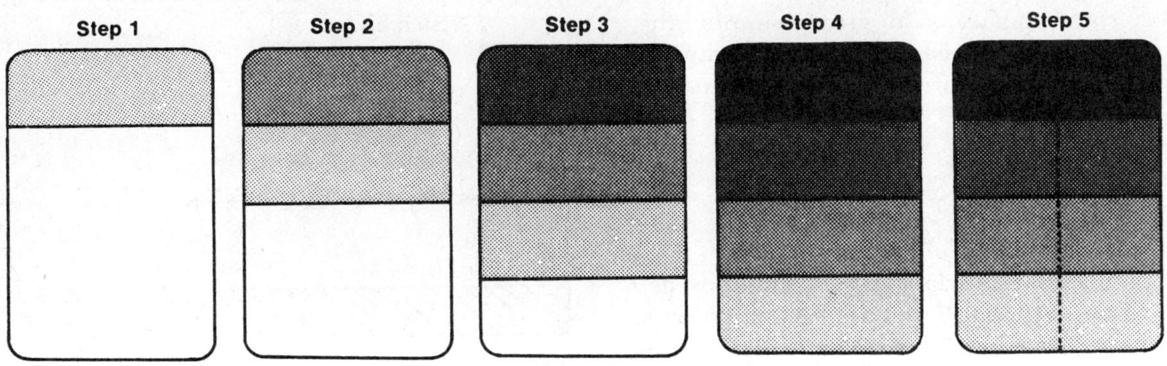

FIGURE 14-25 This example shows the step-by-step creation of a letdown panel. A letdown panel is always recommended when spraying tri-coat finishes.

clipboard. Paint is distributed onto the panel and the drawdown bar is drawn through the paint, spreading it into a uniform paint thickness. After the first drawdown flashes, the process is repeated until the black and white on the sprayout panel is no longer visible through the paint film. Then the sprayout panel is compared to the vehicle.

When working with a base coat/clear coat, make the drawdown over a strip of clear film (placed over a check hiding panel so the painter knows when hiding is achieved). Turn the clear film over and the paint will have a base coat/clear coat appearance. Do not use reduced material with a drawdown bar.

If additional tinting is needed, another drawdown should be made on another sprayout panel. This allows the painter to see the direction in which the tint is moving by comparing the panels to each other. When the desired color is achieved, the material can be sprayed and further adjusted through gun techniques and refinishing skills. The drawdown bar eliminates time, material, and labor needed to adjust the color.

SHOP TALK

When using a tinting system such as the one described in this chapter, the painter should, for the first few times, learn what effect each step or adjustment has on the paint. Once the system is fully understood, then he/she can judge the steps of the procedure. Look for the most obvious problem first, fix it, then check the other adjustments.

Keep good records of each job that is tinted. A copy of the tinting process should be kept with the job paperwork. The vehicle make and model year should be attached to the test panels and then stored. These can prove valuable if the vehicle returns for another repair. A complete tinting record of a given job will speed up any future refinishing.

Always follow the manufacturer's recommended procedures and guidelines for working with their colors. These recommendations can greatly simplify the task of choosing which tint base to use. Some manufacturers have very detailed information on tinting and the use of tinting systems.

OTHER FACTORS AFFECTING COLOR

Acid rain and industrial air pollutants, as described in Chapter 10, can have an adverse effect on a vehicle's finish. These problems are particularly acute in certain areas. How can one tell if a vehicle has been damaged by acid rain? Typically, the damage looks like water droplets that have dried on the paint and caused discoloration. Sometimes the damage appears as a white ring with a clear, dull center. Severe cases of acid rain damage show pitting. Discoloration varies depending on the color. For example, acid rain damage to a yellow finish might appear as a white or dark brown spot. Medium blue can have a whitened look. White might be discolored pink, and medium red turned purple.

Metallic finishes are damaged because the acidic solution reacts with the aluminum flakes and etches away the finish. A new finish is more easily damaged than an aged finish. Lacquers and uncatalyzed enamel finishes are most susceptible to acid rain damage, followed closely by catalyzed enamels.

Clear coat finishes add a layer of protection against acid rain, so late-model vehicles with two- and three-stage finishes are less susceptible to damage. A clear coat protects the paint pigments from discoloration, but it is still possible for acid rain to create a peripheral etch, or ring, on the clear coat.

The procedure for restoring acid rain damage varies depending on the level of the damage. The following steps outline repair procedures according to the level of damage, as illustrated in Figure 14-26. When the problem has been corrected, stop at that stage. Remember that polishing or compounding removes part of the original finish and, thereby, reduces its overall life.

If the surface damage is like that shown in Figure 14-26A, proceed as follows:

1. Wash with soap and water.
2. Clean with wax and grease remover.
3. Neutralize the area by washing with baking soda solution (1 tablespoon baking soda to 1 quart of water) and rinse thoroughly.

If the damage is embedded in the surface coat (Figure 14-26B), proceed as follows:

1. Follow the cleaning and neutralizing steps listed above.

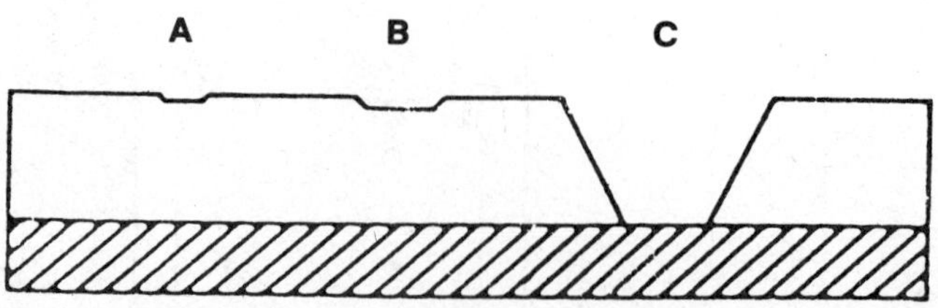

FIGURE 14-26 Levels of acid rain damage

2. Hand polish the problem area. Inspect and continue, if necessary.
3. Buff with a polishing pad. Inspect frequently and remove as little of the original finish as possible.
4. Use rubbing compound. Inspect and continue, if necessary.
5. Wet sand with No. 1500 or No. 2000 grit sandpaper and compound. If damage is still visible, repeat with No. 1200 grit. Do not use grits coarser than 1000.

If the damage is through to the undercoat (Figure 14-26C), proceed as follows:

1. Follow the cleaning and neutralizing steps used for surface damage.
2. Sand with No. 400 to No. 600 grit sandpaper.
3. Reclean and reneutralize prior to refinishing.

Generally speaking, damage from industrial fallout is caused when small, airborne particles of iron fall and stick to the vehicle's surface. The iron eventually eats through the paint, causing the base metal to rust. Sometimes the damage is easier to feel than see. Sweeping a hand across the affected area will likely reveal a gritty or bumpy surface. Rust-colored spots might be visible on light-colored finishes.

The steps for repairing damage caused by industrial fallout are similar to those used when repairing acid rain damage, but with the following exception. After washing the vehicle, treat the repair area with a "fallout remover," a chemical treatment product made especially for industrial fallout damage. Do not buff the damaged area before removing the treatment; buffing will only drive the particles into the paint surface. In addition, if the particles break loose and become lodged in the buffing pad, deep gouges can occur.

REVIEW QUESTIONS

1. ______________ refers to the degree of darkness or lightness of a color.
   a. Chroma
   b. Value
   c. Tint
   d. Hue

2. When using the heat application method of identifying a paint finish, a gloss returns to the surface after the heat is applied. Technician A says this means the paint is an acrylic lacquer. Technician B says it means the paint is a urethane lacquer. Who is correct?
   a. Technician A
   b. Technician B
   c. Both A and B
   d. Neither A nor B

3. Technician A tints only small amounts of paint at a time. Technician B tints an entire can of paint at once. Who is correct?
   a. Technician A
   b. Technician B
   c. Both A and B
   d. Neither A nor B

4. ______________ is a condition unique to metallic and pearl finishes.
   a. Metamerism
   b. Sand scratch swelling
   c. Bleed-through
   d. Flip-flop

5. If modifying the spraying technique does not correct flip-flop, adding a small amount of ______________ will give a more uniform, subdued reflection through the paint film.
   a. solvent
   b. black
   c. white
   d. bleeding agent

6. After the preliminary steps, the next phase of the I-CAR finish matching system is ______________.
   a. prepping the vehicle
   b. prepping the equipment and paint area
   c. plotting the color
   d. selecting the refinish system

7. Which of the following statements is incorrect?
   a. Fluorescent light does not alter the appearance of a paint's color.
   b. A refinish paint can match the vehicle color under shop lighting conditions, but show up as less than a perfect match when viewed in daylight.
   c. Chroma refers to a color's level of intensity and richness.
   d. In some cases, the refinish paint formula is not made of the same pigments as the OEM finish.

8. In which method of determining paint type is the surface wet sanded to dull the paint film?
   a. solvent application method
   b. hardness method
   c. visual inspection method
   d. heat application method

9. Over time, colors can go ____________.
   a. darker
   b. lighter
   c. both a and b
   d. neither a nor b

10. On the General Motors' paint code identification label, what do the first two numbers indicate?
   a. area covered
   b. date of the original finish
   c. basic color
   d. manufacturer's tinting formula

11. ____________ is the volatile ingredient in paint.
   a. Binder
   b. Pigment
   c. Resin
   d. Solvent

12. Which of the following statements concerning tinting is incorrect?
   a. Tinting should never be used to adjust the color on an aged or weathered finish.
   b. Thoroughly mix the tinting color every time any color is added.
   c. Keep the tint on the light side until the final match has been determined.
   d. Both a and b

13. Which of the following affects the appearance of a metallic color?
   a. air pressure
   b. spray technique
   c. solvent evaporation
   d. all of the above

14. Technician A says that an indication of acid rain damage on a vehicle finish is a white ring with a clear, dull center. Technician B says that acid rain damage on a vehicle finish can look like water droplets that have dried and caused discoloration. Who is correct?
   a. Technician A
   b. Technician B
   c. Both A and B
   d. Neither A nor B

15. When making a test panel, Technician A always uses the same spray technique that will be used on the vehicle. Technician B says that it is not necessary to duplicate the spraying technique. Who is correct?
   a. Technician A
   b. Technician B
   c. Both A and B
   d. Neither A nor B

CHAPTER 15

Shop and Vehicle Preparation

OBJECTIVES

- Explain the operation of spray booths and drying rooms.
- Describe the recommended maintenance program for a spray booth.
- List the refinishing materials that are part of everyday painting operations.
- Explain the process of atomization and how it relates to spray gun operation.
- Identify the three types of spray guns and explain how each type operates.
- Measure the viscosity of paint using a Ford cup and a Zahn cup.
- Identify the various types of spray coats.
- Operate and maintain a spray gun.
- Describe the differences between conventional air and HVLP spraying operations.
- Prepare existing paint films and bare metal substrates for refinishing.
- Describe the three methods of stripping paint from metal surfaces.
- Explain how conversion coating enhances primer adhesion to bare metal.
- Determine when to apply primer, primer-sealer, primer-surfacer, sealer, and glazing putty.
- Mask various parts of a vehicle in preparation for painting, using both conventional and liquid masking procedures.

This chapter includes the important subjects of preparing the finishing equipment and paint area (Figure 15-1) and final preparation of the vehicle for refinishing. Starting first in the shop, there are a number of variables that affect the refinishing operation. These variables include the painting environment, as well as the painting equipment and equipment adjustments. They are all important. The technician must pay attention to these variables, because they can affect the quality of his/her work.

The proper painting environment must address:

- Cleanliness
- Temperature
- Light
- Compressed Air

SPRAY BOOTHS

The primary purpose of a spray booth is to provide a clean, safe, well-illuminated enclosure for painting. It

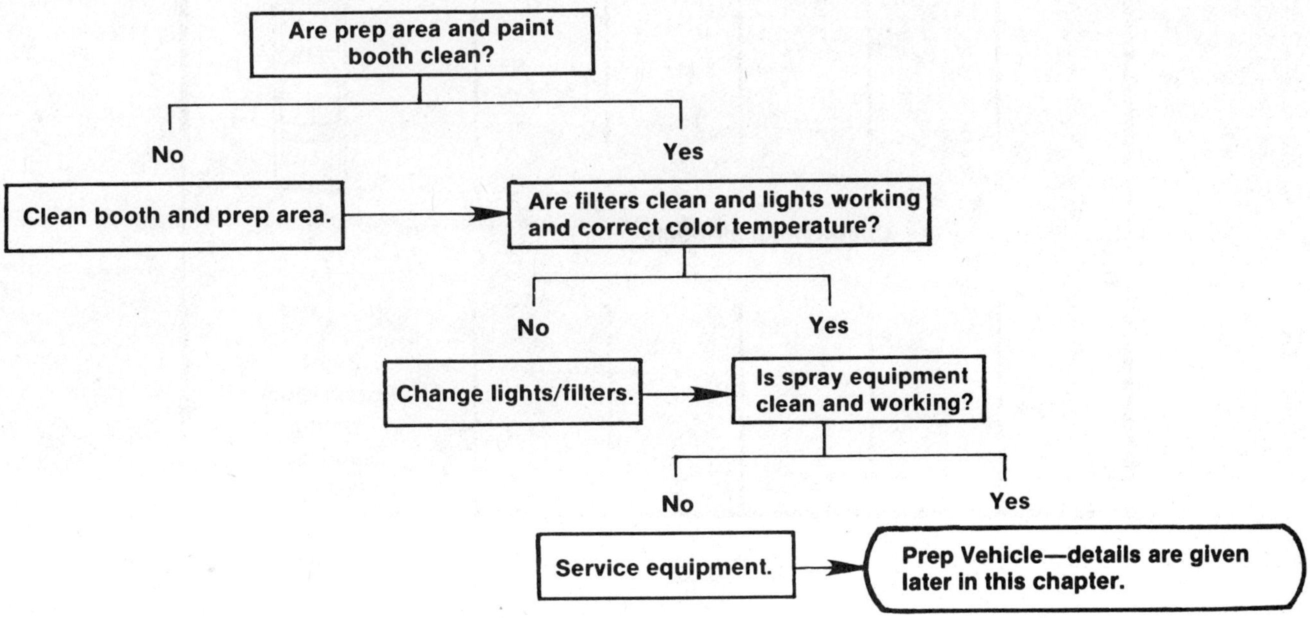

FIGURE 15-1 Steps in prepping the equipment and paint area

isolates the painting operation from dirt and dust and confines the volatile fumes created by spraying. Because colors appear different under different types of light (for instance, incandescent light tends to highlight reds), it is very important to utilize "color corrected" lights in the spray booth. This will provide a perfectly neutral light that makes accurate color matching possible.

Modern spray booths are scientifically designed to conform to federal, state, and local safety codes, not to mention those of insurance underwriters. In some areas, automatically operated fire extinguishers are required because of the highly explosive nature of the refinishing materials. The spray booth should be as far removed as possible from areas where dust and dirt are present. Therefore, it should be isolated from the mechanical and collision repair areas of the shop. This can be accomplished with partitions, walls, or a separate building arrangement. A workbench should be handy in the spray booth for reducing the paint and filling the cup. Paint storage should always be outside the booth.

When a spray booth must be located in the same room with collision repair stalls or other areas of excessive dust, the air can be drawn from the outdoors by means of an air replacement system. This arrangement greatly reduces both the number of filter changes required and the chances of ruined paint jobs.

If the volume of paint work is sufficient, a straight line work flow shop layout is recommended (Figure 15-2). Such a layout is designed for maximum efficiency of personnel and equipment. As mentioned in Chapter 1, jobs are started in the collision repair stalls in the normal manner. From this point, the work flows in a production line manner through each of the various stages, all the way to final cleanup. The cleaning preparation described later in this chapter should be done outside the booth.

After the vehicle is in the spray booth, close the booth doors tightly and tack-rag the entire vehicle before proceeding with the painting operation. The booth doors must be kept tightly closed during painting. If it becomes necessary to open a door, be sure the exhaust fan and air supply are turned off. In fact, many spray booths are equipped with door switches that automatically shut off the air supply and exhaust fan when the doors are opened. The air compressor should be outside the booth, with the air delivery pipes slanting back toward the compressor.

An air makeup or replacement system is important because air must be exhausted from spray booths con-

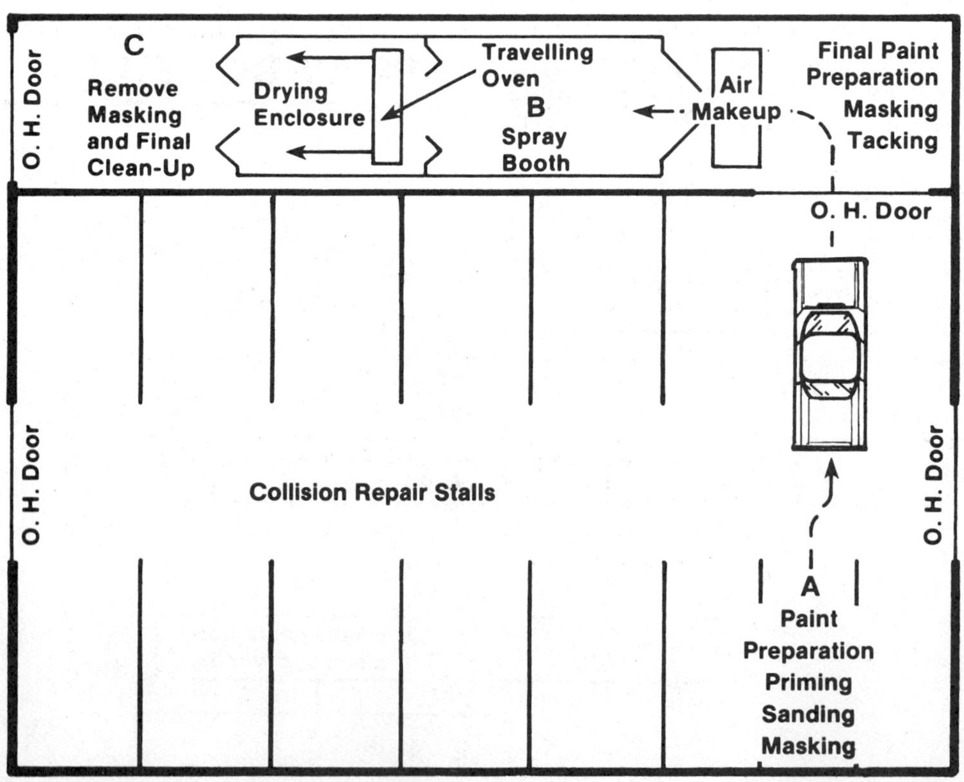

FIGURE 15-2 Typical collision repair and refinishing shop showing straight line work flow layout

tinuously during the spraying process. Under such conditions in winter, the spray area can become cold and uncomfortable, and finish problems can arise from spraying cold materials on cold vehicles in cold air. An air makeup system will provide even temperatures and clean filtered air. Some shops employ an independent air replacement system specifically designed for the spray booth (Figure 15-3). This provides clean, dry, filtered air from the outside, and even heats the air in colder weather. Replacement air can be delivered to the general shop area or directly into the booth for a completely closed system.

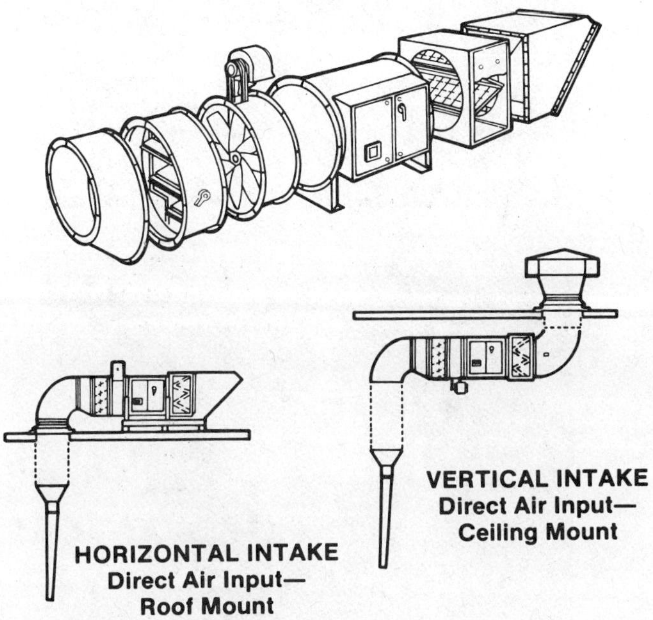

FIGURE 15-3 Air replacement is available in various configurations to fit building needs.

BOOTH DESIGNS

There are four spray booth designs in use today:

- Regular flow booth
- Reverse flow booth
- Crossdraft booth
- Downdraft booth

Both the regular and reverse flow booths were once considered standards in spray booth construction. However, they have been replaced to a great degree in recent years by crossdraft and downdraft types.

The downdraft system is the single most popular air movement system employed today. The downward directional flow of air from the ceiling of the booth to the exhaust pit creates an envelope of air passing by the surface of the vehicle. This process of taking clean, tempered air and directing it downward past the vehicle prevents contamination and overspray from settling on the freshly painted surface of the vehicle (Figure 15-4). But just as important, this air movement helps to remove toxic vapors and harmful overspray from the breathing zone of the painter, providing a safer working environment (Figure 15-5).

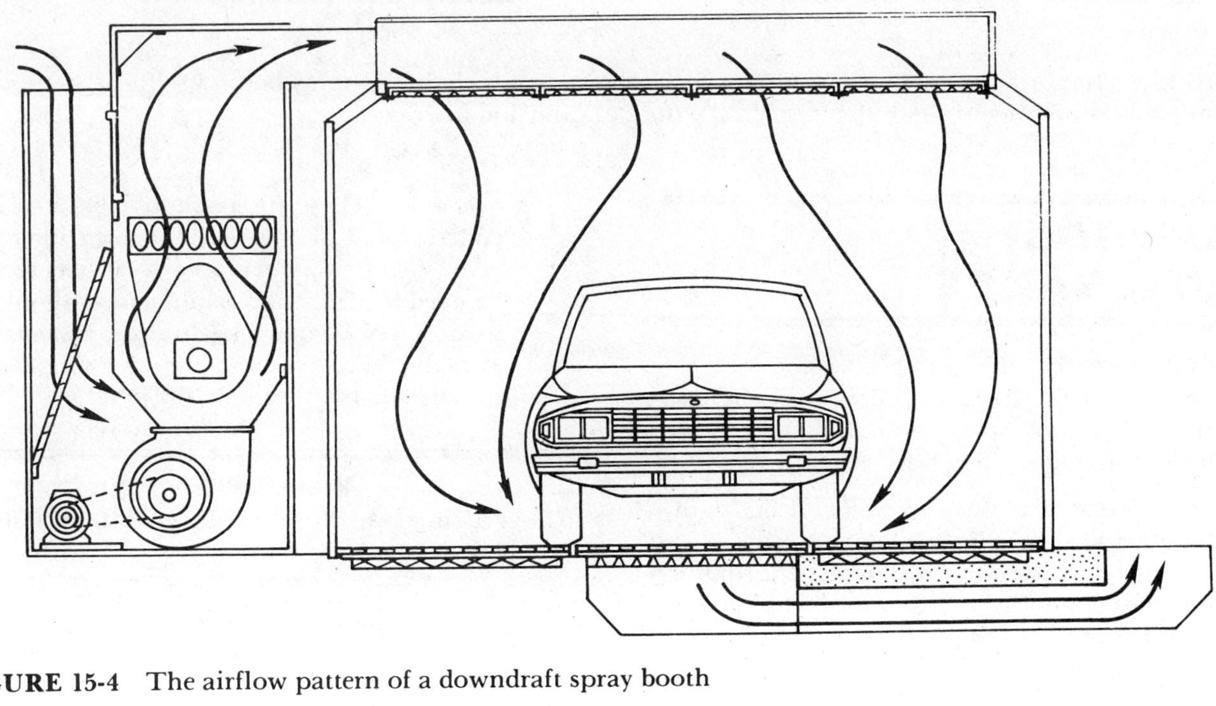

FIGURE 15-4 The airflow pattern of a downdraft spray booth

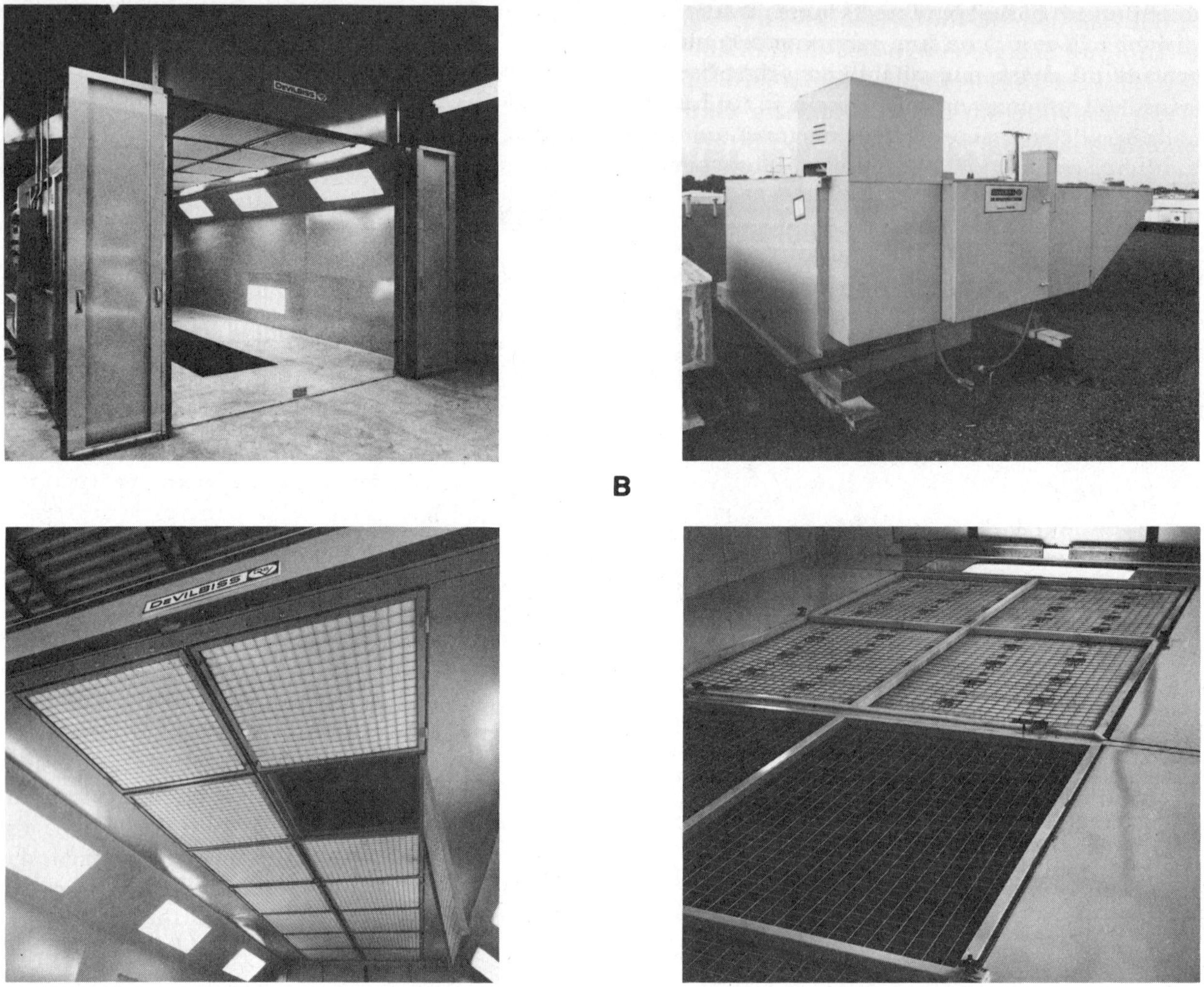

A B C D

FIGURE 15-5 Important parts of a downdraft refinishing system include: (A) spray booth itself; (B) air replacement unit; (C) ceiling plenum and filter system; (D) floor grating and filter.

SPRAY BOOTH MAINTENANCE

All too often, the spray booth is one of the most neglected areas of the shop. In order to get the best results from any type of spray booth, follow a routine maintenance program such as this:

- Be sure that all collision repair work and paint preparation procedures are done outside of the booth. Make certain no sanding or grinding operations are performed in the booth.
- The continuous flow of air through the booth will eventually load the filters with dirt and overspray. These filters should be inspected and replaced at the recommended intervals. And when they are replaced, the filters recommended by the booth manufacturer should be used. Several types of filtration systems are available (Figure 15-6), but the single-stage furnace filters will not do the job.

WARNING

Clogged filters are a fire hazard because they could ignite under certain conditions.

- Monitor the manometer readings daily and know what a normal reading should be. The

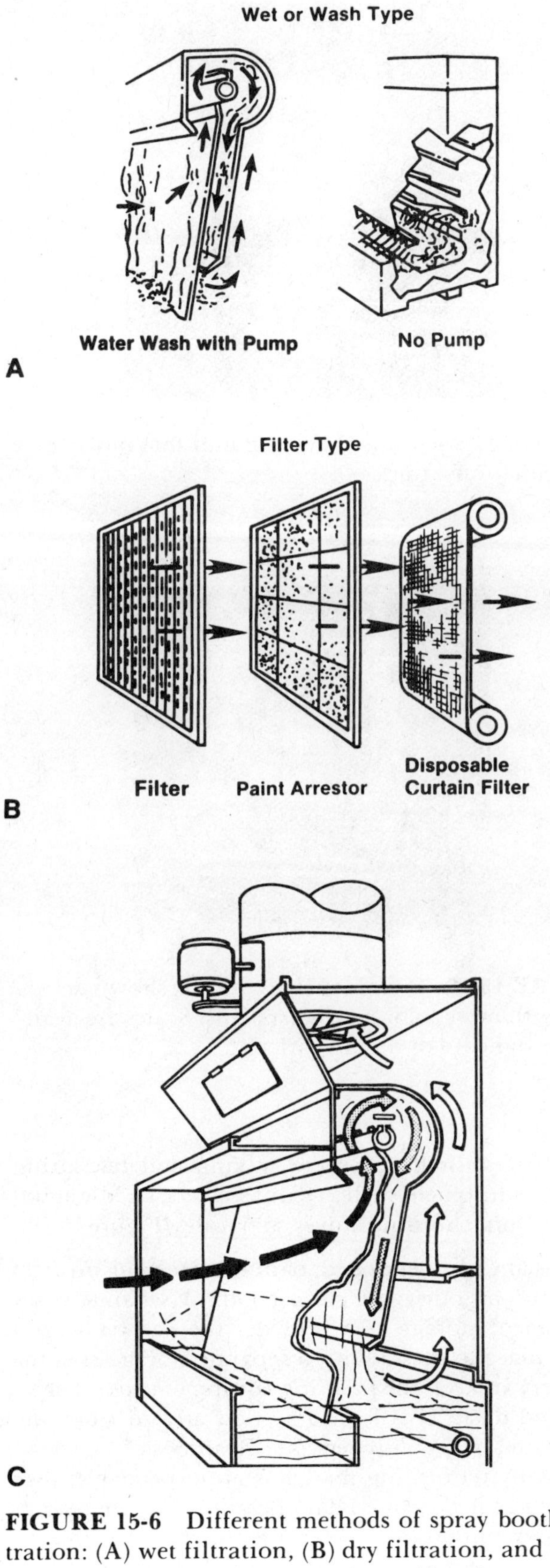

FIGURE 15-6 Different methods of spray booth filtration: (A) wet filtration, (B) dry filtration, and (C) a combination of water spray and centrifugal separation

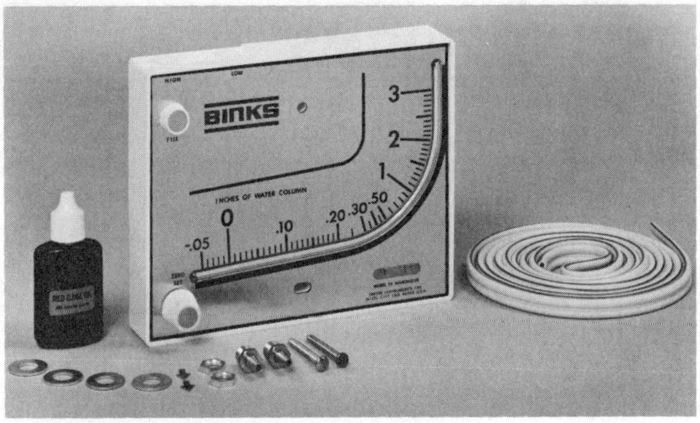

FIGURE 15-7 A typical manometer

manometer (Figure 15-7) indicates when the intake filters are overloaded. Some booths have a pressure switch that shuts off the air supply and exhaust fan when the intake filter is clogged.

- Keep the booth free of dirt and overspray at all times. Floors and walls should be wiped down after every job. Clean up scrap, masking paper, rags, and so forth. Maintain and clean all equipment used in the booth.
- To help keep any stray dust down, wet the spray booth floor before spraying.
- Clean all air hoses several times a day; accumulations of dirt and dust on their outer surface might drop off onto the freshly painted surfaces of the vehicle.
- Coat the inside of the booth with a strippable spray-on coating. When the overspray on it becomes too thick, strip and recoat.
- Periodically check the lighting inside the booth and replace weak or burned out bulbs. Improper lighting can lead to poor finishes.
- Personal cleanliness cannot be overemphasized. Never enter the booth wearing dirty clothes or shoes; the dirt can be easily transferred to the vehicle and into the finish. All clothing and gloves should be clean and lint-free.
- Caulk all seams and cracks where dirt might enter.
- Oil the exhaust fan pulley and motor bearings regularly.

WARNING

Always shut off the main exhaust fan switch or power supply before oiling the fan.

DRYING ROOMS

A dust-free drying room will speed up drying, produce a cleaner job, and increase the volume of refinishing work that can be handled. Some drying rooms have forced air or permanent infrared units for the force drying of paint, particularly enamels. These ovenlike units can speed up the drying time as much as 75 percent. Force drying of putties, primers, and sealers will reduce waiting between operations. Drying rooms can also be used for fast drying spot and panel repairs.

Infrared drying equipment is also available in booth systems; portable panels for partial or sectional drying (Figure 15-8) and large heating units capable of moving automatically along a track over the vehicle for overall drying (Figure 15-9). When using a drying room, caution must be taken not to overheat and destroy the finish.

PAINT MIXING AND STORAGE AREA

Dirt can enter the paint as it is being mixed, so a clean, well-lit mixing area is required. Federal and local laws will dictate the type of mixing area the shop must have. In many cases this means an enclosed space with its own ventilation system and explosion-proof lighting, as well as the following:

- The mixing area should be located next to the spray booth.
- It must have adequate ventilation and fire protection (per local and state codes).

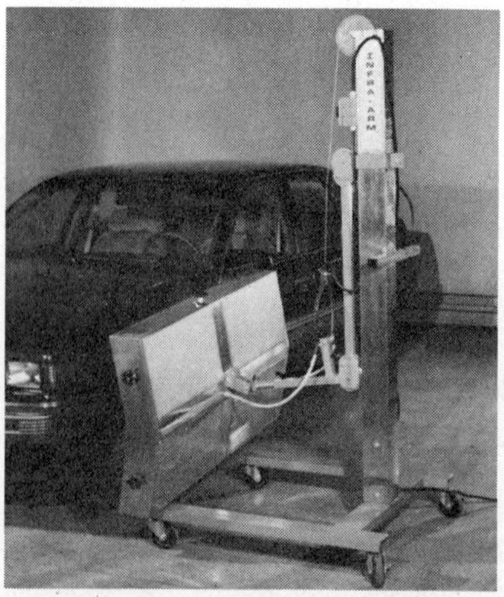

FIGURE 15-8 Typical portable drying equipment

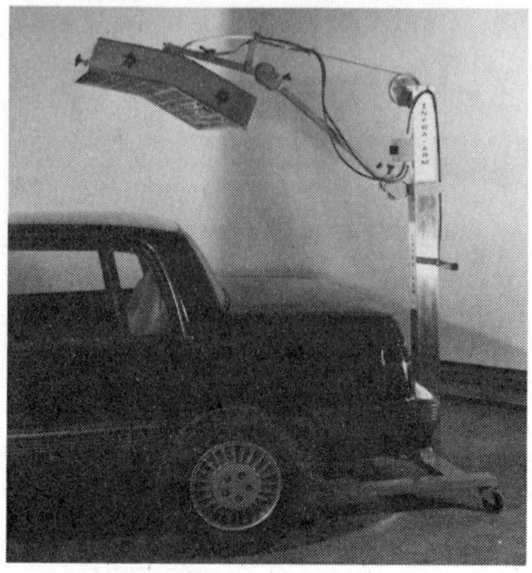

FIGURE 15-9 Infrared heating unit that moves over the vehicle on a track

FIGURE 15-10 A scale such as the one shown here is used with most color mixing systems. Scales are available in standard or digital varieties.

- It must have proper mixing and measuring equipment such as paint shakers, paddle agitators, churning knives, and scales (Figure 15-10).

As already mentioned, painters can bring dirt and dust into an otherwise clean booth. A separate paint equipment storage area provides the painter with a clean place to dress. With a separate storage area the painter can keep the painting equipment free of shop dirt and dust. This is also a good area to store and maintain safety equipment (see Chapter 1).

Many refinishing materials are expendable; they are used up in the day-to-day operation. For this reason, it is very important to keep plenty of such items on hand. It is inexcusable (not to mention very time consuming) to run out of a necessary material while in the middle of a job.

Popular refinishing materials include:

- Abrasive paper and sanding blocks
- Clean wiping cloths or paper towels
- Tack rags
- Paint paddles
- Strainers
- Mixing buckets
- Masking paper and tape
- Squeegees
- Thinner/reducer
- Cleaning solvent

SHOP TALK

By its very nature, spray painting creates a certain amount of overspray, hazardous vapors, and toxic fumes. Even under ideal conditions, there is no way to avoid contact with these materials entirely. Anyone who is around a refinishing operation must wear the safety equipment described in Chapter 1.

PAINT SPRAY EQUIPMENT

Spraying equipment for any refinishing system consists of:

- Spray guns and cups
- Air compressor
- Oil and water extractor and air regulator combination
- Air hose
- Heater (optional)

The spray gun is the key component in any refinishing system. Even though all spray guns have many parts and components in common, each gun type or size is suited for a certain, defined range of jobs. As in most other areas of refinishing work, having the right tool for the job goes a long way toward doing a professional job in minimum time.

The compressed air system, an important part of the refinish operation, can be a source of problems for the painter. The compressor can introduce dirt, moisture, and oil into the air supply, and from there into the paint. A variety of aftercoolers is available to reduce moisture. Traps can be installed which will keep oil, water, and dirt out of the paint. These must be in proper working order to perform at their best.

ATOMIZATION

A thorough understanding of atomization is the key to using a spray gun correctly. Atomization breaks paint into a spray of tiny, uniform droplets. When properly applied to the vehicle's surface, these droplets flow together to create an even film thickness with a mirrorlike gloss. Proper atomization is essential when working with today's base coat/clear coat finishes.

Atomization takes place in three basic stages (Figure 15-11):

- In the first stage, the paint siphoned from the fluid tip is surrounded by air streaming from the annular ring. This turbulence begins the breakup of the paint.
- The second stage of atomization occurs when the paint stream is hit with jets of air from the containment holes. These air jets keep the paint stream from getting out of control and aid in the breakup of the paint.
- In the third phase of atomization, the paint is struck by jets of air from the air cap horns. These jets hit the paint from opposite sides, causing it to form into a fan-shaped spray.

PARTS OF A SPRAY GUN

The principal components of a typical spray gun are illustrated in Figure 15-12. Some guns are equipped with a removable spray head unit containing the air

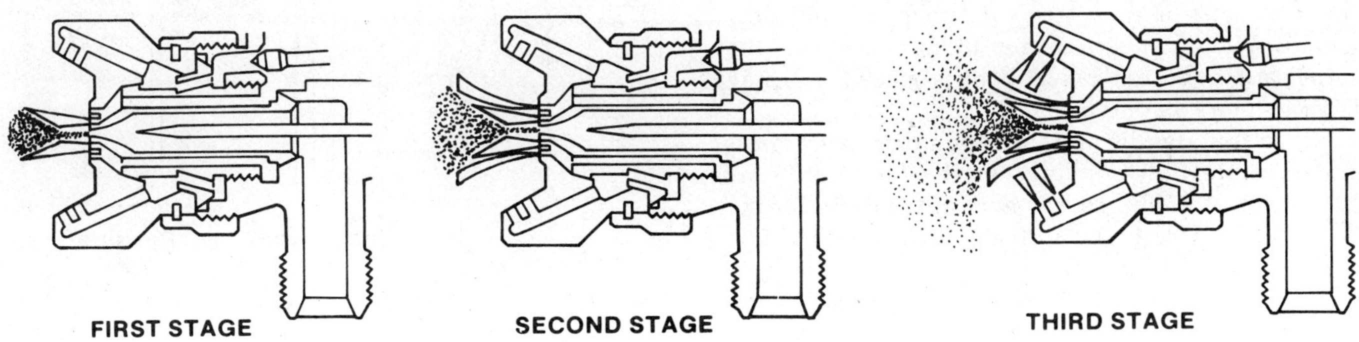

FIGURE 15-11 The three stages of atomization

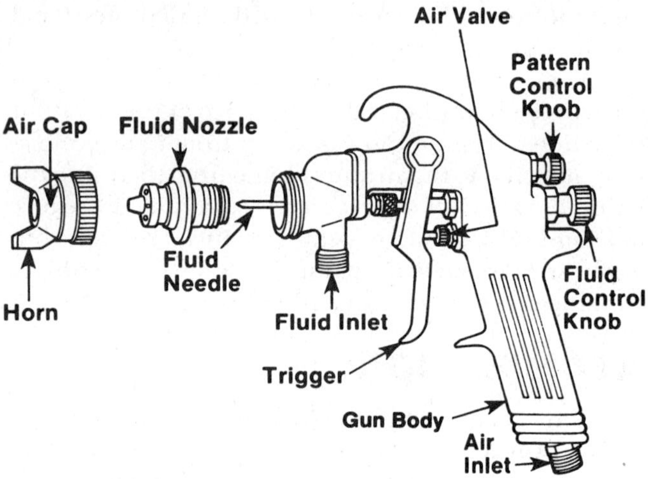

FIGURE 15-12 Parts of a typical spray gun

cap, fluid nozzle, and fluid needle. The parts of a spray gun are as follows:

- Air cap
- Fluid nozzle
- Fluid valve
- Trigger
- Fluid control (or spreader) knob
- Air valve
- Pattern control (or fan adjustment) knob
- Gun body (or handle)

The air cap directs the compressed air into the paint to atomize it and form the spray pattern. Each of the orifices has a different function. The center orifice located at the fluid nozzle creates a vacuum for the discharge of the paint. The side orifices determine the spray pattern by means of air pressure. The auxiliary orifices promote atomization of the paint. There is a definite relationship between the auxiliary orifices and gun performance. Many or larger orifices increase the ability to atomize more material for painting large objects with great speed. Fewer or smaller orifices require less air, produce smaller spray patterns, and deliver less material to conveniently paint smaller objects or apply coatings at lower speeds. Air also flows through the two side orifices; this flow forms the shape of the spray pattern. When the pattern control knob is closed, the spray pattern is round. As the knob is opened, the spray becomes more oblong in shape.

The fluid needle and the fluid nozzle both meter and direct the flow of paint from the gun into the air stream. The fluid nozzle forms an internal seat for the fluid needle, which shuts off the flow of paint. The amount of paint that actually leaves the front of the gun depends on the size of the fluid nozzle opening when the needle is unseated from the nozzle. Fluid nozzles are available in a variety of sizes to properly handle paints of various types and viscosities. The fluid control knob changes the distance the fluid needle moves away from its seat in the nozzle when the trigger is pulled.

Like the fluid valve, the air valve is opened by the trigger. The distance the air valve opens is regulated by a screw adjustment. When the trigger is pulled partway, the air valve opens. When it is pulled a little farther, the fluid nozzle opens.

TYPES OF SPRAY GUNS

There are three basic types of spray guns (Figure 15-13):

- Siphon (or suction) feed
- Pressure (or compressor) feed
- Gravity feed

The siphon feed spray gun is by far the one most commonly used in refinishing. The paint is held in a 1 quart cup attached to the gun. When the trigger is partially depressed, the air valve opens and air rushes

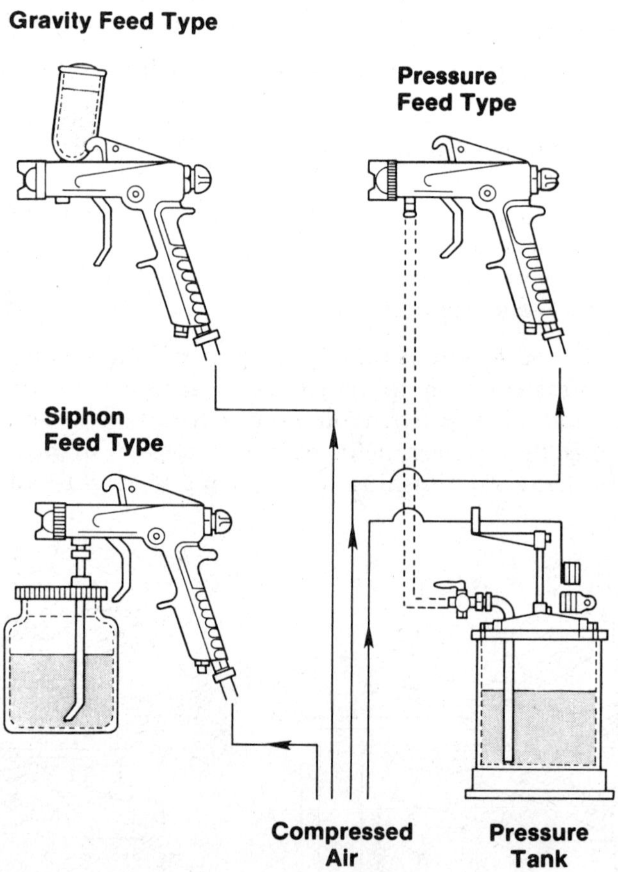

FIGURE 15-13 Basic types of spray guns

through the gun. As the air passes through the openings in the air cap, a partial vacuum is created at the fluid nozzle. Further squeezing of the trigger withdraws the fluid needle from the fluid nozzle. The vacuum sucks the paint from the cup, up the fluid inlet, and out through the open fluid nozzle. Air enters through the air hole and replaces the siphoned paint.

In the design of a pressure feed spray gun, the fluid nozzle is flush with the face of the air cap and no vacuum is created. The fluid is forced to the air cap by pressure sustained in the system by a separate cup, tank, or pot.

In the gravity feed system, the paint is supplied by gravity and the material is suction forced at the fluid nozzle. This system is ideal for high-solids materials. The handling of the gun is the same as a siphon feed gun.

Several other types of spray guns can be found in some collision repair and refinishing shops, including HVLP (high volume, low pressure), airless, and electrostatic. These guns operate and handle in the same manner as other spray guns. The major difference in these systems, as explained later in this chapter, is the methods of atomizing the paint.

Figure 15-14 illustrates how the pressure feed spraying system is hooked up:

- Connect the regulated air hose from the air control device on the tank to the air inlet on the gun.
- Connect the mainline air hose from the regulator to the air regulator inlet on the tank.
- Connect the fluid hose from the fluid outlet on the tank to the fluid inlet on the gun.

Paint pressure tanks are available in sizes from two quarts to 10 gallons. They are available in single, dual, and nonregulated models.

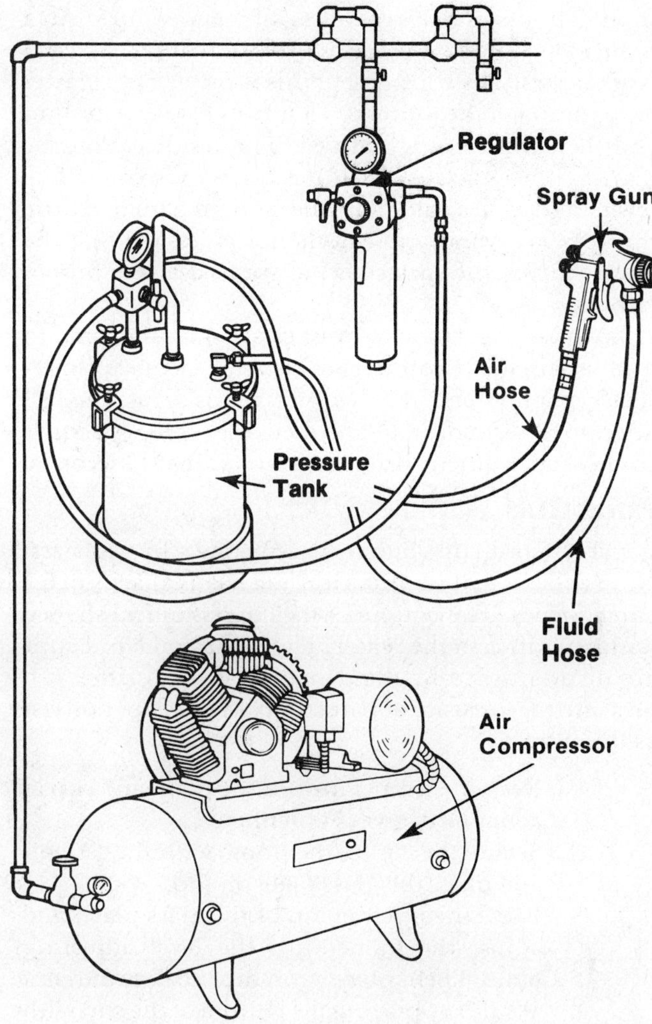

FIGURE 15-14 Hookup for pressure feed spray gun

SPRAYING TECHNIQUES

Spraying a vehicle is a skill that calls for considerably more than just holding down the trigger and hoping that the gun will put the paint where it is supposed to be and in the right amount. Several variables contribute to the quality of the spray finish.

VISCOSITY

The viscosity of a paint is its ability to resist flow. Using an incorrect viscosity paint will result in various finish defects. The paint must be thoroughly mixed and properly thinned or reduced. Paint viscosity can be measured by means of a viscosity cup. The two types of viscosity cups used for automobile paint are the Ford cup and the Zahn cup. Although the Ford cup is very accurate, because of its high cost it is not used as much as the Zahn cup, which is less accurate but also less expensive. Although not every shop has access to a viscosity cup, its use is recommended when available.

The flow characteristics of liquids relate directly to the degree of internal friction; therefore, anything that influences the internal friction (such as solvents, thinners, or temperature change) will influence flow. Similarly, it is the flow characteristics that determine how well the paint will atomize, how well it will "flow out" on the work, and the type of equipment needed to move it.

When preparing material for spraying, thin to the proper viscosity according to the directions on the can, using the thinner or reducer best suited for the shop temperature and conditions. The amount of reduction should be the same regardless of temperature. At a higher temperature the viscosity of the reduced material is actually slightly lower, but this is offset by the faster evaporation of the thinner as it travels between the gun and the surface being painted. The result is that the paint reaches the surface at the correct viscosity. The reverse is true in a cold shop; the reduced paint is a little thicker, but evaporation in the air is less, so that the paint reaches the surface being sprayed at the proper viscosity.

Various automotive finishes are manufactured to spray at specific viscosities. For instance, lacquer primer-surfacer sprays best at 15 seconds viscosity, acrylic lacquer at 12 seconds, and alkyd enamel colors, lacquer colors, and synthetic primer-surfacers all at 19 seconds.

Ford Cup

The Ford cup (Figure 15-15) comes in two sizes: No. 3 and No. 4. It is cylindrical and has a capacity of 6.1 cubic inches. The bottom of the cup is conical shaped with an orifice in the center. The No. 3 and No. 4 cups are distinguished by the diameter of this orifice. To measure the viscosity of paint with a Ford cup, proceed as follows:

1. Keep the temperature of the paint and cup at about 68 degrees Fahrenheit.
2. Secure the cup to the frame with the set bolt and place the glass plate on top.
3. Place the level on top of the glass plate, and adjust the frame with the level adjusting bolts. Then, place a container below the cup.
4. While supporting the bottom of the cup with one hand, place a piece of thick rubber over the orifice. Pour in the paint, being careful that no air bubbles enter.
5. Slide the glass plate horizontally over the top of the cup to remove any excess paint, then set it aside.
6. Release the piece of rubber supporting the orifice, and at the same time begin measuring the continuous downward drain of the paint with a stopwatch. Measure until the continuous paint flow stops. This time is used as an indicator of the paint viscosity. For example, if it takes 15.4 seconds, the viscosity of the paint is said to be 15.4 seconds at 68 degrees Fahrenheit.

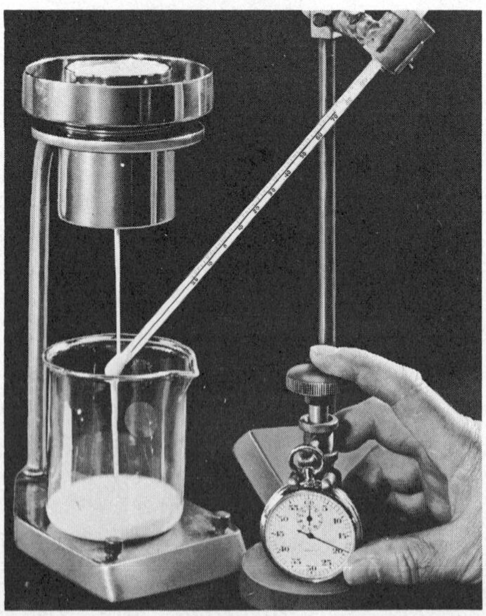

FIGURE 15-15 The Ford cup

Zahn Cup

The Zahn cup is also cylindrical in shape and has an orifice at the bottom. To determine viscosity with a Zahn cup, proceed as follows (Figure 15-16):

1. Prepare the paint to be tested. Mix, strain, and reduce as directed by the manufacturer.
2. Fill the cup by submerging it in the paint.
3. After removing the cup, release the flow of the paint and trigger the stopwatch. Keep all eyes on the flow, not on the watch.
4. When the solid stream of paint "breaks" (indicating air passing through the orifice), stop the watch. As with the Ford cup, the result is expressed in seconds. Manufacturers recommend specific viscosities for their various types of paint.

SHOP TALK

A stopwatch is necessary for measuring paint viscosity with either viscosity cup. Most painters prefer a digital stopwatch to the standard type because it is easily read.

TEMPERATURE

The temperature at which paint is sprayed and dried has a great influence on the smoothness of the finish. This involves not only the temperature of the shop, but the temperature of the work surface as well. A job should be brought into the shop long enough ahead of spraying time so that it achieves approximately the same temperature as the shop. Spraying warm paint on a cold surface or spraying cool paint on a hot surface will completely upset the flow characteristics. Appropriate thinners or reducers should be used for warm and cold weather applications.

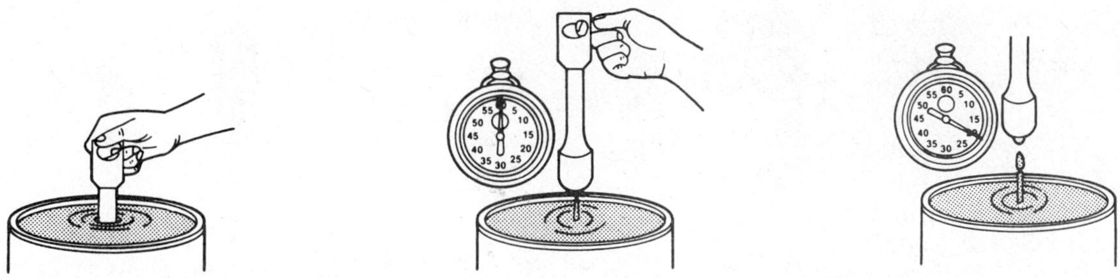

FIGURE 15-16 Using a Zahn cup: (A) Dip the cup into the paint until it is full. (B) Remove the cup, and as it clears the surface of the paint, begin timing the flow of paint from the small hole in the bottom of the cup. (C) Stop the timer when the stream of paint breaks.

ADJUSTING THE SPRAY

A good spray pattern depends on the proper mixture of air and paint droplets, much like a fine-tuned engine depends on the proper mixture of air and gasoline. The paint should go on smoothly in a medium to wet coat, without sagging or running. There are three basic adjustments, which under normal conditions will give the proper spray pattern, degree of wetness, and air pressure.

1. Adjust the air pressure (Figure 15-17). This is generally set at the dryer-regulator (or transformer). But due to friction, as air passes from the dryer-regulator to the gun, pressure will be lost. The difference between the reading at the dryer-regulator and the reading at the gun varies depending on the length and diameter of the hose.

   The surest method to measure this pressure drop is by inserting a regulator between the hose coupler and the gun to allow for checking and setting pressure at the gun (Figure 15-18).
2. Set the size of the spray pattern using the pattern control knob. Turn the pattern control knob all the way in to create a small, round pattern. Backing the pattern control knob all the way out will produce a wide pattern. Figure 15-19 represents the evolution of the spray pattern, from all the way in to all the way out.
3. Use the fluid control knob (Figure 15-20) to regulate the volume of paint according to the selected pattern size; backing the knob out increases the paint flow, and turning it in increases the flow (Figure 15-21).

The optimum spraying pressure is the lowest needed to obtain proper atomization, flow rate, and pattern width. A pressure that is too high results in excessive paint loss through overspray and poor flow due to high solvent evaporation before the paint reaches the surface. A pressure that is too low produces poor drying characteristics due to high solvent retention, and makes the paint film prone to bubbling and sagging. The recommended air pressure varies with the kind of material sprayed. Many low volatile organic compound (VOC) regulations require 10 psi or less at the air cap.

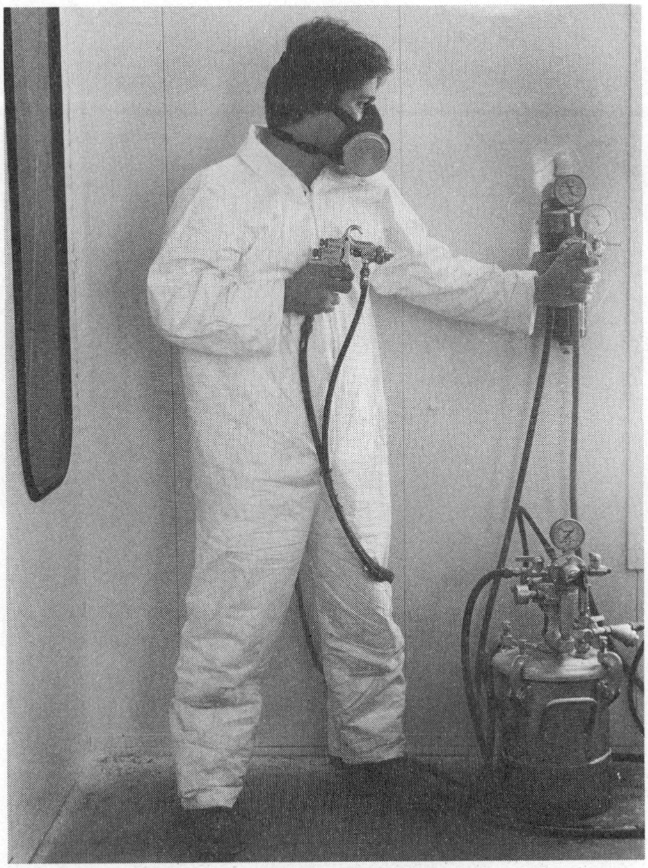

FIGURE 15-17 Adjusting the air pressure

FIGURE 15-18 This gun is equipped with a regulator that allows pressure to be checked and set at the gun.

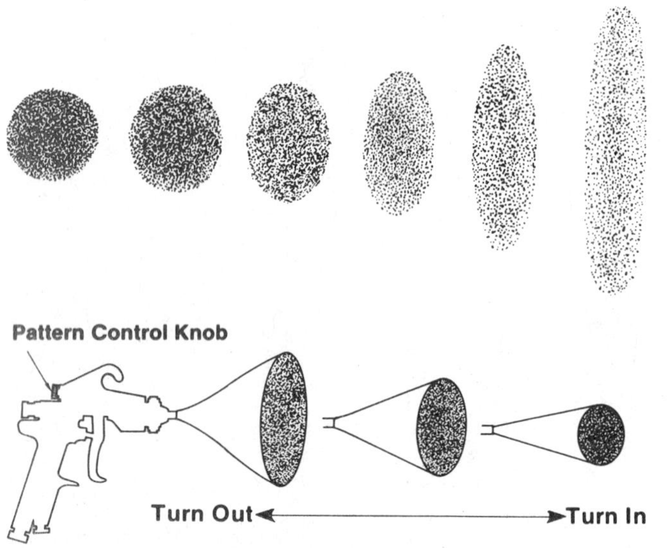

FIGURE 15-19 Setting the size of the spray pattern

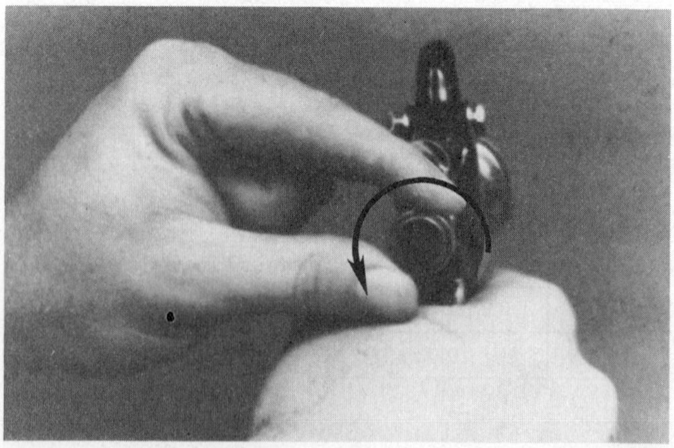

FIGURE 15-20 The fluid control knob is used to regulate paint volume.

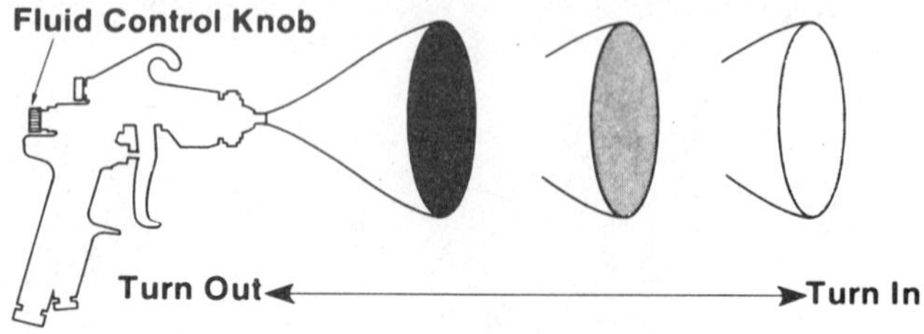

FIGURE 15-21 Paint volume adjustment

TESTING THE SPRAY PATTERN

Before attempting to paint, it is very important to test the spray pattern on a piece of masking paper. Hold the gun 6 to 8 inches away from the paper if spraying lacquer, and 8 to 10 inches away when spraying enamel. Pull the trigger all the way back and release it immediately. This burst of paint should leave a long, slender pattern on the test paper (Figure 15-22).

Spraying primer-surfacer usually requires a small spray pattern. Turn in the pattern control knob until the spray pattern is 6 to 8 inches wide. For spot repair, the pattern should be about 5 to 6 inches from top to bottom. If the paint droplets are coarse and large, close the fluid control knob about one half turn or increase the air pressure 5 pounds. If the spray is too fine or too dry, either open the fluid control knob about one half turn or decrease the air pressure 5 pounds.

Next, test the spray pattern for uniformity of paint distribution. Loosen the air cap retaining ring and rotate the air cap so that the horns are straight up and down. In this position, the air cap will produce a horizontal spray pattern rather than a vertical one. Spray again, but this time hold down the trigger until the paint begins to run. This is known as flooding the pattern. Inspect the lengths of the runs. If all adjustments are correct, the runs will be approximately equal in length (Figure 15-23A).

The uneven runs in the split pattern shown in Figure 15-23B are a result of setting the spray pattern too wide or the air pressure too low. Turn the pattern control knob in one half turn or raise the air pressure 5 pounds. Alternate between these two adjustments until the runs are even in length. If paint runs are longer in the middle than on the edges (Figure 15-23C), too much paint is being discharged. Turn in the fluid control knob until the runs are even in length.

WARNING

Always wear a suitable air respirator when doing any spraying.

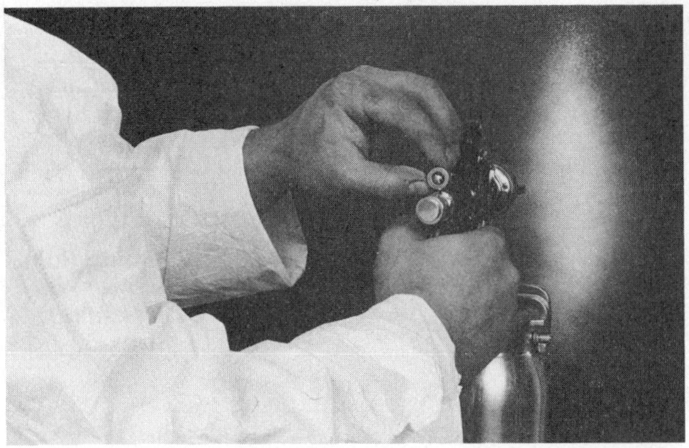

FIGURE 15-22 A long, slender paint pattern should form on the masking paper.

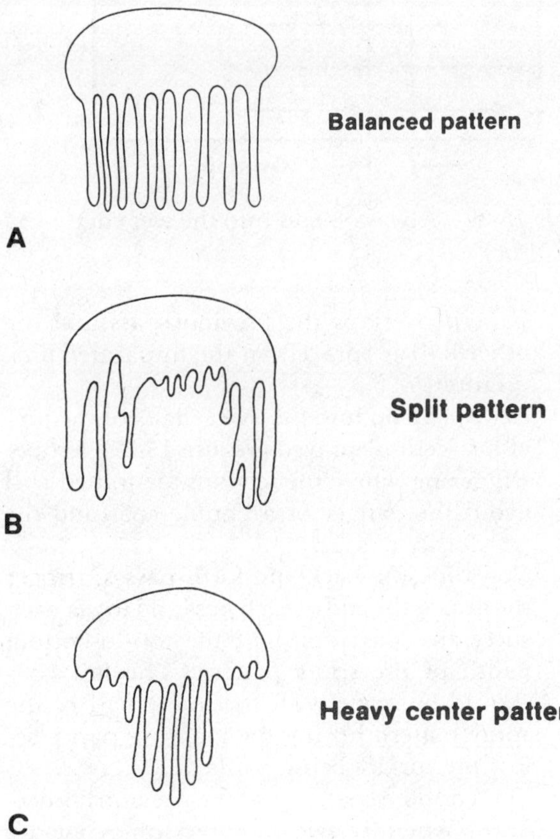

FIGURE 15-23 Spray patterns: (A) balanced, (B) split, and (C) heavy center

THE APPLICATION STROKE

The proper stroke is most important to a successful refinishing job. Practice on masking paper; any problems with the spray pattern must be corrected before painting the vehicle.

1. Hold the gun at the proper distance from the surface—6 to 8 inches for lacquer, 8 to 10 inches for enamel. If the spraying is done from a shorter distance, the high velocity tends to ripple the wet film. If the distance is too great, there will be a greater percentage of thinner evaporated, resulting in orange peel or dry spray. A slower drying thinner will permit more variation in the distance of the gun from the surface, but it will produce runs if the gun gets too close. Excessive spraying distance also causes more overspray.
2. Hold the gun level and perpendicular to the surface. Keep the gun at a right angle to the vehicle even when spraying curves in the body, or an uneven paint film will result (Figure 15-24). On flat surfaces such as the hood or roof, the gun should be pointed almost straight down (Figure 15-25).
3. The gun should be kept in motion until the spraying ceases. This technique gives a fade in and fade out effect, which prevents overloading.
4. Do not fan the gun with wrist motion if a uniform film is desired. The only time it is permissible to fan the gun is on a small spot spray, where the paint film at the edges of the spot should be thinner than the center portion.

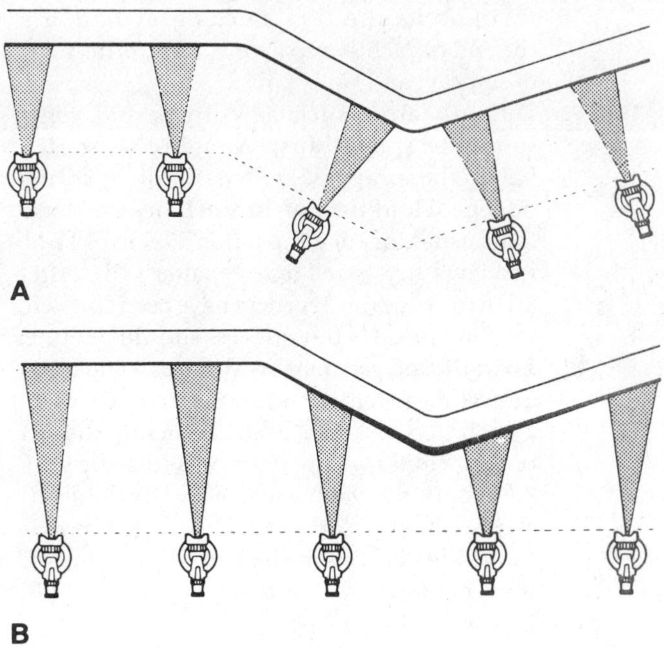

FIGURE 15-24 (A) Proper and (B) improper spray gun movement

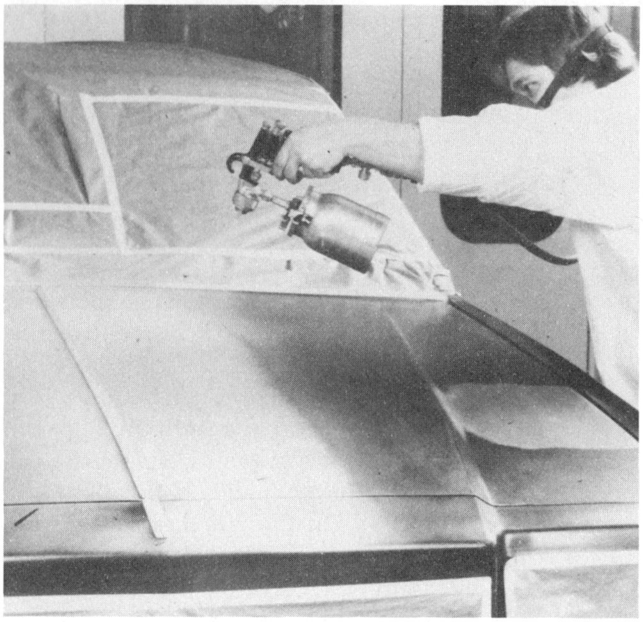

FIGURE 15-25 Position of gun for painting flat surfaces such as the hood or roof

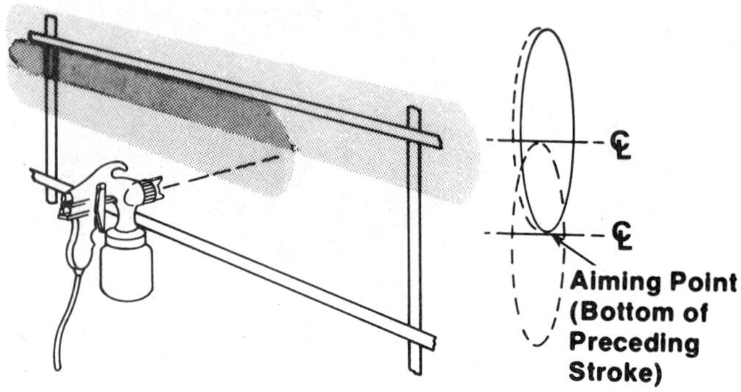

FIGURE 15-26 How to overlap strokes

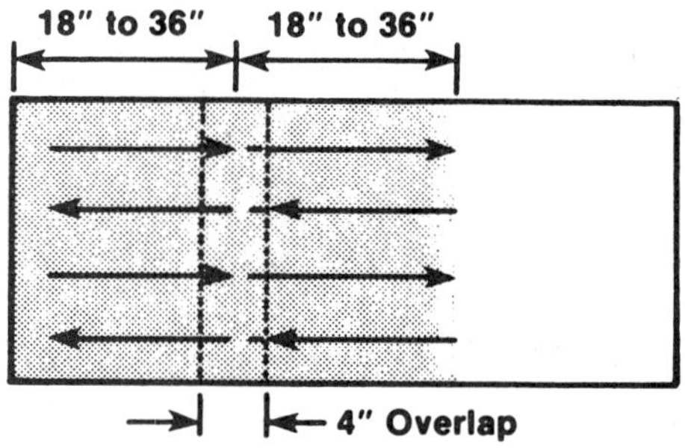

FIGURE 15-27 Always blend into the wet edge.

5. Move the gun with a steady, deliberate pass, about 1 foot per second. The speed must be consistent or it will result in an uneven paint film. Never stop the gun while spraying, or the coat will sag and run.
6. Release the trigger at the end of each pass, then pull it back when beginning the pass in the opposite direction. Turning off the gun at the end of each sweep avoids runs, minimizes overspray, and saves paint.
7. Difficult areas such as corners and edges should be sprayed first. Aim directly so that half of the spray covers each side of the corner or edge. Hold the gun an inch or two closer than usual, or turn the pattern control knob in a few turns to reduce the pattern size. After all of the edges and corners have been sprayed, the flat or nearly flat surfaces should be done.
8. For painting very narrow surfaces, switch to a gun with a smaller spray pattern. This can also be accomplished by reducing the air pressure and spray pattern on a full-size gun.
9. Generally, start at the top of any upright surface such as a door panel. The gun nozzle should be level with the top of the panel so that the upper half of the spray pattern will hit the masking paper.
10. The second pass is made in the opposite direction, with the nozzle level with the lower edge of the previous pass. Thus one half of the pattern overlaps the previous pass and the other half is sprayed on the unpainted area (Figure 15-26).
11. Always blend into the "wet edge" of the previous section sprayed (Figure 15-27). Proper triggering where the sections are joined will avoid the danger of a double coat and the possibility of sags.
12. Continue the back-and-forth passes; trigger the gun at the end of each pass and lower each successive pass one half the top-to-bottom width of the spray pattern. The last pass should be made with the lower half of the spray pattern hitting the masking paper below the surface being painted.
13. For a double coat, repeat the previous procedure. Generally, two or more double coats are required for a lacquer top coat. Allow flash time of several minutes between coats. Two or three single coats are required for enamel top coats. Allow the first coat to set up (become tacky) before applying additional coats.

GUN HANDLING PROBLEMS

The inexperienced painter is prone to making spraying errors. These include the following:

- *Heeling*. This occurs when the gun is allowed to tilt. Because the gun is no longer perpendicular to the surface, the spray produces uneven film thickness, excessive overspray, dry spray, and orange peel.
- *Arcing*. This occurs when the gun is not moved parallel with the surface. At the outer edges of the stroke, the gun is farther away from the surface than at the middle of the stroke. The result is uneven film thickness, excessive overspray, dry spray, and orange peel.
- *Incorrect Stroke Speed*. If the stroke is made too quickly, the paint will not cover the surface evenly. If the stroke is made too slowly, sags and runs will develop. Proper stroke speed is something that comes with experience.
- *Improper Overlap*. Improper overlapping results in uneven film thickness, contrasting color hues, and sags and runs as shown in Figure 15-28.

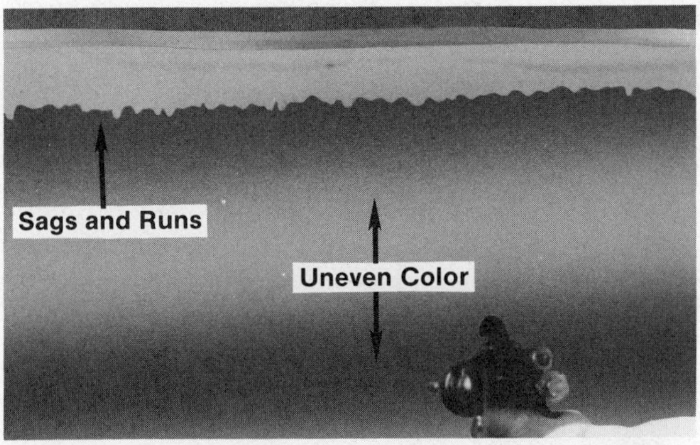

FIGURE 15-28 Results of improper overlapping

FIGURE 15-29 Results of overspraying

FIGURE 15-30 Improper coverage

- *Overspray*. Failure to trigger the gun before and after each stroke results in wasteful overspray and excessive buildup of paint at the beginning and end of each stroke (Figure 15-29).
- *Improper Coverage*. Failure to trigger exactly over the edge of the panel results in improper coverage and uneven film thickness (Figure 15-30).

TYPES OF SPRAY COATS

There are varying degrees of thickness for a spray coat: light, medium, or heavy. The easiest way to control thickness is by the speed with which the gun is moved. That is, the slower the speed, the heavier the coat.

Five different types of spray coats are used:

- Tack
- Full wet
- Dust
- Shading or blending
- Banding

The tack coat allows application of heavier wet coats without sagging or runs. It is a light covering coat applied to the surface and then allowed to flash until it is tacky, which usually takes only a few minutes.

The full wet coats are then sprayed over the tack coat. A full wet coat is a heavy, glossy coat that is applied in a thickness almost heavy enough to run. Spraying this coat requires skill and practice.

A dust coat is a light, dry coating that is similar to a mist coat. It is accomplished by holding the gun a short distance from the surface.

Shading or blending coats are applications of paint on the boundary of spot repair areas, so that a color difference between the new paint and the original

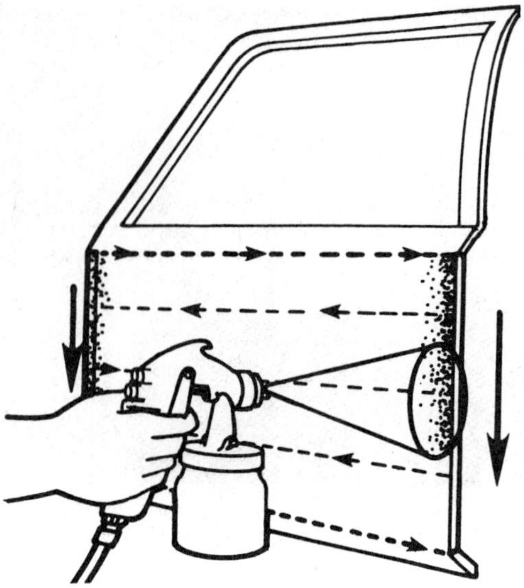

FIGURE 15-31 Spraying the panel ends with a banding coat

paint is not noticeable. They are applied in two coats; the second coat is thinner and sprayed over a wider area than the first. This technique can also be used when repainting a panel so that a color difference is not noticeable between adjacent panels.

A banding coat is a single coat applied in a small spray pattern. This technique assures coverage at the edges without spraying beyond the refinishing area (Figure 15-31). Banding is often used in spraying panel repairs with a primer-surfacer.

HVLP SPRAY SYSTEM

The HVLP spray system uses a *high volume* of air, delivered at *low pressure*, to atomize the paint into a pattern of low-speed particles. The most important way it differs from conventional spraying is its high transfer efficiency, which means that it maximizes the atomization of the paint in order to greatly cut down on overspray. The high pressure of conventional sprays tends to "blast" the paint into small particles. In the process, it creates a fair amount of overspray. Thus, the transfer efficiency suffers as a result of overspray, particle "bounce," and blow back. In contrast, HVLP relies on air delivered at 10 psi or less to break the material into small particles. As the material flows into the air stream, far less is lost in overspray, bounce, and blow back, hence the dramatic improvement in transfer efficiency. HVLP will work with any medium solid material that can be atomized by the gun. These include two-part paints, urethanes, acrylics, epoxies, enamels, lacquers, stains, and primers.

High transfer efficiency is attractive for several reasons. But, perhaps the most important reason is the trend toward legislated transfer efficiency requirements. For example, low VOC regulations require the use of spray methods, which are at least 65 percent transfer efficient. Low pressure spray (from 0.1 to 10 psi) has been approved, along with electrostatic spraying methods. Similar legislation is pending or under consideration in many other states and countries. The forecast is that high transfer efficiency will soon be a requirement everywhere.

The purpose of this legislation is to protect the environment, but there are other good reasons for looking into HVLP. Higher transfer efficiency improves the quality of the workplace and the quality of the finished products. Overspray cuts visibility, which contributes to mistakes and low productivity. Overspray is one of the main causes of paint operation maintenance, so cutting overspray cuts downtime. All paint spraying equipment can be affected by overspray, but the spray booth and its filters are hurt the most.

To illustrate how much of a difference transfer efficiency makes in booth maintenance, consider that HVLP can be two to three times as efficient as conventional spraying. Depending on how it is used, conventional spraying might be as little as 20 to 30 percent efficient. That means for every 3 gallons of paint sprayed, more than 2 gallons are wasted. HVLP is typically between 65 and 90 percent efficient, so as little as 1 pint of paint is wasted for every gallon applied. Thus, a 3:1 difference in transfer efficiency becomes a 16:1 advantage in terms of overspray.

HVLP systems are basically simple, and similar in function to conventional air spray systems (Figure 15-32). They consist of the following:

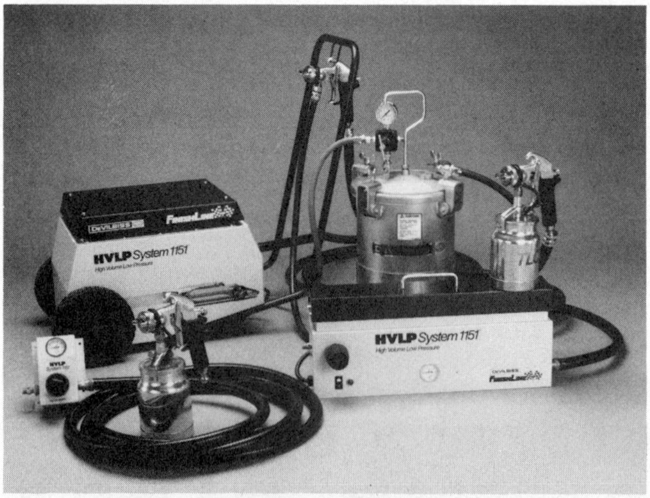

FIGURE 15-32 Basic components of an HVLP system, including a turbine generator, heated and unheated air conversion units, and various spray guns

- High-volume air source
- Material supply system
- Spray gun designed to operate with a high volume of low pressure air

Air sources can be centralized, serving multiple guns, or can be designed for single-gun use. These sources will provide a range of delivery volumes and pressures. As a general rule, the best transfer efficiency is obtained when operating nearer the low end of the 0.1 to 10 psi acceptable atomizing pressure range. Some materials, especially the new low VOC coatings, require atomizing pressure nearer the high end of the range.

Material can be supplied through pressure pots, quart cups, or other standard supply systems. Flow requirements are usually lower than conventional air spray systems for similar applications because of the higher transfer efficiency.

Most HVLP guns are the conventional air spray type (Figure 15-33), with special modifications and nozzles that atomize material at low air pressures and velocities to create a "soft spray" effect for high transfer efficiencies. The air passage in the gun handle incorporates a "sonic venturi" that limits the airflow in the venturi throat. The sonic venturi and the gun nozzle interact to limit the air pressure to a maximum of 10 psi at the nozzle. Most HVLP guns are also nonblur and have no obstructions that could restrict the airflow. In many designs, the airflow through the gun is continuous; the trigger only opens the material flow valve. Special caps and tips are also needed to assure proper atomization.

There are two basic air supply designs. One generates airflow from a turbine generator, the other converts 80 to 100 psi shop air to the necessary 0.1 to 10 psi. Each of these approaches has its pros and cons.

- *Turbine Generators.* In the early days of low pressure technology, most air was turbine generated. This approach offered portability that generally is not available when connecting to shop air lines. Additionally, in some cases existing shop air sources offer insufficient volume. For example, in a shop where the existing air compressor is already working at full capacity, a turbine unit would work well as it would not draw on the shop air. Due to friction, the turbine unit generates enough heat to provide moderate air temperatures. However, the exact temperature depends on turbine design, distance from the gun, and the insulating values of the air conduit, and is not always controllable. Uncontrollable heat in the atomizing air could upset the chemistry of the paint and interfere with the proper leveling of the film. Assuming

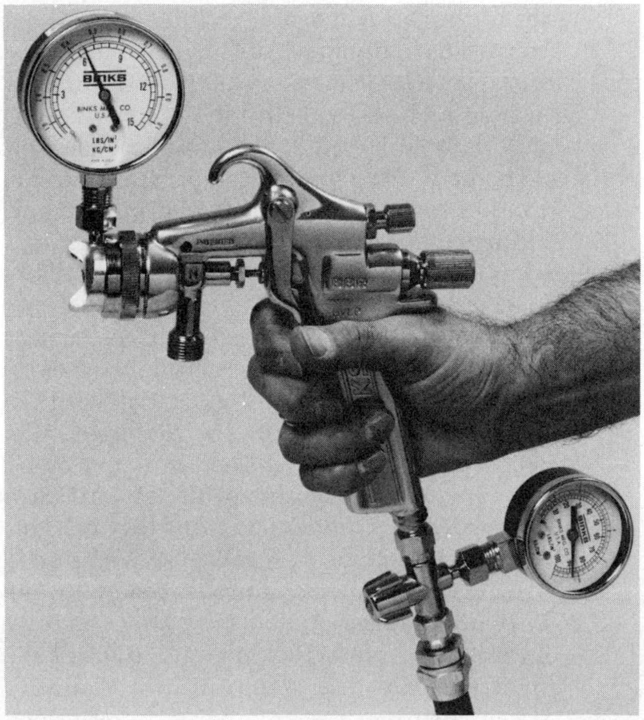

A

B

FIGURE 15-33 (A) Modified conventional air spray guns can be used with the HVLP system. There are gauges for both input and output. (B) Guns are also available with just one gauge.

the air intake system and its filter are well designed and maintained, turbine generators provide relatively clean, oil-free, dry air. For turbine generators to be used inside the spray booth, they must be of costly, explosion-proof design.

- *Air Conversion Units.* A standard shop an air compressor such as described in Chapter 3 and an air conversion unit provide more control over the variables that affect application performance. Heat and pressure are two important HVLP variables. While some air conversion units do have controlled heaters, most operate with the ambient temperature compressed air from the shop air supply. Furthermore, they can usually deliver a consistent 10 psi or be regulated to provide somewhat less pressure. Air conversion units eliminate turbine maintenance and reliability problems. Additional equipment and maintenance procedures could be required to assure that the air is clean, oil-free, and dry; otherwise contaminated air could spoil the paint job. Again, if the air supply system is currently maintained for conventional air spraying, the existing equipment and procedures will do the job.

Except for certain subtle differences, the HVLP spray gun operates in basically the same manner as described earlier. The HVLP gun should be held closer to the surface of the workpiece because of the lower speed of the particles. A rule of thumb would be to hold the gun 6 to 8 inches from the surface when spraying enamel with HVLP, compared to 8 to 10 inches for conventional spraying. Greater distances result in excessive overspray and lack of film buildup.

Many first time HVLP users get the impression that HVLP is slower than conventional spraying, but this is not always the case. Film coverage is often greater than with conventional air spray systems, resulting in fewer total passes for the desired buildup. Sometimes the application is slower, but this is generally because the air source is not delivering sufficient air pressure. To many people, HVLP sounds ineffective. Unlike conventional air spray systems, which sound like a leaking tire when airflow is present, HVLP is very quiet. However, quiet should not be confused with ineffective.

AIRLESS SPRAY GUN SYSTEM

Airless spray equipment (Figure 15-34) uses hydraulic pressure to atomize paint rather than air pressure. With the airless spray method, pressure is applied directly to the paint, which is injected at high speed through small holes in the nozzle and formed into a mist. Unlike the air spray method, there is less mixing of air in the paint and, consequently, less mist dispersion. Also, since the paint is pressurized directly, less energy is used for atomization. Thus, with the same amount of power, a degree of atomization is accomplished that is several times that for air spraying. In fact, the pressure developed in airless equipment ranges from 1,500 to 3,000 psi. Actual pressure depends on the pump ratio of the equipment.

The airless system reduces overspray to a minimum, and application of the finish is much faster than with conventional spraying. Because of the higher pressures involved, the airless system can be used with paints and other materials that have a higher viscosity. However, this system of application cannot be used where a fine finish is required. It is often used to apply

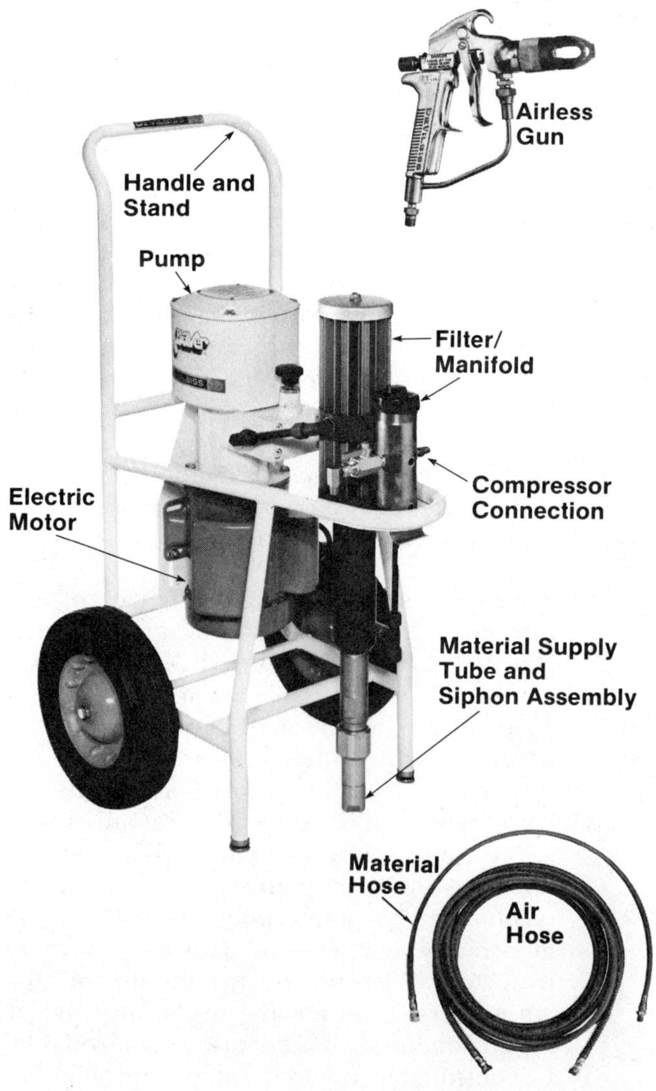

FIGURE 15-34 Typical airless spray equipment

finish coats in the commercial fleet industry. It also has found a place for underbody and corrosion work (see Chapter 10). The air-assisted airless system uses some air to assist in the spraying operation and tends to give a better finish.

WARNING

An airless system maintains pressure after the system is shut down. High pressure can cause serious injury. Before attempting any disassembly of the gun, system pressure must be relieved.

ELECTROSTATIC SPRAYING SYSTEM

Electrostatic spraying (Figure 15-35) utilizes the principle that positive(+) and negative(–) attract each other, but oppose a like charge. Therefore, when paint particles are given a negative charge by a high-voltage generator, the particles oppose each other, causing them to become atomized. On the other hand, because the object being painted is grounded, it is under a positive electrical charge. In this manner, when high voltage is applied between the surface and the electrostatic painting equipment, an electrical field is formed and the air in the field allows the electricity to pass through easily. In other words, electrical passages are formed and the atomized paint passing through these passages is sent to and adheres to the object that is being painted.

As with conventional painting, a spray gun is used for electrostatic painting and the paint is atomized by the force of compressed air. Atomization is further promoted by the application of a negative electrical charge. Therefore, the paint is sprayed onto the surface by both the force of the compressed air and electrical attraction. Adhesion efficiency is better with airless electrostatic spraying (Figure 15-36A), but the air-assisted airless method is better when delicate spray gun manipulation is needed (Figure 15-36B).

Like the normal airless spray method, airless electrostatic spraying utilizes high pressure to atomize the paint by injecting it through small holes in the nozzle, but it also gives the paint a negative electrical charge to further promote atomization. Paint is adhered by means of both injection pressure and electrical attraction. This method provides very good adhesion efficiency and application is faster due to the large discharge volume. However, because compressed air is not used, the injection energy is not as strong and thus air spray prepainting of depressed areas like the underside of the hood and door panels is necessary.

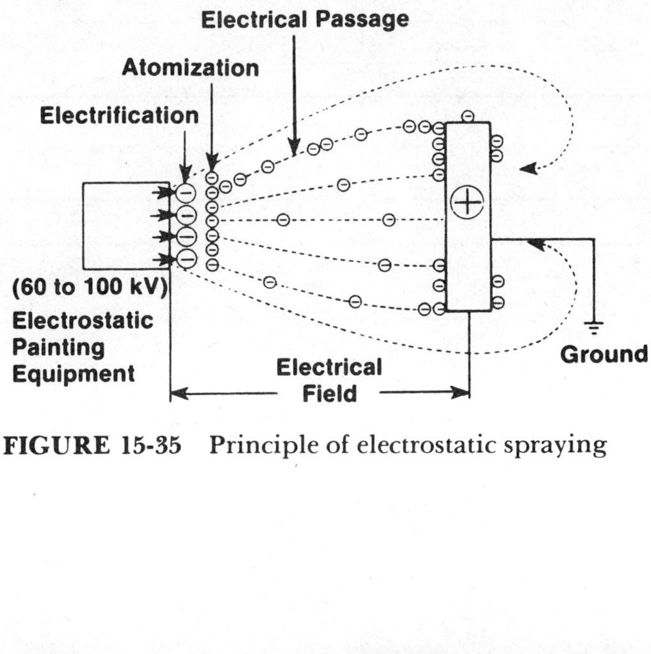

FIGURE 15-35 Principle of electrostatic spraying

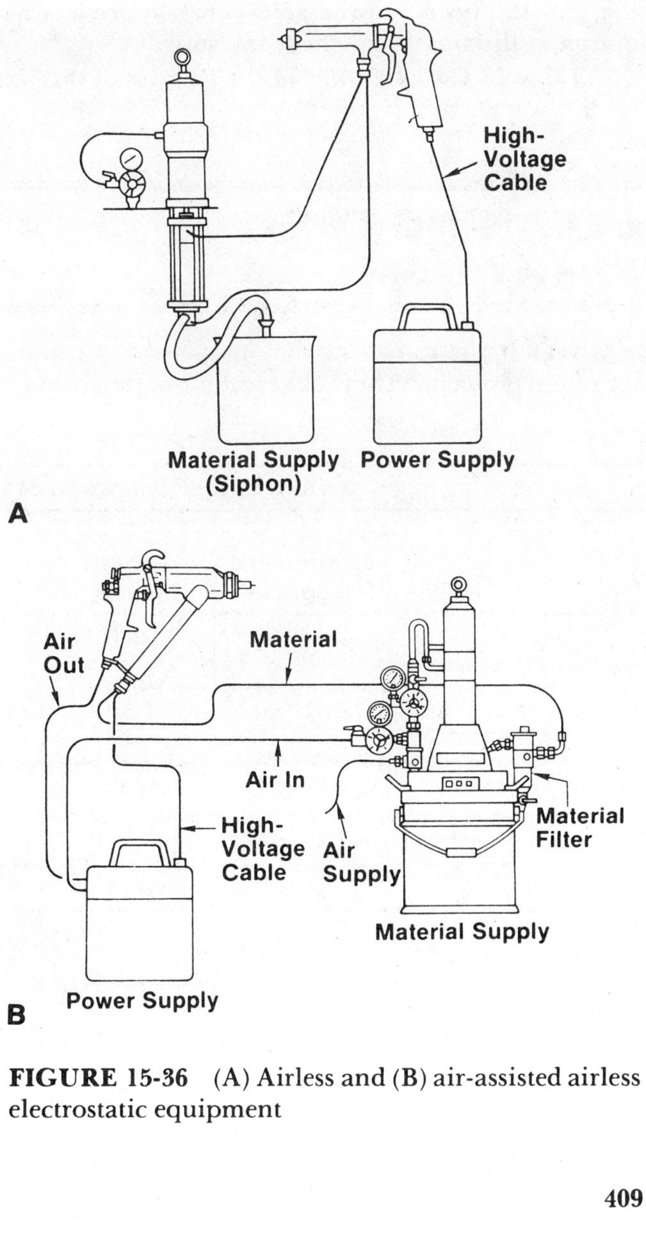

FIGURE 15-36 (A) Airless and (B) air-assisted airless electrostatic equipment

WARNING

Because of the high voltage involved in electrostatic spraying, it is very important to follow the manufacturer's instructions on the use of the equipment and all safety procedures. It is important that the vehicle be grounded. It is a good idea to always ground the vehicle's body (or frame) to a good ground source (such as a water pipe) as soon as it enters any spray booth. Grounding the vehicle will help prevent dust and dirt from being attracted to the paint by static electricity.

The basic operating techniques of airless and electrostatic spray guns are the same as those for conventional spray guns described earlier in this chapter. That is, the gun should be held perpendicular and moved parallel to the surface in order to obtain a uniform coating. Once the best working distance is determined, the gun should be moved across the surface at this optimum distance throughout the stroke.

Table 15-1 details some of the features of the six types of spray painting systems.

CLEANING THE SPRAY GUN

Neglect and carelessness are responsible for most spray gun problems. Thorough cleaning of the gun and accessory equipment immediately after use, as well as lubrication of all bearing surfaces and packings at recommended intervals, are important factors in the care of a spray gun.

To clean a siphon feed gun, first loosen the cup (Figure 15-37A). Keep the fluid tube in the cup. Next, loosen the air cap two or three turns. Hold a folded cloth over the air cap and pull the trigger (Figure 15-37B) to force the paint in the gun back into the cup.

SHOP TALK

Use very low pressure (5 psi). Never attempt to clean a siphon feed gun if the cup is loaded because paint could be splashed from the cup.

Retighten the air cap and pour the paint in the cup back into its original container. Clean the cup and cup cover with solvent and a thin bristle brush (Figure 15-37C). Wipe up any residue with a clean rag soaked with solvent. Then, pour 1 inch of clean solvent into the cup. Spray the solvent through the gun to clean out the fluid passages.

Remove the air cap and soak it in thinner or solvent. Clean out clogged orifices with a round toothpick or broom straw (Figure 15-37D). Remember, never use wire or nails to clean these precision-drilled openings. Next, clean the fluid nozzle with a brush and solvent (Figure 15-37E). With a clean rag soaked in thinner, wipe the outside of the gun to remove all traces of paint (Figure 15-37F).

TABLE 15-1: COMPARISON OF VARIOUS SPRAYING SYSTEMS

	Conventional Air Spraying	Airless Spraying	Air-Assisted Airless Spraying	Air-Assisted Airless Electrostatic Spraying	Airless Electrostatic Spraying	HVLP Spraying
Adhesion of Spray Efficiency	20 to 40%	50 to 60%	40 to 60%	60 to 70%	70 to 80%	65 to 90%
Quality of Finished Surface	Excellent	Poor	Fair to Good	Good	Fair	Good to Excellent
Work Environment (paint mist dispersion)	Poor	Good	Good	Excellent	Excellent	Excellent
Paint Speed	Slow	Fast	Fast	Very fast	Very fast	Slow
Coverage of Depressed Areas	Excellent	Poor	Fair	Good	Fair	Excellent
Gun Handling (partial repainting and touch up)	Excellent	Fair	Fair	Fair	Good	Excellent

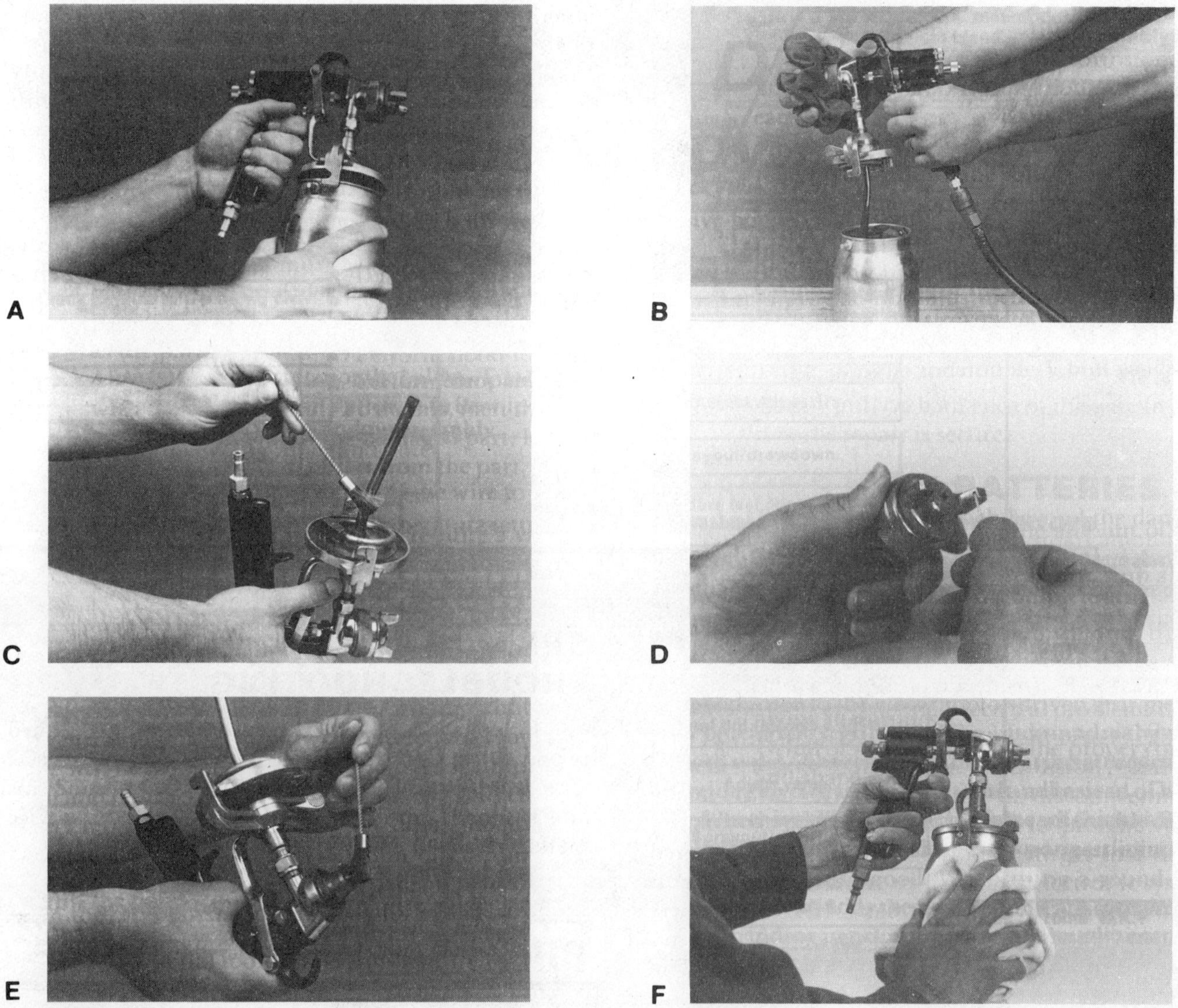

FIGURE 15-37 Steps in cleaning a siphon feed spray gun: (A) loosen the cup; (B) discharge the paint; (C) clean the cup and cup cover; (D) clean out clogged orifices in the air cap; (E) clean the fluid tip with a brush and solvent; and (F) wipe the outside of the gun.

SHOP TALK

If the gun is not cleaned soon after use, the nozzle might partially or completely close, causing the gun to spit (eject pieces of dried paint) or form the wrong spray pattern. This is of particular importance when dealing with enamel paints with additives, because enamel paint can harden in the gun if it is not removed.

On pressure feed guns, release the pressure in the cup, loosen the air cap, and force the paint into the cup by triggering the gun. Take the air cap off and empty the contents into a suitable container, then refill the cup with a clean, compatible solvent. The air cap can be left off. Spray the solvent out of the gun until clean solvent is flowing. If a tank is used, clean as directed by the manufacturer and reassemble.

Areas in the United States with air pollution problems, such as southern California, require the use of enclosed spray gun cleaning equipment. The spraying equipment—guns, cups, mixing sticks, and strainers—is placed in the tub of the gun washer (Figure 15-38). The lid is closed, then the air-operated pump recirculates the thinner into the upper portion of the tub, cleaning the inside and outside of the equipment. In

FIGURE 15-38 Inside a typical gun washer

less than 60 seconds, the equipment is clean and ready for use.

The gun washer saves the painter time. Compared with traditional manual cleaning methods, the gun washer saves 10 minutes on each color change. It offers increased safety for the painter, because the skin no longer is exposed to the drying effects of solvent. The system is designed so sludge from the cleaning action settles to the bottom for easy drainage and disposal with other shop wastes. Check the owner's manual for complete operational details and the proper solvents to use.

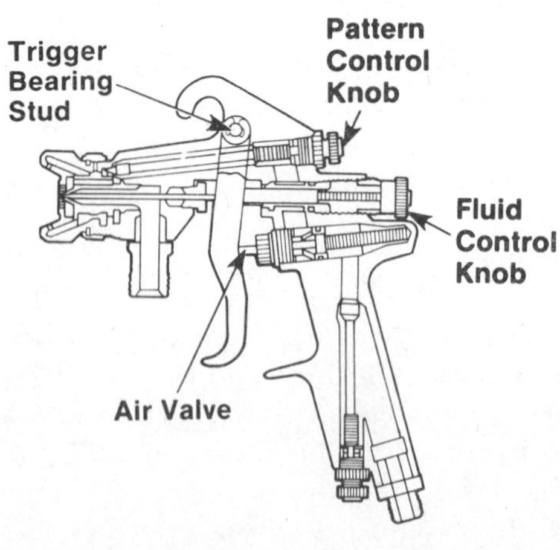

FIGURE 15-39 Parts to lubricate in a spray gun

Most spray gun manufacturers recommend daily lubrication of the parts shown in Figure 15-39 with light machine oil. Packings, springs, needles, and nozzles will still have to be periodically replaced due to normal wear and tear. This should be done only in accordance with the manufacturer's instructions. Extreme care should be taken not to over lubricate; the excess oil could overflow into the fluid passages, mixing with the paint and resulting in a defective finish.

SHOP TALK

Never soak the entire gun in cleaning solvent. Doing this will dry out the packings and neutralize the lubrication. When manually cleaning guns, it is wise to wear rubber gloves and a respirator.

SPRAYING TROUBLESHOOTING

If a spray gun is not adjusted, handled, and cleaned properly, it will apply a defective coating to the surface. The most common spray gun application problems and suggested remedies are given in the gun service manual.

PREPARING THE VEHICLE

Before doing any refinishing work the vehicle must be prepped (Figure 15-40). Keep in mind that it is unwise to apply any kind of finish to a surface that has not been prepared properly. Quality suffers, customer dissatisfaction is inevitable, and costs increase because the job usually has to be done over. A good beginning pays off in a savings of materials and time and in a higher quality refinishing job.

There are two types of automotive refinishing surfaces:

- Previously painted surface
- Bare metal or substrate

Even if the original finish is in good condition, it should be lightly sanded or scuff sanded after washing to remove dead film and to smooth out imperfections. If the surface is in poor condition, all the paint should be removed down to the bare metal. This way, a good foundation is achieved.

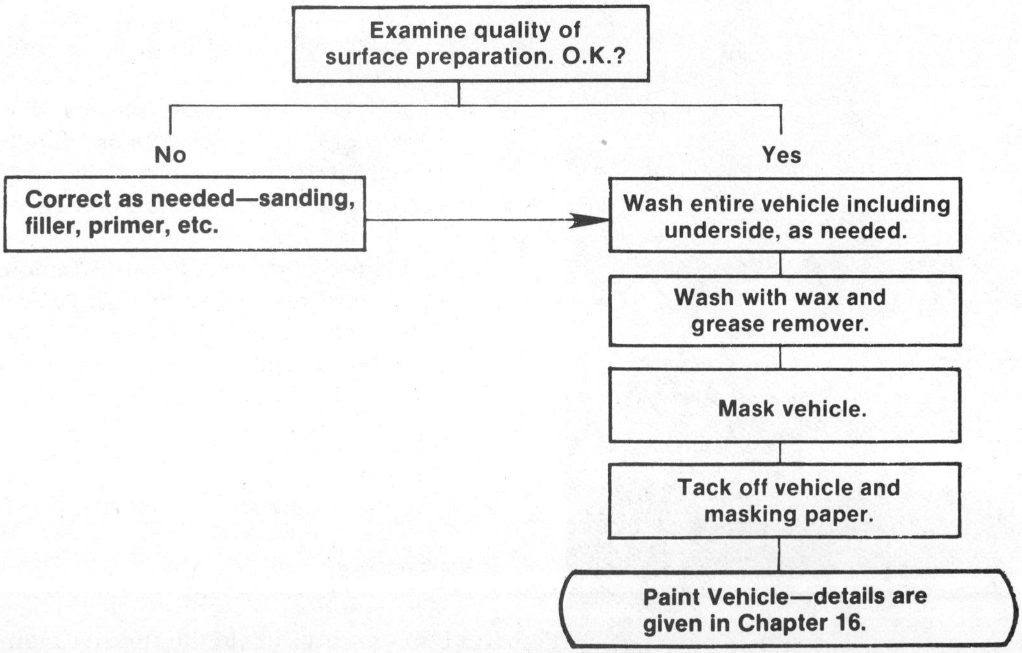

FIGURE 15-40 Steps in prepping a vehicle using the I-CAR system

CLEANING

The vehicle must be washed to remove mud, dirt, and other foreign matter before being brought into the shop (Figure 15-41). Hose it down, sponge with detergent and water, then rinse thoroughly. Wash the top, front, rear, and sides—and let dry. Dirt collects in the doorjamb areas, around the trunk and hood openings, and especially in the wheel wells. Failure to clean these areas can lead to large amounts of foreign matter in the newly painted surface.

FIGURE 15-41 Washing the vehicle

Keep in mind that it is just as important to make sure that all adjacent panels are as clean as the area to be refinished. Be sure there is no wax, grease, or other contaminants imbedded in the finish. Gasoline is a poor wax-removing solvent and can itself deposit contaminating substances on the surface. Do not use synthetic reducers for cleaning up either; they absorb reducer into the paint film and blistering or lifting can result.

Be sure to thoroughly clean areas where heavy wax buildup can be a problem, such as around trim, moldings, door handles, radio antennae, and behind the bumpers. Paint will not adhere properly to a waxy surface. To apply wax and grease remover, fold a clean, dry cloth, soak it with the remover, and wipe the painted surface (Figure 15-42). While the surface is still wet, fold a second clean cloth and wipe dry. Work small areas and wet the surface liberally. Never attempt to clean too large an area; the solvent will dry before the surface can be wiped. Maximum effectiveness will be achieved by wiping up the remover while it is still wet. Always use new wiping cloths because laundering might not remove all oil, wax, or silicone residue.

To remove any last trace of moisture and dirt from seals and moldings, use compressed air at low pressure. Wax and silicone can penetrate beneath the surface. Because this contamination is not easily detectable, it is wise to assume that it is present. Always include some wax and grease remover or detergent in the water when wet sanding. Special attention should be paid to tar,

FIGURE 15-42 Applying a wax and grease remover

gasoline, battery acid, coolant, and brake fluid stains. These can penetrate well beneath the surface of the paint film and must be removed during the sanding operation.

Once the vehicle has been cleaned, the paint surface should be carefully checked. Look for any signs of paint film breakdown—including checking, cracking, or blistering. Pay particular attention to the gloss level; low gloss often indicates surface irregularities. To determine if the old surface has good adhesion, sand through a small part of the finish and featheredge it. If the edge does not break or crumble, the old paint will likely stay on when the refinish color is applied over it. A roughness or pitting of the surface is a sign of developing rust. It must be remembered that most forms of paint failure are progressive. These conditions cannot be stopped by any form of repairing; in fact, repairing will usually accelerate the deterioration of the paint. If the old finish is badly weathered or scarred, it is *not* suitable for refinishing. When this situation occurs, the old finish should be completely removed. There are three common ways of stripping paint from metal surfaces:

- Sanding or grinding
- Chemical stripping
- Sandblasting

SANDING

One of the most important parts of surface preparation is sanding. Sanding prepares the surface for painting in several ways:

- Chipped paint is sanded to taper the sharp edges that would show up as ridges under the new finish.
- Cracking and peeling paint and minor surface rust must be removed by sanding before applying a fresh topcoat. If not, these conditions will continue to deteriorate and will eventually ruin the new finish.
- Primed areas must be sanded smooth and level.
- The entire surface to be refinished must be scuff sanded to improve adhesion of the new finish. Scuff sanding also removes any trace of contaminants on the existing finish.

Because coated abrasives (sandpaper) perform the actual cutting and leveling in the sanding operation, selecting the correct abrasive is critical.

Sandpaper

When sandpaper is constructed, a flexible or semirigid backing attaches to the abrasive grains, which are bonded by an adhesive. Hence, the most efficient results on a particular application depend on the selection of suitable combinations of grains, adhesives, and backings. The painter must select and correctly use the proper sandpaper for optimum productivity, material cost efficiency, and the best finish.

The rough side of the sandpaper is called the grit. Grit sizes vary from coarse to micro fine grades and are ordered by number (Table 15-2). The lower the number, the coarser the grit. For example, a No. 24 grit is used to remove old paint film; while a No. 320, No. 360, or No. 400 grit is used to sand the gloss of an old finish to be refinished. Very fine and ultra fine grits are used primarily for color sanding. Compounding papers, the No. 1250, No. 1500, and No. 2000 grits, are used to solve problems on base coat/clear coat surfaces.

All domestic sandpaper manufacturers conform to the same grading system for uniform consistency of standards. Differences in performance when using the same grit size from different manufacturers can be attributed to differences in manufacturing processes, quality, and/or sanding methods.

Sandpaper is generally manufactured in two types of surface distributions (Figure 15-43):

- Closed coat
- Open coat

A closed coat sandpaper is one in which the abrasive grains completely cover the grit side. An open coat sandpaper is one in which the abrasive grains are spaced to cover between 50 and 70 percent of the grit. As for uses, open coat sandpaper is the popular choice on softer materials such as old paint, body filler, plastic, and aluminum, where premature loading of the abra-

TABLE 15-2: SANDPAPER GRIT SIZES

Grit	Aluminum Oxide	Silicon Carbide	Zirconia Alumina	Primary Use for Collision Repair
Micro fine	—	2000 1500 1250	—	Used for base coat/clear coat paint system
Ultra fine	—	800	—	Used for color sanding.
Very fine	—	600	600	Used for color sanding. Also for sanding the paint before polishing.
	400 320 280 240	400 320 280 240	400 — 280 240	Used for sanding primer-surfacer and old paint prior to painting.
	220	220	—	Used for sanding of topcoat.
Fine	180 150	180 150	180 150	Used for final sanding of bare metal and smoothing old paint.
Medium	120 100 80	120 100 80	— 100 80	Used for smoothing old paint and plastic filler.
Coarse	60 50 40 36	60 50 40 36	60 — 40 —	Used for rough sanding plastic filler.
Very coarse	24 16	24 16	24 —	Used on sander or grinder to remove paint.

sive would otherwise be a problem. Closed coat sandpaper generally provides a finer finish and is most commonly used in wet sanding applications.

The most common abrasive sanding disc sizes for disc and dual action sanders are 5, 6, and 8 inches. Sanding disc grit sizes generally range from No. 50 to No. 400 grit. To apply nonstick sandpaper to the backing pad of a sander, squeeze a few drops of adhesive on the backing pad. Spread the adhesive evenly on the pad (Figure 15-44A), then center the disc on the pad and press it into place (Figure 15-44B). When using a self-

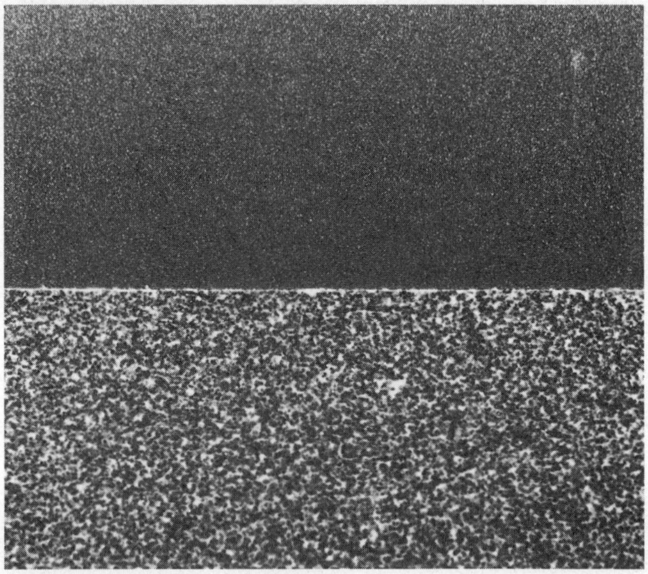

FIGURE 15-43 (Top) Closed coat sandpaper; (bottom) open coat sandpaper

adhesive sanding disc, be sure to center the paper on the pad before pressing it into place. Immediately after finishing the sanding operation, remove the used disc from the backing pad. If it is not removed right away, the adhesive will harden and cause the disc to stick fast to the backing pad. Should this occur, use solvent on a rag to dissolve the adhesive and then remove the disc.

Abrasive grinding discs are used for rough jobs, such as grinding off rust and paint. They are available in grit sizes of No. 16 to No. 50, and in diameters of 3 to 9 inches. A dual action sander can be used to remove light rust, but heavy surface rust must be removed with a grinder (Figure 15-45). The grinding disc is first assembled to the backing plate and then the disc/plate assembly is attached to the grinder.

METHODS OF SANDING

Sanding can be done:

- Using a sanding block
- Using power equipment

Most heavy sanding—such as removal of the old finish—is done by power sanders. But some delicate operations dictate block sanding.

Block Sanding

Block sanding is a simple back and forth scrubbing action with the sandpaper flat against the surface. Use the following general procedure:

1. Use sandpaper that fits the sanding block, or cut the sheet of sandpaper to fit the block.

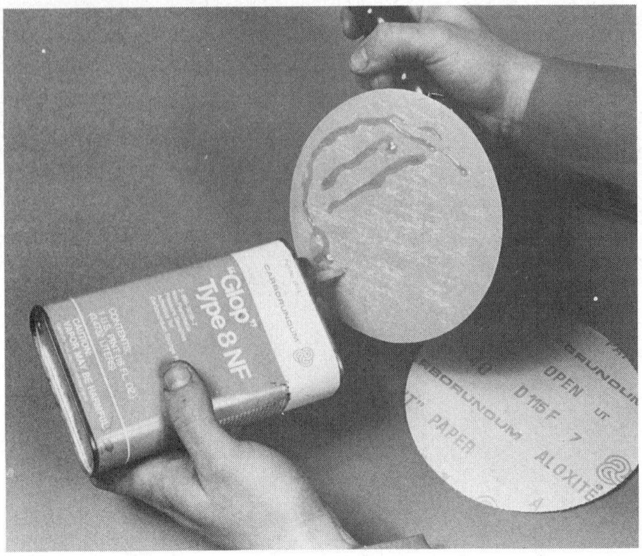

A

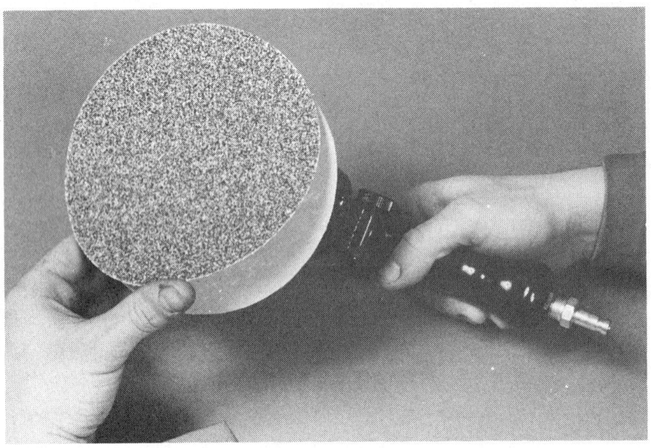

B

FIGURE 15-44 (A) Spread the adhesive on the backing pad, then (B) press the sanding disc in place.

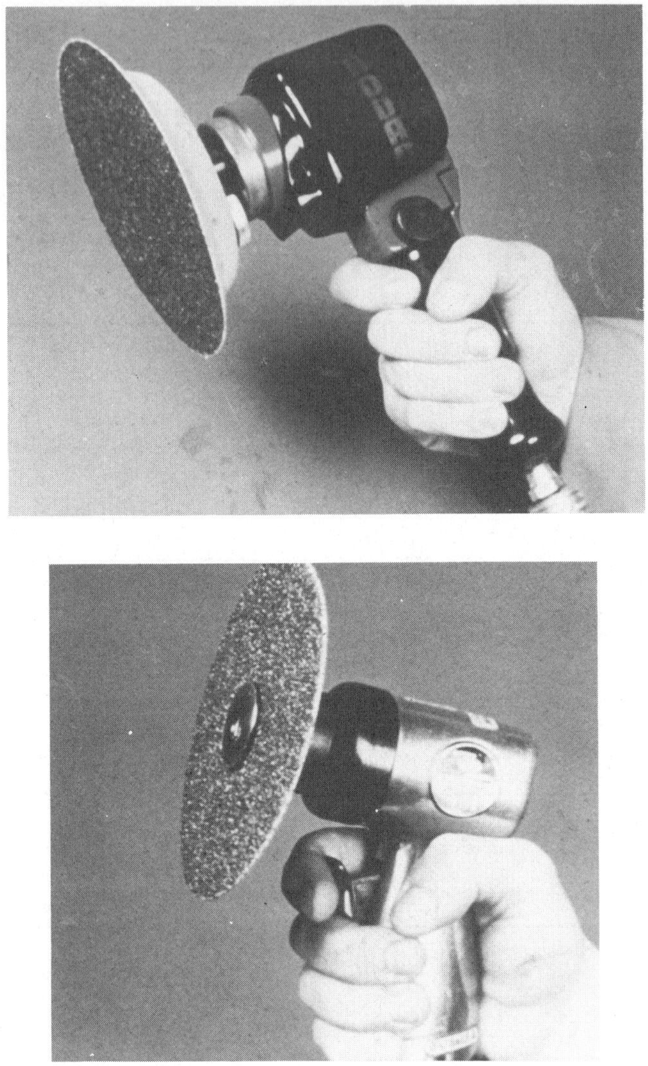

FIGURE 15-45 (Top) Dual action sanders can be used to remove light rust. (Bottom) A grinder must be used for heavy surface rust.

2. Place the paper on the block and hold the block flat against the surface. Apply even, moderate pressure along the length of the block (Figure 15-46). Sand back and forth with long, straight strokes.
3. Do not sand in a circular motion. This will create sand scratches that might be visible under the finish. To achieve the best results, always sand in the same direction as the body lines on the vehicle (Figure 15-47).
4. Be sure to thoroughly sand areas where a heavy wax buildup can be a problem, such as around trim, moldings, door handles, radio antennas, and behind the bumpers. Paint will not adhere properly to a waxy surface.

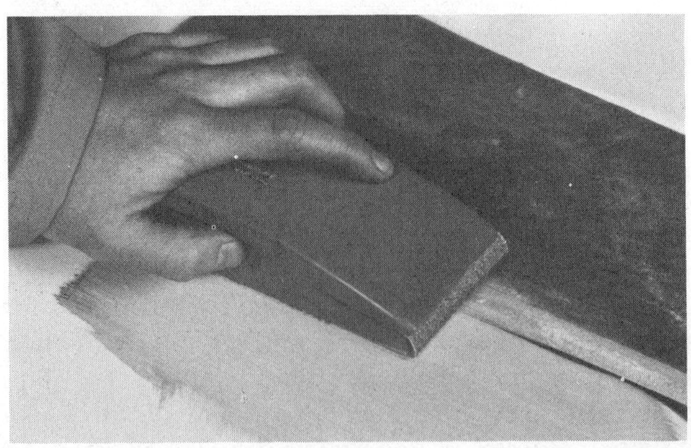

FIGURE 15-46 The sanding block works well on flat surfaces.

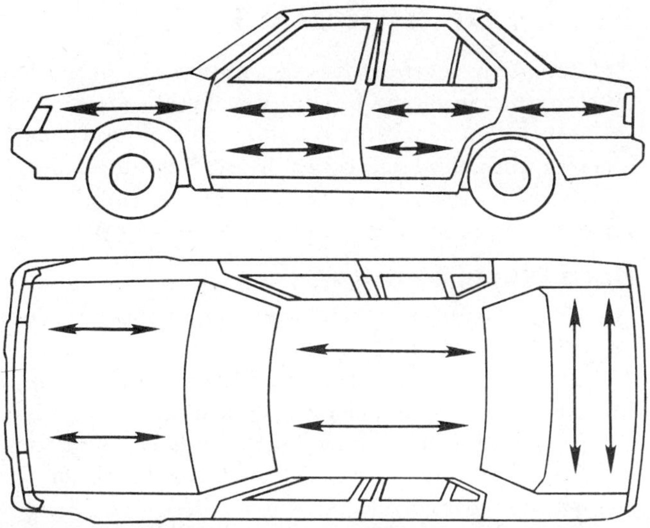

FIGURE 15-47 Sand with the body lines.

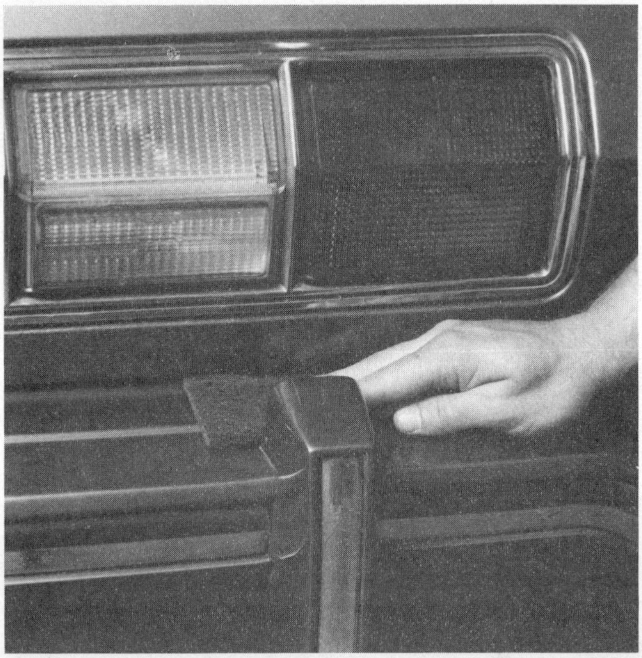

FIGURE 15-49 Using a nylon scuff pad in tight spots

5. To sand convex or concave panels, employ a flexible sponge rubber backing pad (Figure 15-48).
6. Carefully sand coarse grit scratches using a finer grit paper. Hard-to-reach areas are easier to sand with a nylon scuff pad (Figure 15-49).
7. When block sanding primer or putty, make certain to sand the area until it feels smooth and level. Rub a hand or a clean cloth over the surface to check for rough spots.
   —*Dry.* This is basically the back-and-forth procedure just described. But one of the problems with it is that the paper tends to clog with primer or filler dust. Tapping the paper from time to time will remove some of the dust. Another suggestion is to use talc-coated or dry-lubricated sandpaper, hich tends to minimize clogging. Specially treated open coat paper resists clogging for long life. The majority of acrylic primer-surfacers should be sanded dry.
   —*Wet.* Wet sanding also reduces the problem of paper clogging. It is basically the same action as dry sanding except that water, a sponge, and a squeegee are used in addition to the sanding block (Figure 15-50). Sandpapers are available in dry or wet-or-dry types.

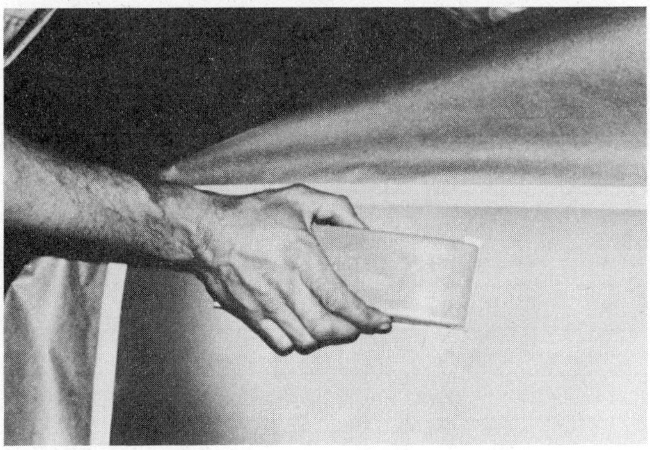

FIGURE 15-48 Use a flexible sponge rubber backing pad on convex and concave panels.

Power Sanding

Four types of power sanders are available:

- Disc sander or grinder
- Orbital or jitterbug pad sander
- Dual action (DA) sander
- Straight line or board sander

WARNING

Always wear a dust mask when sanding, and wear both a dust mask and face shield when grinding.

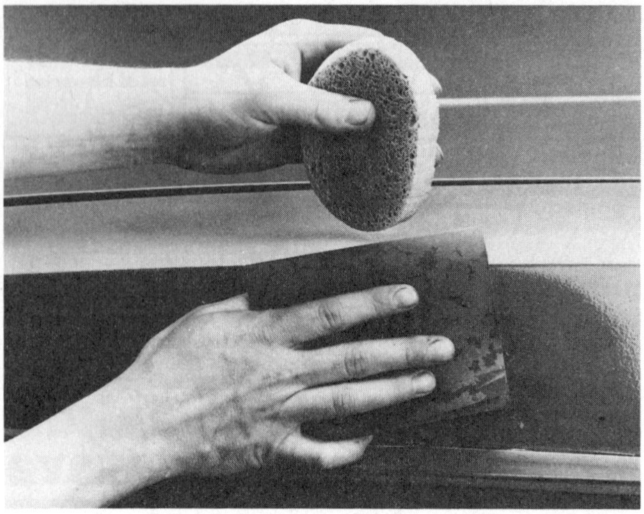

FIGURE 15-50 When wet sanding, water can be applied with a sponge.

When power sanding, replace the sandpaper when paint begins to cake or "ball up." This paint buildup can scratch the surface and reduce the sanding action of the disc. Slowing down the speed of the sander will also help prevent paint buildup on the sanding disc and prolong sandpaper life. Generally, six to eight sanding discs will be required to featheredge the chips and scratches on the average vehicle. Table 15-3 outlines uses for the various power sanders.

TYPES OF SANDING

There are several types of sanding that a painter must master. Some can be completed with power sanders alone, others with a combination of power and block sanding, and still others by block sanding alone.

Most are done during the surface preparation stage, but one—scuffing—is performed after priming.

Bare Metal Sanding

If the metal work has been done properly, little sanding of bare metal should be required. But once in a while the metal arrives very rough from coarse sanding. In such cases, it might be necessary to sand it with No. 50 grit to level out the burrs, nibs, and deep scratches. Remember, the smoother the surface, the easier the refinishing work.

Thorough Sanding

Use this procedure for two specific conditions:

- Where the old finish is rough or in poor shape
- To level and smooth primed areas

Since the primer-surfacer is primarily intended to fill low spots and scratches, thorough sanding must be done in a manner that will leave material in the low spots and cut away the high spots. Block sanding is highly recommended for this purpose. A guide coat is very helpful and assists in pointing out the depressions. A guide coat is applied by spraying a very light coat of a different color material over the primer-surfacer. Sand-

TABLE 15-3: USES OF SANDERS

		Normal Use							
Sander Type	**Normal Area of Operation**	**Sanding old finishes**	**Paint stripping**	**Feather-edging**	**Rough sanding of solder**	**Rough sanding of metal putty**	**Rough sanding of polyester putty**	**Sanding of metal putty**	**Sanding of polyester putty**
Disc Sander	Suitable for narrow areas	A	A	C	B	A	C	C	C
Dual Action Sander		A	C	A	C	A	A	A	A
Orbital Sander		B	C	B	C	A	A	A	A
Straight Line Sander	Suitable for wide open spaces	C	C	C	C	A	A	A	B
Long Orbital Sander		C	C	C	C	A	A	A	B

NOTE: It is important that the correct type of sander and abrasive disc be used for each type of job. Also, always wear a mask or use some sort of dust arrester when using the sander.
A Preferred
B Acceptable
C Least preferred

ing reveals the high spots and the low spots: the guide coat sands off the highs, and remains in the low areas.

For the average wet sanding job, use No. 360 or No. 400 grit when applying an acrylic lacquer or enamel topcoat and No. 320 grit when the topcoat is an alkyd enamel. Dry sanding runs 100 grit coarser because sanding dust partially fills in the grit.

Light Sanding

This procedure should be done on all areas where the old finish is in good condition. The purpose is to partially reduce the gloss and to improve adhesion. Use an orbital or dual action sander, or use a sanding block—but *never* use a disc grinder or sander.

If the new topcoat will be lacquer or enamel, use a No. 360 or No. 400 grit sandpaper. If it will be alkyd enamel, use a No. 320 grit sandpaper. For base coat/clear coat finishes, proper surface preparation is especially critical. It is important to sand all surfaces to be refinished with No. 400 grit or finer paper, followed by 600 grit. Sanding can be done wet or dry.

Featheredging

If a new coat of paint was applied right over the broken areas of the old finish, the broken film would be very noticeable through the topcoat. The broken areas must be featheredged, meaning that the sharp edge of the broken film must be tapered down by sanding (Figure 15-51). Then the bare metal areas are filled with a primer-surfacer and the entire area is sanded smooth and level.

Featheredging with a sanding block is usually a two-step procedure:

1. First, cut down the edges of the broken areas with a coarse No. 80 grit paper and follow with No. 240 grit.

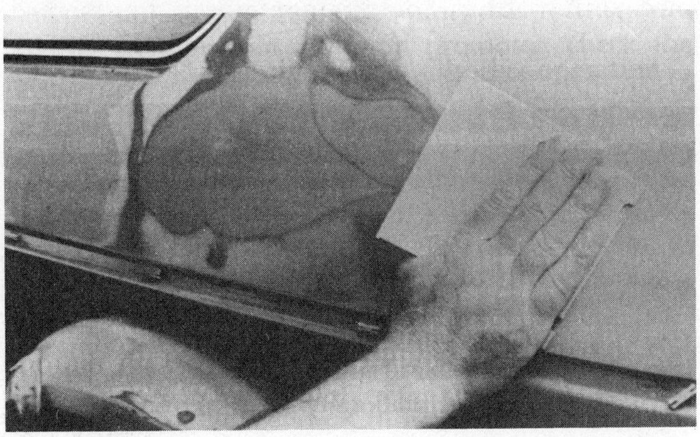

FIGURE 15-51 Featheredging tapers successive coats of paint and primer away from the metal to create a smooth surface.

2. Then, complete the taper of the featheredge with either a No. 360 or No. 400 grit paper and water to produce a finely tapered edge and eliminate coarse paper scratches.

When featheredging with a power sander, an orbital or dual action sander equipped with a flexible backing pad is recommended. Use a No. 80 grit for the rough cut, followed by a No. 180 or No. 240 grit for the fine work. Start by positioning the sanding disc flat against the work surface. Using the outer edge, or approximately 1 inch of the sanding disc, cut away the rough paint edges. Do not hold the sander at an angle greater than 10 degrees from the surface or it will cut a deep gouge in the paint.

After initially leveling the rough paint edges in this manner, lay the sander flat on the panel and finish tapering the paint layers by moving the sander back and forth in a crosscutting pattern. Start over the chipped area and work in an outward direction. Stop frequently and run a hand over the sanded area to feel for rough edges. The total outside edge should be cut new and all original chipped edges should be removed to avoid any lifting during refinishing. When the surface feels smooth, and rings of old paint and primer are visible, the featheredging is complete.

Grinding

Machine grinding is suitable for removing the old finish from small flat areas and gently curved areas. Start with a No. 24 grit open-coat disc. By holding the face of the disc at a slight angle to the surface, work back and forth evenly over the area to remove the bulk of the old finish down to the metal. Follow this with a No. 36 and then a No. 80 close coat disc. Go over the entire repair area and surrounding surface to eliminate the troughs or steps caused by the coarse disc. When using the grinder, care must be taken to prevent gouging or scarring the metal.

After all of the paint is removed, be sure to finish sand the area with the orbital or dual action sander and No. 80 grit paper to remove the metal scratches. Then finish sand using No. 180 grit paper. In this way most of the scratches created by the stripping operation will be eliminated. Remember that any metal that has been scratched with very coarse paper will require filling to the depth of the scratch plus the height of the burr.

The dust from sanding and grinding operations can often cause a major problem. As shown in Figure 15-52, there are several dust control systems available.

CHEMICAL STRIPPING

A chemical paint remover is recommended for stripping large areas of paint. It is very effective in those places that a power sander cannot reach.

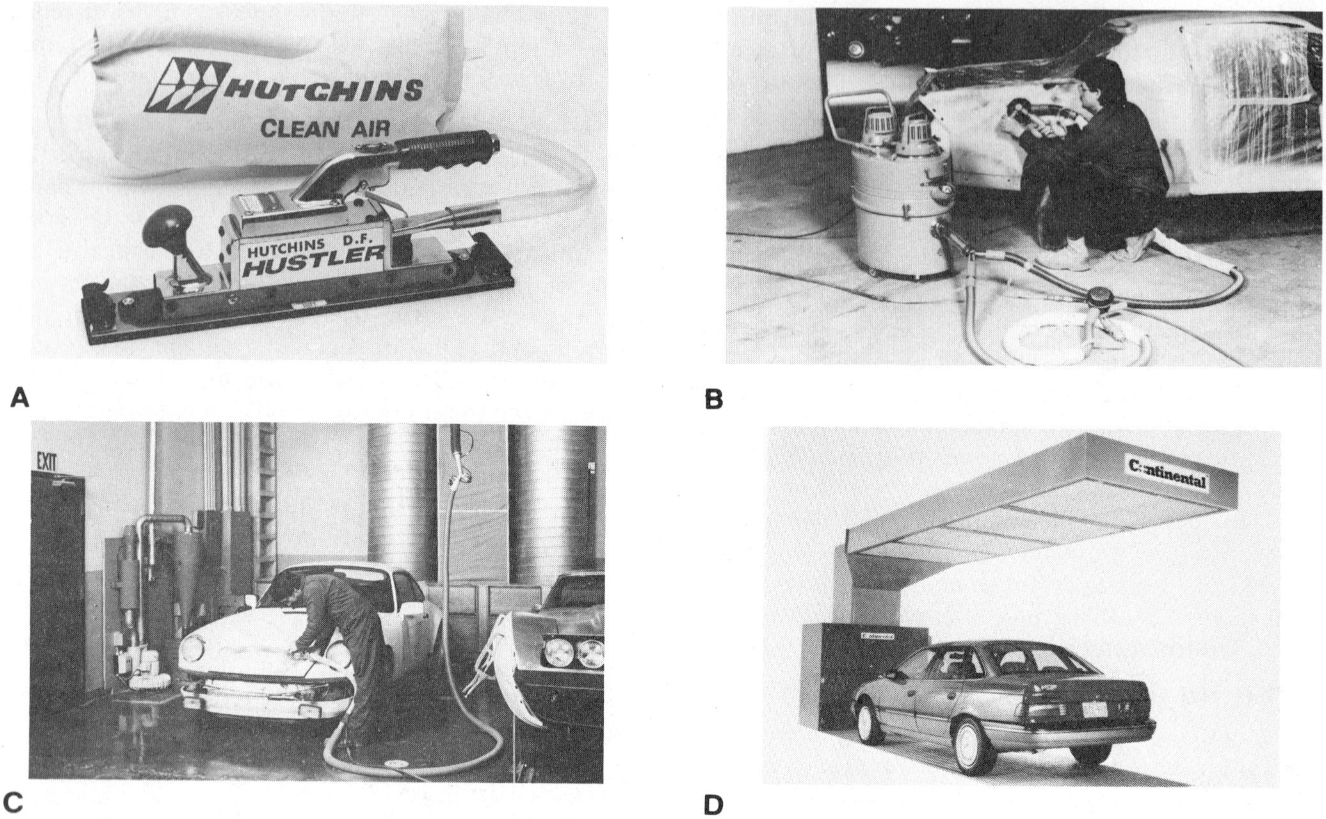

FIGURE 15-52 Various dust control systems: (A) sander equipped with a dust collecting bag; (B) portable vacuum system; (C) central vacuum system; and (D) prep station. With the prep station, all sanding, solvent cleaning, masking, tack cloth cleaning, and sealing can be done in one location.

Before applying paint remover, mask off the area to ensure that the remover does not get on any area that is not to be stripped. Use two or three thicknesses of masking paper to give adequate protection. Cover any crevices to prevent the paint remover from seeping to the undersurface of a panel. Slightly scoring the surface of the paint to be stripped will help the paint remover to penetrate more quickly.

Paint remover should be applied following the manufacturer's instructions. Pay attention to warnings regarding ventilation, smoking, and the use of protective clothing such as PVC or rubber gloves, long sleeve shirts, and safety glasses or goggles. If paint remover comes in contact with the skin or eyes, it will cause irritation and burning.

Traditional paint removers will harm the plastic beneath the paint. Also, the chemical residue is difficult to remove completely, and can cause paint adhesion problems. Today, however, there are a number of paint removers made especially for use on plastics.

To apply paint remover, brush on a heavy coat in one direction only to the entire area being treated. Use a soft bristle brush, but do not brush the material out. Allow the paint remover to stand until the finish is softened. Although paint remover is quickly effective on most topcoats, some surfaces can prove stubborn. (For example, acrylic lacquer becomes sticky, making it more difficult to remove.) More than one application might be needed.

Caution should be taken when removing the loosened paint coatings. Some paint removers are designed to be neutralized by water. Others are more easily removed with a squeegee or scraper (Figure 15-53). Be sure to rinse off any residue that remains using a cleaning solvent and steel wool, followed immediately by wiping with a clean rag. This rinsing operation is essential. Many paint removers contain wax which, if left on the surface, will prevent the paint from adhering, drying, and hardening properly.

SANDBLASTING

Sandblasting leaves a clean, dry surface that is ideal for refinishing. It is very fast and has the further advantage of revealing rusted areas and places where hidden rust can result in scaling after the job has been finished.

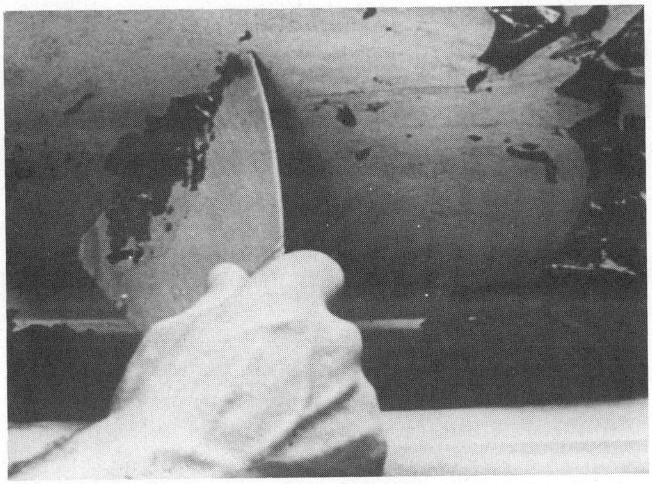

FIGURE 15-53 Removing paint with a scraper after an application of chemical stripper

FIGURE 15-54 Blasting with a "captive" sandblaster

In addition, sandblasting makes hard-to-reach areas accessible. Also this method saves time when compared with sanding/grinding and chemical stripping.

Typically, sandblasting operations concentrate the pressure and flow of air and sand. Usually found in shops in a small version (40 to 300 pound models), the sandblaster can be adjusted to vary the blast volume, focusing on the spot at hand rather than blasting in a wide pattern. But because of possible damage to aluminum, plastic, and other substrates other than sheet steel, sand or hard media blasting is not recommended. With the recent introduction of soft plastic stripping media, however, it is possible to use the sandblasting technique of paint removal on almost any surface. Methods of using a soft plastic stripping media are similar to the conventional sandblasting methods, with the addition of these guidelines:

- Use low (20 to 25 psi) air pressure at the nozzle.
- Hold the nozzle at a shallow (20 to 30 degree) angle to the panel.
- Hold the gun 18 to 24 inches from the panel.
- Use a broad, sweeping back-and-forth motion.
- Stop when the primer is reached.

As mentioned in Chapter 3, some of the newer models provide dust-free, "captive" sandblasting (Figure 15-54). They contain a built-in vacuum and filtration system that vacuums up and recycles the abrasive, while blasting away paint, rust, and other debris.

BARE METAL TREATMENT

Proper bare metal treatment is an important part of every successful automotive refinishing operation. Yet it is often ignored or carried out in a haphazard manner. This can only result in poor adhesion, corrosion, and, as a result, customer dissatisfaction. Several metal treatment systems are available that ensure a good bond. One of the best is metal prep/conversion coating. To apply, proceed as follows:

- *Cleaning to remove contaminants.* Use a wax and grease remover and apply it to the surface. Work small areas, 2 to 3 square feet at a time, wetting the surface liberally. While the surface is still wet, fold a second clean cloth and wipe dry.
- *Cleaning with metal conditioner.* Mix the appropriate cleaner with water in a plastic bucket according to the label instructions. Apply with a cloth, sponge, or spray bottle. If rust is present, work the surface with a stiff brush or abrasive pad. Then while the surface is still wet, wipe it dry with a clean cloth.
- *Applying conversion coating.* Using the proper conversion coating for the metal to which it is applied (check the manufacturer's recommendation), pour the conversion coating into a plastic bucket. Using an abrasive pad, brush, or spray bottle, apply the coating to the surface. Then leave the conditioner on the surface 2 to 5 minutes; it will create a zinc phosphate coating on the steel. Apply only to an area that can be coated and rinsed before the solution dries. If the surface dries before the rinsing, reapply. Flush the coating from the surface with cold water. Wipe dry with a clean cloth and allow to air dry completely. The desired primer or primer-surfacer can then be applied.

Another type of metal treatment is done using a wash primer. A wash primer is a sprayable surface treatment that eliminates the need for a conversion coating. The vinyl resin in the wash primer provides corrosion resistance; the reducer in the wash primer contains phosphoric acid for strong bonding. A wash primer with phosphoric acid reducer cleans and etches the metal and assists in the adhesion of the subsequent paint film. Before applying a wash primer carefully read and follow the manufacturer's directions and special instructions.

Typically, metal treatment and priming are considered separate surface preparation steps. Some new products, however, actually make it possible to combine these steps. Self-etching primers etch the bare metal to improve paint adhesion and corrosion resistance, while providing the priming and filling properties offered by primer-surfacers. Self-etching primers work best on lightly sanded surfaces where a slight-to-moderate amount of filling is required. They must be applied as directed by the manufacturer.

Preparing Metal Replacement Parts

Many manufacturers and parts suppliers prime replacement panels. The function of this primer coat is to protect the metal against corrosion. It does not necessarily provide a firm basis for a paint system. Although most primers do have this dual function, the supplier should always be consulted.

It must be remembered that some replacement parts are provided from the manufacturer with a coating. This coating is *not* intended to serve as the primer. A primer must be applied to these replacement parts or the color coat will not adhere properly. The usual procedure is to clean with wax and grease remover, then examine the part for scratches and other imperfections. Sand any imperfections until smooth but do not try to remove the coating completely. Scuff sand the entire panel, then apply primer before painting. If in doubt about the coating, check with the manufacturer of the part for the recommended finishing procedures.

Any bare metal replacement panels protected with grease should also be cleaned with a wax and grease solvent. They should be washed with liberal amounts of solvent, changing rags frequently. Then treat the panels with a bare metal conditioner, flush them down with water, and dry them off.

UNDERCOAT SYSTEMS

Proper undercoats are the foundation for an attractive, durable topcoat. If the undercoat—or combination of undercoats—is not correct, the topcoat appearance will suffer and might even crack or peel.

Undercoats can be compared to a sandwich filler that holds two slices of bread together. The bottom slice of bread is called the substrate (or surface of the vehicle). That surface can be bare metal or plastic, or a painted or preprimed surface. The undercoat is the sandwich filler applied to the substrate. It makes the substrate smooth and provides a bond for the topcoat. The upper slice is the topcoat, the final color coat that the customer sees.

Undercoats contain pigment, binder, and solvent. There are four basic types of undercoats:

- Primer (Figure 15-55)
- Primer-surfacer (Figure 15-56)
- Primer-sealer
- Sealer (Figure 15-57)

Most surfaces must be undercoated (Table 15-4) before refinishing for several reasons: to fill scratches, to provide a good base for the topcoat, to promote adhesion of the topcoat to the substrate, and to ensure corrosion resistance. A primer alone, however, will not fill sand scratches or other surface flaws. Primer-surfacers are used to provide both priming and filling in one step. Primer-sealers are applied to prevent solvents in the topcoat from being absorbed into the porous primer-surfacer. These three undercoats—primer, primer-surfacer, and primer-sealer—can be used together, singularly, or in various combinations, depending on the surface condition and size of the job (Table 15-5). Sealers are used to improve adhesion between the old and new finishes. To provide good adhesion, a sealer should always be used over an old lacquer finish when the new finish is to be enamel. Under other conditions, a sealer is desirable but not absolutely necessary.

Before applying any undercoat, be sure to treat all bare metal with metal conditioner. Reduce the under-

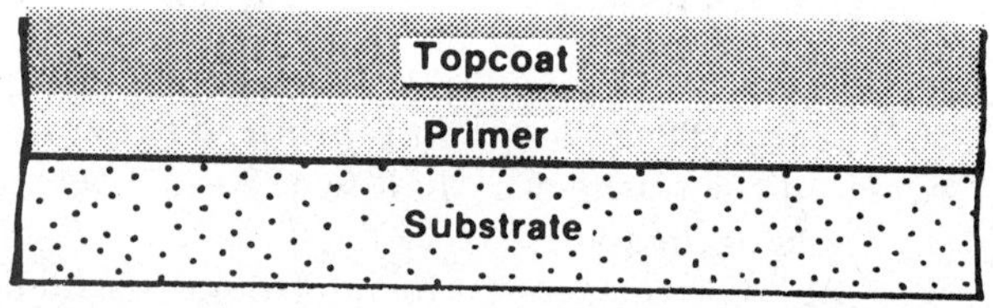

FIGURE 15-55 Primers are generally applied in a thin coat and do not require sanding. Primers are applied over bare metal substrate.

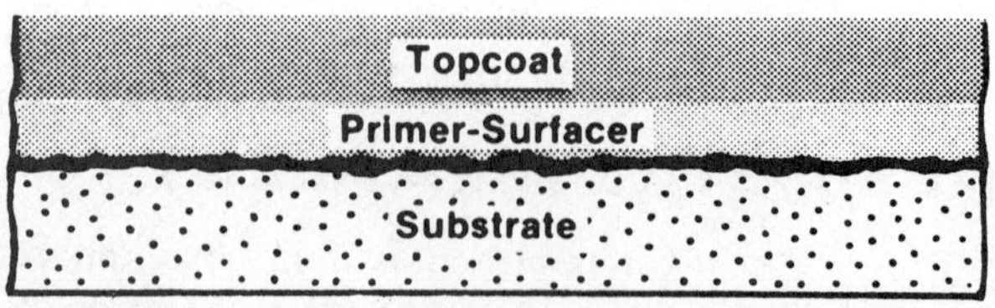

FIGURE 15-56 Primer-surfacers—unlike primers—must be sanded. They can be applied over bare metal or old finishes.

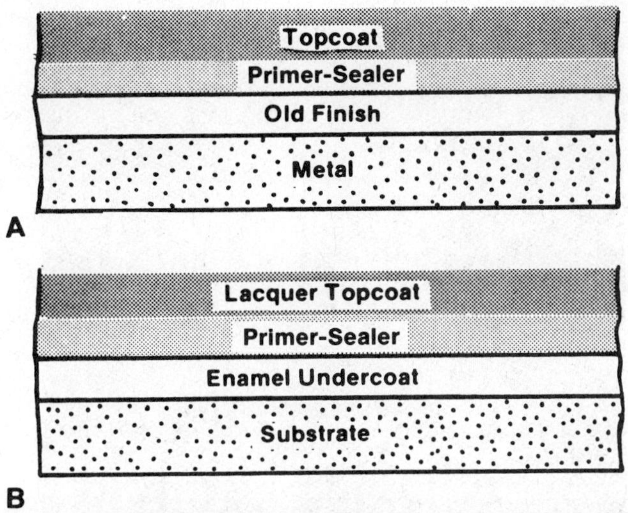

FIGURE 15-57 (A) Primer-sealers improve topcoat adhesion. They can be applied over primers, primer-surfacers, or the old finish. (B) When a topcoat different from the old finish (in quality as well as color) is applied, a primer-sealer should be used.

TABLE 15-4: FUNCTIONS OF UNDERCOATS

Undercoat Function	Primer	Primer-Surfacer	Primer-Sealer	Sealer
Resists rust and corrosion	Yes	Yes	Yes	No
Makes topcoat adhere better	Yes	Yes	Yes	Yes
Fills scratches and nicks	No	Yes	No	No
Provides uniform hold out of the topcoat	No	No	Yes	Yes
Prevents show through of sand scratches	No	No	Yes	Yes

TABLE 15-5: SURFACE APPLICATIONS FOR UNDERCOATS

Undercoat Surface	Primer	Primer-Surfacer	Primer-Sealer	Sealer
Bare substrate (metal, fiberglass, or plastic)	Yes	Yes	Yes	No
Sanded old finish	No	Yes	Yes	Yes

coat according to the manufacturer's instructions. Be careful to select the proper solvent for the weather conditions and mix the materials thoroughly.

Apply the first coat of undercoat and allow it to flash dry, following the recommendations on the label for flash time. Then apply two or three more medium wet coats for additional film buildup, with flash time between each coat. When making a spot repair, extend the additional coats several inches out from the first coat.

Allow the undercoat to dry thoroughly. After the undercoat is dry, block sand the area until it is smooth. For best results, use No. 320 grit sandpaper. If very fine scratches still appear, another coat of undercoat might be all that is required to fill them.

PUTTY APPLICATION

Some sand scratches might require additional filling with a thin glazing putty. This putty is generally used to fill small scratches and pinholes, but use it sparingly. To apply the putty, squeeze a small amount from the tube onto the edge of a squeegee. Quickly press the putty onto the repair area and level it off with a flat scraping motion. Use moderate pressure to assure that the putty fills the depression completely. Make additional passes quickly, if necessary.

Allow the putty to air dry until it is hard. Test with a fingernail for hardness before sanding. If it is sanded too soon, the putty will continue to shrink, leaving part of the scratch unfilled. Once it hardens, the putty should be dry sanded with No. 80 to 180 grit paper or wet sanded with No. 320 grit paper. Wet sanding is carried out in the same manner as dry sanding; however, it requires special paper and plenty of water. After sanding the puttied area, clean the surface and then reprime. If the putty has been wet sanded, make sure to dry the surface thoroughly before applying primer-surfacer.

Lacquer-based glazing putties are being partially replaced by polyester putties, or finishing fillers as they are also called, and by polyester primer/fillers. Both products must be mixed with hardener before starting the application and can be applied to filler, bare metal, or old paint finishes. The use of these high-viscosity, finely textured "fillers" eliminates the traditional primer/putty/primer process. Because they chemically harden, they cure quickly and can be primed and refinished without the worry of sand scratch swelling commonly associated with lacquer-based glazing putties.

MASKING

Masking is a very important step in the vehicle preparation process. It keeps paint mist from contacting areas other than those that are to be refinished.

Masking has become even more important since the use of acrylic urethane and two-component type paints has become popular. Once these paints dry, the overspray cannot be removed with a thinner or other solvent. It must be removed with a rubbing compound or by other time-consuming means.

Masking Materials

The basic materials for any conventional masking job are masking paper and tape. Automotive masking paper comes in various widths—from 3 to 36 inches. It is heat resistant so that it can be used safely in baking ovens. It also has good wet strength, freedom from loose fibers, and resistance to solvent penetration. Never use newspaper for masking a vehicle since it does not meet any of these requirements. Newspaper also has the added disadvantage of containing printing inks that are soluble in some paint solvents and can be transferred to the underlying finish, and cause staining.

Automotive masking tape comes in various widths—from 1/4 to 2 inches. Larger width tapes are used only occasionally because they are difficult to handle. Automotive masking tape should not be confused with tape bought in hardware or paint stores for home use; this will not hold up to the demanding requirements of automotive refinishing. The average size vehicle takes 2 to 2-1/2 rolls of tape to be completely masked. The use of masking paper and tape dispensing equipment makes it easy to pull and tear off the exact amount of paper and tape needed (Figure 15-58). Some dispensers permit tape to adhere to one or both edges of the paper as it is rolled out.

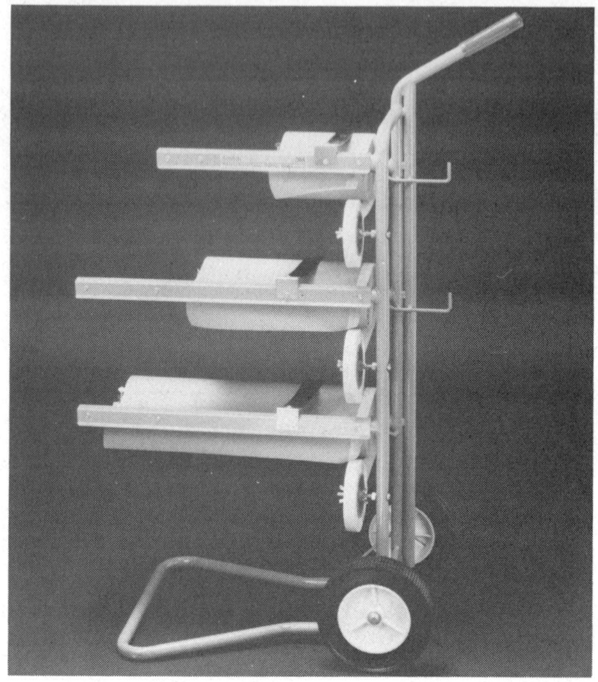

FIGURE 15-58 Typical masking paper and tape dispenser

There are several types of masking covers available. One is the plastic tire cover that eliminates the need for masking off the tire (Figure 15-59). Others include a body cover and a frame cover. Covers are also available in a variety of sizes and shapes to mask headlights and taillights. These covers can be used as the situation demands.

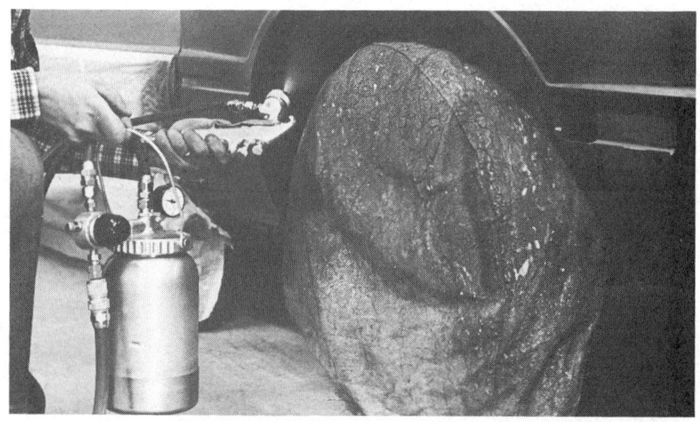

FIGURE 15-59 Plastic tire cover

The newest masking system available to painters is "liquid masking." This effectively seals off the entire vehicle to protect undamaged panels from paint overspray. Liquid masking is used on areas where masking is necessary but difficult to apply, including wheel wells, headlights, grille, underbody chassis, and even the engine compartment.

Before any types of masking materials are applied, the vehicle must be completely cleaned and all dust blown away. Masking tape will not stick to surfaces that are not clean and dry. It is important that the tape is pressed down firmly and adheres to the surface. Otherwise, paint will creep under it. In the case of a two-color job where the color break is not hidden by a stripe or molding, it is vital that the masking tape edge is pressed down firmly.

Conventional Vehicle Masking

There are no clear ground rules on when to mask and when to remove trim, moldings, door handles, etc. This decision depends upon the design of the vehicle, its age and condition, and the expectations of the customer. The objective is good paint adhesion, whether through mechanical or chemical means. Sanding, the use of adhesion promoters, and the ability to clean the area are all variables in the decision-making process.

If it is not possible to adequately sand and clean right up to the part in question, it should be removed. Also, if it will be difficult to mask the piece, it should come off the vehicle. Each part will require an individual decision. Removal of trim is more often necessary when using a base coat/clear coat system than with a single stage finish. The film buildup can be greater with base coat/clear coats, and this additional thickness makes the paint edge more likely to crack or chip.

SHOP TALK

It is wise to completely detail the vehicle before the masking is started and again after the refinishing job is completed. This is because improper masking can cause a dirty paint job if done over a dirty vehicle.

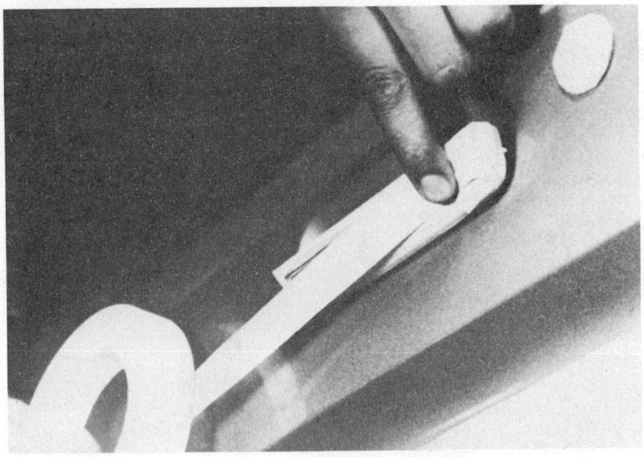

FIGURE 15-60 Method of applying masking tape

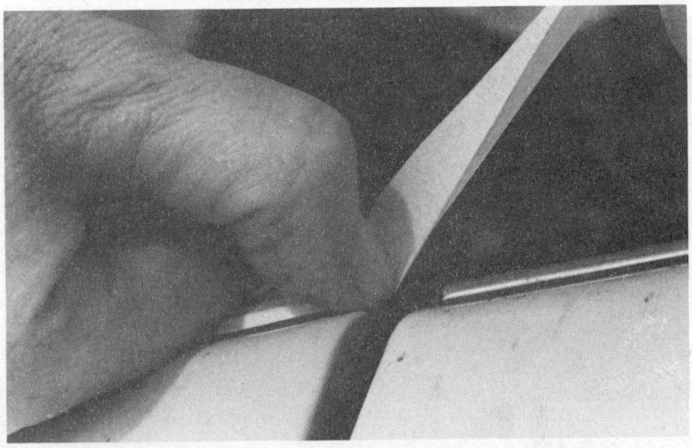

FIGURE 15-61 Cutting tape with a thumbnail

If the painting environment is cold and damp with little air movement, the masking tape probably will not stick to glass or chrome parts because of condensation that has formed on them. The condensation must be wiped off before the tape will adhere properly. When masking doorjambs, be sure to cover both the door lock assembly and the striker bolt.

Although masking tape has an elastic property, it should be stretched only on curved surfaces. This is especially true when masking newly applied finishes that are still soft underneath. It is also wise to avoid stretching the tape because this can increase the degree of tape marking on the finish.

When applying masking tape, it is usually easier to hold and peel the tape in one hand while using the other hand to guide and secure it (Figure 15-60). This provides tight edges for good adherence and also allows you to change directions and go around corners. To cut the tape easily, quickly tear upward against your thumbnail as shown in Figure 15-61. This procedure permits a clean cut of the tape without any stretching.

Be careful that the tape does not overlap any of the areas to be painted. Loop or overlap the inner tape edge to follow curves; the tape will stretch to conform to the curves. Difficult areas such as wheels can be masked using this process, but more often covers are used to save time.

When masking large areas, such as bumpers, it is easier to manage the paper if it is tacked in the middle of the bumper with tape first. Then each side can be secured without the paper dragging on the floor and getting in the way.

Before masking glass areas, remove accessories such as wiper blades. The wiper shafts can be protected in the same manner as radio antennas and door handles. The glass is masked by first applying tape along the very top and along the edges of the moldings. Then, use two pieces of masking paper to cover the glass (Figure 15-62). The tape on the edge of the paper should overlap the tape placed on the molding; the top piece of paper should overlap the bottom piece of paper. If necessary, fold and tape any pleats in the paper to prevent dust and paint seepage.

One-sixteenth and 3/32-inch fine line tape can be used to protect existing stripes from overspray (Figure 15-63). Also use fine line tape for precise color separation in two-color painting and for creating vivid, clean stripes. Its added flexibility makes painting of curved lines easier, with less reworking.

When spraying horizontal surfaces (hood, trunk, and so on), two layers of masking paper should be used to prevent bleed-through or finish-dulling from solvents. Another method to prevent bleed-through is to reverse tape the paper; the paper is taped on the inside and allowed to bellow slightly, which keeps it lifted a bit from the surface.

FIGURE 15-62 Use two pieces of masking paper to cover the glass.

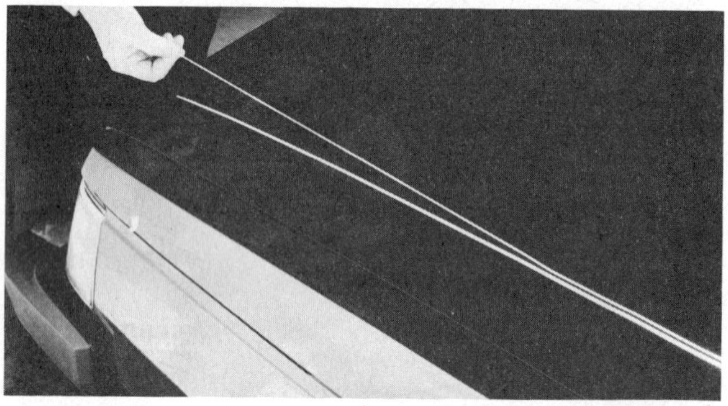

FIGURE 15-63 Fine line tape protects existing stripes from overspray.

A reverse masking method is often used during spot repairs to restrict the shading area and make it less noticeable. A line of definition such as a door edge or a body line next to a refinished panel will sometimes accentuate a slight difference in an otherwise acceptable color match, particularly with metallics. This situation can be avoided by crossing the line and sanding the adjacent panel when preparing for refinishing. Then reverse mask the adjacent panel and refinish as desired. If there is a slight difference after removing the masking paper, just paint across the line and blend in smoothly as with any spot repair. As shown in Figure 15-64, reverse masking can be used for flat surfaces and for curved surfaces.

Inspect the masking very carefully for any overmasked or undermasked areas that will lead to extra work after the vehicle is painted. Overmasked areas mean that the painter must touch up the part of the vehicle that should have been painted. On the other

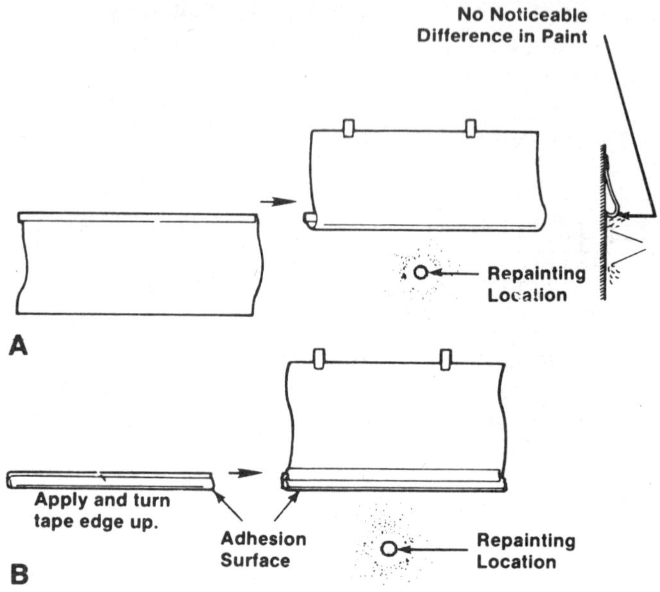

FIGURE 15-64 Reverse masking method used for (A) flat surfaces and (B) curved surfaces

hand, undermasked areas must be cleaned with solvent to remove overspray that detracts from the appearance of an otherwise good job.

Liquid Masking

To mask a vehicle using the liquid masking system, proceed as follows:

1. Mask the area to be painted by outlining it with masking paper. Fold the paper over onto the area to be painted (Figure 15-65A) and secure with masking tape.
2. Apply the liquid masking material with a heavy, single overlapping coat to all surfaces, including bumpers, grilles, doors, windshields, body panels, wheels, wheel wells, door jambs, and even the engine compartment (Figure 15-65B). An airless spray system is generally recommended for applying the material.
3. Fold the masking paper back onto the liquid masking material (Figure 15-65C). Wipe away the material from the area to be painted with a damp sponge.
4. Prepare the surface and apply paint or primer according to the manufacturer's instructions.
5. Allow the paint to dry, then unmask the vehicle. Liquid masking may be used in both air dry or bake conditions.
6. After the paint has cured, wash off the dried liquid masking material with a garden hose or pressure washer (Figure 15-65D).

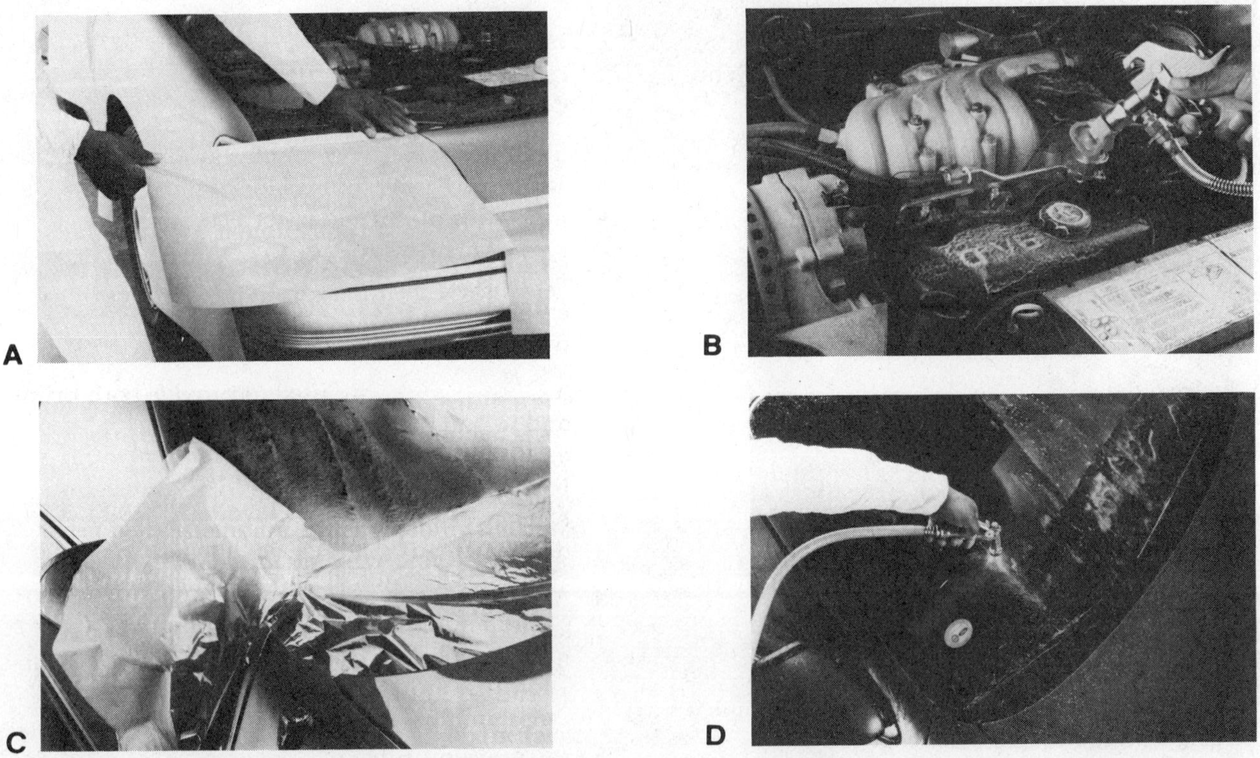

FIGURE 15-65 Steps in liquid masking a vehicle

REVIEW QUESTIONS

1. Which of the following is a result of excessively high spray gun pressure?
   a. bubbling and sagging paint film
   b. poor paint flow
   c. excessive paint loss through overspray
   d. both b and c
2. Upon inspection of a spray pattern, the paint runs are found to be longer in the middle than on the edges. To correct this, Technician A turns the fluid control knob in until the runs are the same length. Technician B turns the pattern control knob in until the runs are the same length. Who is correct?
   a. Technician A
   b. Technician B
   c. Both A and B
   d. Neither A nor B
3. What part of a spray gun directs the compressed air into the paint to atomize it?
   a. fluid valve
   b. air valve
   c. air cap
   d. fluid nozzle
4. On a typical spray gun, which orifice(s) determine(s) the spray pattern by means of air pressure?
   a. auxiliary
   b. side
   c. center
   d. both b and c
5. The ______________ feed spray gun is the one most commonly used in collision repair and refinishing shops.
   a. siphon
   b. gravity
   c. pressure
   d. both a and c
6. Which of the following statements is correct?
   a. A dust coat is applied with the gun a short distance from the surface.
   b. Banding coats are often used when spraying panel repairs with a primer-surfacer.
   c. The tack coat is sprayed over the finish wet coats.
   d. Both a and b

7. Technician A says that the temperature at which paint is sprayed has no effect on the smoothness of the finish, but Technician B says it does. Who is correct?
   a. Technician A
   b. Technician B
   c. Both A and B
   d. Neither A nor B

8. How often should air be exhausted from a spray booth?
   a. daily
   b. twice a day
   c. continuously during the spraying process
   d. at least once a week

9. What is the function of a manometer?
   a. remove toxic vapors from a spray booth
   b. deliver replacement air into the general shop area or directly into the spray booth
   c. indicate when the intake filters of a spray booth are overloaded
   d. indicate when a coating of paint has become tacky enough for a second coat

10. For a heavier coat of paint, Technician A moves the spray gun faster. To achieve the same effect, Technician B moves the gun slower. Who is correct?
    a. Technician A
    b. Technician B
    c. Both A and B
    d. Neither A nor B

11. What sandpaper grit size is used to remove an old paint film?
    a. No. 24
    b. No. 400
    c. No. 1250
    d. No. 2000

12. Technician A uses a circular motion when block sanding. Technician B block sands in a back-and-forth direction. Who is correct?
    a. Technician A
    b. Technician B
    c. Both A and B
    d. Neither A nor B

13. ______________ are used to provide both priming and filling in one step.
    a. Primer-sealers
    b. Sealers
    c. Conversion coatings
    d. Primer-surfacers

14. An HVLP spray system ______________.
    a. uses air delivered at 10 psi or less
    b. works with any medium solid material that can be atomized by the gun
    c. is generally more efficient than conventional spray equipment
    d. all of the above

15. Which of the following statements concerning masking is incorrect?
    a. Newspaper works well as automotive masking paper.
    b. Removal of trim, as opposed to masking, is preferred in some instances.
    c. Masking tape will not stick to surfaces that are not clean and dry.
    d. Using plastic covers on tires eliminates the need for masking them.

CHAPTER

16 Topcoat Application

OBJECTIVES

- Explain the objective of blending a finish and how it is accomplished.
- Understand the difference between spot refinishing, panel refinishing, and overall refinishing.
- Outline the general topcoat application procedures.
- Explain the key points to keep in mind when applying tri-coat finishes.
- Perform topcoat repairs by polishing, using a finesse sanding block or knife, and brushing.
- Finish and/or retexture the various types of automotive plastics.

From the customer's standpoint, the topcoat (or color coat) is the most important step in refinishing because that is all he/she sees. A painter takes special pride in producing a beautiful finish on spot and panel repairs, or in an overall refinishing job that matches both the color (or color effect) and texture of the original finish. Therefore, it is of great importance to fully understand the topcoat materials and how they are applied.

VEHICLE COLORS

Like all colors, automotive refinishing colors are a result of the way they react to light. The color seen depends on the kind and amount of light waves the surface reflects. When these light waves strike the retina, they are converted into electrical impulses that the brain sees as color. As mentioned in Chapter 14, the same color will look very different under natural sunlight, incandescent lights, or fluorescent lights (Figure 16-1). That is why it is so important to check the color match in daylight or under a balanced artificial light.

There are two basic automobile finish types: solid colors and metallic colors.

SOLID COLORS

For many years all vehicles were solid colors, such as black, white, tan, blue, green, maroon, and so on. These colors are composed of a high volume of opaque type pigments. Opaque pigments block the rays of the sun and absorb light in accordance with the type of color they are. That is, the darker the solid color is, the more light it absorbs and the less it reflects. Black will absorb more light and will reflect less; white absorbs less light but reflects a great deal more. When polished, solid colors reflect light in only one direction. Solid colors are still used on vehicles, but to a lesser degree when compared with a few years ago.

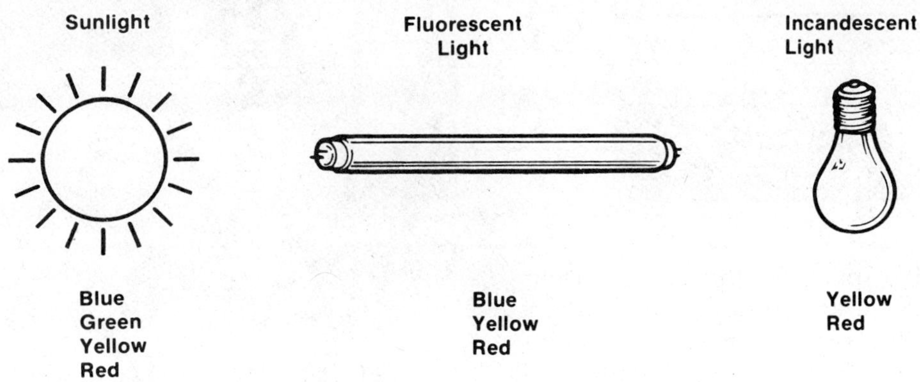

FIGURE 16-1 Types of light vary widely in how they influence the appearance of colors. Daylight or balanced artificial light is best for evaluating the color on vehicles.

METALLIC FINISHES

Metallic paints, as described in Chapter 14, contain small flakes of aluminum suspended in liquid. The position the flakes assume within the paint film and the thickness of the paint affect the overall color of the painted surface. The flakes reflect light, but some light is absorbed by the paint. The thicker the layer of paint, the greater the light absorption.

When metallic paint is sprayed on dry, the aluminum flakes are trapped at various angles near the surface. Light reflection is not uniform, and because the light has less paint film to travel through, little of it is absorbed (Figure 16-2A). The result of nonuniform light reflection and minimum light absorption is a painted surface with a light metallic appearance.

When metallic paint is sprayed on wet, the flakes have sufficient time to settle so they lie parallel to and deeper within the paint film. Light reflection is uniform and, because the light has to go farther into the paint film, light absorption is greater (Figure 16-2B). The result is a painted surface that appears deeper and darker in color.

CAUTION

Metallic colors must be stirred and mixed thoroughly before using. The pigment settles below the binder; the aluminum flakes settle below the pigment. If flakes stay at the bottom of the can, the paint will not match the same color on the vehicle being refinished.

A

B

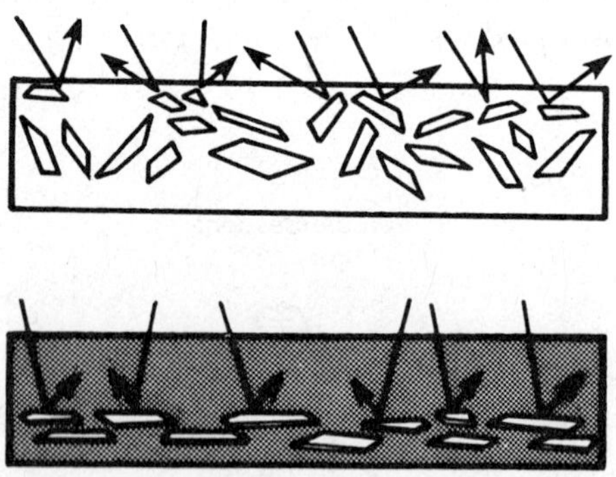

FIGURE 16-2 (A) Dry spray traps the aluminum particles at various angles near the surface and causes a light-metallic color effect. (B) Wet spray allows time for the metallic particles to settle in the paint film and causes a deeper and darker metallic color effect.

PREPARING THE TOPCOAT

After properly masking the vehicle and preparing the surface for refinishing, blow off any remaining dust with an air gun. Give a final touch-up cleaning with wax and grease remover, then clean the surface with a tack cloth. From this point on be careful not to touch the surface being refinished.

Before the topcoat material can be applied, it must be:

- Properly mixed.
- Thinned or reduced with the proper solvent to the desired viscosity.

MIXING PAINT

The failure to properly stir all the settled pigment into the liquid is a principal cause of paint problems. Mixing can be done by hand (Figure 16-3) or by machine (Figure 16-4).

Pigments are the part of the paint that settle and give the paint its color, opacity, and specific performance properties. These pigments vary greatly as to weight. Some of the commonly used pigments are seven to eight times as heavy as the liquid part of the paint. Because of their weight, the heavy pigments settle and it is impossible to keep them in suspension. Other pigments are light and fluffy and have very little tendency to settle. The commonly used pigments that settle quite rapidly are the whites, chrome yellows, chrome oranges, chrome greens, and red and yellow iron oxides.

FIGURE 16-3 Paint sticks are useful in mixing paint by hand.

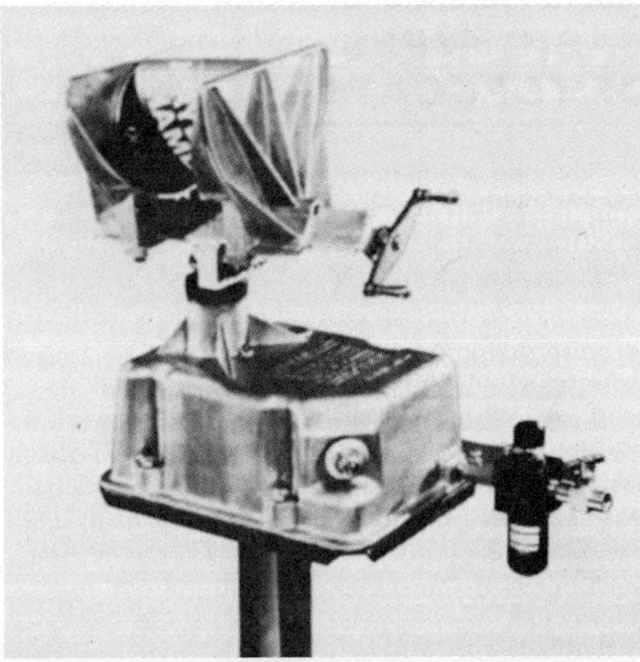

FIGURE 16-4 The paint shaker shown here is operated using compressed air. Paint shakers are also available that are electrically driven.

The consistency or viscosity of the liquid part of the paint has much to do with the rate of settling. The heavier the consistency, the slower the settling. Careful judgment in reducing only enough paint to do the job will lessen the amount of material to be discarded.

If a color that contains one or more of the heavy pigments is reduced to spraying consistency and allowed to stand 10 to 15 minutes without being stirred, it will have settled enough in that time to be off color when sprayed.

After a can of paint has been thoroughly mixed, empty out the contents of the can into another container or the gun cup, wash the can clean with a little solvent, and add this to the paint.

If a paint has settled out very hard, the liquid should be poured off and the residue well broken up. The liquid should then be slowly poured back while vigorously stirring. At least a 1-inch-wide flat bottomed, clean stirring paddle or steel spatula is usually the best choice. When using an intermix system, agitate all base mixing colors for a minimum of 15 minutes.

VISCOSITY

Topcoats are usually manufactured at as high a viscosity as practical to help in slowing down the rate of settling. In order to apply these materials, they must be reduced to a viscosity that can be properly atomized by the spray gun.

Compared with a two-part reaction type paint, the air-dry type has a higher viscosity, so evaporation is not as rapid. Because the initial drying time is faster, more solvent is required to improve the finish; it is also necessary to increase the air pressure. However, increasing the air pressure will cause the solvent to evaporate faster, resulting in an even faster initial drying time and a defective finish. For preventive measures, it is necessary to consider decreasing the distance between the gun and painting surface, increasing the discharge volume, increasing the number of coats, and speeding up the painting operation.

Two-part reaction type paint can be used with less solvent than air-dry type paint. However, more thinner is needed to compensate. To prevent unevenness, metallic colors with a lower viscosity than solid colors are used. The distance between the gun and the surface is greater and there is also more pattern overlap. Application is similar to that of spraying a dry coat of solid color.

Clear coat is applied in a method similar to a solid color but care must be taken to avoid applying a heavy coat. This can cause unevenness or other defects.

TEMPERATURE

The temperatures of concern when refinishing are the room temperature, the surface temperature of the vehicle, and the temperature of the paint. Temperature is especially important in downdraft spray booths. With all the air rushing by, the temperature of the paint is raised 6 or 7 degrees almost immediately. If the paint dries too fast, solvents will be trapped in the base coat. This can cause problems when applying the clear coat. Also be careful when bringing in a vehicle that has been

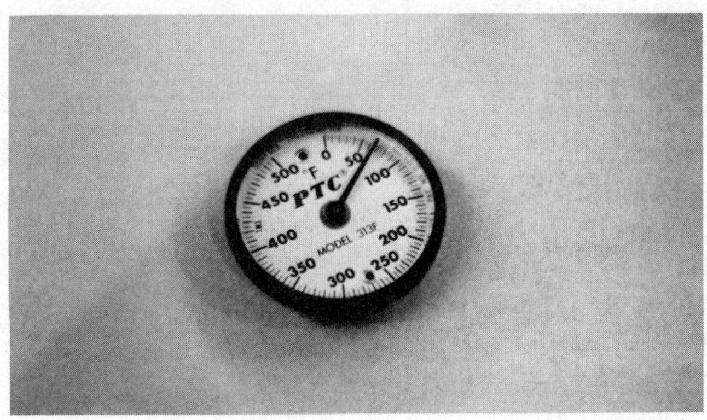

FIGURE 16-5 A surface thermometer is used to determine the surface temperature prior to painting.

stored outside in the cold. If the surface is too cold, the solvents will not evaporate as quickly as they should. Color matching and paint curing problems can result. A surface thermometer should be placed on the vehicle (Figure 16-5) to make sure the surface is at the temperature suggested by the manufacturer before spraying.

COLOR BLENDING

When refinishing, the procedure often will involve color blending for one reason or another. To stay ahead of the game, plan to blend right from the start. When preparing to refinish, always count on using more of the vehicle than you think you will need. That is, always prepare the adjacent panels for blending when preparing the refinishing area. It is easier to do a little extra work before beginning the job than to find out that more preparation is needed after the job is started. Remember that applying base coat/clear coat and tricoat systems requires different procedures than single stage finishes. This chapter covers general application techniques designed to yield professional-looking results. The topcoat application and final detailing of the vehicle are the completion steps of I-CAR's finish matching system.

The objective when blending is to create the illusion that the eye is seeing only one color. To do this, it is necessary to "step-out" the color in successive coats. This allows the original finish to show through the edges of the blend (Figure 16-6). To achieve this, it must be assumed that:

- You have a blend from the start.
- The blend area will be kept as small as possible.
- The adjacent areas are prepared for blending.
- The refinishing area is masked for blending.

Keep in mind, however, that it is impractical to try to get a *perfect* color match. The goal is to get the refinish color close enough to achieve a *blendable* match.

SHOP TALK

If problems arise in matching certain colors, it might be wise to have your eyes checked for color vision. Nearly 10 percent of all men have trouble seeing one or more colors. Blue-greens are most often the problem. However, almost no women have this difficulty. This is one of the reasons women do the touch-up work in auto plants. If a color vision problem is discovered, help can usually be found from someone in the shop for matching colors.

BASIC TYPES OF TOPCOATS

The most common types of topcoats that a painter will work with are:

- Alkyd enamels
- Acrylic lacquers
- Acrylic enamels (including both the hardener and two-part types)
- Polyurethane enamels
- Acrylic urethane enamels

The type of refinish system used for the topcoat ultimately determines the attractiveness of the color, gloss, and overall finish. Table 16-1 is a general summary of the properties of topcoats used for refinishing, and Table 16-2 compares the durability of various topcoat finishes.

SHOP TALK

Polyoxithane enamels are part of the first isocyanate-free base coat/clear coat and single stage systems. They were developed with health considerations in mind as well as to provide the highest quality in color matching. Tests show that polyoxithane enamels provide excellent hiding, durability against industrial pollution and extreme weather conditions, and protection from fading caused by ultraviolet light.

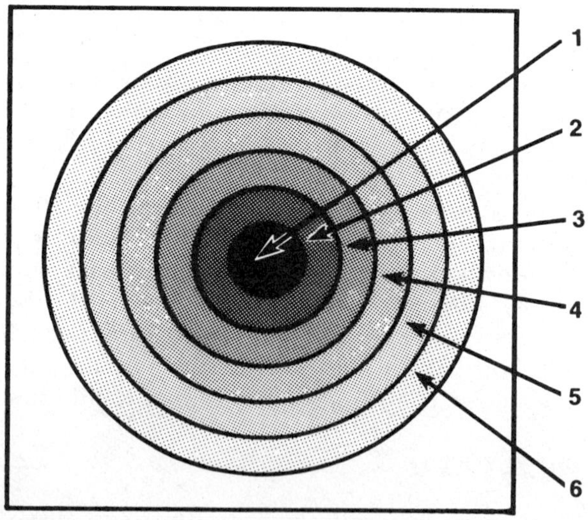

FIGURE 16-6 The "step-out" theory of color blending

TABLE 16-1: SUMMARY OF TOPCOAT FEATURES

Nomenclature		One-Part Type			Two-Part Type	
		Alkyd Enamel	Acrylic Lacquer	Acrylic Enamel	Polyurethane	Acrylic Urethane Enamel
Spray characteristics		Excellent	Excellent	Good	Good	Good
Possible thickness per application		Fair	Fair	Good	Excellent	Excellent
Gloss	without polishing	Fair	Good	Good	Excellent	Excellent
Gloss	after polishing	Good	Good	Good	—	Good
Hardness		Good	Good	Good	Excellent	Excellent
Weather resistance (frosting, yellowing)		Fair	Fair	Good	Excellent	Excellent
Gasoline resistance		Fair	Fair	Fair	Excellent	Good
Adhesion		Good	Good	Fair	Excellent	Excellent
Pollutant resistance		Fair	Fair	Fair	Excellent	Excellent
Drying time	to touch	68°F 5-10 minutes	68°F 10 minutes	68°F 10 minutes	68°F 2-30 minutes	68°F 10-20 minutes
Drying time	for surface repair	68°F 6 hours 140°F 40 minutes	68°F 8 hours 158°F 30 minutes	68°F 8 hours 158°F 30 minutes	—	68°F 4 hours 158°F 15 minutes
Drying time	to let stand outside	68°F 24 hours 140°F 40 minutes	68°F 24 hours 158°F 40 minutes	68°F 24 hours 158°F 40 minutes	68°F 48 hours 158°F 1 hour	68°F 16 hours 158°F 30 minutes

In the past few years, OEMs around the world have increasingly adopted base coat/clear coat systems as the finish of choice for new vehicles rolling off assembly lines. The technology for base coat/clear coat finishes was developed in Europe. The durability and popularity of these finishes prompted Japanese and American manufacturers to begin offering them, too. In fact, most automotive finishing experts agree that the base coat/clear coat system will be used on the vast majority of vehicles before the turn of the century. When the system was first used, it utilized either an acrylic lacquer clear or a polyurethane enamel clear over an acrylic lacquer base coat. Early in 1985, the first acrylic enamel base coat/clear coat system to actually simulate the OEM base coat/clear coat finish was introduced. Like OEM finishes, this system loads more transparent pigments into the base coat and locks the aluminum flakes into a flat arrangement, enabling it to match the brightness and color intensity of an OEM base coat/clear coat finish. This makes it easy to get the best color match available.

An added benefit of this new technology is minimal recoat time with the clear coat. Conventional base coats use "chemical drying" to build solvent resistance. Chemical drying generally requires 4 to 6 hours to prevent the clear coat solvents from redissolving the base coat and allowing the aluminum flakes to streak. Another new base coat/clear coat technology builds the necessary solvent resistance without chemical drying in just 30 minutes at 75 degrees Fahrenheit and 1.5 mils dry film thickness. These qualities enable this new system to be applied quickly and easily with superior results.

More recently, painters have been switching to the new acrylic urethane base coat/clear coat system. Many paint manufacturers now offer this excellent two-stage

TABLE 16-2: COMPARING THE DURABILITY OF TOPCOATS

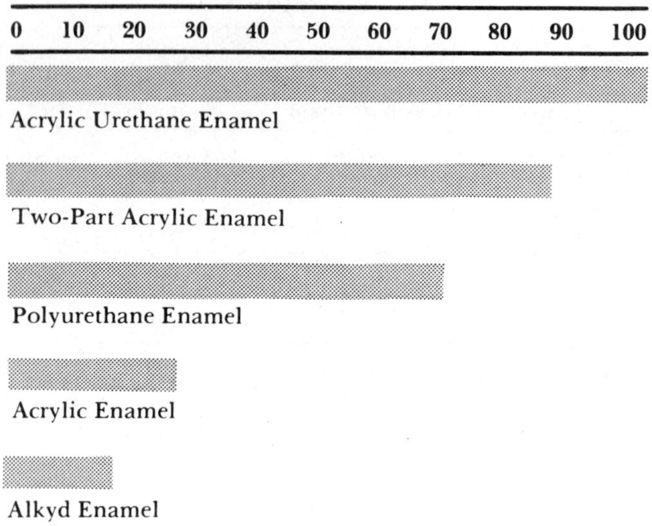

system, and it has quickly achieved a strong following. The reason for its popularity stems from not only meeting the practical needs of the painter, but also because of its advanced technology. Acrylic urethanes are fast, offer the best color matches, and provide better coverage and hiding. Thus, they increase productivity and improve customer satisfaction.

SHOP TALK

Recently introduced fluorine clear coats are designed to provide superior weathering characteristics and paint film durability. This is due to their higher resistance to ultraviolet rays. The result is a finish that requires no waxing. It keeps its gloss and water shedding properties as well as, or better than, conventional clears when waxed regularly. If the vehicle has an OEM fluorine clear topcoat (check the manufacturer's paint manual), the refinish system must also use a fluorine clear coat. These finishes are now becoming available through domestic paint manufacturers. Some require slightly different application and blending procedures than the clears now being used. This might include longer cure time for the base coat before the clear is applied and before the use of adhesion promoter.

In a few years, most vehicles will be finished in base coat/clear coat colors. This trend is certain to have several effects on the collision repair and refinishing shop. Surface preparation and a clean shop environment will be more important than ever, as the higher gloss and the clear coat make imperfections easier to see. A greater emphasis on cleanliness will result in increased use of downdraft booths in progressive shops.

With good surface preparation and a clean shop, however, painters will find that the new acrylic and urethane base coat/clear coats are very easy to apply and result in better looking, more durable finishes. For example, streaking and mottling are eliminated. Spot repair is made easy because base coat/clear coat finishes make it simple to blend in an edge. And the need for buffing is just about eliminated. But most important to painters is the outstanding color match that state-of-the-art base coat/clear coat systems provide.

One challenge the painter faces is becoming well-educated about the many different systems available in order to select the best one. No customer likes a second-rate finish. No shop can afford to lose potential income because time is lost due to difficult and involved application procedures and long dry times. Thus, when choosing a refinishing system, the key factors involved are appearance and ease of application.

REFINISHING METHODS

As mentioned in Chapter 14, refinishing methods are classified according to the condition of the original finish, the size of the area to be refinished, and the location on the vehicle. These methods include:

- Spot repair
- Panel repair
- Overall refinishing

SPOT AND PANEL REFINISHING

Spot repairs are recommended where a complete panel repair is either uneconomical (due to the size of the repair or the amount of masking involved) or impractical (because of difficulty rendering the repair invisible, particularly in the case of metallic finishes). Blending makes slight differences in color or texture of spot repairs less conspicuous.

Panel repairs are is done on complete panels that are separated by a definite boundary, such as a door or a fender. It is not necessary to blend the paint in such cases, unless it is difficult to match or if it is a metallic color. However, blending is routinely done for such areas as between the quarter panel and roof panel.

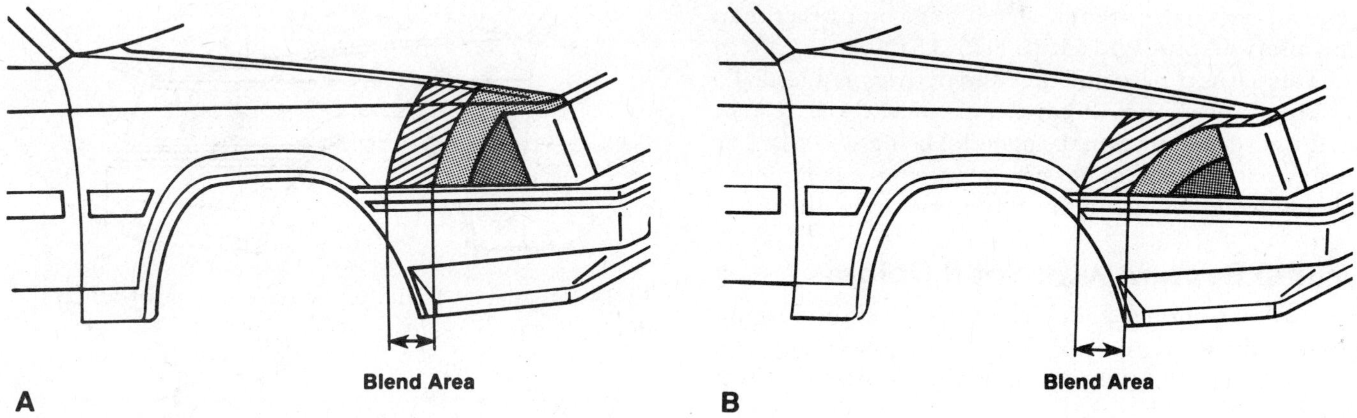

FIGURE 16-7 Methods of spot repair blending

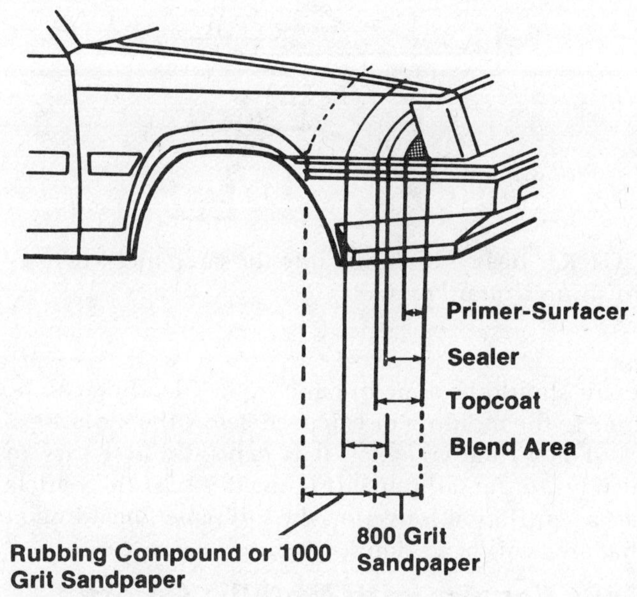

FIGURE 16-8 Spot repair with a solid color

Spot Repairs with Solid Colors

Definite rules cannot be laid down as to the range of blending needed for spot repairs with solid colors. This method is used for situations such as light damage to the fender edge. In this case, there are two methods of blending: at the hood-to-fender gap (Figure 16-7A) and at the body line (Figure 16-7B). If the blending can be done within the range shown in Figure 16-7B, it will not be necessary to paint the upper portion of the fender where shading shows most conspicuously, thereby avoiding problems with color and texture differences. As shown in Figure 16-8, when spot repairs are made with a solid lacquer, the blend area should be treated with rubbing compound or sanded with 1000 grit paper before refinishing.

Apply the topcoat with a fanning motion, working the spray gun from the center outward (Figure 16-9). This allows each coat to blend out a bit further than the previous one. An alternative method is to apply the finish in short strokes from the center outward. Again,

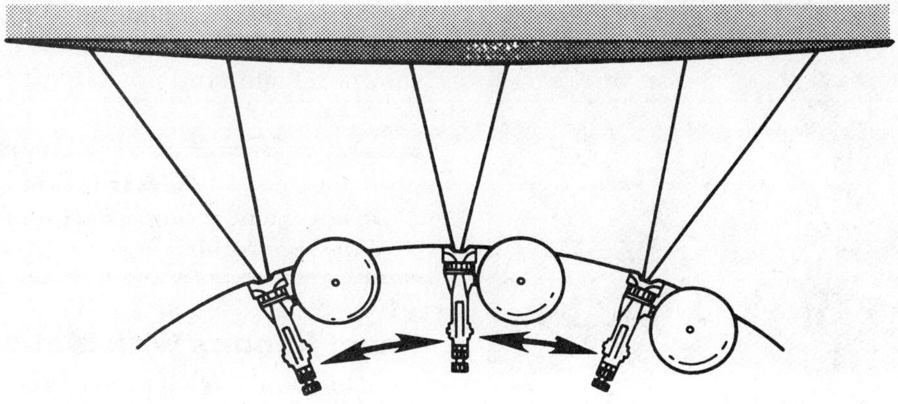

FIGURE 16-9 When spraying a blend area, use a fanning motion, working the gun from the center outward.

extend each coat so that it blends out a bit further than the previous one. With either method, the spray pattern should be narrowed and the fluid delivery reduced. To minimize overspray, the air pressure may also have to be reduced, depending on the material being sprayed. The spray pattern should never be reduced to a completely round jet; otherwise, paint control becomes difficult.

Panel Repairs with Solid Colors

For a complete panel repair, mask off the area not to be painted (Figure 16-10). If a panel has damage at two different locations, as shown in Figure 16-11, the repair should be done in panel style. Blending can be done to the molding or extended below the molding.

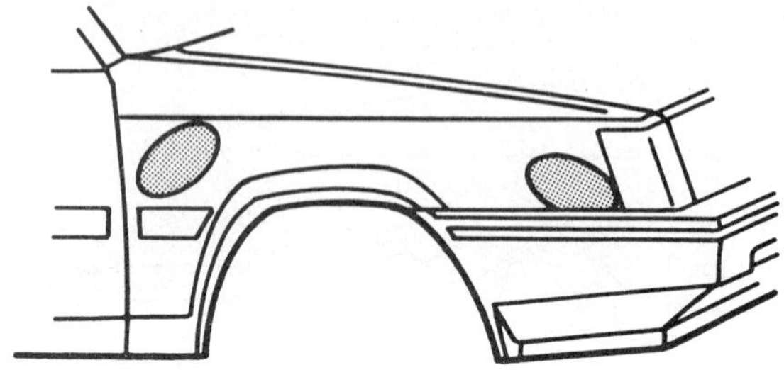

FIGURE 16-11 Panel repairs are called for when the damage is in two areas.

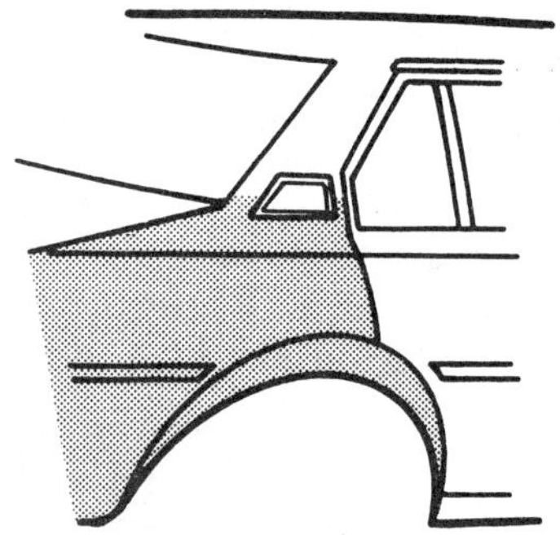

FIGURE 16-12 Blending into the sail panel when refinishing a quarter panel

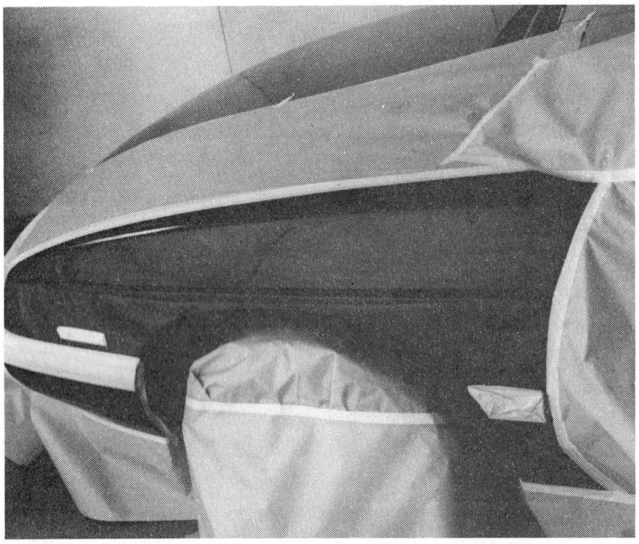

FIGURE 16-10 Masking for panel repairs on the front and rear of the vehicle.

For a quarter panel, it is generally necessary to blend into the sail panel (Figure 16-12). If the vehicle has a ventilation louver on the sail panel, blending at that area will be less noticeable.

Spot Repairs with Metallic Colors

If spot repairs are done with a metallic color as shown in Figure 16-13, skill is required in matching the color tone and bringing out the metallic image through proper distribution of the paint. The blend area will be less noticeable if it is angled away from the body line as shown in Figure 16-14.

CAUTION

Do not extend the topcoat beyond the sanded area of the spot repair.

Panel Repairs with Metallic Colors

Unevenness tends to be very noticeable with bright metallic colors. Therefore, if it is impossible to match the finish exactly with the previous one, extend the

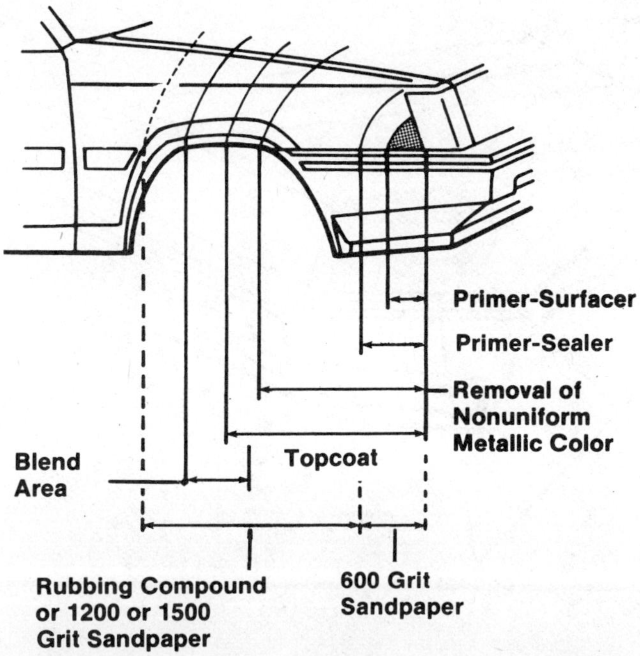

FIGURE 16-13 Spot repair with a metallic color

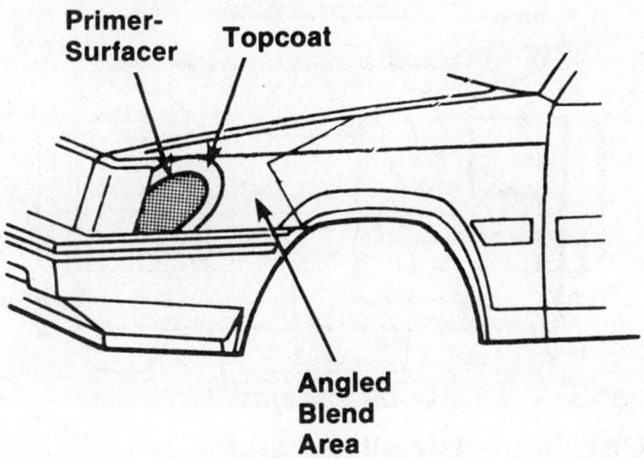

FIGURE 16-14 A blend area will be less noticeable if it is angled away from the body line.

blending over a wide range, as shown in Figure 16-15. This will make it less distinguishable.

If a panel has damage at both ends or if the whole panel is to be refinished, the blending must extend onto the adjacent panels. As shown in Figure 16-16, the clear coat extends beyond the fender onto one of the adjacent door panels and to the second body line of the hood.

OVERALL REFINISHING

For overall refinishing, keep a wet edge while maintaining minimal overspray on the horizontal sur-

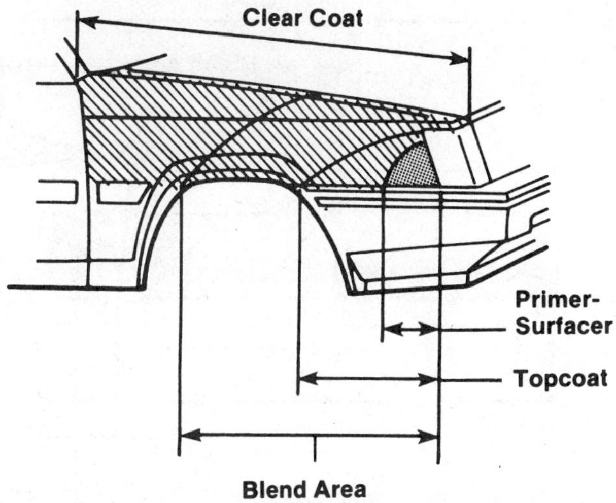

FIGURE 16-15 Panel repair with a metallic color

faces. This prevents spray from settling onto areas that have already dried, which would cause a gritty surface. Avoid sags in the overlap line by changing the point of overlapping as shown in Figure 16-17.

Although there is no single perfect procedure for overall refinishing of a vehicle, most painters will agree that the diagram in Figure 16-18 illustrates the best technique. In the spray booth, by starting with the top of the vehicle and proceeding to the trunk lid, side, and so on, the painter can best keep a "wet edge" while maintaining minimum overspray on the horizontal surfaces. This prevents overspray from settling onto areas that have already dried and causing a gritty or dry surface. If possible, it is better to work in a downdraft booth. The spray pattern is different because of the direction of the airflow (top to bottom). Following the procedure shown in Figure 16-19 allows the three main horizontal surfaces to remain as wet as possible while maintaining minimum overspray. The procedures in Figures 16-18 and 16-19 also allow the painter to continue to apply additional coats as needed without a significant loss of time due to flash time between coats.

TOPCOAT APPLICATIONS

Before applying any topcoat, it is necessary to carefully read the paint manufacturer's directions that appear on the paint container (Figure 16-20). While the basic type of paints—alkyd enamel, acrylic urethane, and so on—have the same general characteristics, each manufacturer has specific formulations for its prod-

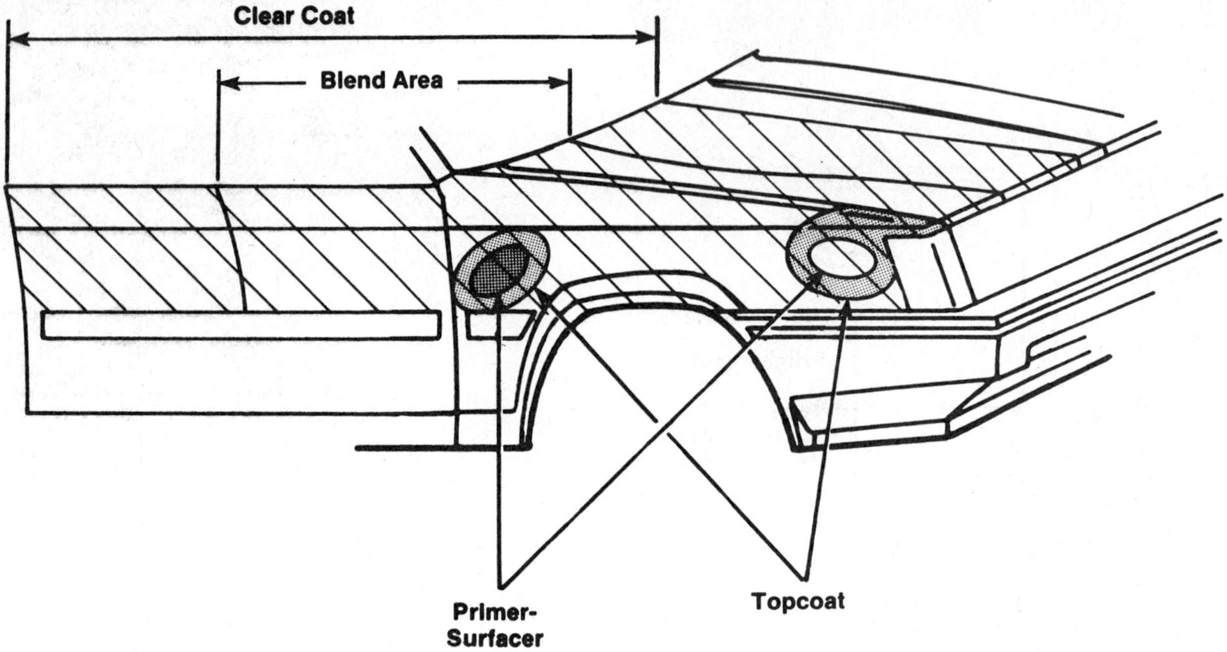

FIGURE 16-16 Blending a metallic color onto adjacent panels

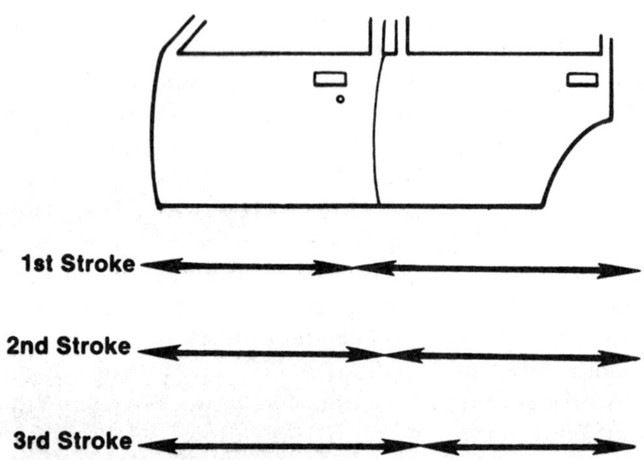

FIGURE 16-17 Changing the overlapping point to avoid sags

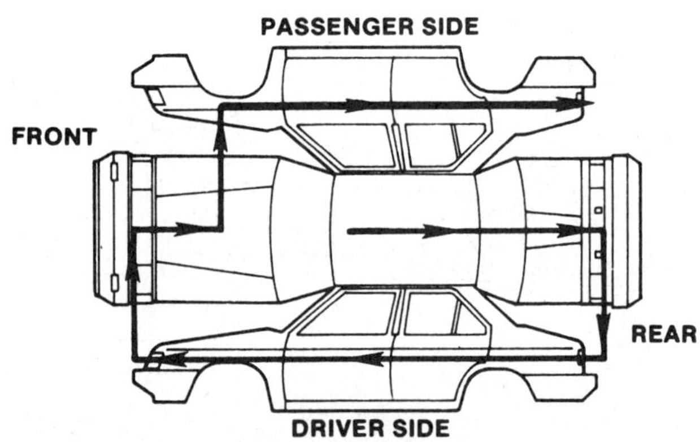

FIGURE 16-18 Overall refinishing procedure in a spray booth

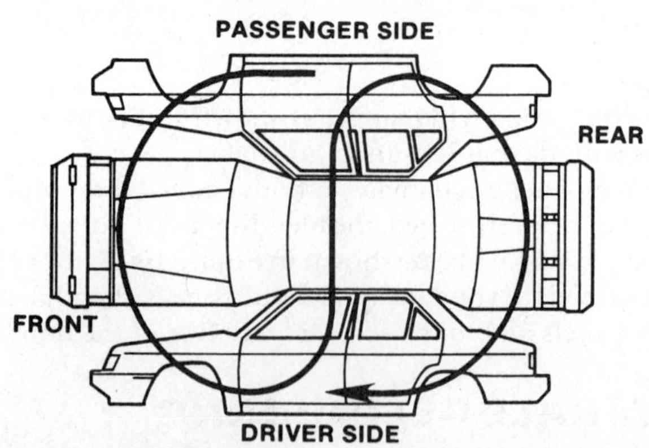

FIGURE 16-19 Overall refinishing procedure in a downdraft spray booth

ucts. For this reason, the best source of data on how to apply a specific brand of paint is the container label. Another source of good information can be found in the manufacturer's literature.

Some of the more important label and literature data that should be checked include:

- Viscosity recommendations
- Air pressure recommendations
- Use of additives, reducers, thinners, and activators
- Application techniques

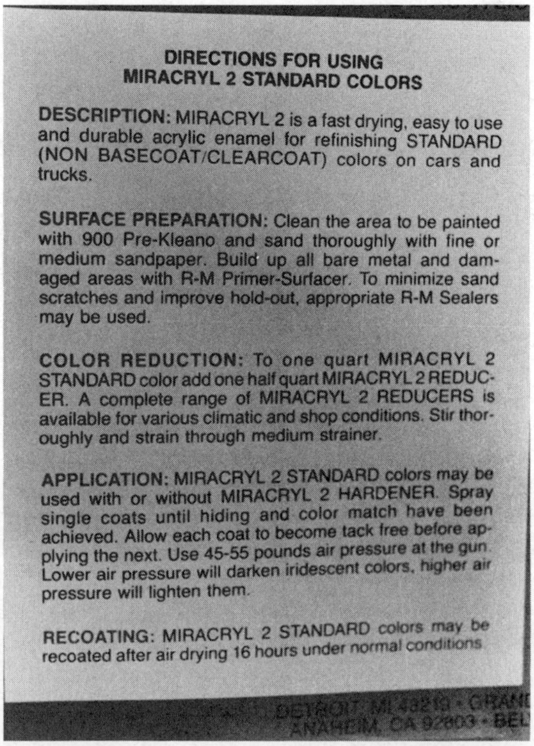

FIGURE 16-20 Carefully read the manufacturer's directions that appear on the paint container.

- Number of coats required
- Blending and mist coat procedures
- Polishing and compounding recommendations
- Cleanup procedures

APPLYING BASE COAT/ CLEAR COATS

Base coat/clear coat systems pose a major challenge to painters. To help match and refinish base coat/clear coat systems, paint manufacturers offer special clear coats for use over base coats. Since more and more vehicles have base coat/clear coat finishes, it is very important to become familiar with them.

When estimating a base coat/clear coat repair, carefully examine the finish on the area adjacent to the damage. If it is chalked, dulled, or otherwise impaired, color matching might prove impossible. Ideally, such repairs should be performed as overall refinishing jobs. This approach will eliminate many problems in repairing base coat/clear coat finishes that are severely weathered.

When spraying, two medium coats of base coat should be applied. The base coat does not need to be glossy, and only enough should be used so as to achieve hiding. Two or three medium wet coats of clear should be applied next, with at least 15 minutes flash time between coats.

Sanding of the base coat should be avoided. If sanding must be done because of dirt or imperfections after the first coat, it can be done after the paint has had time to dry. Wet sanding with 800 to 1200 grit paper will minimize sand scratches. The sanded area must then be given another base coat to prevent streaking and mottling.

Several things can be done to help clear coats wear better:

- Do not load clear coats on heavily. Because these finishes are clear, many painters have a tendency to use too much in an attempt to increase the desired glamour effect. As a result, they "bury" the topcoat.
- Do not use thick clear coats. Contrary to some opinions, clears do not perform better when they are underreduced. Thin or reduce according to the label instructions.
- Do not use economy thinners or reducers with clear coats. These only weaken the performance of clears by trapping solvents and hurting the flow and leveling characteristics. Use a quality thinner/reducer and let each coat flash thoroughly before applying the next one.
- Apply an adhesion promoter over the entire panel to be refinished, or at least past where any color or clear will be applied (Figure 16-21). Apply clear coat over the entire panel within the adhesion promoter area. If necessary, blend the clear as described earlier. Remember to step out the successive coats of clear.

TRI-COAT FINISHES

Although tri-coats require somewhat different refinishing procedures and techniques, they have much in common with base coat/clear coat finishes. Following are some key points to keep in mind when performing a tri-coat repair:

- Follow the recommendations for this type of repair furnished by the paint manufacturer.
- Pay close attention when the instructions call for the use of adhesion promoters, antistatic materials, and so on.
- Make a letdown panel, as described in Chapter 14. A mismatch in the base coat, mica coats, or clear coat can affect the overall finish match.
- Keep the repair area as small as possible.
- Avoid a "halo" effect by applying the first coat of mica to the base coat only.

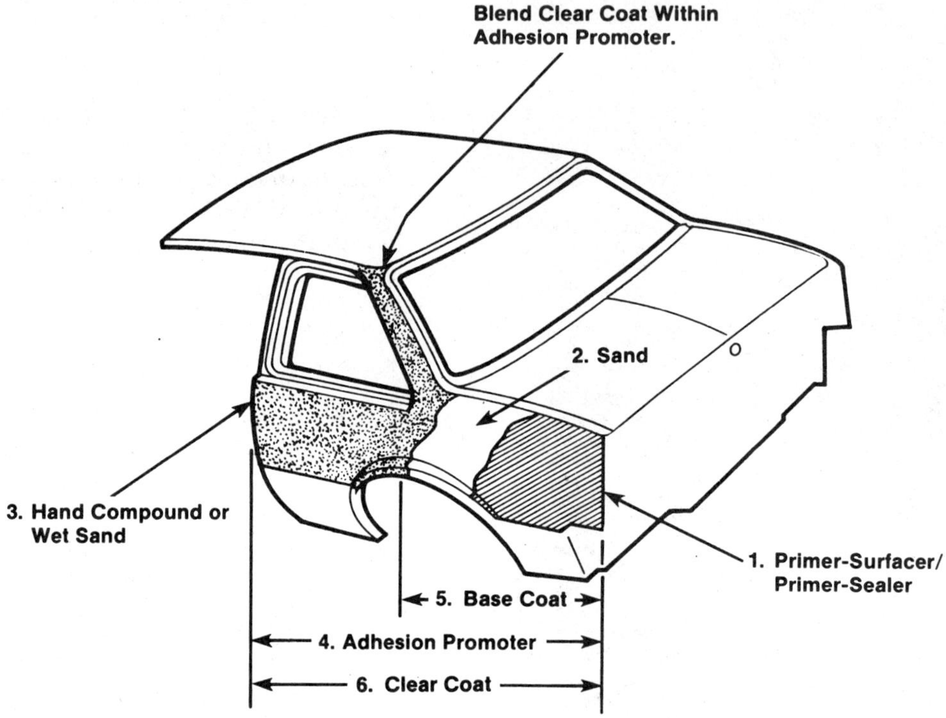

FIGURE 16-21 Typical base coat/clear coat panel repair using adhesion promoter

- The more intermediate mica coats that are applied, the darker the finish will appear.
- Allow a larger area in which to blend the mica intermediate mica coats. They require more room to blend than a standard base coat.
- Do not try to substitute another type of paint for the recommended base coat.
- Always check the base coat color against the OEM base coat. To do this, find an uncleared, mica-free area of the vehicle. Some auto companies leave an exposed portion of base coat beneath the right and left sill plates that is perfect for this (Figure 16-22). Other common locations for the exposed portion of base coat are under the deck lid and under the hood.

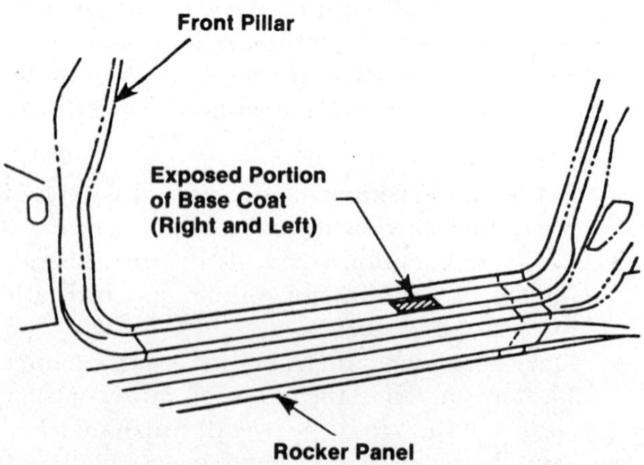

FIGURE 16-22 Testing the base coat color

FINAL DETAILING

Final detailing is the last step in the I-CAR finish matching system (Figure 16-23). The objective of final detailing is to remove any defects in the finish that might cause customer complaints, while at the same time keeping the repair area as small as possible. Final detailing includes:

- Removal of masking tape and paper
- Polishing the surface
- Repairing small surface defects

REMOVAL OF MASKING MATERIALS

If the finish has been force dried, the masking tape should be removed while the finish is still warm. If the

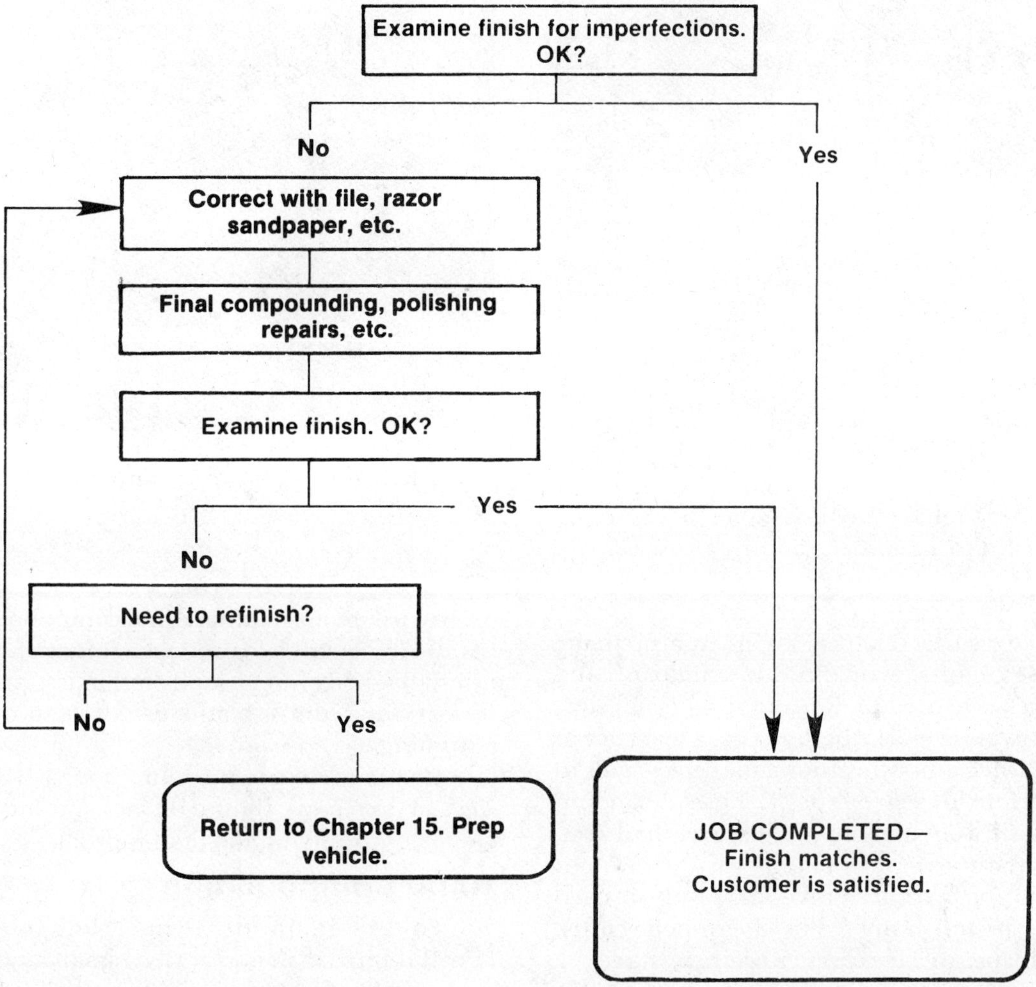

FIGURE 16-23 Steps in final detailing

finish is allowed to cool, the tape is more difficult to remove and leaves adhesive behind. Remove the tape slowly so that it comes off evenly (Figure 16-24). Take care not to touch any painted areas because the paint might not be completely dry; fingerprints or tape marks could result.

Never allow a lacquer paint to dry thoroughly before removing the tape because the paint film might peel off along with the tape. It is best to remove the tape immediately after refinishing. When doing so, be careful not to touch the freshly painted surface.

POLISHING

Acrylic lacquer finishes and most clears need a light compounding or polishing to achieve the desired gloss and to remove any unevenness (Figure 16-25). If slight defects in finished surfaces should occur, they should not be compounded until the paint has had a chance to cure. This could involve a period of several

FIGURE 16-24 Remove masking tape slowly.

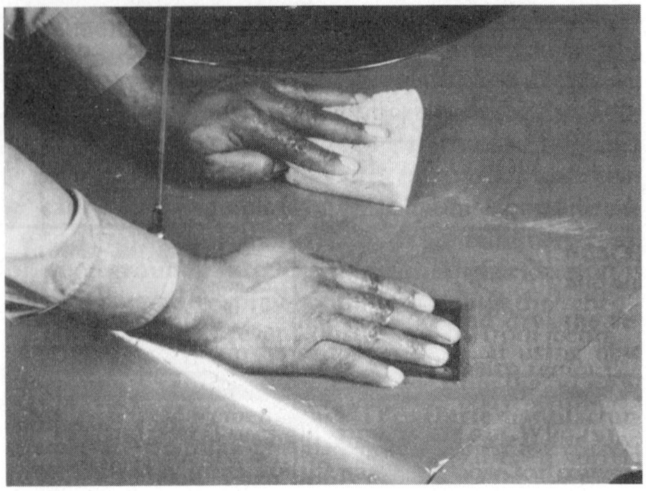

FIGURE 16-25 Light compounding or polishing is sometimes needed to achieve the desired gloss.

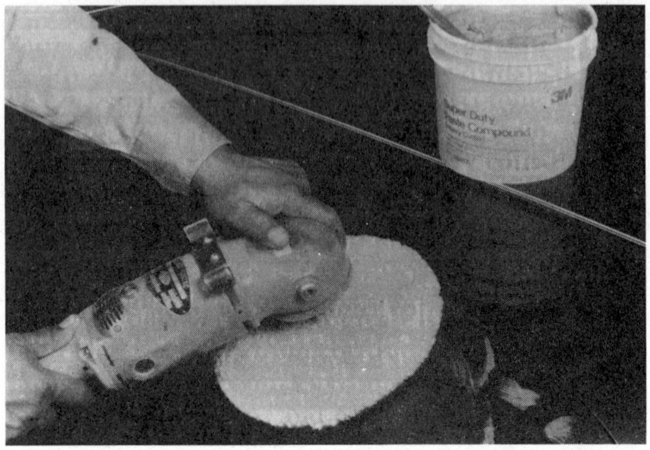

FIGURE 16-26 Machine compounding using a power buffer

days. To remove surface imperfections in a urethane finish, "finesse sanding" with ultra-fine sandpaper and a sanding block is needed. If the sandpaper is not supported with a sanding block, the defect has a tendency to crown; that is, the paint film around the defect tends to wear away before the defect itself. Finesse sanding should be done in a circular motion just over the defect, keeping the sanding area as small as possible. Do not sand back and forth because this will result in deep gouges which are not easily removed. For best results, soak the sandpaper in a clear water before using it.

Rubbing Compound

Rubbing compounds are much like the paste and liquid cleaners used before a vehicle is waxed. They contain an abrasive such as pumice that levels the top of the finish, making it smooth and lustrous. Rubbing compounds are available in various cutting strengths for both hand and machine compounding. Hand compounds are oil based to provide lubrication; machine compounds are water based to disperse the abrasive while using a power buffer (Figure 16-26).

Rubbing compounds are used to:

- Eliminate fine sand scratches around a repair area
- Correct a gritty surface
- Smooth and bring out the gloss of lacquer topcoats

Small areas or blended areas are best done by hand compounding. On large surfaces, however, machine compounding is recommended. To avoid cutting through the paint on body lines, apply a strip of masking tape along the edge. After the compounding is completed, remove the tape and compound the edge by hand—just enough to produce a smooth finish. Keep in mind that body lines usually retain less paint than flat surfaces and thus should get only minimum compounding.

Compounding is sometimes done as a final smoothing step to remove light scratches and small dirt particles *before* applying the final topcoat.

Hand Compounding

Fold a soft, lint-free flannel cloth into a thick pad or roll it into a ball and apply a small amount of hand compound to it. Use straight, back-and-forth strokes and medium-to-hard pressure until the desired smoothness is achieved.

Hand compounding takes a lot of elbow grease and is time consuming. To keep the compounding of lacquer topcoats to a minimum, it is important to apply the finish as wet as possible (without sags or runs) by using the proper thinner for the shop temperature.

Machine Compounding

Apply machine compound over a small area using a medium-to-coarse bristled paintbrush or a squeeze bottle. Then compound using a slow-speed power buffer and pad. It is important to use a slow-speed machine to avoid static buildup and high surface temperatures. Because the compound has a tendency to dry out, do not try to do too large an area at one time. Do not apply too much pressure, and always keep the machine moving to prevent cutting through or burning the topcoat. Figure 16-27 shows how to install both buffing and polishing pads.

The primary advantages of machine compounding—in addition to saving time—are proper cutting

A

B

FIGURE 16-27 (A) A buffing pad is usually secured with a washer nut. (B) The polishing pad can be tied in place over the buffing pad for extra softness.

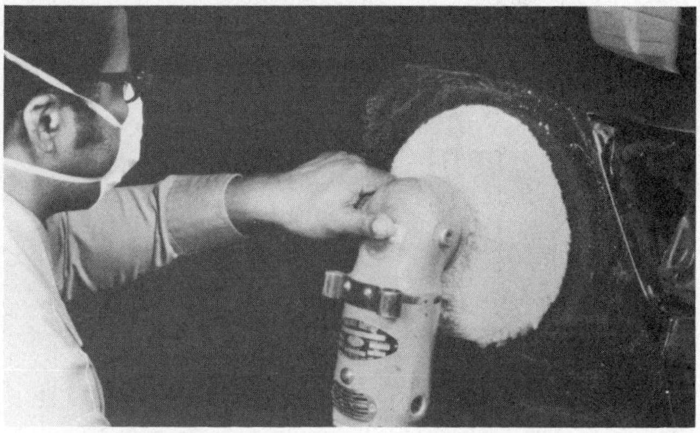

FIGURE 16-28 Using a polishing pad

TABLE 16-3: TOPCOAT REPAIR TECHNIQUES

	Repair Method		
Defect	Polishing	Finesse Sanding Block	Brush
Sags (small)		X	
Dust specks		X	
Orange peel	X		
Repair marks	X		
Tape marks	X		
Paint splashes	X		
Scratches			X
Burn-throughs			X
Adhesion of metal particles	X		

action and the nonclogging of the buffing pad. When the operation is completed, replace the buffing pad with a polishing pad or lamb's wool bonnet and polish (Figure 16-28). Remember to keep the machine moving at all times, and check the finish often. To remove all of the compound without leaving streaks, clean with a solution of 25 percent rubbing alcohol and 75 percent water.

TOPCOAT REPAIRS

The recommended repair methods that can be used on various topcoat defects are discussed below and are outlined in Table 16-3.

Polishing Repairs

Slight defects in the topcoat can be repaired by polishing. The choice of rubbing compound depends on the extent of the damage, but final polishing should always be with an extra-fine compound.

Finesse Sanding Block and Dirt-Nib File Repairs

Small protruding defects in the topcoat such as sags should first be corrected with a finesse sanding block, then polished with a rubbing compound. Use a finesse sanding block as follows:

1. Dress the surface of the block with wet 220 grit paper to make it smooth and flat.
2. Thoroughly soak the block in clear water.
3. Place the block over the protrusion and move it back-and-forth (Figure 16-29). If necessary,

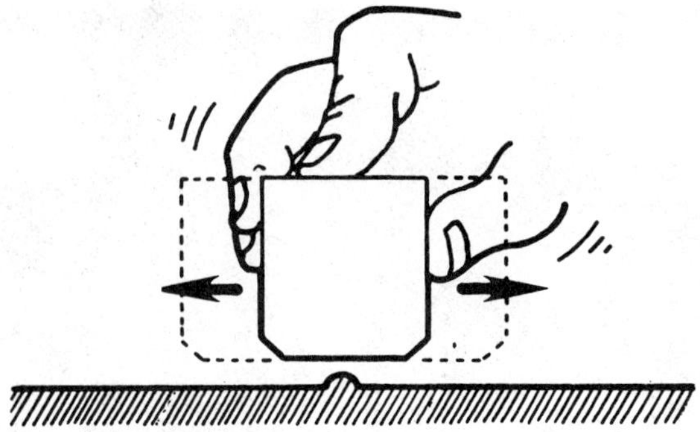

FIGURE 16-29 Removing a small protrusion with a finesse sanding block

FIGURE 16-30 Removing a large protrusion with a knife or razor

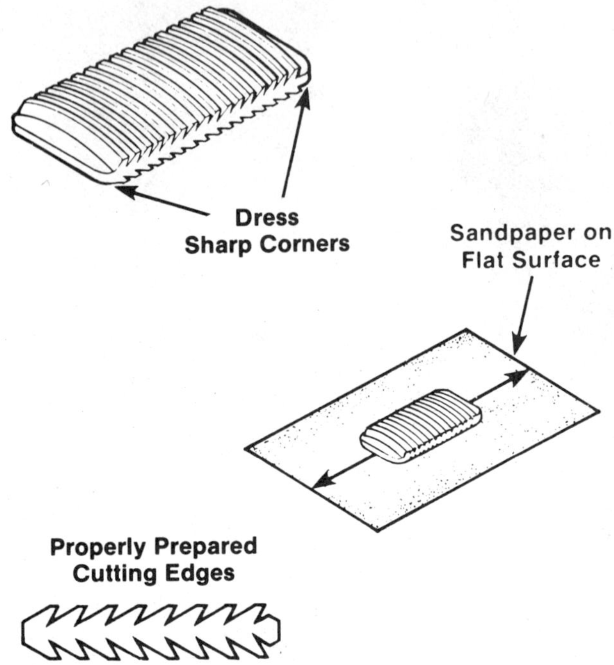

FIGURE 16-31 Making a dirt-nib file

use a little water to help make the movement smoother.

4. When the protrusion has all but disappeared, blow off the loose particles and finish the job with extra-fine rubbing compound.

If the protrusion is larger, repair it with a knife or single-edge razor blade as follows:

1. Being careful not to take off more finish than is necessary, cut off the protrusion (Figure 16-30). The tip of the knife or razor blade should be pointed slightly upward.
2. Smooth the area with a finesse sanding block or 3500 grit paper.
3. Blow off any particles and finish with an extra-fine rubbing compound.

Larger defects can also be removed using a dirt-nib file. This tool will remove the protrusion with minimum damage to the surrounding paint film. Dirt-nib files are available commercially, or you can make your own. To make a dirt-nib file (Figure 16-31), obtain a 1-1/2" to 2"-long piece of a vixen body file. After dressing down the sharp corners on a grinder, place a piece of 400 grit sandpaper on a flat surface and draw the piece of file back and forth over the paper until the teeth become smooth. A light machine oil will speed up the operation.

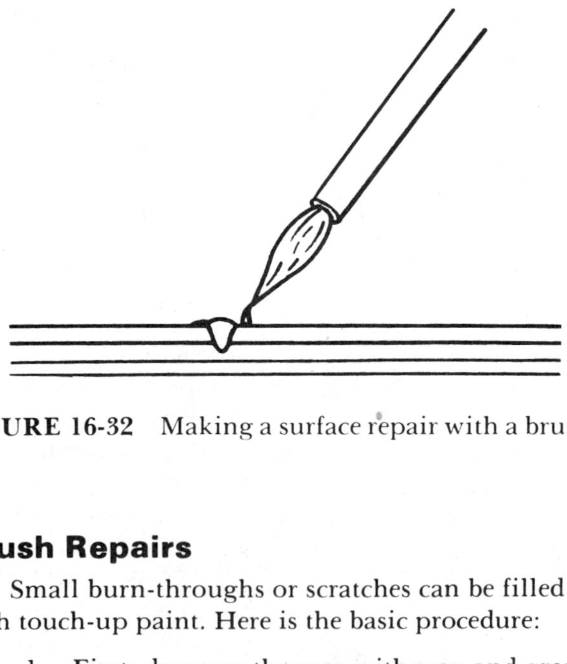

FIGURE 16-32 Making a surface repair with a brush

Brush Repairs

Small burn-throughs or scratches can be filled in with touch-up paint. Here is the basic procedure:

1. First, degrease the area with wax and grease remover.
2. Dip the brush into lacquer thinner to ensure a thin paint film.
3. Lightly dip the brush into the touch-up paint and brush thin coats over the imperfection (Figure 16-32).
4. Let the repair sit until completely dry.

SHOP TALK

Until a new topcoat has cured at least 60 to 90 days, wax should not be applied.

FINISHING AUTOMOTIVE PLASTICS

Follow the manufacturer's recommendations to determine if a particular finishing system can be used on a specific type of automotive plastic. The manufacturer will also point out if a special plastic primer, flex agent, or adhesion promoter is required. Table 16-4 lists the more popular automotive plastics along with their suggested finishing systems.

Most rigid (hard) plastics generally require no primers; the paint will adhere properly to the plastic. Semirigid (flexible) plastics might require the addition of a flex agent to the finishing system. Semirigid plastics expand, contract, and bend more easily than other substrates. The flex agent keeps the paint film flexible so it can accommodate the movement of the plastic without cracking.

TABLE 16-4: FINISHING SYSTEMS FOR AUTOMOTIVE PLASTICS

KEY

I Interior
E Exterior
P Primer
NP No primer
SP Special primer/adhesion promoter
NA None approved
* Flexible primer and/or additive recommended

		Standard Lacquer System	Flexible Lacquer/ Enamel System	Polypropylene System	Vinyl System	Urethane System
ABS	Acrylonitrile-Butadiene-Styrene	I/NP E/NP				
ABS/PVC	ABS/Vinyl (Soft)		I/NP E/NP		I/NP	
EP I, EP II, or TPO	Ethylene Propylene			E/SP*		
PA	Nylon	E/P				
PC	Lexan	I/NP				
PE	Polyethylene	NA	NA	NA	NA	NA
PP	Polypropylene			I/SP		
PPO	Noryl	I/NP				
PS	Polystyrene	NA	NA	NA	NA	NA
PUR, RIM, or RRIM	Thermoset Polyurethane		E*			E
PVC	Polyvinyl Chloride (Vinyl)		E/NP I/NP		E/NP	E I/NP
SAN	Styrene Acrylonitrile	I/NP				
SMC	Sheet Molded Compound (Polyester)	E/P				
UP	Polyester (Fiberglass)	E/P				
TPUR	Thermoplastic Polyurethane		E*			E
TPR	Thermoplastic Rubber		E*			E

Some manufacturers require that different flex additives be used in the various finishing systems. Others offer a universal flex additive that can be used in many systems. As always, it is best not to mix manufacturers' products. The flex additive, the topcoat, the undercoat, and the reducer or thinner should all be provided by the same manufacturer whenever possible. Mixing labels or using different manufacturers' products on the same job can yield poor results.

SHOP TALK

Plastic parts are normally painted before they are installed. However, if painting is done with the part already in the vehicle, it is important that the adjacent surfaces are properly masked off using the procedures discussed in Chapter 15.

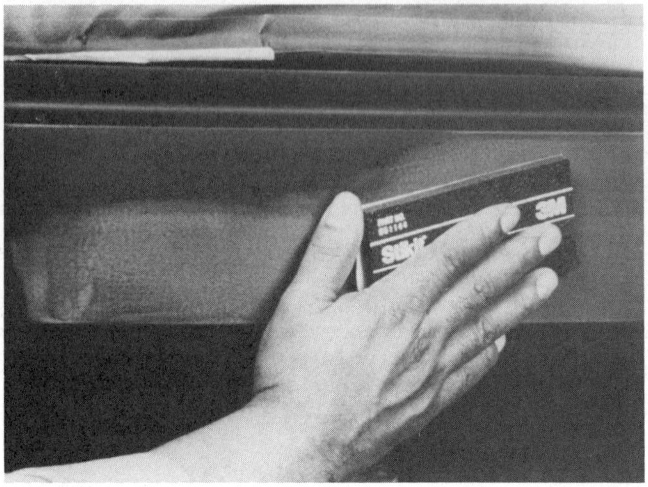

FIGURE 16-33 Sanding the area to be clear coated with 600 grit paper

FINISHING RIGID INTERIOR PLASTIC PARTS

Rigid or hard interior plastic parts generally require no primer-surfacer or primer-sealer. The finishing procedure is as follows:

1. Wash the part thoroughly with the recommended solvent.
2. Apply suitable interior acrylic lacquer (consult the paint supplier color chart). To avoid washout of the grain effect, apply only enough color to obtain hiding.
3. Let dry according to the label directions, then reinstall the part.

FINISHING RIGID EXTERIOR PLASTIC PARTS

While most rigid exterior plastics do not require a primer, some paint manufacturers nevertheless recommend a primer application before the color coat. Conversion coatings, metal conditioners, self-etching primers, and flex agents should not be used. The finishing procedure is as follows:

1. Wash the part thoroughly with the recommended solvent. Use mild cleaning solutions rather than harsh chemical cleaners.
2. The area that will receive the base coat should be sanded with 400 grit paper. The blend area that will be clear coated should be sanded with 600 grit or finer paper (Figure 16-33).
3. Color coat using the appropriate acrylic lacquer, acrylic enamel, urethane, or base coat/clear coat system. Specific reduction ratios supplied by the paint manufacturer should be followed.
4. Only enough film thickness to achieve full hiding is needed; usually two or three medium wet coats are sufficient. The base coat should be allowed to dry completely before the clear coat is applied.

When working with SMC, if any of the reinforcing fibers are exposed, more repair work is needed. Do not apply finish over exposed SMC fibers. The actual SMC finishing procedure is the same as for steel; use primer or primer-surfacer as required.

FINISHING INTERIOR/ EXTERIOR FLEXIBLE PLASTIC PARTS

As previously mentioned, most flexible or semirigid plastics require the addition of an additive to the paint to allow it to flex without cracking. There are several such systems available; in most cases, it is a matter of personal preference as to which one is used. Manufacturers do not recommend the use of flex additives in their base coat when using a base coat/clear coat system, but do recommend their use for lacquer and enamel clear coats.

To apply a base coat/clear coat flexible (elastomeric) finish, proceed as follows:

1. Thoroughly sand the entire part with 400 grit paper. Clean the surface with a cleaning solvent.

FIGURE 16-34 Mix the base coat and solvent according to the manufacturer's instructions.

2. Following the manufacturer's instructions, mix the base coat with the recommended solvent (Figure 16-34).
3. Using the recommended air pressure at the gun, apply a sufficient number of coats to achieve complete hiding and the proper color match. Be sure to allow flash time between coats.
4. Allow the base coat 30 to 60 minutes drying time before applying the clear coat. Do not sand the base coat before applying the clear coat.
5. If sanding of the base coat is necessary to remove imperfections, such as dirt or sags, sand with 400 grit or finer sandpaper and reclean the area. Apply one additional coat of base coat and let dry.

CAUTION

As the paint film builds, it will be necessary to allow more flash time between coats to avoid trapping the solvent. If solvent is trapped, pinholes and bubbles might result.

Apply the clear coat in the following manner:

1. Mix the clear coat with reducer and flex additive as per the label instructions.
2. Strain the mixture and apply 2 to 3 coats using the recommended air pressure at the gun.
3. Allow each coat to flash completely before applying the next one. Allow at least 4 hours air drying time, or force dry for 30 minutes at 180 degrees Fahrenheit. Follow the label instructions for the amount of flash time needed before force drying.

SHOP TALK

Compounding is not necessary when a flex additive is used; the mixture will dry with acceptable gloss. Compounding dulls the gloss of elastomeric finishes, causing a flat appearance. In this case, the finish cannot be brought back to the same gloss level without applying more paint.

Flexible replacement panels are factory primed with an elastomeric enamel-based primer. The only preparation required prior to finishing is cleaning with solvent, sanding with 400 grit paper, and a second cleaning after the sanding is completed. In the event the OEM primer is scratched and has left the plastic substrate exposed, or the part has been repaired with a flexible filler material, it is necessary to cover the exposed area with a flexible primer-surfacer prior to finishing. If the exposed surface is not primed, the area will be highlighted after the topcoats are applied. A fast evaporating solvent should be used to reduce the primer-surfacer and prevent swelling of the base material caused by absorption of the solvent into the plastic.

FINISHING INTERIOR/ EXTERIOR POLYPROPYLENE PARTS

Finishing polypropylene (PP) parts involves the use of a special primer. Since polypropylene is hard, it can be color coated after priming with conventional interior acrylic lacquer.

The most common exterior use of polypropylene is for bumpers (Figure 16-35). The surface preparation on PP bumpers is different from that of steel bumpers; flex agents are required. Therefore, a special PP primer must be used for the undercoat and a flex agent added to the topcoat. If not, peeling will result.

If a PP bumper has major structural damage, it must be replaced. Replacement bumpers of this type are usually primed and ready to be painted. If they are not primed, the special PP primer must be applied over the entire bumper. Before finishing, be sure to wash the surface with solvent. When applying an acrylic enamel topcoat, proceed as follows:

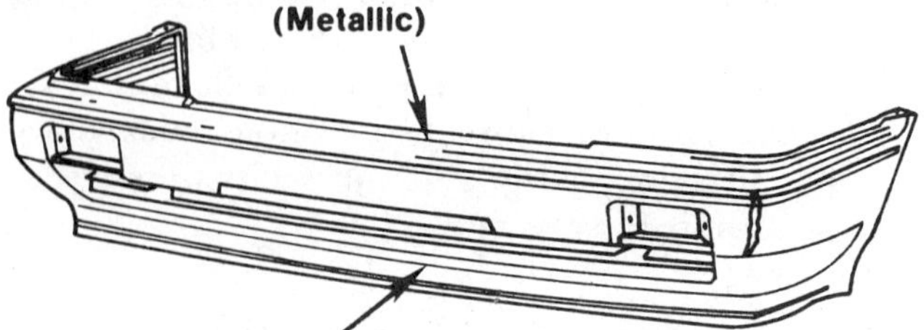

FIGURE 16-35 Typical polypropylene bumper

1. Apply properly reduced and mixed PP primer and flex additive as directed by the manufacturer. Allow 1 to 2 hours drying time before applying any color coats.
2. Apply reduced and mixed acrylic enamel and hardener additive. Flex additive should not be used in the topcoat.
3. Allow 8 hours (overnight if possible) drying time.

If an acrylic lacquer base coat/clear coat is being used, proceed as follows:

1. Apply properly reduced and mixed PP primer. Allow 1/2 to 1 hour drying time before the application of acrylic lacquer. Flex additives should not be used with base coat/ clear coat finishes.
2. Apply properly thinned acrylic lacquer. Allow 15 to 30 minutes drying time before applying the clear coat.
3. Apply properly thinned acrylic enamel clear coat. Allow 8 hours (overnight if possible) drying time.

CAUTION

Because polypropylene bumpers are made of thermoplastic resin, force drying at more than 212 degrees Fahrenheit could result in deformation.

Minor surface scratches can usually be repaired by following the same procedures used for finishing replacement PP bumpers, with the following changes:

1. If the scratches do not penetrate the substrate, follow the entire procedure but do not apply primer.
2. If the scratches penetrate the substrate, use a thin coating of special plastic body filler and prime the repair area only with special PP primer.

Finishing Urethane Bumpers

Following is the procedure for finishing a colored urethane bumper:

1. Mask off the area to be finished and clean it with a wax and grease remover. Keep in mind that insufficient cleaning will result in peeling or blistering.
2. Apply a coat of primer-surfacer over the entire bumper (Figure 16-36). Because it is extremely difficult to match the topcoat for spot refinishing, the entire bumper should be done. Prepare the surface by wet sanding with 600 grit paper.
3. Clean the surface again.
4. Apply the topcoat over the entire bumper (Figure 16-37). Use a two-part acrylic urethane with a softening agent added. For metallics, allow a flash time of approximately 5 minutes after application, then apply a clear coat.
5. Follow the dry time recommended by the manufacturer.

Finishing Interior Vinyl and Soft Plastic Parts

The rigid outer surface of instrument panels is generally made of ABS plastic blended with PVC or vinyl. The same is true of many padded door trim panels. The soft cushion padding under the rigid outer surface is urethane foam.

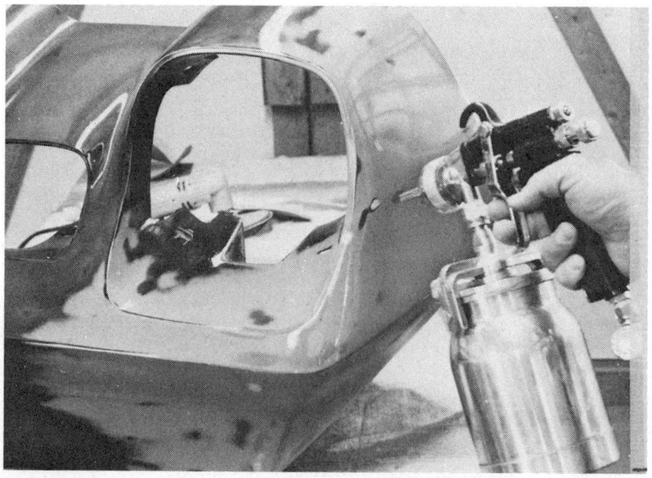

FIGURE 16-36 Applying primer-surfacer to a urethane bumper

FIGURE 16-37 The finished bumper

The most common use of flexible vinyls such as polyvinyl chloride are the coated fabrics used in seat trim, some door trim assemblies, headliners, and sun visors. Examples of hard vinyls are seat back handles, coat hooks, and exterior molding inserts.

The finishing system used for soft plastics is vinyl lacquer. Originally, this heavy-bodied finish was used over painted steel tops to simulate vinyl fabric. By changing reduction formulas and air pressures, vinyl lacquer will dry to a leatherlike texture, similar in appearance to textured vinyl. This product is frequently used to restore faded vinyl tops.

Vinyl lacquer is also used as a flat black topcoat to produce accent stripes and nonglare hood trim. Once dry, most vinyl lacquer can be recoated with acrylic lacquer or acrylic enamel to match the vehicle color. Vinyl lacquer is usually available in a wide array of colors. No primer, undercoat, or thinning is necessary. The finishing procedure is as follows:

1. Make sure the area to be finished is free of oils, waxes, and all other contaminants. An appropriate cleaning solvent should be used to clean vinyl. If an extremely soiled condition exists, detergent and water can be used for a first washing before solvent is used. Be sure all moisture has completely evaporated before any coatings are applied.
2. Apply vinyl lacquer in wet coats. Allow flash time between coats according to the label directions. To avoid washing out the grain effect, apply only enough color for proper hiding, and use an air pressure of 20 to 25 psi.
3. Before the color flashes completely, apply one wet double coat of clear vinyl. This is necessary to control the gloss and to prevent crocking (rubbing off) of the color coat after drying. Instrument panel covers require a final nonglare topcoat.
4. Allow to dry according to the label directions before installing the part and putting the vehicle back in service.

FINISHING EXTERIOR VINYL TOPS

Exterior vinyl tops can be finished using the following procedure:

1. Wash the top with a bleach-type detergent, a brush, and plenty of water. Rinse thoroughly with clean water.
2. Clean the top with cleaning solvent or vinyl prep conditioner.
3. Blow out all gap spacing and crevices around the top and tack-wipe as required.
4. Mask carefully and cover the entire hood and deck lid. The adhesion property of the vinyl system will make overspray difficult to remove.
5. Start the color application with a banding coat using low air pressure (20 to 25 psi) and a narrow fan. Spray into drip rails and cracks where the windshield and rear window molding meet the top.
6. Increase the air pressure and open the fan to a normal spray pattern. Apply the finish while working toward the center.
7. On the opposite side of the vehicle, start at the center and maintain a wet application over the remainder of the top. Use a full and uniform 50 to 75 percent stroke overlap. Keep the spray gun as perpendicular to the surface as possible.
8. Apply a second full wet coat for complete hiding and uniformity of wetness. Adjust the gun distance and speed for the desired texture.
9. For the final coat, many manufacturers recommend an application of one wet coat of 200 percent thinned vinyl lacquer (1 part vinyl color to 2 parts lacquer thinner) over the entire top.
10. After 1 hour of drying, remove the masking. Let dry a minimum of 4 hours before putting the vehicle back into service.
11. To protect the top and other vinyl surfaces, use vinyl preserver. It can be applied with a sponge or clean, soft cloth (Figure 16-38).

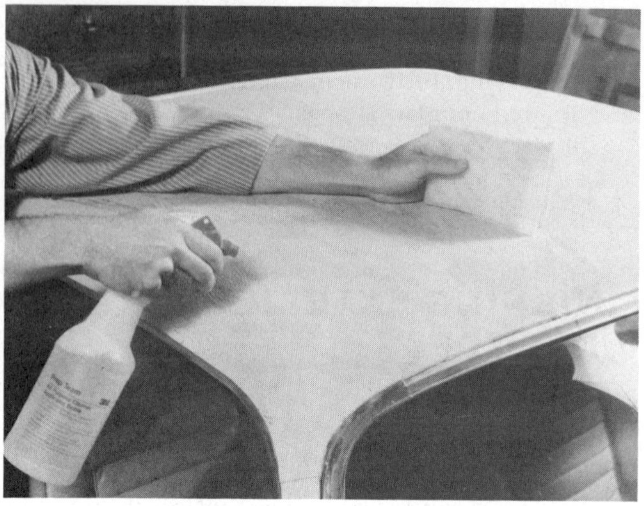

FIGURE 16-38 Applying vinyl preserver

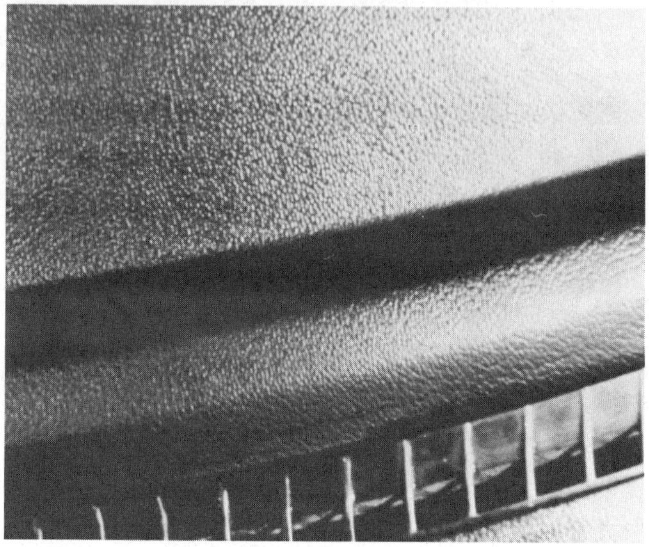

FIGURE 16-39 A retextured plastic surface

RETEXTURING INTERIOR PLASTIC PARTS

Many different textures or grains are found in the average automobile interior. When retexturing a repaired plastic part, it is important to keep in mind that the existing texture does not have to be duplicated. There is no need to spend time and effort trying to get the repair area to look exactly like the rest of the piece; a variation in the grain is acceptable. Only the coarseness of the texture must be duplicated in order to achieve professional results (Figure 16-39).

Retexturing can be done one of three ways:

- Blend the new texture out into the old
- Retexture to a natural break line
- Retexture the entire part

Use a refillable aerosol sprayer instead of a spray gun; the lower pressure will prevent the material from atomizing. To achieve a coarse texture, use the texture material unthinned. For a finer texture, use a small amount of lacquer thinner. A typical retexturing procedure is as follows:

1. Mix the texture material as per the manufacturer's instructions. Direct the first coat toward the repair area only. Hold the sprayer 18 to 24 inches from the surface, and always use a dry spray. Wet spraying will destroy the grain effect.
2. Allow flash time between coats. As many as eight to ten light coats might be needed to achieve the required buildup; this is by no means a one-coat application.
3. When buildup has been achieved, begin blending the texture out and away from the repair area, similar to a color coat. Force-drying between coats speeds up the process.
4. When the surface has dried, nib sand with 220 grit paper. This will blend the newly textured area into the original texture of the part.
5. If not satisfied with the texture, apply additional light coats of material and repeat the sanding.

After retexturing, the part should be blown dust-free in preparation for refinishing. Do not use any type of cleaner on a newly retextured area. Since texture material is usually lacquer based, a conventional interior acrylic lacquer is considered the best topcoat.

ADDING THE FINISHING TOUCHES

Before returning the vehicle, the painter must be aware of all those little things that make a big difference to the customer. These include:

- Vacuum the interior of the vehicle carefully. It should be cleaner than when the customer brought it in. To achieve this it might be necessary to use an automobile fabric reconditioner on the upholstery and carpet surfaces. Reduce the reconditioner with water according to the ratio on the label.

After all areas to be reconditioned have been vacuumed, apply the solution with a soft brush to all soiled spots on the seats, door panels, seat belts, and carpets. No masking is required because leather, vinyl, and metal surfaces are not affected by most reconditioners. (Check the manufacturer's instructions.) It is important to saturate stains thoroughly and brush them vigorously with the brush. After an application of reconditioner, use a highly absorbent towel or cloth and, with heavy pressure, rub and remove the excess. Be sure to remove all excess reconditioner from the seat crevices and folds. Stubborn stains remaining after the application can be removed by brushing undiluted reconditioner directly on them. Under normal conditions, the entire interior should dry within 12 hours.

- Thoroughly clean all the glass, including windows, mirrors, and lights (Figure 16-40).
- Carefully remove any overspray that may have been left on windows or chrome.
- Clean and polish chrome, moldings, and bumpers. Steel wool should not be used to polish chrome because pieces of the wool can easily become embedded in the new finish. Instead, use a commercial chrome polisher.
- Use a brush with soap and water to clean the tires and wheels. Do not let dirty wheels spoil the appearance of an otherwise quality job.
- Use chassis black to blacken wheel openings and any other exposed undercarriage parts, since overspray often gets on these areas. The customer generally will not notice this, but it certainly will be noticed if it is not done.
- Replace wipers, moldings, and emblems that were removed before finishing. Take the time to clean off these items and be certain that everything is replaced.
- If the vehicle has a vinyl top, do not forget to wipe it with a damp cloth or a commercial vinyl cleaner.
- As a finishing touch, clean the engine compartment. The easiest way to do this is to spray it with a heavy-duty engine cleaner, then flush it out with high-pressure water. A clean engine compartment usually makes a big impression on the customer.

FIGURE 16-40 Clean all glass, including headlights.

FIGURE 16-41 Always try to send the customer home satisfied.

Finally, inspect the vehicle with a careful eye for details. Make the extra effort. If a window is smeared, clean it again. If a piece of masking tape remains, remove it. If an emblem is missing, replace it before the customer asks where it is. If the vehicle gets dirty while waiting to be picked up, wipe it down with a clean cloth. The number one objective should always be a satisfied customer (Figure 16-41).

REVIEW QUESTIONS

1. To achieve a lighter color with a metallic paint. Technician A sprays it on dry. To achieve the same effect, Technician B sprays it on wet. Who is correct?
   a. Technician A
   b. Technician B
   c. Both A and B
   d. Neither A nor B

2. Which part of the paint settles in the can?
   a. binder
   b. solvent
   c. resin
   d. pigment

3. What can be done during overall refinishing to prevent a gritty surface?
   a. maintain minimal overspray on the horizontal surfaces
   b. keep a wet edge
   c. both a and b
   d. neither a nor b

4. Which of the following statements concerning base cloat/clear coats is incorrect?
   a. Sanding of the base coat should be avoided.
   b. Clear coats should not be loaded on heavily.
   c. Clears perform better when underreduced.
   d. All of the above

5. When spot refinishing with solid colors, Technician A applies the topcoat with a fanning motion from the center outward. Technician B applies the topcoat in short strokes from the center outward. Who is correct?
   a. Technician A
   b. Technician B
   c. Both A and B
   d. Neither A nor B

6. Technician A says that acrylic urethane enamel is more durable than polyurethane enamel. Technician B says polyurethane enamel is more durable than acrylic urethane enamel. Who is correct?
   a. Technician A
   b. Technician B
   c. Both A and B
   d. Neither A nor B

7. After force drying the topcoat, Technician A removes the masking tape while the finish is still warm. Technician B waits to remove it until after the finish has cooled. Who is correct?
   a. Technician A
   b. Technician B
   c. Both A and B
   d. Neither A nor B

8. Which type of compounding is best on blended areas?
   a. machine
   b. hand
   c. both a and b
   d. neither a nor b

9. How long should a new topcoat be allowed to cure before automotive finish wax is applied?
   a. 24 hours
   b. 1 week
   c. 30 to 45 days
   d. 60 to 90 days

10. When refinishing a hard interior plastic part, Technician A always uses primer. Technician B always uses primer-sealer. Who is correct?
    a. Technician A
    b. Technician B
    c. Both A and B
    d. Neither A nor B

11. When painting a polypropylene part, ______________.
    a. a flex agent must be added to the topcoat
    b. a special PP primer is required for the undercoat
    c. both a and b
    d. neither a nor b

12. Texture material should be applied with the spray gun ______________.
    a. 18 to 24 inches from the surface and using a wet spray
    b. 6 inches from the surface and using a wet spray
    c. 6 inches from the surface and using a dry spray
    d. 18 to 24 inches from the surface and using a dry spray

13. Technician A finishes an exterior vinyl top by starting with a banding coat and wide spray fan. Technician B starts with a banding coat and narrow spray fan. Who is correct?
    a. Technician A
    b. Technician B
    c. Both A and B
    d. Neither A nor B

14. Technician A retextures by blending the new texture out into the old. Technician B retextures to a natural break line. Who is correct?
    a. Technician A
    b. Technician B
    c. Both A and B
    d. Neither A nor B

15. If necessary, a small amount of ______________ can be used to smooth the movement of a finesse sanding block when removing finish defects.
    a. oil
    b. water
    c. grease
    d. none of the above

APPENDIX A
GLOSSARY

Abrasive Material such as sand, crushed steel grit, aluminum oxide, silicon carbide, or crushed slag used for cleaning or surface roughening.

Accessible Area An area that can be reached without parts being removed from the vehicle.

Access Time Time required to remove extensively damaged collision parts by cutting, pushing, pulling, and so on.

Actual Cash Value (ACV) Current market value of a standard production vehicle and its accessory options as determined by used car guidebook listings or car dealer assessments.

Adhesion The ability of one substance to stick to another.

Adjuster The insurance company representative who is responsible for approving the collision repair bid and satisfying a customer's vehicle damage claim; same as appraiser.

Air Drying Allowing paint to dry at ambient (surrounding) temperatures without the aid of an external heat source.

Airless Spraying A method of spray application in which atomization is affected by forcing the paint under high-pressure through a very small orifice in the spray gun cap. On emerging, the paint instantly expands and breaks into very fine particles.

Align To adjust to a line or predetermined relative position.

Anchor To hold in place.

Assembly A number of parts that are either bolted or welded together to form a single unit.

Atomize The extent to which the air at the spray gun nozzle breaks up the paint and solvent into fine particles.

Automatic Welding Welding with equipment that performs the entire welding operation without constant observation and adjustment of the controls by an operator. The equipment may or may not perform the loading and unloading of the work.

Back Light Vehicle window located behind the passengers.

Baffle Panel used to direct air to the radiator.

Bar Gauges Gauges used to accurately measure and diagnose body and frame collision damages for all conventional and unitized vehicles.

Base Coat The coat of paint on which the final coats are applied.

Base Coat/Clear Coat A paint system in which the color effect is given a highly pigmented base coat. Gloss and durability are given by a subsequent clear coat. The base coat can be either solid color or metallic.

Belt Line The horizontal molding or crown along the side of the vehicle.

Belt Molding A trim strip that goes around the vehicle slightly below the lower portion of the vehicle's windows.

Bench System A vehicle underbody measurement device. The vehicle is set on fixtures to check its frame and suspension dimensions for damage.

Betterment Depreciation of vehicle parts, due to age or wear, determined by the insurance adjuster.

Binder The ingredient in a paint that holds the pigment particles together.

Bleeding Original color showing through after a new topcoat has been applied.

Body The apparent viscosity of a paint as assessed when stirring it.

Body Filler A heavy-bodied plastic material that cures very hard and is used to fill small dents in metal.

Body Panels Sheets of steel that form the car body.

Body Trim Upholstery or body ornamentation.

Bounce-back Atomized particles of paint that rebound from the surface being sprayed and contribute to overspray.

Braze Welding A method of welding using a filler metal. Unlike brazing, the filler metal is not distributed in the joint by capillary attraction.

Brazing A group of welding processes that produces coalescence of materials by heating them to a suitable temperature and by using a filler metal. The filler metal is distributed between the closely fitted surfaces of the joint by capillary attraction.

Bulge High crown or area of stretched metal.

Burnishing To polish or buff a finish by hand or machine using a compound or liquid manufactured for this purpose.

Butt Weld Weld made along a line where two pieces are put edge to edge.

Camber The inward or outward tilt of the wheel at the top from true vertical. It is the tire wearing angle in degrees.

Caster The backward or forward tilt of the kingpin or spindle support arm at the top from true vertical. It is the directional control angle measured in degrees.
Catalyst A substance that causes or speeds up a chemical reaction when mixed with another substance but does not change by itself.
Centering Gauge Used in sets of four, this frame gauge locates horizontal datum planes and centerlines on a vehicle.
Chassis The assembly of mechanisms that makes up the major operating systems of the vehicle; usually assumed to include everything except the car body.
Clearance The amount of space between adjacent panels.
Clear Coat A top coating on a painted surface that is transparent so that the color coat beneath it is visible.
Closed Structural Members Boxed-in sections, typically accessible only from the outside, such as rails, pillars, and so forth.
Coatings Covering material used to protect an area.
Compounding The action of using an abrasive material—either by hand or machine—to smooth and bring out the gloss of the applied lacquer topcoat.
Contaminants Foreign substances on the surface to be painted (or in the paint) that would adversely affect the finish.
Control Points Points on a vehicle, including holes, flats, or other identifying areas, used to position panels and rails during manufacturing and redimensioning of vehicles.
Conventional Frame Construction type in which the engine and body are bolted to a separate frame.
Conventional Points Points on the unibody used as a reference to make a repair.
Conversion Coating A special metal conditioner or primer used on galvanized steel, uncoated steel, and aluminum to prevent rust.
Corrosion The chemical reaction of air, moisture, or corrosive materials on a metal surface; usually referred to as rusting or oxidation.
Coverage The area a given amount of paint will cover.
Cowl Panels Panels forward of the passenger compartment to which the fenders, hood, and dashboard are bolted.
Crush Zones Buckling points designed into certain structural components for absorbing the impact energy in a collision.
Cure The process of drying or hardening of a paint film.
Datum Line An imaginary line that appears on frame blueprints or charts to help determine correct height.
Dedicated Fixture Measuring Systems A bench with fixtures that are set in specific points for body measurement.
Degreasing Cleaning a substrate (usually metal) by removing greases, oils, and other surface contaminants.
Diamond A damage condition where one side of the vehicle has been moved to the rear or front causing the frame/body to be out of square.
Direct Damage The damage that occurs to the area that is in direct contact with the damaging force or impact.
Dog Tracking Off-center tracking of rear wheels as related to the front wheels.
Double Coat Spray first pass left to right, spray second pass right to left directly over first pattern.
Drier A catalyst added to a paint to speed up the cure or dry.
Drivetrain Engine, transmission, and drive shafts/axles.
Drying The process of changing a coat of paint from a liquid to a solid state due to evaporation of solvent, chemical reaction of the binding medium, or a combination of these causes.
Dry Spray An imperfect coat, usually caused by spraying too far from the surface being painted or on too hot a surface.

Edge Joint A joint between the edges of two or more parallel or nearly parallel members.
Elastomer A man-made compound with flexible and elastic properties. The resin system of elastomeric enamels and lacquers is made of these elastomeric compounds.
Enamel A type of paint that dries in two stages: first by evaporation of the solvent and then by oxidation of the binder.

Factory Specifications Data as supplied by the individual car manufacturer covering all measurement and areas of the vehicle.
Fan The spray pattern of a spray gun.
Fatigue Failure A type of metal failure resulting from repeated stress that finally alters the character of the metal so that it cracks.
Featheredge Tapering the edges of the damaged area with sandpaper or special solvents.
Filler Any material used to fill (level) a damaged area.
Film A very thin continuous sheet of material. Paint forms a film on the surface to which it is applied.

Finish A protective coating of paint; to apply a paint or paint system.
Fixtures A set of accessories for a dedicated measuring system designed to attach to a bench to fit the reference points for a specific family of body styles.
Flash The first stage of drying where some of the solvents evaporate, which dulls the surface from an exceedingly high gloss to a normal gloss.
Flat Lacking in gloss.
Frame The heavy metal structure that supports the auto body and other external components.
Frame Gauges Gauges that can be hung from the car frame to check its alignment.
Frame Straightener A pneumatic-powered machine used to align and straighten a distorted frame or body.
Front-Wheel Drive A vehicle that has its drive wheels located on the front axle.
Full Body Section Simultaneous section repairs to both rocker panels, windshield pillars, and floor pan; required to join the undamaged front half of one vehicle to the undamaged rear half of another vehicle.

Galvanized Metal coated with zinc.
Gap The distance between two points.
Gas Metal Arc Cutting An arc cutting process used to sever metals by melting them with the heat of an arc between a continuous metal (consumable) electrode and the work. Shielding is obtained entirely from an externally supplied gas or gas mixture.
Gas Metal Arc Welding An arc welding process that produces coalescence of metals by heating them with an arc between a continuous filler metal (consumable) electrode and the work. Shielding is obtained entirely from an externally supplied gas or gas mixture.
Gauge A measure of the thickness of sheet metal.
Goggles A device with colored lenses or clear safety glass that protects the eyes from harmful radiation during welding and cutting operations.
Grille Decorative panel in front of the radiator.
Grit A measure of the size of particles on sandpaper or discs.
Gun In semiautomatic, machine, and automatic welding, a manipulating device to transfer current and guide the electrode into the arc. It can include provisions for shielding and arc initiation.

Hardener A curing agent used in plastics.
Header Bar The framework or inner construction that joins the upper sections of the windshield and pillars; forms the upper portion of the windshield opening and reinforces the turret top panel.
Headlining The cloth or plastic material covering the ceiling area in the passenger compartment.
High-Strength Steel A low-alloy steel that is much stronger than hot-rolled or cold-rolled sheet steels; normally used in the manufacture of vehicle structural parts.
Hinge Pillar The framework or inner construction to which the door hinges fasten.
Hood Panel The large metal panel that generally fills in the space between the two fenders and closes off the engine compartment so that rain cannot fall on the engine.
Hot Spot An unprotected area subject to corrosion.
HSLA High-strength low-alloy steel used in unibody construction.
HSS Steel High tensile strength steel whose strength is derived from heat treatment. This steel will tear or fracture if the collision stresses exceed the tensile strength.

Independent Front Suspension The conventional front suspension system in which each front wheel moves independently of the other.
Independent Rear Suspension Rear suspension that has no cross axle shaft. Each wheel and related suspension is allowed to act individually.
Interior Trim All the upholstery and moldings on the inside of the vehicle.
Isocyanate Resins This is a principal ingredient in urethane hardeners. Because this ingredient has toxic effects on the painter, he or she is always advised to wear a respirator that is approved by NIOSH.

Jig Mechanical device for holding work in its exact position while it is being welded.
Joint The point or line at which two pieces are connected.
Jounce The compression of the springs, or the upward movement of the wheel and downward movement of the frame.

Lacquer Type of paint that dries by solvent evaporation; can be rubbed to improve appearance.
Lap Weld A weld is made along the edge of an overlapping piece.
Laser System Type of universal measuring system that uses laser optics in part or total vehicle redimensioning.

Manual Welding Welding in which the entire operation is performed and controlled by hand.

Mash Vehicle body damage condition in which the length of any section or frame member is shorter than factory specifications.
Mechanical Components Powertrain, accessories, and suspension system.
Member Any essential part of a machine or assembly.
Metal Conditioner Chemical cleaner that removes rust and corrosion from bare metal and helps prevent further rusting.
Metallic Paint Finish-paint colors that contain metallic flakes in addition to pigment.
Misaligned Having uneven spacing, as between body panels.
MIG (Metal Inert Gas) Gas metallic arc welding.
Model Year Production period for new model vehicles or new engines, designated by the calendar year in which the period ends.
Monocoque Unibody vehicle construction in which the sheel metal of the body supports most of the structural strength of the vehicle.
Multiple-Pull Systems Pulling in two or more directions to correct damage.

Nonferrous Metals that contain no iron. Aluminum, brass, bronze, copper, lead, nickel, and titanium are nonferrous.

OEM Original Equipment Manufacturer.
Opaque Impervious to light; not transparent.
Open Structural Members Flat panels, typically accessible from both sides, such as floor pans and trunk floors.
Outer Panels Sheet metal sections that, when attached to the inner panels, form the exterior of a vehicle.
Oven A piece of equipment used to bake finishes.
Overlap The amount of the spray pattern that covers the previous spray stroke.
Overpulling Stretching the metal too much, resulting in the need to replace the part.
Overspray Paint that falls on the area next to the one being painted.
Oxyacetylene Welding A welding process that produces coalescence of metals by heating them with a gas flame or flames obtained from the combustion of acetylene with oxygen. The process can be used with or without the application of pressure and with or without the use of filler metal.

Paint Film The actual thickness of the paint on a surface.
Panel Repair A type of refinish repair job in which a complete section (door, hood, rear deck, and so on) is repainted.
Panel Spotting The means by which both lap and flange joints can be made with a spliced or full panel installation.
Parallelogram Linkage Steering system in which a short idler arm is mounted on the right side so that it is parallel to the pitman arm.
Parent Metals Original metals in the unibody.
Penetration Depth of fusion into the metal being welded.
Pigment Material in the form of fine powders; used to impart color, opacity, and other effects to paint.
Plasma A gas that has been heated to at least a partially ionized condition, which enables it to conduct an electric current.
Plasma Arc Cutting Arc cutting process in which the metal is severed by melting a localized area with a constricted arc and removing the molten material with a high velocity jet of hot, ionized gas issuing from the orifice.
Plastic Filler A compound of resin and fiberglass used to fill dents on car bodies.
Plug Weld Adding metal into a hole and fusing all the metal.
Polychromatic Term used by some paint manufacturers for color coats that contain aluminum powder in flake form.
Porosity Gas pockets or voids in metal.
Pot Life The amount of time a painter has to apply a plastic or paint finish to which a catalyst or hardener was added.
Primary Damage Damage that occurs at the point of impact on the vehicle.
Prime Coat The first coat in a paint system. Its main purpose is to improve adhesion and provide corrosion protection.
Primer-Sealer An undercoat that improves adhesion of the topcoat and seals old painted surfaces that have been sanded.
Primer-Surfacer A high-solids primer that fills small imperfections in the substrate and usually must be sanded.
Pulling Applying force to a part to make a change.
Putty A material made for filling small holes or sand scratches.

Quarter Panel The side panel, which in four door sedans, is generally a quarter of the total length of the automobile and extends from the rear door to the end of the car.

Rack-and-Pinion Steering Gear Steering gear in which a pinion gear on the end of the steering shaft meshes with a rack gear on the steering linkage.
Reducer The solvent combination used to thin enamel is usually referred to as a reducer.

Refinish To remove or seal the old finish and apply a new topcoat; to repaint.
Resistance Welding Welds made by passing an electric current through metal between the electrodes of the welding gun.
Respirator A device worn over the mouth to filter particles and fumes out of the air being breathed.
Retarder A slow evaporating thinner used to retard drying.
Rocker Panel The narrow, outer panel attached below the car door.
Roof Rails Framework or inner construction that reinforces and supports the sides of the roof panel or turret top.

SAE Society of Automotive Engineers.
Sag A type of frame damage in which one or both side rails bend and sag at the cowl, causing buckles to be formed on the top of the side rails.
Scuff To rough up a surface by rubbing lightly with sandpaper to provide a suitable surface for painting.
Sealer An intercoat between the topcoat and the primer or old finish, giving better adhesion.
Secondary Damage Damage that occurs due to misplaced energy that causes stresses in the suspension and/or body dimensions at areas other than the primary impact zone.
Sectioning A means of replacing partial areas of a vehicle.
Service Manual A book published annually by each vehicle manufacturer that lists the specifications and service procedures for each make and model of vehicle; also called a shop manual.
Shim Small metal piece used behind panels to bring them into alignment.
Show Through Sand scratches in the undercoat that are visible through the topcoat.
Single Pull Systems Pulling one way at a time.
Solvents That which puts a paint into solution for application; a single such ingredient. Thinners and reducers are composed of one or several solvents.
Specifications Information provided by the manufacturer that describes each automotive system and its components, operation, and measurement.
Spot Repair Type of refinish repair job in which a section of a car smaller than a panel is refinished (often called ding and dent work).
Spot Weld Weld in which the arc is directed to penetrate both pieces of metal, while triggering a timed impulse of wire feed.
Spray Paint is atomized in a spray gun and the stream of atomized paint is directed at the part to be painted.
Spray Gun A painting tool powered by air pressure that atomizes liquids.
Steering System Mechanism that enables the driver to turn the wheels to change the direction of the vehicle's movement.
Stitch Welding Using intermittent welds to join two or more parts.
Straight-In Damage Damage resulting from a direct hit, a collision for example, rather than from a glancing hit.
Stress Relieving Taking tension off a part.
Structural Member Any primary load-bearing portion of the body structure that affects its over-the-road performance or crashworthiness.
Structural Panels Panels used in the unibody that become part of the whole unit and are vital to the strength of the body.
Strut Suspension MacPherson strut attached to the tower and camber reference angle.
Stub Frame Unitized body frame with only the front and rear stub sections of the frame rails and no side rail portions.
Stud Welding General term for the joining of a metal stud or similar part to a workpiece. Welding can be accomplished by arc, resistance, friction, or other suitable process with or without external gas shielding.
Subassembly Several parts put together before the "whole" is attached.
Substrate The surface that is to be finished (painted). It can be anything from an old finish or primer to an unpainted surface.
Suspension System The springs and other components that support the upper part of a vehicle on its axles and wheels.
Symmetrical Design Design in which both sides of a unibody are identical in structure and measurement.

Tack Cloth A cheesecloth that has been treated with nondrying varnish to make it tacky. Used to pick up dust and lint from the surface to be painted.
Thinner The solvent combination used to thin lacquers and acrylics to spraying viscosity is usually called thinner.
Three-Stage Consists of three distinct paint layers that produce a pearlescent appearance: base coat, mid coat, and clear coat.
Tint A very light color, usually a pastel. To add color to another color or to white.
TIG (Tungsten Inert Gas) Gas Tungsten arc welding.
Toe The straight-ahead position of the wheels.
Toe-In The distance the front of the wheels is closer together than rear of the front wheels.

Toe-out The distance the front of the front wheels is farther apart than the rear of the front wheels.
Tolerances Acceptable alteration limit of the vehicle dimensions as provided by the manufacturer (for example, plus or minus 3 mm is typical).
Topcoat The last or final color coat.
Toxicity Pertaining to a poisonous effect.
Trim Decorative metal pieces on a car body.
Turning Radius A tire wearing angle measured in degrees; the amount one front wheel turns sharper than the other one turns.
Two-Part A paint or lacquer supplied in two parts that must be mixed together in the correct proportions before use. The mixture remains usable for a limited time.
Two-Stage Consists of two distinct layers of paint: base coat and clear coat.

Ultrasonic Plastic Welding Method of repairing auto plastics in which the welding time is controlled by the power supply. This method is best suited to rigid plastics.
Ultrasonic Stud Welding A variation of the shear joint used to join plastic parts in which the weld is made along the circumference of the stud.
Undercoat A first coat; primer, sealer, or surfacer.
Undercut A groove melted into the base metal adjacent to the toe or root of a weld and left unfilled by weld metal.
Unibody Vehicle in which the parts for the body structure serve as the support for the overall vehicle components.
Universal Measuring System A measuring system that has devices mounted on a frame that can be adjusted for various vehicle bodies. This system can utilize laser systems, universal benches, or mechanical gauges.

Vehicle Identification Number (VIN) The number assigned to each vehicle by its manufacturer, primarily for registration and identification purposes.
Viscosity Consistency or body of a paint; thickness or thinness.
VOC Volatile organic compound found in most paint solvents.

Weld To join two metal or plastic pieces together by bringing them to their melting points, often involving the use of a welding rod to add metal to the joint.
Wheel Alignment Positioning the suspension and steering components to assure a vehicle's proper handling and optimum tire wear.
Wheel Balancing Proper distribution of weight around a tire and wheel assembly to counteract centrifugal forces acting on the heavy areas to maintain a true running wheel perpendicular to its rotating axis.
Wheelbase Distance between the centerlines of the front and rear axles.
Wheelhouses The deep curved panels that form the compartments in which the wheels rotate. They are generally bolted to the front fenders and spot welded to the rear quarter panels.

Zoning Method of systematically observing a damaged vehicle. It includes checking primary damage, secondary damage, then the suspension, steering and mechanics, the interior components and finally, the exterior paint and trim of the vehicle.

APPENDIX B

ABBREVIATIONS USED BY BODY REPAIR TECHNICIANS AND ESTIMATORS

The estimator and collision repair technician must be able to communicate verbally as well as in writing. Both in estimates and work procedure reports most estimators use abbreviations. Generally, these abbreviations are the same as those used in crash estimating guides. Some abbreviations are even used verbally. For example, three of the most commonly used abbreviations in a body shop are:

- **R&I: Remove and Reinstall.** The item is removed as an assembly, set aside, and later reinstalled and aligned for a proper fit. This is generally done to gain access to another part. For example, R&I bumper would mean that the bumper assembly would have to be removed to install a new fender or quarter panel.
- **R&R: Remove and Replace.** Remove the old parts, transfer necessary items to new part, replace, and align.
- **O/H: Overhaul.** Remove an assembly from the vehicle, disassemble, clean, inspect, replace parts as needed, then reassemble, install and adjust (except wheel and suspension alignment).

In addition to these abbreviations, the following terms are those accepted by most estimating guides, shop manuals, and estimators. They are the ones used in most written forms.

A	Manufacturer has no list price for the part.
ABS	Anti-lock Brake System
A/C	Air Conditioner
ACRS	Air Cushion Restraint System
adj	adjuster or adjustable
A/F	Air Fuel Ratio
AIR	Air Injector Reactor
alt	alternator
alum	aluminum
amp	ampere
approx.	approximately
assy	assembly
A/T	Automatic Transaxle
AT	Automatic Transmission
auth	authority
auto	automatic
aux	auxiliary
B+	Battery Voltage
bat.	battery
bbl	barrel
bk	back
blwr	blower
bmpr	bumper
brg	bearing
brk	brake
brkt	bracket
Bro	Brougham
btry	battery
btwn	between
B-U	Back-Up
bush	bushing
Calif	California
chnl	channel
c/mbr	cross member
cntr	center
col	column
comp	compressor
compt	compartment
conn.	connector
cond	conditioning or conditioner
cont	control
conv	converter or convertible
cor	corner
cov	cover
Cpe	Coupe
C/R	Customer Requested
crossmbr	cross member
c/shaft	crankshaft
ctl	control
Ctry	Country
Cust	Custom
CV	Constant Velocity
cyl	cylinder
D	Discontinued part
d	drilling operational time
da	dash
dbl	double

def	deflector
dehyd	dehydrator
desc	description
dia	diameter
diag	diagonal
dist	distributor
div	division
Dlxe	DeLuxe
DP	Dash Pot
dr	door
ea	each
ECU	Electronic Control Unit
EFI	Electronic Fuel Injection
EGR	Exhaust Gas Recirculation
elec	electric
emiss	emission
eng	engine
EP	Exhaust Purging
equip	equipment
evap	evaporator
exc	except
exh	exhaust
extn	extension
FED	Federal (All States Except Calif.)
flr	floor
Fndr	Fender
Fr & Rr	Front & Rear
frm	from
fr or rr	front or rear
ft	foot
ft-lb	foot-pounds
FWD	Front Wheel Drive
gal	gallon
gen	generator
GND	Ground
grds	guards
grv	groove
harn	harness
H'back	Hatchback
hd	heavy duty
HDC	Heavy Duty Cooling
hdr	header
HEI	High Energy Ignition
Hi Per	High Performance
horiz	horizontal
H.P.	High Performance
hp	horsepower
hsg	housing
HSLA	High Strength Alloy Steel
HSS	High Strength Steel
HT	Hard Top
H'Top	Hard Top
hyd	hydraulic
Hydra	Hydramatic
ID	Inside Diameter
ign	ignition
in	inch
incl	includes
INJ	Injection
in-lb	inch-pounds
inr	inner
inst	instrument
inter	intermediate
IP	Instrument Control Panels
L	Left
LED	Light Emitting Diode
LF	Left Front
LH	Left Hand
lic	license
lp	lamp
LR	Left Rear
LS	Left Side
lwr	lower
max	maximum
mdl	model
mldg	molding
MT	Manual Transmission
mtd	mounted
mtg	mounting
muff	muffler
NAGS	National Auto Glass Specification
neg	negative
OD	Outside Diameter
OD, O/D	Overdrive
OEM	Original Equipment Manufacturer
opng	opening
orna	ornament
O/S	Oversize
otr	outer
p	paint operational time
pass	passenger
PCV	Positive Crankcase Ventilation
pkg	package
plr	pillar
pnl	panel
pos	positive
PS	Power Steering
psi	pounds per square inch
pt	pint
Pwr	Power
qtr	quarter
R	Right
rad	radiator
R-L	Right or Left
rec	receiver

refl	reflector
reg	regulator
reinf	reinforcement
reson	resonator
RH	Right-Hand
Sed	Sedan
ser	serial or series
shld	shield
sidembr	side member
sig	single
SIR	Supplemental Inflatable Restraint
spd	speed
spec	special
SRS	Supplemental Restraint System
Sta	Station
stab	stabilizer
stat	stationary
Std	Standard
stl	steel
strg	steering
Sub	Suburban
sup	super
supt	support
surr	surround
susp	suspension
SW	Station Wagon
tach	tachometer
t & t	tilt & telescope or tilt & travel
TE	Thermactor Emission
tel	telescope
trans	transmission
TWC	Three-Way Catalyst
U/D	Underdrive
upr	upper
U/S	Undersize
vent	ventilator
vert	vertical
vib	vibration
VIR	Valve-In-Receiver
w/	with
w/b	wheel base
WB	Wheelbase
WD	Wheel Drive
Wgn	Wagon
w'house	wheelhouse
whl	wheel
whlse	wheelhouse
wndo	window
w/o	without
wo/	without
wshd	windshield
w'strip	weatherstrip

APPENDIX C

MEASUREMENT EQUIVALENTS

CONVERSION FACTORS		
Multiply	**By**	**To obtain**
Length		
Millimeters (mm)	0.03937	Inches
	0.1	Centimeters (cm)
Kilometers (km)	0.6214	Miles
	3281	Feet
Inches	25.4	Millimeters (mm)
Miles	1.6093	Kilometers (km)
Area		
$Inches^2$	645.16	$Millimeters^2$ (mm^2)
	6.452	$Centimeters^2$ (cm^2)
$Feet^2$	0.0929	$Meters^2$ (m^2)
	144	$Inches^2$
Volume		
$Centimeters^3$ (cc)	0.06102	$Inches^3$
	0.001	Liters (L)
Liters (L)	61.024	$Inches^3$
	0.2642	Gallons
	1.0567	Quarts
$Inches^3$	16.387	$Centimeters^3$ (cc)
$Feet^3$	1728	$Inches^3$
	7.48	Gallons
	28.32	Liters (L)
Fluid ounces (oz)	29.57	Milliliters (mL)
Mass		
Gram (g)	0.03527	Ounce
Kilograms (kg)	2.2046	Pounds
	35.274	Ounces
Force		
Ounce	0.278	Newton (N)
Pound	4.448	Newton (N)
Kilogram	9.807	Newton (N)
Torque		
Foot-pounds	1.3558	Newton-meters (Nm)
	0.1383	Kilogram/meter (kg/m)
Inch-pounds	0.11298	Newton-meters (Nm)
	0.0833	Foot-pounds

CONVERSION FACTORS (CONTINUED)

Multiply	By	To obtain
Torque		
Kilogram-meters (Kg/m)	7.23	Foot-pounds
	9.80665	Newton-meters (Nm)
Pressure		
Atmospheres	14.7	Pounds/square inch (psi)
	29.92	Inches of mercury (In. Hg)
Inches of mercury (In. Hg)	0.49116	Pounds/square inch (psi)
	13.1	Inches of water
	3.377	Kilopascals (kPa)
Bars	100	Kilopascals (kPa)
	14.5	Pounds/square inch (psi)
Kilogram/cu² (Kg/cm²)	14.22	Pounds/square inch (psi)
	98.07	Kilopascals (kPa)
Kilopascals (kPa)	0.145	Pounds/square inch (psi)
	0.2961	Inches of mercury (In. Hg)
Pascal	1	Newton/square meter (N/m²)
Fuel Performance		
Miles/gallon	0.4251	Kilometers/liter (km/L)
Velocity		
Miles/hours	1.467	Feet/second
	88	Feet/minute
	1.6093	Kilometers/hour (km/h)
Kilometers/hour	0.27778	Meters/second (m/s)

TORQUE CONVERSIONS
(FOOT-POUNDS, NEWTON-METERS, METER-KILOGRAMS)

ft/lb	0	1	2	3	4	5	6	7	8	9
0	0	1.35	2.70	4.05	5.40	6.75	8.10	9.45	10.8	12.1
10	13.5	14.9	16.2	17.6	18.9	20.3	21.6	22.9	24.3	25.6
20	27.0	28.3	29.7	31.0	32.5	33.7	35.1	36.4	37.8	39.1
30	40.5	41.8	43.2	44.5	45.9	47.2	48.6	49.9	51.3	52.6
40	54.0	55.3	56.7	58.0	59.4	60.7	62.1	63.4	64.8	66.1
50	67.5	68.8	70.2	71.5	72.9	74.2	75.6	76.9	78.3	79.6
60	81.0	82.3	83.7	85.0	86.4	87.7	89.1	90.4	91.8	93.1
70	94.5	95.8	97.2	98.5	99.9	101	102	103	105	106
80	108	109	110	112	113	114	116	117	118	120
90	121	122	124	125	126	128	129	130	132	133
100	135	136	137	139	140	141	143	144	145	147
110	148	149	151	152	153	155	156	157	159	160
120	162	163	164	166	167	168	170	171	172	174
130	175	176	178	179	180	182	183	184	186	187

Note: The following formulas can be used:

$$\text{ft-lb} \times 1.35 = \text{Nm}$$
$$\text{ft-lb} + 7.23 = \text{mkg}$$

For meter-kilograms within 5%, divide Newton-meters by 10 (move the decimal point one place left). For greater accuracy, divide by 9.81.

DECIMAL AND METRIC EQUIVALENTS

Fractions	Decimal (in.)	Metric (mm)	Fractions	Decimal (in.)	Metric (mm)
1/64	.015625	.397	33/64	.515625	13.097
1/32	.03125	.794	17/32	.53125	13.494
3/64	.046875	1.191	35/64	.546875	13.891
1/16	.0625	1.588	9/16	.5625	14.288
5/64	.078125	1.984	37/64	.578125	14.684
3/32	.09375	2.381	19/32	.59375	15.081
7/64	.109375	2.778	39/64	.609375	15.478
1/8	.125	3.175	5/8	.625	15.875
9/64	.140625	3.572	41/64	.640625	16.272
5/32	.15625	3.969	21/32	.65625	16.669
11/64	.171875	4.366	43/64	.671875	17.066
3/16	.1875	4.763	11/16	.6875	17.463
13/64	.203125	5.159	45/64	.703125	17.859
7/32	.21875	5.556	23/32	.71875	18.256
15/64	.234275	5.953	47/64	.734375	18.653
1/4	.250	6.35	3/4	.750	19.05
17/64	.265625	6.747	49/64	.765625	19.447
9/32	.28125	7.144	25/32	.78125	19.844
19/64	.296875	7.54	51/64	.796875	20.241
5/16	.3125	7.938	13/16	.8125	20.638
21/64	.328125	8.334	53/64	.828125	21.034
11/32	.34375	8.731	27/32	.84375	21.431
23/64	.359375	9.128	55/64	.859375	21.828
3/8	.375	9.525	7/8	.875	22.225
25/64	.390625	9.922	57/64	.890625	22.622
13/32	.40625	10.319	29/32	.90625	23.019
27/64	.421875	10.716	59/64	.921875	23.416
7/16	.4375	11.113	15/16	.9375	23.813
29/64	.453125	11.509	61/64	.953125	24.209
15/32	.46875	11.906	31/32	.96875	24.606
31/64	.484375	12.303	63/64	.984375	25.003
1/2	.500	12.7	1	1.00	25.4

APPENDIX D

TAP AND DRILL BIT DATA

DECIMAL EQUIVALENTS AND TAP DRILL SIZES		
Drill Size	**Decimal**	**Tap Size**
1/64	.0156	
1/32	.0312	
60	.0400	
59	.0410	
58	.0420	
57	.0430	
56	.0465	
3/64	.0469	0-80
55	.0520	
54	.0550	1-56
53	.0595	1-64, 72
1/16	.0625	
52	.0635	
51	.0670	
50	.0700	2-56, 64
49	.0730	
48	.0760	
5/64	.0781	
47	.0785	3-48
46	.0810	
45	.0820	3-56, 4-32
44	.0860	4-36
43	.0890	4-40
42	.0935	4-48
3/32	.0937	
41	.0960	
40	.0980	
39	.0995	
38	.1015	5-40
37	.1040	5-44
36	.1065	6-32
7/64	.1093	
35	.1100	
34	.1110	6-36
33	.1130	6-40
32	.1160	
31	.1200	
1/8	.1250	
30	.1285	
29	.1360	8-32, 36
28	.1405	8-40
9/64	.1406	

DECIMAL EQUIVALENTS AND TAP DRILL SIZES		
Drill Size	**Decimal**	**Tap Size**
27	.1440	
26	.1470	
25	.1495	10-24
24	.1520	
23	.1540	
5/32	.1562	
22	.1570	10-30
21	.1590	10-32
20	.1610	
19	.1660	
18	.1695	
11/64	.1719	
17	.1730	
16	.1770	12-24
15	.1800	
14	.1820	12-28
13	.1850	12-32
3/16	.1875	
12	.1890	
11	.1910	
10	.1935	
9	.1960	
8	.1990	
7	.2010	1/4-20
13/64	.2031	
6	.2040	
5	.2055	
4	.2090	
3	.2130	1/4-28
7/32	.2187	
2	.2210	
1	.2280	
A	.2340	
15/64	.2344	
B	.2380	
C	.2420	
D	.2460	
E, 1/4	.2500	
F	.2570	5/16-18
G	.2610	
17/64	.2656	
H	.2660	

DECIMAL EQUIVALENTS AND TAP DRILL SIZES			DECIMAL EQUIVALENTS AND TAP DRILL SIZES		
Drill Size	**Decimal**	**Tap Size**	**Drill Size**	**Decimal**	**Tap Size**
I	.2720	5/16-24	33/64	.5156	9/16-18
J	.2770		17/32	.5312	5/8-11
K	.2810		35/64	.5469	
9/32	.2812		9/16	.5625	
L	.2900		37/64	.5781	5/8-18
M	.2950		19/32	.5937	11/16-11
19/64	.2968		39/64	.6094	
N	.3020		5/8	.6250	11/16-16
5/16	.3125	3/8-16	41/64	.6406	
O	.3160		21/32	.6562	3/4-10
P	.3230		43/64	.6719	
21/64	.3281		11/16	.6875	3/4-16
Q	.3320	3/8-24	45/64	.7031	
R	.3390		23/32	.7187	
11/32	.3437		47/64	.7344	
S	.3480		3/4	.7500	
T	.3580		49/64	.7656	7/8-9
23/64	.3594		25/32	.7812	
U	.3680	7/16-14	51/64	.7969	
3/8	.3750		13/16	.8125	7/8-14
V	.3770		53-64	.8281	
W	.3860		27/32	.8437	
25/64	.3906	7/16-20	55/54	.8594	
X	.3970		7/8	.8750	1-8
Y	.4040		57/64	.8906	
13/32	.4062		29/32	.9062	
Z	.4130		59/64	.9219	
27/64	.4219	1/2-13	15/16	.9375	1-12, 14
7/16	.4375		61-64	.9531	
29/64	.4531	1/2-20	31/32	.9687	
15/32	.4687		63/64	.9844	
31/64	.4844	9/16-12	1	1.000	
1/2	.5000				

PIPE THREAD SIZES

Thread	Drill	Thread	Drill
1/8-27	R	1 1/2-11 1/2	1 47/64
1/4-18	7/16	2-11 1/2	2 7/32
3/8-18	37/64	2 1/2-8	2 5/8
1/2-14	23/32	3-8	3 1/4
3/4-14	59/64	3 1/2-8	3 3/4
1-11 1/2	1 5/32	4-8	4 1/4
1 1/4-11 1/2	1 1/2		

METRIC TAP DRILL SIZES

Diameter and Pitch	Metric Drill	Inch Drill
5 × .80	4.20	11/64
6 × 1.00	5.00	13/64
7 × 1.00	6.00	15/64
8 × 1.25	6.75	17/64
10 × 1.50	8.50	11/32
12 × 1.75	10.25	13/32

Note: Nominal outside diameter minus the pitch equals the tap drill size for 75% thread contact area.

DECIMAL EQUIVALENTS OF NUMBER SIZE DRILLS

No.	Size of Drill (Inches)	No.	Size of Drill (Inches)	No.	Size of Drill (Inches)	No.	Size of Drill (Inches)
1	.2280	21	.1590	41	.0960	61	.0390
2	.2210	22	.1570	42	.0935	62	.0380
3	.2130	23	.1540	43	.0890	63	.0370
4	.2090	24	.1520	44	.0860	64	.0360
5	.2055	25	.1495	45	.0820	65	.0350
6	.2040	26	.1470	46	.0810	66	.0330
7	.2010	27	.1440	47	.0785	67	.0320
8	.1990	28	.1405	48	.0760	68	.0310
9	.1960	29	.1360	49	.0730	69	.0292
10	.1935	30	.1285	50	.0700	70	.0280
11	.1910	31	.1200	51	.0670	71	.0260
12	.1890	32	.1160	52	.0635	72	.0250
13	.1850	33	.1130	53	.0595	73	.0240
14	.1820	34	.1110	54	.0550	74	.0225
15	.1800	35	.1100	55	.0520	75	.0210
16	.1770	36	.1065	56	.0465	76	.0200
17	.1730	37	.1040	57	.0430	77	.0180
18	.1695	38	.1015	58	.0420	78	.0160
19	.1660	39	.0995	59	.0410	79	.0145
20	.1610	40	.0980	60	.0400	80	.0135

DECIMAL EQUIVALENTS OF LETTER SIZE DRILLS

Letter	Size of Drill (Inches)	Letter	Size of Drill (Inches)
A	.234	N	.302
B	.238	O	.316
C	.242	P	.323
D	.246	Q	.332
E	.250	R	.339
F	.257	S	.348
G	.261	T	.358
H	.266	U	.368
I	.272	V	.377
J	.277	W	.386
K	.281	X	.397
L	.290	Y	.404
M	.295	Z	.413

HELICOIL TAP DRILL SIZES

Inches		Metric	
Thread	Tap Drill Size	Thread	Tap Drill Size
1/4–20	17/64	5 × 0.8	13/64
5/16–18	21/64	6 × 1.0	1/4
3/8–16	25/64	7 × 1.0	9/32
7/16–14	29/64	8 × 1.25	21/64
1/2–13	17/32	10 × 1.50	13/32
		12 × 1.75	31/64

APPENDIX E
REFERENCE TABLES

DRY TORQUE RECOMMENDATIONS

Diameter	Grade 5		Grade 6		Grade 8	
	Coarse	Fine	Coarse	Fine	Coarse	Fine
1/4	108[a]	120[a]	132[a]	156[a]	17	19
5/16	17	20	23	25	34	37
3/8	31	35	40	45	60	68
7/16	50	55	64	72	96	108
1/2	75	85	98	110	145	165
9/16	110	120	140	160	210	235
5/8	150	170	195	220	290	330

[a]Any torque less than 15-lb is given in in-lb.

LUBRICATED TORQUE RECOMMENDATIONS (REDUCED 33 PERCENT)

Diameter	Grade 5		Grade 6		Grade 8	
	Coarse	Fine	Coarse	Fine	Coarse	Fine
1/4	72[a]	80[a]	88[a]	104[a]	136[a]	152[a]
5/16	135[a]	160[a]	15	17	22	25
3/8	21	23	26	30	40	45
7/16	33	37	43	48	64	72
1/2	50	57	65	73	96	110
9/16	73	80	93	107	140	157
5/8	100	113	130	147	193	220

[a]Any torque less than 15-lb is given in in-lb.

METRIC DRY TORQUE RECOMMENDATIONS

Diameter (mm)	Pitch	Grade 8.8	Grade 12.9
6	1.00	84[a]	132[a]
7	1.00	132[a]	20
8	1.25	18	29
8	1.00	20	32
10	1.50	33	58
10	1.25	35	61
10	1.00	38	64
12	1.75	59	100
12	1.50	65	110

[a]Any torque less than 15 ft-lb is given in in.-lb.

METRIC LUBRICATED TORQUE RECOMMENDATIONS (REDUCED 33 PERCENT)

Diameter (mm)	Pitch	Grade 8.8	Grade 12.9
6	1.00	56[a]	88[a]
7	1.00	88[a]	160[a]
8	1.25	144[a]	19
8	1.00	160[a]	21
10	1.50	22	39
10	1.25	23	41
10	1.00	25	43
12	1.75	39	67
12	1.50	43	73

[a]Any torque less than 15 ft-lb is given in in.-lb.

APPENDIX F
TORQUE VALUES

The following charts give the standard torque values for bolts, nuts, and taperlock studs of SAE Grade 5 or better quality.

GENERAL TIGHTENING TORQUE FOR BOLTS, NUTS, AND TAPERLOCK STUDS

Thread Diameter		Standard Torque	
Inches	**Millimeters**	**ft.-lb.**	**N·m***
Standard Thread 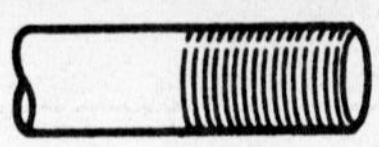		**Use these torques for bolts and nuts with standard threads (conversions are approximate).**	
1/4	6.35	9 ± 3	12 ± 4
5/16	7.94	18 ± 5	25 ± 7
3/8	9.53	32 ± 5	45 ± 7
7/16	11.11	50 ± 10	70 ± 15
1/2	12.70	75 ± 10	100 ± 15
9/16	14.29	110 ± 15	150 ± 20
5/8	15.88	150 ± 20	200 ± 25
3/4	19.05	265 ± 35	360 ± 50
7/8	22.23	420 ± 60	570 ± 80
1	25.40	640 ± 80	875 ± 100
1-1/8	28.58	800 ± 100	1100 ± 150
1-1/4	31.75	1000 ± 120	1350 ± 175
1-3/8	34.93	1200 ± 150	1600 ± 200
1-1/2	38.10	1500 ± 200	2000 ± 275
		Use these torques for bolts and nuts on hydraulic valve bodies.	
5/16	7.94	13 ± 2	20 ± 3
3/8	9.53	24 ± 2	35 ± 3
7/16	11.11	39 ± 2	50 ± 2
1/2	12.70	60 ± 3	80 ± 4
5/8	15.88	118 ± 4	160 ± 6

GENERAL TIGHTENING TORQUE FOR BOLTS, NUTS, AND TAPERLOCK STUDS (CONTINUED)			
Thread Diameter		Standard Torque	
Inches	Millimeters	ft.-lb.	N·m*
Taperlock Stud 		Use these torques for studs with taperlock threads.	
1/4	6.35	5 ± 2	7 ± 3
5/16	7.94	10 ± 3	15 ± 5
3/8	9.53	20 ± 3	30 ± 5
7/16	11.11	30 ± 5	40 ± 10
1/2	12.70	40 ± 5	55 ± 10
9/16	14.29	60 ± 10	80 ± 15
5/8	15.88	75 ± 10	100 ± 15
3/4	19.05	110 ± 15	150 ± 20
7/8	22.23	170 ± 20	230 ± 30
1	25.40	260 ± 30	350 ± 40
1-1/8	28.58	320 ± 30	400 ± 40
1-1/4	31.75	400 ± 40	550 ± 50
1-3/8	34.93	480 ± 40	650 ± 50
1-1/2	38.10	550 ± 50	750 ± 70

*1 Newton-meter (N·m) is approximately the same as 0.1 mkg.

The torques shown in the charts that follow are to be used on the part of 37° flared, 45° flared and inverted flared fittings (when used with steel tubing), O-ring plugs, and O-ring fittings.

TORQUE FOR FLARED AND O-RING FITTINGS

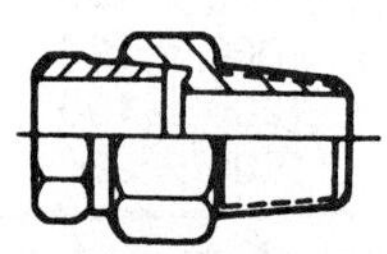

Inverted 45° Flared

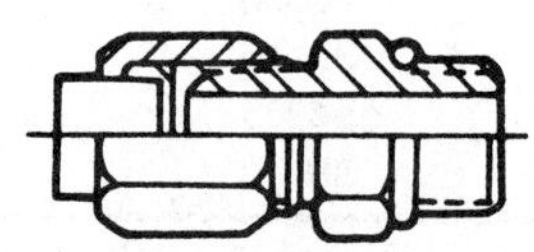

37° Flared

Tube Size (O.D.)	mm	3.18	4.78	6.35	7.92	9.52
	in.	.125	.188	.250	.312	.375
THREAD SIZE (in.)		5/16	3/8	7/16	1/2	9/16 5/8
TORQUE N·m		5 ± 1	11 ± 1	16 ± 2	20 ± 2	25 ± 3
TORQUE lb.-in.		45 ± 10	100 ± 10	145 ± 20	175 ± 20	225 ± 25

TORQUE FOR FLARED AND O-RING FITTINGS

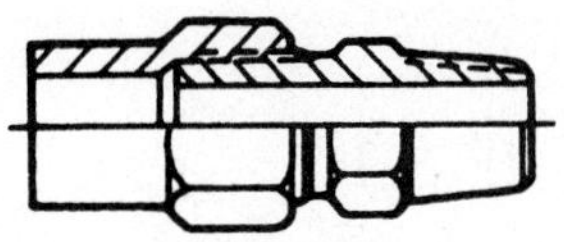

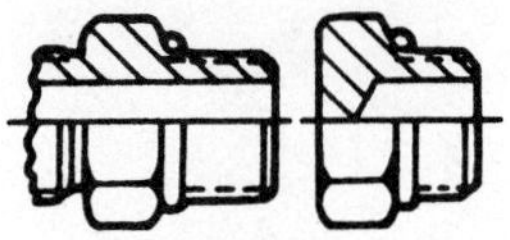

		45° Flared		O-Ring Fitting—Plug			Swivel Nuts		
Tube Size (O.D.)	mm	12.70	15.88	19.05	22.22	25.40	31.75	38.10	50.80
	in.	.500	.625	.750	.875	1.000	1.250	1.500	2.000
THREAD SIZE (in.)		3/4	7/8	11/16	1-3/16 1-1/4	1-5/16	1-5/8	1-7/8	2-1/2
TORQUE N·m		50 ± 5	75 ± 5	100 ± 5	120 ± 5	135 ± 10	180 ± 10	225 ± 10	320 ± 15
TORQUE lb.-ft.		35 ± 4	55 ± 4	75 ± 4	90 ± 4	100 ± 7	135 ± 7	165 ± 7	235 ± 10

INDEX

A

B

C

D

E

F

G

H

I

J

K

L

M

N

O

P

R

S

Z